R.G. Mitchel

San Ramon, Feb 1999

Carbonate Seismology

Edited by
Ibrahim Palaz
Kurt J. Marfurt

Series Editor
Leon A. Thomsen

Geophysical Developments Series, No. 6
Society of Exploration Geophysicists

SEG thanks Amoco Production Company for its contribution of typesetting and graphics services toward the publication of this book.

Carbonate Seismology / edited by Ibrahim Palaz, Kurt J. Marfurt
p. cm. -- (Geophysical developments series; no. 6)
Includes bibliographical references.
ISBN 1-56080-038-0 (hc)
1. Carbonate reservoirs. 2. Stratigraphic traps (Petroleum geology). 3. Seismic prospecting. I. Palaz, Ibrahim. II. Marfurt, K. J. III Series: Geophysical development series; v. 6.
TN870.57.C37 1997
622'. 1828--cd20

ISBN 0-931830-41-9 (Series)
ISBN 1-56080-038-0 (Volume)

Society of Exploration Geophysicists
P. O. Box 702740
Tulsa, OK 74170-2740

Published 1997
Printed in the United States of America

Contents

Preface

We first collaborated in the area carbonate seismology in 1990 while mapping Cretaceous and Tertiary carbonate reservoir facies from neighboring seismic data surveys gathered in the Pelagic Sea of Tunisia and Malta. Both areas, one shallow water (<50 m) and one deep water (>500 m), were plagued by a "penetration" problem through shallow carbonates and by a resolution problem of low-relief stratigraphic targets at depth. While the geologists on our teams had an ample supply of up-to-date sources devoted to the details of carbonate sedimentology and sequence stratigraphy, those of us working on the seismic data were left to our own devices. With considerable effort, we were able to come up with a handful of technical papers, some good notes from continuing education courses, and a thick pile of expanded abstracts from diverse sources to help us understand the seismic expression of carbonates. We augmented this sparse material with expert advice from our Amoco colleagues, our contractors, and our partners.

It was at this point when we first saw the need for an integrated reference book on carbonate seismology, and we vowed that once we were finished with our assignments, we would attempt to put such a book together. The years of 1992–1994 were tumultuous in the petroleum industry, with most of the major oil companies downsizing and the competing service companies consolidating. During this period, we saw many of our experienced colleagues who had provided us with expert advice leave the oil industry. With this additional lack of available "folk" wisdom in the area of carbonate seismology, we found it more imperative than ever to capture the current state-of-the-art before it was lost to posterity.

Our goal was to produce a book that would integrate the principles of carbonate geology with its seismic expression and would be readily understandable to the practicing geologists, geophysicists, and engineers that form the exploration and exploitation teams in the petroleum industry. The result is a single integrated volume, written in plain language by acknowledged experts in their fields, that illustrates the interrelationships of carbonate geology, petrology, sequence stratigraphy, rock properties, seismic data acquisition, seismic data processing, and integrated interpretation.

We have taken care in the editing process to ensure that every concept is explained clearly and concisely without getting lost in domain-specific terminology. Our hope is that this volume will sit dog-eared on the desk of every practicing geoscientist, to help the seismic data processor determine parameters to enhance the fidelity of carbonate images, to help the seismic interpreter better recognize the expression of sequence stratigraphy, to help the engineer understand patterns of permeability and fractures, and to help the carbonate geologist understand the expression of the rock record at the seismic scale and differentiate it from common seismic acquisition and processing artifacts.

We have provided ample examples on the application of carbonate AVO and acoustic logging. Tying acoustic logs to seismic is a common theme throughout the book. We have included two chapters by Fischer et al. and by D'Angelo et al. that show how, with the aid of careful seismic modeling, AVO can be calibrated and used to map porosity in carbonate rocks.

We wish to thank all the contributing authors for their hard work, perseverance, and patience. We also want to thank those authors who had hoped to contribute to this volume and did much of the work but, through the turmoil in the oil industry, found themselves severed from their data and ultimately unable to contribute.

Ibrahim Palaz
Kurt J. Marfurt

Acknowledgments

Each of these chapters underwent a rigorous peer review process. All but two of the reviewers have served as associate editors for journals of the Society of Exploration Geophysicists or the American Association of Petroleum Geologists. We therefore wish to thank Edith Wilson, J. P. Blangy, Norm Neidell, R. Loucks, Ron Nelson, Don Rice, Mike Graul, David Farel, S. Mallick, Ken Nibbelink, Don Winterstein, Randy Keller, H. Karazincir and Greg Riley. The early design of this book was heavily influenced by Prof. Cahit Coruh at Virginia Tech, while our SEG Series Editor, Leon Thomsen, guided and encouraged us through the SEG publications process. We are most grateful for the careful editing and design provided by Kathy Walker of Editorial Technologies, as well as to SEG Publication Manager Ted Bakamjian, who managed production of the book and proofread the final version.

Most important of all we wish to thank our colleagues at Amoco—our fellow geoscientists who helped us with the preliminary reviews, Rita Langford who helped with the communications with the authors, and the workers in the Amoco Westlake Media Center, who helped in translation and drafting of several of the documents. In particular, we wish to acknowledge the care and dedication exercised in the typesetting of this book by Julie Youngblood, Vicki Wilson and Charlie Johnson of Amoco's Tulsa Document Services and Graphic Services.

I. P.
K. J. M.

Chapter 1

Carbonate Seismology: An Overview

Ibrahim Palaz
Amoco Production Company
Houston, Texas, U.S.A.

Kurt J. Marfurt
Amoco EPTG
Tulsa, Oklahoma, U.S.A.

Introduction

Carbonate rocks make up only 20% of the sedimentary rock record yet account for more than 50% of the world's proven hydrocarbon reserves. Therefore, it should be no surprise that carbonate reservoirs are very important targets for oil and gas exploration. While surface mapping, paleontology, palynology, oxygen and carbon isotope studies, petrography, X-ray diffraction, gravity, magnetics, and core and well log analyses all play essential roles in oil and gas exploration, a vast majority of the data is obtained by reflection seismology. In addition to being relatively inexpensive, reflection seismology provides the only practical method of obtaining a 3-D image of the earth's subsurface. After more than 70 years, the intricacies of seismic data acquisition, data processing, and data interpretation are well established. However, much of this methodology and expertise has been developed for siliciclastic terranes, where the sand and clay have been transported by streams, rivers, wind, or waves or have been mechanically reworked by other processes usually understood by most practicing seismologists.

Contrasts Between Carbonates and Siliciclastics

Carbonates differ significantly from siliciclastics in that they are not transported, but are largely organically grown, organically precipitated, or geochemically precipitated in place. Carbonate generation has specific environmental requirements. For example, organically grown carbonates require clear water with little or no silt contamination, specific water temperatures, and food supply. Carbonate generation occurs along long slopes, platforms, and margins extending for hundreds of kilometers, in contrast to the point source influx of siliciclastics from streams and rivers. Environments favorable to the precipitation of carbonates are often also favorable to the precipitation of evaporites (salt and anhydrite) such that an already difficult seismic interpretation problem is often complicated by salt diapirism, withdrawal, and dissolution.

Once generated, carbonates are subject to radical and rapid diagenetic changes, creating wild variations in porosity and permeability that have little, if anything, to do with the original stratigraphic boundaries. Such diagenetically altered and organically bound carbonate rocks are able to maintain steep marginal slopes and wave-resistant structures not seen in siliciclastic terranes. While siliciclastics are also subject to diagenetic changes, the chemical reaction rate of calcium carbonate is on the order of tens of thousands of years, whereas that of silica is on the order of a million years. While carbonates may undergo diagenesis subaerially, silica dissolution and recrystallization may require temperatures and pressures corresponding to depths of at least 300 m (e.g., in the Gulf of Mexico) to as much as 2 km (e.g., in the North Sea). Timing can be critical, with porosity often being locked in due to the early accumulation of hydrocarbons.

Carbonate stratigraphy shows greater variation over geologic time than does siliciclastic stratigraphy. Siliciclastic depositional textures and styles have changed little after the evolution of soil-retaining land plants. In contrast, carbonate deposition is subject to changes in flora and fauna throughout evolution, with some geologic periods showcasing reef builders, mound builders, and other organic precipitators. This evolution of carbonate generators is also coupled to the evolution of the ocean's chemistry over geologic time. While climatic changes such as glaciation, increased rainfall, and aridity affect both siliciclastic and carbonate deposition, carbonates are also influenced by changes in ocean temperature and salinity. Finally, since carbonates are generated only in shallow water, their origination depends on platform subsidence, driven either by sedimentary loading or plate tectonic activity, which results in carbonates being drowned by rapid subsidence or poisoned by terrigenous sediments associated with rapid uplift.

From a seismic viewpoint, carbonate velocities are faster and carbonate densities greater than those of most siliciclastics. Associated anhydrite velocities may be faster still. These faster velocities result in lower vertical and spatial resolution in carbonate strata versus siliciclastics of comparable thickness. When overlain by less dense siliciclastics with slower velocities, the top-of-carbonate reflection can be easily recognized on a seismic section from an observer across the room. However, when the target horizon or carbonate reservoir is embedded in yet more carbonate, the reflectivity between formations is notoriously low, giving a transparent appearance. This is particularly irksome where the carbonate section continues up to the water bottom, where the reflection coefficient may be 60% or more, such that strong multiples overwhelm the 1% reflection coefficient of internal carbonate-on-carbonate reflections.

To complicate matters further, such a simple vertical stack of homogeneous carbonate layers is quite rare. With the rise and fall of sea level, carbonates are often exposed subaerially, where they undergo further diagenesis and karsting. While the generation of macroscopic porosity associated with karsted terranes (e.g., vugs, caves, and collapse features) may be of significant interest at the reservoir level, such karsting greatly denigrates seismic waves passing through it. For example, beneath the central Mediterranean waters of Malta and Tunisia, there are at least two well-documented levels of karsting for seismic waves to pass through: the Messinian and Cretaceous unconformities. The relatively shallow Messinian unconformity is further complicated by these karsted features being filled or overlain with anhydrite. Laterally back-scattered seismic noise from these features plagues seismic exploration in the Mediterranean from the Nile Delta to the Balearic Islands.

The effects of such karsting are even worse in a terrestrial versus marine setting, such as in Papua New Guinea, Madagascar, and Croatia. Here, each air- or water-filled inhomogeneity back-scatters and converts surface waves to compressional waves and shear waves in all three dimensions, significantly reducing seismic data quality. Since waves are scattered according to the relationship between the size of the scatterer and the seismic wavelength, only the longest wavelengths have sufficient signal-to-noise ratio at the target depth, giving rise to what many geophysicists call the "penetration problem."

In more structurally complex terranes where mixed lithologies intersect the earth's surface, significant problems are encountered in equalizing the seismic signal generated and measured in soft soil and hard carbonate outcrops, giving rise to what seismic acquisition workers call the "coupling problem." If the carbonates are differentially weathered, the seismic "statics problem" is encountered.

In the most extreme case, the differential weathering between lithologies gives rise to the rough topography and significant seismic "imaging problems" seen in the Dinaric Alps and the Canadian Rockies. Even thin layers of deeply buried carbonates can wreck havoc on the best thought-out seismic data processing schemes, such as the difficulties of using the workhorse Kirchhoff (ray trace driven) seismic migration schemes to image below the thin but high-velocity chalks of the North Sea, giving rise to what seismic imagers call the "lensing problem." Finally, since carbonate velocites are significantly faster than those of siliciclastics, the wavelengths for a given frequency are significantly longer, giving rise to what is called the "resolution problem."

Overview of the Volume

Given the lower quality seismic data than what we would expect at corresponding depths in a siliciclastic terrane, it is more important than ever to integrate the seemingly disjointed scraps of geologic, geophysical, and engineering data that we have available to construct an integrated picture. A good detective integrates disjointed data by discovering the *modus operendi* of the suspect. For this reason, we begin the first half of our book by describing the *geologic framework* of carbonate seismology with Wilson's description of the inner workings and interrelationships of carbonate depositional environments and diagenesis in Chapter 2. Here, Wilson explains the where and how of nine juxtaposed environments (called "Wilson zones" by other workers): basinal pelagic, open-marine platform, toe of slope/basin margin, rimmed platform, organic ecological reef, platform edge, restricted platform interior, sabkha, and supratidal. The temporal and spatial framework of the various depositional zones is the key to the heterogeneous data integration that is a recurrent theme throughout this book, enabling us to map or infer lithologic facies, porosity, and permeability. This chapter not only clarifies where but also at what scale and with which geometrical configuration a postulated facies may occur, allowing us to predict beyond the limits of seismic resolution.

High seismic velocities giving rise to lower resolution seismic images is only one manifestation of carbonate rock properties. In contrast to siliciclastics, there appears to be little written on carbonate seismic rock properties. Although one might imagine that the seismic behavior of rocks composed almost entirely of calcite, aragonite, and dolomite would be quite simple, variations in texture, fabric, porosity, fractures, cementation, and fluid fill lead to highly variable carbonate rock properties. Though carbonate permeability is not readily measured by the seismic method, explorationists would like to be able to *infer* permeability patterns using well control and 3-D seismic data as a guide. We therefore use Chapter 3, 4, and 5 to lay the necessary foundation on which the seismic method and much of the remainder of this book is constructed.

Wang lays the cornerstone of this foundation by describing the seismic properties of carbonate rocks in Chapter 3. He begins by defining what geophysicists mean by seismic velocity, attenuation, and anisotropy, and he summarizes modern laboratory core analysis techniques with an eye toward quantifying the strengths and weaknesses of each kind of measurement. While surface seismic methods cannot image the individual pore configuration so important to reservoir capacity and performance, Wang compares and contrasts the more common models and theories used to map microscopic grain, pore, and pore fluid properties to the macroscopic carbonate properties we measure with surface seismic techniques. Carbonate pore types range from intercrystalline and interparticle porosity common to siliciclastics, through the more carbonate-specific moldic, intraparticle, vug, channel, and fenestral porosities, to the formation of fracture and breccia porosity and stylolites. Searching for other factors affecting seismic velocities, Wang evaluates the role of mineralogy, porosity, saturation, fluid type, depth of burial, pressure, and bulk density on seismic velocities as a function of frequency. Finally, he integrates these findings through two short case studies of enhanced oil recovery.

In a complementary study, Anselmetti and Eberli look at sonic velocities in carbonate sediments and rocks in Chapter 4 through a comprehensive study of 295 shallow carbonate minicores from Florida Bay, the Great Bahamas Bank, and Montagna della Maiella in central Italy. This chapter contains an outstanding variety of samples from diverse depositional environments, ranging from shallow-water platform to deeper slope deposits and from ancient outcrop to modern seafloor. The authors augment and correlate their measurements of velocity, density, permeability, and porosity with petrographic studies. They find that factors affecting velocity include mechanical compaction, burial depth, age, effective pressure, depositional lithology, mineralogy, porosity, pore type, density, and pore fluid. Of particular interest is how seismic velocity varies with different states of diagenesis, perhaps allowing us to use velocity as a tool to identify where porosity has been enhanced by diagenesis and to identify carbonate seals where porosity has been diminished by diagenesis.

In Chapter 5, the third of three chapters emphasizing rock properties, Marion and Jizba discuss acoustic properties of carbonate rocks and their use in quantitative interpretation of sonic and seismic measurements. They demonstrate that seismic data can be used not only to image the structure of reservoirs and determine their depositional and geologic framework but also to assess the petrophysical properties necessary for exploitation. While porosity plays the largest role in determining carbonate velocity and may be sufficient for mapping reservoirs in the exploration mode, in reservoir development, it is necessary to understand how velocity depends on mineralogy, pore shape, and fluid type. After quantifying these three important secondary components to the velocity equation, the authors dedicate the bulk of their chapter to correlating core and well control from some 30 wells with 3-D surface seismic data over a Tertiary carbonate oil field. Through calibration and sensitivity analysis, they demonstrate the correlation of surface-derived seismic inversion for acoustic impedance to reservoir porosity, saturation, and bed thickness.

Carbonate seismologists often have no well control at all and must infer carbonate lithologies from seismic data alone. In Chapter 6, which covers carbonate seismic facies analysis, Macurda shows how the carbonate depositional environment described by Wilson in Chapter 2 is expressed by the reflection seismic technique. By using careful but simple concepts of pattern recognition and texture analysis, he demonstrates how an interpreter can obtain significant information about carbonate depositional environments, diagenesis, and lithofacies from seismic reflection data. Macurda breaks seismic facies analysis into the mapping of four seismic attributes: configuration or shape of reflections, amplitude or relative strength of peaks and troughs, frequency or number of reflections per unit traveltime, and reflection continuity or lateral continuity. Reflection configuration is the most varied of these seismic facies attributes, now commonly classified as parallel-subparallel-divergent, progradational, mounded-draped, or onlap-fill. Macurda illustrates each of these reflection configurations from carefully selected seismic lines over carbonate terranes. Having defined his terminology, he draws examples from around the world to demonstrate how carbonate facies can be mapped from various parts of the carbonate depositional environment. Even in the absence of well control, this chapter coupled with Chapter 2 puts the interpreter well on the road to mapping prospective areas for exploration.

When there is access to disparate, but high-resolution well log, core, outcrop, biostratigraphic, petrographic, and geochemical data, it is possible to integrate them with seismic data using the principles of sequence stratigraphy, such as described by Fitchen in Chapter 7 on carbonate sequence stratigraphy and its application to hydrocarbon exploration and reservoir development. The seismic method makes it possible to resolve sequences around the platform margin and slope where strata are thickest and stratal geometries and discordances are best developed. However, in inner platform settings, all strata are flat-lying, such that there may be no stratal discordances at sequence boundaries. Integration of high-resolution methods allows the mapping of these boundaries, as well as those where there is little to no contrast in acoustic impedance. By interpreting genetically related sedimentary facies within the framework of chronostratigraphically significant surfaces, it is possible to make meaningful predictions about lithology and porosity.

The vocabulary of sequence stratigraphic nomenclature is overwhelming for many geophysicists and engineers. Fitchen takes great care in defining his terms and relating them to nomenclature used in state-of-the-art literature. Likewise, he carefully describes the sedimentologic and stratigraphic processes of carbonate systems, emphasizing the controlling factor of sea level change. After familiarizing the reader with all the necessary components, he outlines the methods and tools of high-resolution sequence stratigraphic analysis using seismic, well log, core, and outcrop data. Having built this framework, he concludes the chapter by presenting three case studies to illustrate the application of carbonate sequence stratigraphic concepts to sequences that are thick enough to be at or above seismic resolution: the Miocene Terumbu Formation of the Indonesian Natuna field, the Lower Cretaceous Shuaiba Formation of the Abu Dhabi Bu Hasa field, and the Middle Permian San Andres Formation of the Texas/New Mexico Permian Basin. He demonstrates that data integration in a carbonate sequence stratigraphy context make it possible to map porosity and seals accurately. These practices, developed by Fitchen and co-workers, can be used to apply the tools of sequence stratigraphy to mapping chronostratigraphic boundaries that can only be partially imaged by the seismic method.

Porosity, whether primary or diagenetically induced, is one means of storing large oil and gas volumes in a reservoir. Lorenz and his co-authors discuss the characteristics of natural fractures in carbonate strata in Chapter 8. Such naturally fractured rocks can be first-order exploration targets even when they do not have primary or diagenetically induced porosity. Fractures, like sequence boundaries, are often at or below seismic resolution. Instead, we infer fractures from traveltime anomalies, birefrigent effects on shear waves, and subtle effects on seismic attributes, such as seismic coherency. These inferences are often best calibrated to ground truth by use of geostatistics. Only after we have obtained a clear understanding of the phenomena of natural fractures within the carbonate strata can we effectively design a seismic survey, properly process and image the data, and make realistic predictions. To this end, the authors study fractures in mildly, moderately, and highly deformed strata using examples from Oklahoma, Texas, New Mexico, Wyoming, Alaska, and the North Sea. These examples show that the most common cause of natural fractures is extension and that the single most important factor controlling natural fractures is lithology, with dolomite and dolomitic mudstones being much more susceptible to natural fractures than limestones. Fractures are more intense along fold hinges than along fold limbs, while fractures caused by faults alone tend to be limited to within a few hundred meters of the fault. This completes the first half of the book describing the geologic framework of carbonate seismology.

The second half of the volume on the *seismic expression* of carbonates begins with a paper by Fischer and his co-authors on integrated exploration strategy applied to carbonate environments in Chapter 9. This chapter, previously published by the Society of Petroleum Engineers, is the only reprint contribution to this volume. Given our goal of providing a well integrated reference book on carbonate seismology for use by geophysicists, geologists and engineers, we felt it appropriate for inclusion. The authors develop an image of the carbonate depositional setting presented in Chapters 2 through 7 by designing an interconnected structural and stratigraphic road map that carefully integrates all seismic, well log, gravity, and magnetic data. Milestones along the route include mapping structural closure and sequence boundaries, classification of stratigraphic intervals, and generation of facies trends using A-B/C encoding. Having thus generated a geologic working hypothesis of the earth's subsurface, the authors test their hypothesis with full waveform seismic models which are then compared to the processed seismic data. In their example, the authors show that, by replacing their initial hypothesis of a grainstone facies with boundstones and by replacing the adjacent packstone–grainstone facies with grainstones, it is possible to both honor their geologic depositional model and fit the seismic data accurately.

Up to this point, our focus has been on carbonates acting as source rocks, reservoirs, and seals. Carbonates can also form part of the overburden. In Chapter 10, on the design of seismic data processing flows in areas of shallow carbonates, Taner demonstrates how carbonate outcrops at the base of the weathering zone, in shallow water, or elsewhere in the shallow part of the seismic section, can greatly degrade seismic data quality. Because of high velocity, high acoustic impedance, and high variability to erosion, carbonates create challenging problems in seismic data acquisition and processing. Taner's experience is that it is best to first identify these problems. If a flat, relatively thick carbonate layer is overlain by slow-velocity sediment or water, only the small cone of seismic rays below the critical angle of reflection will be able to penetrate through the carbonate to illuminate deeper, slow-velocity clastic sediments. Of those few rays that do penetrate the carbonate layer, those that reflect off steeply dipping deeper structures will on their return hit the bottom of the carbonate layer beyond the critical angle and never make it back to the earth's surface.

Taner shows how these complicated velocity problems, exacerbated by the high impedance contrast encountered at shallow carbonate boundaries, give rise to strong surface and interbed multiples. High impedance and fast, shallow velocities also give rise to strong, fast surface waves on land and in shallow water, and to strong reverberating refractions both on land and in marine environments, further exacerbating the already poor signal-to-noise ratio. High-velocity carbonates encased in lower velocity siliciclastics behave as transverse isotropic media, causing problems with normal moveout (NMO) and seismic imaging algorithms. Confronted with this myriad of problems, Taner asks the simple question, "How do we determine an ultimate processing flow when carbonates are present?" He works from the desired end-product, step by step backward, to ensure that everything is done to achieve the best possible results. In this process, it becomes clear that what appears to be trivial processing steps, such as gain correction and multiple elimination, are critically important in obtaining the best possible section. Taner illustrates each step of the processing flow with a set of carefully chosen figures that will be invaluable to geophysicists in communicating their processing needs to their geologic and engineering teammates.

High-quality seismic processing and interpretation are based on the acquisition of high-quality seismic data in the field. While back-scattered energy from surface topography and irregular weathering zones is present in nearly all geologic terranes, it is particularly a problem over differentially weathered surface carbonates. In Chapter 11, on the measurement and identification of 3-D coherent noise generated from irregular surface carbonates, Regone quantifies these problems through careful 3-D field tests performed on Edward's Plateau in West Texas and on a karsted surface in Croatia. Rayleigh waves over surface carbonates can be quite fast, such that an "inside seismic mute" applied to the data leaves little information outside the noise cone. In addition, shallow scatterers convert outgoing direct waves, ground roll, and head waves into back-scattered head waves having apparent velocities approaching those subsurface reflections of interest. Because of this velocity ambiguity, it is difficult to differentiate side-scattered noise from deeper primary diffractions on conventional 2-D acquisition wave tests. To address this problem, Regone uses a 3-D grid of receivers to create a "box wave test" that quantifies both the magnitude and directionality of the signal-to-noise ratio in the data. He analyzes these data using a display that looks much like a commercial radar screen. Only after such analysis can economical 3-D source and receiver arrays be designed to address the problem of side-scattered noise in a site-specific way.

In his paper on seismic signatures of carbonate diagenesis in Chapter 12, Wagner builds on Wang's discussion of porosity presented in Chapter 3. Diagenetic processes such as karsting, mechanical compaction, chemical compaction, burial leaching, cementation, and dolimitization can change carbonate seismic rock properties severely. Wagner points out that carbonate reservoirs can be created, degraded, or rejuvenated if the correct conditions exist. There are several important factors in recognizing diagenetic changes on seismic data. First is the scale factor. A large-scale, basinwide acoustic impedance contrast may be caused by diagenetically induced changes in the mineralogy phase. However, a small-scale change of an individual facies or bed may be of primary importance to entrapment and later production of oil and gas accumulations. These small-scale changes may be difficult or impossible to image with surface seismic data. Using a case study from China, Wagner discusses in detail the limits of seismic resolution, as well as the use of 3-D compression mapping, compressional velocity, shear velocity, and acoustic impedance inversion to identify and delineate a diagenetically altered section.

Gravity flow redeposition of chalks by submarine slides, slumps, debris flows, and turbidites forms the basis of many North Sea reservoirs. In Chapter 13 on porosity detection and mapping in a basinal carbonate setting in offshore Norway, D'Angelo and co-workers integrate specialized geophysical studies with rock property data from drilling operations in producing fields to map stratigraphically thickened redeposited chalks. Petrophysical analysis of 41 plugs from chalk cores show a strong correlation between porosity and *P*-wave velocity, *S*-wave velocity, and bulk density. The authors perform geophysical studies ranging from nonroutine high-density stacking velocity analysis, through seismic trace impedance inversion and amplitude versus offset analysis, to full waveform tau-*p* inversion and elastic parameters. As in Chapter 9, the authors calibrate their results by forward ray tracing and amplitude variation with offset (AVO) modeling. Three-dimensional impedance maps and 2-D AVO and tau-*p* inversion cross-sections predict anomalies thought to be caused by high-porosity zones in the chalk. Based on this analysis, an exploration well was drilled in 1994 to test the concept; the well was a discovery.

In addition to gravity flow induced porosity, chalks are also subject to fracture. Li and Mueller, in Chapter 14, discuss several case studies of multicomponent seismic data for fracture delineation using examples from the Austin Chalk. Since individual fractures are below seismic resolution, they must be detected in a less direct way using seismic attributes. Li and Mueller find that the most effective attributes are obtained by using multicomponent shear wave data, in particular those of polarization, time delay, and reflectivity at normal incidence. The Austin Chalk trend of Texas is a natural laboratory for using shear wave anisotropy as a tool because it has little matrix porosity and it has vertically aligned fractures that are major oil migration pathways. Here the challenge is not only to map the fractured reservoir, but also to delineate zones of more intense fracturing in order to enhance production from existing fields.

To achieve this goal, Li and Mueller describe the necessary components of data acquisition, processing, and interpretation. They demonstrate a good correlation between fractures seen on mudlogs of three horizontal wells and dim spots observed on the surface-gathered slow split shear wave section. In addition to six multicomponent surface lines, the authors present the results of three multicomponent vertical seismic profiles (VSPs) from wells having different production rates. The VSP from the nonproductive well shows minimal shear wave splitting, while the VSP from the water-producing well shows shear wave splitting but no amplitude anomaly. In contrast, the VSP from the oil-producing well shows both shear wave splitting and amplitude anomalies. The authors present the strengths and limitations as well as the common pitfalls of using multicomponent shear wave technology in the delineation of fractured reservoirs.

Many of the papers in this unit on seismic expression begin with low-resolution surface seismic images that must be reconciled with high-resolution well control, after which the high-resolution inferences can be carried away from the well. In Chapter 15, Anselmetti and co-workers further our understanding of the relationship of geologic processes, rocks, and seismic measurements through a discussion of seismic modeling of a carbonate platform margin in Montagna della Maiella, Italy, and its variations in seismic facies and implications for sequence stratigraphy. Unlike the seismic models generated by Fischer et al. in Chapters 9 and 16 for hypothesis testing, Anselmetti et al. model an outcrop of a carbonate platform margin that is 14 km in length by 1.5 km in height. By generating the corresponding synthetic seismic section, the results of detailed sequence analysis performed on the Montagna della Maiella can be compared to sequence analysis based largely on seismic sections. The measurement and analysis of acoustic impedance corresponding to each of the 75 lithologic units is described in Chapter 4. Given these impedances and the structural cross-section obtained by directly mapping the outcrop, the authors generate their synthetic seismic cross-section using normal incidence ray tracing. This synthetic section shows a largely transparent platform that is onlapped along the escarpment by a succession of high-amplitude slope reflections. This difference in reflectivity is directly correlated to higher acoustic impedance contrasts in the slope zone and correspondingly smaller acoustic impedance contrasts on the platform.

In terms of sensitivity analysis, synthetics offer several distinct advantages over real data. First, we can quantify the effect of a variable signal-to-noise ratio. In addition, we can analyze the effect of geologic resolution with seismic frequency content. To this end, Anselmetti and co-workers mapped seven unconformity-bounded supersequences and modeled them with various levels of noise and signal frequency content. They demonstrate that, if the acquired seismic data has a dominant frequency of 20 Hz, then only five of the seven supersequences can be resolved. In contrast, all seven supersequences are imaged at 60 Hz. These studies reveal another significant problem: the appearance of pseudo-unconformities where no unconformity actually exists, but rather tuning artifacts. Such limitations on seismic resolution in a platform environment are further discussed in a real data case by Stiteler et al. in Chapter 18.

In Chapter 16, Fischer et al. study the development of an exploration concept for the Shuaiba Formation (Thamama Group) using seismic sequence and facies analysis in combination with forward modeling for a carbonate stratigraphic play in Abu Dhabi by applying the methodology he and his co-workers outlined in Chapter 9. The Early Cretaceous Shuaiba/Bab Formation, which was deposited in a shallow-marine environment, presents several lithologic units comprising lateral and vertical variations in reservoir development. They begin by mapping the relevant sequence boundaries from a grid of 2-D seismic data along with two wells. They next generate a detailed cross-section derived from a high-resolution (L_1 norm deconvolved) version of their seismic data. This is followed by cross-correlating velocity and porosity data obtained from core measurements for each of the stratigraphic units, after which they block the sonic and density logs. These blocked velocity and density values, now correlated to each of the stratigraphic units, are then transferred to the 2-D seismic cross-sections. The validity of this stratigraphic model is tested by the generation of a normal incidence full waveform finite difference seismic section, which is then time migrated and compared to the real processed seismic data. Finally, discrepancies between the modeled and real seismic data are fixed by adjusting the external geometries of the lithologic units to obtain anintegrated solution that honors the carbonate depositional model yet fits both the seismic and well control data.

In Chapter 17, on imaging of a stratigraphically complex carbonate reservoir with crosswell seismic data, Langan and his co-authors address how complexities in field production of a carbonate reservoir can be better mapped than by simply using surface seismic data or VSPs. This study, on the Midland Basin of West Texas, was initiated to better understand subtle 3-D surface seismic anomalies correlated with production from the carbonate Wolfcamp Formation. These anomalies were at the resolution threshold of the surface seismic data. While the drilling of one well proved the presence of reservoir-quality massive carbonate rock, an adjacent well only 195 m away penetrated a nonproducing siliciclastic-rich carbonate section instead. By use of a crosswell survey, the authors are able to bridge the gap between low-resolution 3-D surface seismic and high-resolution but only point control well logs. The tomography/migration algorithm they use produced three high-resolution images: a *P*-wave velocity cross-section, a migrated reflectivity image, and an anisotropy image. Generation of a transverse isotropy image is particularly intriguing in that it allows the differentiation between isotropic massive carbonates, strongly anisotropic shales, and mildly anisotropic intercalated carbonates and siliciclastics. By tying these three images to the wells, it becomes possible to map lithology and lithological changes more definitively between the wells.

The most difficult data interpretation problem is saved for the last chapter, Chapter 18, on carbonate platform seismic sequence attributes in the Maracaibo Basin of Venezuela, by Stiteler et al. Here in the inner platform, there is little change in gross acoustic impedance and monotonous substratal parallel seismic reflection patterns predominate. While the seismic data contain 5–50 Hz energy at the shallow objective, attenuation lowers this to only 5–30 Hz at the deeper objective. This low-frequency content is exacerbated by a strong top-of-carbonate Colon shale/Socuy carbonate reflector whose wavelet sidelobes strongly interfere and often mask the subtle low-amplitude carbonate-on-carbonate reflections that follow. To make matters worse, this large impedance change leads to transmission losses, as previously discussed by Taner in Chapter 10, and further degradation of the quality of the deeper section.

By carefully generating synthetics at the seven wells in the survey, Stiteler and his co-authors are able to quantify the effect of wavelet sidelobes and interbed multiples, thus preventing them from misinterpreting changes in these seismic artifacts with frequency and depth as real changes in the seismic reflectivity due to variations in stratigraphy. In a theme constant throughout this book, high-resolution interpretation is achieved by integrating the results of detailed X-ray diffraction of clays, micropaleontology, palynology, carbon and oxygen isotopes, petrography, and well logs with low-resolution seismic data, such that sequence boundaries can be inferred in a 3-D sense. In addition to mapping subtle stratigraphic changes in the Cogollo Group carbonates, 3-D seismic coverage allows the authors to map fault-related fracture zones associated with the development of good secondary porosity.

The extensive reference lists provided by the chapter authors will prove to be invaluable to workers who wish to immerse themselves more fully in specialized areas of carbonate deposition, rock properties, seismic stratigraphy, sequence stratigraphy, seismic data acquisition, seismic processing, and seismic modeling, or to search for further case studies more closely analogous to their problem at hand.

Chapter 2

Carbonate Depositional Environments and Diagenesis

James Lee Wilson
Consultant
New Braunfels, Texas, U.S.A.

Abstract

Carbonates are calcareous sediments composed of shredded and reworked remains of shelly invertebrate animals and calcareous algae, microscopic lime-secreting organisms (foraminifera and algae), and coated fecal pellets mixed with lime mud matrix. In addition to these shallow-water sediments, deep ocean bottoms are covered with limey organic microscopic debris (pelagic lime mud). Carbonate facies that make hydrocarbon reservoirs have been formed in clear, shallow, subtropical waters within the photic zone. They consist mainly of aragonite and Mg-rich calcite, which are stable only in marine saline water. Present-day depositional environments of shallow carbonates have been studied worldwide, e.g., Bahamas, Florida shelf, Arabian Gulf, Yucatan, and Red Sea.

Deposition of carbonate sediment follows a consistent sequential pattern. A halo of about eight sediment types is formed as a carbonate ramp, and then a rimmed platform is established. This occurs in a clear, warm marine environment on an evenly downsloping sea floor around a gentle uplift that furnishes a minimum amount of siliciclastics. The environments range in sequence from deep offshore muds and turbidites, to debris flows on the slope, reef tracts or oolitic lime sand shoals at the platform margin, to deposition of lime mud sediments in low-energy, open-marine, backreef shelf or interior shelf lagoons with restricted circulation, and finally to shoreward tidal flats or evaporitic salt marshes and ponds. Both interior shelf deposits and their subsequent alteration are controlled by climate (arid versus humid). This predictable pattern of depositional environments is valuable in discerning carbonate facies progression. Since slope debris flow, reef rock, and lime sand grainstones often retain some of their originally high porosity, it is important to map them in sequence and to locate them with reference to structures. Even the poorly permeable lime muds of the shelf lagoons can be improved by preferential dolomitization and can engender the Mg-rich fluids necessary for subsequent dolomite formation.

Porosity in carbonates is thus depositionally controlled but is also related to diagenetic alteration to which carbonate sediment is susceptible, including (1) cementation in marine environments, (2) presence of fresh meteoric water, (3) presence of warm, hypersaline burial fluids, (4) replacement of original aragonite and high-Mg calcite with crystals of low-Mg calcite, (5) leaching of aragonitic shells, and (6) brecciation, fracturing, and vein and stylolite formation. Most carbonate diagenesis reduces original porosity. Limestones are good reflecting horizons and have high velocity when well cemented. Reservoir discernment and prediction of trends on noncemented sediment or dissolution and fracturing of limestone and dolomite is at present challenging. Isotope geochemistry and fluid inclusion studies appear to help geoscientists in exploration of carbonate reservoirs.

Principles of Carbonate Sedimentation

Formation of Carbonates

Although limestones and dolomites constitute only about 20% of the Phanerozoic stratigraphic record, they contain nearly half of the world's discovered hydrocarbons. For this reason, they have been assiduously studied by petroleum geologists for almost 50 years. Modern textbooks on sedimentology outline principles of carbonate deposition and diagenesis in some detail (e.g., Laporte, 1969; Bathurst, 1971; Flugel, 1982; Wilson, 1975; Reading, 1978; Blatt et al., 1980; Tucker and Wright, 1990).

Most calcium carbonate in the geologic record is produced biochemically from shallow depths of either marine or fresh lake water. The distribution of calcium carbonate on modern shallow sea floors is by far highest in subtropical latitudes, in water shallow enough for light penetration, which infers a linkage between photosynthesis and carbonate production. There is some geochemical evidence based on strontium trace element content that lime muds may, in places, form by direct chemical precipitation (Milliman et al., 1993), but much is produced either by algal photosynthesis or by breakdown and transportation of finely comminuted, organically precipitated shells and tests.

Today, about 90% of carbonate marine sediment lies within the oceanic basins in the southern hemisphere, except in the Atlantic, and resides in deep water (shallower than the carbonate compensation depth of 5000 m or more). This pelagic sediment is composed of debris of microscopic plants (coccolithophores) and other tiny organisms such as foraminifera and pteropods which originally lived within the warm water of the photic zones, shallower than about 100–200 m deep.

Most carbonate strata of interest to petroleum explorationists, however, are formed in shallow shelfal environments in clear, warm marine water. Remains of contained organisms and sedimentary structures impressively attest to this. Hence, shallow-water accumulations of carbonate sediments have been studied around the world in some detail (Figure 1). Such areas include the Florida Bay and reef tract, Bahama Banks, Batabano Bay in Cuba, Caribbean coasts of Yucatan and Belize, Gulf of Mexico coast off Veracruz Bay, Persian-Arabian Gulf, Thousand Islands on the Sunda shelf of Indonesia, Great Barrier Reef of northeastern Australia, and Shark Bay complex of Western Australia.

From these studies of Holocene shallow-water environments, the following sedimentary principles have been formulated:

1. Modern carbonates are formed mostly in the photic zone of warm, clear, tropical to subtropical marine water.

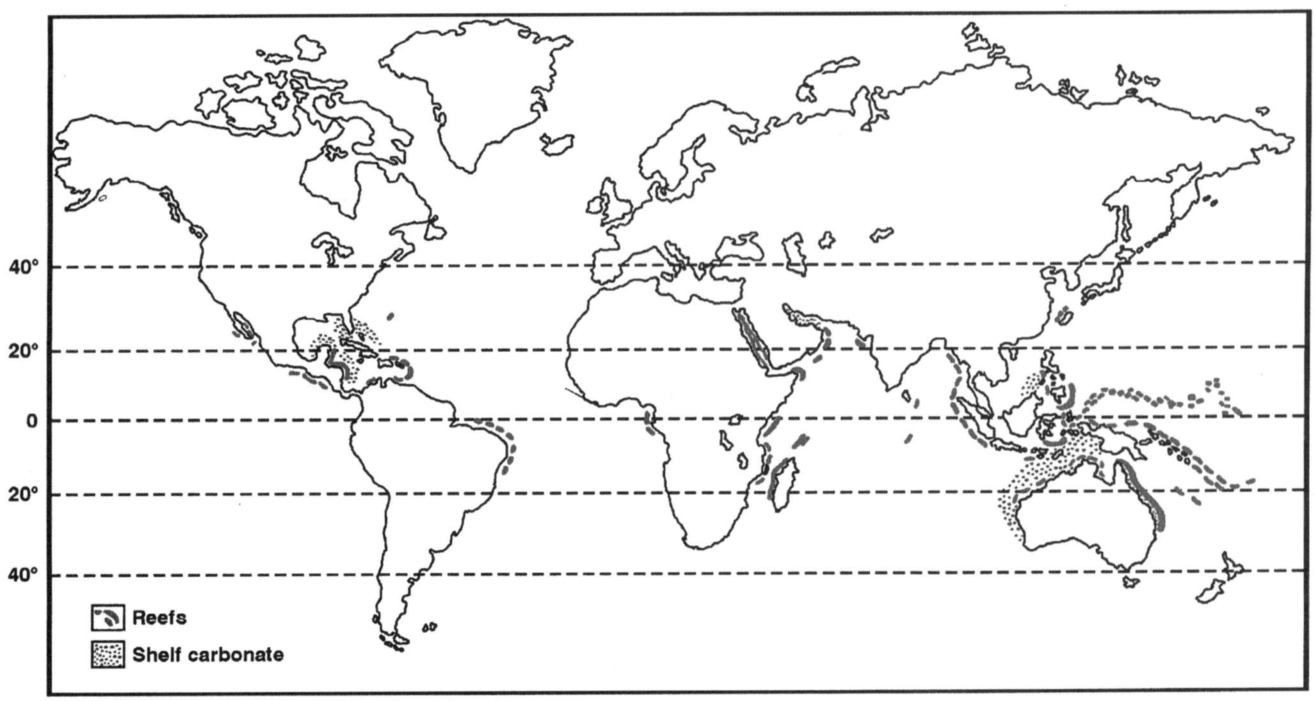

Fig. 1. Worldwide distribution of present-day shallow-water carbonate sediments. From Wilson (1975), after J. Rodgers.

2. Carbonate sediment is essentially formed in place, and intrabasinal factors control facies development. This contrasts with sandstone and shale sediments which are delivered into the basin from outside and form from point sources (e.g., deltas).
3. Basin shape and depth (controlling water energy) are the prime factors controlling sedimentary patterns. Mechanical processes can shape bodies of grainy carbonates (lime sands) as in clastics, but in addition, upwelling and water agitation on shelf breaks can induce massive organic carbonate accumulations (reefs). Likewise, vast low-relief areas of restricted circulation can trap fine-grained carbonate muds and lead to tidal flat and sabkha evaporite deposition.
4. Biological activity along with the shape and fragility of organic tests are of great importance in creating, sizing, and modifying carbonate particles. Organic diversity and abundance reflect original environmental conditions despite a bias caused by differences in preservation.
5. The controls on and significance of the texture of carbonate versus detrital clastic sediments are different. Carbonate particles are created in various sizes and shapes and are not necessarily mechanically sorted by depositional agents. Modern classifications of carbonate textures are based principally on the presence or absence of interpreted original lime mud matrix and grain types. The presence and amount of lime mud is considered a better guide to water energy and circulation than grain size or shape. Lime texture is more dependent on the nature of contributing skeletal producers than on external depositional agents. This realization has resulted in special terms for carbonate textures proposed by R. L. Folk (1959), R. J. Dunham (1962), and Embry and Klovan (1971) (Figures 2 and 3).
6. The rate of sedimentation can be moderately rapid for ancient carbonates, from 10 to 200 m/m.y. The growth potential of ancient carbonate platforms has been estimated by Schlager and Bosscher at 300–1000 m/m.y. (Schlager, 1992). For comparison, the rate of Mississippi delta Miocene sedimentation is 680 m/m.y. and the rate of deep-sea carbonate and siliceous deposition is 4–5 m/m.y. Modern rates for coral growth and tidal flat sedimentation are generally about 3000 and 1000 m/m.y., respectively, an order of magnitude higher than estimates for ancient rates. This leads to the conclusion that the most favorable environment for carbonates can produce vast amounts of sediment but that these environments are very sensitive and their sedimentation can be easily interrupted. The rate of modern carbonate deposition is much more rapid than that of tectonic subsidence or eustatic marine rise and fall (Figure 4). As a result, carbonate sedimentation is easily stopped by cooling down or muddying up the water, or by poisoning the environment with hypersalinity, but difficult to drown through normal long-term subsidence alone (Wilson, 1975; Schlager, 1992).

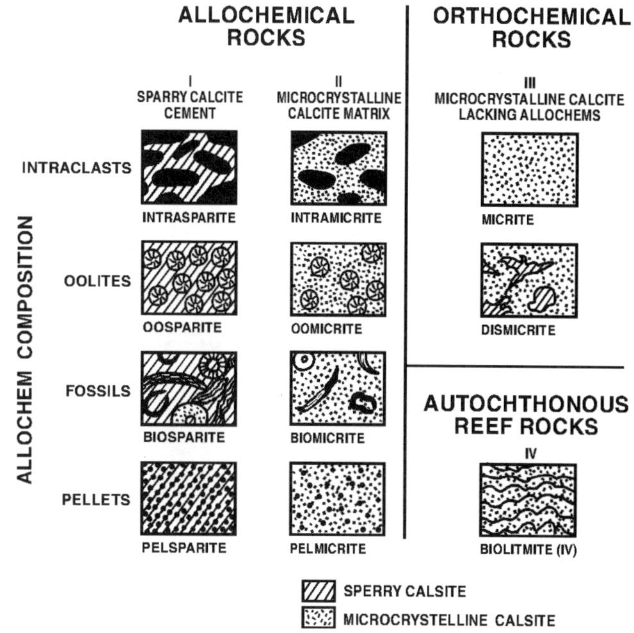

Fig. 2. R. L. Folk's (1959) classic textural classification of carbonate sediments.

Rimmed Platforms and Facies Model

The above rates of sedimentation result in two special characteristics of carbonate stratigraphy: the development of rimmed carbonate platforms and a useful model for facies prediction (Figure 5). These characteristics and platform architecture are discussed by Wilson (1975), Schlager (1981), Read (1982), Bosellini (1984), and Handford and Loucks (1993).

Rimmed platforms initially form on a gently inclined sea floor around a mildly positive tectonic element. This setting possesses a high-energy zone shoreward with a gradation outward to finer sediment, a pattern termed *carbonate ramp sedimentation.* On such a gently inclined sea bottom that deepens gradually away from land, a zone exists offshore where normal wave action drags the sea floor. Where the sea is of such depth and far enough from land to be uncontaminated by fine muddy sediment and fresh water, organic growth can be

<table>
<tr><th colspan="6">ALLOCHTHONOUS LIMESTONES
ORIGINAL COMPONENTS NOT ORGANICALLY BOUND DURING DEPOSITION</th><th colspan="3">AUTOCHTHONOUS LIMESTONES
ORIGINAL COMPONENTS ORGANICALLY BOUND DURING DEPOSITION</th></tr>
<tr><td colspan="4">LESS THAN 10% > 2mm COMPONENTS</td><td colspan="2">GREATER THAN 10% > 2mm COMPONENTS</td><td rowspan="4">BY ORGANISMS WHICH ACT AS BAFFLES</td><td rowspan="4">BY ORGANISMS WHICH ENCRUST AND BIND</td><td rowspan="4">BY ORGANISMS WHICH BUILD A RIGID FRAMEWORK</td></tr>
<tr><td colspan="3">CONTAINS LIME MUD (<.03mm)</td><td>NO LIME MUD</td><td rowspan="3">MATRIX SUPPORTED</td><td rowspan="3">> 2mm COMPONENT SUPPORTED</td></tr>
<tr><td colspan="2">MUD SUPPORTED</td><td colspan="2" rowspan="2">GRAIN SUPPORTED</td></tr>
<tr><td>LESS THAN 10% GRAINS (>.03mm–< 2mm)</td><td>GREATER THAN 10% GRAINS</td></tr>
<tr><td>MUD - STONE</td><td>WACKE - STONE</td><td>PACK - STONE</td><td>GRAIN - STONE</td><td>FLOAT - STONE</td><td>RUD - STONE</td><td>BAFFLE - STONE</td><td>BIND - STONE</td><td>FRAME - STONE</td></tr>
</table>

Fig. 3. Carbonate textural classification of Embry and Klovan (1971), amplifying Dunham's (1962) classification.

very rapid. Here the beginning of a raised organic rim is built. This may develop with gradual subsidence into a vast carbonate platform with a rimmed border, an extensive backreef, and a seaward-facing slope. The bordering basin or shelf has much slower sedimentation rates and thus may be "starved" (Figures 4 and 5).

Another consequence of platform construction and rapid sedimentation potential is that many carbonates produced during high stands of sea level are regressive in pattern. The construction of paleotopography by the normal process of carbonate sedimentation also results in a well-established facies model that is useful in predicting changes in rock types and is thus important in exploration. This facies model is illustrated in Figure 5 (from Wilson, 1970, 1974, 1975) and consists of nine belts, not all of which may be present, but which always follow the same order:

1. **Deep basin**—euxinic or evaporitic
2. **Open sea shelf**—offshore of rimmed platform
3. **Deep shelf margin**—toe of slope or distal slope off platform
4. **Foreslope**—proximal part of slope off platform
5. **Organic buildup**—ecological reef along platform margin
6. **Winnowed sands**—lime grainstone on platform edge
7. **Open shelf lagoon**—normal marine environment across platform top or middle shelf
8. **Restricted circulation shelf and tidal flats**—on platform or cratonic interior
9. **Evaporites on sabkhas and salinas**—evaporites and saline ponds in platform interior

Diagenesis

Carbonate sediment, once formed, is very susceptible to *diagenesis,* the term for low-temperature rock-fluid reactions under overburden load, excluding the higher temperature and pressure of metamorphism. Three aqueous diagenetic environments can be distinguished. As it lithifies, carbonate sediment can be affected by all of them in turn: (1) Shallow-marine depositional diagenesis, (2) meteoric or groundwater diagenesis in the vadose and phreatic zones, and (3) postburial diagenesis where water movement is slower but temperature and pressure are higher.

Changes begin in loose grains of sediment before and during deposition and continue on the sea floor and within the shallow substrate (Longman, 1980; James and Choquette, 1983). Significant solution and mineralogic changes occur as soon as carbonate sediment is removed from the influence of sulfate and magnesium in marine water and saturated with fresh water under sub-

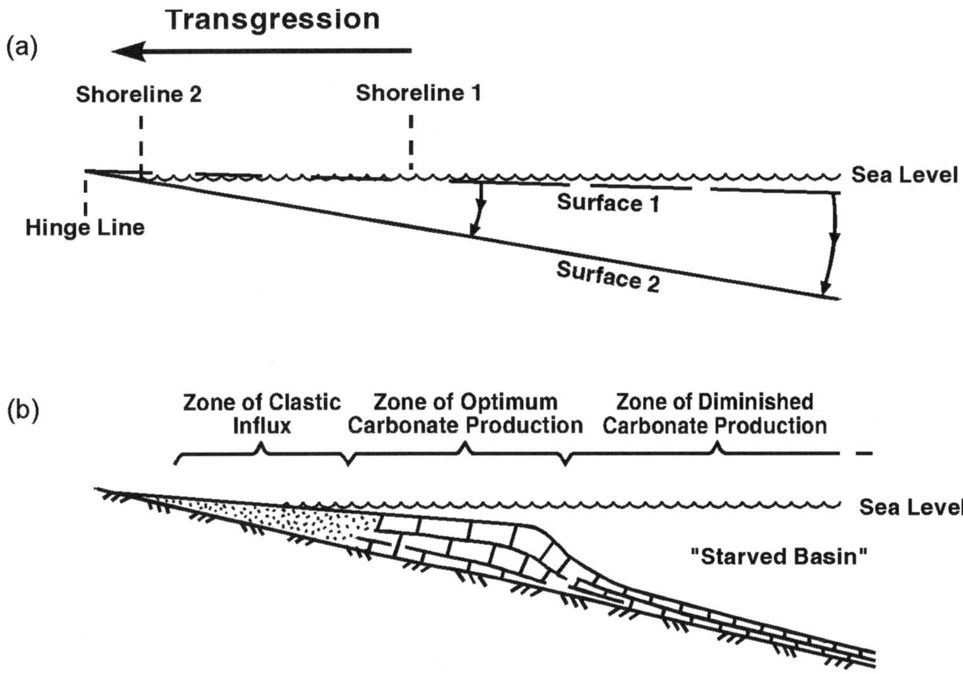

Fig. 4. Schematic cross-sections showing (a) transgression caused by differential subsidence and its relationship to depositional surfaces, and (b) creation of shelf to basin topography. From Wilson (1975), after Meissner (1972).

XIII-3

	Basin	Open sea shelf	Deep shelf margin	Foreslope	Organic build up	Winnowed edge sands	Shelf lagoon open circulation	Restricted circulation shelf and Tidal Flats	Evaporites on Sebkhas-salinas
	1	2	3	4	5	6	7	8	9

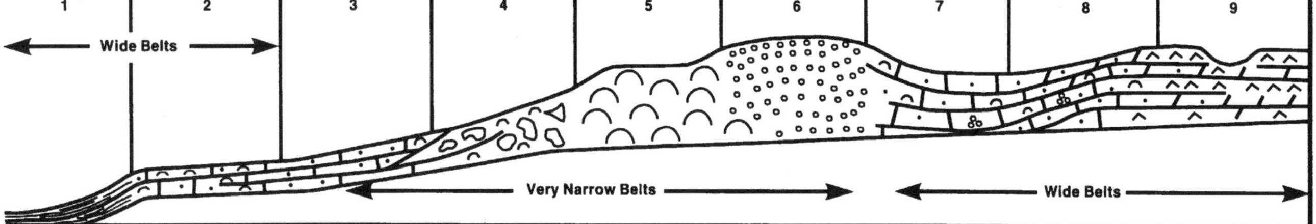

	1	2	3	4	5	6	7	8	9
2nd Order Sedimentary Bodies		Ball and flow	Debris flows & turbidites in fine laminate strata Mounds on slope toe	Giant talus blocks Infilled large cavities Downslope mounds	Downslope mounds Reef knolls Boundstone patches Fringing and barrier frame work reef Spur & groove	Islands Dunes Barrier Bars Passes and channels	Tidal deltas Lagoonal ponds Typical shelf mounds Columnar algal mats Channels and tidal bars of lime sand	Tidal flats Channels Natural levees Ponds Algal mat belts	Anhydrite domes Tepee structures Laminated crusts of gypsum Salinas (evap. ponds) Sebkhas (evap. flats)
Standard Microfacies	1 Spiculite 2 Microbio-clastic calcisilt 3 Pelagic micrite Radiolarite shale	2 Microbioclastic calcisilt 8 Whole shells in micrite 9 Bioclastic wackestone 10 Coated grains in micrite	2 Microbio-clastic calcisilt 3 Pelagic micrite 4 Bioclastic-Lithoclastic microbreccia	4 Bioclastic-Lithoclastic microbreccia Lithoclastic conglomerate 5 Bioclastic grainstone-packstone Floatstone 6 Reef rudstone	7 Boundstone 11 Coated-worn bioclastic grainstone 12 Coquina (shell hash)	11 Coated-worn bioclastic grainstone 12 Coquina (shell hash) 13 Onkoidal bioclastic grainstone 14 Lag breccia 15 Oolite	8 Whole shells in micrite 9 Bioclastic wackestone 10 Coated grains in micrite 16 Pelsparite 17 Grapestone onkoids in micrite 18 Foram-Dasy-cladacean grainstone	16 } 17 } 18 } 19 Fenestral peloidal laminate micrite 20 Stromat-olitic micrite 24 Rudstone in channels 21 Spongiostrome micrite 23 Non laminate pure micrite 22 Onkoidal micrite	20 Stromatolitic micrite 23 Non laminate pure micrite Nodularpearl enterolithic anhydrite Selenite blades in micrite

Fig. 5. Descriptive synopsis of idealized facies belts of a rimmed carbonate platform. From Wilson (1975).

aerial and meteoric conditions (Longman, 1980; James and Choquette, 1984). Original marine carbonate sediment is composed principally of aragonite and high-Mg calcite (as high as 15 mol %) and to a lesser degree of low-Mg calcite (<4 mol %). Composition is largely controlled by organic composition. Aragonite and high-Mg calcite are unstable in fresh water and quickly dissolve or alter to ordinary calcite. Lake water is usually fresh and deposits low-Mg calcite. Other changes occur in carbonates when evaporitic brines and other types of connate waters migrate through limy strata both before and after deep burial and during various stages of lithification (Longman, 1980; Choquette and James, 1987). One of these processes is *dolomitization,* an important metasomatic replacement adding Mg to the $CaCO_3$ and generally forming rhombic crystals. Dolomitization rearranges and enlarges pore space because of attendant spaces between rhombs and solution of residual calcitic shell debris. When the rock is completely cemented by dolomite, its density increases.

Other diagenetic changes include organic rotting of particles, solution and enlargement of existing cavities, cementation of pore space in lime sands, crystal rearrangement (neomorphism), and metasomatic replacement by anhydrite, which makes the sediment more susceptible to later dissolution. All these processes affect to some degree the reservoir quality, rock density, and velocity of sound waves passing through the rock. Cemented grainstone or reef rock is a better conductor than porous uncemented sediment (Eberli and Ginsburg, 1989).

Research is still ongoing in the prediction of diagenetic sequences and geographic trends of diagenesis in carbonate strata. The application of isotope geochemistry, trace element content, cathodoluminescence, fluorescence, and fluid inclusion studies to carbonate petrography is increasingly important in attempting to relate real rock to petrophysical logs and seismology. The ultimate goal is to predict more accurately reservoir character, shape, trend, and geologic history using these indirect tools (Asquith, 1982).

Description of Depositional Environments of Carbonate Strata

A massive compendium, *Carbonate Depositional Environments* (Scholle et al., 1983, AAPG Memoir 33), describes and illustrates in color all the depositional environments known for carbonate rocks. The following summary is derived chiefly from this volume and also from Wilson (1975) and Jordan and Wilson (1993). The following are given for each environment: (1) rock types, (2) sedimentary structures, (3) organic composition, (4) rate of sedimentation, (5) geometry and scale, and (6) some idea of porosity and density resulting from diagenesis.

Deep Basin (Facies 1)

Basinal pelagic sediments are deposited in oceanic waters or in deep, narrow basins within orogenic belts. Their composition depends partially on depth because calcium carbonate is partially soluble in cold, deep water, leaving only siliceous sediment (such as radiolarite or red mud) on the sea floor. In areas of great carbonate productivity by pelagic organisms and of moderate depth (less than 5000–5500 m), ultra fine-grained calcitic organic carbonates (coccolith ooze, pteropods, foraminifers, and calcispheres) can be deposited and preserved on the ocean floor. In anoxic troughs and within intrashelf basins, where bottom water is stagnant and abundant silty or argillaceous mud is introduced, organic-rich black shales may be formed in this deep-water environment.

Texturally, these sediments are mud or silt (particles <62 mm in diameter). Strata characteristically have a uniform rhythmic bedding of thin, compacted marls or shale and less compacted carbonate layers that are three to four times thicker. The carbonate layers range from 10 to 40–50 cm thick (Figures 6, 7, and 8). The definitive sedimentary structure is planar fine lamination of millimeter thickness or less. These rocks commonly possess layers of bedded chert derived from solution and reprecipitation of silica from organic opaline tests.

Basinal pelagic sediments are deposited slowly in trough or oceanic environments, from 1 to 4 m/m.y. Organisms are almost exclusively pelagic (floating) or nektonic (swimming), and in carbonate areas shallower than the carbonate compensation depth, coccolith oozes containing scattered larger particles such as foraminifers and pteropods are common. Radiolarites made of altered siliceous radiolarians also occur. These black shaly lime mudstones deposited in nonoxidizing environments are known to be petroleum source rocks. They contain from 1 to 15 wt. % organic carbon consisting of several types capable of producing oil or gas. Source rocks are developed on intrashelf basins such as in the Upper Jurassic of the Arabian Shield, the Upper Jurassic of the northwestern Gulf of Mexico, the Cretaceous of Maverick Basin on the south Texas Gulf Coast, and La Luna of Venezuela. The lack of permeability in such strata also makes them good seals, inhibiting both vertical and lat-

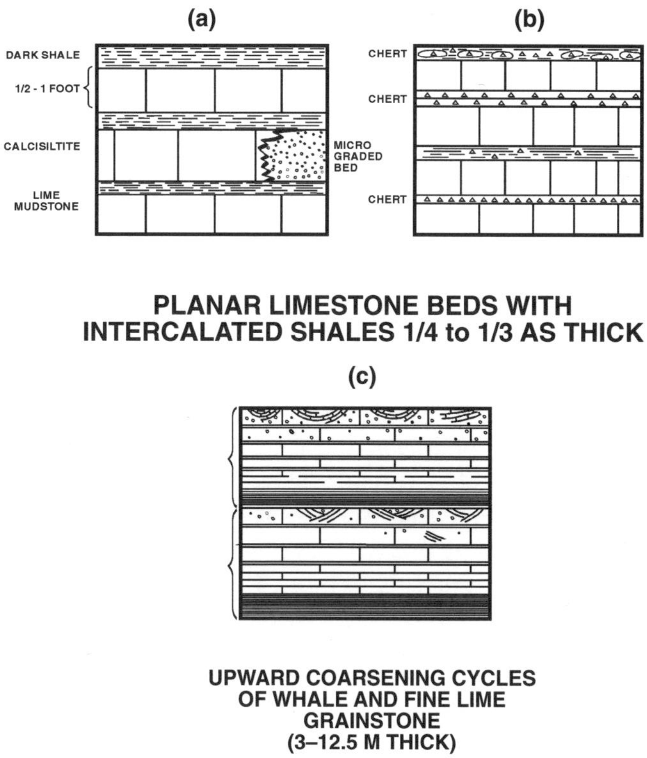

Fig. 6. Core sketches of various types of bedded rhythmites with normal carbonate bed thickness of 25–75 cm. (a) and (b) Planar limestone beds with intercalated thin shales. (c) Upward-coarsening cycles of shale and fine lime grainstone. From Wilson (1969).

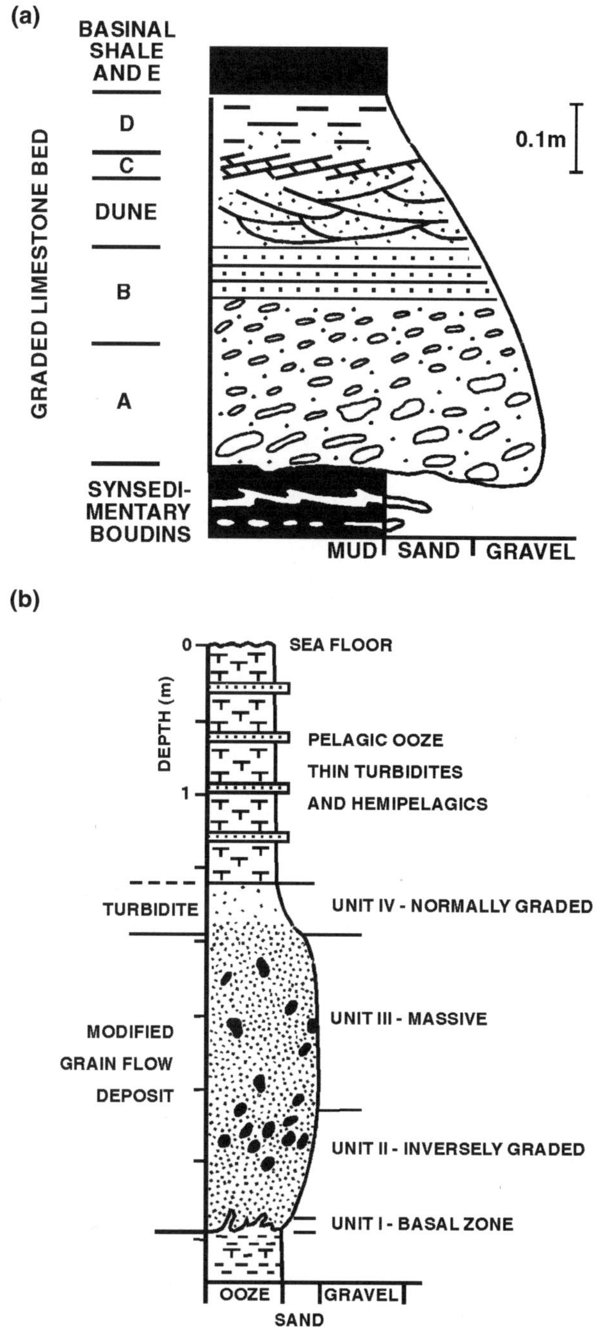

Fig. 7. Sedimentary sections of vertical turbidite sequences. From Tucker and Wright (1990). (a) Idealized Bouma sequence. (b) Holocene core off Little Bahama Bank, described by Mullins and Van Buren (1979).

eral petroleum migration. We thus recognize carbonates as key components of several facets of the petroleum system (see conclusions and Figure 17).

These sediments can be extremely dense, but chalk-textured mudstones with finely disseminated microporosity, when preserved from compaction by overpressuring and early oil migration, make excellent reservoirs. This is particularly true when the homogeneous fine-grained rock is fractured, such as in the Ekofisk Danian of the North Sea and the Austin Chalk of Texas.

Note that Facies 2 is discussed with Facies 7 because they have the same lithology.

Deep Shelf Margin (Facies 3): Toe-of-Slope or Basin Margin

Toe-of-slope, or deep shelf margin, environments lie at the foot of large carbonate platforms or rimmed shelves that rise from relatively deep basins or from open marine, moderately deep water covering extensive shelves behind them. This environment is complex and includes displaced resedimented material as well as sediments formed *in situ* or in the water column directly above. Pelagic or semipelagic beds (fine sediment filtered down from near the water surface), debris flows and rock slides (gravity deposits), carbonate turbidites (products of moving, dense, soupy mass flows of suspended material), and grain flow deposits (movement of sand-sized grainstones caused by dispersive pressure)

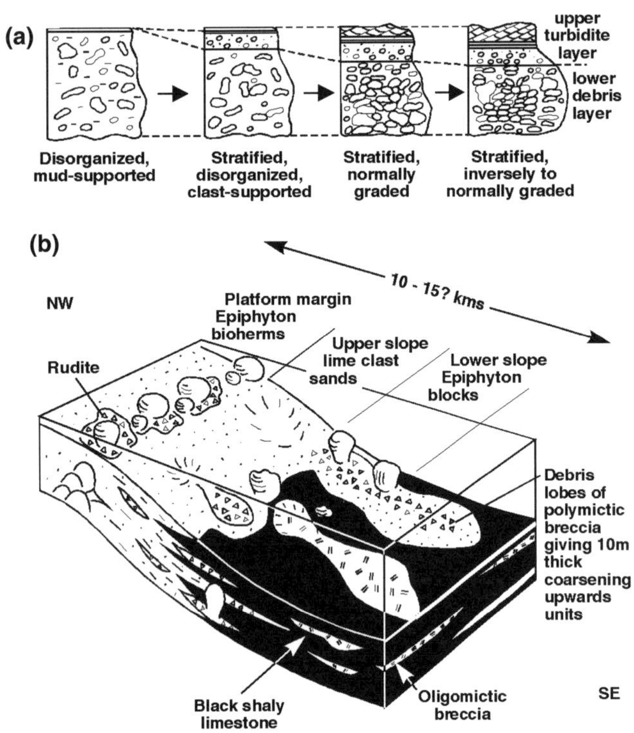

Fig. 8. (a) Sketches of composite lateral sequences of debris flows and overlying turbidites. From Tucker and Wright (1990). (b) Model of a Cambrian shelf margin from the Appalachians. After Pfeil and Read (1980).

are all present (Figure 8). In addition, certain organic buildups are known far down slopes off platforms. These may be organic accumulations of deeper water organisms forming knoll reefs or current-shaped bodies of sediment (lithoherms).

Textures and sedimentary structures in toe-of-slope deposits are varied, ranging from fine-textured, dark to reddish, laminated pelagic muds, to graded turbidite beds with Bouma sequences (Figure 7), inversely graded sand-sized grain flow sediment, and to very coarse conglomerates and breccias representing debris flows (Figure 8). Individual exotic blocks occasionally rest within thin-bedded finer sediments.

The biota is mixed with displaced fossils from organisms living upslope and with some deeper water bottom-dwelling forms such as bryozoans and crinoids (in places as reddish coquinas). There may also be layers of pelagic microorganisms and in-situ buildups (bioherms or knoll reefs) composed of colonies of echinoderms, sponges, and certain corals. Characteristic assemblages of trace fossils occur, and it is possible to distinguish basin assemblages from those of sedimentary slopes.

Bedding is rhythmic and cyclic in places, jumbled in other localities, and may be interrupted by major channels both cut into and filled with rhythmic-bedded limestone. In addition, major slide blocks may occur within such rhythmites. Abrupt changes in dip occur between the individual slide blocks without evidence of any soft sediment deformation.

The best chance for reservoir porosity and permeability is afforded by large voids that have been only partially filled with cement. These occur between clasts in debris flows and grain flow deposits. Preferential solution and dolomitization of the lower portions of slope deposits can also occur. Major downslope debris reservoirs exist in the Mesozoic of Mexico (Poza Rica field, Enos, 1985; and fields in offshore Campeche) and in the Lower Permian of the West Texas Permian Basin (Mazzullo and Reid, 1989).

Foreslope off Rimmed Platform (Facies 4)

The foreslope environment off the reef or rimmed platform is found farther upslope than that of Facies 3. These sediments have many of the characteristics of downslope deposits, but lack the interstratified pelagic rhythmites and have abundant fossil debris of normal marine organisms. The sediments vary in texture, but are commonly composed of bioclastic debris, mostly grainstone–packstone, but in places mixed with abundant lime mud matrix (wackestone). Lithoclasts are present in places. Most sediment is gravity flow in origin, and much mixing with fine material in suspension occurs. Exotic blocks, debris flows, and proximal turbidite structures are characteristic.

Bedding tends to be thick, irregular, and steeply dipping. The angle of repose seems to be about 25–30° for forereef sediments, which are commonly (but not always) progradational (Figure 9) and formed on the seaward side of rimmed carbonate platforms. Sedimentary structures include coarser phases of Bouma sequences, extensive channels cut out of slopes, preserved submarine cliffs cut into the slopes, and truncation surfaces that may result from sliding of large masses of lithified sediment. Thus, three types of platform foreslopes are known: progradational, by-pass, and channelized (Read, 1982).

Fossil debris is a major lithologic component of foreslope sediments. It characteristically represents abundant normal marine organisms living on the upper slope and outer fringes of the rimmed platform.

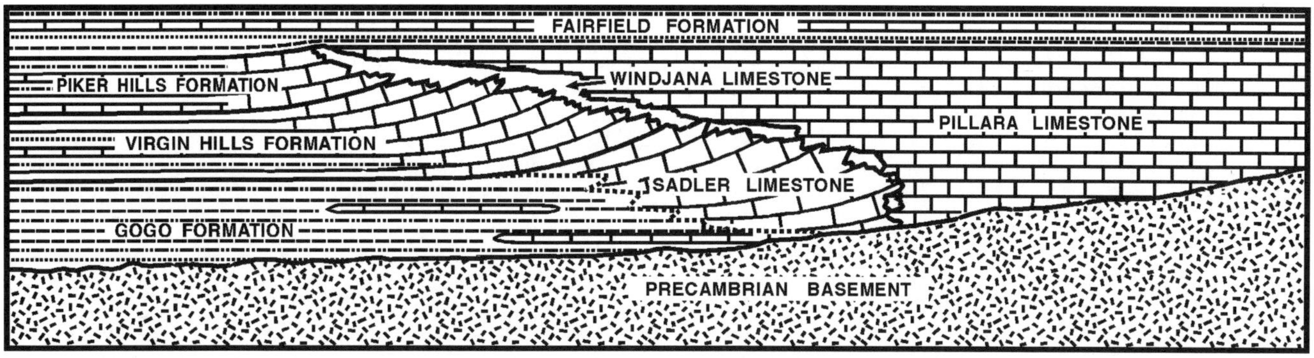

Fig. 9. Schematic section of a shelf margin in the Middle–Upper Devonian Canning Basin reef trend, Hull-Emanuel Range, showing expanding platform with reef front rising above strata on the prograding slope. From Playford and Lowry (1966).

The rate of sedimentation is relatively high (estimated to be about 20–50 m/m.y.) and is indicated by the common occurrence of prograding clinoforms off the shelf margin. Relief on such platforms may approach 1000 m.

As for the contiguous belt of ecologic reef located higher upslope, the upper foreslope may possess some enlargement of already existing intergrain porosity by meteoric solution and enhancement of permeability through dolomitization. The large amount of aragonite in reef and perireef deposits is important because of its solubility in fresh water. Conversely, surficial marine cementation is also common in upper slope deposits. Early lithification in the shallow parts of forereef deposits is widely recognized in the Holocene of Belize (James and Ginsburg, 1979). The cemented, highly dense layer at the outer surface of foreslope deposits should make for good velocity contrast (impedance). Major petroleum reservoirs are known in forereef deposits of Poza Rica and Campeche offshore in Cretaceous strata of Mexico. Porosity in deposits results from fracturing and both dolomitization and meteoric solution, enhancing original pore space.

Organic Buildup (Facies 5)

Organic buildups or ecological reefs are composed of a substantial amount of calcium carbonate framework formed in place within the tissues of various marine organisms. As such, the carbonate buildup it produces creates its own environment and influences surrounding sedimentation. Organic reefs can thus be distinguished from other types of carbonate buildups such as lime sand bars, piles of lime mudstone, and mud mounds caused by the baffling action of organisms accumulating piles of lime mud. Some of these mounds are cemented and stabilized by large amounts of marine aragonite and calcite.

Reefs have been studied extensively both in modern environments and in the geologic record because they are hosts for a disproportionately large amount of petroleum and economically significant metallic sulfides. Reefs make good reservoirs because (1) they commonly build up to form their own structural traps, (2) they are commonly composed of aragonite, which is subject to leaching in the meteoric environment, and (3) they are initially porous enough to be easily dolomitized. They form massive bodies of rock that are prone to fracturing instead of taking stress along bedding planes. Some reefs are completely dense, particularly when filled with lime mud or secondarily cemented with halite, anhydrite, or calcite.

Reefs are commonly classified geographically into fringing, patch, barrier, and deeper water knoll reefs. They also have internal subdivisions, including immediate backreef, reef flat, reef crest, reef front (facing the swells and waves), and forereef slope (Figures 10 and 11). Many reef complexes in the geologic record consist of pods of organic framework with much interreef debris (Figure 12) or flanking beds (Figures 10 and 11). Many modern reefs worldwide contain a near-surface surf zone fringe of the fast-growing coral *Acropora.* Ancient reefs consist more often of pods of framework growing in slightly deeper water just at wave base. One can distinguish the various subenvironments of reef complexes by slightly irregular, wavy bedding in reef flank and interreef beds and by the lack of bedding in massive pods of reef framework.

Because ecologic (framework) reefs are huge colonies of colonial organisms, their biological composition changes during geologic time so that their position, abundance, internal make up, and diagenetic susceptibil-

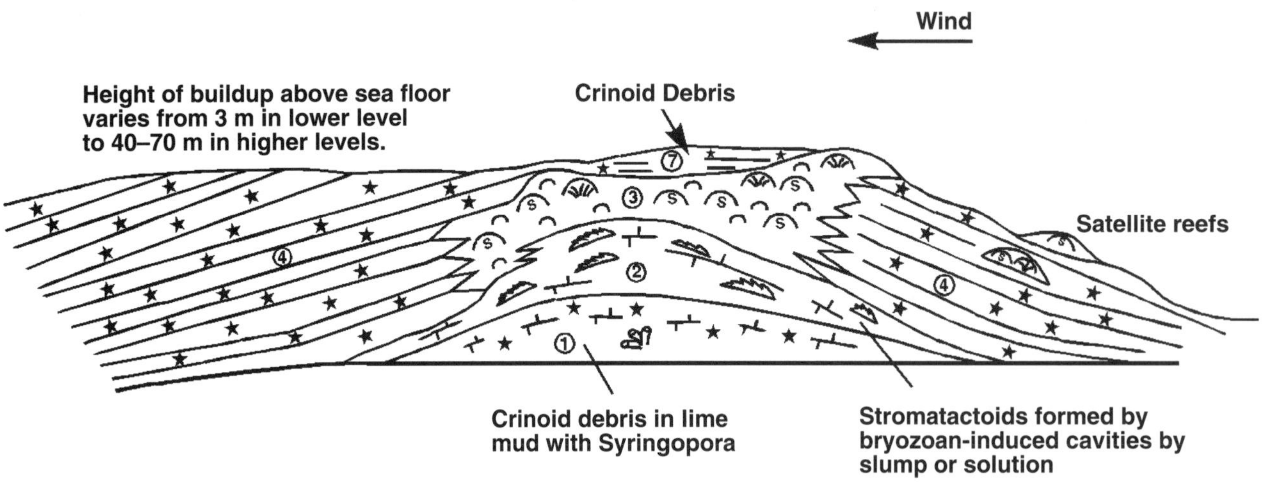

Fig. 10. Idealized cross-section of a Middle Silurian isolated reef. From Wilson (1975). Key to circled numbers: (1) Basal bioclastic pile of initial accumulation; (2) micrite bafflestone core, mud mound buildup below wave base; (3) crestal boundstone and frame-built organic reef; (4) flank beds of crinoids derived from reef top dwellers; (5 and 6) talus and organic veneer not recognized; (7) capping crinoidal grainstone.

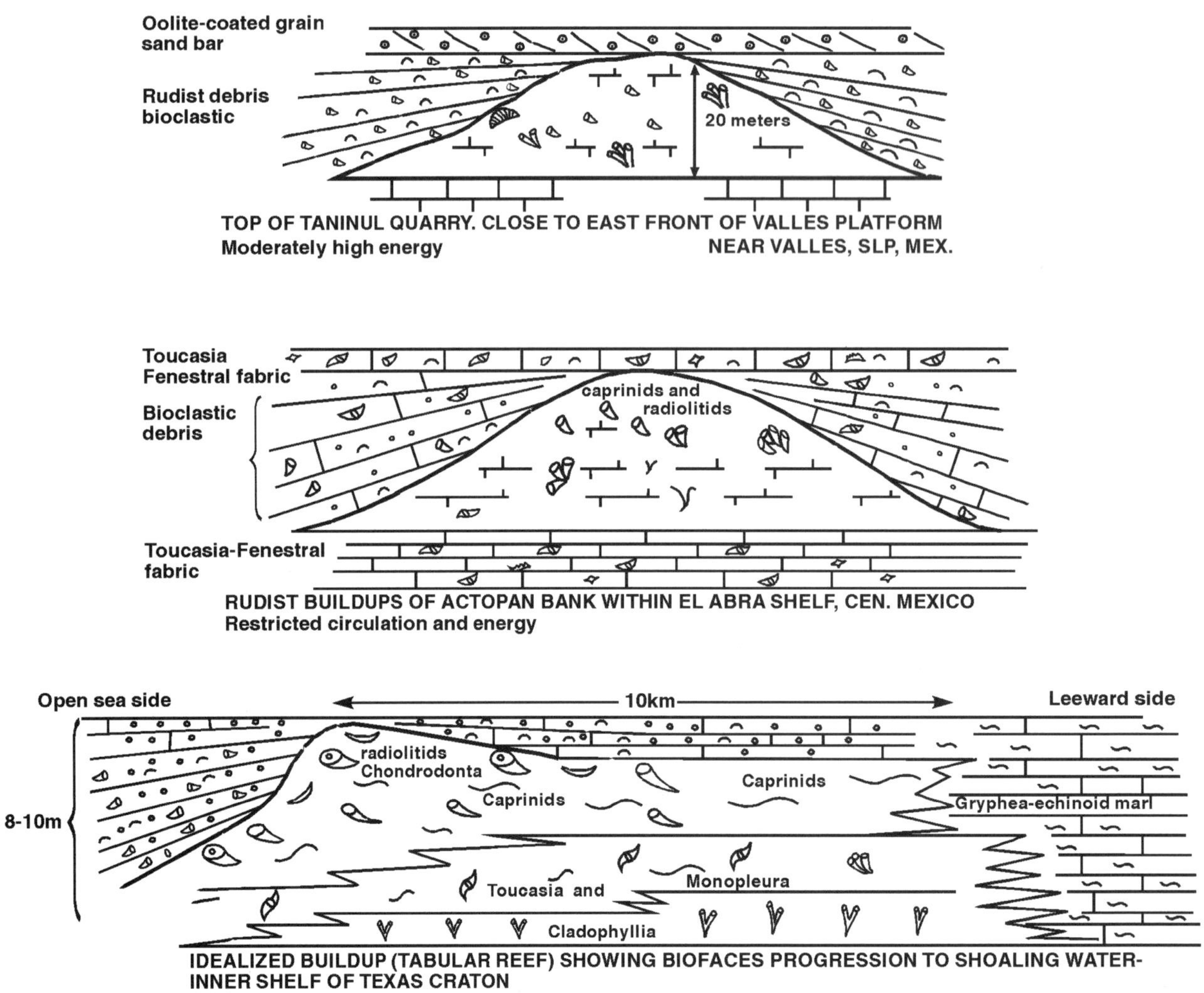

Fig. 11. Facies in three rudist mounds on shelves or along gently dipping shelf margins in the middle Cretaceous of Texas and Mexico. From Wilson (1975).

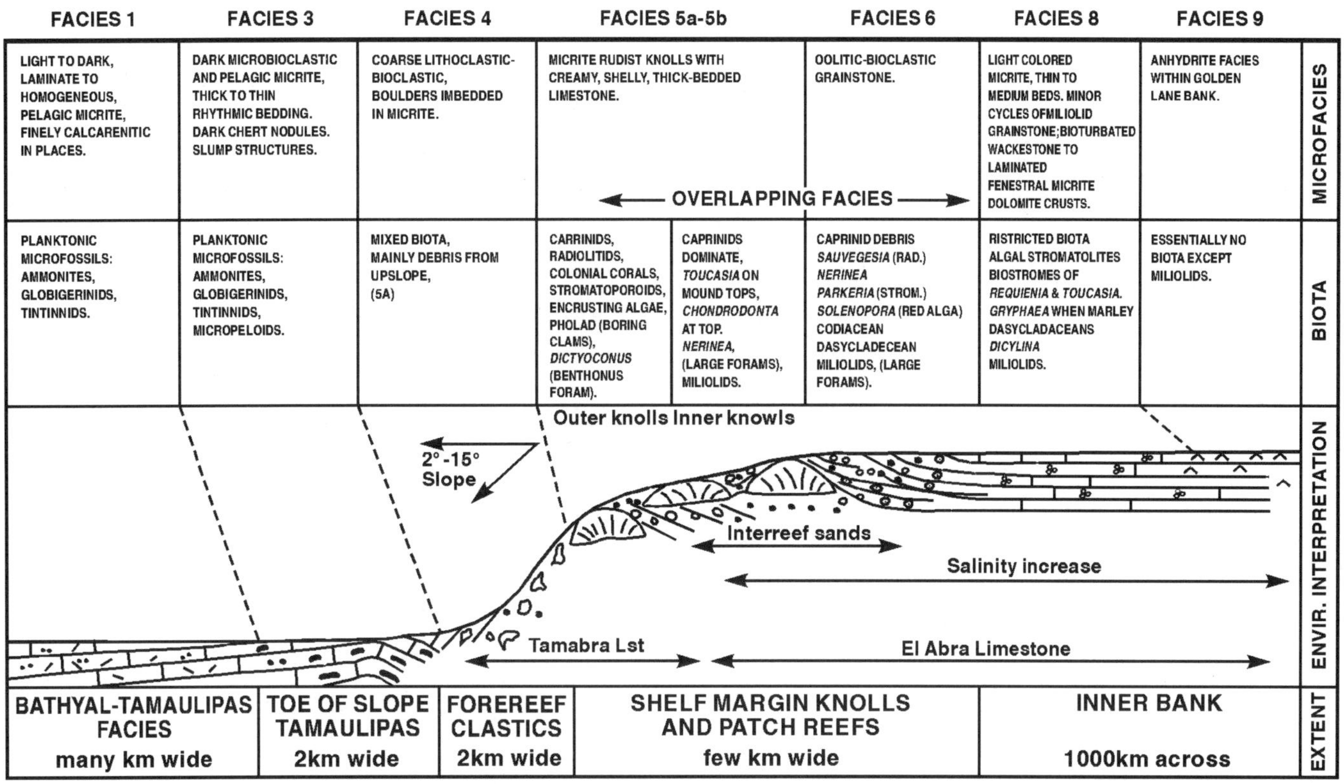

Fig. 12. Idealized cross-section of middle Cretaceous facies across large offshore banks in central Mexico, such as the Golden Lane and Poza Rica areas. From Wilson (1975).

ity also vary in different periods of geologic history. Growth habits of organisms also vary and have different responses to varying marine conditions, thus creating organisms that are encrusting, massive, branching, or platy in growth form depending on depth, water energy, and turbidity. The principal framebuilders of reefs are corals, sponges, stromatoporoids, bivalves, crinoids, and even brachiopods, which may contribute to the organic frame. Red algae such as the modern *Lithothamnion* are important binders in the framework.

Natural cavities overgrown by reef frames are in places filled with drusy, fan-shaped aragonite cement or with internal laminated sediment visible in both modern and ancient reefs. The constant surge of water through reef frames brings finer sediment that becomes trapped along with organic debris. This material, along with large amounts of marine cement, helps stabilize the reef mass. The rate of such sedimentation in carbonate buildups is the fastest known. It is about 3000 m/m.y. in modern reefs and approaching 100–300 m/m.y. in ancient examples.

Reef masses at shelf margins and large isolated bioherms are comparatively easy to recognize seismically. Foreset clinoforms seaward of shelf margin buildups are often visible at seismic scale, and the impedance contrast of the tops of massive carbonate reef rock, coupled with irregular or shattered patterns within biohermal buildups, are well known, such as in the Silurian bioherms of the Michigan Basin (Caughlin et al., 1976).

Winnowed Sands on Platform Edge (Facies 6)

Lime grainstones (carbonate "sands") are common on the edges of carbonate platforms where they form belts of sand bars shaped primarily by tidal action. They can be found in place of the reef trend (Facies 5) along the shelf margin crest forming both submarine bars and dune islands, or they may occur just behind a fringing or barrier reef belt. Coated grainstones (ooids) form best where tidal action is strongest, enhanced by upwelling against a rise in submarine topography and in places where both geographic and topographic constriction causes stronger tides. Individual bars within the belt of tidal bars are oriented normal to the direction of the shelf margin, but the belt as a whole parallels it. The grains are made in these tidal "factories" by persistent but sporadic coating of smaller particles that can be moved regularly by tidal currents, buried, and then reexcavated. Such ooid particles can form up to 1.5 mm in diameter in less than 1000 years.

Other settings for grainstones include (1) coastal dunes and spits shaped by a combination of wave action and longshore drift, (2) flanking beds to shelf interior patch reefs, (3) sand accumulations associated with uplifted structures on shallow shelves, (4) sand banks parallel to and interspersed with shelf marginal reefs, and (5) great spillover masses of lime sand swept across a carbonate bank and deposited downslope on its leeward side.

All these sand bodies consist of well washed, often beautifully sorted grainstones with rounded, coated, and partially rotted skeletal and peloid particles, but the best formed ooids are those subjected to constant tidal movement, such as the belts of bars at platform edges. Particle size is strikingly uniform, ranging on average from 1 to 2 mm. The dominant sedimentary structure is cross-bedding, which can form large-scale foresets from tidally induced spillover lobes at ends of bars or, at a smaller scale, from strong ripples caused by persistent wave action modifying bar surfaces. Modern oolite bars can be as high as 7 m. Grains may be partially cemented at bar surfaces, which are intermittently in and out of water. Since bars build up to sea level and migrate, they result in vertical sequences with upward coarsening texture.

Most sand bars are biological deserts because the substrate of shifting sand makes it impossible for most bottom-dwelling lifeforms to exist on the bars. It can survive only within the intervening channels. Rates of sedimentation for modern lime sands are as rapid as for tidal flat sedimentation, about 1000 m/m.y., with rates for the ancient record being more than an order of magnitude less.

Porosity and permeability in lime sands depends chiefly on the types and amount of cementation under marine, burial, or meteoric conditions. Cementation occurs in stages. Initial porosity is often modified by marine cementation in the form of circumgrain fringes to about 20–25%. Initial porosity of uncompacted spherical grains is about 40%. A pack of a variety of irregularly shaped, multi-sized grains can have porosity of 75–80% for a given volume. Upon uplift, this porosity may be either enhanced by meteoric leaching within and between grains or completely obliterated by a later stage of cementation if the sediment has remained too long in the phreatic (water table) zone. Dolomitization of a grainstone usually enhances permeability significantly. The largest and most productive reservoirs in the world, the Arab D and C zones of Saudi Arabia, produce from skeletal, peloidal, and oolitic grainstones that have undergone little modification by late-stage cementation. About 45% of the hydrocarbons produced from carbonates comes from this type of rock, mostly in Saudi Arabia (Wilson, 1980) (Figures 13 and 14).

Open Sea Shelf and Shelf Lagoon with Open Circulation (Facies 2 and 7)

Rimmed carbonate platforms with well-differentiated facies can rise from deep-water basins present in orogenic or rifted areas or from shallow (~100 m deep) neritic shelves developed well within the craton when the latter is covered with water of sufficient depth to permit open-marine circulation. These environments constitute Wilson's Facies 2 and are often called open sea (or neritic) shelves. Similar conditions of moderate depth and open-marine circulation can also be found over isolated carbonate platforms if the marginal barrier rim has passes through it that allow open sea water to cover the platform at sea level highstands. This is Wilson's Facies 7, shelf lagoon with open circulation, which is also referred to as open marine platform by Wilson (1975) or as middle shelf by Wilson and Jordan (1983). Although geographically different, sediments of open sea shelves and open marine platform interiors have essentially identical lithologies. Such sediments have the following general characteristics:

1. Deposition in sea water, mostly of normal salinity
2. Depths from a few meters to tens of meters, but rarely more than 100–200 m (within photic zone)
3. Temperatures of 10–30°C (tropical to subtropical)
4. Well-oxygenated water

Open shelf (Facies 2) sediments generally have lime mud textures with dominant wackestone and packstone, although patch reefs and sand shoals also occur, producing local accumulations of boundstone and skeletal grainstone. Bedding is of variable thickness with lenses or wedge-shaped strata common. The sediments are commonly argillaceous, and thin gray or green shale beds may interrupt sequences of shelf limestone or dolomite. Sedimentary structures such as extensive bioturbation (mostly by arthropods), preserved burrow forms, nodular beds, and wispy flaser bedding are typical. Hardground surfaces are also common.

The biota of open shelf seas is generally adapted to normal marine salinity (30–38 ppm). Such animals and plants are said to be stenohaline. These include red algae, corals, echinoderms, cephalopods, many bivalves, and certain types of foraminifera, brachiopods, and bryozoans.

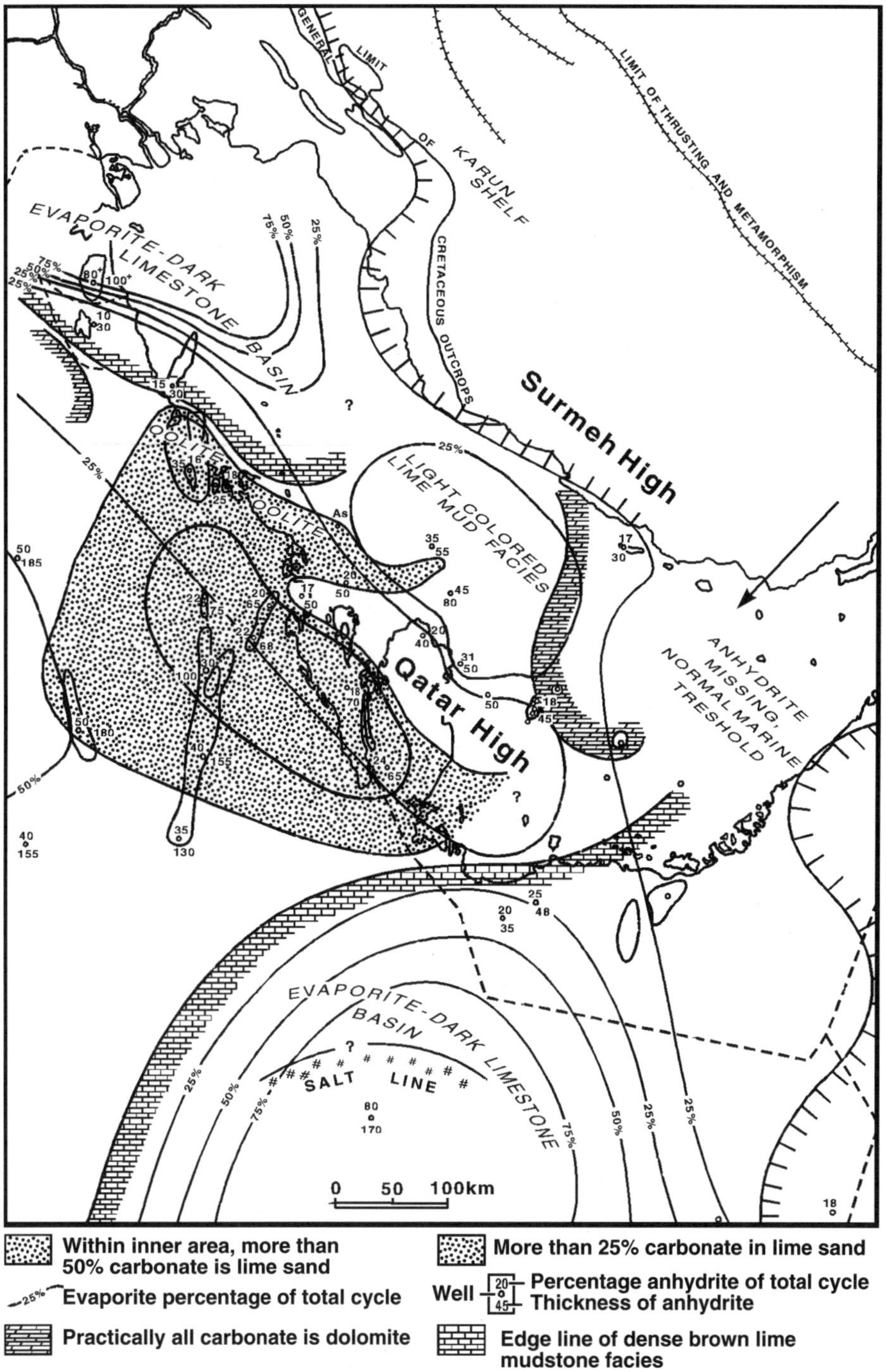

Fig. 13. Facies and isopachous map of Upper Jurassic Arab D carbonate-evaporite and grainstone-rich cycle in Saudi Arabia. From Wilson (1975). Note extensive thickness of lime sand across shelf in large area separating two basins to the north and south and bordered on the west by Arabian shield and on the east by Qatar-Surmeh highs. Oolite is best developed facing north into Basrah Basin.

Sedimentation rates are moderate in the open sea shelf environment, but not as rapid as on the shelf margin which often forms an island rim. As in restricted marine shoreward tidal flats, modern rates of 1000 m/m.y. have been measured. Examples from the geologic record are one or two orders of magnitude slower. Two examples illustrate the minimal rates possible in the open sea shelf environment. First, several banks in the Bahamas are rimmed by islands with Holocene eolianite as much as 10 m above sea level, whereas areas of open circulation in the bank interior are about 10 m deep and covered with thin peloidal sands all deposited within the last 5000 years. The second example is the western Florida shelf of the eastern Gulf of Mexico, which has been covered by the postglacial sea for at least 10 000 years. Water is as deep as 200 m. Although much fine sediment (both clays and carbonates) is piled against the shore, the shelf has so little Holocene sediment that Pleistocene

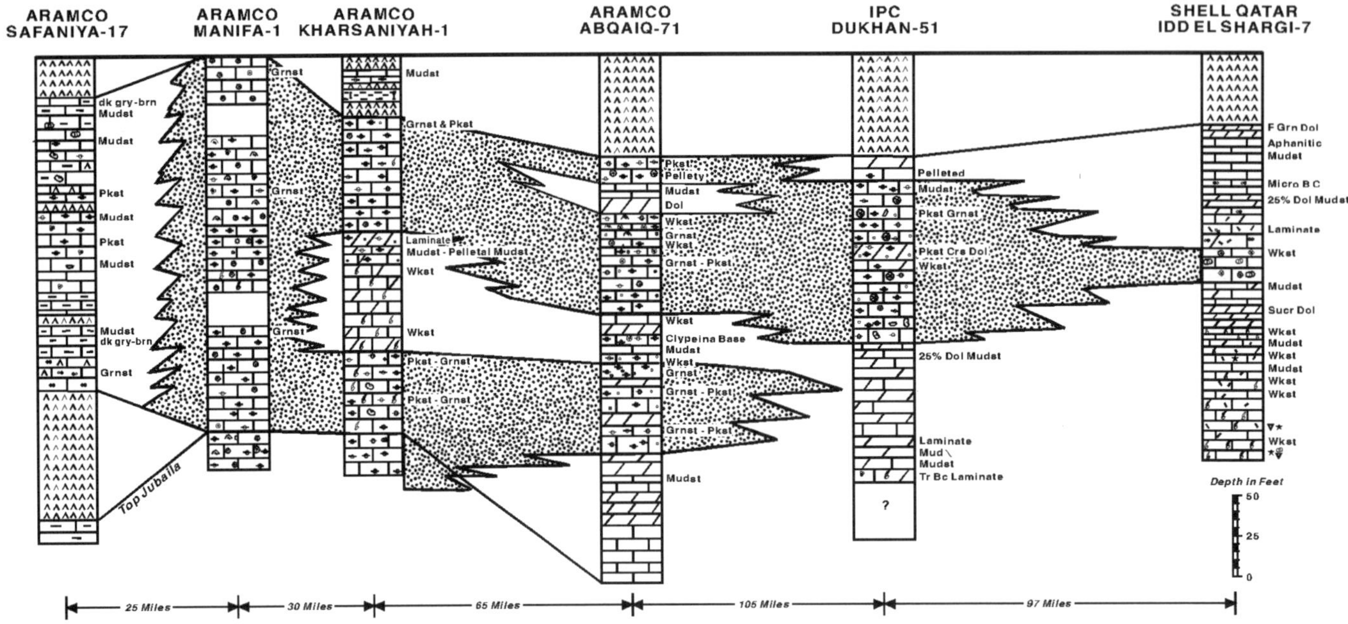

Fig. 14. Cross-section of Arab D facies along the Hasa coast from the Neutral Zone to Qatar and east into the Persian-Arabian Gulf, about 550 km. It shows northern grainstone buildup and eastward disappearance of grainstone into tidal flat dolomites of Qatar-Surmeh highs. From Wilson (1975).

bedrock is widely exposed beneath the covering sea water.

The open marine middle shelf on carbonate platforms commonly grades shoreward into restricted marine environments and thence to evaporitic salt flats (sabkhas). These three major environments commonly follow Walther's law, which states that vertical sequences are representative of geographically arranged facies patterns. Consequently, these three environments are seen many times in cyclic arrangement across platform interiors (e.g., Wilson, 1975; James, 1977; Enos, 1983). Widely extending, uniform, thin stratal units are developed in these cycles, apparently as a result of extensive progradation of the environments. This is presumed by the fact that shorelines in these modern environments are only a few kilometers wide, but in the geologic record they have been traced over much wider areas.

Except for some secondary porosity developed through dolomitization, leaching, and fracturing, the lithology of middle shelf sediments seldom forms good or widespread reservoir rock.

Restricted Circulation Shelf and Tidal Flats (Facies 8)

Platform interiors of both large isolated carbonate banks and shoreward portions of continent-fringing platforms form vast areas of restricted marine water from a few meters to only centimeters in depth. Many such areas are sediment traps for lime mud that is caught where rising diurnal or wind-driven tides frequently flood. The adjacent shallow bays may be chiefly subtidal, but are nonetheless filled with hypersaline to brackish marine-derived water with poor circulation.

Tidal flats are probably the best known carbonate environment, having been extensively studied in the Bahamas, the southeastern Persian-Arabian Gulf, and Shark Bay, Western Australia, in both humid and arid climates. Subenvironments include tidal channels, natural levees, ponds, and mangrove or palmetto islands and associated bays and beaches. In arid climates, extensive belts of algal stromatolites mark mean high tide level (Figure 15).

Tidal flats and associated subenvironments are complex because they are the meeting place of land, fresh water, marine water, and air. Each subenvironment area has its own suite of sedimentary structures, various types of laminae, ripple cross-lamination, and winnowed sediments forming migrating bars and beaches. The most characteristic sedimentary structure consists of varied forms of lamination caused by periodic flooding by storms and the stranding of lime mud over gluey cyanobacterial ("blue-green algae") mats. Other sedimentary structures include mud cracks, tracks of many animals from snails to dinosaurs, minor channeling, flat pebble conglomerates, and preserved algal stromatolites in various forms.

Biota is restricted to plants and animals that can withstand fluctuating conditions of heat, subaerial exposure, and water salinity and can survive on or within a

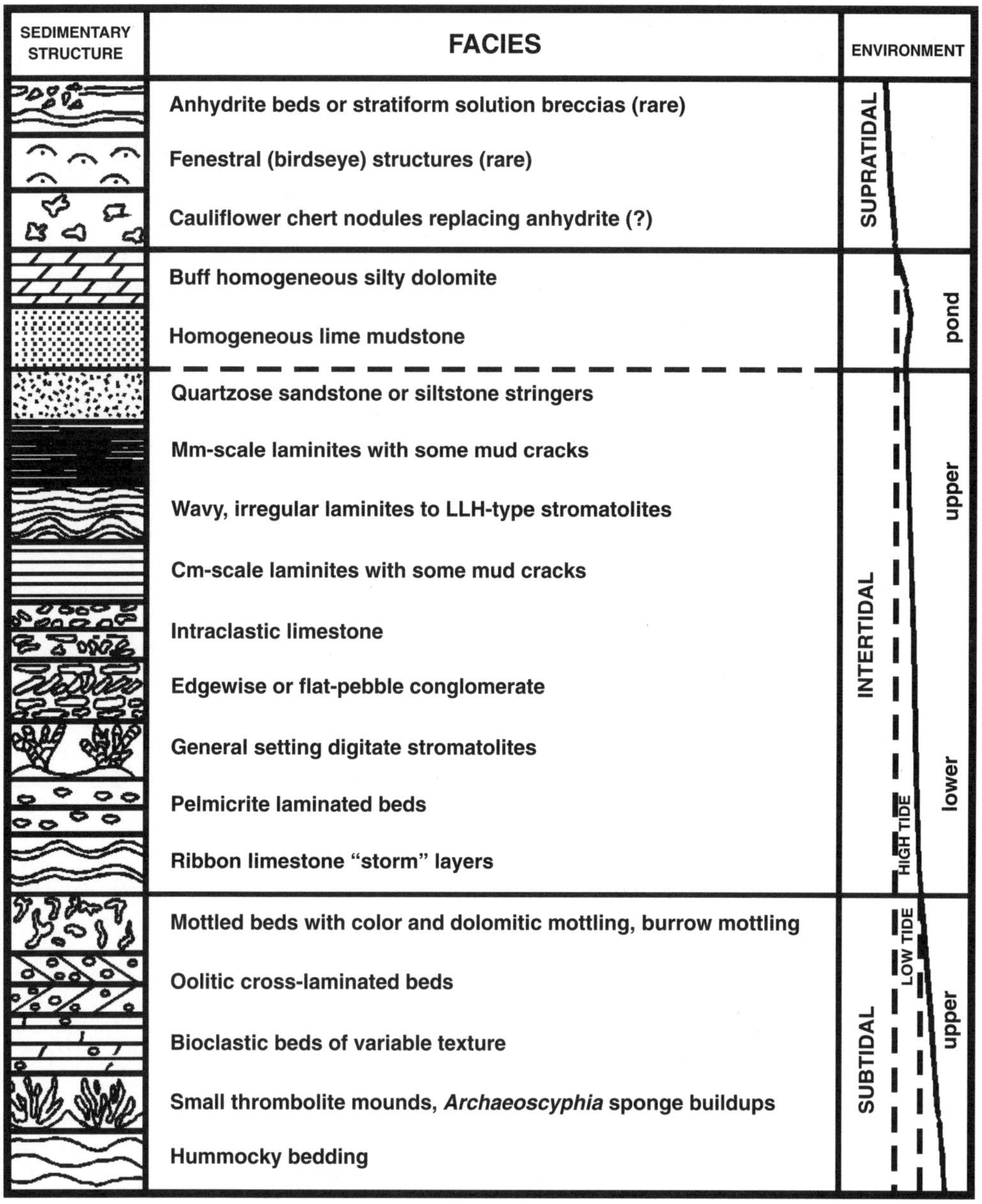

Fig. 15. Sedimentary structures of shoaling-upward tidal flat sequences in the Arbuckle Group of southern Oklahoma. Courtesy of Masera Group of Tulsa and the Oklahoma Geological Survey.

mud substrate. These euryhaline forms are limited in kind, but thrive in vast numbers in certain places, forming monospecific coquinas of cerithid snails, small bivalves, foraminifera such as peneropolids and miliolids, and crustaceans. Ubiquitous cyanobacteria leave their stromatolitic laminations everywhere, chiefly where there are few cerithid snails on the surface and where burrowing and bioturbation do not occur in the substrate.

Modern tidal flats have accumulated at the rate of about 1000 m/m.y., which is moderately fast and sufficient to fill in embayments and protected areas in lee of barrier islands. The Abu Dhabi tidal flats of the Persian-Arabian Gulf have prograded 5–20 km in the last 5000 years forming a wedge of sediment 5 m thick at its distal end. Ancient strata record much lower rates of sedimentation, but are nevertheless thick in platform interiors and often appear to form prograding upward-shoaling cycles (Figure 16).

The discovery of poorly ordered dolomite in several modern tidal flat complexes and its relationship to gypsum-anhydrite diagenetic replacement and precipitation in the Abu Dhabi tidal flats enhanced interest in the recognition of this particular environment and the largely dolomitized strata equivalent to it in the geologic record. Roehl (1967) first called attention to the number of carbonate reservoirs present in dolomitized cyclic tidal flat strata. Major fields developed in these kinds of rocks include the Middle Permian of West Texas, the Asmari of Iran, the Red River (Ordovician) of the Williston Basin,

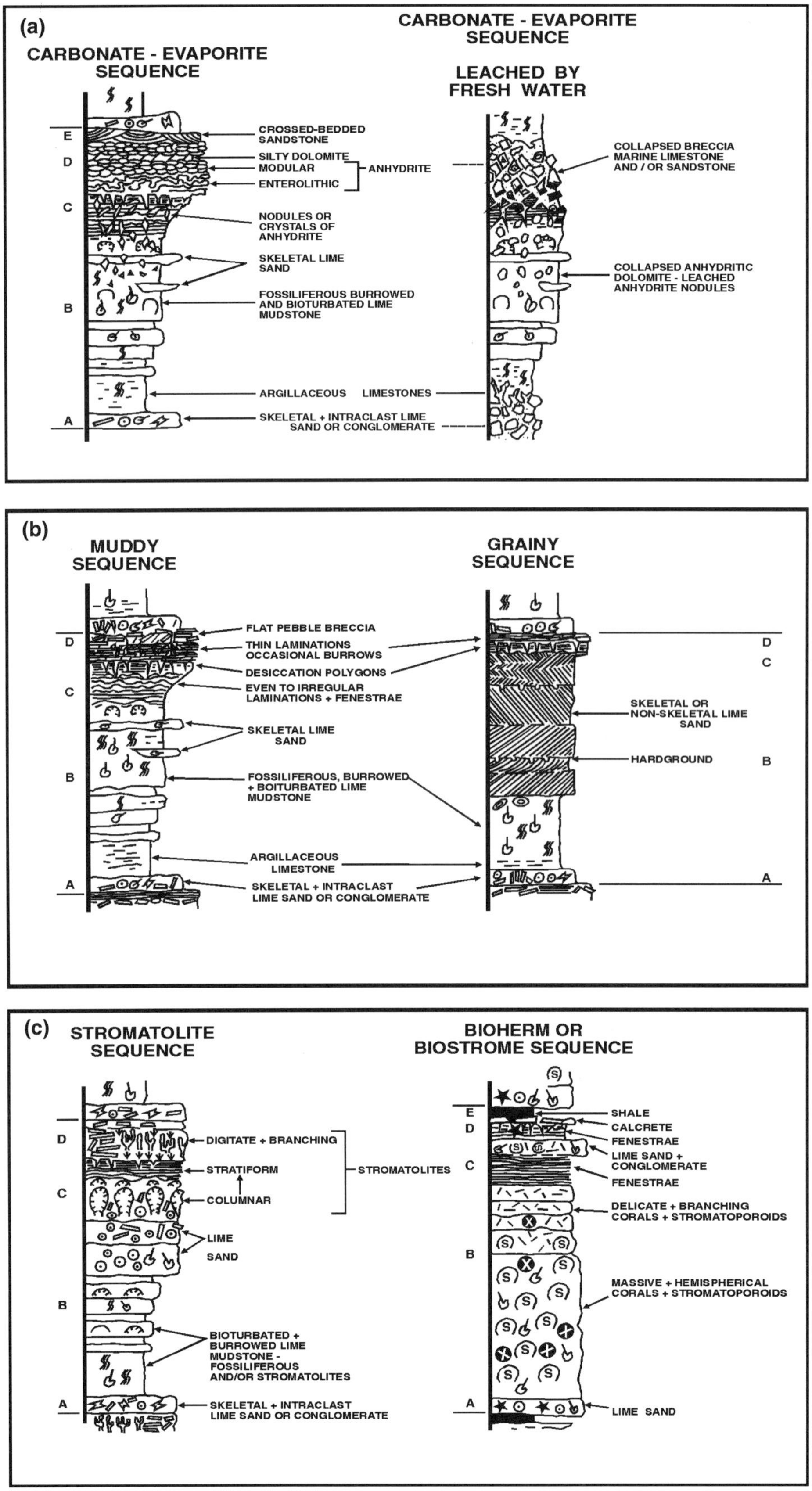

Fig. 16. Various types of restricted marine and tidal flat sedimentary cycles. From James (1977). (a) A low-energy intertidal unit (left) and supratidal evaporite unit leached by fresh water (right). (b) Contrasting muddy and grainy sequences, both beginning with argillaceous subtidal limestone. (c) Low-energy intertidal unit passing upward into algal stromatolites (left) and a biohermal unit with a thin tidal flat and supratidal cap (right).

and the Ellenburger-Arbuckle of southern United States. The best dolomite reservoirs, formed by ongoing processes under both depositional and later burial conditions, are in strata that were originally of coarser textures (e.g., channels, bars, beaches, and bays). Lower permeability and susceptibility to anhydrite plugging makes the finer grained sediment at the tops of cycles much poorer reservoir rock (see Lucia, 1972, for a thorough review).

Evaporites on Sabkhas and Salinas (Facies 9)

The evaporitic supratidal environment consists of sediment deposited just above mean high tide in flat platform areas covered only occasionally by high tides and mostly associated with storms. Modern supratidal deposits are divided into humid and arid types depending on climate. The humid type is represented in the Holocene by the areas west of Andros Island of the Great Bahama Bank, and the arid type by the evaporitic sabkha coastal areas of the Persian-Arabian Gulf in the United Arab Emirates.

The humid supratidal environment consists of vegetated high areas at the heads of channel distributaries interspersed with brackish to fresh water ponds where water remains about half the time. Crusts of extremely fine-grained poorly ordered dolomite form around the high areas surrounding the ponds. The pelleted mud found in higher parts of levees on channel sides or pond edges often has fenestral fabric caused by escaping gas from organic decay pushing apart the half-soft pellets as the fabric hardens. This is a definitive structure for this environment.

In arid regions, dolomite forms in supratidal areas of tidal flats. These areas are covered by water infrequently, but water is sucked in from surrounding bays when onshore winds and high tides prevail. This water mingles to some degree with shoreward-derived groundwater percolating from below. Tidal flat waters become highly evaporative under subaerial desert conditions. Calcium sulfate is precipitated when interstitial water reaches sufficient concentration, raising the magnesium-calcium ratio of the water to the extent that some of the aragonite and high-Mg calcite is replaced with micron-grained poorly ordered dolomite. The extent and rate to which the heavy, evaporated, dolomitizing brines seep down and reflux through the tidal flat is debated, but the wide extent of thin beds of dolomite in ancient tidal flat strata indicates that this may be an effective process of dolomitization.

The preserved sequence in Abu Dhabi tidal flats show about 5000 years of progradation above the bay sediment by the stromatolitic tidal flat facies. In turn, above the stromatolitic layers is a zone of nodular and bladed calcium sulfate crystals mixed with dolomite and unreplaced aragonitic mud capped by a cover of desert wind-blown dust which is about half dolomitic carbonate and half siliciclastic silt. Such carbonate-evaporite cycles are common in the geologic record in shoreward parts of many lime mud sequences.

Sedimentary structures are important in recognition of subenvironments. Crinkly and dome-shaped millimeter-scale laminae mark the stromatolite zone, and nodular ("chicken-wire") forms of gypsum-anhydrite pods are found in the sabkha. Internal deformation due to expansion under load caused by anhydrite transformation to gypsum creates enterolithic structures (accordion-pleated folds) to form in some evaporite layers. All of these features make it possible to distinguish sabkha evaporites from those formed in standing bodies of water; the latter chiefly contain planar millimeter-scale laminae and few or none of the above structures. Biota is essentially lacking in the arid supratidal environments except for cyanophyte bacteria and a few hardy mollusks such as cerithid snails.

No reservoirs exist in these sediments, but when evaporites are well-developed, they form excellent seals for trapping hydrocarbons. The cyclicity exhibited often enables prediction of facies changes both vertically and geographically and thus is a valuable concept for both exploration and field development.

Minor Environments

Other environments known in carbonate strata are less prevalent and less significant as reservoirs; many of these are subaerial. They include eolianite dunes, beaches, exposure surfaces (unconformity and hardground surfaces), paleokarst (collapsed cavern breccia), and caliche. There are many types of unconformity surfaces in the geologic record. With the advent of sequence stratigraphic concepts, it has become necessary to assemble more useful and significant criteria for condensed zone marine firmgrounds (layers hardened under submarine conditions) and subaerial hardgrounds. Typical hardground surfaces show considerable lithification, and the increased density of such surfaces may be of sufficient contrast to show seismically.

Eolianite is the term applied to carbonate coastal sand dunes. These are common worldwide in the upper Pleistocene and have been carefully studied, but such strata are difficult to distinguish from marine sands in the subsurface and almost no petroleum reservoirs have been recognized in eolianites. The best criteria for recognition are steeply dipping foreset strata (30–35°) with dipping surfaces from 7 m to more than 20 m in length. Common preservation of calcareous insect nests, land snails, and impressive root casts is typical. Bodies of such carbonate dunes, often with caliche crusts, are generally elongate parallel to the coast. Grains may be peloidal, oolitic or skeletal. Cement configurations typical of the vadose zone include dripstone below grains, meniscus cement between grains, and blocky calcite or needle-fiber crystal forms. Crystalites making up tiny globular bodies are termed *microcodium.*

Lacustrine carbonate sediments are also known and are well described in Scholle et al. (1983). Calcium carbonate becomes saturated in fresh water and precipitates as low-Mg calcite in both cool, temperate lakes and hot, saline waters. Lake facies tend to be ephemeral and complex, changing quickly. They can be divided into littoral and deep (profundal) types. Deep facies are often anoxic in deeper lakes unless lake water turns over seasonally and extensively. Anoxic conditions are also known in evaporitic bodies of water. Such sediments are dark and may be varved (with annual laminations). They can form important source beds, as in the Green River Eocene oil shales of the North American Rocky Mountains. Shoreward littoral facies can be reservoirs, but these beds are mostly littoral sandstones. Lake shore facies are noted for spectacular algal stromatolites. Lacustrine deposits are well studied in northern lakes of the United States. Interior basins in Africa and oil-producing basins in China are also important hydrocarbon sources.

Paleokarst is a diagenetic phenomenon rather than a depositional environment, although it may be depositionally controlled. It results from solution of preexisting rock associated with subaerially formed unconformities in the geologic record. It is important because such dissolution may result in coarse collapse solution breccias with associated dolomite and good porosity and permeability. Various rock types are included in paleokarst profiles and are controlled by the amount and incidence of rainfall, elevation (affecting depth and thickness of the water table and efficiency of drainage), presence or absence of soil and plant cover, purity of bedrock (absence of argillaceous content), and fracturing of bedrock. Paleokarsting may cause considerable surface relief at the unconformity or a single sharp planar contact, depending on the maturity of the surface and the intensity of the process. Meteoric water may be an effective solvent at depths of hundreds or even thousands of feet below the unconformity.

Paleokarst strata are commonly arranged in profiles, such as a sequence of typical caliche soil with a cemented hardpan under it, followed downward by platy and nodular caliche, and underlain by chalky microporous beds. The position of the hardpan layer can vary in the vertical profile. The result of thorough solution and oxidation at an exposure surface in the tropics or in arid regions is a reddish terra rosa soil, in places rich in bauxite. Paleokarst breccias also form from cavern collapse, which has a typical profile (Kerans, 1990) of a mosaic of cave roof fractures, a cave-fill zone where phreatic moving water has washed erosion residue into the cavern, and an underlying cave floor rubble of blocks derived from cavern roof collapse (Figure 17). This telescoping of rock types through collapse and compaction of dissolved caverns is common and enhanced by the presence of interbedded evaporites within the carbonate section.

Major reservoirs exist in these rocks, including the Arbuckle-Simpson in the Oklahoma City field, the Elk Basin dome in the Mississippian Madison Group of Wyoming, and the Lisbon field of Mississippian age in the Paradox basin.

Conclusions

Conclusions and a summary of carbonate facies types are presented in Figure 18 which is based on the rimmed platform model. Although this chapter has em-

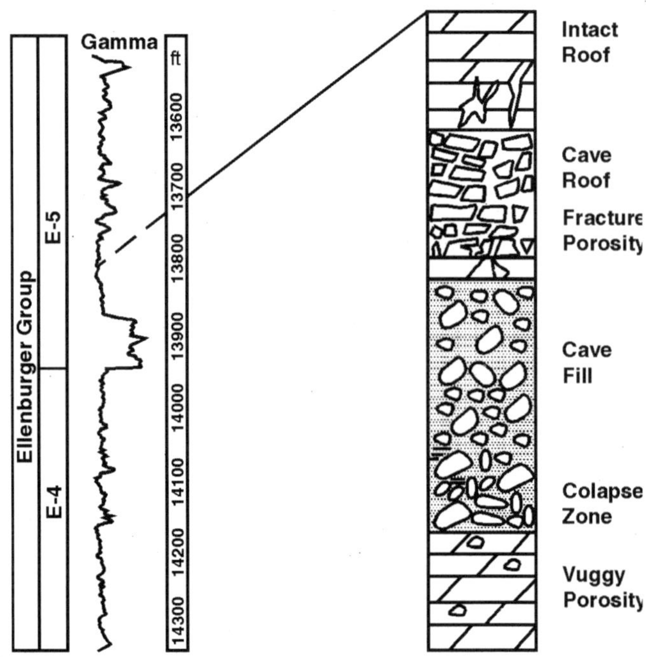

Fig. 17. Log signature and sedimentary profile of paleokarst facies resulting from cavern collapse in the Ellenburger section. After Kerans (1990); from R. Geesaman and L. Canter, Coyote, Val Verde Basin report.

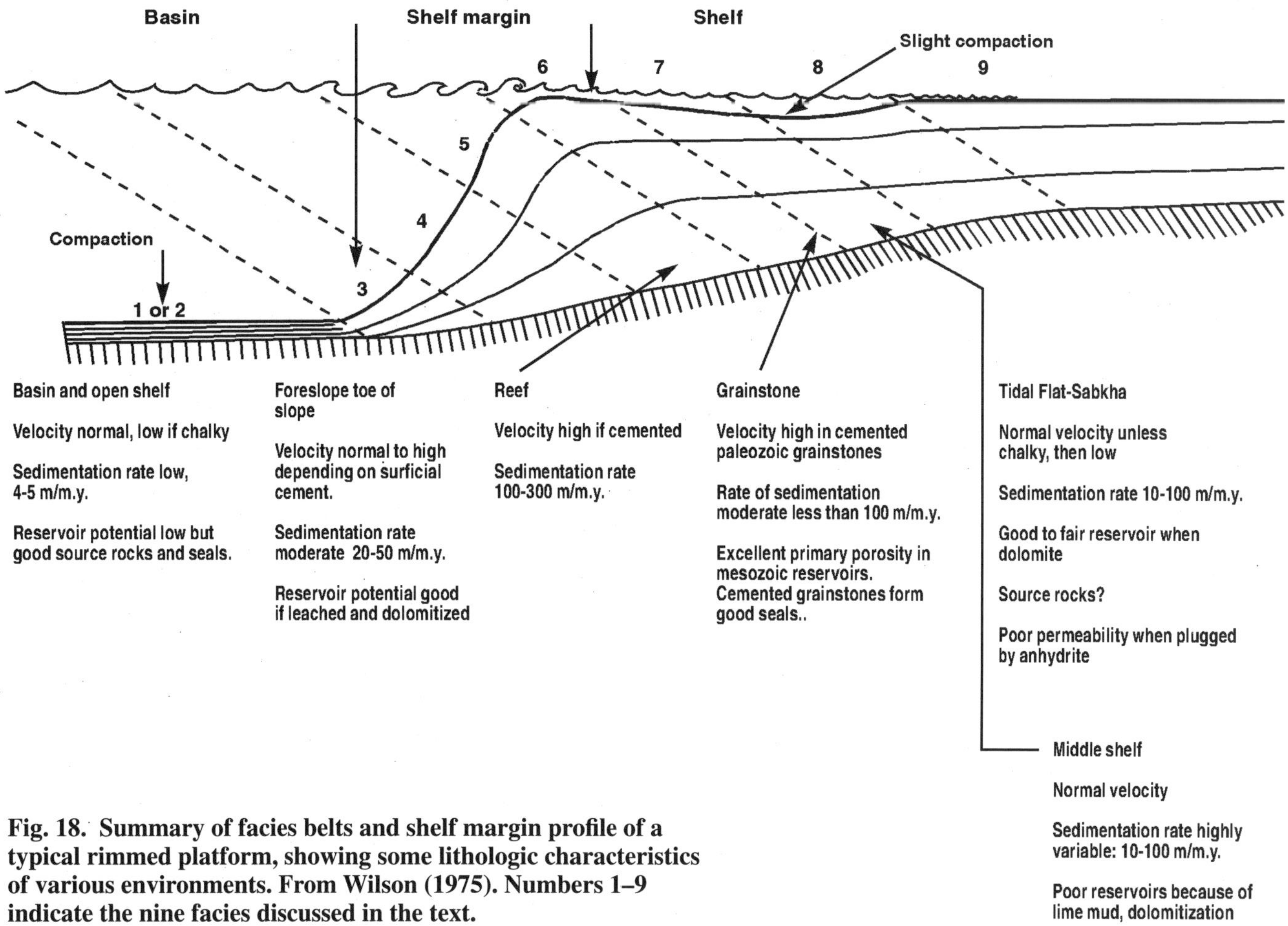

Fig. 18. Summary of facies belts and shelf margin profile of a typical rimmed platform, showing some lithologic characteristics of various environments. From Wilson (1975). Numbers 1–9 indicate the nine facies discussed in the text.

phasized carbonates as reservoir rock, it is clear that they also play important roles as source rocks and seals.

Acknowledgments

The author is particularly grateful to his colleague Clif Jordan, who cooperated in writing the chapter on middle shelf sedimentation in AAPG Memoir 33, and to Edith Wilson (Amoco Production), who kindly critiqued this chapter. Word processing was done efficiently by Della Wilson, my incomparable wife and secretary. Many of the illustrations were taken from the writer's book, *Carbonate Facies in Geologic History* (Springer-Verlag, 1975).

References

Asquith, G. B., 1982, Basic well log analysis for geologists: AAPG Methods in Exploration **3**.

Bathurst, R. G. C., 1971, Carbonate sediments and their diagenesis: Developments in Sedimentology **12**: Elsevier.

Blatt, H., Middleton, G., and Murray, R., 1980, Origin of sedimentary rocks: Prentice-Hall.

Bosellini, A., 1984, Progradation geometries of carbonate platforms: example from the Triassic of the Dolomites, northern Italy: Sedimentology, **31**, 1-24.

Caughlin, W. C., Lucia, F. J., and McIver, N. L., 1976, The detection and development of Silurian reefs in northern Michigan: Geophysics, **41**, 646-658.

Choquette, P. W., and James, N. P., 1987, Diagenesis 12, limestones—the deep burial diagenetic environment: Geoscience, Canada, **14**, No. 1, 3-35.

Dunham, R. J., 1962, Classification of carbonate rocks according to depositional texture, *in* Ham, W. E., Ed., Classification of carbonate rocks: AAPG Memoir **1**, 108-121.

Eberli, G. P., and Ginsburg, R. N., 1989, Cenozoic progradation of northwestern Great Bahama Bank, *in* P. Crevelo et al., Eds., Controls on carbonate platform and basin development: SEPM Special Publication **44**, 339-351.

Embry, A. F., and Klovan, J. E., 1971, A Late Devonian reef tract on northeastern Banks Island, N.W.T.: Bull. Can. Petrol. Geol., **19**, 730-781.

Enos, P., 1985, Cretaceous debris reservoirs, Poza Rica field, Veracruz, Mexico, *in* Roehl, P. O., and Choquette, P. W., Eds., Carbonate petroleum reservoirs: Springer-Verlag.

Enos, P., 1983, Shelf environments, *in* Scholle, P. A., Bebout, D. G., and Moore, C. H., Eds., Carbonate depositional environments: AAPG Memoir **33**, 267-295.

Flugel, E., 1982, Microfacies analysis of limestones: Springer-Verlag.

Folk, R. L., 1959, Practical petrographic classification of limestones: AAPG Bulletin, **43,** 1-38.

Handford, R., and Loucks, R. G., 1993, Carbonate depositional sequences and system tracts, responses of carbonate platforms to sea level changes, *in* Handford, C. R., and Loucks, R. G., Eds., Carbonate depositional sequences and system tracts: AAPG Memoir **57**, 3-42.

James, N. P., 1977, Shallowing upward sequences in carbonates: Geoscience Canada, **4**, 126-136.

James, N. P. and Choquette, P. W., 1983, Diagenesis 6, limestones—the seafloor diagenetic environment: Geoscience Canada, **10**, 162-180.

James, N. P., and Choquette, P. W., 1984, Diagenesis 9, limestones—the meteoric diagenetic environment: Geoscience Canada, **11**, 161-195.

James, N. P., and Ginsburg, R.N., 1979, The seaward margin of Belize barrier and atoll reefs: International Sedimentologists Assoc. Spec. Publ. **3**, Blackwell Scientific Publ.

Jordan, C., and Wilson, J. L., 1993, Carbonate reservoir rocks, *in* Magoon, L. B., and Dow, W. C., Eds., The petroleum system—from source to trap: AAPG Memoir **60**, 141-158.

Kerans, C., 1990, Depositional systems and karst geology of the Ellenburger Group (Lower Ordovician), subsurface, West Texas: Univ. Texas Bureau Econ. Geol., Rept. of Invest. **193**.

Laporte, L. F., 1969, Recognition of a transgressive carbonate sequence within an epeiric sea: Helderberg Group (Lower Devonian) of New York State, *in* Friedman, G. M., Ed., Depositional environments in carbonate rocks: SEPM Special Pub. **14**, 98-119.

Longman, M., 1980, Carbonate diagenetic textures from nearshore diagenetic environments: AAPG Bull., **64**, 461-487.

Lucia, F. J., 1972, Recognition of evaporite-carbonate shoreline sedimentation, *in* Rigby, J. K., and Hamblin, W. K., Eds., Recognition of ancient sedimentary environments: SEPM Spec. Publ. **16**, 160-191.

Mazzullo, S. J., and Reid, A. M., 1989, Lower Permian platform and basin depositional systems, northern Midland Basin, Texas, *in* Crevello, P. D., Wilson, J. L., Sarg, J. F., and Read, J. F., Eds., Controls on carbonate platform and basin development, SEPM Spec. Publ. **44**, 305-320.

Meissner, F., 1972, Cyclic sedimentation in Middle Permian strata of the Permian basin, *in* Elam, J. C., and Chuber, S., Eds., Cyclic sedimentation in the Permian Basin, 2nd edition: West Texas Geol. Soc. Pub., 203-232.

Milliman, J. D., Freile, D., Steinen, R. P., and Wilber, R. J., 1993, Great Bahama Bank aragonitic muds: mostly inorganically precipitated, mostly exported: Jour. Sed. Petrol., **63**, 589-595.

Mullins, H. T., and Van Buren, H. M., 1979, Modern modified carbonate grain flow deposit: Jour. Sed. Petrol., **48**, 747-752.

Pfeil, R. W., and Read, J. F., 1980, Cambrian carbonate platform margin facies, Shady Dolomite, southwestern Virginia, USA: Jour. Sed. Petrol., **50**, 91-115.

Playford, P. E., and D. C. Lowry, 1966, Devonian reef complexes of the Canning Basin, Western Australia: Geol. Surv. Western Australia Bull., **118**.

Read, J. F., 1982, Carbonate platforms of passive (extensional) continental margins: types, characteristics, and evolution: Tectonophysics, **81**, 195-212.

Reading, H. G., 1978, Sedimentary environments and facies: Elsevier.

Roehl, P. O., 1967, Stoney Mountain (Ordovician) and Interlake (Silurian) facies analogs of Recent low-energy marine and subaerial carbonates, Bahamas: AAPG Bull., **51**, 1979-2032.

Schlager, W., 1981, The paradox of drowned reefs and carbonate platforms: Geol. Soc. Amer. Bull., **92**, 197-211.

Schlager, W., 1992, Sedimentology and sequence stratigraphy of reefs and carbonate platforms: AAPG Continuing Education Course Notes **24**.

Scholle, P. A., Bebout, D. G., and Moore, C. H., Eds., 1983, Carbonate depositional environments: AAPG Memoir **33**.

Tucker, M. E., and Wright, V. P. 1990, Carbonate sedimentology: Blackwell Scientific Publ.

Wilson, J. L., 1969, Microfacies and sedimentary structures in "deeper water" lime mudstones, *in* Friedman, G. M., Ed., Depositional environments in carbonate rock: SEPM Special Publ. **14**, 4-19.

Wilson, J. L., 1970, Depositional facies across carbonate shelf margins: Trans. Gulf Coast Assoc. Geol. Soc., **20**, 229-233.

Wilson, J. L., 1974, Characteristics of carbonate platform margins: AAPG Bull., **58**, 810-824.

Wilson, J. L., 1975, Carbonate facies in geologic history: Springer-Verlag.

Wilson, J. L., 1980, A review of carbonate reservoirs, *in* Miall, A. D., Ed., Facts and principles of world petroleum occurrence: Can. Soc. Petrol. Geol. Memoir **6**, 95-115.

Wilson, J. L., and Jordan, C., 1983, Middle shelf, *in* Scholle, P. A., Bebout, D. G., and Moore, C. H., Eds., Carbonate depositional environments: AAPG Memoir **33**, 297-343.

Chapter 3

Seismic Properties of Carbonate Rocks

Zhijing Wang
Chevron Petroleum Technology Company
La Habra, California, U.S.A.

Abstract

The study of seismic reservoir properties is an important aspect of reservoir characterization and management using seismic technology because it directly links seismology to rock properties. Results of such a study, both experimental and theoretical, provide a fundamental understanding of seismic responses to reservoir properties and their dynamic changes so that seismic attributes can be converted to reservoir parameters. Unlike siliceous clastic reservoir rocks, carbonate rocks tend to have complicated pore systems, such as vugs, molds, channels, and fractures. Different types of pores have different compressibilities so they affect seismic wave velocities differently. Furthermore, pore types also control seismic responses to pore fluid changes in the reservoir. Besides pore types, factors that influence seismic wave velocities in carbonate rocks include mineralogy, porosity, pore fluid type and saturation, and reservoir parameters such as pressure and temperature. Studies of seismic properties of carbonate rocks play an important role in seismic reservoir characterization and monitoring. Recent laboratory results show that carbonate pore type and pore fluid compressibility contrast are the two most important factors in the feasibility of seismic monitoring of fluid movement in carbonate reservoirs.

Introduction

Seismic properties of carbonate rocks are important to petroleum exploration and production because more than half of the world's hydrocarbon reserves are found in carbonate rocks. In seismic exploration, the velocity and density contrast between two rock layers causes reflection of seismic waves at the interface. Through the seismic properties of rocks and their relationship to other rock properties, the reflected seismic waves reveal information about the structure of the earth's crust so that potential hydrocarbon traps can be identified from seismic surveys.

In addition to identifying potential structural traps, seismic methods have been successfully applied in recent years to direct hydrocarbon detection, production assessment and monitoring, reservoir characterization, and enhanced oil recovery (EOR) monitoring. In seismic monitoring applications, the impedance contrast caused by various EOR and production processes is the physical basis of the method. Laboratory results have shown that large velocity decreases are found in heavy oil sands as temperature increases (Wang and Nur, 1990a), in oil-saturated rocks when flooded with carbon dioxide (CO_2) (Wang and Nur, 1989), and in oil-saturated rocks when flooded with gas and hydrocarbon solvent (Hirsche et al., 1990a). Furthermore, laboratory results show large differences between the compressional velocities in rocks saturated with gas and liquid (water or oil). These results are the physical foundation for reservoir and production seismology.

Little has been published on the seismic properties of carbonate rocks. Rafavich et al. (1984) measured 93 Mission Canyon Formation carbonate cores from four wells in the Williston Basin field of North Dakota. Wang et al. (1991) measured 80 carbonate core samples from 11 carbonate reservoirs in Alberta and performed systematic studies on the seismic velocities of these carbonate rocks, as well as petrographic image analysis, porosity and permeability, electrical properties, and capillary pressure. Anselmetti and Eberli (Chapter 4, this volume) studied velocities in 89 core samples of shallow carbonates from two wells drilled on the modern Great Bahamas Bank.

The objective of this paper is to provide the reader with a reference on the seismic properties of carbonate rocks, including laboratory handling and measurement of carbonate core samples, some basic theories, and discussion of factors that influence seismic velocity, especially pore type. Some applications of seismic property studies to carbonate reservoirs are also included.

Laboratory Measurements

Laboratory measurements of rock properties play a vital role in exploration and development of oil fields. Many exploration and development decisions are based on laboratory results. For example, laboratory measurements of the seismic properties of reservoir rocks have provided the physical basis for exploration seismology and interpretation, direct hydrocarbon detection using bright-spot and amplitude versus offset (AVO), seismic reservoir surveillance, and monitoring of production and enhanced oil recovery (EOR) processes. Here we briefly introduce some of the most commonly used laboratory methods for measuring seismic properties of reservoir rocks.

Heterogeneity and Sample Selection

Many carbonate rocks are heterogeneous at the laboratory scale (2.5–10.2 cm, or 1.0–4.0 inches). This heterogeneity can be caused by depositional features such as shell fragments, diagenesis, and secondary porosity development. In laboratory measurements, the selected samples must represent larger scale (e.g., log and seismic) formation rocks. Features such as large vugs, cavities, fractures, and stylolites should be avoided in the selected samples. Furthermore, different laboratory measurements require different sample selection criteria. For example, fractures may not make much difference to porosity and bulk density measurements but are critical to permeability and velocity measurements.

One way to avoid heterogeneity is to use as large a core as available and low-frequency laboratory measurements. In general, any heterogeneity (e.g., vug or lens) larger than 5 mm should be avoided in ultrasonic measurements using 1-MHz acoustic transducers. For carbonate cores, full-diameter samples (8.9–10.2 cm, or 3.5–4.0 inches) of 13–25 cm (5.0–10 inches) in length and low-frequency (200–500 kHz) ultrasonic transducers are recommended for petrophysical and seismic property measurements.

Reservoir-scale heterogeneity can be minimized by measuring as many samples as possible from different wells at different depths. Statistical analysis can be performed to look for averaged rock properties that can be used in large-scale applications such as reservoir simulation and seismic interpretation.

Sample Preparation

Carbonate cores should be handled with care because they are fragile. The cores are first cleaned using toluene and/or methanol to extract the oil and salt. In most cases, the extraction procedure should be done at low temperature to prevent thermo-fracturing of the core and dehydration of possible gypsum lenses. For example, dolomite cores from a West Texas field contain substantial amounts of gypsum that were deposited after dolomitization. At high temperature, the gypsum dehydrates, resulting in higher measured porosity and lower seismic velocities (Nolen and Hoeksema et al., 1995).

In laboratory measurements, full cores or plugs are prepared by drilling and cutting. The samples are prepared in a regular (usually cylindrical) shape so that the dimensions and bulk volume can be measured precisely. For irregularly shaped samples, the bulk volume can be measured using Archimedes principle (buoyancy method). For seismic property measurements, the two end-surfaces of a cylindrical core have to be precisely ground to high flatness and parallel so that the wave path can be measured accurately.

If the sample is vuggy, surface vugs should be filled with an epoxy. Otherwise, the measured porosity may be uncharacteristically high and difficulties will occur during seismic property measurements at high pressures. Because pore fluid pressure is separated from overburden pressure by a plastic jacket in laboratory seismic measurements, the overburden pressure may punch holes in the jacket through surface vugs, resulting in communication between the pore fluid and the hydraulic oil used for overburden pressure and thus experiment failure.

Laboratory Velocity and Attenuation Measurements

Laboratory measurements of core samples provide important data on reservoir rocks. Systematic laboratory studies of seismic properties in relation to other rock properties are the foundation for converting seismic attributes to reservoir properties such as porosity, fluid saturation, density, and fracture orientation. Several methods of measuring the seismic properties of rocks do exist, however. It is important to understand the measurement principles and procedures in interpreting and applying the laboratory data. In this section, we briefly introduce some of the most common laboratory methods for measuring seismic properties of carbonate rocks.

Pulse Transmission Method

The pulse transmission method (Birch, 1960) is used by most rock physics laboratories around the world. In this method, ultrasonic pulses generated by the pulse generator are converted to mechanical vibration by the transmitting piezoelectric transducer. After traveling through the rock sample, the mechanical (elastic) waves are converted back to an electrical signal by the receiving transducer. The traveltime of the acoustic waves can be measured on a digital oscilloscope so that the velocity is calculated by

$$V = \frac{L}{\Delta t}, \tag{1}$$

where V is either the compressional or shear wave velocity, L is the pressure- and temperature-corrected sample length, and Δt is the traveltime in the sample.

Spectrum Ratio Method

The spectrum ratio method (Toksöz et al., 1976) uses the recorded waves through pulse transmission to calculate the quality factor or attenuation. In this method, a wave, either compressional or shear, traveling in a medium can be expressed mathematically by

$$A = A_0 G e^{-\alpha x + i\omega t}, \tag{2}$$

where A is the amplitude, A_0 is the initial amplitude, G is the geometrical spreading factor, α is the attenuation coefficient, x is the distance, ω is the angular frequency, and t is time. A reference material, usually an aluminum block, is used to obtain the reference waveform amplitudes and traveltimes. Thus, a wave traveling through the aluminum block is

$$A_\mathrm{a} = A_\mathrm{ao} G_\mathrm{a} e^{-\alpha_\mathrm{a} x_\mathrm{a} + i\omega t}, \tag{3}$$

where subscript "a" represents aluminum. Similarly, for a rock sample,

$$A_\mathrm{r} = A_\mathrm{ro} G_\mathrm{r} e^{-\alpha_\mathrm{r} x_\mathrm{r} + i\omega t}, \tag{4}$$

where subscript "r" represents a rock sample.

In these measurements, the same set of acoustic transducers are used for both the aluminum and the rock sample at a fixed power level. Also, the sizes of the aluminum and the rock sample are the same, such that

$$A_\mathrm{a0} \approx A_\mathrm{r0} \quad \text{and} \quad G_\mathrm{a} \approx G_\mathrm{r}. \tag{5}$$

The ratio of equations (3) and (4) gives

$$\frac{A_\mathrm{a}}{A_\mathrm{r}} = e^{\alpha_\mathrm{r} x_\mathrm{r} - \alpha_\mathrm{a} x_\mathrm{a}}. \tag{6}$$

The attenuation coefficient α is defined as

$$\frac{1}{Q} \approx \frac{\alpha V}{\pi f}, \tag{7}$$

where Q is the quality factor and f is the frequency. Substituting equation (7) into (6), we have

$$\ln\left(\frac{A_\mathrm{a}}{A_\mathrm{r}}\right) = \left(\frac{1}{Q_\mathrm{r}}\frac{x_\mathrm{r}}{V_\mathrm{r}} - \frac{1}{Q_\mathrm{a}}\frac{x_\mathrm{a}}{V_\mathrm{a}}\right)\pi f. \tag{8}$$

Equation (8) shows a linear relationship between $\ln(A_\mathrm{a}/A_\mathrm{r})$ and f, with the slope being

$$\begin{aligned} \text{Slope} &= \left(\frac{1}{Q_\mathrm{r}}\frac{x_\mathrm{r}}{V_\mathrm{r}} - \frac{1}{Q_\mathrm{a}}\frac{x_\mathrm{a}}{V_\mathrm{a}}\right)\pi \\ &= \left(\frac{1}{Q_\mathrm{r}}\Delta t_\mathrm{r} - \frac{1}{Q_\mathrm{a}}\Delta t_\mathrm{a}\right)\pi, \end{aligned} \tag{9}$$

where Δt_r and Δt_a are the traveltimes in the rock sample and the aluminum block, respectively. The quality factor Q_r of the rock sample is thus

$$Q_\mathrm{r} = \frac{\Delta t_\mathrm{r}}{(\text{Slope}/\pi) + (\Delta t_\mathrm{a}/Q_\mathrm{a})}. \tag{10}$$

Since $Q_\mathrm{a} \approx 150\,000$ in standard aluminum blocks, Q_r can be obtained by measuring the traveltimes in the aluminum block and the rock sample and $\ln(A_\mathrm{a}/A_\mathrm{r})$ as a function of frequency.

Figure 1 shows the compressional (V_P) and shear (V_S) wave velocities and the quality factors Q_P and Q_S versus pressure measured in a limestone core sample using the pulse transmission and spectrum ratio methods. In using the spectrum ratio method, the transducer frequency should be kept as low as possible because the higher frequency components of the wave may be scat-

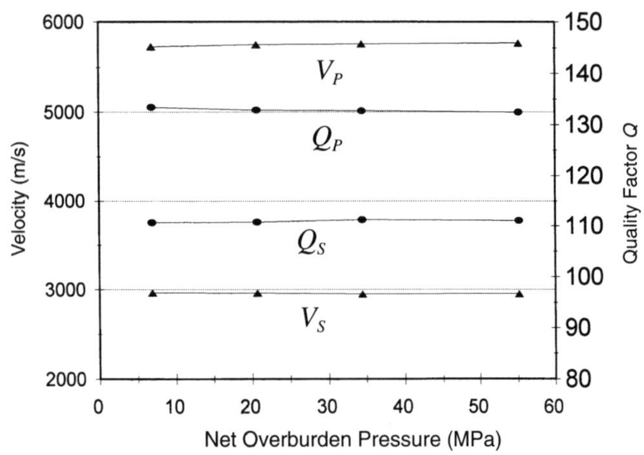

Fig. 1. Seismic velocities (V_P and V_S) and quality factors (Q_P and Q_S) measured in a brine-saturated limestone (porosity, 0.7%; diameter, 3.8 cm; length, 5.1 cm) using the pulse transmission and spectrum ratio methods. The net overburden pressure is the difference between the overburden pressure and the pore pressure and is synonymous with the effective pressure and net pressure.

tered by the rock grains and pores. Since the reference aluminum block is nonporous, the higher frequency portion of the spectrum ratio may be too high if scattering occurs in the rock sample, resulting in uncharacteristically low Q. The pulse transmission measurement, however, requires that the sample length be at least half a wavelength, thus the transducer frequency has to be greater than $V/2L$, where V and L are the velocity and sample length, respectively.

Another problem that may arise in attenuation measurements using the spectrum ratio method is the contact between the sample and the transducers. In general, couplants of the same material with the same thickness have to be used in both the reference and core measurements.

Pulse Echo or Reflection Method

The pulse echo or reflection method is similar to the pulse transmission method except that only one ultrasonic transducer is used. The transducer is attached to one end-surface of the core sample and excited by a pulse generator. The electrical pulses are converted to an elastic wave that travels through the rock sample and is then reflected back and forth at the two end-surfaces. The velocity can therefore be calculated by

$$V = \frac{2L}{\Delta t_2}, \tag{11}$$

where L is the sample length and Δt_2 is the difference between the arrival times of the first and second reflected waves.

Assuming plane wave propagation, the ratio of the spectral amplitudes of the first (A_1) and second (A_2) reflected waves, according to equation (6), is

$$\frac{A_1}{A_2} = e^{2L\alpha}. \tag{12}$$

The quality factor is thus

$$Q = \frac{2L\pi f}{V \ln\left(\frac{A_1}{A_2}\right)}. \tag{13}$$

The pulse echo or reflection method works poorly in highly attenuative rock samples because most energy is absorbed inside the sample. Furthermore, it requires plane wave propagation or else geomechanical spreading comes into play. The advantage of this method is that it does not require measurements of a reference sample so the potential contact problem between the transducer and the sample is eliminated.

Resonant Bar Method

Unlike the ultrasonic pulse transmission and pulse echo methods, the resonant bar method determines velocities and attenuation at the kilohertz frequency range, which is comparable to those frequencies used in acoustic logging. In this method, a rock is cut into a bar-shaped sample with uniform cross-sectional areas. The cross-section of a bar can be circular, square, or rectangular. The bar sample is excited by a periodic signal (usually sine or cosine) through piezoelectric or elastostatic transducers, causing it to vibrate at one of its resonance frequencies (usually at the fundamental resonance frequency). Once the resonance frequency is found, the signal is shut off so the resonance decays exponentially. The bar sample can be excited at both extensional and shear modes using different configurations of the transducers. The extensional (V_E) and shear (V_S) velocities are

$$V_E = \sqrt{\frac{E}{\rho}} = 2Lf_E, \tag{14}$$

$$V_S = \sqrt{\frac{\mu}{\rho}} = 2Lf_S, \tag{15}$$

where μ and E are the shear and Young's moduli of the sample, ρ is the bulk density, L is the sample length, and f_E and f_S are the resonance frequencies at the extensional and shear modes, respectively.

The quality factor is obtained by

$$Q = \frac{f}{\Delta f}, \tag{16}$$

where Δf is the width of the resonance peak between 70.7% $(1/\sqrt{2})$ of the peaks maximum height. Once the extensional and shear velocities (V_E and V_S) and quality factors (Q_E and Q_S) are calculated, they can be converted to compressional velocity (V_P) and the quality factor (Q_P):

$$V_P = V_S\sqrt{\frac{4V_S^2 - V_E^2}{3V_S^2 - V_E^2}}, \tag{17}$$

$$\frac{(1-\sigma)(1-2\sigma)}{Q_P} = \frac{(1+\sigma)}{Q_E} - \frac{2\sigma(2-\sigma)}{Q_S}, \tag{18}$$

where σ is Poisson's ratio.

The advantage of the resonant bar method is that it measures velocities and quality factors at relatively low frequencies. The disadvantages are that (1) extensive system calibration and data correction are needed, (2) high-pressure, high-temperature data are hard to obtain, and (3) pure shear mode is difficult to achieve.

Stress/Strain Wave Method

Instead of exciting the rock sample at the resonance frequency in the resonant bar method, the stress/strain wave method excites the sample at much lower frequencies. In this method, a sinusoidal stress is applied to the rock sample and the resulting strain is measured so that extensional and shear moduli and thus velocities can be calculated. The quality factor Q is calculated through the measured phase-angle difference (δ) between the stress and strain as follows:

$$Q = \frac{1}{\tan\delta}. \tag{19}$$

The advantages of this method are that the measured velocities and quality factors are low frequency (0.1 H to a few hundred hertz) and it is suitable for measurement of low Q materials such as soft sediments. The disadvantages are that (1) it is difficult to implement, (2) extensive system calibration is required, and (3) it is difficult to measure fluid-saturated samples because of the boundary problem. Figure 2 shows compressional and shear velocities measured in a gas-saturated Alberta limestone using the stress/strain wave method at a frequency of about 50 Hz.

Anisotropy

Because transversely isotropic rocks have five independent elastic stiffness constants, at least five mutually independent velocities must be measured to characterize the rock. In most laboratory measurements, three core plugs are drilled adjacently from a full-diameter (usually 3.5-inch) core at 0°, 45°, and 90° to the symmetry axis (Figure 3). Each core plug is fitted into a plastic jacket and attached to the top and bottom transducers. The jacket separates the overburden pressure

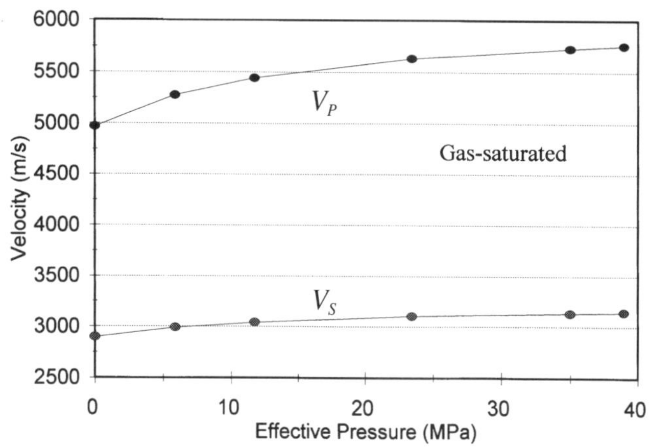

Fig. 2. Seismic velocities (VP and VS) measured in a gas-saturated limestone (porosity, 1.9%; diameter, 3.8 cm; length, 10.2 cm) from Alberta, Canada, using the stress/strain wave method at about 50 Hz.

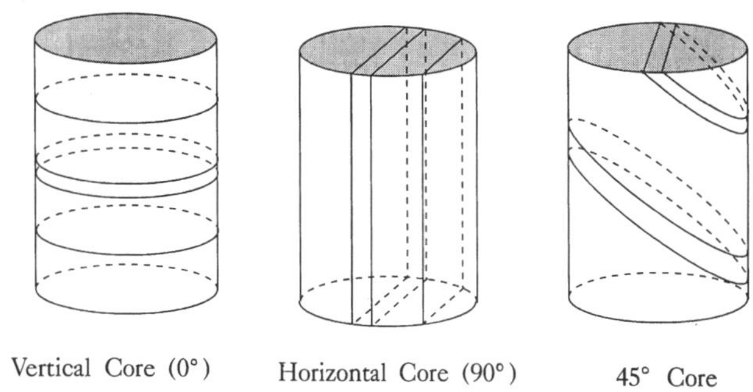

Fig. 3. The three core plug orientations used in velocity measurements in transversely isotropic rocks needed to characterize the rock: a vertical plug (perpendicular to bedding plane or parallel to symmetry axis), a horizontal plug (parallel to bedding plane or perpendicular to symmetry axis), and a third plug at a 45° angle to both the bedding plane and symmetry axis.

from the pore pressure. Each transducer has one piezoelectric crystal for generating and receiving P-waves and two perpendicularly polarized piezoelectric crystals for generating and receiving S-waves. In this three-plug method, the velocities and densities measured on the 0°, 90°, and 45° cores yield the stiffness constants C_{33} and C_{44}, C_{11} and C_{66}, and C_{13}, respectively.

One advantage of this method is that all the measurements can be done under simulated reservoir condition (pore and overburden pressures, temperature, and saturation). The other advantage is that the measured velocities are redundant for calculating the five independent stiffness constants because each core plug measurement yields three velocities (one P and two S). The most obvious disadvantage is that it requires measurements of three core plugs. Because shale measurements are extremely time consuming, the cost for measuring three core plugs is high. The three samples may not represent the same rock in terms of rock properties. Another disadvantage is that it can be frustrating to prepare three core plugs because most shales are friable and fissured.

Limitations of Laboratory Measurements

Although laboratory measurements provide fundamental data and understanding of reservoir rock properties, they also have limitations. First, laboratory samples are usually too small to be representative of large-scale reservoir rock properties, especially when the rock is heterogeneous. Second, it is vital to preserve and prepare laboratory samples correctly so they are not altered. The first limitation can be minimized by measuring full cores and large suites of samples from various locations in the formation so that systematics can be developed. Scale-up schemes such as using statistics can be applied to laboratory data sets. The second problem can be solved by developing a set of best-practice procedures in laboratory measurements and through quality assurance.

Laboratory measurements are complicated. Many laboratory results are process and procedure dependent. The cores should be preserved and measurements made as close to the in-situ condition as possible. During core handling and sample preparation, extreme care must be taken to not alter the rock sample (e.g., in terms of porosity, permeability, grain contact, fracturing, and wettability).

Some Theories and Models Applied to Carbonates

Numerous theories and models on elastic properties of effective media have been published, many of which can be directly applied to carbonate rocks. Watt et al. (1976) reviewed and summarized some of the important theories and models. Wang and Nur (1992) also gave an extensive review of the existing theories and models. In this section, we review four commonly used theories or models which can be readily applied to carbonate rocks.

Time-Average Equation

Wyllie et al. (1956) discovered that velocities measured in piles of alternating Lucite (a plastic material) and aluminum discs fit the time-average equation

$$\frac{1}{V_P} = \left(\frac{\varphi}{V_{\text{Luc}}} + \frac{1-\varphi}{V_{\text{Al}}}\right), \qquad (20)$$

where V_P is the time-average compressional velocity based on the summation of traveltimes in aluminum and Lucite, φ is the fractional volume of Lucite in the sample, and V_{Luc} and V_{Al} are the compressional velocities in Lucite and aluminum, respectively. In fluid-saturated rocks, equation (20) becomes

$$\frac{1}{V_P} = \frac{\phi}{V_{\text{f}}} + \frac{1-\phi}{V_{\text{m}}}, \qquad (21)$$

where ϕ is the porosity, V_{f} is the compressional velocity in saturating fluid, and V_{m} is the compressional velocity in the solid rock matrix.

Since its inception, the simple time-average relationship has been used in sonic log analysis to derive porosity from formation velocity. Two problems that arise in using equation (21) are generally related to (1) estimation of V_{m} and V_{f} and (2) the range of applicability. Velocities in the rock matrix and pore fluid are normally assumed because detailed lithology and pore fluid composition are not available at the time of log interpretation. This assumption often generates errors in the derived porosity. Equation (21) is often applied beyond its range of applicability. Because it is based on an oversimplified model, it does not work for rocks containing fluids of low velocities such as gases and live oils (oils with gas in solution), rocks with vugular pores or fractures (e.g., some carbonate rocks), and rocks with loose matrix (e.g., soft and unconsolidated sands).

In rocks saturated with low-velocity fluids, V_P in equation (21) is dominated by the effect of pore fluids. In general, this equation should not be used whenever $V_{\text{f}} \le 1.0$ km/s. For carbonates with vugs, equation (21) underestimates the porosity because vugular pores tend to be big, spherical, and incompressible and V_P is insensitive to pore fluid saturation. As pointed out by Wyllie et al. (1958), equation (21) should not be applied to determine the porosity of carbonate rocks that are vugular or fractured. Because fractures tend to be small in volume, elongated, and highly compressible, V_P is sensitive to the presence of fractures and pore fluid saturation, so that equation (21) substantially overestimates the porosity.

Figure 4 shows compressional velocities measured on 66 carbonate core samples saturated with a 35° API oil in the laboratory versus the calculated velocities using the time-average equation (Wang et al., 1991). In general, the time-average equation underestimates the oil-saturated V_P.

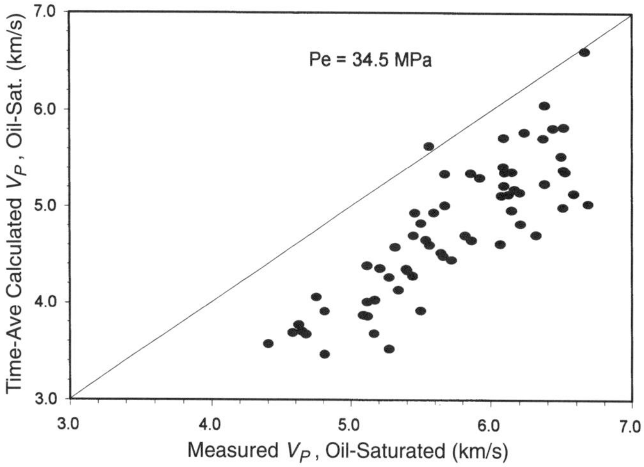

Fig. 4. Calculated V_P using the time-average equation versus measured V_P in 66 carbonate core samples saturated with 35° API oil at an effective pressure of 34.5 MPa (5000 psi). The time-average equation severely underestimates the V_P and thus overestimates the porosity in these carbonate rocks.

Gassmann Equation

Gassmann (1951) derived an equation to calculate the bulk modulus of a fluid-saturated porous medium using the known bulk moduli of the solid matrix, the frame, and the pore fluid. For a rock, the solid matrix is the rock-forming minerals, the frame refers to the rock sample with empty pores (dry rock), and the pore fluid can be gas, oil, water, or a mixture of these. The equation is

$$K = K_d + \frac{\left(1 - \frac{K_d}{K_m}\right)^2}{\frac{\phi}{K_f} + \frac{1-\phi}{K_m} - \frac{K_d}{K_m^2}}, \tag{22}$$

where K is the bulk modulus of a rock saturated with a fluid of bulk modulus K_f, K_d is the frame (dry) bulk modulus, K_m is the matrix bulk modulus of the same rock, and ϕ is the porosity. The shear modulus of the rock is not affected by fluid saturation, such that

$$\mu = \mu_d, \tag{23}$$

where μ_d is the frame (dry) shear modulus of the rock. The density of the saturated rock is simply given by

$$\rho = \rho_d + \phi\rho_f \tag{24}$$

where ρ_d and ρ_f are the frame (dry) and fluid densities, respectively.

To use equations (22) and (23), K_d and μ_d can be obtained through laboratory velocity measurements on dry rocks or through formation evaluation. The porosity and densities can be measured in the laboratory or obtained from density logs. The value for K_m can be calculated by effective medium theories, such as Reuss-Voigt-Hill (Hill, 1962) and Hashin-Shtrikman (1963), when the mineralogic analysis data are available. Otherwise, K_m can be assumed to be 76 GPa for limestones and 95 GPa for dolostones. The calculated compressional velocities are sensitive only in low-porosity rocks to the uncertainty in the assumed K_m values.

Figure 5 shows the Gassmann calculated V_P versus the measured V_P and V_S in 80 carbonate core samples from 11 reservoirs in Alberta, Canada, at an effective pressure of 6.9 MPa (Wang et al., 1991). The velocities were measured on the sample when both dry and saturated with a 35° API oil at different reservoir temperatures. The Gassmann calculated V_P in Figure 5a is lower than the measured V_P in all the carbonate samples by as much as 23%. The Gassmann calculated V_S (Figure 5b) is also generally lower than the measured V_S, but it fits mostly within 10%. The difference between the calculated and the measured velocities decreases as effective pressure increases. For example, at an effective pressure of 34.5 MPa, the Gassmann calculated V_P is still lower than the measured V_P, but it fits mostly within 10% (Figure 6a). The calculated V_S at 34.5 MPa generally fits the measured V_S within 4% (Figure 6b).

The misfits between the Gassmann calculated and measured velocities are caused by four factors: (1) the presence of cracks or microfractures in the rock samples, (2) uncertainties in the input parameters to the Gassmann equation, (3) velocity dispersion, and (4) nonconformance of real rocks to Gassmann's assumptions. The presence of cracks or microfractures

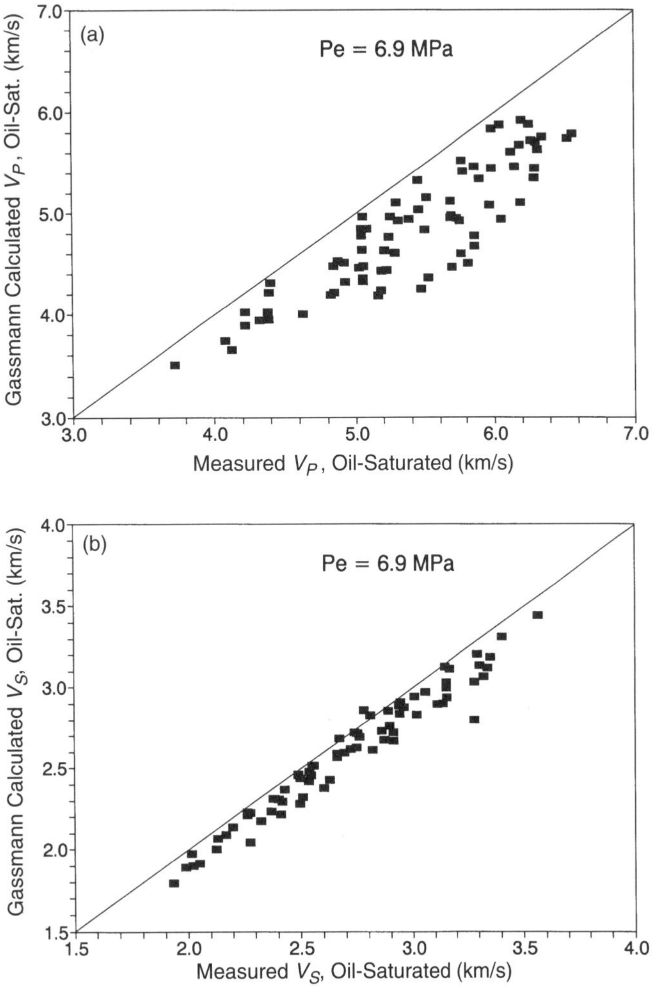

Fig. 5. Calculated (a) V_P and (b) V_S using the Gassmann equation versus measured V_P in 80 carbonate core samples saturated with 35° API oil at an effective pressure of 6.9 MPa (1000 psi). The Gassmann-calculated velocities are almost always lower than the measured velocities.

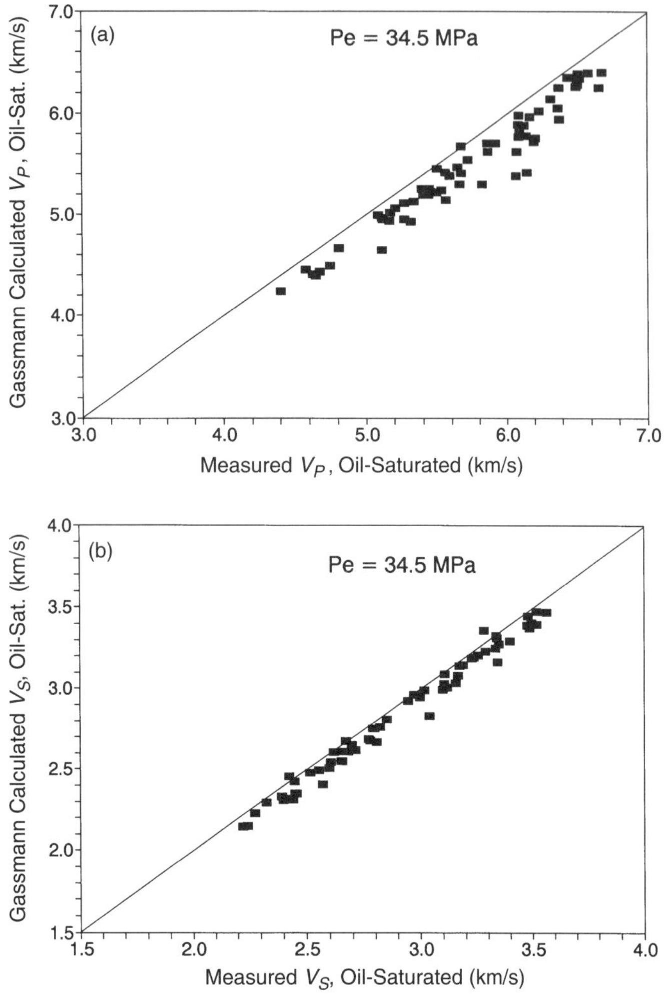

Fig. 6. Calculated (a) V_P and (b) V_S using the Gassmann equation versus measured V_P in 80 carbonate core samples saturated with 35° API oil at an effective pressure of 34.5 MPa (5000 psi). Because most cracks, fractures, and thin pores close at high effective pressures, the Gassmann-calculated velocities are closer to, but still mostly lower than, the measured velocities.

lowers the velocities, especially the V_P, in dry rocks (the frame). As the cracked rock becomes saturated with a relatively incompressible liquid (water or oil), the bulk modulus of the rock increases (compressibility decreases) dramatically if the rock-fluid system is closed. In this case, the fluid saturation effect on V_P is higher than that predicted by the Gassmann equation. As the effective pressure increases, the cracks are closed so that the Gassmann calculated velocities are closer to the measured velocities.

Uncertainties exist in the input parameters to the Gassmann equation. The real K_m values are essentially unknown for a rock sample, the measured frame velocities have about 1% uncertainty, and the measured density, porosity, and fluid velocity all have uncertainties. Furthermore, the measured fluid-saturated velocities also have about 1% uncertainty. All these uncertainties have to be accounted for when the measured velocities are compared to the Gassmann calculated velocities. For practical purposes, a 2% difference should be considered as an excellent fit between the Gassmann calculated and measured velocities because of these uncertainties.

Since the Gassmann equation is a low-frequency equation, some of the misfits may be caused by velocity dispersion because the laboratory data are usually measured at high frequencies (0.5–1.0 MHz). The magnitude of velocity dispersion can be high according to the "local flow" mechanism introduced by O'Connell and Budiansky (1977) and Mavko and Nur (1978). At low effective pressures, the magnitude of velocity dispersion is high because the cracks are still open. As the effective pressure increases, velocity dispersion decreases.

It is also worth noting that equation (22) can be applied to suspensions by setting $K_d = 0$. In this case, it is reduced to Wood's equation:

$$\frac{1}{K} = \frac{1-\phi}{K_m} + \frac{\phi}{K_f} . \tag{25}$$

In practice, K_d in equation (22) is often unknown while K is known. If the Gassmann equation is assumed to model adequately the effective elastic properties of a fluid-saturated porous medium, K_d can be inverted from equation (22):

$$K_d = K - \frac{K_f\left(1 - \frac{K}{K_m}\right)^2}{\left(1 - \frac{K_f}{K_m}\right)\phi - \frac{K_f}{K_m}\left(1 - \frac{K}{K_m}\right)} . \tag{26}$$

The practical use of this equation is that when both V_P and V_S are available (e.g., from full-wave sonic logs or laboratory measurements), equations (26), (23), and (24) can be used to derive the frame (dry) or gas-saturated velocities in the same rock or formation. This inversion for frame velocities must be used with caution because of the misfits between the Gassmann calculated and measured velocities (Figure 5 and 6). In general, the inversion works best at high effective pressures (deep formations) and better in sandstones than in carbonates.

Biot Theory

As mentioned above, the Gassmann equation is a low-frequency approach to elastic wave propagation in porous media. At higher frequencies, some of its assumptions break down so that it becomes inadequate to describe the wave phenomenon in fluid-saturated porous media. Biot (1956, 1962) developed a theory that covers the whole frequency range.

The Biot equations are identical to the Gassmann equations as the wave frequency approaches zero. This is not surprising because at zero frequency, there is no relative motion between the pore fluid and the rock matrix. In general, whenever the wave frequency $f << 0.1f_c$, Biot's results should agree with Gassmann's results (White, 1983), where

$$f_c = \frac{\eta\phi}{2\pi\kappa\rho_f} \tag{27}$$

is defined in the Biot theory as a characteristic frequency. In this equation, η and ρ_f are the viscosity and density of the pore fluid, respectively, and ϕ and κ are the porosity and permeability of the rock. Equation (27) shows that f_c is high in low-permeability rocks or when the pore fluids viscosity is high.

In the Biot theory, when the wave frequency approaches infinity ($f \rightarrow \infty$), a set of Biot high-frequency equations can be derived for shear velocity ($V_{S\infty}$) and compressional velocity ($V_{P\infty}$):

$$V_{S\infty} = \frac{\mu_d}{\rho_{11} - \frac{\rho_{12}^2}{\rho_{22}}} = \frac{\mu_d}{\rho_d + \left(1 - \frac{1}{\alpha}\right)\phi\rho_f} . \tag{28}$$

$$V_{P\infty} = \frac{A + [A^2 - 4B(PR - Q^2)]^{1/2}}{2B} , \tag{29}$$

where

$$A = P\rho_{22} + R\rho_{11} - 2Q\rho_{12} , \tag{30}$$

$$B = \rho_{11}\rho_{22} - \rho_{12}^2 , \tag{31}$$

$$P = \frac{(1-\phi)\left[1 - \phi - \frac{K_d}{K_m}\right]K_m + \phi\frac{K_d K_m}{K_f}}{D} + \frac{4}{3}\mu_d , \tag{32}$$

$$R = \frac{\phi^2 K_m}{D} , \tag{33}$$

$$Q = \frac{\left(1 - \phi - \frac{K_d}{K_m}\right)\phi K_m}{D} , \tag{34}$$

$$D = 1 - \phi - \frac{K_d}{K_m} + \phi\frac{K_m}{K_f} , \tag{35}$$

$$\rho_{11} = \rho_d - (1-\alpha)\phi\rho_f , \tag{36}$$

$$\rho_{22} = \alpha\phi\rho_f , \tag{37}$$

$$\rho_{12} = (1-\alpha)\phi\rho_f , \tag{38}$$

and α is the tortuosity factor, defined as a nondimensional parameter dependent on the pore geometry of the rock. For pores of parallel tubes, $\alpha = 1$, while for most rocks, $\alpha = 2$ to 3 (Johnson and Plona, 1982; Winkler, 1986; Wang and Nur, 1990b). In the Biot theory, the relative motion is implicitly caused by the sharp contrast in shear moduli between the pore fluid and the rock matrix, which is reflected by the coupling densities of equations (36) through (38).

Although Biot's theory proves that an elastic wave propagating through a fluid-saturated porous medium is both dissipative and dispersive, it generally underestimates the magnitude of energy dissipation and velocity dispersion (Winkler, 1986; Han, 1987; Wang, 1988; Wang and Nur, 1990b). Figure 7 shows the Biot V_P and V_S dispersions in 80 carbonate core samples versus porosity at two effective pressures, where the Biot dispersion is defined as the fractional difference between the velocities calculated by the Gassmann equation (V_G) and by the Biot high-frequency limit (V_B):

$$\text{Biot dispersion} = \frac{V_B - V_G}{V_G} \times 100\% . \tag{39}$$

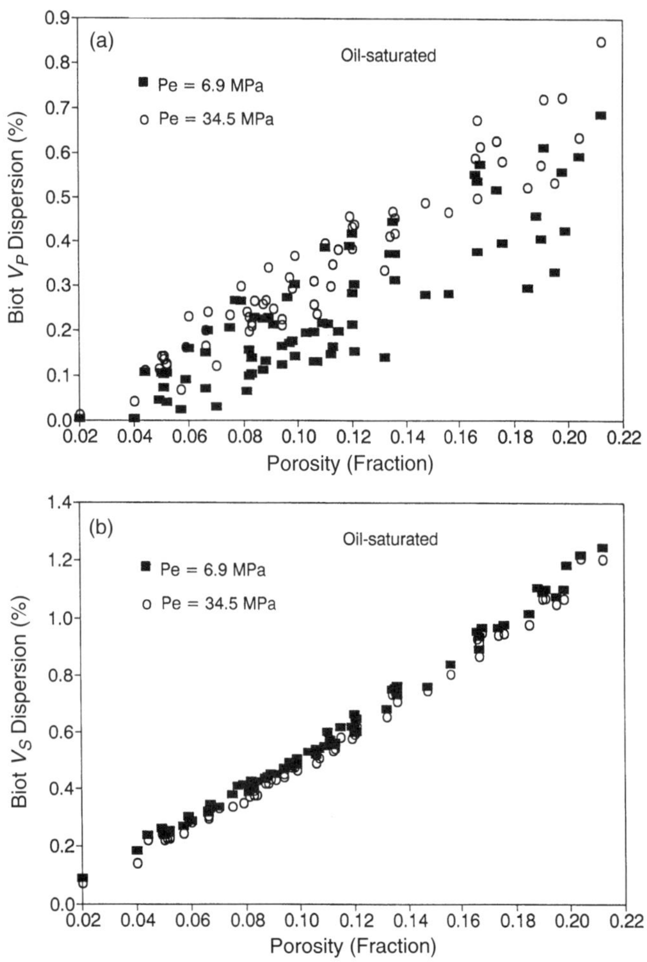

Fig. 7. Biot dispersion in (a) V_P and (b) V_S in 80 carbonate core samples saturated with 35° API oil at two effective pressures. The Biot theory predicts little velocity dependence on frequency.

Here α is assumed to be a constant of 3 for all the carbonate rock samples. In these carbonate core samples, V_P dispersion is below 1%, while V_S dispersion can be as large as 3–4%. Because, in reality, α is higher at higher effective pressures but is assumed to be constant in the calculations, the Biot dispersions are slightly dependent on the effective pressure.

Biot's theory does not predict much frequency dependence of compressional velocities. For practical purposes, the Biot high-frequency limit equations are considered the same as the low-frequency limit (Gassmann) equations for compressional velocities. For shear velocities, the Biot dispersion is only related to the tortuosity factor α (equation 28). For high-porosity, high-permeability rocks, α is small ($1 < \alpha < 2$) and has more impact on the V_S dispersion. Otherwise, the Biot V_S dispersion is negligible.

Kuster-Toksöz Model

Kuster and Toksöz (1974) derived a model (here referred to as the KT model) of wave velocity and attenuation in a two-phase medium based on the scattering theory. The KT model for a fluid-saturated rock with spheroidal inclusions gives the following equations:

$$K = \frac{3K_m(3K_m + 4\mu_m) + 4\mu_m(K_f - K_m)A}{3(3K_m + 4\mu_m) - 3(K_f - K_m)A} , \tag{40}$$

$$\mu = \frac{25\mu_m^2(3K_m + 4\mu_m) - \mu_m^2(9K_m + 8\mu_m)B}{25\mu_m(3K_m + 4\mu_m) + 6\mu_m(K_m + 2\mu_m)B} , \tag{41}$$

where K_m and μ_m are the bulk and shear moduli of the solid matrix, K_f is the bulk modulus of the pore fluid,

$$A = \sum_{n=1}^{N} c(\alpha_n)\, T_{iijj}(\alpha_n) , \tag{42}$$

$$B = \sum_{n=1}^{N} c(\alpha_n) \left[T_{ijij}(\alpha_n) - \frac{1}{3} T_{iijj}(\alpha_n) \right] . \tag{43}$$

Here, α_n is the nth aspect ratio of the inclusions, defined as the ratio between the short and long axis for a given pore (e.g., a round pore has an aspect ratio of 1, while a crack has a small aspect ratio), $c(\alpha_n)$ is the fractional concentration of α_n,

$$\sum_{n=1}^{N} c(\alpha_n) = \phi \text{ (porosity)},$$
$$T_{iijj} = \frac{3F_1}{F_2}, \quad (44)$$

and

$$T_{ijij} - \frac{1}{3}T_{iijj} = \frac{2}{F_3} + \frac{1}{F_4} + \frac{F_4F_5 + F_6F_7 - F_8F_9}{F_2F_4}. \quad (45)$$

The parameters in equations (44) and (45) are defined as follows:

$$F_1 = 1 - \left[\frac{3}{2}(g+\theta) - R\left(\frac{3}{2}g + \frac{5}{2}\theta - \frac{4}{3}\right)\right], \quad (46)$$

$$F_2 = 1 - \left[\frac{3}{2}(g+\theta) - R\left(\frac{3}{2}g + \frac{5}{2}\theta - \frac{4}{3}\right)\right] - \frac{1}{2}(3B-1)(3-4R)\left[g + \theta - R(g - \theta + 2\theta^2)\right] \quad (47)$$

$$F_3 = 1 - \frac{1}{2}\left[R(2-\theta) + \frac{1+\alpha^2}{\alpha^2}g(R-1)\right], \quad (48)$$

$$F_4 = 1 - \frac{1}{4}\left[3\theta + g - R(g-\theta)\right], \quad (49)$$

$$F_5 = B\theta(3-4R) - R\left(g + \theta - \frac{4}{3}\right) + g, \quad (50)$$

$$F_6 = B(1-\theta)(3-4R) - g + R(g+\theta), \quad (51)$$

$$F_7 = 2 - \frac{1}{4}\left[9\theta + 3g - R(5\theta + 3g)\right] + B\theta(3-4R), \quad (52)$$

$$F_8 = B(1-\theta)(3-4R) - 1 + 2R - \frac{g}{2}(R-1) - \frac{\theta}{2}(5R-3), \quad (53)$$

$$F_9 = B\theta(3-4R) - g(R-1) + R\theta, \quad (54)$$

where

$$B = \frac{K_f}{3K_m}, \quad (55)$$

$$R = \frac{3\mu_m}{3K_m + 4\mu_m} = \left(\frac{V_S}{V_P}\right)_m^2, \quad (56)$$

$$g = \frac{\alpha^2}{1-\alpha^2}(3\theta - 2), \quad (57)$$

$$\theta = \frac{\alpha}{(1-\alpha^2)^{3/2}}\left[\cos^{-1}\alpha - \alpha(1-\alpha^2)^{1/2}\right], \quad (58)$$

and α is the aspect ratio.

The KT model must be used with caution. Equations (40) through (58) are based on the assumption that the inclusion concentration is small and noninteracting, that is,

$$\sum_{m=1}^{M} \frac{c(\alpha_m)}{\alpha_m} < 1. \quad (59)$$

Whenever this assumption is violated, the KT model breaks down. Furthermore, the distribution of the aspect ratio is difficult to obtain in a rock sample. In fact, one can adjust the pore aspect ratios and their distribution to fit any laboratory velocity data. Figure 8 shows both the measured and calculated V_P and V_S using the KT model versus net overburden pressure (equivalent to effective pressure) in a limestone sample. In the calculation, we forced the pore aspect ratios and their distribution to fit the measured velocity data.

Limitations of Theories

Although theoretical models play an important role

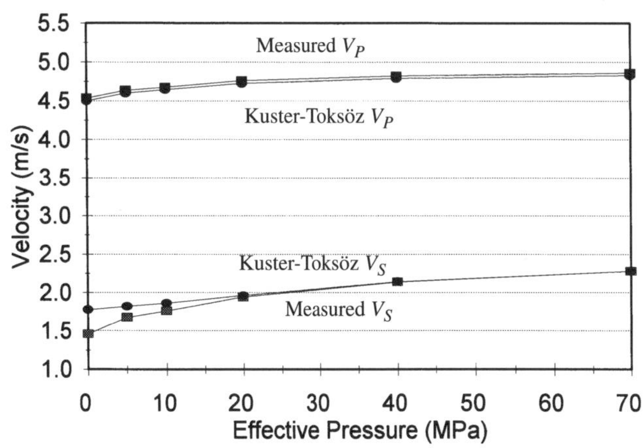

Fig. 8. Measured and calculated velocities using the Kuster-Toksöz model in a brine-saturated limestone (porosity, 14%). Because the real values are unknown, we maneuvered the pore aspect ratios and their distributions until the calculated velocities fit the measured data.

in our understanding of rock properties, they do have serious limitations. First, all theories have to make substantial assumptions about the rock, fluid, and rock-fluid interactions. These assumptions set the boundaries within which the theories are valid. Theories are often used outside their assumption boundaries, causing erroneous results. Second, rocks are so complicated that simple mathematics are inadequate to describe them. In more complicated models attempting to describe rocks with more accuracy, the mathematics are often so complicated that the models are of little use. Third, most models require many input parameters that are difficult and sometimes impossible to obtain.

All theoretical models should be used in light of a full understanding of their assumptions and limitations. Theoretical results are often different from the corresponding laboratory measurements. Further work is needed to better understand such differences.

Carbonate Pore Types and Their Effects on Seismic Velocity

Choquette and Pray (1970) classified carbonate porosities into 15 basic types which, combined with other elements (modifiers), provide a detailed geologic characterization of almost all types of porosity in sedimentary carbonates. The most important and frequently observed types of porosity are interparticle, intercrystalline, moldic, vug, intraparticle, fracture, and fenestral. The porosity type not only determines hydrocarbon migration, storage, and reserve economics but also the seismic properties of carbonate rocks.

Seismic velocities in sedimentary rocks are affected not only by porosity but more importantly by pore type and shape. Intuitively, seismic velocities are nothing but a measure of the deformability (compressibility and rigidity) of the rock, which in turn depends on the shape and amount of pores. A flat, thin pore is easy to deform so seismic velocities are low in rocks containing this type of pore. In contrast, round cavities and circular holes are hard to deform so seismic velocities are high, even though the porosity may be high. In this section, we briefly summarize seismic velocities in carbonate rocks in relation to their major porosity types.

Intercrystalline and Interparticle Porosity

Intercrystalline porosity means porosity between crystals of approximately equal size. It can be of either primary or secondary origin. Interparticle denotes porosity between particles of any size. It is generally of primary origin (depositional), although occasionally it is of secondary origin (postdepositional).

The common feature of intercrystalline and interparticle porosity is the high internal surface area to porosity ratio, where internal surface area means the surface area of the pores. Such pores are usually irregular and angular in shape so they are easy to deform (Figure 9a). Figure 9b shows a dolomite sample (diameter, 3.8 cm; porosity, 10.8%) with intercrystalline porosity. In terms of seismic velocities, both V_P and V_S are relatively low at a given porosity and are strongly dependent on the effective or net overburden pressure (overburden pressure minus pore pressure) because higher pressures create better contacts among the crystals or particles. The better contacts, however, affect the bulk modulus (incompressibility) more than the shear modulus so that V_P is affected more by pressure than V_S. As a result, V_P/V_S increases as pressure increases (Figure 9c).

Moldic and Intraparticle Porosity

Moldic pores are created mostly by the selective solution of individual constituents in carbonate rocks. Such individual constituents can be shells, grains, salt, or plant roots. Moldic pores usually have regular shapes and are of secondary origin. Intraparticle and intracrystal pores are the pores within individual particles, grains, or crystals. Most of these pores are of primary origin. Primary intraparticle pores are seen mostly as internal chambers or other openings within individual or colonial skeletal organisms. Secondary intraparticle pores form chiefly by solution and borings (Choquette and Pray, 1970).

Rocks with moldic or intraparticle pores usually have strong frames and low internal surface area to porosity ratios. Figure 10a shows some moldic pores in a dolomite sample shown in Figure 10b (diameter, 3.8 cm; porosity, 15.5%). The pores in this dolomite are mostly in the shape of tubes with circular cross-sections, perhaps as a result of solution of fusulinid-type organisms. Velocities in rocks with moldic or intraparticle pores are relatively insensitive to pressure changes and are generally high at a given porosity because both these pores are difficult to deform. Figure 10c shows that V_S in this gas-saturated core sample is less dependent on pressure than V_P, so V_P/V_S increases as pressure increases. Also, the velocity values are uncharacteristically high for a rock with 15.5% porosity.

Vug and Channel Porosity

Choquette and Pray (1970) classified a vug as a pore that is (1) approximately equant or not markedly elon-

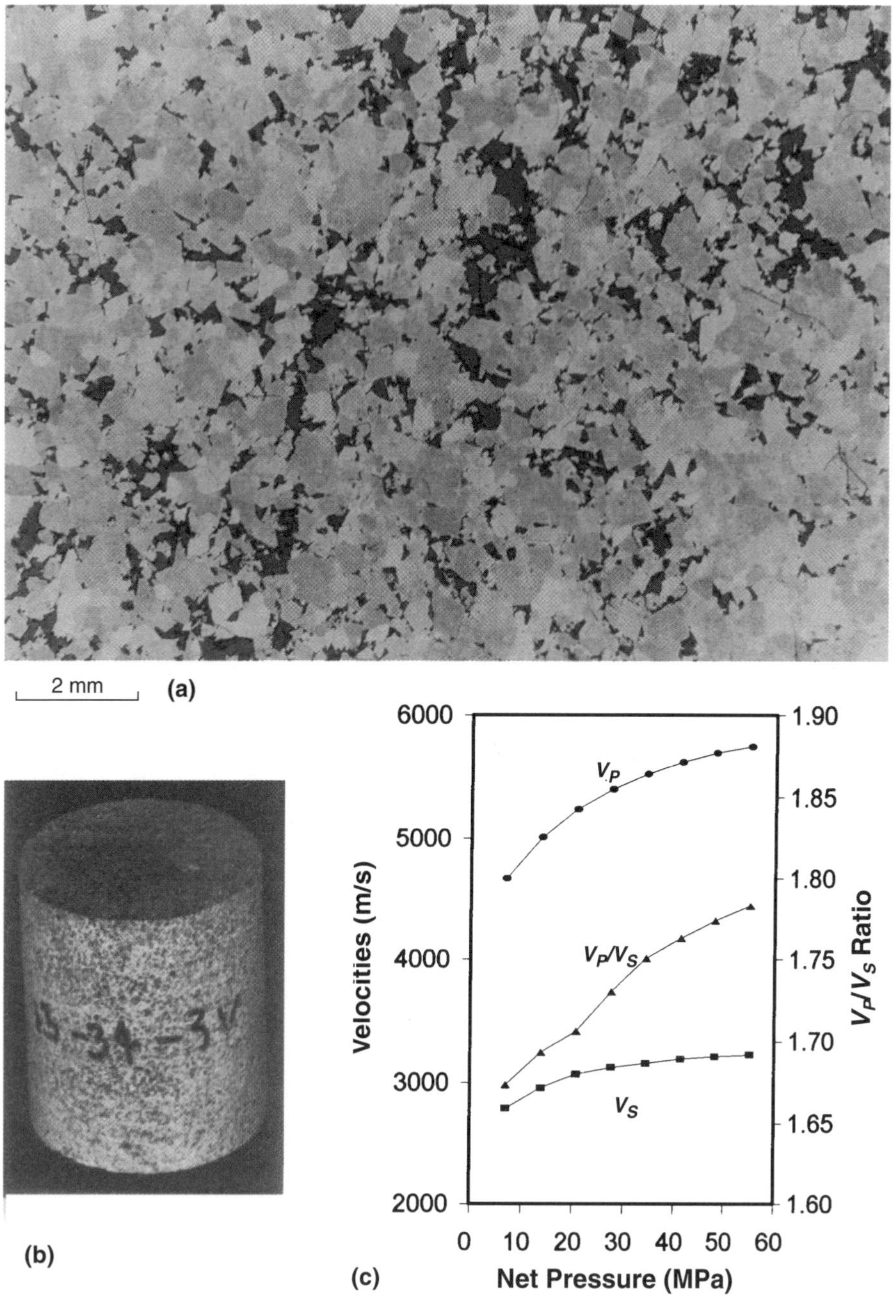

Fig. 9. (a) Thin section of (b) a dolomite core sample with intercrystalline porosity (porosity, 10.8%; diameter, 3.8 cm). (c) Velocities in rocks with interparticle or intercrystalline pores are usually low for the porosity they have and sensitive to pressure changes.

gate, (2) large with a diameter greater than 1/16 mm, and (3) not fabric selective. Vugs occur mainly as a result of solution and are of secondary origin. Channel pores are defined on the basis of their shape. This term is used to describe those markedly elongate pores or irregular openings with elongation or continuity in one or two dimensions relative to the third dimension. Channels are usually a result of solution along fractures or enlargement of other types of pores.

Rocks with vuggy pores also have strong frames and low internal surface area to porosity ratios. In terms of velocity, these rocks behave similarly to those with moldic pores. That is, both V_P and V_S are relatively high at a given porosity and are insensitive to pressure changes because of their strong frames and low internal surface area to porosity ratios. Figure 11a shows a thin section of a large vug in a dolomite sample that is shown in Figure 11b (diameter, 8.9 cm, porosity, 7.9%). The velocities measured in this gas-saturated core sample are shown in Figure 11c; note that V_P, V_S, and V_P/V_S all increase slowly as pressure increases.

Rocks with channel porosities are usually easy to deform, especially when the channels result from fracture enlargement by solution or are formed by connected

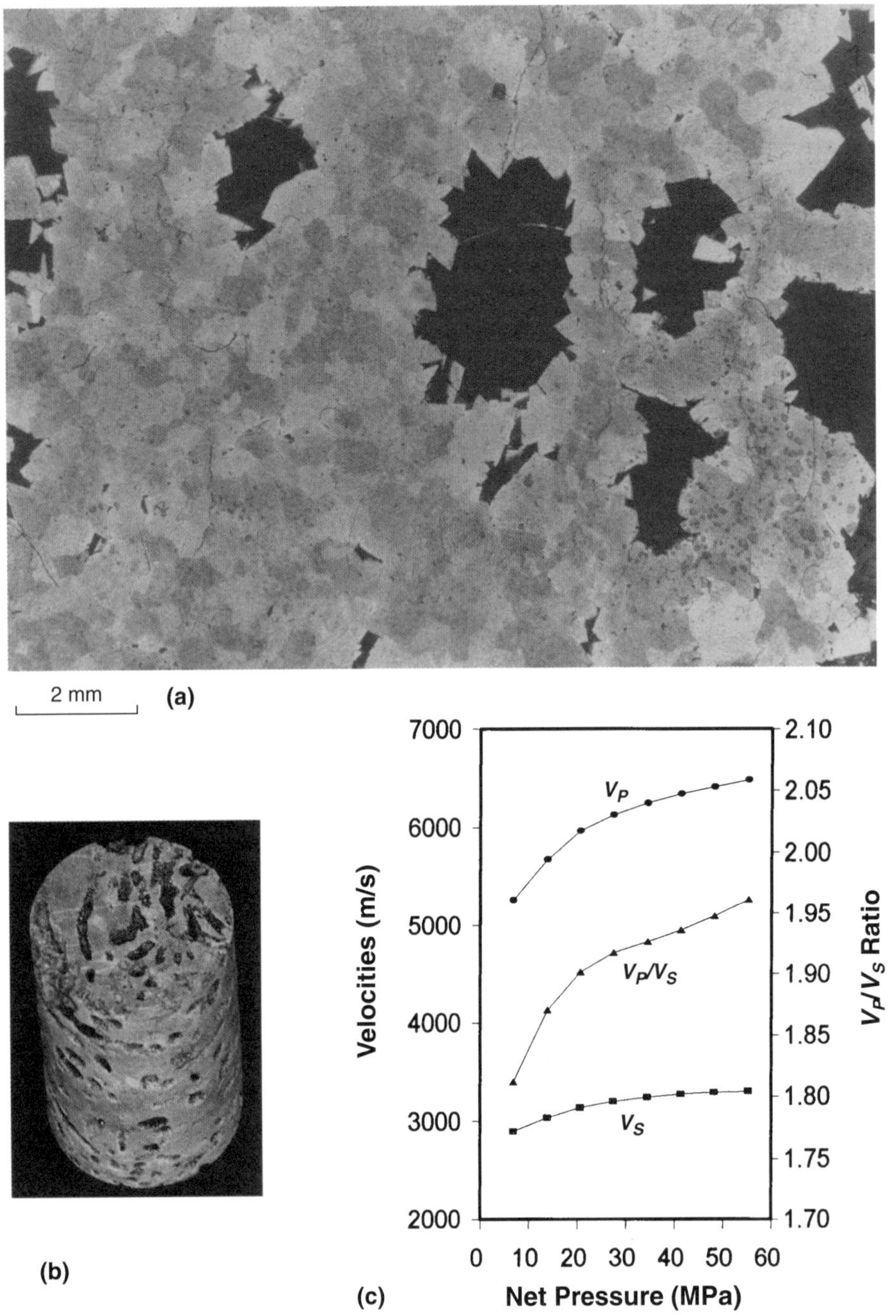

Fig. 10. (a) Thin section of (b) a dolomite core sample with moldic porosity (porosity, 15.5%; diameter, 3.8 cm). (c) Velocities in rocks with moldic pores are usually uncharacteristically high for the porosity they have due to the relatively low compressibility of moldic pores.

vugs. Consequently, these rocks usually show relatively low velocities at a given porosity. Figure 12a shows a dolomite core sample with channel pores or connected vuggy pores (diameter, 3.8 cm; porosity, 14.1%). The velocities in this gas-saturated core sample are relatively low (Figure 12b). As pressure increases, both V_P and V_S change slope a few times, reflecting that some pores may have been crushed during the process.

Fenestral Porosity

Fenestral pores generally mean openings that are larger than the grain-supported interstices. They are sometimes synonymous with birdseye pores. Their shapes can be rounded, lenticular, or irregular. They are basically of secondary origin and formed by decay of sediment-covered algal mats, shrinkage during drying, and accumulation of pockets of gas or water shortly after deposition (Choquette and Pray, 1970). The thin section in Figure 13a shows fenestral porosity in a dolomite core shown in Figure 13b (diameter, 3.8 cm; porosity, 11.7%).

In terms of seismic velocities, rocks with fenestral pores behave similarly to those with intercrystalline

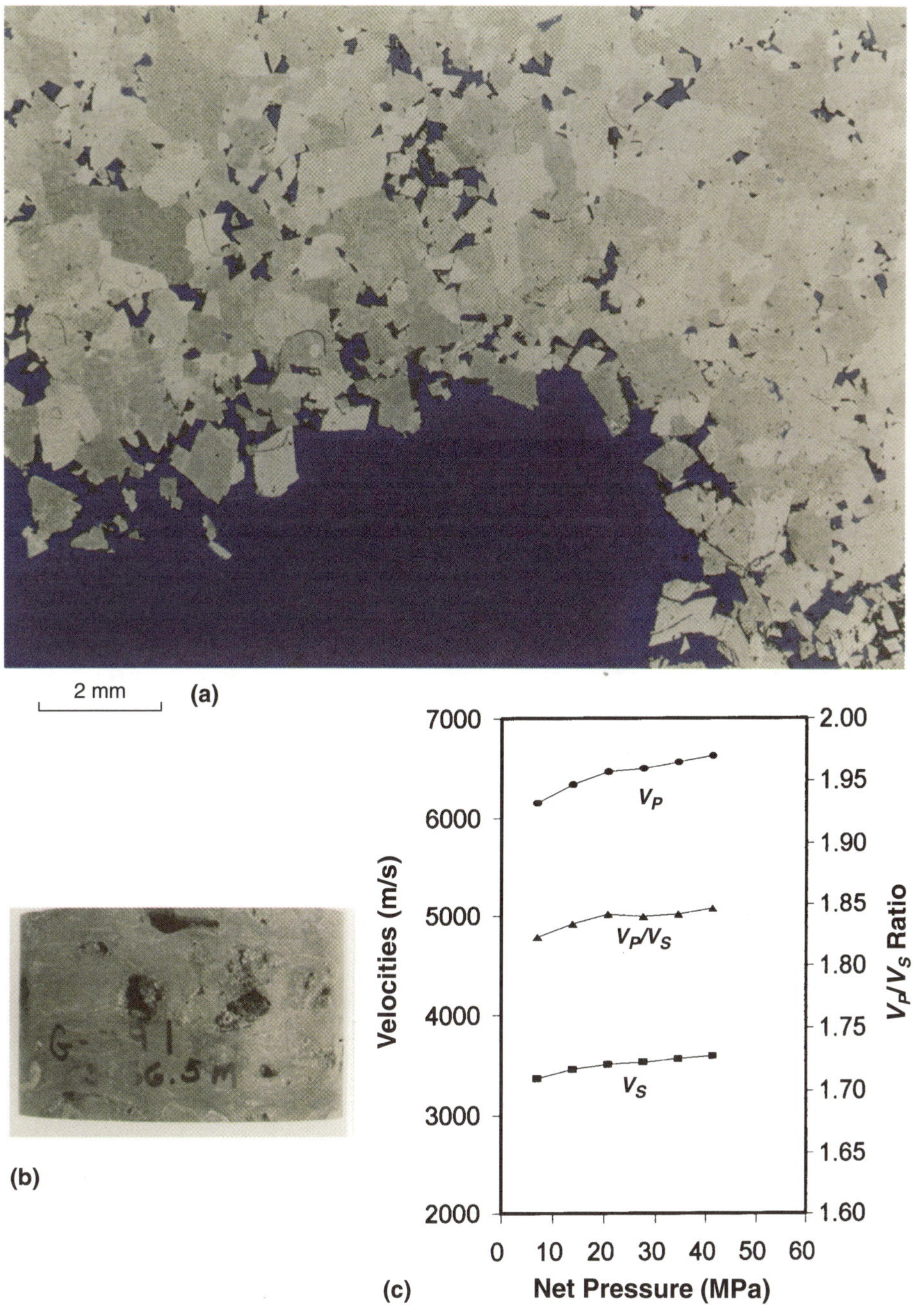

Fig. 11. (a) Thin section of (b) a dolomite core sample with vug porosity (porosity, 7.9%; diameter, 8.9 cm). (c) Velocities in rocks with isolated vugs are usually uncharacteristically high for the porosity they have due to the relatively low compressibility of vugs.

pores. At a given porosity and pressure, they have relatively low V_P and V_S, but as pressure increases, these velocities increase due to the deformation of the pores. Since V_P increases faster than V_S, V_P/V_S also increases in this gas-saturated dolomite sample (Figure 13c).

Fracture and Breccia Porosity and Stylolites

Fracture porosity is formed by fracturing. Fracture porosity evolves to breccia porosity with increasing dislocation between the opposing fracture walls. A unique feature of a fracture is that both V_P and V_S depend on the direction of wave propagation relative to the fracture's orientation (anisotropic) because the rock is most compliant in the direction perpendicular to the fracture. If fractures are abundant and randomly oriented in a rock, the rock will behave isotropically in terms of seismic properties.

Figure 14a shows a thin section taken from a dolomite sample (diameter, 3.8 cm; porosity, 3.1%) with vertical fractures (Figure 14b). Figure 14c shows that V_P was measured vertically so it is not affected by the fractures; V_S was also measured vertically but polarized across the fractures. At low pressures, V_S in this gas-sat-

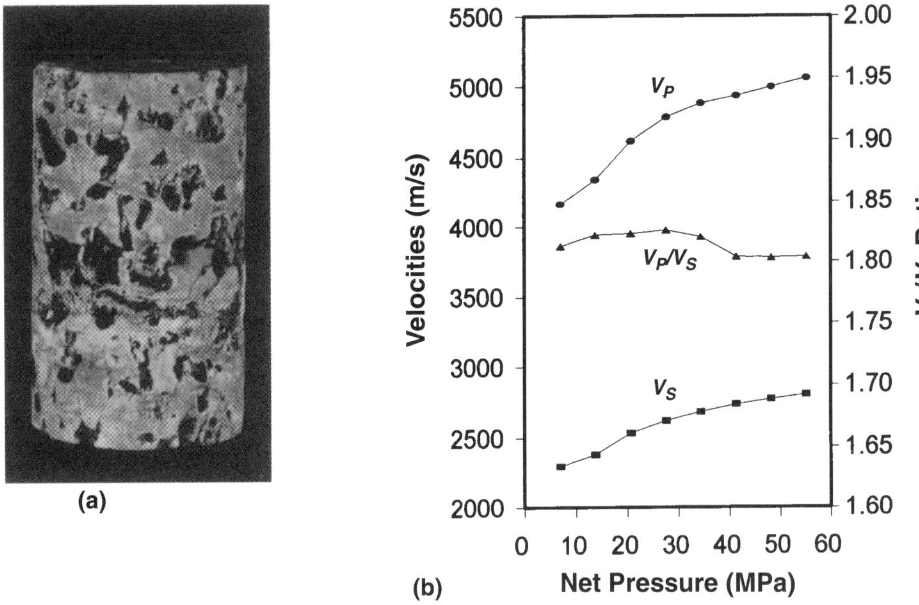

Fig. 12. (a) A dolomite core sample with vug and channel porosity (porosity, 14.1%; diameter, 3.8 cm). (b) Velocities in rocks with connected vugs or channel porosity are usually low.

urated dolomite sample is low because of the fractures. As pressure increases, the fractures are closed so V_S increases. This translates to a decreasing V_P/V_S in the low-pressure range and an increasing V_P/V_S in the high-pressure range when V_P/V_S is plotted versus pressure.

Fractures contribute little to the total porosity of the rock, but they can greatly decrease the velocities because of their high compliance. In deep reservoirs (greater than 1.5 km), however, most thin fractures are probably closed, with minimal effect on *P*-wave velocities.

Stylolites are depositional features of thin layers of organic or clay materials. Figure 15a shows a thin section of some stylolites in a dolomite sample shown in Figure 15b (diameter, 3.8 cm; porosity, 1.3%). Rocks with stylolites usually have abnormally low velocities, especially at low pressures because stylolites are soft and easy to deform. Both V_P and V_S increase substantially as pressure increases and the rock sample is better compacted. Figure 15c shows the velocities versus pressure in the gas-saturated dolomite sample.

Although we have discussed the effect of various types of pores on seismic velocities, rocks in nature almost always contain more than one type of pore. The most compliant pores control the magnitude of and pressure effect on seismic velocities.

Other Factors Affecting Seismic Velocities

Seismic velocities in rocks are affected in a complex way by many factors. In this section, we discuss some of these factors and their relationship to seismic velocities.

Mineralogy

Carbonate reservoir rocks are mostly formed by dolomite [$CaMg(CO_3)_2$] and calcite ($CaCO_3$) which are sometimes accompanied by evaporites such as halite (NaCl), anhydrite ($CaSO_4$), and gypsum ($CaSO_4 \bullet 2H_2O$). Figure 16 shows both V_P and V_S versus bulk density in these nonporous minerals. For comparison purposes, V_P and V_S in pure quartz are also shown. Dolomite has the highest seismic velocities, while halite has the lowest among the minerals associated with carbonate rocks.

Porosity

With all other parameters fixed, seismic velocities should decrease as porosity increases. Such a velocity-porosity relationship is the foundation of sonic logging, which derives porosity from seismic wave traveltimes. Figure 17 shows V_P and V_S versus porosity in 173 gas-saturated carbonate core samples measured in the laboratory under fixed pressure and temperature conditions. The data are scattered, suggesting that porosity is not the only parameter that controls velocity. Dolomite and limestone are separated in terms of velocity values. In general, however, both V_P and V_S decrease with increasing porosity.

In reality, the relationship between seismic velocity and porosity is complicated by geometric pore types and mineral content. As discussed in the previous section, rocks with thin or flat pores have low velocities even though porosity may be low. In contrast, seismic velocities can be high in some high-porosity carbonate rocks if

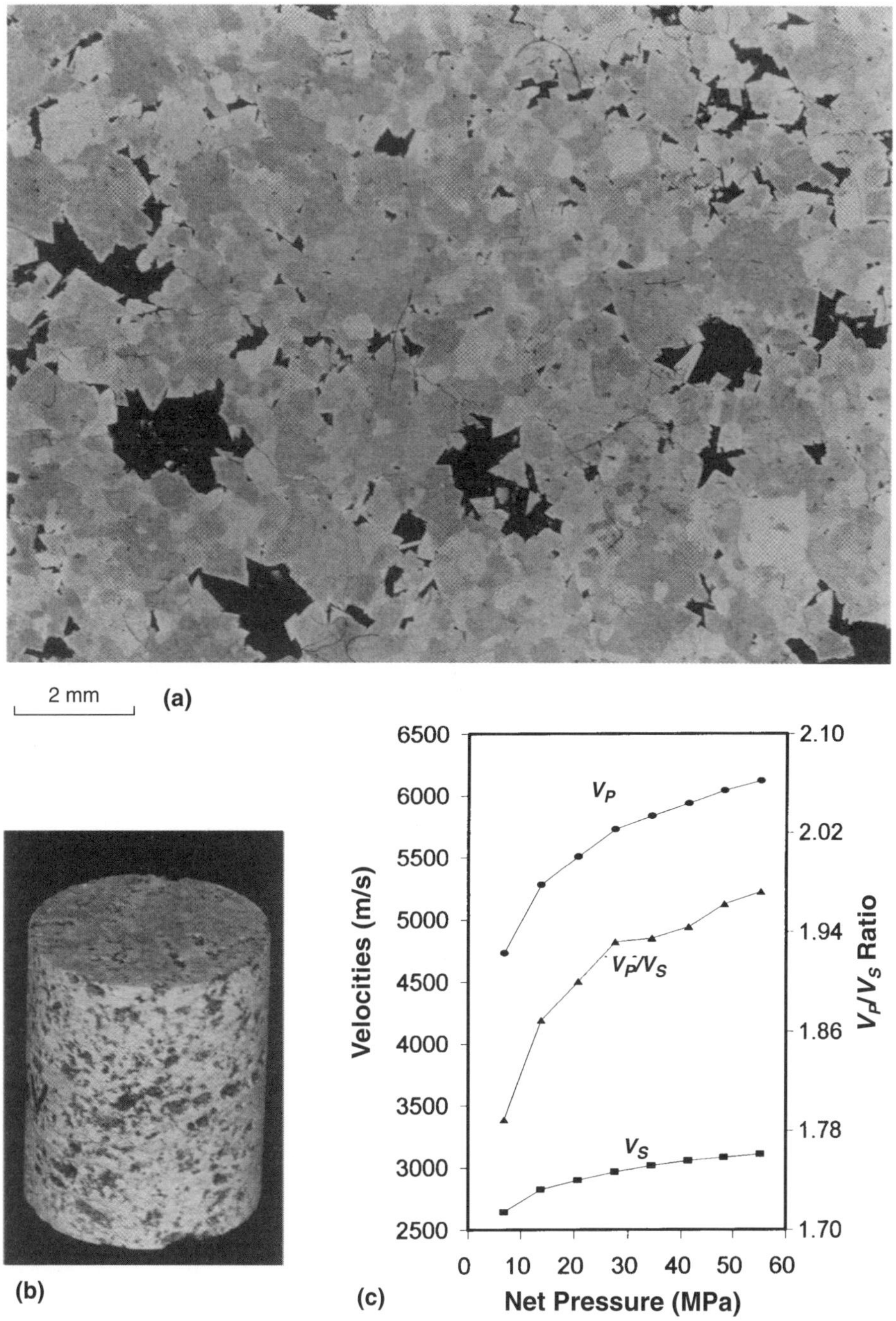

Fig. 13. (a) Thin section of (b) a dolomite core sample with fenestral porosity (porosity, 11.7%; diameter, 3.8 cm). (c) Like velocities in rocks with intercrystalline pores, velocities in rocks with fenestral pores are usually low for the porosity they have.

the pores are in the shape of round vugs or molds. Figure 18a shows V_P and V_S versus porosity in dolomites for some typical pore types at a pressure of 6.9 MPa. The dolomite sample with 8% vug porosity has velocities higher than those for a nonporous dolomite sample with stylolites. Figure 18a also suggests that the scatter in the velocity-porosity relationship may be caused mostly by the pore type or by differences in pore shape. At higher pressures, some thin, flat pores or fractures close, causing the velocity scatter to decrease (Figure 18b).

Saturation, Pore Fluid Type, and Velocities in Pore Fluids

Theoretically, the shear modulus of a rock is not affected by the pore fluid substitution so that the shear velocity is only affected by the bulk density change. The compressional velocity is affected by the effect of pore fluid on both the bulk density and the bulk modulus. In most reservoir rocks, a few percent of gas saturation can decrease the V_P substantially because gases are much more compressible than other reservoir fluids (oil and brine). In carbonate rocks, experimental results do not

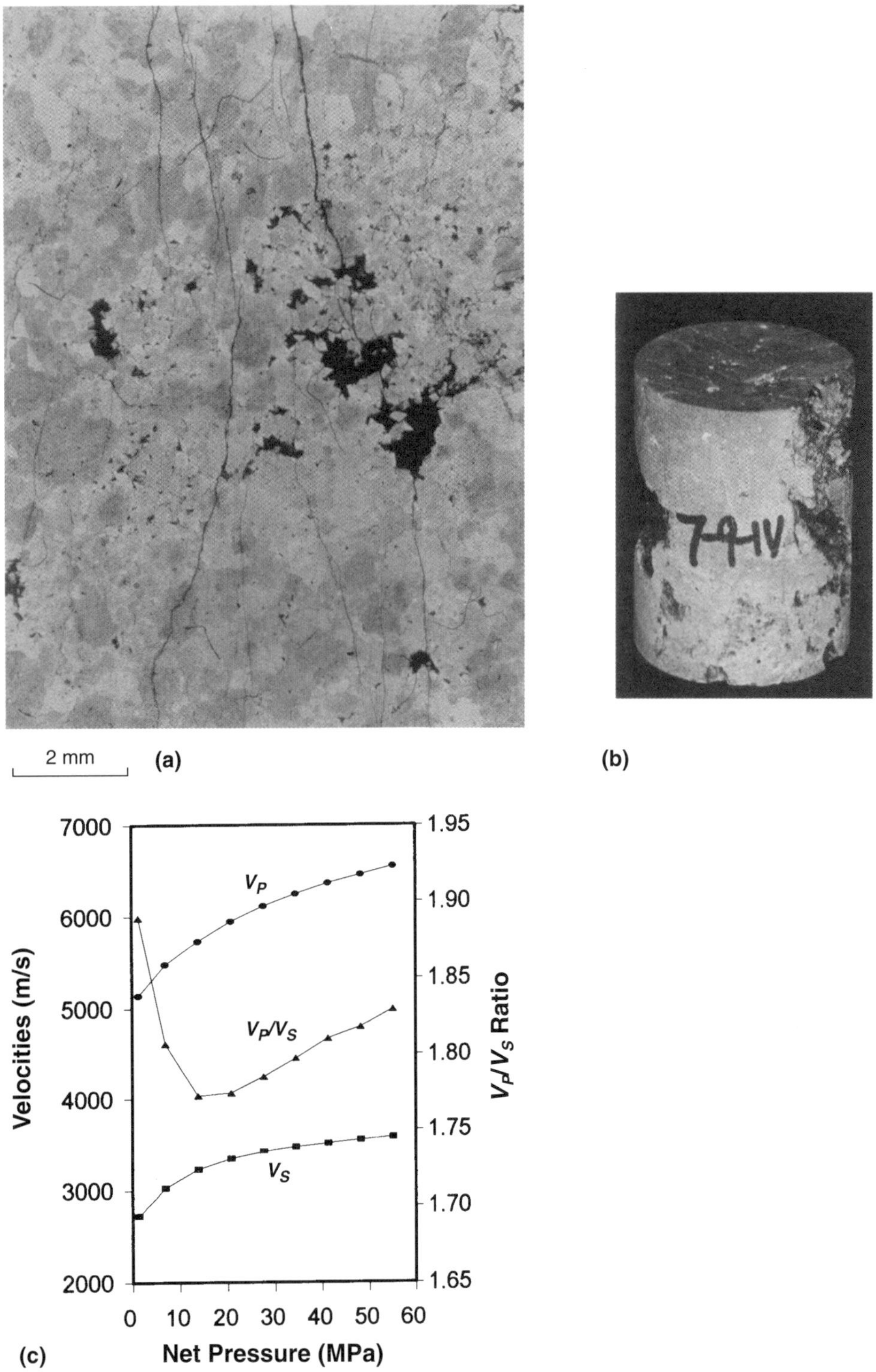

Fig. 14. (a) Thin section of (b) a dolomite core sample with fractures (porosity, 3.1%; diameter, 3.8 cm). (c) Velocities in rocks with fractures are usually low and sensitive to pressure changes and are anisotropic if the fractures are aligned.

always follow theoretical predictions made by the Gassmann equation or Biot theory (Figures 5 and 6) because of the vast variation in pore types, shapes, and connectivity.

Although theoretical calculations are not always reliable due to built-in assumptions, care must also be taken in using laboratory results. For example, laboratory results of V_P and V_S as functions of fluid saturation in rocks are dependent on the wettability of the rock and on whether the saturation is due to imbibition or drainage. As a fully liquid-saturated rock is drained, the large, well-connected pores are drained first due to the capillary effect, causing the rock to be stiff before the small, thin, and flat pores are drained, resulting in a higher than

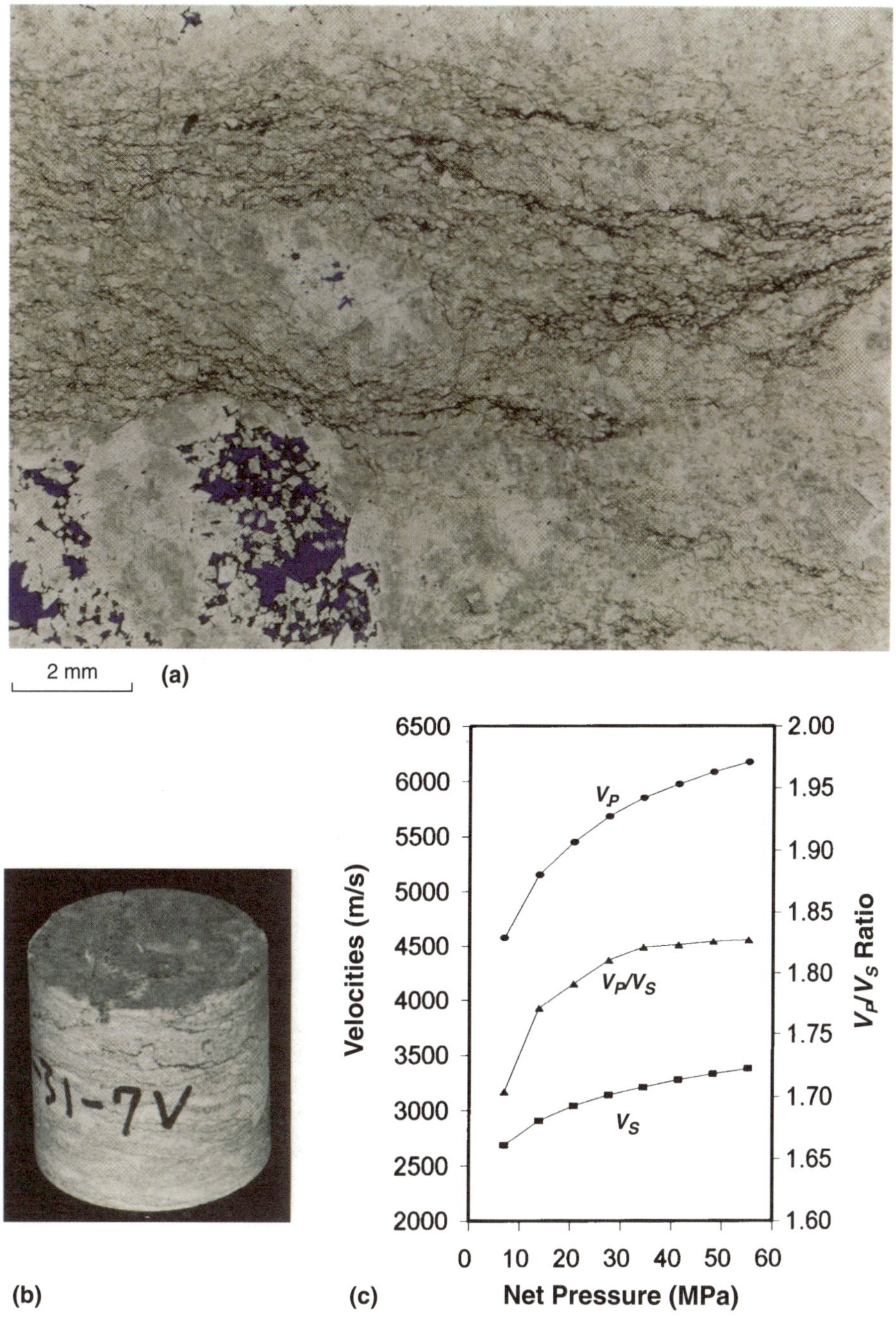

Fig. 15. (a) Thin section of (b) a dolomite core sample with stylolites (porosity, 1.3%; diameter, 3.8 cm). (c) Because stylolites are usually more compressible than the host rocks, velocities in rocks with stylolites are usually low and sensitive to pressure changes.

expected V_P. In contrast, as a liquid is imbibed into a gas-saturated rock, the large, well-connected pores are filled up first, so that the rock's moduli and V_P are barely affected until the thin pores are filled up. This velocity hysteresis phenomenon is especially common in carbonate rocks because of their wide spectrum of pore sizes and shapes. Furthermore, the hysteresis is frequency dependent. At extremely low frequencies (subseismic in some cases), the hysteresis should disappear.

The effect of total fluid substitution on seismic velocities is further complicated by pore types. If a rock contains mostly round pores such as vugs or molds, substituting the gas for brine or oil does not change the V_P much because the pores are relatively incompressible (Figure 19). In contrast, fluid substitution should have a larger effect on the V_P in rocks with flat pores or grain contacts because of the higher compressibility of these pores and contacts (Figure 20).

The effect of total fluid substitution on seismic velocities is also influenced by the pore fluid type. For example, high-salinity brines are less compressible than low-salinity brines and heavy oils are less compressible

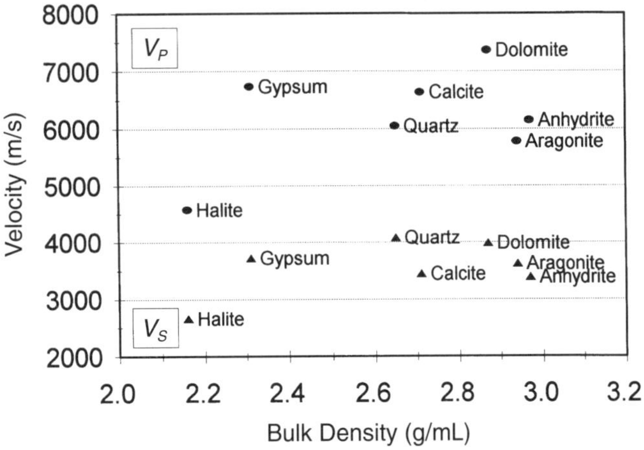

Fig. 16. Seismic velocities in carbonate-forming minerals. The velocities in pure quartz are also plotted for comparison. Dolomite has the highest velocities among the minerals.

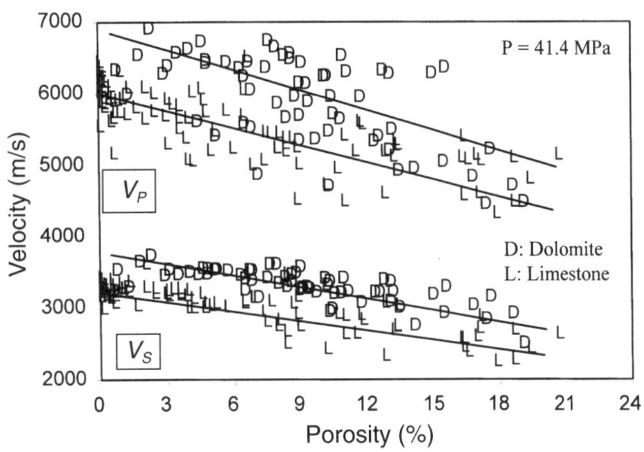

Fig. 17. Velocities (V_P and V_S) versus porosity in 173 gas-saturated carbonate core samples. The severe scatter in the velocity-porosity relationship suggests that porosity is not the only parameter affecting seismic velocities. Overall, dolomites show higher velocities than limestones at a given porosity.

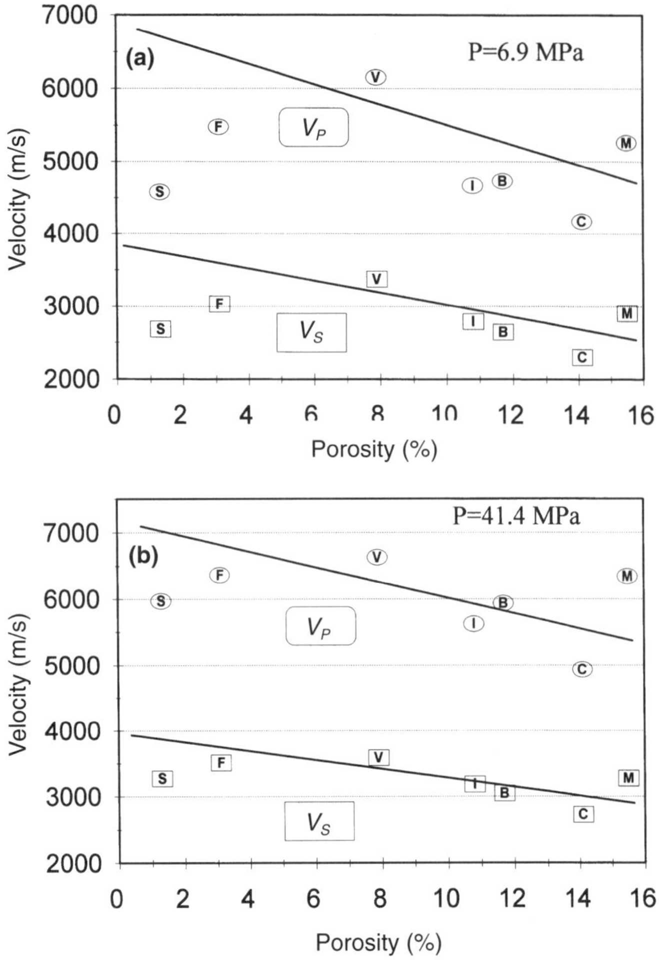

Fig. 18. Velocities (V_P and V_S) versus porosity in dolomite core samples for some typical pore types at (a) 6.9 MPa and (b) 41.4 MPa effective pressures. Key to pore types: S, stylolite; F, fracture; V, vug; I, intercrystalline or interparticle; B, birdseye or fenestral; C, channel; M, moldic. The solid lines represent the velocity-porosity trends shown in Fig. 17.

than light oils. Gases are extremely compressible, with bulk moduli ranging from a few megapascals to 500 MPa, depending on pressure, temperature, and gas gravity. In most practical applications, the bulk modulus and density of gases can be calculated using equations of state (Batzle and Wang, 1992).

The bulk density and velocity of oils can be calculated by the following equations given by Batzle and Wang (1992) and Wang (1988):

$$\rho_{oil} = \frac{\rho_0 + (0.00277P - 1.71\times10^{-7}P^3)\,(\rho_0 - 1.15)^2 + 3.49\times10^{-4}P\ (\text{g/cm}^3)}{0.972 + 3.81\times10^{-4}\,(T + 17.78)^{1.175}}, \quad (60)$$

$$V_{oil} = \frac{15540}{\sqrt{77.1 + \text{API}}} - 3.71T + 4.64P + 0.0115\,(0.36\sqrt{\text{API}} - 1)\,TP\ (\text{m/s}), \quad (61)$$

where P is pressure in MPa, T is temperature in °C, and ρ_0 is the bulk density at 15.6°C and 1 atm and is related to API gravity by

$$\text{API gravity} = \frac{141.5}{\rho_0} - 131.5\ . \quad (62)$$

The bulk modulus of brine and water changes less dramatically than that of oil as pressure and temperature change. In general, both the density and bulk modulus of

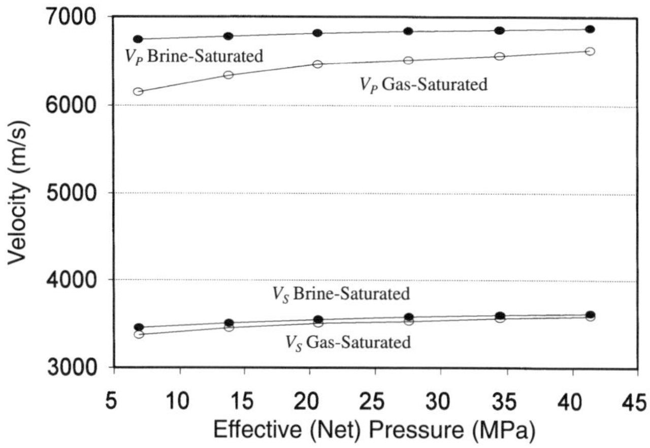

Fig. 19. Velocities (V_P and V_S) in a dolomite core sample (porosity, 7.9%) with vug and moldic porosity. Because most vugs and moldic pores are relatively incompressible, full substitution of a brine for the gas in these pores changes V_P only a little.

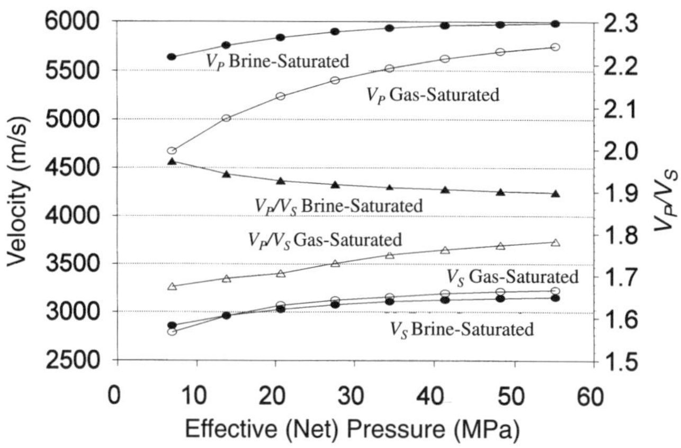

Fig. 20. Velocities (V_P and V_S) in a dolomite core sample (porosity, 10.8%) with intercrystalline porosity. Because intercrystalline pores are relatively compressible, full substitution of a brine for the gas in these pores increases V_P dramatically. As a result, V_P/V_S also increases.

brine are influenced by pressure, temperature, salinity, and gas in solution. For a set of equations calculating the velocities and bulk densities in water and brine, see Batzle and Wang (1992) (note that the last term in their equation 29 should be $-1820S^2$ which was incorrectly typed as $-820S^2$).

Depth and Pressure

The overburden pressure of a rock is directly proportional to the burial depth. Normally, overburden pressure increases by about 23 kPa/m, or 1.0 psi/ft. Pore fluid pressure is also proportional to the depth or hydraulic column with a gradient of about 9.72 kPa/m, or 0.43 psi/ft. Seismic velocities in general increase as the burial depth or overburden pressure of the rock increases because at depth, rocks are usually better compacted, fractures may be closed, and rock grains, if any, are in better contact. The effect of pore pressure is to counteract that of overburden pressure. Therefore, it is the difference between overburden pressure and pore pressure that affects the seismic velocities. Such a pressure difference is often called differential pressure, net overburden pressure, net pressure, or effective pressure.

The sensitivity of seismic velocities to depth or pressure in carbonate rocks depends on the pore type or shape. For rocks with interparticle and intercrystalline pores, fractures, or stylolites, both V_P and V_S increase rapidly in the low-pressure range (<20 MPa) as the net overburden pressure increases because most fractures and thin, flat pores close at a modest pressure. At higher pressures, the velocities become insensitive to further pressure changes (Figure 20). For rocks with mostly round vugs or molds, both V_P and V_S increase slowly as the net overburden pressure increases due to low compressibility of the pores (Figure 19). However, the velocity-pressure relationship in rocks is dominated by the presence, not necessarily the amount, of thin pores, cracks, and fractures.

If a rock is fully saturated with a liquid such as brine, V_P may become much less sensitive to pressure changes than in the same rock saturated with a gas, especially in rocks with thin pores. As the pores are fully filled with liquids, they become much less compressible due to the incompressibility of the liquid, resulting in lower compressibility of the overall rock (Figures 19 and 20).

Bulk Density

From their definitions, it would seem that both V_P and V_S decrease as the bulk density increases. In reality, as the bulk density increases, the elastic moduli usually increase faster than the bulk density so that velocities actually increase. In fully gas-saturated rocks, the pore fluid affects the bulk density little so that the velocities are affected by the bulk density through the mineral content (e.g., dolomite in limestones) and the porosity. Therefore, both V_P and V_S should increase with increasing bulk density because lower bulk density means higher porosity or higher content of lighter minerals, which are more compressible.

Figure 21 shows both V_P and V_S versus bulk density in 173 gas-saturated carbonate samples. The solid lines in the figure represent the best fit to the data according to the following linear relationships:

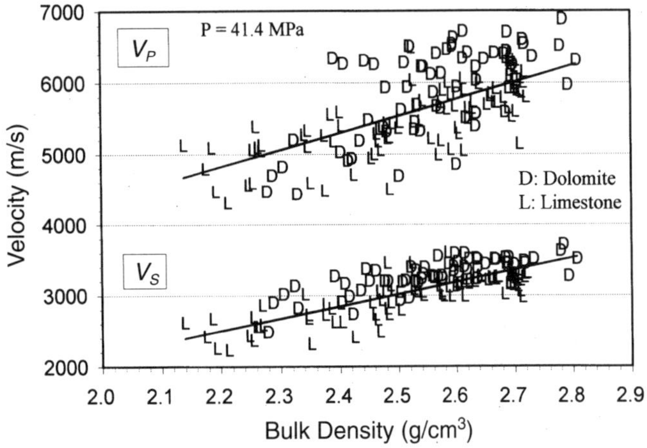

Fig. 21. Velocities (V_P and V_S) versus bulk density in 173 gas-saturated carbonate core samples. Although a trend clearly exists that both V_P and V_S increase with increasing bulk density, the correlation between the velocities, especially V_P, and the bulk density is poor.

$$V_P = -409 + 2377\rho_b \text{ (m/s)}, \quad R^2 = 0.27, \tag{63}$$

$$V_S = -1275 + 1717\rho_b \text{ (m/s)}, \quad R^2 = 0.62, \tag{64}$$

where ρ_b is the bulk density (in g/cm^3) and R^2 is the correlation coefficient. Figure 21 shows that systematic relationships exist between both V_P and V_S and bulk density with severe scatter. The scatter is caused mostly by differences in pore type and shape.

As the rocks become fully saturated with a liquid, V_P increases while V_S decreases slightly. The relationship between the velocities and bulk density still exists, but the correlations in equations 63 and 64 change slightly.

Wave Frequency

An elastic wave is said to be dispersive if the velocities are dependent on the frequency. As mentioned before, most laboratory measurements of seismic velocities are performed at ultrasonic frequencies. In fully gas-saturated or dry rocks, seismic velocities are essentially independent of wave frequency (Spencer, 1981). Our laboratory measurements on gas-saturated rocks, including carbonates, have proven that seismic waves are indeed nondispersive in dry or fully gas-saturated rocks. In both partially and fully liquid-saturated rocks, evidence shows that seismic waves are dispersive so that much of the wave dispersion is caused by pore fluids.

Two important dispersion mechanisms exist for liquid-saturated rocks: the Biot mechanism and the local flow mechanism. The Biot (1956) mechanism accounts for the inertia effect of pore fluid relative to the solid rock frame. Because fluids do not support the applied shear stress, Biot assumes that the sharp contrast in shear moduli causes an inertia effect in which the pore fluid moves behind the solid rock frame as elastic waves propagate through the rock. The inertia effect in turn causes wave dissipation and dispersion. Intuitively, the higher the rock's permeability, the higher the inertia effect and the dispersion. In reality, the Biot dispersion for both V_P and V_S in most rocks is small to negligible, as shown in Figure 7.

The local flow mechanism (O'Connell and Budiansky, 1977; Mavko and Nur, 1978; Winkler, 1986; Wang and Nur, 1990b) accounts for the effect of pore-scale fluid flow and pressure equilibrium on the elastic moduli. Because different types of pores have different compressibilities, the passing wave in this mechanism induces fluid flow from compliant pores to the stiff, less compliant pores. If such a pore-scale fluid flow cannot equilibrate within a half-period of the passing wave, the compliant pores appear to be less compressible due to the support of the pore fluid. As a result, the velocities appear to be higher. In this theory, the lower the rock's permeability or the higher the pore fluid's viscosity, the more time it takes the pore fluid to equilibrate.

One way to estimate the magnitude of total velocity dispersion is to calculate the difference between the laboratory-measured and Gassmann-calculated velocities in liquid-saturated rocks (Winkler, 1986; Wang and Nur, 1990b). This method gives the upper limit of wave dispersion in fluid-saturated rocks. As discussed before, the Gassmann equation is only approximate because of its many assumptions. Because of the difficulties in directly measuring velocities as a continuous function of frequency in liquid-saturated rocks in the laboratory, the problem of wave dispersion remains unsolved at this time.

Applications of Rock Physics to Carbonate Seismology

Rock physics plays a fundamental role in the development of many new seismic technologies, such as production and development seismology, seismic monitoring of EOR processes, and direct hydrocarbon and porosity detection. Rock physics not only provides the physical and feasibility basis for developing and using seismic technologies in characterizing reservoirs and reservoir processes, but also bridges exploration, production, and management of reservoirs through interpreting seismic results. In this section, we briefly summarize two applications of rock physics to carbonate seismology.

Seismic Monitoring of Hydrocarbon Miscible Flood

In many carbonate reservoirs, light hydrocarbons composed of mostly methane and ethane are injected into the reservoir formation to drive the in-situ oil out through expansion and dissolution. Because the light hydrocarbons injected are compressible, V_P should decrease as the formation is flooded.

Because most carbonate reservoirs are deep and the seismic signal-to-noise ratio degrades with increasing depth, seismic modeling (Hirsche et al., 1990b) has shown that to use surface seismology to map the injected light hydrocarbons, a V_P contrast of about 5% or more is needed. Hirsche et al. (1990a) measured 80 carbonate core samples from 11 carbonate reservoirs in Alberta under reservoir conditions and found that V_P decreased by 2–14% after the rock samples, initially saturated with 35° API oil, were flooded with the light hydrocarbons. In the majority of the samples (about 75%) measured, the V_P decrease was more than 5%.

Both Hirsche et al. (1990a) and Wang et al. (1991) concluded that there is an excellent chance of success in using high-resolution seismic methods (e.g., surface 3-D and crosswell) to map injected light hydrocarbons and monitor the injection and flooding process, especially in fractured or shallow reservoirs. The 2-D test line shot at an Alberta reservoir by Hirsche et al. (1990b) clearly shows the position of the injected light hydrocarbons. The test line also matched seismic modeling results using laboratory rock physics data.

Seismic Monitoring of Carbon Dioxide Flood

Another often-deployed EOR method in carbonate reservoirs is carbon dioxide (CO_2) injection. This recovery process can be either miscible or immiscible, depending on the reservoir pressure and temperature. In miscible CO_2 flooding, the injected CO_2 is mostly in the liquid phase so oil is recovered through expansion and dissolution. In the immiscible process, the injected CO_2 is mostly in the gas phase and oil is recovered by expansion.

Because CO_2 is much more compressible than most reservoir oils (Wang and Nur, 1989), CO_2 injection causes V_P to decrease and V_S to increase slightly. Nolen and Hoeksema et al. (1995) measured both V_P and V_S in 21 dolomite core samples from a West Texas field which is undergoing a CO_2 flood. They then used the Gassmann equation to predict the effect of CO_2 on the seismic velocities. An average of about 2% V_P decrease was found as the oil-saturated dolomites were flooded by CO_2. This small change in V_P may be too small for surface seismic methods to detect the injected CO_2, but it may be resolved by high-resolution crosswell seismic methods. In reality, the V_P change may be higher than that predicted by the Gassmann equation, which usually underestimates the pore fluid effects on seismic velocities in hard rocks such as carbonates.

A joint university and industry pilot project is ongoing in this West Texas carbonate field to use crosswell seismology to monitor the CO_2 injection process and to map the injected CO_2.

Conclusions

The study of seismic reservoir properties is an important aspect of reservoir characterization and management using seismic technology because it directly links seismology to rock properties. Results of such a study, both experimental and theoretical, provide a fundamental understanding of seismic responses to reservoir properties and their dynamic changes so that seismic attributes can be converted to reservoir parameters. Because carbonate rocks are extremely complex and little has been published about seismic carbonate properties, more effort is needed in further understanding seismic properties and their interaction with other properties of carbonate rocks.

Seismic properties of carbonate rocks are affected in complex ways by many parameters, such as pore type and shape, porosity, pore fluid, and saturation, especially pore type. In reality, most carbonate rocks contain more than one type of pores. The most compliant pores affect seismic properties the most.

Results of seismic properties of carbonate rocks have many potential applications, such as lithology, porosity, and fracture detection; pore fluid identification and monitoring; and seismic surveillance of EOR processes. As seismic technology further improves in terms of spatial resolution and signal-to-noise ratio, detailed descriptions of reservoirs and detailed monitoring of the dynamic changes and movement of reservoir fluids using seismic methods will become routine in the near future. Seismic rock property studies will undoubtedly play a vital role in such technology advancement because they provide the keys to interpretation of the seismic results.

Acknowledgments

I thank Chevron Petroleum Technology Company for approving the publication of this paper, Ibrahim Palaz and Kurt Marfurt for inviting me to contribute to their book, W. D. Wiggins for preparation of the thin sections, and Clint Frasier, Gary Graf, Ken Szata, and Jeff Wright for their contributions.

References

Batzle, M., and Wang, Z., 1992, Seismic properties of pore fluids: Geophysics, **57**, 1396-1408.

Biot, M. A., 1956, Theory of propagation of elastic waves in a fluid saturated porous solid, I. Low frequency range; II. High frequency range: J. Acoust. Soc. Am., **28**, 168-191.

Biot, M. A., 1962, Generalized theory of acoustic propagation in porous dissipative media: J. Acoust. Soc. Am., **34**, 1254-1264.

Birch, F., 1960, The velocity of compressional waves in rocks to 10 kilobars, part 1: J. Geophys. Res., **65**, 1083-1101.

Choquette, P. W., and Pray, L. C., 1970, Geologic nomenclature and classification of porosity in sedimentary carbonates: AAPG Bulletin, **54**, 207-250.

Gassmann, F., 1951, Elastic waves through a packing of spheres: Geophysics, **16**, 673-685.

Han, D., 1987, Effects of porosity and clay content on wave velocities in sandstones and unconsolidated sediments: Ph.D. thesis, Stanford University.

Hashin, Z., and Shtrikman, S., 1963, A variational approach to the theory of the elastic behavior of multiphase materials: J. Mech. Phys. Solids, **11**, 127-140.

Hill, R., 1962, The elastic behavior of a crystalline aggregate: Proc. Phys. Soc. London, **A65**, 349-354.

Hirsche, W. K., Wang, Z., and Sedgwick, G., 1990a, Seismic monitoring of miscible/immiscible floods, part I—petrophysical studies: SEG Annual Conference, San Francisco, Extended Abstracts, 816-819.

Hirsche, W. K., Sedgwick, G., and Wang, Z., 1990b, Seismic monitoring of miscible/immiscible floods, part II—model studies and field test: SEG Annual Conference, San Francisco, Extended Abstracts, 233-236.

Johnson, D. L., and Plona, T. J., 1982, Acoustic slow waves and the consolidation transition: J. Acoust. Soc. Am., **72**, 556-565.

Kuster, G. T., and Toksöz, M. N., 1974, Velocity and attenuation of seismic waves in two-phase media, part I—theoretical formulations: Geophysics, **39**, 587-606.

Mavko, G. M., and Nur, A., 1978, The effect of nonelliptical cracks on the compressibility of rocks: J. Geophys. Res., **83**, 4459-4468.

Nolen-Hoeksema, R. C., Wang, Z., Harris, J. M., and Langan, R. T., 1995, High-resolution crosswell imaging of a west Texas carbonate reservoir, part 5—core analysis: Geophysics, **60**, 712-726.

O'Connell, R. J., and Budiansky, B., 1977, Viscoelastic properties of fluid-saturated cracked solids: J. Geophys. Res., **82**, 5719-5725.

Rafavich, F., Kendall, C. H. St. C., and Todd, T. P., 1984, The relationships between acoustic properties and the petrographic character of carbonate rocks: Geophysics, **49**, 1622-1636.

Spencer, J. W., 1981, Stress relaxation at low frequencies in fluid-saturated rocks: attenuation and modulus dispersion: J. Geophys. Res., **86**, 1803-1812.

Toksöz, M. N., Cheng, C. H., and Timur, A., 1976, Velocities of seismic waves in porous rocks: Geophysics, **41**, 621-645.

Wang, Z., 1988, Wave velocities in hydrocarbons and hydrocarbon saturated rocks—with application to EOR monitoring: Ph.D. thesis, Stanford University.

Wang, Z., and Nur, A., 1989, Effect of CO_2 flooding on the velocities in rocks with hydrocarbons: SPE Res. Eng., **4**, 429-436.

Wang, Z., and Nur, A., 1990a, Wave velocities in hydrocarbon saturated rocks: experimental results: Geophysics, **55**, 723-733.

Wang, Z., and Nur, A., 1990b, Dispersion analysis of acoustic velocities in rocks: J. Acoust. Soc. Am., **87**, 2384-2395.

Wang, Z., and Nur, A., 1992, Elastic wave velocities in porous media: a theoretical recipe, *in* Wang, Z., and Nur, A., Eds., Seismic and acoustic velocities in reservoir rocks: Soc. Explor. Geophys., 1-35.

Wang, Z., Hirsche, W. K., and Sedgwick, G., 1991, Seismic velocities in carbonate rocks: J. Canadian Petro. Tech., **30**, 112-122.

Watt, J. P., Davies, G. F., and O'Connell, R. J., 1976, The elastic properties of composite materials: Rev. Geophys. Space Phys., **14**, 541-563.

White, J. E., 1983, Underground sound: Elsevier.

Winkler, K. W., 1986, Estimates of velocity dispersion between seismic and ultrasonic frequencies: Geophysics, **51**, 183-189.

Wyllie, M. R. J., Gregory, A. R., and Gardner, G. H. F., 1958, An experimental investigation of factors affecting elastic wave velocities in porous media: Geophysics, **23**, 459-493.

Wyllie, M. R. J., Gregory, A. R., and Gardner, L. W., 1956, Elastic wave velocities in heterogeneous and porous media: Geophysics, **21**, 41-70.

Chapter 4

Sonic Velocity in Carbonate Sediments and Rocks

Flavio S. Anselmetti
Gregor P. Eberli
University of Miami
Rosenstiel School of Marine and Atmospheric Sciences
Miami, Florida, U.S.A.

Abstract

Compressional wave and shear wave velocities (V_P and V_S) of 295 carbonate minicores from different areas and different ages were measured under variable confining and pore fluid pressures. The samples range from unconsolidated carbonate mud to completely lithified limestones. The results of the measurements show that, unlike siliciclastic and shaly sediments, pure carbonate rocks show little direct correlation between acoustic properties (V_P and V_S) and depth of burial or age. Velocity inversions with increasing depth are thus common. Sonic velocity in carbonates is more controlled by the combined effect of depositional lithology and diagenetic processes, such as cementation and dissolution.

At 8 MPa effective pressure, V_P ranges from 1700 to 6500 m/s and V_S from 700 to 3400 m/s. These ranges are caused mainly by variations in the amount of porosity and porosity type and not by variations in mineralogy. Measured velocities generally show a positive correlation with density and an inverse correlation with porosity, but departures from general trends can be as high as 2500 m/s. These deviations can be explained by the occurrence of different pore types that form during specific diagenetic phases. Commonly used correlations such as Gardner's law (V_P versus density) and the time-average equation (V_P versus porosity) should be modified because they often result in velocities that are too low for carbonates.

Velocity measurements in unconsolidated carbonate mud at different stages of experimental compaction show that velocity increases due to compaction are lower than observed velocity increases at decreasing porosities in natural rocks. This difference shows that diagenetic changes such as cementation and dissolution, which may predate or accompany compaction, influence velocity more than simple compaction at increasing burial depth.

Introduction

Knowledge of the relationship between sonic velocity in sediments and lithology is one of the keys to interpreting data from seismic sections or from acoustic logs of sedimentary sequences. Reliable correlations of rock velocity with other petrophysical parameters, such as porosity and density (Gregory, 1977; Bourbié et al., 1987), are essential for calculating impedance models for synthetic seismic sections (Fagin, 1991) or for identifying the origin of reflectivity on seismic lines (Sellami et al., 1990; Christensen and Szymanski, 1991). Velocity is thus an important parameter for correlating lithologic data with geophysical data.

Recent studies have increased our understanding of elastic rock properties in siliciclastic and shaly sediments. The causes of variations in velocity have been investigated for siliciclastic rocks (Han et al., 1986; Vernik and Nur, 1992), mixed carbonate-siliciclastic sediments (Christensen and Szymanski, 1991), synthetic sand-clay mixtures (Marion et al., 1992), and claystones (Japsen, 1993). The concepts derived from these studies are, however, only partially applicable to carbonates.

Carbonates do not have large compositional variations that are responsible for velocity contrasts, as in other sedimentary rocks. Pure carbonates are characterized by lack of clay and siliciclastic content and are mostly produced and deposited on the top or the slope of isolated or detached carbonate platforms, which have no hinterland as a source of terrigeneous material (Wilson, 1975; Eberli, 1991). They consist mainly of the carbonate minerals calcite (both low-Mg and high-Mg), dolomite, and aragonite. These minerals have very similar physical properties, which excludes compositional variation as a major reason for the large variability of velocity in carbonates.

Theories that describe sonic wave propagation in porous media (Gassmann, 1951; Biot, 1956; Toksöz et al., 1976) are hard to apply to the complex system of carbonates because they form a variety of unique diagenetic rock fabrics with specific elastic properties. The characteristics of velocities in carbonates has been studied on drowned Cretaceous guyots (Kenter and Ivanov, 1995), highly lithified, low-porosity carbonate rocks (Rafavich et al., 1984; Wang et al., 1991) and pelagic, deep-water carbonates (Schlanger and Douglas, 1974; Milholland et al., 1980; Urmos and Wilkens, 1993). The present study (see also Anselmetti and Eberli, 1993) includes carbonates from several depositional environments ranging from shallow water platform to lower slope at all stages of diagenetic alteration.

Sonic velocity measurements were performed in combination with a thorough lithologic and diagenetic examination of thin sections and X-ray defraction analysis. Porosities in the samples ranged from 0 to 60%, and depositional environment varied from a protected shallow water platform over reefal and platform marginal sediments to deeper slope deposits. The correlation of velocity measurements with lithologic and mineralogic data enabled characteristic velocities to be assigned to depositional and diagenetic stages. It also allowed the tracing of diagenetic evolution and velocity development from the time of deposition through various burial stages, recognizing that each diagenetic process alters velocity in a characteristic fashion.

Sample Areas

This study relies on velocity analyses of 295 discrete samples from three different settings: (1) modern carbonate mud from Florida Bay, (2) Cenozoic cores from deep drillholes on the Great Bahama Bank, and (3) Mesozoic–Cenozoic outcrops from an exhumed carbonate platform margin in central Italy.

Modern Unconsolidated Carbonate Sediments from Florida Bay

The velocities of 20 carbonate mud samples were measured at various stages of artificial compaction to determine the increase of velocity caused by porosity reduction during pure mechanical compaction. The mud was collected with push cores of about 70 cm length from the interior pond of Cluett Key (Figure 1), a mangrove island in Florida Bay, south Florida. Initial porosities of the sampled carbonate mud ranged from 55 to 70%, depending on depth and lithology (Swart and Kramer, in press). The mineralogic composition averaged 55% aragonite, 25% high-Mg calcite, and 20% low-Mg calcite, with less than 2% insolubles. Only traces (1–5%) of dolomite were observed in the sampled upper parts of the cores, although pore water chemistry indicated that dolomite may be forming in small quantities (Burns and Swart, 1992; Swart and Kramer, in press).

Cores of Deep Drillholes on Great Bahama Bank

Two continuous core borings, taken during the Bahamas Drilling Project (Ginsburg, in press), were located along the multichannel Western seismic line across the Great Bahama Bank (Figures 1 and 2) (Eberli and Ginsburg, 1989). They provided an excellent opportunity to correlate the physical properties of Miocene–Pleistocene carbonate sediments with their depositional lithologies and levels of diagenetic alteration. The two drillholes (named Unda and Clino; Figures 1 and 2) penetrated to depths of 442 and 662 m below seafloor,

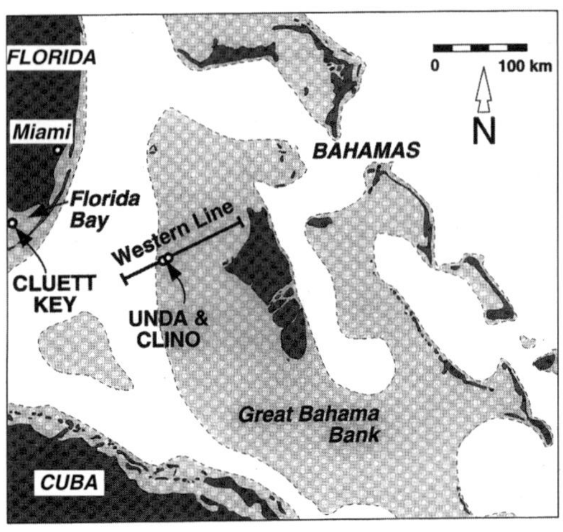

Fig. 1. Location map showing the positions of the drillholes Unda and Clino along the Western seismic line on the Great Bahama Bank and the location of Cluett Key in Florida Bay. Areas outlined in dashed lines are less than 20 m deep.

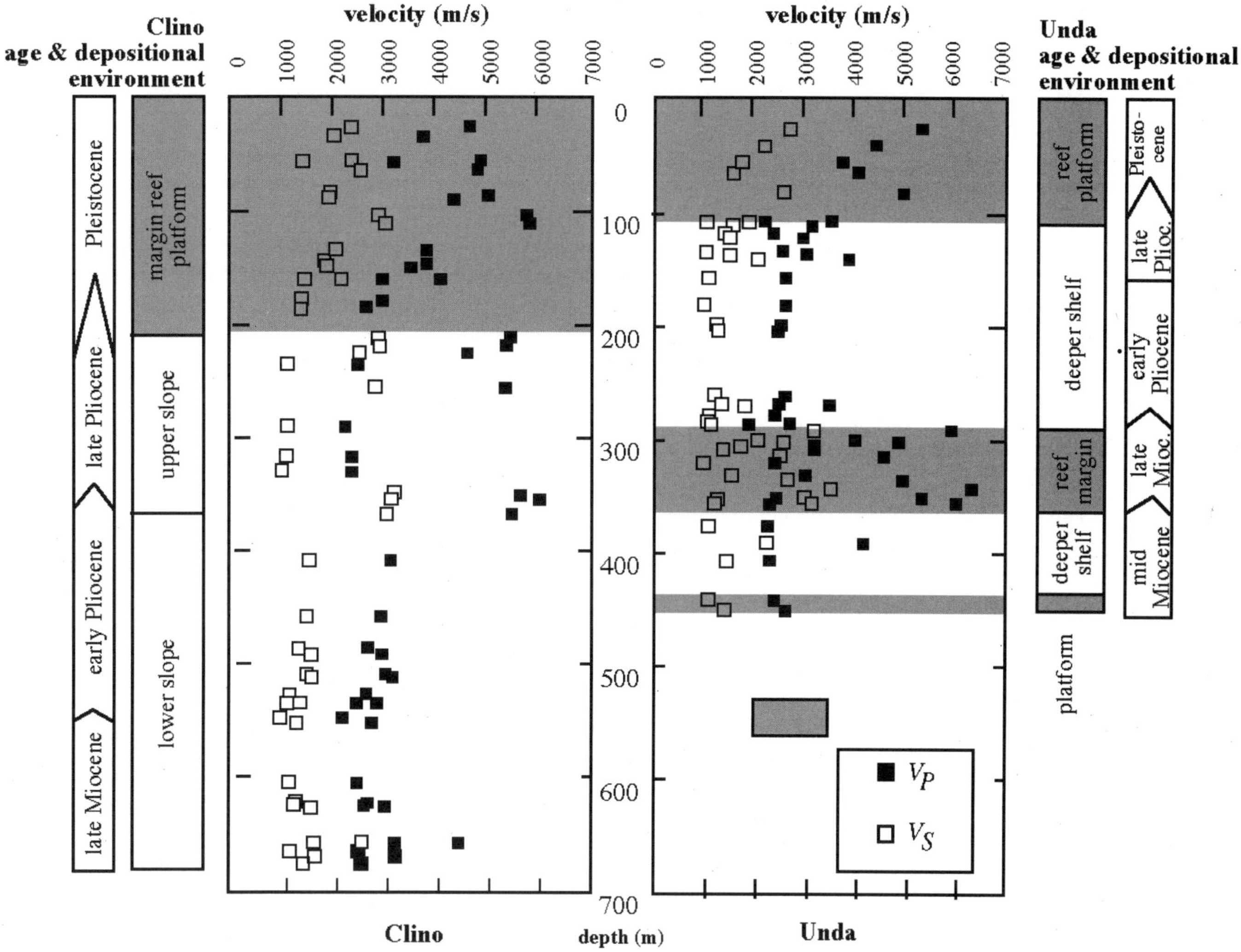

Fig. 2. Correlation of V_P and V_S (at 8 MPa effective pressure) with depth, depositional environment, and age of the drilled sediments from the drillholes Unda and Clino on the Great Bahama Bank. Velocity inversions are common in both holes and show that diagenetic alterations and sediment type are more influential than depth. Velocities of carbonates deposited on the shallow-water platform (shaded areas on graphs) have more variability and higher velocities than deeper water samples.

respectively. The continuous cores had an average recovery of over 80%. A total of 89 samples from the two drillholes were analyzed. Unlike older and exhumed outcrop samples, the young age of the drilled sediments allowed for the measurement of sonic velocities of carbonates that are partially unconsolidated and not in their final stage of post-depositional alteration. The variety of diagenetic processes encountered has enabled the velocity evolution of different carbonate sediments to be traced under different diagenetic conditions through burial history and time.

The top of the rock section in both holes is Pleistocene in age (Kenter and Ginsburg, in press). The oldest drilled sediments are dated as middle Miocene at the bottom of Unda, whereas the bottom of Clino reaches carbonates of late Miocene age (Figure 2). The retrieved lithologies range from platform interior to platform margin and slope carbonates; there is no siliciclastic sediment on this isolated carbonate platform. Not only the depositional lithology but also the diagenetic overprinting changes several times downhole (Melim et al., 1994). The abundance of intense diagenetic alterations, such as dissolution, cementation, and dolomitization, make these rocks an ideal case to study how diagenetic processes affect the petrophysical characteristics of carbonate sediments and rocks.

Outcrop Samples from Montagna della Maiella, Central Italy

Velocity measurements were made on 186 samples collected in the Montagna della Maiella in the Abruzzi in central Italy, which is part of the Apulian carbonate platform. It is an uplifted and exhumed carbonate platform margin exposed in several valley flanks. The exposed platform and slope carbonates range in age from Late Jurassic to late Miocene (Bernoulli et al., 1992; Eberli et al., 1993). The velocities of these carbonates, which underwent burial diagenesis, were determined for two purposes. First, they provide information about the controls on sonic velocity in carbonates because they are examples of completely altered and cemented rocks. Second, knowledge of physical properties, combined with an assessment of large-scale geometric patterns of outcropping formations, are used to construct synthetic seismic sections (Anselmetti et al., Chapter 15, this volume) to investigate the seismic responses of such a carbonate platform setting.

Unlike the Great Bahama Bank, the Maiella Platform margin shows almost no signs of dolomitization. This explains why, despite their older age, original lithologies are generally better preserved than many dolomitized Bahama carbonates. Some of the bioclastic sands of the Upper Cretaceous still have porosities of more than 30%, whereas most of the platform deposits are densely cemented (Mutti, 1995).

Methods of Laboratory Analyses

Sampling Techniques

The samples used for velocity determinations were cylindrical miniplugs 2.5 or 3.8 cm in diameter. The miniplugs of unconsolidated mud from Florida Bay were sampled from short push cores 7.6 cm in diameter. To avoid compaction and fabric destruction during sampling, a thin-walled tube with a diameter of 3.8 cm was used to cut the miniplugs vertically out of the cores. The 2–4-cm-long cylinders were compacted longitudinally by a hydraulic press with a uniaxial pressure of up to 170 MPa. The velocities of the mud samples were measured at variable degrees of compaction. The maximum compaction attained was about 50%, so that the initial porosity of about 60% was reduced to 26%. The samples from the deep Bahama drillholes were cored horizontally, or occasionally vertically, into 7.6-cm diameter-cores. Plugs from the Maiella Platform were cored from hand samples collected from outcrops. All rock cores were trimmed to a length of 1.5–5 cm.

Velocity Measurements

The velocities of all Bahama and 114 (out of 186) Maiella samples were measured in the Petrophysics Laboratory at the University of Miami using a pulse transmission technique (Birch, 1960). The same pair of transducers generates one compressional wave signal (V_P) and two orthogonally polarized shear wave signals (V_{S1}, V_{S2}) with a center frequency of 0.6–1.2 MHz. The V_S used for the calculation of the V_P/V_S ratio is the mean V_S of the two measurements. The samples were water saturated and jacketed in rubber or heat shrink tubing, which seals the pore fluid from the confining oil in the pressure vessel (Figure 3). Confining and pore fluid pressures were chosen independently to simulate in-situ conditions of a buried sediment. Pore fluid pressures as high as 50 MPa can be obtained, but most experiments

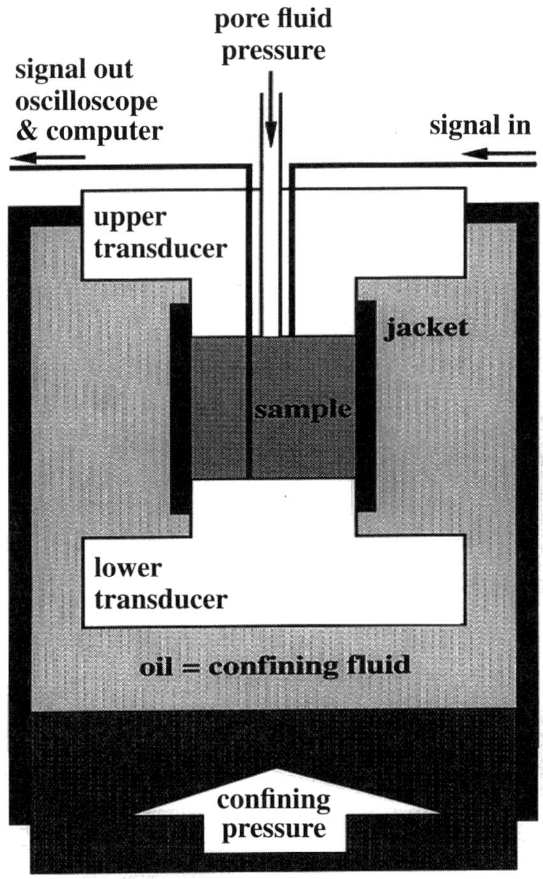

Fig. 3. Diagram of ultrasonic velocity meter at the petrophysics laboratory, University of Miami. Confining pressure and pore fluid pressure can be adjusted independently to simulate true in-situ conditions of buried sediment at any depth. An elastic jacket seals the pore space with the pore fluid from the confining oil. Maximum confining pressure is 100 MPa (1 kbar).

were run at 2 MPa. Confining pressure was varied between 3 and 100 MPa, resulting in an effective maximum pressure (confining pressure minus pore fluid pressure) of up to 98 MPa. However, many samples collapsed and failed at pressures below 100 MPa.

The velocities of the compacted mud from Florida Bay were measured with the same set of transducers but with a benchtop measuring system without confining pressure. This measuring system allowed recognition of compaction due to the axial pressure of the transducers during the measurement. The two transducers were pressed together with a piston at an axial pressure of 0.1–1 MPa. This relative low pressure allowed the transmission of the signal from the transducers into the mud, but did not drastically compact the still deformable mud samples. Corrections for length changes were made in the velocity determination, where the minimum required transducer pressure reduced the sample length by a few percent.

A suite of the Maiella miniplugs (72 samples) was measured with a similar apparatus in the petrophysics laboratory at the University of Geneva, Switzerland. The transducer pair of this machine creates only a compressional wave signal. These measurements were performed under dry conditions without pore fluid pressure and under confining pressures varying up to 400 MPa.

Petrography

X-ray diffraction analyses were performed on the cut-offs of the drilled plugs from the Bahama samples. Calibration of these results with carbonate standards allowed the percentage of calcite, dolomite, and aragonite to be determined for each sample. The cut-offs of the miniplugs were also used to make thin sections from all Bahama samples and from most Maiella samples. Thin sections were examined to determine sediment type, composition, grain size, porosity type, and diagenetic alterations. These observations enabled the physical properties to be correlated with the lithologic parameters.

Density

Dry bulk densities were calculated by weighing the oven-dried rock plugs and calculating the volume by measuring the diameters and lengths. Because these almost pure carbonates consist dominantly (>95%) of three carbonate minerals, the percentage of each mineral (determined by X-ray diffraction) leads to a theoretical grain density (matrix density) that can be calculated using individual mineral densities (Carmichael, 1989):

$$\rho_{grain} = [(\%_{calcite} \times \rho_{calcite}) + (\%_{dolomite} \times \rho_{dolomite}) + (\%_{aragonite} \times \rho_{aragonite})] \div 100 \quad (1)$$

where

$$\rho_{calcite} = 2.71 \text{ g/cm}^3,\ \rho_{dolomite} = 2.87 \text{ g/cm}^3 \quad (2)$$

and

$$\rho_{calcite} = 2.93 \text{ g/cm}^3. \quad (3)$$

The grain density of the pure Maiella carbonates is close to the plain calcite value (2.71 g/cm^3) because no significant dolomite was found. However, the hemipelagic carbonates have a slightly reduced grain density, probably due to some clay content. Several bauxitic samples from a major exposure horizon (middle Cretaceous unconformity) show significantly higher grain densities, depending on the content of the iron and aluminum hydroxides, whose densities were measured with a helium pycnometer using powdered samples.

Porosity

Rock porosity was calculated by comparing the grain density (matrix density) with the measured dry or saturated bulk density:

$$\rho_{bulk} = \frac{\%_{porosity}}{100} \times \rho_{fluid} + \left(1 - \frac{\%_{porosity}}{100}\right) \times \rho_{grain} \quad (4)$$

or

$$\frac{\%_{porosity}}{100} = \frac{\rho_{grain} - \rho_{bulk}}{\rho_{grain} - \rho_{fluid}}. \quad (5)$$

Velocity Data

Compressional and Shear Wave Velocities

Velocities at a confining pressure of 10 MPa and a pore fluid pressure of 2 MPa were used in the following correlations. The resulting effective pressure of 8 MPa (10 MPa for dry samples) was high enough to allow good signal transmission, but did not cause significant fracturing in the highly porous samples. The V_P measurements on dry samples are also presented because major differences between dry and saturated V_P are expected only in rocks with a dominant crack porosity

(Nur and Simmons, 1969), whereas the saturation of round-shaped pores, abundant in the samples of this study, does not influence V_P drastically.

Despite the limited variability in mineralogy, the measured carbonates have an extraordinarily wide range in velocities. V_P varies between 1700 and 6500 m/s, and V_S between 700 and 3400 m/s. The three different datasets show different ranges of V_P and V_S (Figure 4). The unconsolidated mud samples from Florida Bay have the lowest velocities, with minimum V_P and V_S of 1700 and 700 m/s, respectively. The Bahama and Maiella samples reach velocities of up to 6500 m/s (V_P) and 3400 m/s (V_S). Unlike siliciclastic sediments, where variations in mineralogy (e.g., clay content) can cause

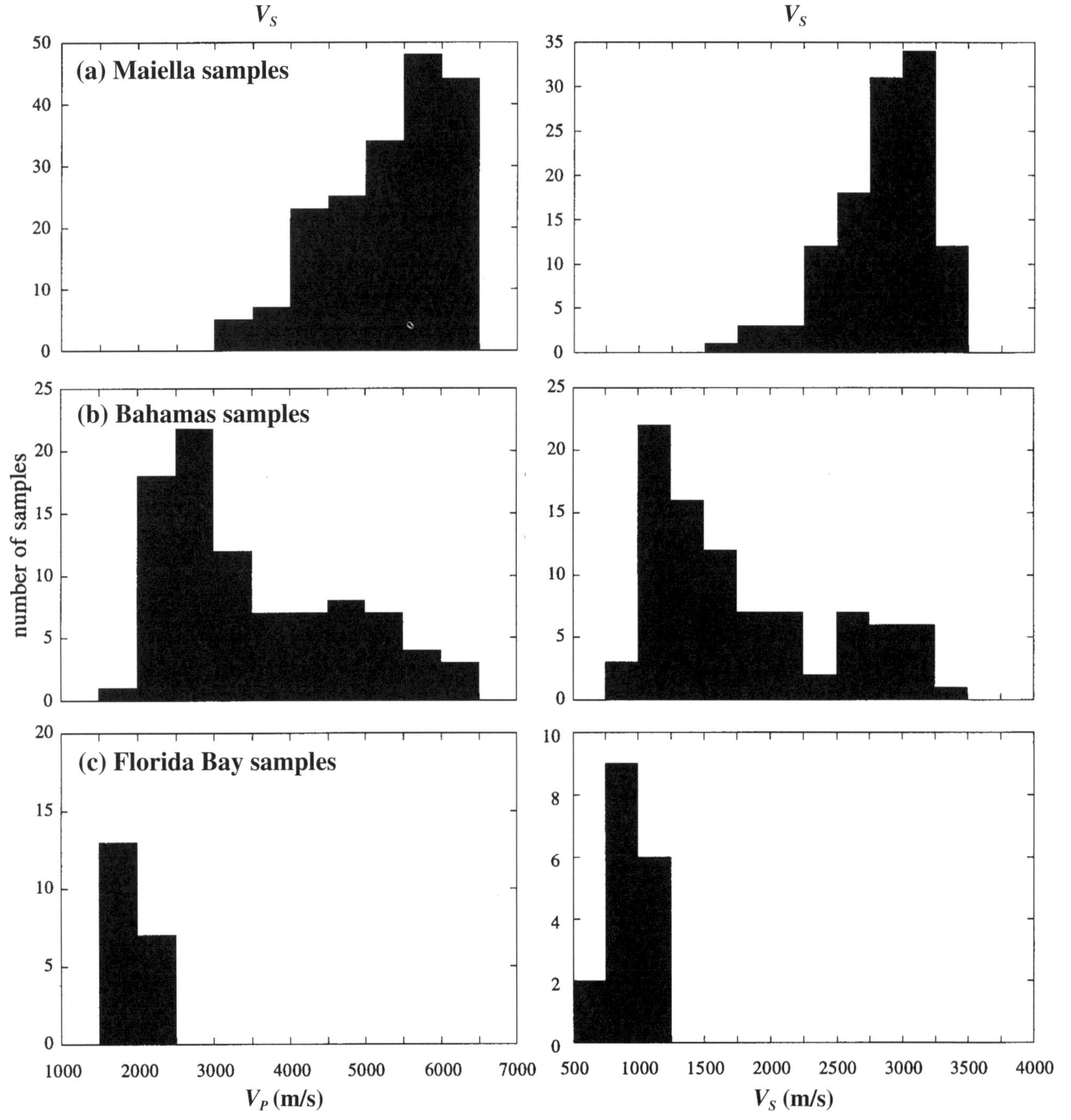

Fig. 4. Range of V_P and V_S (at 8 MPa effective pressure) for all samples from the three investigated areas. (a) The Maiella samples have higher average velocities than (b) the Bahama samples. (c) The artificially compacted mud samples from Florida Bay have low velocities despite compaction of up to 50%. The large velocity ranges of all samples is remarkable for the similar mineralogies of these pure carbonate sediments.

large velocity contrasts, the different carbonate minerals (calcite, dolomite, and aragonite) have very similar physical properties, and thus differences between them are not responsible for the large variability in velocities.

Acoustic Impedance

The observed ranges of V_P and density, and thus the large contrasts in acoustic impedance (Figure 5), can explain the excellent seismic reflectivity of many seismic sections recorded over pure carbonates. For example, the two drillholes from the Great Bahama Bank have acoustic impedance values that range from 3.8 to 17.4 10^6 kg/m²·s. The observed good reflections on seismic sections, often believed to be caused by intercalations of carbonate and noncarbonate sediments, can be explained solely by variations in the fabric of the carbonates.

V_P/V_S Ratios

Compared to other studies of V_P/V_S ratios in sedimentary rocks (Tatham, 1982; Rafavich et al., 1984; Wilkens et al., 1984; Johnston and Christensen, 1992, 1993), the V_P/V_S ratios of this study, which were measured only under saturated conditions, have a very wide range and reflect the large variety of encountered carbonate fabrics. The cross plot of V_P/V_S with V_P in Figure 6 shows that V_P/V_S in indurated, low-porosity rocks with high V_P normally falls between 1.8 and 2, which is in good agreement with the measurements of limestones and dolostones by Johnston and Christensen (1992). At lower V_P, the V_P/V_S ratio can reach much higher values of up to 2.8, reflecting that V_S is generally more affected by the highly porous fabric of low-velocity carbonates than is V_P.

Some readings of shear wave velocity, and thus the V_P/V_S ratio, might have large errors due to poor V_S signal quality, such as when low shear wave amplitudes are combined with high background noise. Therefore, some of the extreme V_P/V_S values might be unreliable. These few values, however, do not change the general pattern of V_P/V_S values. The artificially compacted samples from Florida Bay have an extreme range of V_P/V_S values, varying from 1.7 to 2.8 over a narrow range of V_P from 1700 to 2300 m/s. The uncompacted mud, with a porosity of about 60%, is not strong enough to sustain shear stresses (Laughton, 1957), thus inhibiting the transmission of shear wave signals. The slightly compacted mud samples have high V_P/V_S values of up to 2.8. With increasing compaction, the V_P/V_S of these Florida Bay samples approach more normal values of about 2.2.

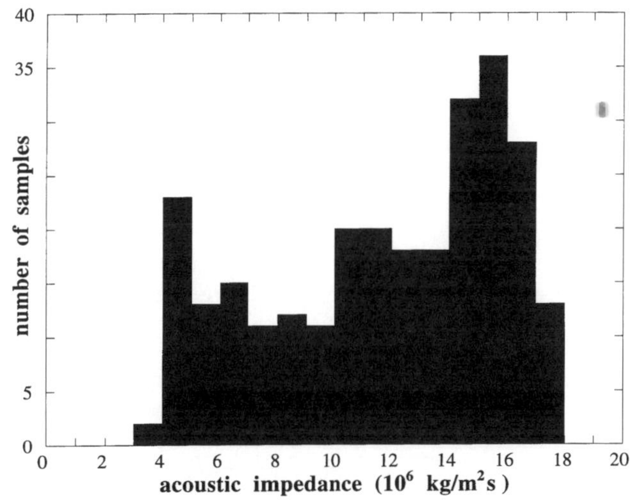

Fig. 5. Range of acoustic impedance of the Bahama and Maiella samples (at 8 MPa effective pressure). Impedances in high-velocity, high-density rocks are more than five times higher than in low-velocity rocks. Impedance variations in pure carbonates are caused by differences in fabric and not in composition. The large range explains the good reflectivity seen in many carbonate seismic sections.

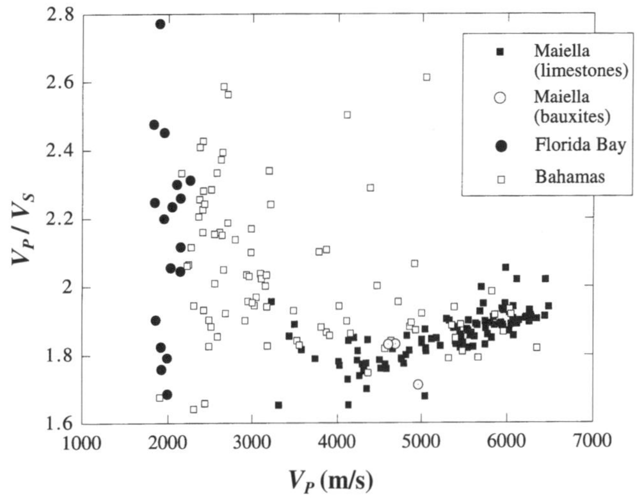

Fig. 6. V_P/V_S as a function of V_P (at 8 MPa effective pressure). The wider range of V_P/V_S toward higher values in low-velocity rocks shows that shear waves are more affected by high-porosity fabrics than compressional waves. Some of the extreme low and high values might be caused by a questionable registration of the arriving V_S signal, in particular in high-porosity rocks.

Factors Affecting Velocity

In many sedimentary rocks, the concept of a grain- and matrix-supported fabric with a critical porosity explains variations in velocity and relates them to changes in sediment composition (Nur et al., 1991; Yin et al.,

1993). However, diagenetic changes in carbonates cause cementation, dissolution, and recrystallization that form unique rock fabrics with associated velocity patterns that do not simply reflect compositional variations in the sediment.

Velocity in carbonates is a complex function of several factors. It is possible to distinguish between rock intrinsic and rock extrinsic parameters. *Rock intrinsic parameters*, such as porosity, pore type, composition, and grain size, are factors related to the lithology and thus to the physical properties of the rock. *Rock extrinsic parameters* are factors not physically related to the rock fabric but rather determined by external boundary conditions. Examples of rock extrinsic parameters include burial depth, confining pressure, and age. It can be shown that, for carbonates, rock intrinsic factors are more important than extrinsic ones.

Velocity as a Function of Rock Extrinsic Parameters

Compaction

Compaction experiments and associated velocity measurements were performed on modern carbonate mud from Florida Bay to determine the manner in which velocity changes in response to porosity reduction from purely mechanical compaction. With a uniaxial pressure of 170 MPa, the porosity could not be reduced to less than 29%, indicating that the microfabric of the rock (55% aragonite needles) was close to the densest packing that can be reached by mechanical compaction. This demonstrated that samples with porosities of 0–25% require diagenetic closing (cementation) of part of the pore space.

It is known that porosity has a major influence on velocity, with porosity decreases usually producing velocity increases. The Florida Bay samples showed a relatively subtle increase in velocity with increasing compaction or decreasing porosity (Figure 7). At porosities close to 60%, the poorly compacted samples had a V_P of 1700 m/s and no measurable V_S. The samples had to be compacted by 10–15% to allow the transmission of a V_S signal. This observation corresponds well with the measurements of Laughton (1957), who detected shear waves in unconsolidated ocean sediments only at a compacting pressure of over 5 MPa. At maximum compaction with remaining porosities of 29%, V_P increased to 2250 m/s and V_S was about 900–1200 m/s.

The gradient of measured increases in velocity of the compacted mud was significantly lower than observed gradients in the other Bahama and Maiella sam-

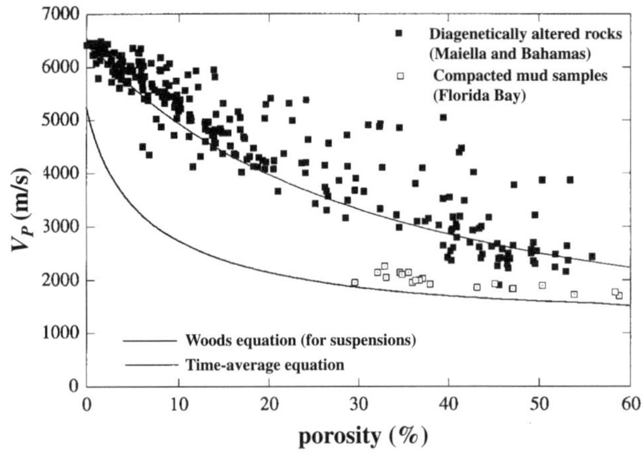

Fig. 7. Graph of velocity as a function of porosity for all datasets. V_P and porosity show an inverse correlation, but the gradient of increasing V_P with decreasing porosity is lower for the compacted Florida Bay samples than for the natural rocks of the Bahamas and Maiella. Mud sample velocities are only slightly higher than those of media without rigidity (suspensions, Wood equation), indicating that porosity reduction due to mechanical compaction has a minor effect on V_P. However, porosity reduction due to diagenetic processes can increase rigidity, which results in higher velocities than predicted by the time-average equation.

ples. Due to their low shear modulus, the mud samples behaved similarly to materials that have no rigidity (liquids), as suggested by Hamilton (1971). He showed that most unconsolidated marine sediments possess little rigidity but nevertheless have a shear modulus of greater than zero and a definite structure. The Wood equation (Wood, 1941; Nafe and Drake, 1963), valid for media without shear modulus or rigidity, can be used to compare the observed porosity to velocity relationship of the artificially compacted sediments with the values of liquids:

$$V_P = \{ [\phi\beta_{\mathrm{fluid}} + (1-\phi)\,\beta_{\mathrm{solid}}]\,[\phi\rho_{\mathrm{fluid}} + (1-\phi)\,\rho_{\mathrm{solid}}] \}^{-1/2} \quad (6)$$

where ϕ is the fractional porosity, β is the compressibility, and ρ is the density.

The Wood equation curve in Figure 7 was derived using compressibility values for water (4.06×10^{-10} m^2/N) and for calcite (1.34×10^{-11} m^2/N) because the elastic properties of aragonite are not well known. Comparisons of calculated and measured velocities show that the measured velocities of the mud samples are only slightly above values predicted by the Wood equation, whereas all other samples (altered during diagenesis) show much higher values (Figure 7). The lower gradient of the mud samples can be explained by the absence of "additional" processes (e.g., diagenesis) that increase the rigidity of the rock and normally accompany the compaction of sediments. The artificial compaction experiments happen so quickly that no diagenetic alterations are initiated. In nature, however, the effects of such diagenetic processes as recrystallization and cementation are superimposed on the effect of porosity reduction. The velocity, therefore, represents the combined effect of these different processes. The difference between the velocity–porosity relationship in natural rocks and that in artificially compacted rocks demonstrates how much diagenesis contributes to the observed velocity increases. The results in Figure 7 show that diagenetic alteration is more effective in increasing velocity than compaction because this alteration significantly increases the rigidity of the rock.

Burial Depth and Age

Samples taken from the two core borings in the Bahamas Drilling Project from the western margin of the Great Bahama Bank clearly show that velocity is neither primarily a function of sediment age nor of burial depth. In contrast to the usual assumption that velocity increases with depth (Hamilton, 1980; Japsen, 1993), the depth plots of V_P and V_S (Figure 2) from the two drillholes display velocity inversions that make velocity predictions, based only on depth, impossible. Both holes display a pattern of high variability of velocity in the carbonates that were deposited in shallow-water environments (e.g., reef, platform margin, and platform interior). Several zones with trends of decreasing velocity with greater depth indicate that diagenetic processes other than simple compaction substantially control velocity evolution. In the young sediments from the shallow parts of drillholes Clino and Unda, high velocities result from intense diagenetic alterations that happen much faster than compaction.

It is concluded that velocity predictions cannot be made from knowledge that a carbonate sediment has a certain age or is at a certain depth. The acquisition of additional intrinsic rock parameters is necessary to give a reliable velocity estimation.

Effective Pressure

To investigate the pressure dependence of V_P and V_S in consolidated rocks, sonic velocities of the Bahama and Maiella samples were measured under varying confining and pore fluid pressures. Sonic velocity in rocks is a function of the differential or effective pressure, which is the difference between the confining and pore fluid pressure. Minor departures from this relationship are caused by changing pore fluid properties at different pressures (Coyner, 1984).

At low pressures, all samples show an increase in velocity with increasing effective pressure due to better grain contacts, changing pore shapes, and closing of microcracks (King, 1966; Gardner et al., 1974). This increase is large for unconsolidated samples with low velocities (Figure 8), whereas the velocities of indurated, dense samples are usually less affected by higher pressures. The measured velocity-pressure traces display a systematic pattern with higher gradients for low-velocity samples and lower gradients for high-velocity samples. A minor part of the observed velocity increase is an artifact because compaction at increasing pressures reduces the sample length. At low pressures, velocity increases significantly, resulting in a steep gradient. Higher confining pressure usually results in stagnating velocities because most cracks in the rock were already closed at pressures below 100 MPa. A characteristic of many samples is that both V_P and V_S reach a maximum during increasing pressure and suddenly decrease above a critical pressure. This velocity decrease is caused by

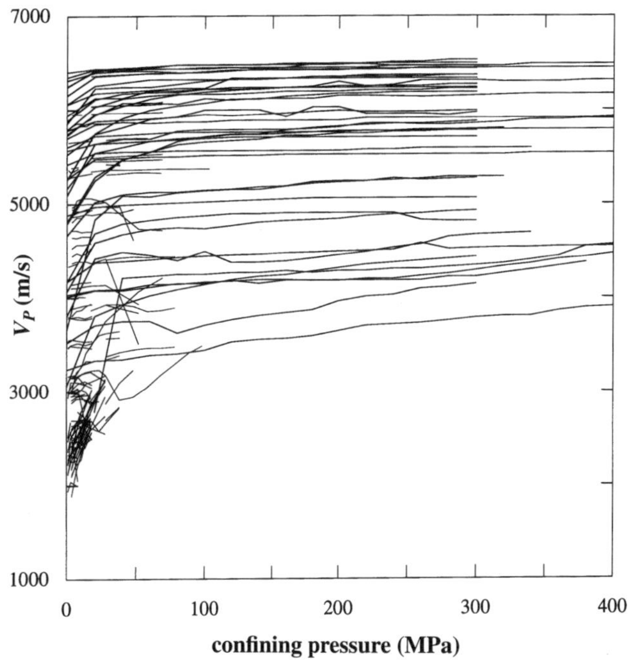

Fig. 8. Graph showing the velocity evolution of the Bahama samples and some Maiella samples at increasing effective pressure. Each trace represents the velocities for one sample at different pressures. The gradients of low-velocity samples are higher than those of high-velocity samples. Decreasing velocities at higher pressures mark a critical pressure at which the samples are intensely fractured and collapse. At pressures above 100 MPa, velocities generally plateau.

disintegration and collapse of the sample in the pressure vessel, with progressive destruction of the partially cemented grain contacts that support the transmission of the acoustic signal. Eventually, velocity may increase with further increasing pressure as the newly formed fractures that reduced the velocity close.

The critical pressure at which the first velocity decrease occurs varies with lithology. Dense, indurated rocks display no evidence of fabric destruction up to the highest measured pressures of 400 MPa, whereas soft, unconsolidated samples show signs of velocity decrease at pressures as low as 5 MPa. Under hydrostatic conditions, an effective pressure of 5 MPa is equivalent to a burial depth of less than 500 m, which is about the depth of the two drillholes on the Great Bahama Bank. The low critical pressures, at which fracturing occurs, demonstrate that some carbonates, especially slope deposits or sucrosic dolomites, became buried without being progressively indurated. Also, many samples show no signs of destruction of the fabric up to high pressures, demonstrating that relatively high porosity can be preserved even at pressures greater than 100 MPa. This pressure is equivalent to a depth of 5 km, assuming an average rock density of 2 g/cm^3 (as measured in the Bahama samples). If the pore water network is open to the surface, an effective pressure of 100 MPa is equivalent to a burial depth of up to 10 km (since effective pressure equals confining or lithostatic pressure minus pore fluid pressure).

Velocity as a Function of Rock Intrinsic Parameters

Depositional Lithology

The lithology of a carbonate sediment at the time of deposition has a strong influence on the evolution of velocity because it controls future alterations of the rock. At the time of deposition, all unconsolidated sediments have velocities between 1550 and 1800 m/s. The different lithologies, however, have different susceptibilities to diagenetic alteration that will change the physical properties and thus the velocities of the sediments over time. The diagenetic susceptibility of certain sediments causes fast or slow alterations of the rock fabric, depending on the diagenetic potential of the sediment (Schlanger and Douglas, 1974) and the diagenetic regime.

In the Bahama cores, the velocities of the shallow and deeper water deposits have different ranges, with an overlap of values in the low-velocity range (Figure 2). Most of the Bahama slope carbonates are unconsolidated and thus have low velocities, whereas the platform carbonates range from low to relatively high velocities. Therefore, not all platform deposits reach high velocities, but both the average velocity and overall velocity range are much higher for the platform samples than for the slope samples. Only turbidites in the slope section, which contain platform-derived material, have high velocities resulting in similar velocity variability as in the platform deposits (Figure 2).

Comparison of velocity ranges with depositional environments in the Maiella Platform confirms the relationships just described. For example, if the Maiella samples are grouped into two categories of depositional environment—(1) platform deposits and (2) basin, slope, and deeper shelf deposits (Figure 9)—then the velocity range of the two categories form two distinct clusters that overlap only in the high-velocity range. The platform carbonates have a significantly higher average velocity than the basin, slope, and deeper shelf carbonates.

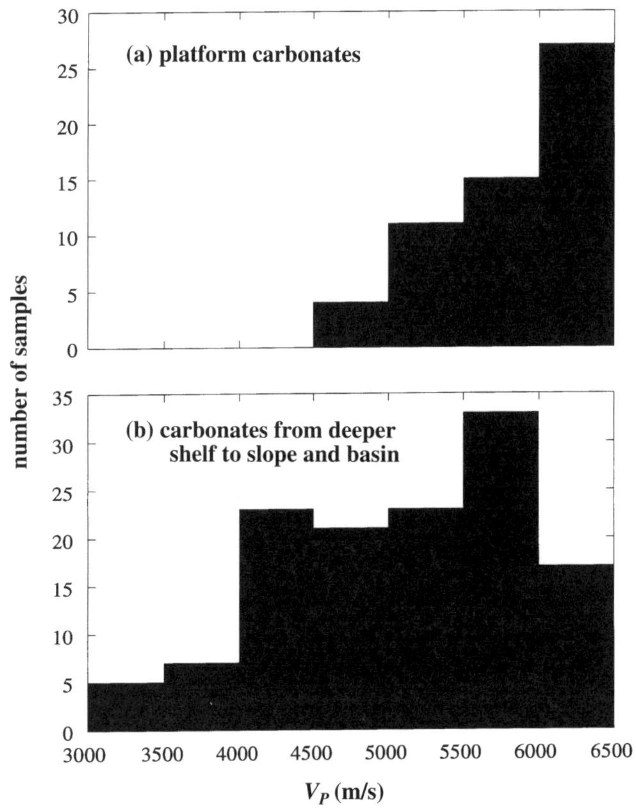

Fig. 9. Histograms of V_P (at 8 MPa effective pressure) in different depositional environments from Maiella platform. (a) Carbonates deposited on the shallow-water platform have a narrow range with generally high velocities. (b) Sediments from the deeper shelf, slope, and basin show a higher variability toward lower V_P, but maximum velocities are the same as for platform rocks.

The velocity signatures of the different depositional environments can be explained by the diagenetic potential of a sediment. This potential is a factor mainly of the grain size and the amount of metastable minerals (Schlanger and Douglas, 1974). A high content of fine-grained micritic material, as found in mudstones and wackestones, results in a low permeability. The resulting low rate of fluid flow inhibits or slows diagenetic alterations that rely on the transport of chemicals in water. In contrast, sediments with grain-supported fabric and a low content of micrite (grainstone to packstone) have a higher permeability and, as a consequence, a higher fluid flow. This accelerates diagenetic processes such that the sediment is quickly altered and consolidated. Thus, original coarse-grained rocks can reach higher velocities after a relatively short time of burial, whereas fine-grained rocks tend to preserve their unaltered fabrics and their low velocities to deeper burial conditions. In addition to grain size, the amount of metastable minerals, such as aragonite and high-Mg calcite, controls the diagenetic potential. A high amount of metastable components causes a high diagenetic potential, leading to rapid dissolution or recrystallization of the sediment. This alteration can enhance the elastic properties and increase permeability, resulting in accelerated lithification.

In the Bahamas and Maiella, the platform carbonates are usually high in coarse skeletal or nonskeletal grains (ooids and peloids). They consist mainly of aragonite (ooids, peloids, corals, *Halimeda,* and bivalves), which is metastable in seawater. In contrast, slope and deeper water deposits are normally characterized by high micritic grain fractions, higher contents of pelagic calcareous organisms (globigerinids and coccoliths), and more stable low-Mg calcite shells. Shallow-water carbonates thus fulfill both conditions for fast diagenetic alterations: coarse grain size and a high amount of metastable minerals. Some turbidites deposited on the slope contain many skeletal aragonitic fragments from the platform top and are therefore similar to platform deposits and different from normal slope sediments (Eberli, 1988). This explains the fast velocities in the generally low-velocity slope section of Clino (Figure 2).

In the Maiella samples, most of the platform carbonates had enough time for diagenetic alteration, allowing the sediments to reach their stable mineralogy and thus their final high velocities. As a consequence, the platform sediment velocities cluster at higher values than those of the slope (Figure 9). In the Bahama cores, the velocities of shallow-water carbonates are also higher than those of the slope sections and may reach the same maximum velocities as samples from the Maiella. However, due to their younger age and partially incomplete diagenetic alteration, they have a broader range that includes lower velocities.

These data strongly suggest that the depositional environment of a carbonate sediment determines the starting conditions for a sediment to undergo diagenetic alterations. This indirect influence controls rock intrinsic parameters such as porosity, pore type, density, and mineralogy.

Mineralogy

More than 95% of the minerals in the Bahama core samples are calcite, dolomite, and aragonite, and the Maiella carbonates consist almost totally of calcite. The difference among the velocities of calcite (6.53 km/s), dolomite (~7.4 km/s), and aragonite (~5.8 km/s) (Simmons and Wang, 1971; Carmichael, 1989) are high enough to expect some control of mineralogic composition on velocity. In siliciclastics, the physical variety of minerals is very large (e.g., quartz and clay) and mineralogy has a strong influence on sonic velocity (Christensen and Szymanski, 1991). The data of this study, however, suggest that changes in mineralogic composition in pure carbonates have no major influence on velocity.

A plot of dolomite content versus velocity of the Bahamas samples clearly shows no correlation (Figure 10). Fully dolomitized rocks can have completely different

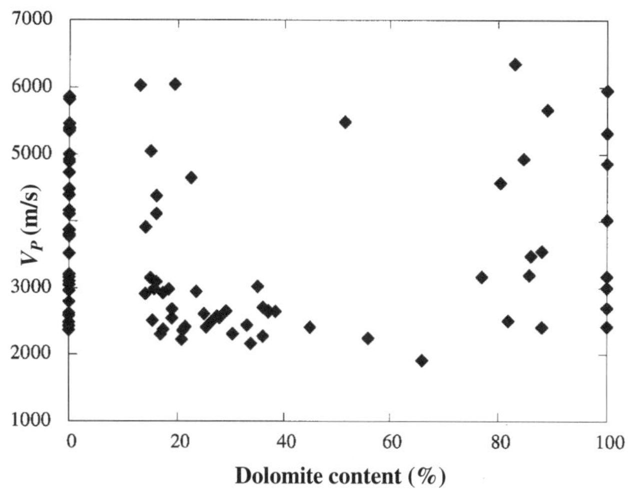

Fig. 10. Graph of velocity (at 8 MPa effective pressure) as a function of dolomite content in the Bahama samples. No correlation is seen between these two parameters; thus, dolomite content alone cannot be used as an indicator of velocity. However, dolomite fabrics, such as sucrosic dolomite and dolomite cement, have profound effects on velocity.

velocities: fabric-preserving dolomitization usually yields extremely high velocities, whereas sucrosic dolomitization results in very low velocities. This suggests that the effect of mineralogy, in this case the higher dolomite crystal velocity, is insignificant compared to the effect of porosity and fabric. Velocity depends more on the type of dolomite and thus more on the associated fabric, porosity, and pore type.

While the mineralogic composition of carbonates has little direct influence on velocity, the processes that alter mineralogy do have a strong influence. Such processes as sucrosic dolomitization and dolomitic cementation, influence velocity, along with mineralogy, porosity, and porosity type.

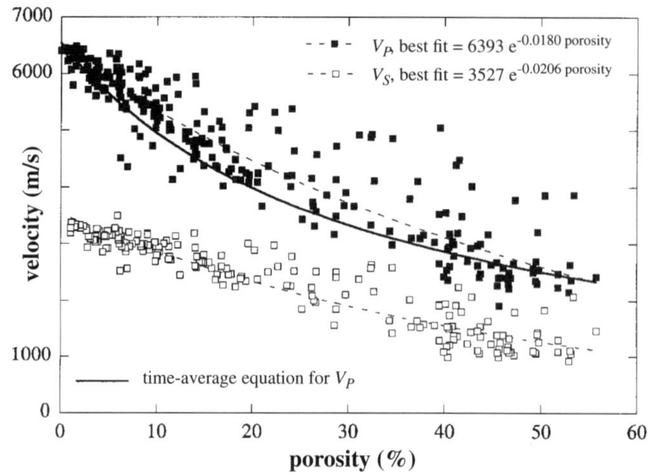

Fig. 11. Graph of V_P and V_S from the Bahama and Maiella samples (at 8 MPa effective pressure) as a function of porosity with exponential best-fit equations. Both V_P and V_S show a trend of decreasing velocity with increasing porosity. However, scattering occurs around the best-fit lines (dashed), especially at higher porosities as a result of certain fabrics and pore types that enhance the elastic moduli of the rock without filling pore space. As a consequence, measured V_P is higher than V_P predicted by the time-average equation (solid line; Wyllie et al., 1956).

Porosity and Pore Types

Velocity is strongly dependent on rock porosity (Rafavich et al., 1984; Wang et al., 1991). A plot of porosity versus velocity clearly shows an inverse trend: an increase in porosity produces a decrease in velocity (Figure 11). The following relationships show that the general trend of Bahama and Maiella samples have correlation coefficients of 0.94 for V_P and 0.92 for V_S:

$$V_P\,(\text{m/s}) = 6393e^{-0.0180 \times \text{porosity (\%)}} \tag{7}$$

$$V_S\,(\text{m/s}) = 3527e^{-0.0206 \times \text{porosity (\%)}} \tag{8}$$

Nevertheless, the measured values display a large scatter around these best-fit exponential curves. Velocity differences at equal porosities can exceed 2500 m/s, especially at high porosities. For example, rocks with porosities of 40% can have velocities ranging from 2100 to 5000 m/s, which is an extraordinary range for rocks having the same chemical composition and porosity. It is shown that this variation is caused by the ability of carbonates to form cements and particular fabrics that, although highly porous, can enhance the elastic properties of the rock without filling the pore space. The high elastic moduli result in velocities that are higher than predicted by theoretical equations, such as the time-average equation (Wyllie et al., 1956), shown in Figure 11.

In other datasets, such as from artificial sand-clay mixtures (Marion et. al, 1992) and siliciclastic sediments (Vernik and Nur, 1992), similar scattering in the velocity versus porosity diagram is observed. But unlike carbonates, the scattering in these rocks can be explained by compositional variations, in particular by changes in clay content. Wilkens et al. (1991) noticed that velocities of low-porous basalts are dependent on pore shapes. Samples containing pores with low aspect ratios (cracks) are associated with lower velocities than samples with round pores or high aspect ratios. As a result, high velocity contrasts without large variations in total porosity are observed. The pores in the measured high-porosity carbonates generally have high aspect ratios; thus, other factors must cause the velocity variations.

In this case, the high velocity contrasts between rocks with similar total porosity can be related to specific pore types resulting in characteristic and different elastic properties. Based on thin section observations, the Bahama samples can be grouped into five categories of predominant pore types, all of which have characteristic clusters in the velocity–porosity diagram (Figure 12). Five dominant pore types can be distinguished:

1. **Interparticle and intercrystalline porosity** (Figures 12b and 13a, d): Interparticle porosity predominates after deposition of a sediment when grains form a loose package with little cementation. Intercrystalline porosity develops at a later stage during diagenesis, when newly crystallized minerals such as dolomite rhombohedra form a loose aggregate. It has a petrophysical behavior similar to interparticle porosity. The accumulation of unconnected grains without cement or matrix results in low velocity because, without a rigid

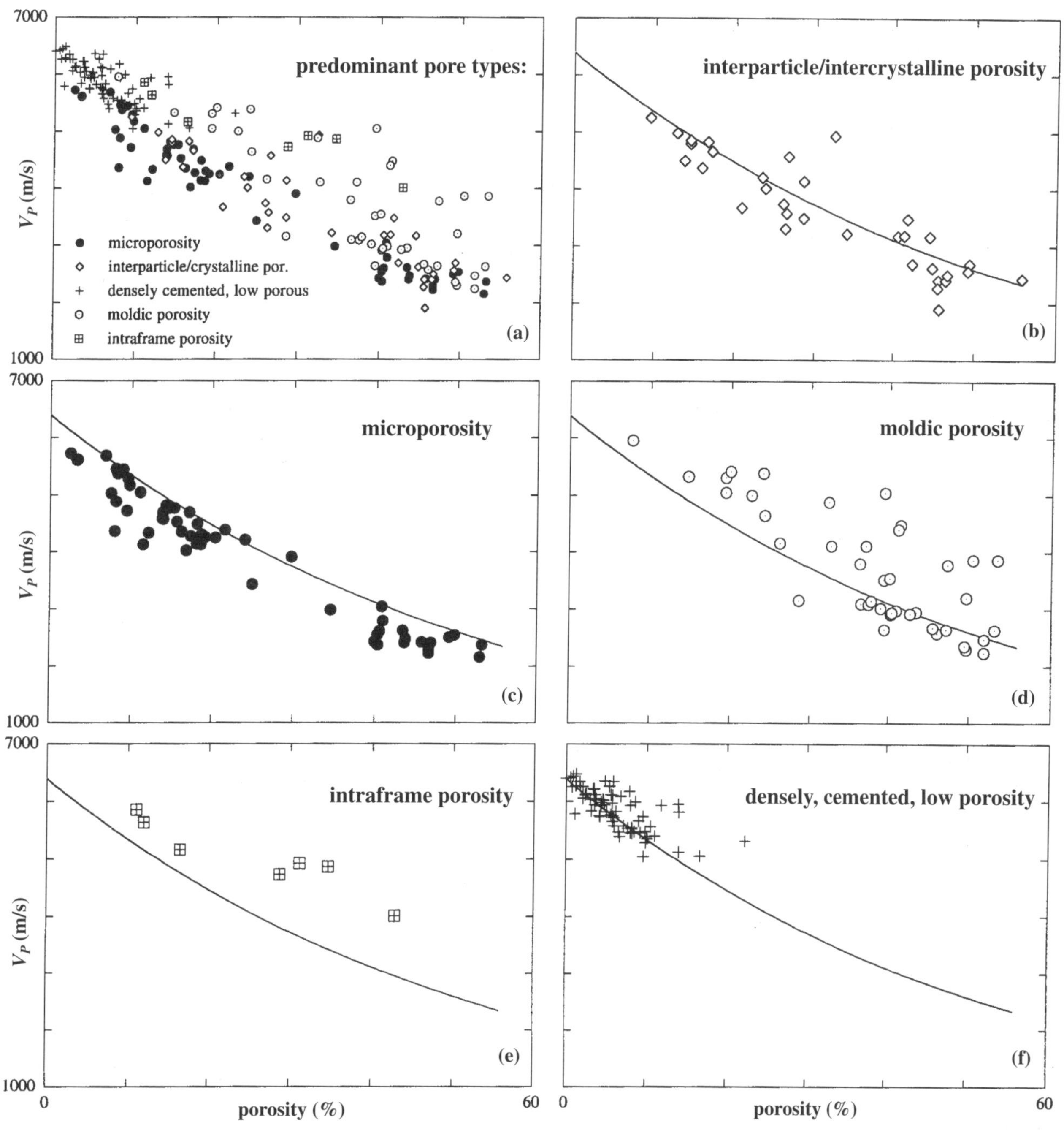

Fig. 12. Graphs of velocity (at 8 MPa effective pressure) versus porosity for various pore types in carbonates. The large scattering of velocities at the same porosity is a result of different predominant pore types in the samples. (a) Values of all Bahama and Maiella samples for which thin sections were analyzed for pore type. (b–f) Graphs of five predominant pore types with exponential best-fit curves through all samples as a reference. (See Figure 13 for photomicrographs of pore types.) Rocks with moldic or intraframe porosity have positive departures from the general trend, whereas rocks with interparticle, intercrystalline, or microporosity have relatively low velocities and thus show negative departures.

framework, the rock has low elastic moduli. Most of these samples show a slightly negative departure from the average velocity–porosity exponential curve (Figure 12b).

2. **Microporosity** (Figures 12c and 13b): Micropores (<10 μm) are abundant in carbonate mud, either within micrite grains or in the micritic matrix. High microporosity is thus expected in carbonates with high micritic content. Due to the lack of cementation that results in an unconnected grain fabric, microporosity has a similar effect on velocity as fine-grained, interparticle porosity. It shows a

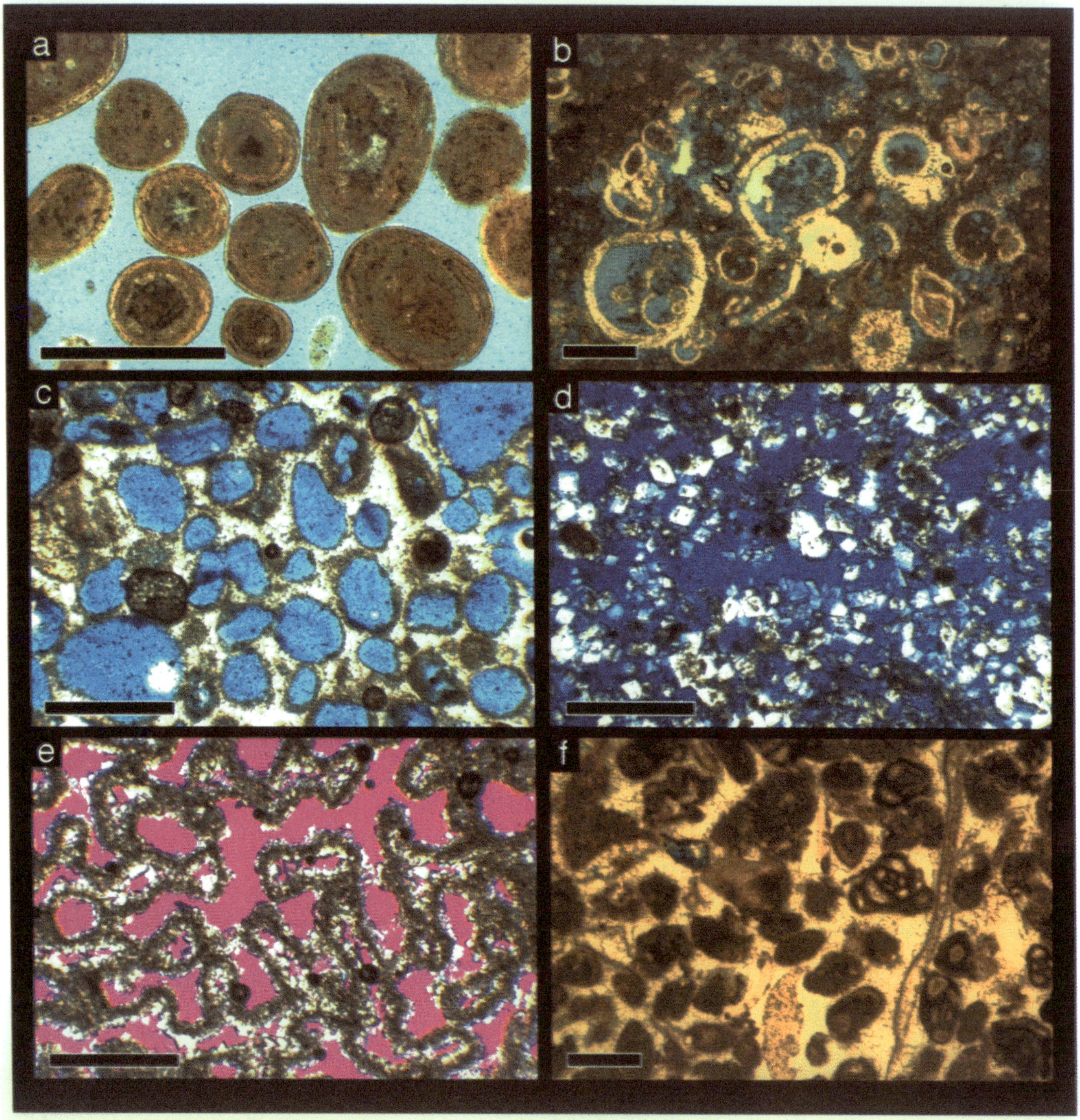

Fig. 13. Photomicrographs of observed pore types in carbonates from Great Bahama Bank. Porosity is shown as blue or purple. Scale bars are 0.5 mm. (a) Interparticle porosity in a modern ooid sand from a sandbar near Cat Cay, Bahamas. Porosity of unconsolidated sediment is 45%. (b) High microporosity in a slope deposit that is rich in globigerinids and micritic matrix; little diagenetic alterations and compaction are visible. Clino drillhole, 510 m, plug porosity 35%. (c) Moldic porosity in a grainstone in which all components were dissolved during and after cementation of interparticle pore space. Unda drillhole, 65 m, plug porosity 37%. (d) Intercrystalline porosity in a sediment completely altered to microsucrosic dolomite. Unda drillhole, 286 m, plug porosity 49%. (e) Intraframe porosity in a coral; the porosity is integrated into the construction of the framestone. Unda drillhole, 302 m, plug porosity 43%. (f) Low porosity in a skeletal grainstone densely cemented with blocky calcite that completely fills the former interparticle pore space. Drillhole Unda, 358 m, plug porosity 3%.

negative departure from the average velocity–porosity trend.

3. **Moldic porosity** (Figures 12d and 13c): Moldic porosity develops by dissolution of grains with a metastable mineralogy (e.g., grains of aragonite and high-Mg calcite). Selective dissolution can occur before, during, or after cementation of the interparticle pore space. After dissolution, the rock consists mainly of molds and the partially cemented former interparticle pore space, which is a fabric type with high values of elastic parameters. Samples in which moldic porosity predominates have higher velocities than expected from their total porosities and therefore a positive departure from the best-fit curve (Figure 12d). These high velocities are caused by the self-supporting framework comprising cement and micrite surrounding the molds. The traveltime through this framework is faster than through grains that are only connected by point contacts, as in rocks with interparticle porosity. Rocks with larger molds tend to have higher velocities than samples with fine-grained moldic porosity. This is mainly the result of increased permeability in coarser grainstones due to larger pores and the absence of micrite in the interparticle pore space. These coarse grainstones usually get cemented quickly and, after dissolution of the original components, form a clean, high-velocity cement frame.
4. **Intraframe porosity** (Figures 12e and 13e): Framestones and boundstones, formed by organisms such as corals and bryozoans, consist of a constructional framework with a porosity that is embedded in the solid frame. These samples thus show a similar velocity–porosity pattern to moldic porous rocks that also have a framework with high elastic rigidity, resulting in high velocities. The samples with predominant intraframe porosity all show positive departures from the general trend in the velocity–porosity diagram.
5. **Low-porosity samples with dense cementation** (Figures 12f and 13f): These samples show extensive blocky cementation with porosities of 20% or less. They are close to the final stage of diagenetic evolution. Velocities are high and close to the intrinsic velocities of calcite (6530 m/s). These samples form the upper part of the velocity–porosity correlation line.

The variety of pore types explains why rocks with the same porosity can have extremely different velocities. The most significant velocity contrasts at equal porosities are measured between moldic rocks and rocks in which interparticle porosity predominates (Figures 12b, d). Moldic rocks with 40–50% porosity can have velocities up to 5000 m/s, whereas rocks with interparticle or intercrystalline porosity have velocities of about 2500 m/s with the same porosities. As a consequence, velocity estimation for a given carbonate sample should not be attempted using only the porosity values, but in combination with an assessment of the pore type. The observed complicated velocity–porosity patterns, which cause matching impedance–porosity patterns, imply that an impedance contrast between two layers can occur even without a porosity change, due only to different pore types.

Density

Seismic reflection patterns are a function of acoustic impedance contrasts. In many case studies, only one parameter, either velocity or density, is known and the other has to be estimated using empirical functions or formulas. Because density is closely related to porosity, velocity shows a good correlation with density (Figure 14). Despite the good correlation coefficients of 0.94 for V_P and 0.93 for V_S, the data in a plot of velocity versus den-

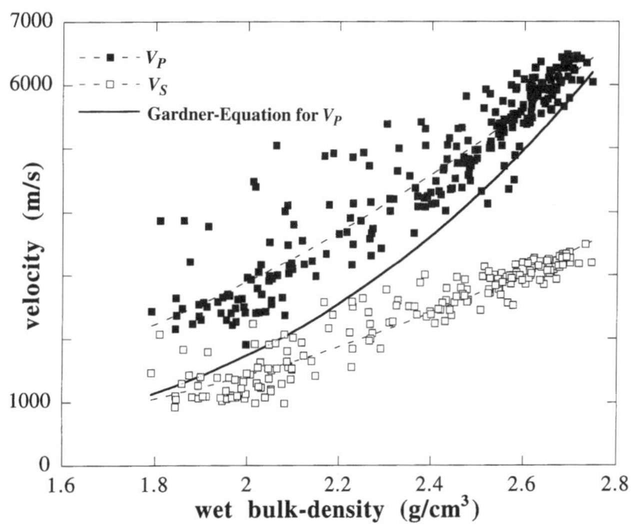

Fig. 14. Velocity (at 8 MPa effective pressure) as a function of density in the Bahama and Maiella samples. The solid line is V_P calculated by Gardner's law [density (g/cm^3) = 0.23 V_P (ft/sec)$^{1/4}$], an empirical formula used to calculate impedance values in sedimentary rocks from only velocity or density data (Gardner et al., 1974). Because the measured velocities are higher than the Gardner equation velocities, the equation must be modified toward higher velocities to produce more reliable velocity-density pairs in carbonates.

sity scatter around a best-fit curve that reflects the carbonate specific pore types. The velocity–density trend from our dataset can be used to calculate a velocity–density formula for pure carbonate rocks. Up to now, most empirical formulas, such as the general Gardner's law (Gardner et al., 1974), were applicable mainly to siliciclastic rocks. Gardner's law is an empirical equation for sedimentary rocks relating V_P to density as follows:

$$\text{Density (g/cm}^3\text{)} = 0.23\,[V_P\,\text{(ft/sec)}]^{1/4}\,; \qquad (9)$$

thus,

$$V_P\,\text{(m/s)} = 108.9\,[\text{density (g/cm}^3\text{)}]^4\,. \qquad (10)$$

All of the laboratory measurements in this study resulted in velocities that are higher than those predicted by Gardner's law (Figure 14). This deviation implies that Gardner's law, which is an average formula for all sedimentary rocks, requires a modification for carbonates toward higher velocities. Based on data from the Bahama and Maiella samples, the following empirical formulas are suggested:

$$V_P\,\text{(m/s)} = 524\,[\text{density (g/cm}^3\text{)}]^{2.48} \qquad (11)$$

and

$$V_S\,\text{(m/s)} = 199\,[\text{density (g/cm}^3\text{)}]^{2.84}\,. \qquad (12)$$

The proposed formula for V_P fits well the dataset shown by Kenter and Ivanov (1995) and fits well the special Gardner correlation for limestones [density (g/cm^3) = 1.493 V_P (km/s)$^{0.324}$] plotted by Gardner et al. (1974) and graphically depicted by Kenter and Ivanov (1995).

Pore Fluid

Experimental and theoretical studies have shown that the state of saturation and type of pore fluid strongly influence the acoustic properties of porous rocks (Nur and Simmons, 1969; Wang et al., 1990; Batzle and Wang, 1992). The investigated samples with maximum porosity values of over 50% in this study are expected to have velocities that are sensitive to changes in pore fluid properties. The effect of different pore fluids, however, has not been addressed in this study. Except for a few measurements of selected Maiella samples, which were measured under dry conditions, all velocities were determined on minicores that were saturated with brine or water and set under independent confining and pore fluid pressures.

Velocity Evolution During Diagenesis

Diagenesis is an important process in carbonates that can transform a lithified sediment into a rock with completely different physical properties. A number of postdepositional processes alter porosity and cause transformations to different fabric and pore types that produce characteristic patterns in velocity evolution. The different ages of analyzed samples, which cover almost continuously the period from Cretaceous to Recent time, offer the opportunity to measure velocities at a number of stages of various diagenetic processes. In particular, the samples from the Bahama cores document in detail the evolution of the porosity–velocity relationship for carbonate sediments under varying diagenetic stages because most of these samples were caught in the act of ongoing diagenetic alterations. The development of porosity and pore types can be determined by analyzing the original composition and subsequent suite of diagenetic events. This development can then be correlated to the measured physical properties.

Concept of Velocity–Porosity Paths

The correlation of lithology with velocity and porosity shows that different diagenetic processes affect the evolution of porosity and velocity of a sediment much more than compaction does. In the case of the measured Bahamas samples, several stages of dissolution and cementation can be distinguished during the diagenetic history (Melim et al., 1994). These stages lead to different pore types that have characteristic patterns in the velocity–porosity diagram. The effect of diagenetic evolution, in particular the effect of the transformation of pore types, can be described by a velocity–porosity path (see Figures 15–18). During burial, every sediment follows a velocity–porosity path, which starts at deposition and ends when velocity–porosity values are measured. This diagenetic path can be simple or complicated, depending on the number of diagenetic stages and the related pore types. Rather than straight lines, the velocity–porosity paths are curved paths that depend on the timing of diagenetic events, which give rise to a history of different pore types for a given rock. By examining the petrography, especially the paragenesis and cement stratigraphy of a rock in thin section, a velocity–porosity path can be determined.

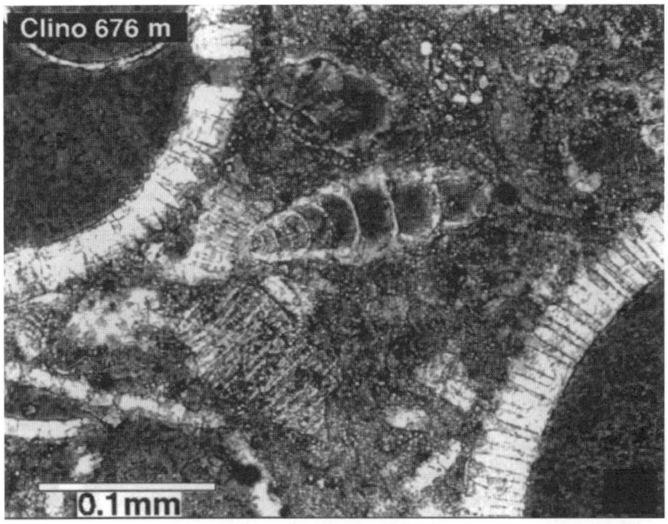

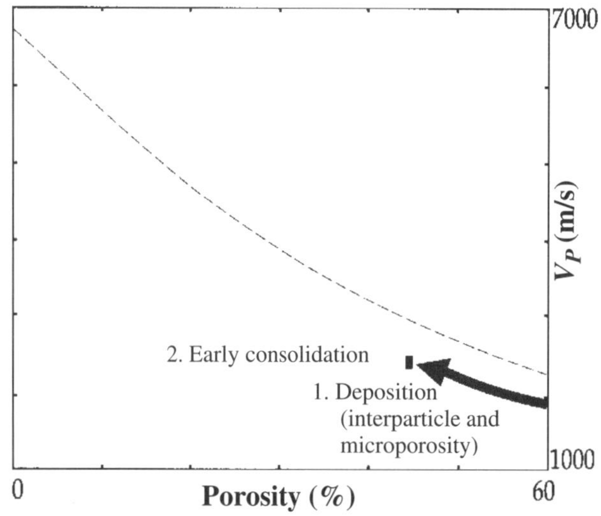

Fig. 15. (Left) Photomicrograph of a periplatform slope sediment containing globigerinids (Great Bahama Bank, drillhole Clino, 676 m). (Right) V_P–porosity diagram for the sample showing a simple path pointing from deposition to current values. The black square marks the measured values (porosity 44%, V_P 2478 m/s, V_S 1356 m/s). The dashed curve is exponential best-fit line through all samples. Despite the deep burial of this sample, little compaction and matrix recrystallization are observed. The velocity–porosity path is thus a short straight line from conditions at deposition to the present.

Figure 15 shows a simple path produced by early compaction, which is usually one of the first processes affecting a sediment. Early compaction results in initial consolidation, dewatering, and grain rearrangement, but no cementation, cracking, or breaking of the components. Initial values of about 50–60% porosity and 1600 m/s for V_P change during this first consolidation stage to values close to 40–50% and 2000 m/s, respectively. At this early stage, the sediments are characterized by interparticle porosity (grainstone) or high microporosity (mudstone to packstone). The sample in Figure 15 is an example of a carbonate sediment that shows hardly any diagenetic alteration other than a little compaction. Interestingly, it is the deepest sample from both drillholes in the Bahamas, but its velocity–porosity path is a simple, short, straight line. The great burial depth of this sample and the absence of any major alteration document again the insignificant effect of burial depth on the velocity of carbonates. In this case, the fine-grained, low-aragonite lithology inhibited the initiation of diagenetic reactions that would have indurated the rock and increased the elastic properties.

A second example displays the velocity–porosity path of the fabric inversion of a grainstone to a coarse moldic rock (Figure 16). A clean, unconsolidated ooid sand from Cat Cay (Bahamas) with a depositional, mainly interparticle porosity of 40–50% (Enos and Sawatsky, 1981) was measured in its unconsolidated and unaltered state (Figure 13a). The resulting V_P equals 1779 m/s at 8 MPa effective pressure. This rock type

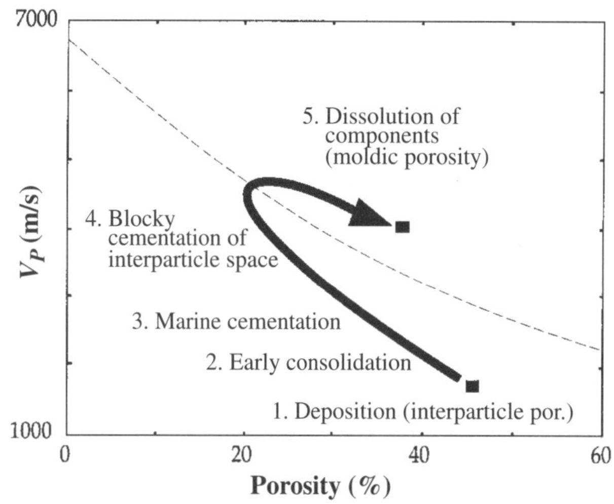

Fig. 16. Reconstruction of a complicated velocity–porosity path from an unconsolidated ooid sand with interparticle porosity as the depositional starting point (Cat Cay, Bahamas, Figure 13a) to a rock with moldic porosity as the diagenetic endpoint (Figure 13c). The ooid sand has a porosity of 45% and a V_P of 1779 m/s. The sample at the end of the path (Drillhole Unda, 65 m, porosity 37%, V_P 4105 m/s, V_S 1640 m/s) is a former ooid(?) grainstone with coarse moldic porosity. Transformation of pore types and dissolution of grains, which occurred after cementation, led to a fabric inversion (Figure 13a, c). The velocity–porosity path has a characteristic loop shape caused by the succession of diagenetic processes. The moldic framework provides high elastic rigidity and thus high velocity at high porosity.

can be considered as the potential precursor of a sample encountered in the Bahama cores, which is characterized by interparticle cement and moldic porosity (Figure 13c). Only the former interparticle pore space is cemented, but the molds are still open and not filled with cement. This indicates that the interparticle pore space became cemented mainly before the dissolution of the ooids or peloids. As a consequence, the rock has a completely inverted fabric compared to the original ooid sand. It has a porosity of 37% (mainly moldic) and a V_P of 4105 m/s.

During fabric inversion, the rock must have gone through a stage with much lower porosity. This stage marks a turning point in the loop in the velocity–porosity path (Figure 16) documenting the interference of two processes with different effects on porosity and sonic velocity. This evolution shows how diagenetic processes can invert the fabric and change the elastic properties of the rocks, even without changing total porosity. The velocity–porosity path, deduced from petrographic examination of the rock, is useful for describing such an evolution, which would not be recognized by laboratory measurements alone.

After an early dissolution stage, some samples undergo an intense cementation of the newly created moldic pore space (Figure 17a). The relics of the stage with a dominant moldic porosity can be recognized by the abundance of former molds that are filled with a blocky spar and surrounded by a micritic rim that mimics the former grain shape. The beginning of the velocity–porosity path toward the moldic stage is the same as in the example of Figure 16 and describes a clockwise loop. The additional stage of cementation reduces porosity, increases velocity, and creates a much denser rock fabric. The resulting velocity–porosity path is a Z-shaped curved line. If the rock had not undergone a stage of dominant moldic porosity, the final high velocity and low porosity values would have been reached along a relatively straight-line path.

In contrast to the constructive process of cementation, fabric destructive processes, such as sucrosic dolomitization, can decrease velocity and increase porosity of an already altered sediment. Figure 17b shows the characteristic evolution of such a destructive diagenetic process. In the thin section of this sample, no original components or textures (except for few relics of red algae) can be recognized. Therefore, it is difficult to reconstruct the beginning and intermediate stages of the velocity–porosity path, but it is likely that some marine or freshwater environment led to cementation and dissolution before sucrosic dolomitization started. As a result, the path has a loop shape which, due to the destructive nature of this dolomitization process, points downward toward lower velocities and higher porosities.

All described velocity–porosity paths can be summarized in a graph (Figure 18) that shows the effects of the various diagenetic processes. Each process has its characteristic direction in the velocity–porosity diagram, but the absolute position within the diagram is dependent on the succession and timing of the different processes. The path for all sediments starts at deposition. The subsequent evolution is mainly dependent on the depositional lithology (diagenetic potential) and the succession of diagenetic processes. It is remarkable that dissolution has only a minor effect on impedance because dissolution roughly follows the lines of equal impedance. This leads to preservation of elastic properties by selective dissolution, during which the rock maintains a framework with relatively high elastic stiffness.

Apart from purely calcitic pelagic oozes (Schlanger and Douglas, 1974), most carbonate sediments are characterized by high diagenetic potential. All the described diagenetic processes associated with them happen much faster than compaction. Consequently, carbonates are quickly dissolved, cemented, and recrystallized.

In siliciclastic or shaly sediments, the diagenetic potential is much lower than in carbonates, and increasing burial pressure leads to an increase in velocity at greater depth (Hamilton, 1980; Japsen, 1993). In carbonates, the diagenetic processes control and alter the velocity before compaction can play a significant role. The dominance of lithification by chemical diagenesis over lithification by compaction is the reason why the velocities of carbonates show no clear correlation with increasing depth. Before burial pressure compacts the rock fabric, the sediment is already altered and the cemented fabric and some of the porosity can survive the increasing overburden to considerable depth.

Conclusions

Laboratory measurements of velocities from 295 carbonate rock samples gave a remarkably wide range of more than 4500 m/s for V_P and 2500 m/s for V_S. The maximum velocities (V_P = 6500 m/s, V_S = 3400 m/s) are about four times higher than the minimum values (V_P = 1700 m/s, V_S = 700 m/s). Together with the density variations, these velocity contrasts cause large impedance differences that explain the excellent seismic reflectivity of pure carbonate rocks observed in seismic sections.

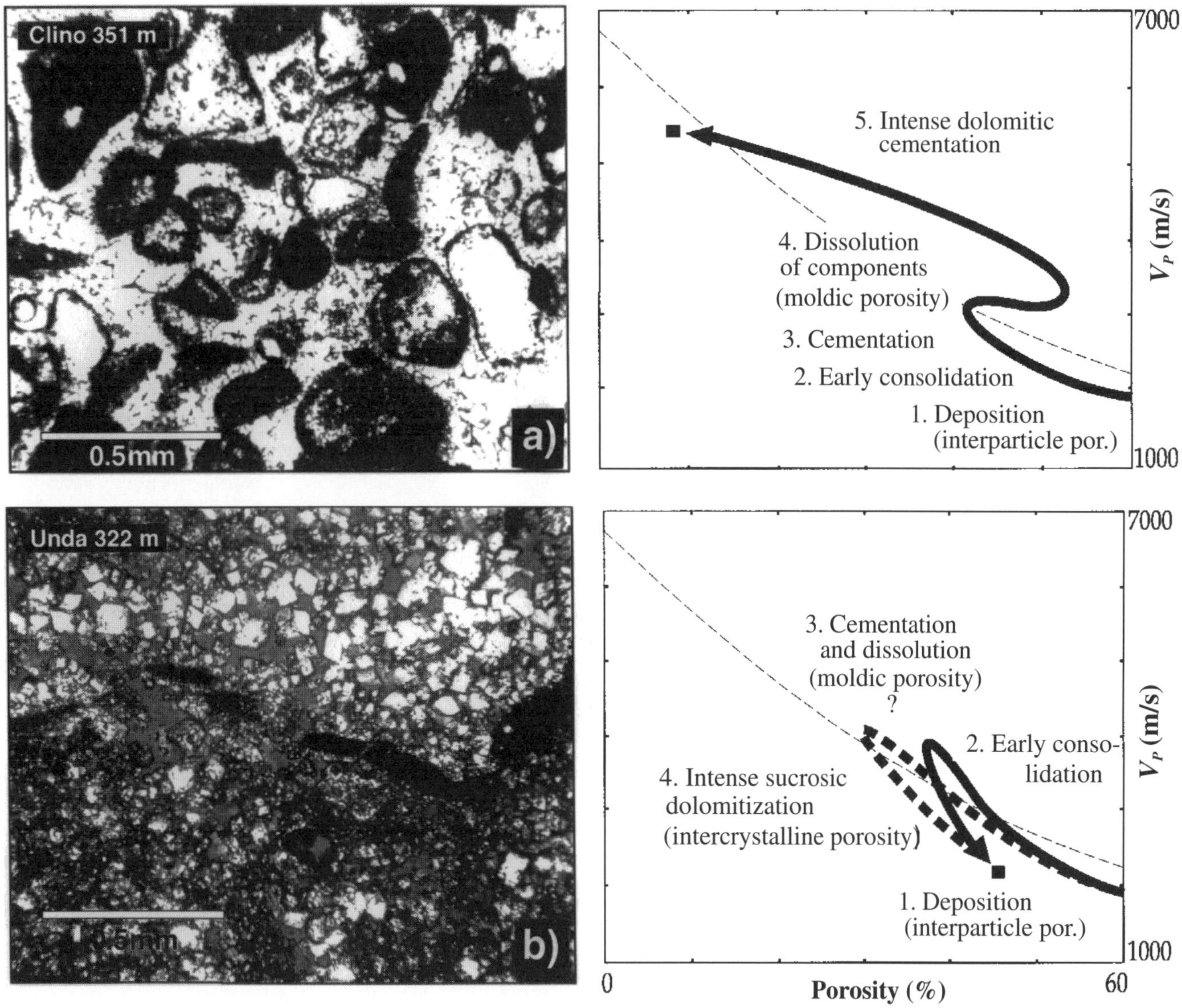

Fig. 17. Photomicrographs and plots of inferred velocity–porosity paths for two samples from Bahama drillholes. (a) Densely cemented grainstone with a similar evolution as in Figure 16, but an additional cementation after dissolution filled the earlier moldic pore space, resulting in reduced porosity and increased velocity. Drillhole Clino, 351 m, porosity 9%, V_P 5661 m/s, V_S 3158 m/s. (b) Sucrosic dolomite with a few relics of red algae. The sedimentary fabric was destroyed by sucrosic dolomitization that created an intercrystalline porosity and caused velocity to decrease and porosity to increase. Unda drillhole, 322 m, porosity 46%, V_P 2405 m/s, V_S 991 m/s.

The performed laboratory analyses document that the variability of velocity in carbonates is a complex product of several factors that have different relevances and effects:

- Changes in carbonate mineralogy are not responsible for the large variability in velocities, as most measured samples contain only carbonate minerals that have similar physical properties. Fully dolomitized rocks can have very fast or very slow velocities, demonstrating the low significance of mineralogic composition.
- Carbonate sediments deposited in shallow water generally have higher average velocities than carbonates from the deeper shelf, slope, or basin. This relationship can be explained by the relatively high diagenetic potential of shallow-water carbonates resulting from their high content of metastable aragonite and coarser grain size and their exposure to freshwater diagenesis.

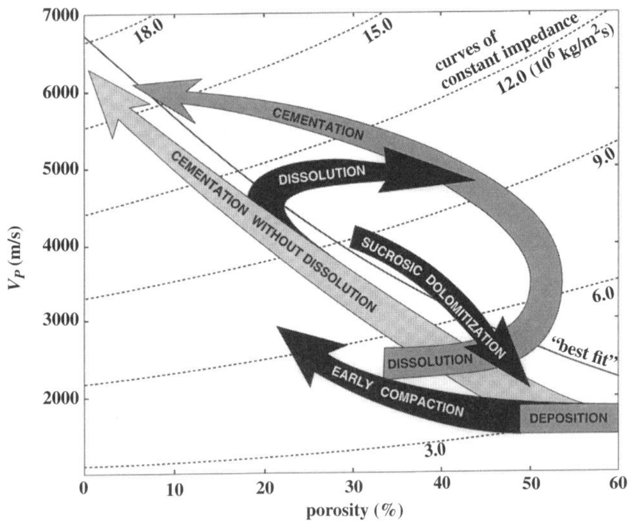

Fig. 18. Velocity–porosity paths of various diagenetic processes in carbonate sediments. Any given process may have a different position on the diagram depending on the timing of the diagenetic events, but the trend (direction of arrow) should be the same. The solid line is the exponential best-fit line through all data points measured in this study; dashed curves are lines of equal impedance (calculated using the density of calcite).

- Comparison of velocity with burial depth and age shows that neither factor on its own exerts a significant control on velocity evolution. In fact, velocity inversions with increasing depth or age are common. For instance, a velocity increase caused by pure mechanical compaction, as measured in compacted carbonate mud, is lower than the observed velocity increase with decreasing porosities in natural rocks. This difference in the manner of velocity increase clearly shows that velocity is influenced primarily by postdepositional diagenetic processes other than pure compaction during increasing burial depth and time.
- Porosity is the most important physical factor that influences velocity. V_P and V_S increase with decreasing porosity, but large departures from this general trend are possible.
- Different velocities in rocks with equal porosities are the result of different pore types. Rocks with frame-forming pore types, such as moldic porosity, can have very high velocities even with high-porosity fabrics, whereas rocks with interparticle, intercrystalline, or high microporosity have much lower velocities at the same porosities.
- Increasing effective pressure usually increases velocity until it reaches a plateau at pressures of about 100 MPa. However, an increase in pressure can also lead to fracturing of the rock and thus to a dramatic decrease in velocity. The critical pressure at which fractures form varies substantially and can be as low as 5 MPa.
- The high values of elastic properties of self-supporting fabrics within a framework enable rocks to maintain extensive porosity even at relatively high overburden pressures. In particular, cementation and dissolution processes, which occur during early stages of burial, create stable rock fabrics with high velocities. As a consequence, commonly used velocity–density or velocity–porosity functions, such as the general Gardner equation or the time-average equation, must be modified toward higher velocities to produce reliable velocity data in carbonate rocks.

Most physical properties are the combined result of (1) the initial sediment type and (2) diagenetic alterations. The depositional lithology determines the diagenetic potential and, together with the succession of diagenetic processes, controls the postdepositional history of the carbonate sediment. The timing of the diagenetic events determines the porosity–velocity evolution and therefore the velocity development.

In summary, it is impossible to make velocity predictions for carbonate rocks based solely on depth or age data. High susceptibility toward diagenetic alteration distinguishes carbonate sediments from other sedimentary rocks. Diagenetic processes, such as dissolution, cementation, and recrystallization, can occur so rapidly and at such shallow depths of burial that the rock fabric is completely altered before compaction reduces porosity. The sediment is lithified mainly by early diagenesis, and to a lesser degree by increased overburden, which explains the poor correlation between velocity and age or depth.

Acknowledgments

Many of the results of this study have been published in the "Schreiber volume" (Liebermann and Sondergeld, 1994). We were encouraged by Dominique Marion and the editors to summarize and update these earlier results for this publication. The Bahama samples for this study were collected under NSF grant OCE-89117295 and the Industrial Associates of the Comparative Sedimentology Laboratory of the University of Miami. The Maiella samples were collected during a project funded by the Swiss NSF grant 20-27457.91. We thank the collaborators of these studies for providing the geologic background information. G. Eberli was supported by DOE grant DE-FG05-92ER14253, and F. Anselmetti was supported by Shell Research and Geological Institute, ETH Zürich, Switzerland.

References

Anselmetti, F. S., and Eberli, G.P., 1993, Controls on sonic velocity in carbonates: Pure Appl. Geophys., **141**, 287-323.

Batzle, M., and Wang, Z., 1992, Seismic properties of pore fluids: Geophysics, **57**, 1396-1408.

Bernoulli, D., Eberli, G. P., Pignatti, J. S., Sanders, D., and Vecsei, A., 1992, Sequence stratigraphy of Montagna della Maiella: Quinto simposio di ecologia e paleoecologia delle comunita' bentoniche, Paleobenthos V, libro-guida delle escursioni, Dip. di Scienze della Terra, Univ. "La Sapienza," 85-109.

Biot, M. A., 1956, Theory of propagation of elastic waves in a fluid-saturated porous solid, I. Low frequency range, II. Higher frequency range: J. Acoust. Soc. Am., **28**, 168-191.

Birch, F., 1960, The velocity of compressional waves in rocks to 10 kilobars, part 1: J. Geophys. Res., **65**, 1083-1102.

Bourbié, T., Coussy, O. and Zinszner, B., 1987, Acoustics of porous media: Editions Téchnip.

Burns, S.J. and Swart, P.K., 1992, Diagenetic processes in Holocene carbonate sediments: Florida Bay mudbanks and islands: Sedimentology, **39**, 285-304.

Carmichael, R. S., Ed., 1989, Practical handbook of physical properties of rocks and minerals: CRC Press.

Christensen, N. I., and Szymanski, D. L., 1991, Seismic properties and the origin of reflectivity from a classic Paleozoic sedimentary sequence, Valley and Ridge province, southern Appalachians: Geol. Soc. Am. Bull., **103**, 277-289.

Coyner, K. B., 1984, Effects of stress, pore pressure, and pore fluids on bulk strain, velocity and permeability in rocks: Ph.D. thesis, Mass. Inst. Tech.

Eberli, G. P., 1988, Physical properties of carbonate turbidite sequences surrounding the Bahamas: implications for slope stability and fluid movements, *in* Austin, Jr., J. A., and Schlager, W., Eds., Proc. Ocean Drill. Prog., Scient. Results, **101,** 305-314.

Eberli, G. P., 1991, Growth and demise of isolated carbonate platforms: Bahamian controversies, *in* Müller, D. W., McKenzie, J. A., and Weissert, H., Eds., Controversies in modern geology: Academic Press, 231-248.

Eberli, G. P., and Ginsburg, R. N., 1989, Cenozoic progradation of northwestern Great Bahama Bank, a record of lateral platform growth and sea-level fluctuations, *in* Crevello, J. L., Wilson, J. L., Sarg, J. F., and Read, J. F., Eds., Controls on carbonate platform and basin development: Soc. Econ. Paleont. Mineral. Spec. Pub. **44**, 339-351.

Eberli, G. P., Bernoulli, D., Sanders, D., and Vecsei, A., 1993, From aggradation to progradation: the Maiella platform, Abruzzi, Italy, *in* Simo, J. T., Scott, R. W., and Masse, J.-P., Eds., Cretaceous carbonate platforms: Am. Assn. Petr. Geol. Mem. **56**, 213-232.

Enos, P., and Sawatsky, L. H., 1981, Pore networks in Holocene carbonate sediments: J. Sed. Pet., **51**, 961-985.

Fagin, S. W., 1991, Seismic modeling of geologic structures: Soc. Expl. Geophys.

Gardner, G. H. F., Gardner, L. W., and Gregory, A. R., 1974, Formation velocity and density: the diagnostic basics for stratigraphic traps: Geophysics, **39**, 770-780.

Gassmann, F., 1951, Elastic waves through a packing of spheres: Geophysics, **16**, 673-685.

Ginsburg, R. N., in press, The Bahamas Drilling Project: summaries of background and acquisition of cores and logs: Soc. Econ. Paleont. Mineral. Concept Series.

Gregory, A. R., 1977, Aspects of rock physics from laboratory and log data that are important to seismic interpretation, *in* Payton, C. E., Ed., Seismic stratigraphy—applications to hydrocarbon exploration: Am. Assn. Pet. Geol. Mem. **26**, 15-46.

Hamilton, E. L., 1971, Elastic properties of marine sediments: J. Geophys. Res., **76**, 579-604.

Hamilton, E. L., 1980, Geoacoustic modeling of the sea floor: J. Acoust. Soc. Am., **68**, 1313-1340.

Han, D., Nur, A., and Morgan, D., 1986, Effects of porosity and clay content on wave velocities in sandstones: Geophysics, **51**, 2093-2107.

Japsen, P., 1993, Influence of lithology and Neogene uplift on seismic velocities in Denmark: implications for depth conversion of maps: Am. Assn. Pet. Geol. Bull., **77**, 194-211.

Johnston, J. E., and Christensen, N. I., 1992, Shear wave reflectivity, anisotropies, Poisson's ratios and densities of a southern Appalachian Paleozoic sedimentary sequence: Tectonophysics, **210**, 1-20.

Johnston, J. E., and Christensen, N. I., 1993, Compressional to shear velocity ratios in sedimentary rocks: Int. J. Rock. Mech. Min. Sci. and Geomech. Abstr., **30**, 751-754.

Kenter, J. A. M., and Ginsburg, R. N., in press, Lithologic and sedimentologic observations on two core borings through the western prograding Neogene–Quaternary margin of Great Bahama Bank: Soc. Econ. Paleont. Mineral. Concept Series.

Kenter, J. A. M., and Ivanov, M., 1995, Parameters controlling acoustic properties of carbonate and volcaniclastic sediments at sites 866 and 869, *in* Winterer, E. L., Sager, W. W., et al., Eds., Proc. Ocean Drill. Prog., Scient. Results, **143**, 287-303.

King, M. S., 1966, Wave velocities in rocks as a function of changes in overburden pressure and pore fluid saturants: Geophysics, **31**, 50-73.

Laughton, A. S., 1957, Sound propagation in compacted ocean sediments: Geophysics, **22**, 233-260.

Liebermann, R. C., and Sondergeld, C. H., Eds., 1994, Experimental techniques in mineral and rock physics—the Schreiber volume: Birkhäuser Verlag.

Marion, D., Nur, A., Yin, H., and Han, D., 1992, Compressional velocity and porosity in sand-clay mixtures: Geophysics, **57**, 554-563.

Melim, L. A., Swart, P. K., and Maliva, R. G., in press, Diagenesis of carbonates from the Bahamas Drilling Project, western margin Great Bahama Bank: meteoric versus marine burial diagenesis: Soc. Econ. Paleont. Mineral. Concept Series.

Milholland, P., Manghani, M. H., Schlanger, S. O., and Sutton, G. H., 1980, Geoacoustic modeling of deep-sea carbonate sediments: J. Acoust. Soc. Am., **68**, 1351-1360.

Mutti, M., 1995, Porosity development and diagenesis in the Orfento supersequence and its bounding unconformities, Upper Cretaceous, Montagna della Maiella, Italy, *in* Budd, D., Saller, A., and Harris, M., Eds., Unconformities in carbonate strata: their recognition and the significance of associated porosity: Am. Assn. Pet. Geol. Mem. **63**, 141-158.

Nafe, J. E., and Drake, C. L., 1963, Physical properties of marine sediments, *in* The Sea 3: Interscience Publishers, 794-815.

Nur, A., and Simmons, G., 1969, The effect of saturation on velocity in low porosity rocks: Earth and Planet. Sci. Lett., **7**, 183-193.

Nur, A., Marion, D., and Yin, H., 1991, Wave velocities in sediments, in Hovem, J. M., Richardson, M. D., and Stoll, R. D., Eds., Shear waves in marine sediments: Kluwer Academic Publishers, 131-140.

Rafavich, F., Kendall, C. H. St. C., and Todd, T. P., 1984, The relationship between acoustic properties and the petrographic character of carbonate rocks: Geophysics, **49**, 1622-1636.

Schlanger, S. O., and Douglas, R. G., 1974, The pelagic ooze-chalk-limestone transition and its implications for marine stratigraphy, in Hsu, K. J., and Jenkyns, H. C., Eds., Pelagic sediments on land and under the sea: Internat. Assn. Sed. Spec. Publ. **1**, 117-148.

Sellami, S., Barblan, F., Mayerat, A.-M., Pfiffner, O. A., Risnes, K., and Wagner, J.-J., 1990, Compressional wave velocities of samples from the NFP-20 East seismic reflection profile: Mém. Soc. Géol. Suisse, **1**, 77-84.

Simmons, G., and Wang, H., 1971, Single crystal elastic constants and calculated aggregate properties: A handbook: M.I.T. Press.

Swart, P. K., and Kramer, P. A., in press, The geology of mud-islands in Florida Bay, *in* Vacher, H. L., and Quinn, T. M., Eds., Geology and hydrogeology of carbonate islands: Elsevier.

Tatham, R. H., 1982, *VP/VS* and lithology: Geophysics, **47,** 336-344.

Toksöz, M. N., Cheng, C. H., and Timur, A., 1976, Velocities of seismic waves in porous rocks: Geophysics, **41**, 621-645.

Urmos, J., and Wilkens, R. H., 1993, In situ velocities in pelagic carbonates: new insights from Ocean Drilling Program Leg 130, Ontong Java Plateau: J. Geophys. Res., **98**, 7903-7920.

Vernik, L., and Nur, A., 1992, Petrophysical classification of siliciclastics for lithology and porosity prediction from seismic velocities: Am. Assn. Pet. Geol. Bull., **76,** 1295-1309.

Wang, Z., Batzle, M. L., and Nur, A., 1990, Effect of different pore fluids on seismic velocities in rocks: Can. J. Expl. Geophys., **26**, 104-112.

Wang, Z., Hirsche, W. K., and Sedgwick, G., 1991, Seismic velocities in carbonate rocks: J. Can. Petr. Tech., **30,** 112-122.

Wilkens, R., Simmons, G., and Caruso, L., 1984, The ratio *VP/VS* as a discriminant of composition for siliceous limestones: Geophysics, **49**, 1850-1860.

Wilkens, R. H., Fryer, G. F., and Karsten, J., 1991, Evolution of porosity and seismic structure of upper oceanic crust: importance of aspect ratios: J. Geophys. Res., **96**, 17981-17995.

Wilson, J. L., 1975, Carbonate facies in geologic history: Springer Pub. Co.

Wood, A. B., 1941, A textbook of sound: Macmillan.

Wyllie, M. R., Gregory, A. R., and Gardner, G. H. F., 1956, Elastic wave velocities in heterogeneous and porous media: Geophysics, **21**, 41-70.

Yin, H., Nur, A., and Mavko, G., 1993, Critical porosity—a physical boundary in poroelasticity: Int. J. Rock. Mech. Min. Sci. and Geomech. Abstr., **30**, 805-808.

Chapter 5

Acoustic Properties of Carbonate Rocks: Use in Quantitative Interpretation of Sonic and Seismic Measurements

D. Marion
Elf Gabon
Port-Gentil, Gabon

D. Jizba
Elf Aquitaine Production
Pau, France

Abstract

Four main groups of parameters influence the seismic response in carbonate rocks: (1) petrophysical properties, such as porosity, pore compressibility, fluid saturation, and fluid type; (2) lithologic parameters, such as mineralogy and shaliness; (3) measured conditions, such as effective stress, temperature, and frequency; and (4) geometric factors such as bed thickness and anisotropy. Knowledge of porosity variations over any carbonate reservoir has significant engineering and economic implications for production development activities. Because porosity is the most important factor determining the seismic response in carbonate rocks, quantification of its variation over a reservoir becomes a critically important task. Thus, before any velocity change is used for quantifying reservoir porosity, relative influence of other factors must be identified and accounted for.

An integrated approach to achieve this objective starts by building a rock physics model using core data, validating it using log data, and then applying it to the seismic interpretation. A laboratory study of selected cores from several wells over a field should be carried out to determine *P*- and *S*-wave velocities, porosity, permeability, and petrophysical parameters. This information is then used to build a rock physics model that accounts for the influence of porosity, pore shape, mineralogy, saturation, and measured frequency on seismic responses. The next step is to calibrate the rock physics model to log data which involves interpretation of logs. Finally, this rock physics model, established for core data and validated by logs, is applied to the seismic interpretation.

A case study of a Tertiary carbonate reservoir confirmed the quantitative interpretation of seismic impedance traveltimes in terms of total reservoir thickness. It also confirmed that the validity of the method strongly relies on three factors: (1) an accurate relationship between seismic attributes and rock properties; (2) coherency among data acquired by cores, logs, and the seismic model; and (3) high-quality seismic data, reliable seismic interpretation, and adapted seismic processing and inversion techniques.

Introduction

The efficient development and production of carbonate oil and gas reservoirs relies on an accurate vertical and lateral description of reservoir properties. Because 2-D and 3-D seismic techniques provide extensive measurement coverage, they have been widely used to image the structure of reservoirs and to provide a framework for depositional and structural geologic models. In addition, seismic attributes can also be used to assess reservoir petrophysical properties and provide constraints in building reservoir simulation models.

When seismic amplitudes and traveltimes are used in a such a quantitative manner, it is important that the relationship used to transform seismic attributes into rock properties is calibrated correctly. The need for such calibration is motivated by two main considerations:

1. One must be able to distinguish reservoir parameters that are identifiable on the seismic response. Many petrophysical properties (porosity, pore compressibility, fluid saturation, and fluid type), lithologic parameters (mineralogy and shaliness), measurement conditions (effective stress, temperature, and frequency), and geometric factors (bed thickness, dip, and anisotropy) can affect seismic wave propagation. Hence, in an attempt to interpret seismic attributes quantitatively, it is first necessary to identity those parameters that may vary significantly, laterally or vertically for a given field, and to estimate their relative impact on seismic properties.
2. One must be able to use calibrated rock physics relationships for a given field. Empirical relationships from the literature commonly used to relate seismic velocities and impedances to porosity, mineralogy (Wyllie et al., 1956; Raymer et al., 1980), and fluid type (Krief et al., 1990) may not be relevant or sufficiently precise for a given field. Furthermore, theoretical models such as Gassmann's relationships for fluid substitution modeling (Gassmann, 1951) require calibration of several inputs.

In this paper, we describe a method in which core and log measurements are used to establish a calibrated rock physics model for seismic interpretation. The first part reviews several aspects of the relationships between petrophysical, lithologic, and acoustic properties in carbonate rocks that justify a need for calibration of seismic interpretation. Using a case study on a carbonate oil reservoir, we demonstrate in the second part how the calibration method is applied to sonic and seismic measurements and used in a quantitative manner to characterize the reservoir pay zone.

Elastic Properties of Carbonate Rocks

Many studies have shown that in carbonate rocks, porosity is the rock parameter that has the most impact on velocity (Rafavitch et al., 1984; Wang et al., 1991; Anselmetti an Eberli, 1993). Hence, characterizing lateral variations in porosity within a field from seismic velocities or impedances is potentially feasible and represents a major contribution of seismic techniques for reservoir appraisal. However, because velocities in carbonate rocks also depend on other parameters, such as mineralogy, pore shape, and fluid type, the velocity–porosity trend may be too scattered to permit precise determination of porosity from velocity measurements.

To gain insight as to how petrophysical and lithologic parameters can affect the velocity–porosity relationship, one can use the following physically based relationship between the elastic modulus of the skeleton frame of a rock and porosity (Walsh, 1965):

$$1/K_{\mathrm{fr}} = 1/K_{\mathrm{'m}} + \phi C_{\mathrm{'p}}\,, \tag{1}$$

where K_{fr} and K_m are the bulk moduli of the skeletal frame and mineral phase of the rock, respectively, ϕ is the porosity, and C_p is the compressibility of the pores. K_{fr} is related to the *P*- and *S*-wave velocities and bulk density (ρ) of an isotropic dry rock according to

$$K_{\mathrm{fr}} = \rho\,(V_P^2 - 4/3\,V_S^2)\,. \tag{2}$$

From equation (1), we note that the bulk modulus varies with porosity, ϕ, but that it also depends on the bulk modulus of the mineral phase, K_m, which varies with mineralogy, and the compressibility of the pores, C_p, which depends mostly on pore shape and aspect ratio. Furthermore, for fluid-saturated rocks, the bulk modulus also depends on fluid compressibility (Gassmann, 1951) and measurement frequency (Biot, 1956; O'Connell and Budiansky, 1974; Mavko and Jizba, 1991).

Hence, precise porosity determination from elastic properties relies on an accurate assessment of the influence of mineralogy, pore shape, and fluid type on elastic properties. This issue is addressed in the following section, which illustrates the possible impact of such parameters on velocity–porosity relationships.

Influence of Mineralogy

The minerals that are commonly found in carbonate rocks—calcite, dolomite, and aragonite—have a wide range of elastic properties and density, as illustrated in Table 1.

From equation (1), the effect of mineralogy on bulk modulus should correspond to a change in the intercept in the relationship between porosity and the inverse bulk modulus. For calcite and dolomite, the influence of mineralogy on bulk modulus is expected to be negligible, considering the narrow range of mineral bulk modulus between calcite and dolomite. This is illustrated in Figure 1a where compressibility (the inverse of bulk modulus) is plotted against porosity for cores composed

Table 1. Elastic Properties of Common Carbonate Minerals

Mineral	ρ (g/cm³)	K_m (GPa)	G_m (GPa)	V_P/V_S ratio	V_P (km/s)	V_S (km/s)	Reference
Calcite	2.71	70	29	1.94	6.34	3.27	Peselinick and Robie (1963)
Calcite	2.71	77	32	1.93	6.64	3.44	Simmons (1965)
Dolomite	2.88	69	52	1.64	6.93	4.23	Schlumberger (1984)
Aragonite	2.92	45	39	2.05	5.75	3.65	Ellis et al. (1965)

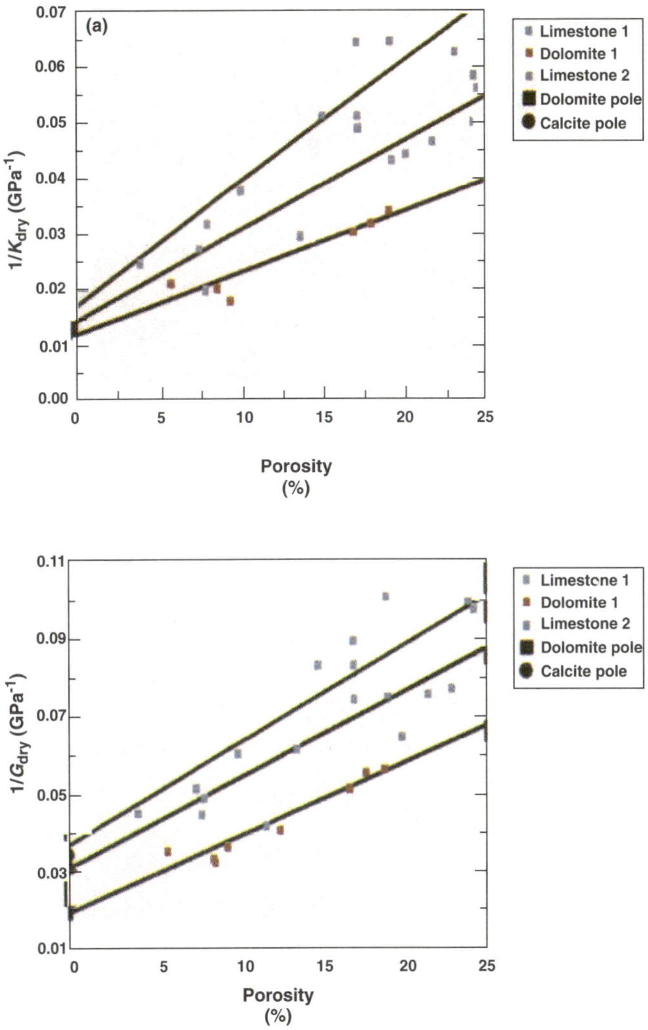

Fig. 1. Graphs of (a) the inverse bulk modulus and (b) the shear modulus versus porosity showing the sensitivity of mineralogy for three carbonate rock suites. Elastic moduli were derived from *P*- and *S*-wave velocity and density measured on cores in dry conditions at 25 MPa effective pressure. The three lines are best fits to the corresponding data points.

of at least 90% dolomite and calcite each. In this data set, bulk modulus data were derived from ultrasonic *P*- and *S*-wave velocities measured on dry samples at an effective stress of 25 MPa. Note in Figure 1a that the intercept describing the influence of mineral compressibility on bulk compressibility is comparable for dolomite and limestone and is coherent with values in Table 1. However, the slope, which according to equation (1) is a measure of pore compressibility, is extremely variable from one set of rocks to another.

If we consider as a first-order approximation that equation (1) also applies to the shear modulus, a greater separation between calcite and dolomite should be observed at low porosity according to Table 1. This is indeed observed on Figure 1b where the inverse shear modulus is plotted against porosity for the dolomite and limestone samples shown in Figure 1a. We note from linear fits to the data that the intercept values for dolomite and limestone range from 0.02 to 0.033 GPa^{-1}. These results are consistent with values from Table 1.

Of the parameters involved in wave propagation, density is also influenced by mineralogy. Figure 2 illustrates the influence of mineral density on the relationship between bulk density and porosity for the dolomite and calcite samples shown in Figure 1. For reference, theoretical lithologic relationships for pure dolomite and pure calcite are shown for dry rocks.

From this data set, we conclude that mineralogy (calcite-dolomite) has a negligible impact on the bulk modulus value, whereas shear modulus and density are influenced to greater extent by calcite-dolomite content. Consequently, compressional and shear velocities (V_P and V_S) and impedances (I_P and I_S) are related to density and elastic moduli through the following equations and should be sensitive to mineralogy because shear modulus (G) and bulk density (ρ) depend on mineral content:

$$V_P = \sqrt{(K + 4/3G)/\rho}\ , \qquad (3)$$

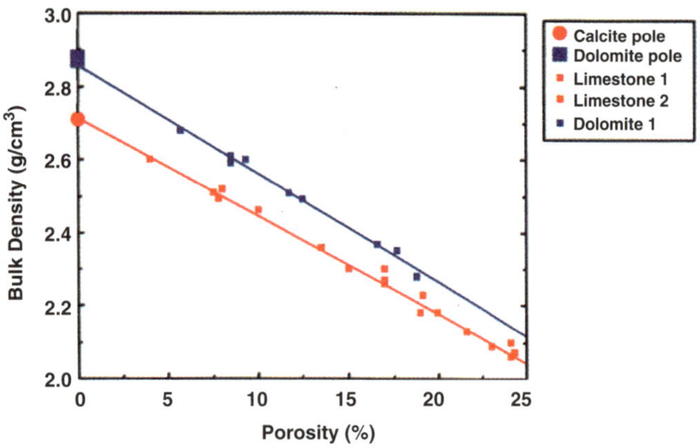

Fig. 2. Graph of bulk density versus porosity showing the influence of mineralogy for three carbonate rock suites. The two lines are the best fits to the data points.

$$V_S = \sqrt{G/\rho}\ , \tag{4}$$

$$I_P = \sqrt{\rho\,(K + 4/3G)}\ , \tag{5}$$

$$I_S = \sqrt{G\rho}\ . \tag{6}$$

Figure 3 shows the velocity–porosity and impedance–porosity crossplots of core data. Similar trends can be observed at a log scale, as illustrated in Figure 4. From these core and log data, we see that dolomites typically have 4–15% higher values than limestones for the elastic properties of velocity, impedance, and density. We also see that variations in mineral composition generate apparent scatter in the velocity–porosity and impedance–porosity crossplots. We find that for a given velocity or impedance, porosity can vary as much as 10%, depending on lithology.

Hence, in the process of estimating porosity from velocity or impedance measurement, information about the mineralogy is essential. Note that for seismic applications, the ratio of compressional to shear velocity, V_P/V_S can be combined with impedance data to distinguish dolomites from limestones, as shown in the well log data in Figure 5.

Influence of Pore Shape

The effect of pore compliance and shape on the elastic properties of rocks has been investigated from both theoretical (Eshelby, 1957; Hill, 1965; Kuster and Toksoz, 1974) and experimental standpoints (Marion and Zinszner, 1991; Anselmetti and Eberli, 1993). Figure 6 shows an example of the impact of pore shape

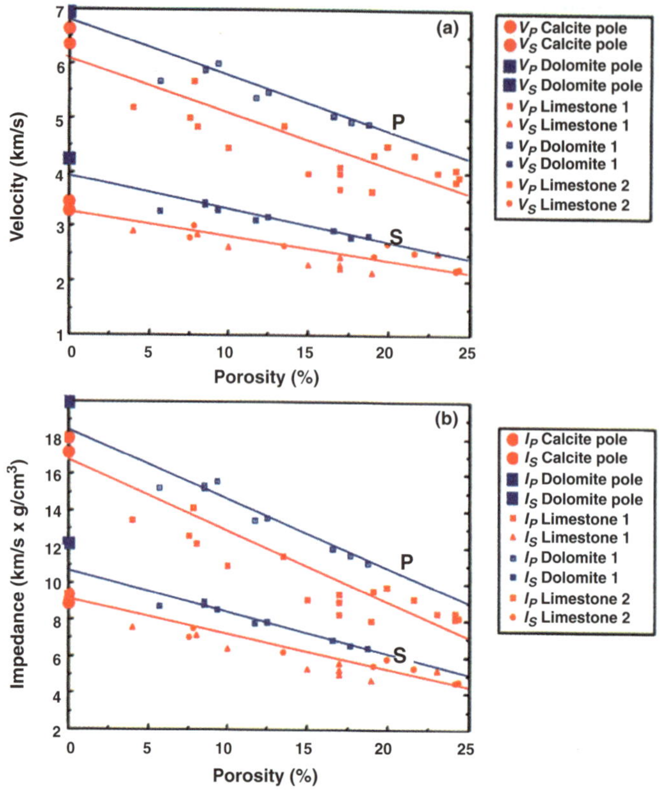

Fig. 3. Graphs of (a) compressional and shear velocities and (b) compressional and shear impedance versus porosity showing the influence of mineralogy for three carbonate rock suites. The four lines are the best fits to the data points.

on velocity. A self-consistent modeling technique (Watt, 1976) was used to estimate the relationship between velocity and porosity for two media composed of spherical and elliptical inclusions embedded in a solid elastic matrix. This numerical example illustrates how, for a given velocity, porosity can vary depending on pore shape.

Such influence of pore shape on velocity can also be observed qualitatively in core data. Figure 7 illustrates the relationship between velocity and porosity for dolomite core samples containing varying amount of vugs. The proportion of vugs was estimated visually, and rocks were classified into three groups: V_0 = no vugs, V_1 = small number of vugs, and V_4 = large number of vugs. For this data set, it is clear that the presence of vuggy porosity strongly influences the velocity–porosity behavior of *P*- and *S*-waves. For example, we note that for a given velocity of 5 km/s, the porosity ranges from 5 to 15% depending on whether or not vugs are present. Also note that the lack of dispersion in the density–porosity relationship for these samples confirms that scatter in the velocity–porosity relationship is due essentially to pore shape and not lithology.

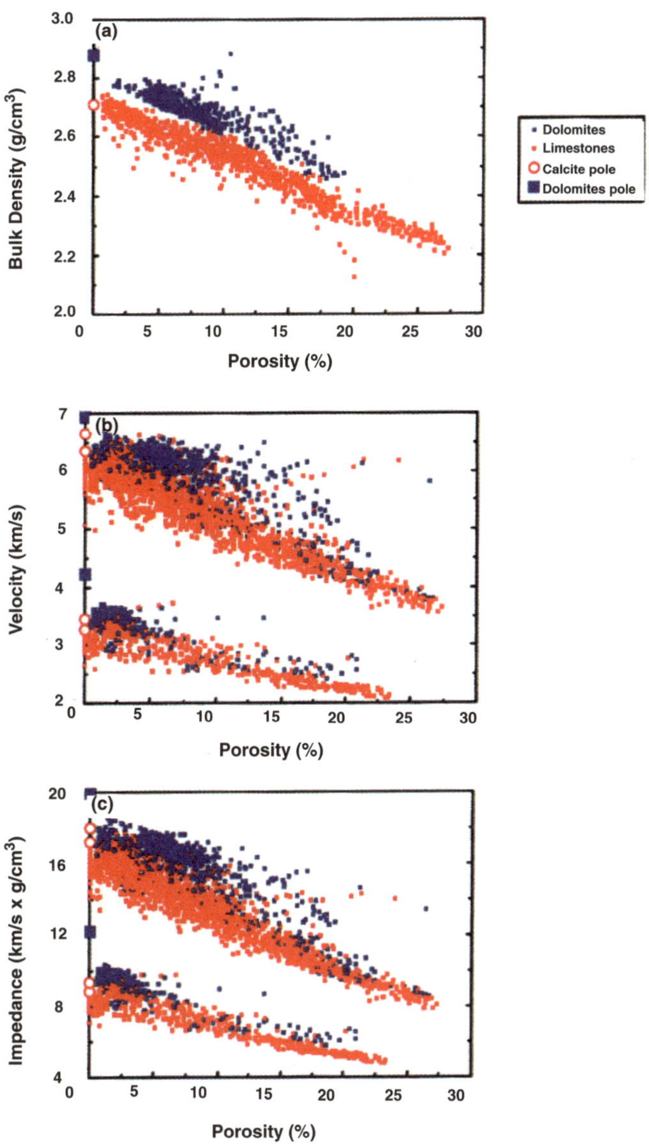

Fig. 4. Graphs showing the influence of mineralogy on log-derived elastic properties versus porosity: (a) bulk density, (b) compressional and shear velocities, and (c) compressional and shear impedance.

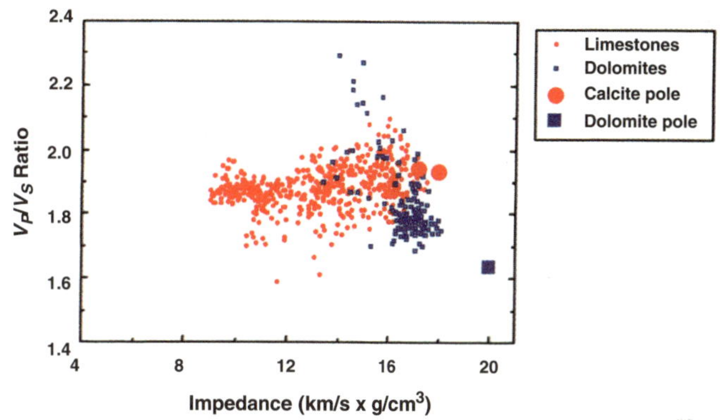

Fig. 5. Graph of V_P/V_S ratio versus *P*-wave impedance, which can be used to identify lithology.

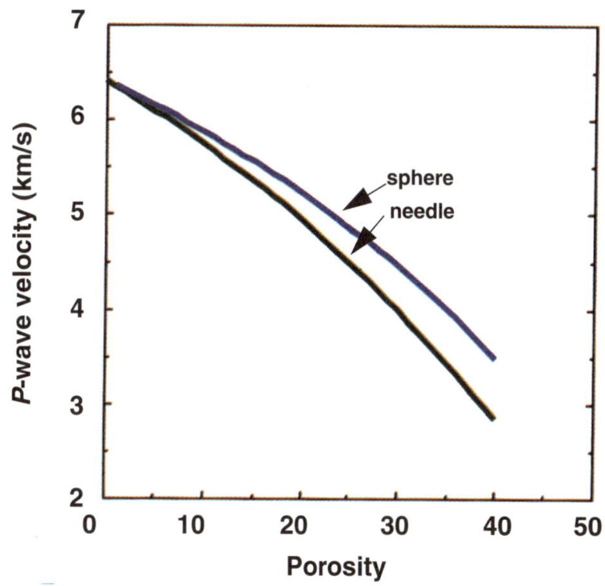

Fig. 6. Graph of *P*-wave velocity versus porosity showing the influence of pore geometry on the velocity-porosity relationship using self-consistent modeling.

Influence of Fluid Type

In addition to mineralogy and pore type, saturation also influences velocity measurements and hence can effect the velocity–porosity trend. Figure 8 illustrates the dependence of the ultrasonic velocity of dolomite and limestone on fluid type for gas- and brine-saturated samples. The results are compared with Gassmann's model, which is commonly used for fluid substitution modeling at seismic frequencies. We find that changes in velocity due to fluid type from gas to brine average about 400 m/s at ultrasonic frequencies. Gassmann's model predicts that at low frequencies, the velocity changes from 0 to 200 m/s depending on the porosity and dry rock velocity. Samples 18–23 show negligible dependence of compressional velocity on saturation. These samples correspond mainly to low-porosity dolomites for which the small amount of pore space and the rigid matrix tend to mask fluid effects. The discrepancy between ultrasonic measurements and Gassmann's predictions increases for low-porosity, high-velocity rocks. This is consistent with observations of sandstone (Wang and Nur, 1988) and can be explained by local grain-scale flow effects (O'Connell and Budiansky, 1974; Mavko and Jizba, 1991).

Note that ultrasonic frequency measurements and Gassmann's predictions can be considered as upper and

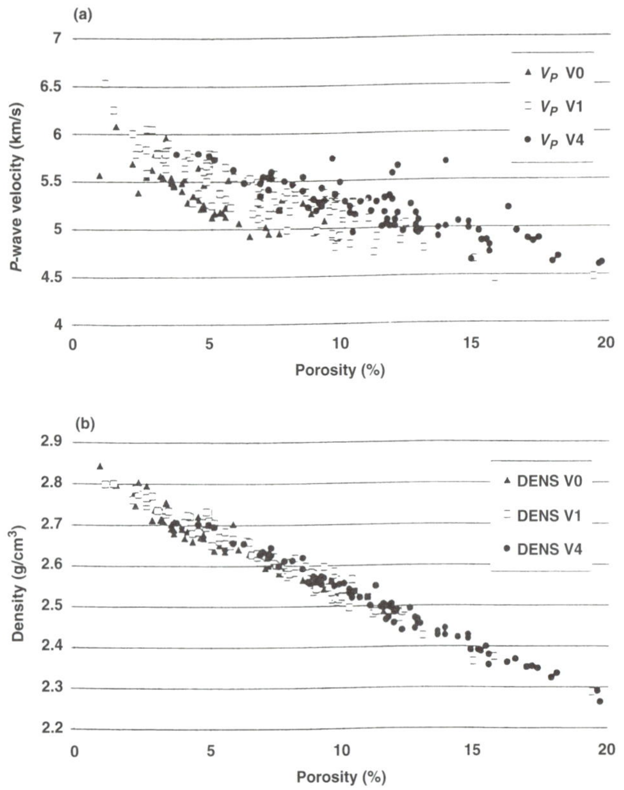

Fig. 7. Graphs of (a) *P*-wave velocity and (b) density versus porosity from core measurements on dolomites showing the influence of vuggy porosity.

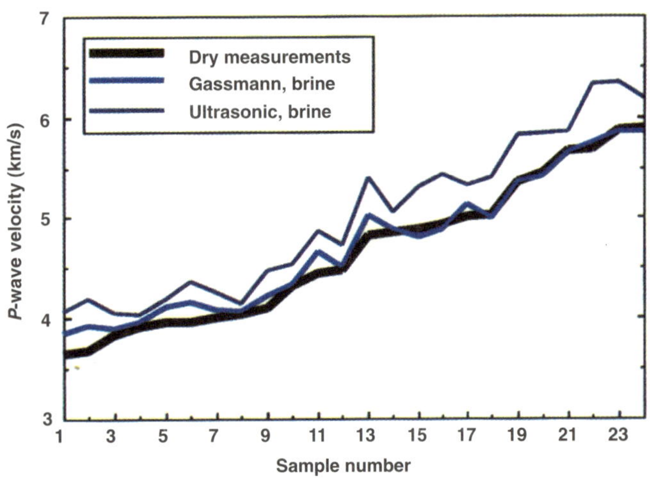

Fig. 8. Graph of *P*-wave velocity in carbonates showing the influence of fluid type. Ultrasonic measurements and Gassmann's low-frequency predictions are compared for 24 carbonate samples.

lower boundaries, respectively, for estimating the influence of fluid type on elastic properties or seismic and well-log applications.

Discussion

This section has illustrated that in addition to porosity, other factors such as mineralogy, pore shape, and fluid type can also have a significant impact on the elastic properties of carbonate rocks. Hence, in attempts to estimate porosity from velocity measurements, it is necessary to quantify the relative influence of these parameters using core and log measurements. The case study in the following section demonstrates a procedure to calibrate seismic interpretation using core and log data.

Case Study: A Carbonate Oil Field

This case study on a Tertiary carbonate oil field illustrates a method that uses combined information acquired at core, log, and seismic scales to characterize petrophysical properties from seismic measurements. The oil field under consideration (location is proprietary at this time) has estimated reserves of 110 million tons of oil. The 30 wells drilled to date, along with 3-D seismic data, indicate a mixed stratigraphic and structural trap for this Tertiary carbonate reservoir. The main problem confronting development of this field is the need to characterize reservoir limits and internal reservoir properties more accurately. Information from wells indicates a large variability of porosity, oil saturation. and reservoir thickness across the field. Therefore, this case study examines in detail the influence of these parameters on velocity and impedance in an attempt to determine the possible contribution of seismic measurements to the reservoir model revision.

From Core to Log to Seismic: An Integrated Approach

To calibrate the seismic attributes, a three-step method based on core and log measurements was developed (Figure 9). First, a rock physics model was derived using core measurements which related compressional and shear wave velocities and impedances to rock properties (porosity and saturation fluid type) and to measurement conditions (frequency and stress). This approach involved controlled experimental conditions and core selection and was thus suited to systematic investigation of the relative importance of each rock parameter to elastic properties. Second, a comparison of core, log, and seismic measurements was performed to check that the rock physics model could be applied with confidence to well log and seismic interpretation. This step required upscaling procedures to bring the various measurements to comparable scales of resolution.

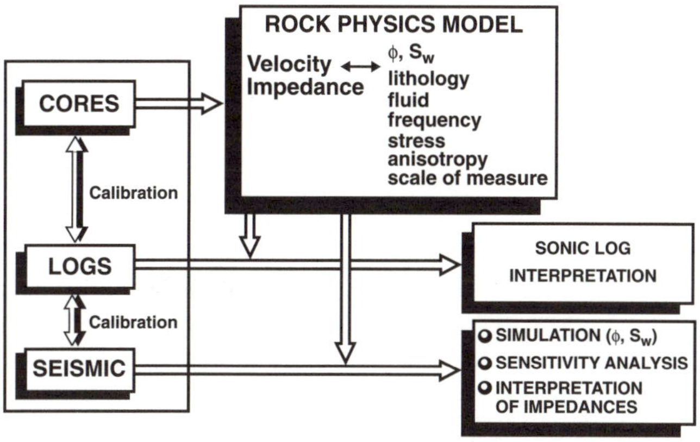

Fig. 9. Graph of the relationships of core, log, and seismic data, showing that the quantitative interpretation of seismic attributes requires integration of all available data.

Third, the rock physics model was used for quantitative interpretation of sonic and seismic data. In seismic applications, it is necessary to first ascertain whether the sensitivity of acoustic properties persists at seismic scales of resolution. This is performed through modeling and sensitivity analysis on log data. If this step is completed successfully and if data quality permits, seismic attributes can be interpreted in terms of quantitative reservoir parameters. The approach (described above) was applied to this carbonate field to characterize porosity and reservoir thickness from seismic measurements using core and log information.

Core Measurement Results

A laboratory study was conducted on cores selected from three wells. Measurements included *P*- an *S*-wave velocity, porosity, and gas permeability as a function of stress mineral density, and a quantitative petrographic description for each plug. Using these measurements, we were able to establish a rock physics model for the reservoir rocks that could account for the influence of porosity, pore shape, mineralogy, saturation, and measurement frequency.

Influence of Porosity, Pore Shape, and Mineralogy

The samples considered in our analysis were representative of various facies encountered in this field. Figure 10 shows thin section images of typical samples. The reservoir facies are predominantly nummulitic limestone with dolomite content ranging from 0 to 30%. In addition, cores were taken in the adjacent reservoir cover and base. From thin section analysis, we were able to identify two types of porosity: intraparticle porosity corresponding to chamber cells within the nummulitie shells and interparticle porosity related to vugs between the shells (Figure 10).

To quantify the influence of total porosity, pore shape, and mineralogy on *P*- and *S*-wave velocities, we constrained the empirical fitting of elastic moduli data using equation (1), in which elastic moduli were computed from velocity and density measurements. The resulting relationships, equations (7) and (8) in Table 2, are compared to the actual laboratory measurements in Figure 11.

Although a good fit of the data is obtained, the correlation between elastic moduli and porosity can be improved slightly when the effects of the two different pore types are considered separately. More specifically, we find that the correlation coefficient is higher when pore type is accounted for, as shown in equations (9) and (10) in Table 2. This is a statistical confirmation that the

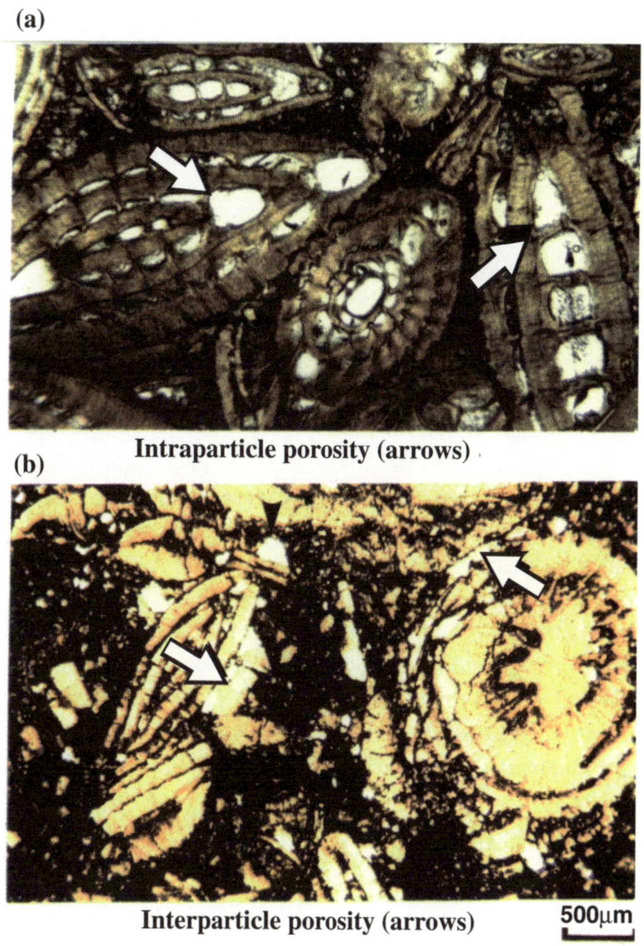

Fig. 10. Thin section photographs of typical reservoir samples composed predominantly of nummulitic limestones with (a) intraparticle porosity and (b) interparticle porosity.

Table 2. Model Fitting to Elastic Moduli Laboratory Data[1]

Equation	R_2 Value	Equation Number
Porosity term:		
$1/K_{fr} = 0.0198 + 0.0198\phi$	$R_2 = 0.94$	(7)
$1/G_{fr} = 0.0374 + 0.250\phi$	$R_2 = 0.91$	(8)
Porosity term with pore shape term considered:		
$1/K_{fr} = 0.0199 + 0.0285\phi_{inter} + 0.104\phi_{intra}$	$R_2 = 0.97$	(9)
$1/G_{fr} = 0.0378 + 0.407\phi_{inter} + 0.082\phi_{intra}$	$R_2 = 0.98$	(10)
$\phi_{inter} + \phi_{intra} = \phi$		(11)
Porosity term with mineralogy term considered:		
$1/K_{fr} = 0.002 + (1.194/K_m) + 0.289\phi_{inter} + 0.117\phi_{intra}$	$R_2 = 0.97$	(12)
$1/G_{fr} = 0.024 + (0.452/K_m) + 0.417\phi_{inter} + 0.079\phi_{intra}$	$R_2 = 0.98$	(13)

[1] Elastic moduli are computed from velocities measured at four confining pressure levels between 23 and 35 MPa.

two pore types identified in thin section analysis (intraparticle and interparticle) have a distinctly different influence on elastic properties. However, it is worth noting that in these limestones, ϕ_{intra} and ϕ_{inter} are not really independent variables. As total porosity increases, the proportion of intraparticle porosity also generally increases.

To estimate the influence of mineralogy on elastic properties, we first computed the bulk (K_m) and shear (G_m) moduli of the mineral phase for each sample using the Hill (1963) average based on the sample's mineralogic composition. Then the new variables K_m and G_m were introduced into the regression analysis. We found that mineralogy had little impact on the quality of the regression fit between the elastic moduli and the porosity (see equations (9) and (10), Table 2) because of the predominance of calcite in these samples. The influence of effective stress on elastic moduli was also investigated. We found the impact of stress variations to be negligible in the stress range of 27–35 MPa encountered in this field. Hence, for practical purposes, information about mineralogy and effective pressure is apparently not necessary for porosity estimation in this particular carbonate reservoir.

A good first-order approximation of porosity from elastic properties can be achieved using equations (7) and (8). When porosity is available through an independent measurement, equations (9) and (10) can be used to characterize the relative amount of intra- and interparticle porosity. Note, however, that these relationships are valid only for dry rocks. For oil- and brine-saturated rocks, the influence of fluid type and measurement fre-

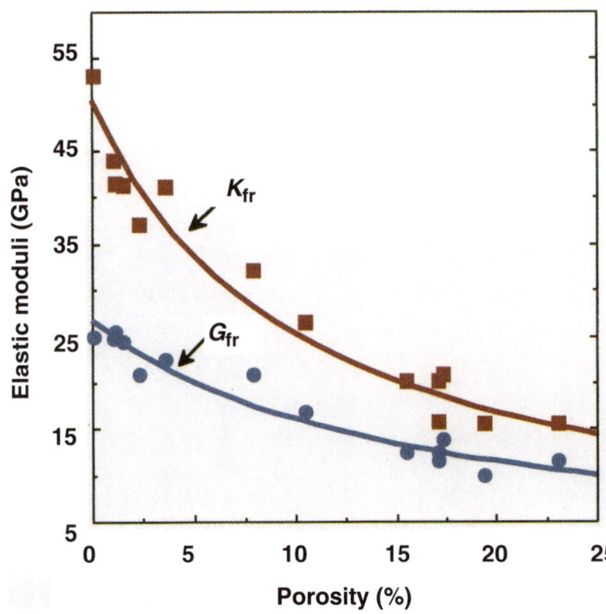

Fig. 11. Graph of elastic moduli versus porosity measured on cores and fitted using equations (7) and (8) in Table 2. The two curves are the best fit to the data.

quency must be accounted for prior to any porosity determination.

Influence of Fluid Type

To investigate the influence of fluid type on elastic properties, measurements of *P*- and *S*-wave velocities were taken under stress on dry and fully brine-saturated samples at ultrasonic frequencies. Results were then compared with Gassmann's model. We found that shear velocity measurements are nearly unaffected by fluid type, as predicted by Gassmann's equations (Figure 12). In contrast, compressional velocities are greater by about 10% in brine-saturated rocks than in dry rocks. For comparison, Gassmann's model predicts only an average 5% increase in *P*-wave velocity from a dry to a brine-saturated rock. The remaining 5% discrepancy in brine-saturated rocks between ultrasonic measurements and Gassmann's low-frequency prediction (Figure 12) can be considered as a measurement of velocity dispersion (Winkler, 1986).

Influence of Frequency

In an attempt to quantify velocity dispersion among ultrasonic laboratory measurements, sonic frequencies, and seismic frequencies, *P*- and *S*-wave velocity measurements were made under uniaxial loading on dry and fully brine-saturated samples at ultrasonic (500 kHz) and sonic (30 kHz) frequencies. For dry samples, no significant discrepancies were found between measurements at sonic and ultrasonic frequencies, suggesting that dispersion is negligible. In contrast, ultrasonic velocities in brine-saturated samples are about 4% higher than sonic velocities (Figure 13). Furthermore, when comparing these results with Gassmann's low-frequency prediction, we find that ultrasonic velocities are greater than Gassmann's predictions by about 5%. Hence, assuming that Gassmann's model applies to low-frequency seismic data and using a Geertsma-Smit (1961) dispersion curve, we can summarize the dispersion correction for compressional velocities (*V*) as follows:

For dry rocks,

$$V_{\text{seismic}} = V_{\text{sonic}} = V_{\text{ultrasonic}},$$

and for fluid-saturated rocks,

$$V_{\text{seismic}} = V_{\text{gassmann}},$$

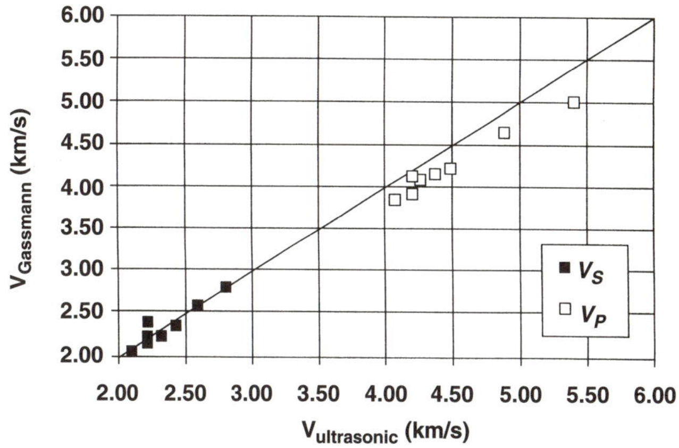

Fig. 12. Graph of velocities predicted by Gassmann versus those measured on brine-saturated samples (ultrasonic) showing a comparison between V_P and V_S.

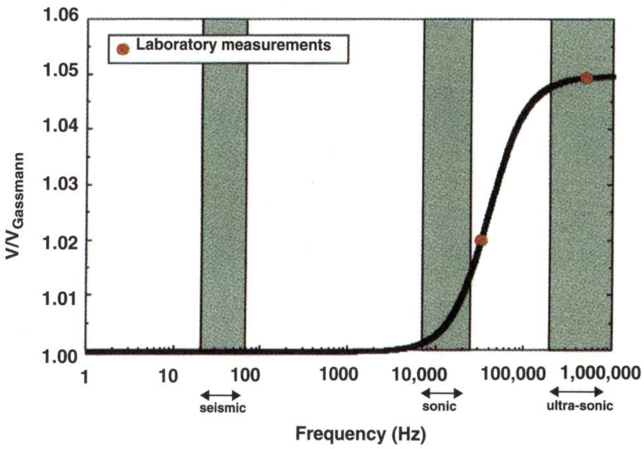

Fig. 13. Graph of V/V_{gassmann} versus frequency showing the influence of frequency on compressional velocity measurements and extrapolation to other frequencies using a Geertsma-Smit dispersion curve.

$$V_{\text{sonic}} = V_{\text{gassmann}} \times 1.01,$$

$$V_{\text{ultrasonic}} = V_{\text{gassmann}} \times 1.05,$$

$$V_{\text{ultrasonic}} = V_{\text{sonic}} \times 1.04.$$

Rock Physics Model

The relationships previously described can be summarized in a rock physics model that describes the influence of porosity, pore shape, mineralogy, fluid type, saturation, and measurement frequency on velocity and impedance. A comparison between the model's prediction and the experimental velocity data for ultrasonic brine-saturated samples is shown in Figure 14.

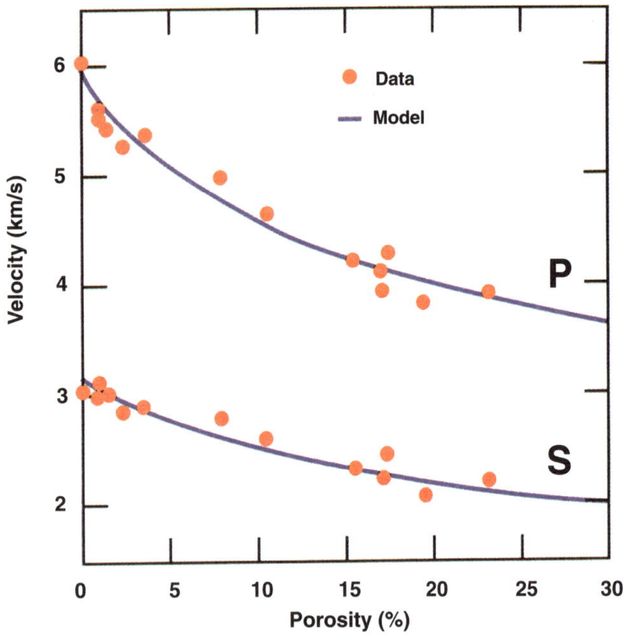

Fig. 14. Comparison of velocity–porosity relationships of values measured on cores (dots) with values predicted using the rock physics model (curves).

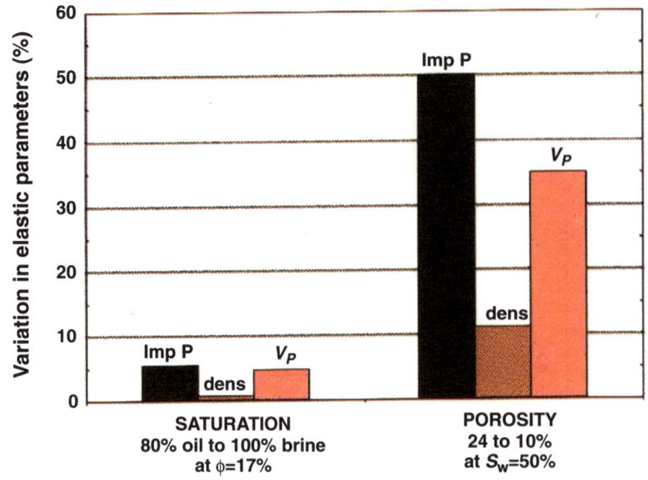

Fig. 15. Graph of elastic properties versus reservoir parameters showing the sensitivity of impedance, density, and velocity to variations in saturation and porosity.

The role of the rock physics model is twofold. First, the relationships can be used in quantitative sonic and seismic interpretation once calibration of core, log, and seismic measurements is performed. This is described next. Second, the model can be used as a feasibility tool to investigate the sensitivity of velocity and impedance to variations in rock and fluid parameters. Using this analysis, we found that acoustic properties were primarily influenced by porosity, as seen in Figure 15. Other parameters such as oil and brine saturation, pore shape, and mineralogy can be considered as second-order terms for seismic applications.

Core to Log Seismic Comparison

In the general approach described in Figure 9, it is necessary to verify the coherency of acoustic measurements at the core, log, and seismic scales if we wish to apply the relationships established on cores to sonic and seismic interpretation.

An example of a core-log comparison is shown in Figure 16. In this figure, *P*-wave velocities measured on dry cores at atmospheric pressure (V_{P0}) were corrected for stress (V_{P300}), saturation (V_{Psat}), and frequency using the rock physics model. Furthermore, to allow direct comparison between core and well log measurements, core data were brought to the scale of resolution of sonic measurements (V_{Poy}) using an up-scaling procedure described by Marion and Pellerin (1994). To compare log and seismic measurements, the classic method of comparing synthetic seismograms and seismic traces was used. In this procedure, impedances measured from log data are sampled every 4 ms in the time domain, then converted to reflectivities (derivatives of impedance), and finally are convolved with seismic wavelet extracted from seismic data. The resulting synthetic seismogram can then be compared to seismic traces (Figure 17). In this calibration step, in which we compared core with log data and log with seismic data, no systematic discrepancies were observed between measurements obtained at various scales. This is encouraging for subsequent use of the rock physics model at sonic and seismic scales.

Sonic Log Interpretation

Quantitative interpretation of sonic logs using calibrated relationships is an important application of the rock physics model for well log formation evaluation. In this integrated approach, it is essential to ensure that the rock physics model is predictive at the log scale before applying it to the seismic scale. Hence, the rock physics model was used in this case study to estimate porosity from sonic measurements on a multiwell basis. In the calculations, we neglected variations in mineralogy and pore shape (use of equations 7 and 8). We assumed an average oil saturation of 50% and used a dispersion factor of 1% to Gassmann's model for application to sonic frequencies. Results in Figure 18 show good agreement between sonic porosity and core porosity on a multiwell

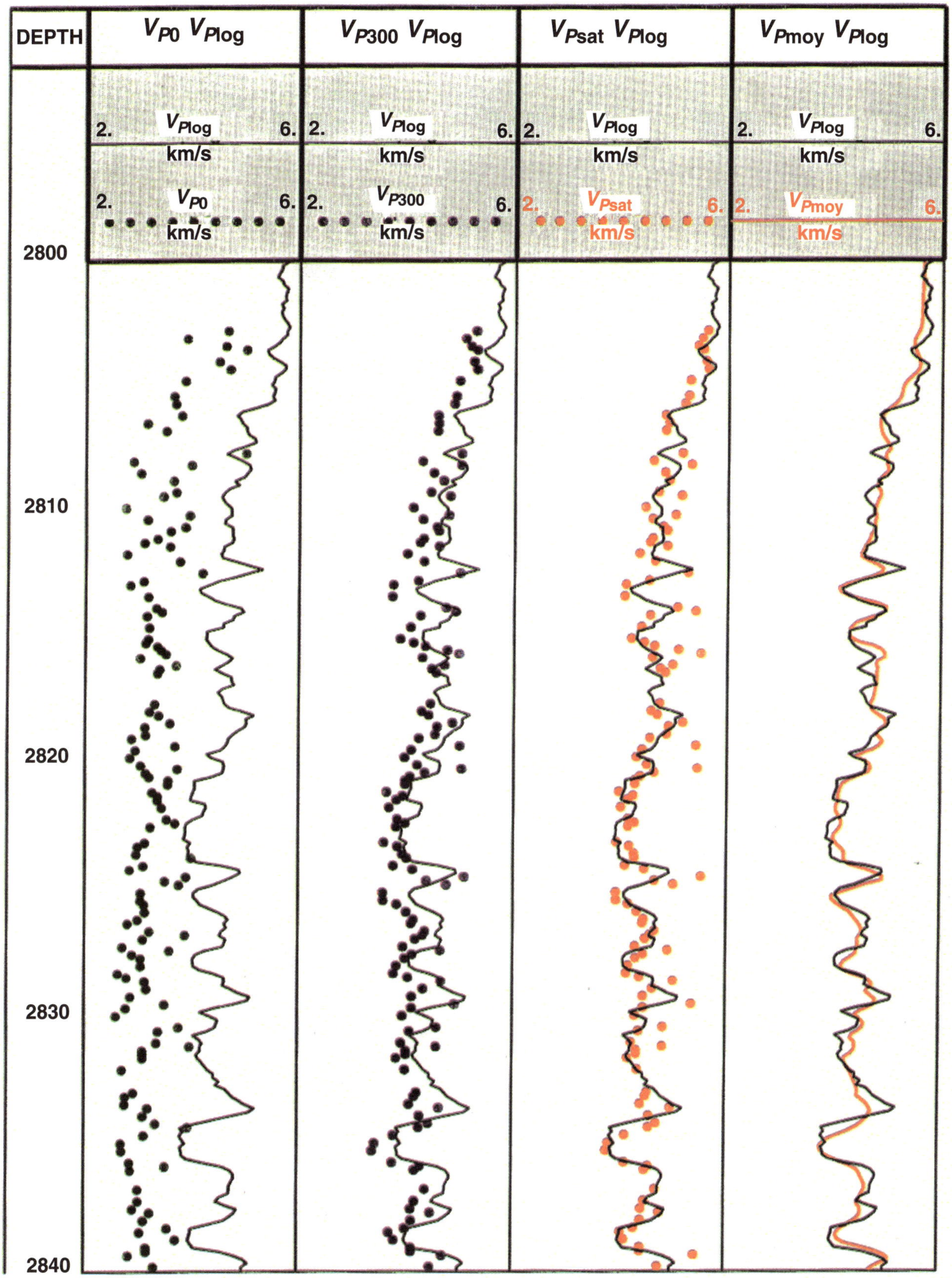

Fig. 16. Logs of sonic data (solid lines) overprinted with upscale core measurements (dots) showing a comparison of the two data sets.

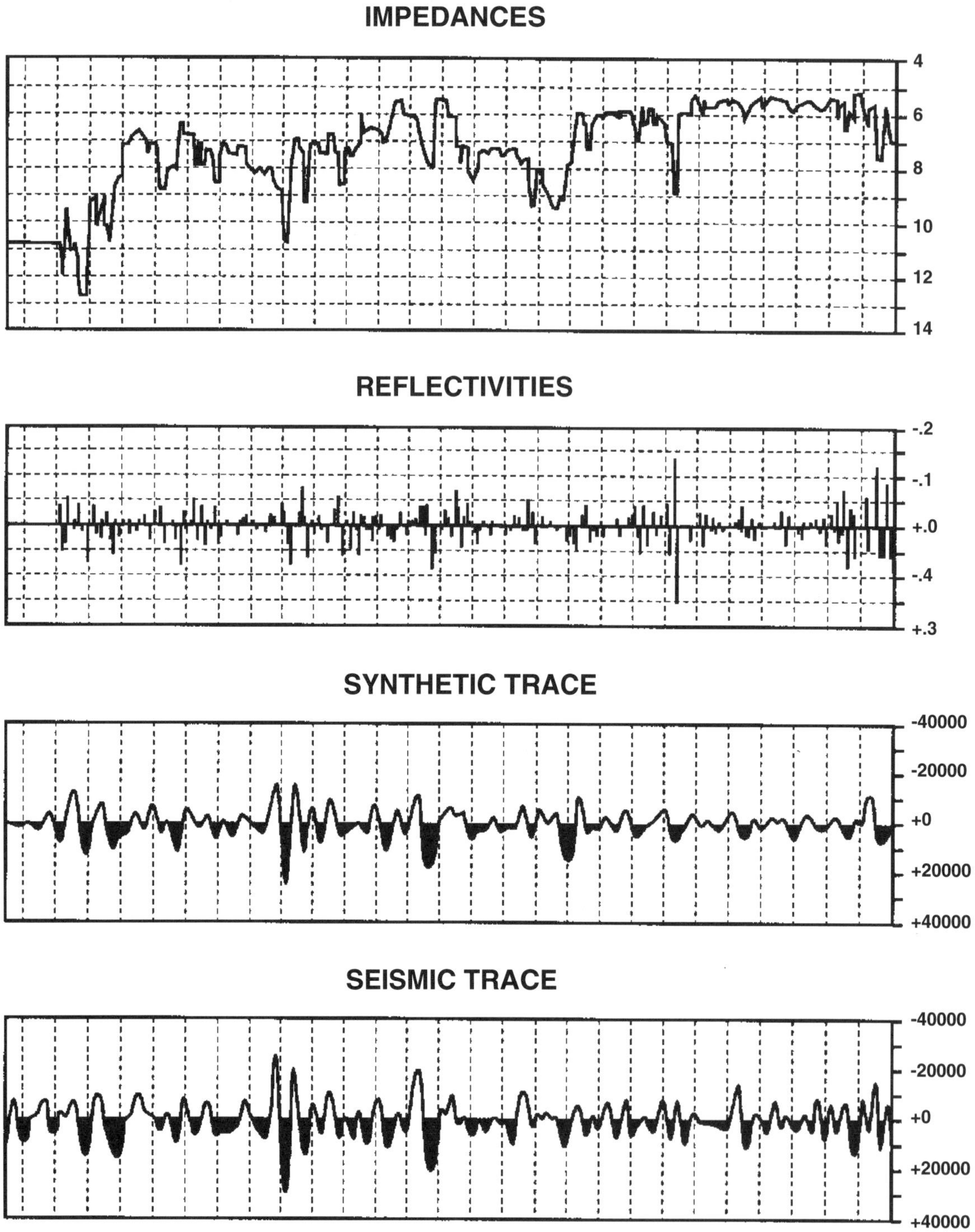

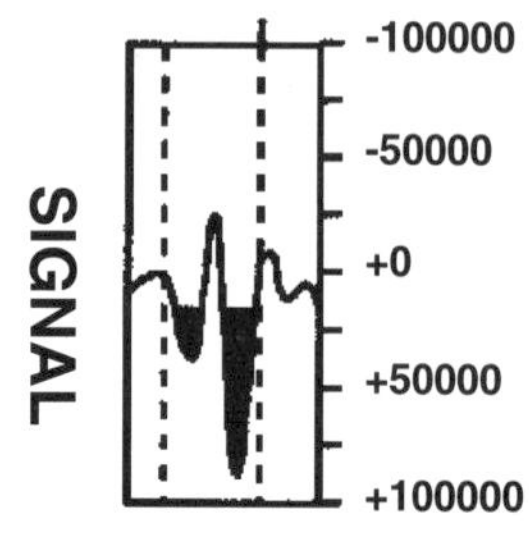

Fig. 17. Comparison between seismic trace (bottom) and synthetic seismogram derived from log data.

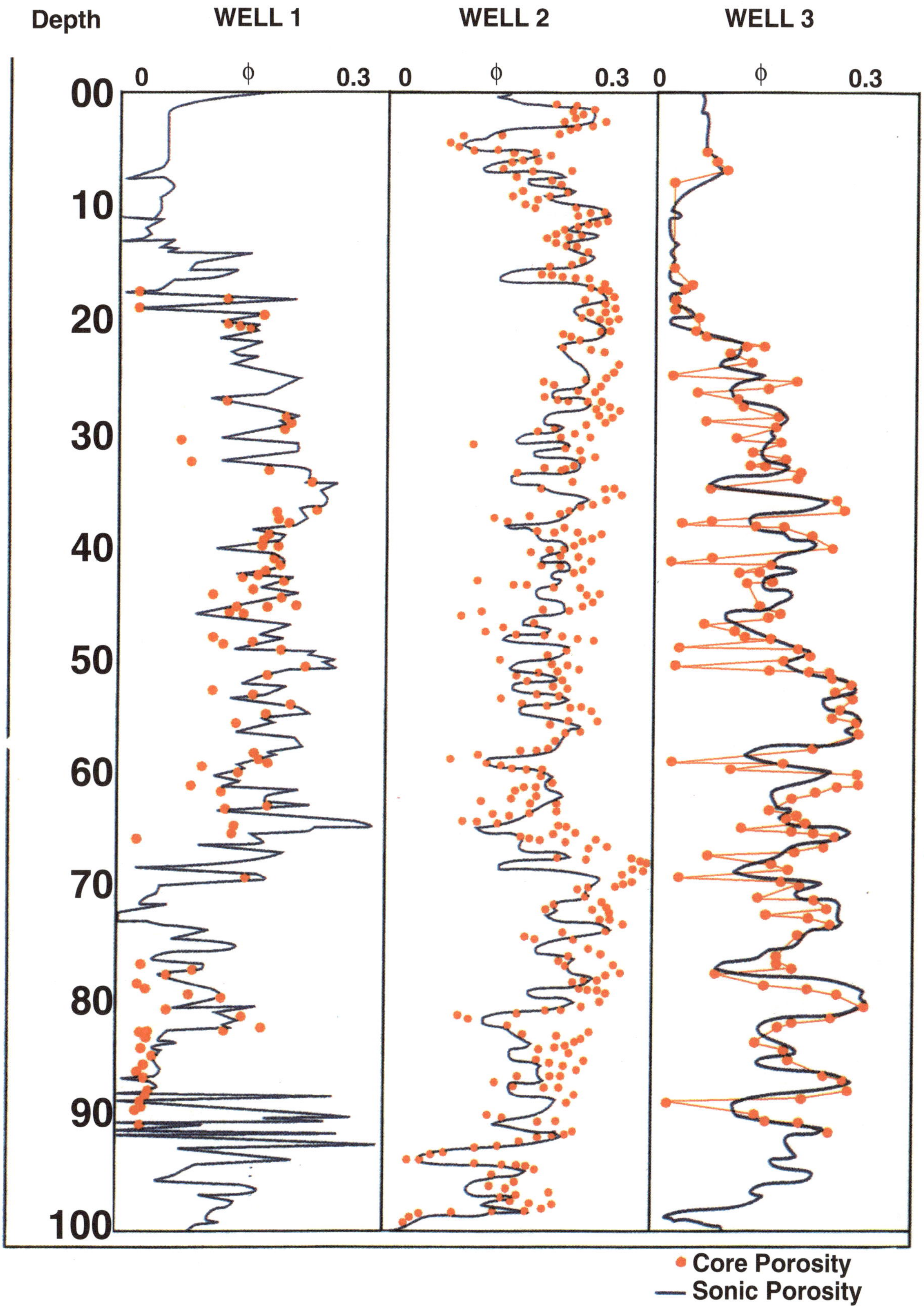

Fig. 18. Comparison of porosity values calculated using sonic logs (solid lines) with values measured on cores (dots) from three different wells.

basis. These results confirm (1) the predictive value of the rock physics model at the sonic scale and (2) the small impact of saturation, mineralogy, and pore shape on porosity determination from velocity measurements. Finally, this study shows that sonic measurements, once properly calibrated, can be used quantitatively for porosity estimation.

Seismic Interpretation

In a final step, the rock physics model, established for cores and validated for logs, was applied to seismic interpretation. A sensitivity analysis to reservoir parameters was first performed to check if the elastic sensitivity to porosity observed at the core and log scales persisted at the seismic scale. Finally, a seismic line was inverted, and seismic impedances were interpreted in terms of reservoir parameters.

Sensitivity Analysis of Seismic Response

As seen from core and log results, measurable changes in impedance and velocity can be caused by porosity variations. To assess whether porosity variations are seismically observable, we investigated their impact in the seismic frequency band. The impact of variable reservoir parameters on the seismic response was first investigated at well locations using synthetic seismograms. On the four wells in Figure 19, note that changes in seismic attributes from one well to another can arise from porosity, saturation, and bed thickness variations. To study in more detail the influence of these parameters, a systematic sensitivity analysis was conducted using synthetic seismograms (Figure 20). These seismograms were generated by varying one parameter at a time, such as porosity changes at fixed saturation and bed thickness, saturation changes at fixed porosity and bed thickness, and bed thickness changes at a given porosity and saturation.

The seismic modeling shown in Figure 20 confirms that variations in the average reservoir porosity observed in the field (11–21% from well observations) are detectable in the seismic bandwidth. More specifically, an increase in porosity corresponds to an increase in reflection amplitude at the base of the reservoir (labeled R_2 in Figure 20) and to an increase in traveltime between reflectors from the reservoir top and base (R_1 and R_2). These simulations also confirm that variations in oil saturation from 0 to 70% are not seismically detectable. Finally, these simulations show that variations in cover thickness from 7 to 24 m influence reflection amplitudes at the reservoir top due to the limited seismic frequency band (30 Hz) and tuning effects.

Seismic Inversion

To extend seismic interpretation in terms of reservoir properties and to minimize wavelet effects, an inversion was conducted on a seismic line passing through the four wells shown in Figure 19. Figure 21a shows the initial seismic line extracted from 3-D time-migrated seismic data. Well data were then used to establish a lay-

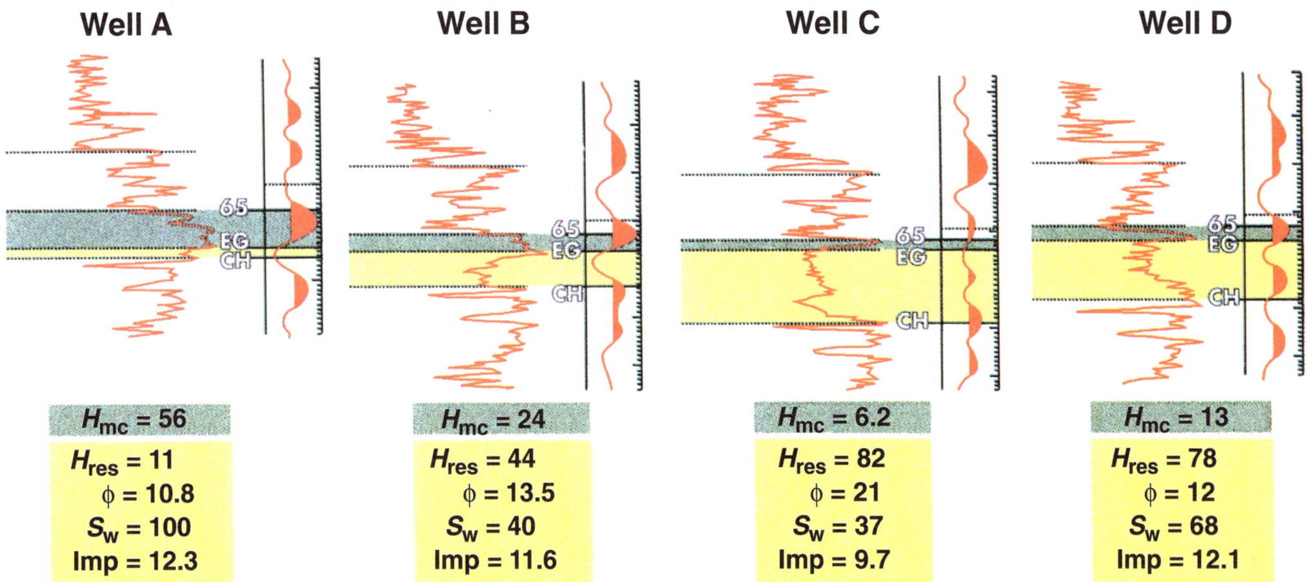

Fig. 19. Synthetic seismograms computed for four wells showing the changes in seismic attributes due to a wide range of porosities, ϕ; saturations, S_w; reservoir thickness, H_{res}; and micrite seal thickness, H_{mc}. Wells A–D correspond to those in Figures 21 and 22.

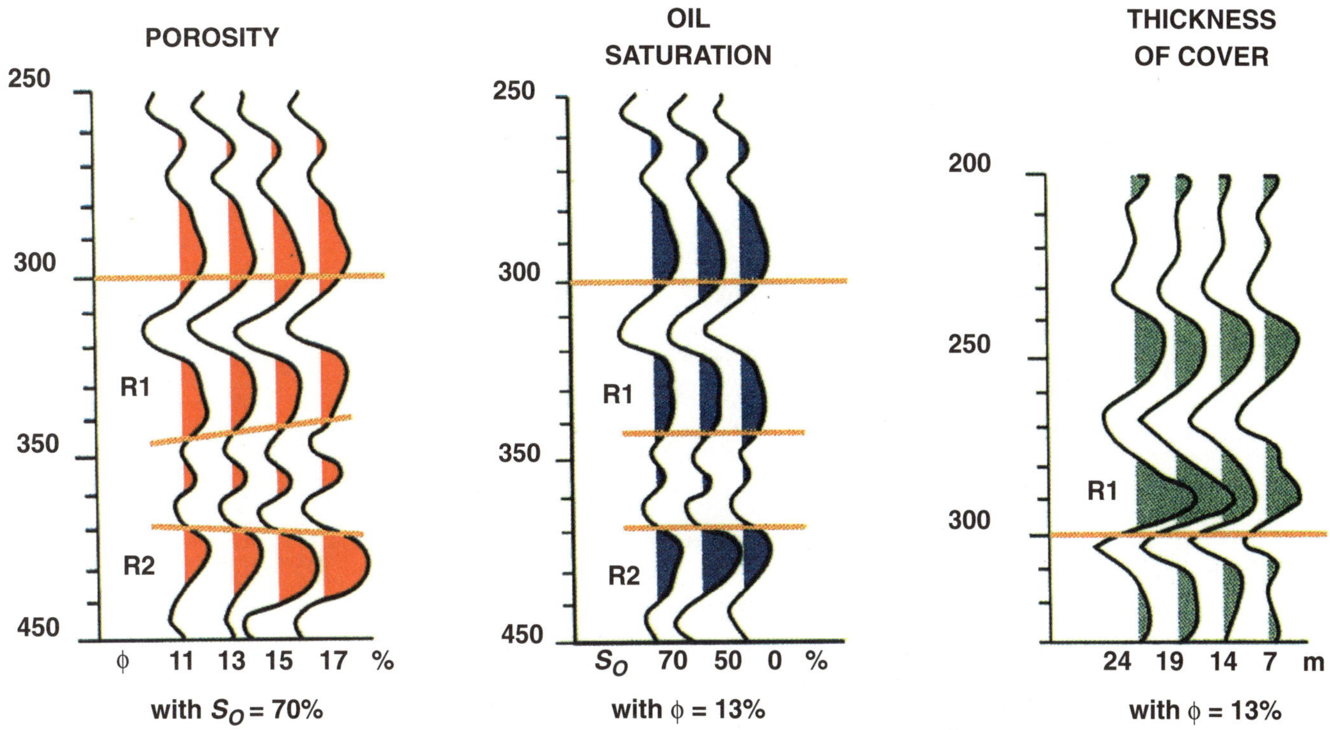

Fig. 20. Three synthetic seismograms showing the sensitivity of the seismic response to porosity (ϕ), oil saturation (S_o), and bed thickness.

ered initial impedance model (Figure 21b). This was done by blocking the impedance logs and extrapolating laterally along interpreted seismic horizons. This initial layered model was constructed using only two wells (B and D). Wells A and C were blind wells that were subsequently used to validate the inversion results.

The reservoir model for inversion is comprised of three layers plus a compact micrite layer that corresponds to the reservoir cover. The choice of three layers in the reservoir allowed the inversion to converge to a more stable solution, but it has no significance in terms of reservoir parameters. The acoustic impedance in this initial model ranges from 11 to 13 m/s × g/cm^3 in the reservoir zone. In the inversion results (Figure 21c), we see a greater variability of impedance from 9 to 13 km/s × g/cm^3. This suggests that there are stronger variations in porosity than estimated from the initial model, especially in the central zone of the reservoir (vicinity of well C).

Interpretation of Seismic Impedance

From the inversion results, we can access two seismic attributes: the average impedance and the total traveltime in the reservoir. These attributes are used to estimate porosity and reservoir thickness according the following procedure. Porosity is calculated directly from the relationship between impedance and porosity from the rock physics model (Figure 22a). Traveltimes are used to compute the reservoir thickness, knowing the average velocity in the formation. The average velocity is calculated from impedance at each trace using the relationship between velocity and impedance from the rock physics model (Figure 22b).

The relationship between impedance and porosity was applied to the inverted seismic impedance to generate a profile of porosity, shown in Figure 23a. To obtain results comparable to the reservoir model dimensions, porosity and thickness profiles were averaged over 10 and 40 traces corresponding to 120 and 500 m, respectively. Similarly, using the information contained in impedance and traveltimes, a total reservoir thickness profile was obtained (Figure 23b). Also, Figure 23b shows the product porosity–thickness that characterizes the pay zone. For comparison, we superposed the average reservoir porosity values measured at the wells. The seismically derived properties are coherent at well sites with log observations. In particular, seismic inversion successfully predicts porosity and reservoir thickness at the blind well locations (wells A and C).

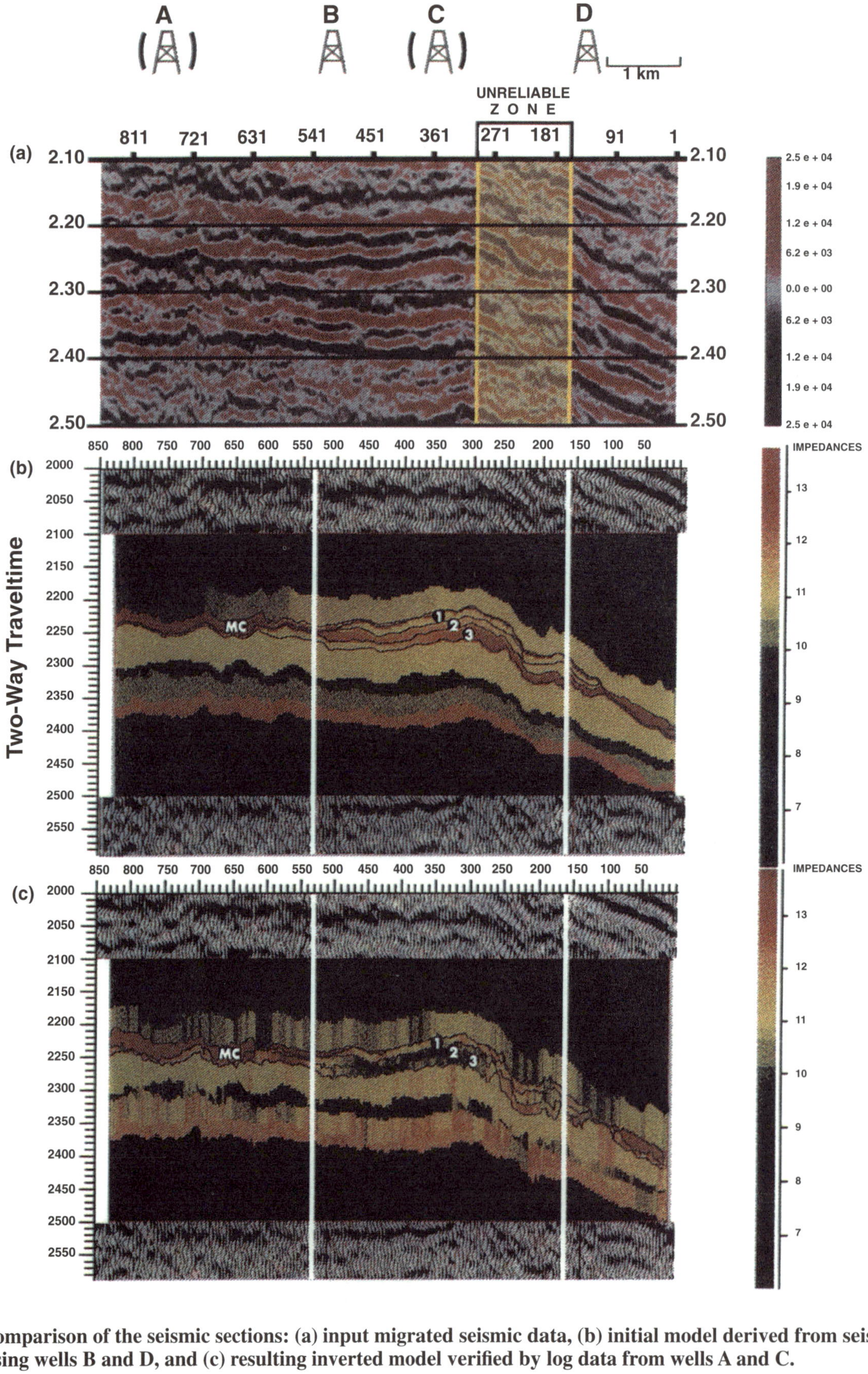

Fig. 21. Comparison of the seismic sections: (a) input migrated seismic data, (b) initial model derived from seismic picks and logs using wells B and D, and (c) resulting inverted model verified by log data from wells A and C.

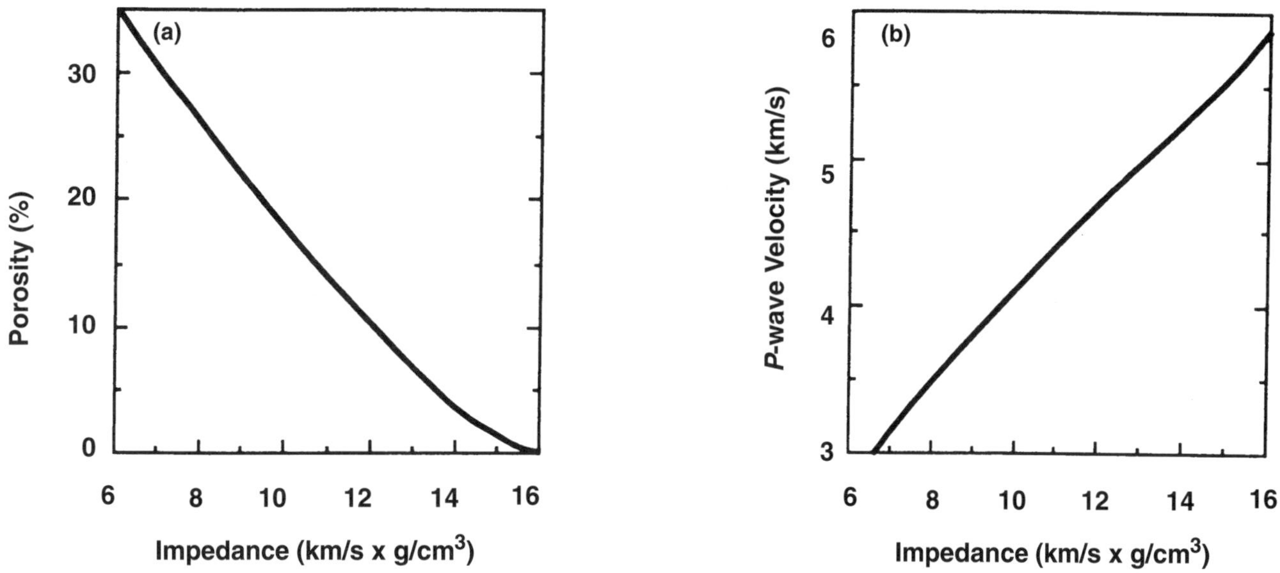

Fig. 22. Graphs showing (a) porosity and (b) *P*-wave velocities versus impedance relationships.

These results provide confidence in estimating porosity and pay zone away from the well using seismic measurements, with the exception of the zone of poor-quality seismic data near well D (Figure 21a). We also compared the seismically derived properties with the previous reservoir model, which was based on well interpolation. There is an overall agreement between the initial reservoir model and the seismic results in the pay zone. However, in a more detailed comparison, we find a greater lateral extension of the pay zone estimated seismically. These new results are consistent with production data.

Conclusions

This integrated study confirms that quantitative interpretation of seismic impedance traveltimes in terms of total reservoir and thickness is feasible. The validity of the method strongly relies on three factors:

1. Accurate relationships between seismic attributes and rock properties that can be assessed from a detailed laboratory study,
2. Coherency among data acquired at core, log, and seismic scales, and
3. Good-quality seismic data, interpretation, and adapted seismic processing and inversion techniques.

Acknowledgments

The authors thank Elf Aquitaine Production for permission to publish this work. Special thanks go to Pierre Coudin for his detailed laboratory study, Jean Marrauld for performing the seismic inversion, and the Aramis Reconciliation des Facies research teams.

References

Anselmetti, F. S., and Eberli, G. P., 1993, Controls on sonic velocity of carbonates: Pure Applied Geophysics, **141**, 287-323.

Biot, M. A., 1956, Theory of propagation of elastic waves in a fluid-saturated porous solid, II. Higher frequency range: J. Acoust. Soc. Am., **28**, 179-191.

Ellis, D., Howard, J., Flaum, C., McKeon, D., Scott, J., Serra, O., and Simmons, G., 1988, Mineral logging parameters: nuclear and acoustic: Schlumberger Tech. Rev., **36**, 38-51.

Eshelby, J. D., 1957, The determination of the elastic field of an ellipsoidal inclusion and related problems: Proc. Roy. Soc. London, Ser. A, **241**, 376-396.

Gassmann, F., 1951, Elastic waves through a packing of spheres: Geophysics, **16**, 673-685.

Geertsma, J., and Smith, D. C., 1961, Some aspects of elastic wave propagation in fluid-saturated porous solids: Geophysics, **26**, 169-181.

Hill, R., 1963, Elastic properties of reinforced solids, some theoretical principles: J. Mech. Phys. Solids, **11**, 357-372.

Hill, R., 1965, A self-consistent mechanics of composite materials: J. Mech. Phys. Solids, **13**, 213-222.

Krief, M., Garat, J., Stellingwerff, J., and Ventre, J., 1990, A petrophysical interpretation using the velocities of *P*- and *S*-waves (full-waveform sonic): The Log Analyst, 355-369.

Kuster, G., and Toksöz, N., 1974, Velocity and attenuation of seismic waves in two-phase media, 1. Theoretical formulations: Geophysics, **39**, 587-606.

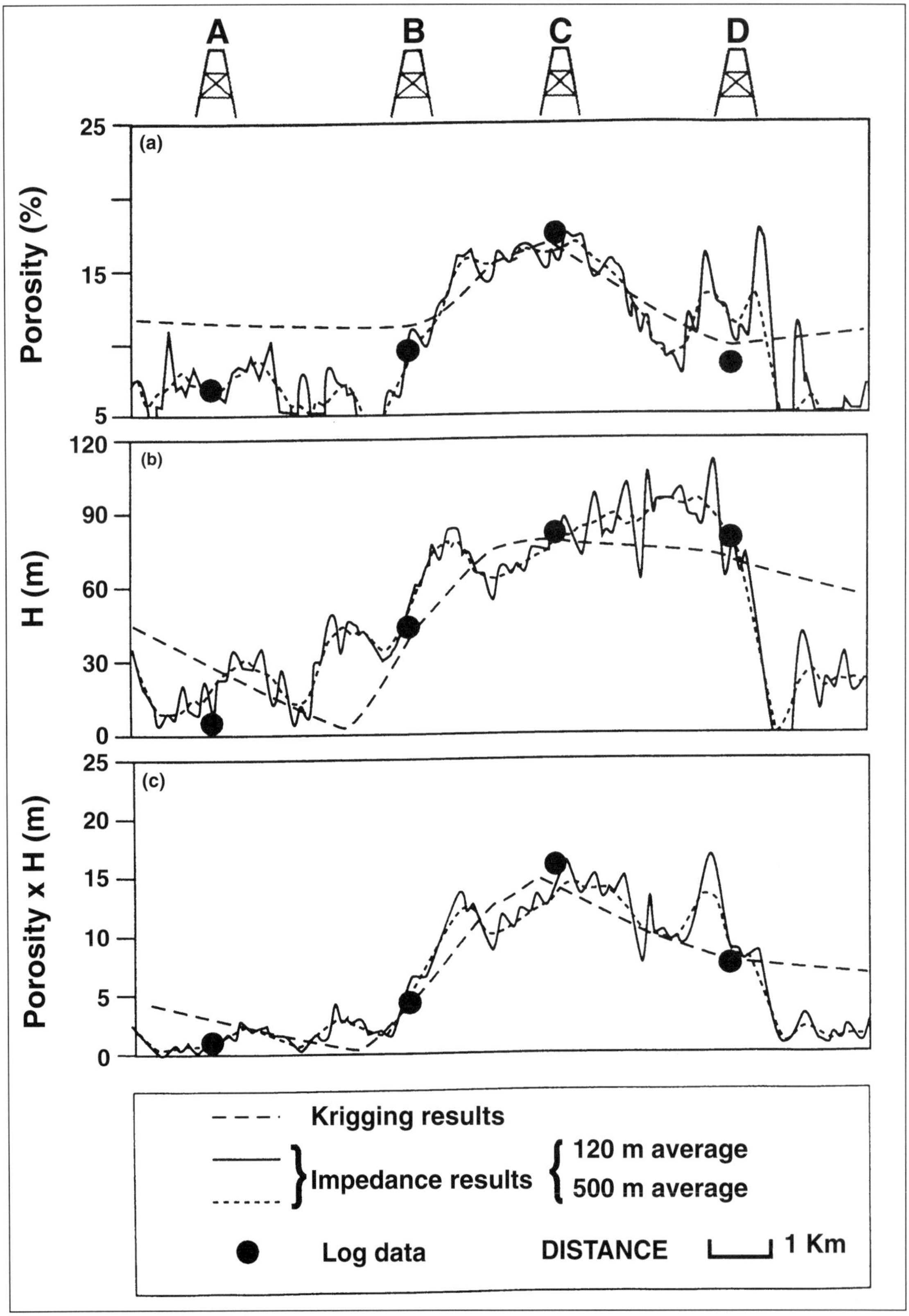

Fig. 23. Graphs showing the accuracy of the seismic inversion technique. (a) Porosity profile generated by applying the relationship in Figure 22 to the inverted seismic model. (b) Bed thickness, H, profile predicted using traveltime and impedance data. (c) Pay zone profile generated by multiplying the values of porosity by bed thickness.

Marion, D., and Pellerin, F. M., 1994, Acoustic measurements on cores as a tool for calibration and quantitative interpretation of sonic logs: Soc. Ptr. Eng. Form. Eval., 100-104.

Marion, D., and Zinszner, B., 1991, Core analysis to calibrate geophysical interpretation, *in* Proc., 2nd Society of Core Analysis Symp., 17-34

Mavko, G., and Jizba, D. L., 1991, Estimating grain-scale fluid effects on velocity dispersion in rocks: Geophysical J., **56**, 1940-1949.

O'Connell, R. J., and Budiansky, B., 1974, Seismic velocities in dry and saturated cracked solids: J. Geophys. Res., **79**, 5412.

Peselinick, L., and Robie, R. A., 1963, Elastic constants of calcite: J. Appl. Phys., **34**, 2494-2495.

Rafavitch, F., Kendall, C. H. St. C., and Todd, T. P., 1984, The relationship between acoustic properties and the petrographic character of carbonate rocks: Geophysics, **49**, 1622-1636.

Raymer, L. L., Hunt, E. R., and Gardner, J. S., 1980, An improved sonic transit time-to-porosity transform: SPWLA 21st Ann. Logg. Symp.

Schlumberger, 1984, Log interpretation charts: Schlumberger Ltd., publ. SMP-7006.

Simmons, G., 1965, Single crystal elastic constants and calculated aggregate properties: J. Grad. Res. Center, Southern Methodist Univ., **34**, 1-269.

Walsh, J., 1965, The effect of cracks on the compressibility of rocks: J. Geophys. Res., **70**, 381-389.

Wang, Z., and Nur, A., 1988, Velocity dispersion and the local flow mechanism in rocks: 58th Ann. Internat. Mtg., Soc. Explor. Geophys., Exp. Abst., 928-930.

Wang, Z., Hirsche, W. K., and Sedgwick, G., 1991, Seismic velocities in carbonate rocks: J. Can. Petr. Tech., **30**, 112-122.

Watt, J. P., Davies, G. F., and O'Connell, R. J., 1976, The elastic properties of composite materials: Rev. Geophys. Sp., Phys., **14**, 541-563.

Winkler, K., 1986, Estimates of velocity dispersion between seismic and ultrasonic frequencies: Geophysics, **51**, 183-189.

Wyllie, M. R. J., Gregory, A. R., and Gardner, G. H. F., 1956, Elastic wave velocities in heterogeneous and porous media: Geophysics, **21**, 41-70

Chapter 6

Carbonate Seismic Facies Analysis

D. Bradford Macurda, Jr.
The Energists
Houston, Texas, U.S.A.

Abstract

The appearance of carbonate rocks in seismic data contains information about their original depositional environments, lithofacies, diagenesis, and source rock and reservoir potential. Carbonate seismic facies can be interpreted through analysis of the configuration, amplitude, frequency, and continuity of the seismic reflections. The geometries of the reflection configurations are the most varied aspect of seismic data, including parallel, prograding, mounded, draped, onlap, and onlap fill. Each geometry suggests different depositional processes that can occur in different depositional environments. Amplitude and frequency can also be diagnostic of carbonate rocks; a lack of continuity is often characteristic of carbonate buildups. Carbonate platforms, platform margin, slope environments, and basinal deposits have characteristic combinations of seismic facies that aid in their recognition and in the evaluation of their hydrocarbon potential.

Introduction

The term *facies* is used in many different ways in geology, either alone or in combination with modifiers or prefixes. Examples include biofacies, metamorphic facies, lithofacies, and seismic facies. The word itself is derived from the Latin word *facies*, which means aspect or appearance. To characterize seismic facies, we need to know what aspects or appearances we can examine. These include the configuration or shape of the reflections, the amplitude (relative strength of the peaks and troughs), the frequency (number of reflections per unit traveltime), and the continuity (lateral persistence of the reflections). Of these four, the greatest variation is found in the configuration, although the other three do play an important role in differentiating carbonate environments.

The identification of seismic facies is basically a study in pattern recognition. What has rich detail to one observer may seem nondescript to another. Observational skills are usually acquired, not inborn. Interpreting depositional environments and predicting carbonate lithofacies from carbonate seismic facies is a two-step process. One must first recognize the facies variation and then ask what depositional environments these facies could have formed in. Then, when one has decided among the possible environments, one must predict the lithofacies. Given that carbonates usually form in the basin where they are found, are temperature dependent (Arctic versus tropical), depth dependent (supratidal versus abyssal), time dependent (due to evolution of species), and chemically reactive (causing diagenetic modification to occur readily), this interpretation is no simple task.

Fontaine et al. (1987) have applied seismic facies analysis to the interpretation of carbonate depositional environments. Their work should be consulted for more examples of carbonate environments.

Reflection Configurations

There are four basic reflection configurations found in seismic data. These are parallel or subparallel, prograding, mounded or draped, and onlap fill. Mitchum et al. (1977) has published a series of tables characterizing both siliciclastic and carbonate facies, which have been amended by Brown and Fisher (1980). My initial discussion of carbonate facies is drawn from these works plus numerous case histories, various authors, and personal projects over the intervening years.

It is fortunate that many outcrop studies of carbonates are at a scale that approaches that of seismic data, including Mesozoic rocks of the Dolomites in northern Italy (Bosellini, 1984), the Cretaceous of northern Mexico (Goldhammer and Lehmann, 1991), and Permian outcrops along the northwestern edge of the Permian Basin in West Texas and southeastern New Mexico (Sarg and Lehmann, 1986; Sonnenfield, 1991). In this last example, there is comparative seismic data along strike across the same platform and slope. Other important areas that provide comparisons between the surface and subsurface are the Alberta Basin and the Rocky Mountains in western Canada, the Pennsylvanian of the Paradox Basin in Utah, the Jurassic in the Atlas Mountains of Morocco, the Miocene of the Gulf of Suez, the Devonian of the Canning Basin in western Australia, the Tertiary of the Nullabor Plain and South Australian Bight, the Miocene of southeastern Asia, and the Pleistocene–Recent of the Bahama Platform. These areas provide calibration for our observations and conclusions.

Parallel, Subparallel, and Divergent Reflections

When one looks at a seismic reflection configuration such as parallel, subparallel, or divergent, the mental query should be, "What depositional process could be responsible for this reflection configuration and what are the environments in which it could occur?" *Parallel* or *subparallel reflections* (Figure 1) imply that space is available or being made available vertically; there is a regional increase in accommodation potential. *Divergent reflections*, in contrast, imply that there is a regional difference in the rate at which the accommodation potential is changing (such as in a half-graben being filled or in front of a growing reef trend) (Figure 2).

There is nothing inherent in the reflections to tell us if the accumulation of parallel or subparallel reflections is fast or slow. Biostratigraphic information is needed to calibrate the data. Two examples at the ends of the spectrum are the Cretaceous platform carbonates of the Gulf of Mexico rim and the pelagic chalks and marls of the foundered Blake Plateau east of Florida and the Carolinas. Furthermore, the configuration does not directly tell us whether the accumulations formed under high- or low-energy conditions. Low-energy environments with subparallel mudflats may be more common, such as those in Abu Dhabi or the carbonate grapestones in the lee of Andros Island in the Bahamas. High-energy environments are also possible, however, as the oolites of Joulter's Keys in the northern part of the Great Bahama Bank or the bryozoan grainstones in the shelf environments of Australia.

Divergent reflections (Figure 2) usually imply synsedimentary deformation or differential supply, such as on a carbonate margin that is growing rapidly while the basinal area (with lower rates of sedimentation) fails to keep up. Carbonate slopes can become highly oversteepened, resulting in even greater bathymetric differentiation as the rimmed platform aggrades. Divergent reflections can be caused by initial rifting, particularly in arid areas where siliciclastic pollution is not an overwhelming factor, such as in the Pennsylvanian of the Paradox Basin. Divergence can also be caused by halokinesis, such as in the Jurassic of the northeastern Gulf of Mexico. The differential movement in divergence results in a more rapid increase in accommodation potential in the subsiding areas, and if bathymetric conditions are appropriate, carbonate sediments can accumulate rapidly. Given differential subsidence, faster rates of deposition are more likely than slower rates in the more rapidly subsiding parts of a depression.

Carbonate environments that are characterized by aggradation on a regional scale include carbonate ramps and platforms, from shallow subtidal to intertidal to supratidal. The landforms that result from the uplift of such environments are strong testimony to the regional extent of the parallel bedding. One example is the Cretaceous platform carbonates west of the Balcones Escarpment in central Texas, which provide a flat, monotonous landscape for hundreds of kilometers. Another example is the Nullabor Plain of South Australia, which is an uplifted Tertiary platform environment. The railroad west of Adelaide runs perfectly straight for more than 320 km. It is the longest straight track in the world, with the same scenery all the way.

Another regional-scale carbonate environment is the basin floor. The pelagic chalks and marls of the abyssal ocean depth are widespread. These have been well documented through the Deep Sea Drilling Program and its successor, the Ocean Drilling Program. One factor affecting these accumulations is the carbonate compensation depth (CCD), which varies according to global climate regimes. Conditions in the Cretaceous were much more favorable for deep-water carbonate accumulation than today. The depth need not be abyssal. Bathyal depths are found on the Blake Plateau. Lacustrine systems that are saturated with respect to CO_2 ions also have parallel basin floor deposits, such as the lower Tertiary Green River Formation of the Uintah Basin in Utah.

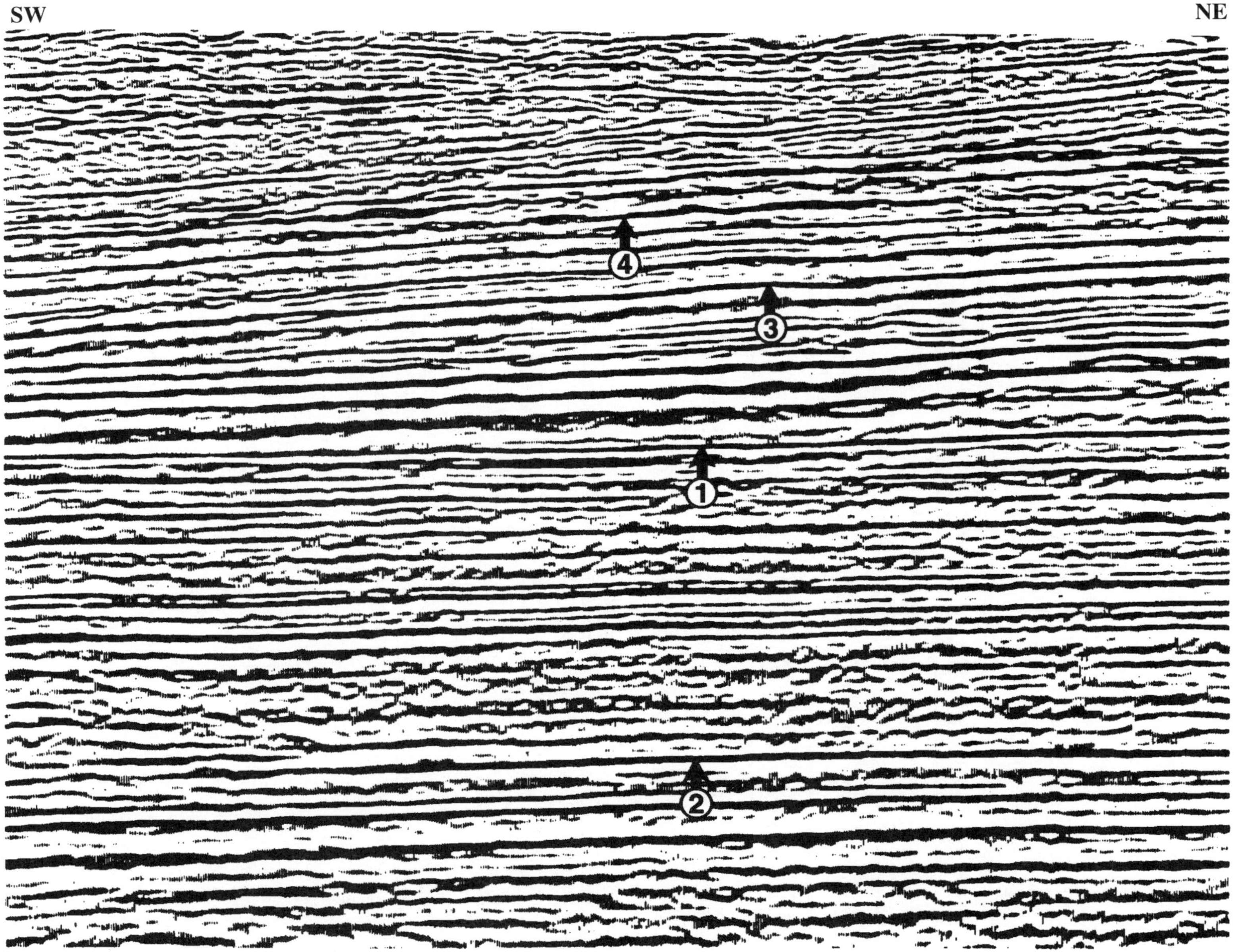

Fig. 1. Seismic profile from the northeastern Gulf of Mexico, perpendicular to the Early Cretaceous shelf edge, showing parallel and subparallel reflections. The tops of the Cretaceous Sligo Formation (2) and the Edwards (or Stuart City) Formation (1) are shown. After deposition of the Edwards, the Cretaceous carbonate megabank foundered and was followed by Late Cretaceous–early Tertiary deposition of chalks and marls (1–4); there are numerous unconformities within (3). The boundary at (1) represents drowning on a carbonate platform; erosion occurred at its edge and at the base of slope.

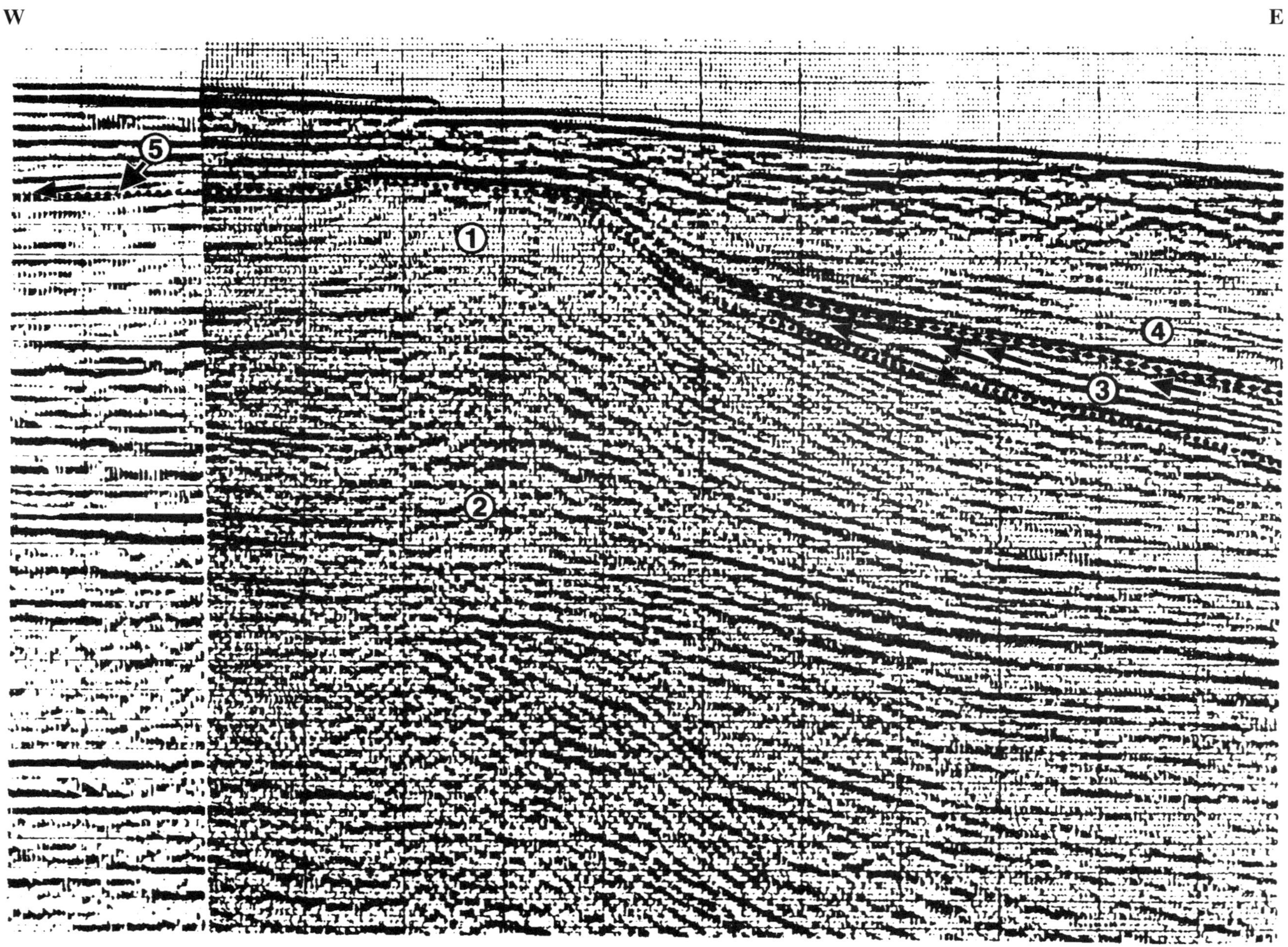

Fig. 2. Seismic profile from the southeastern United States. The large reef (1) is Aptian–Albian in age. Debris derived from it prograded shelfward (5) and basinward (3); other episodes of infilling also occurred (4). Prior to the growth of the reef, the environment was a ramp (2) of Barremian age. Beds at (3) are oblique; those below it show a divergent pattern.

Another carbonate environment where parallel reflections occur is the lower slope. Carbonate margins are line sources, and sediments shed by the margins travel downslope as turbidites or debris flows. If sufficient fluidization of the latter occurs, the resulting deposits will be sorted. Deposition occurs from the traction and traction carpet portion of the turbidity current as opposed to simple settling by pelagic or hemipelagic sedimentation.

Carbonates are often intermixed with evaporites, thus parallel reflections involving carbonates can also be encountered in evaporitic basins. Most evaporitic environments are interpreted as shallow water, such as the Permian of northwestern Europe, but some evaporite deposits are thought to be of a deep water origin. For example, in the Michigan Basin, which was almost circular during the Silurian, there are intermixed carbonates and evaporites. The evaporites increase toward the basin center while the carbonates decrease.

Progradational Reflections

Progradation is one of the reflection patterns most likely to be differentiated quickly by an observer. Progradation takes many forms, and each has a consequence for reservoir prediction. One common prograding configuration is *sigmoid*, which infers that depositional space is available or being made available both vertically and laterally. This pattern is characteristic of carbonate margins in basins where the rate of subsidence is fairly rapid. Spectacular examples have been recorded by Eberli and Ginsburg (1987) in the Bahamas where rapid supplies of sediment fill in the lows adjacent to the highs (Figure 3). Other examples include the Cretaceous of the northeastern Gulf of Mexico and the Permian Capitan reef complex of the Permian Basin. Prediction of lithofacies and porosity from the sigmoid pattern has some interesting consequences. We might expect the lithofacies to be muddier because of lower energy environments; we would also expect the porosity to be poorer than in a higher energy oblique configuration. However, the sediments in higher energy environments become penecontemporaneously cemented because of the enhanced circulation of seawater through it, while the lower energy sediments remain uncemented for a longer period of time (Sarg, 1988).

Oblique prograding reflections, in contrast to sigmoid, imply a strict control on the space available vertically but unrestricted control laterally. Such patterns occur along the periplatform margin west of Florida where currents sweep the sediments over the shelf edge or on the Campeche Bank north of the Yucatan Peninsula (Figure 4). These patterns can also occur in carbonate sediments generated in the high-energy environment on the crest of a reef, but they are limited by high energy and a static sea level, which prevent vertical accumulation. Outcrop studies of this kind of configuration in the Silurian of Indiana (Pipe Creek Quarry) show that carbonate producers can grow on inclined bedding planes, acting as contributors to the forward progradation of the oblique reflections. Oblique reflections are usually thought to form under higher energy conditions, have more of a grainstone or rudstone composition, and cement more rapidly because of the penecontemporaneous circulation of seawater through the sediments. Spectacular examples of oblique prograding carbonates have also been observed in lacustrine basins, such as in the Tertiary of the northwestern United States.

Both the oblique and sigmoid reflection configurations are thought to characterize "keep-up" margins in which space is infilled as it becomes available. This is in contrast to "catch-up" margins which backstep or are drowned due to increased rates of subsidence or the rise of sea level.

The third prograding pattern is *complex sigmoid-oblique*. Usually this appears as alternating sets of oblique and sigmoid clinoforms. There are several processes that, alone or in combination, can produce this alternation. One is varying rates of subsidence in a tectonically active area. A second is fluctuations in current intensity along a margin. Another may relate to climatic fluctuations that affect the carbonate producers. Complex sigmoid-oblique is a rare pattern. One area where it may be seen is in the Jurassic of the eastern margin of the United States (Figure 5). The controlling factors here are not known.

The fourth prograding pattern is *shingled* (Figure 6). It resembles the oblique pattern in its configuration but is distinguished by being only one reflection thick in two-way traveltime (TWT), while oblique shows more than one inclined reflection. It is differentiated because it is characteristic of environments with limited vertical space but high energy along the active prograding edge. It is particularly characteristic of carbonate ramp environments where grainstones or oolites may develop. It can also develop in distal carbonate slope environments in which the supply of sediment has diminished as a turbidity current moved downslope or in the distal portion of a carbonate contourite where the sediment supply was diminishing.

The fifth prograding pattern is *hummocky clinoforms*. This pattern was not clearly differentiated from hummocky by Mitchum et al. (1977), with the same illustration being used for both hummocky clinoform and hummocky. The simple hummocky pattern has a

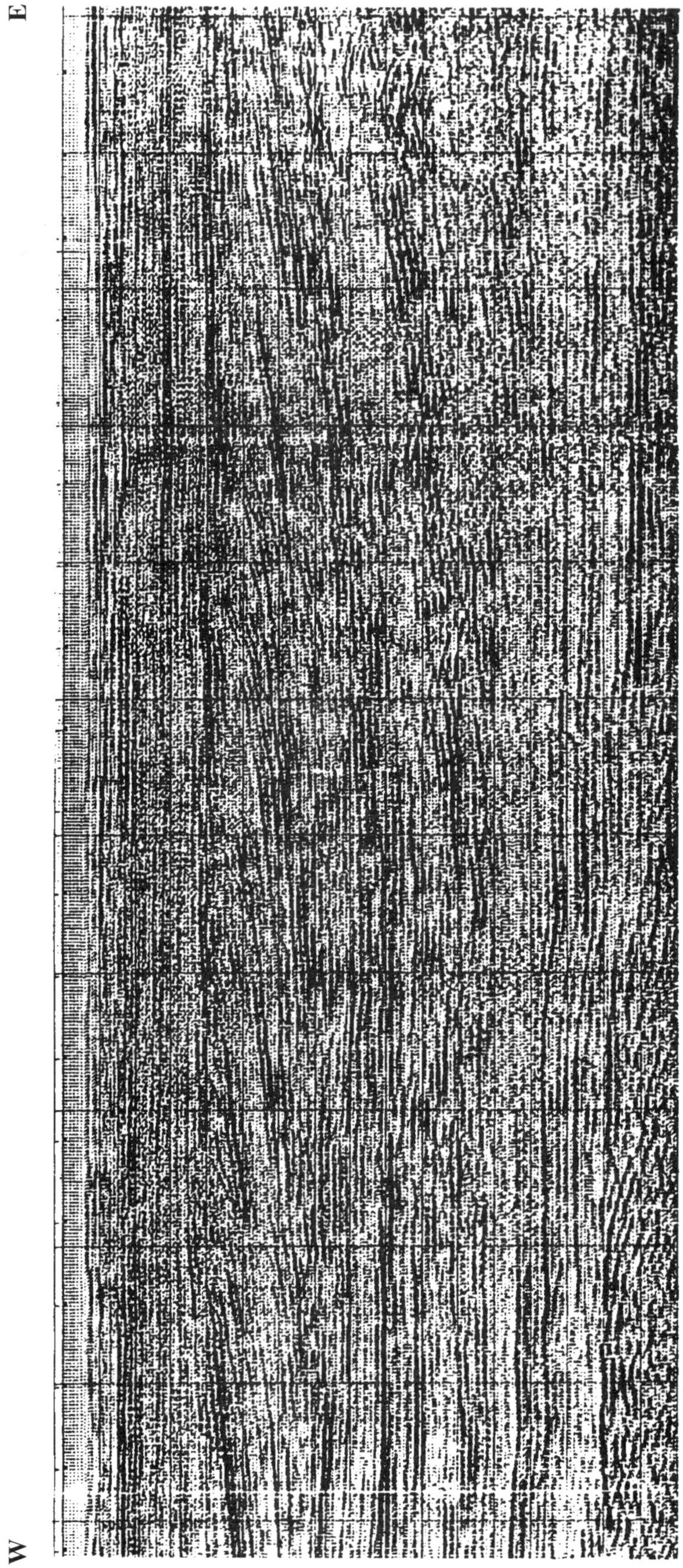

Fig. 3. Seismic profile of prograding carbonate sediments, Great Bahama Bank. The platform is to the right. From Eberli and Ginsburg (1987).

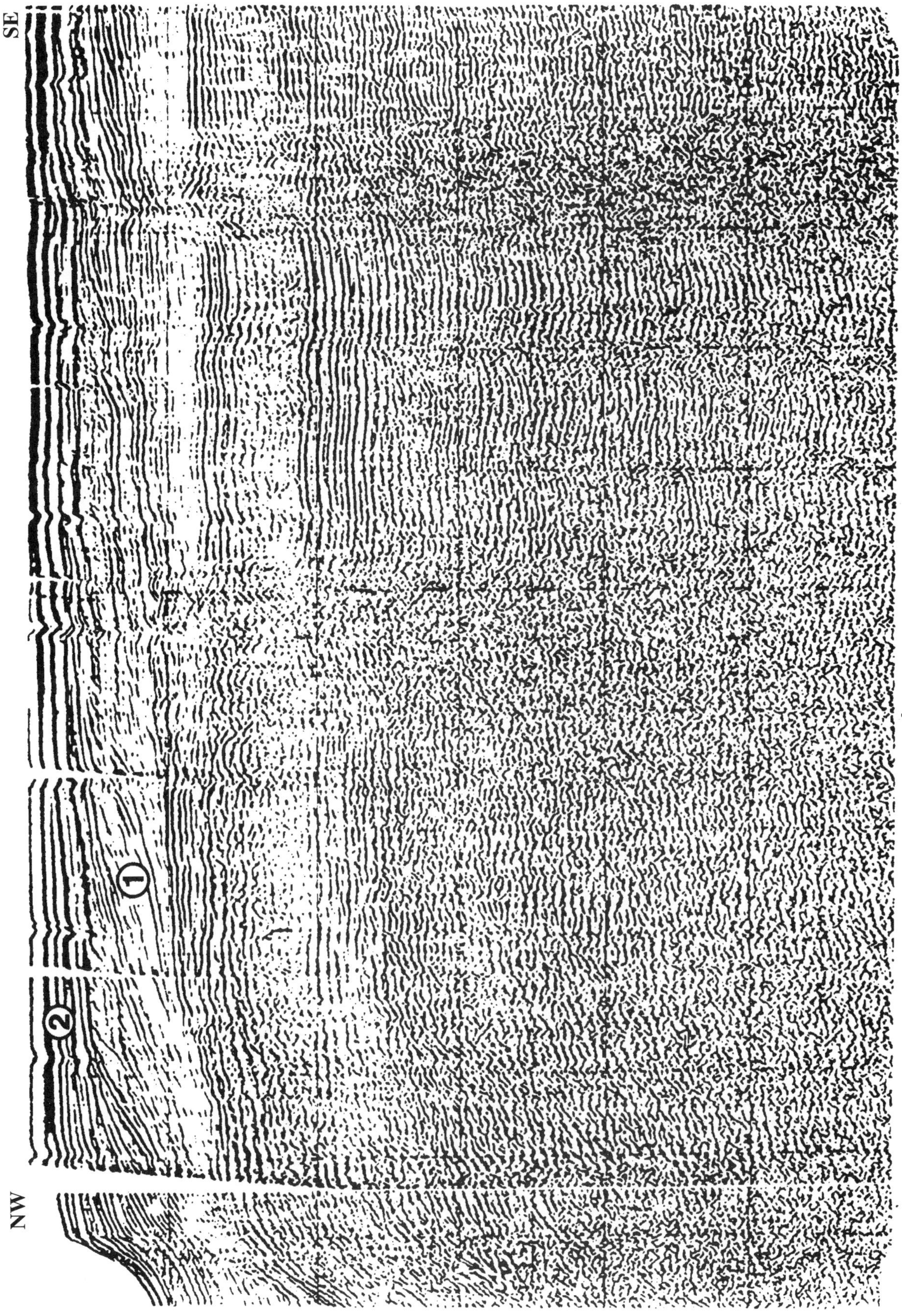

Fig. 4. Northwest-southeast seismic profile across the Campeche Bank north of the Yucatan Peninsula, Mexico. The shelf margin is to the left. A series of inclined oblique reflections (1) record northwestward progradation of this carbonate shelf platform. Unconformities within the sloping reflectors show that progradation was episodic. Flat reflections (2) mark shelfal carbonate sediments. It is difficult to record deeper events in the platform carbonates because of signal absorption.

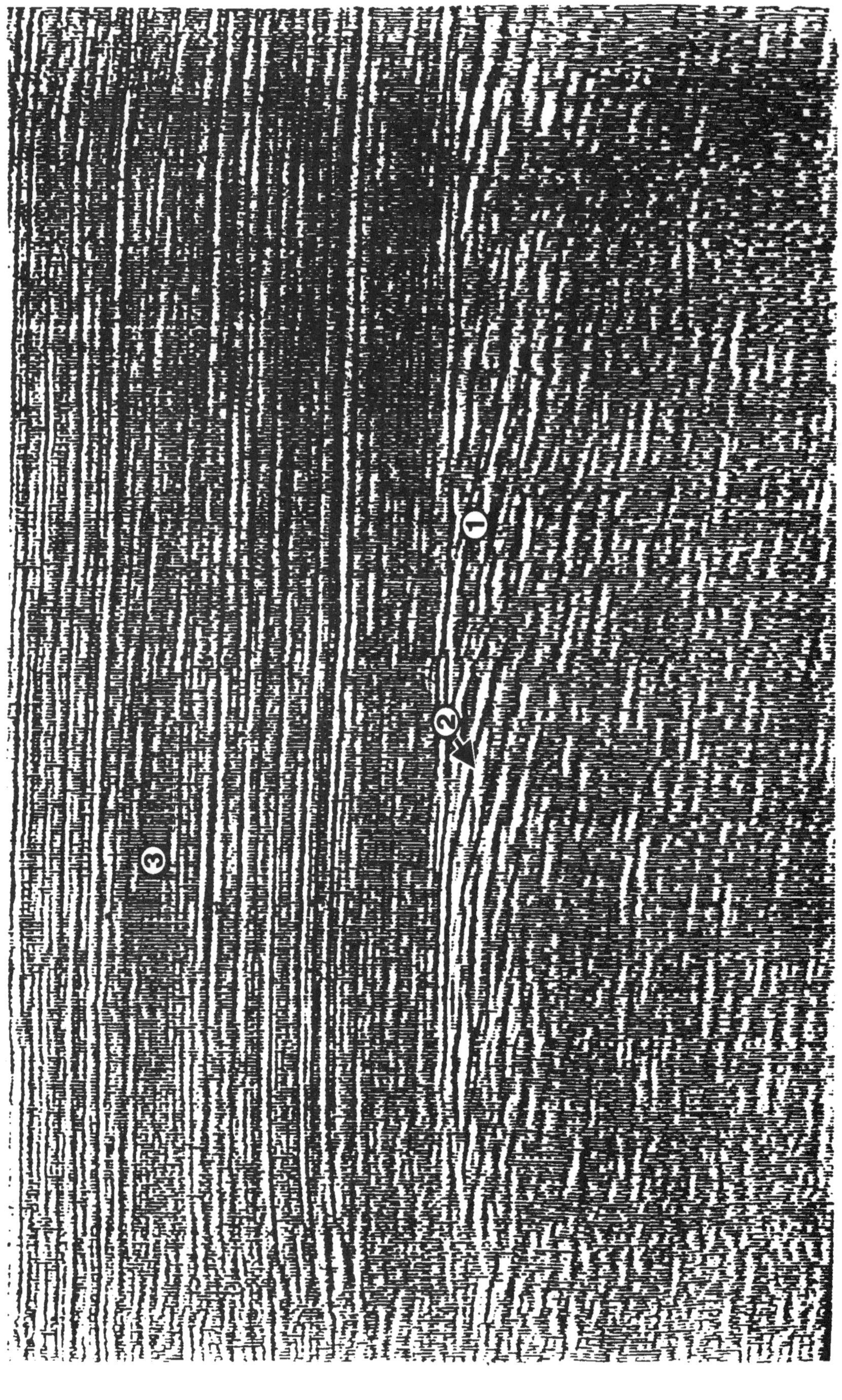

Fig. 5. East-west (left) seismic profile across the continental shelf off New Jersey on the western Atlantic margin. The inclined (complex sigmoid-oblique) reflectors (1) in the lower part of the section are a prograding bank of Mesozoic carbonates. The sloping surfaces (2) represent a continental carbonate slope environment associated with a prograding carbonate shelf (1). The overlying flat reflections (3) represent sediments deposited in a siliciclastic shelf platform environment, with the shelf margin lying farther to the east.

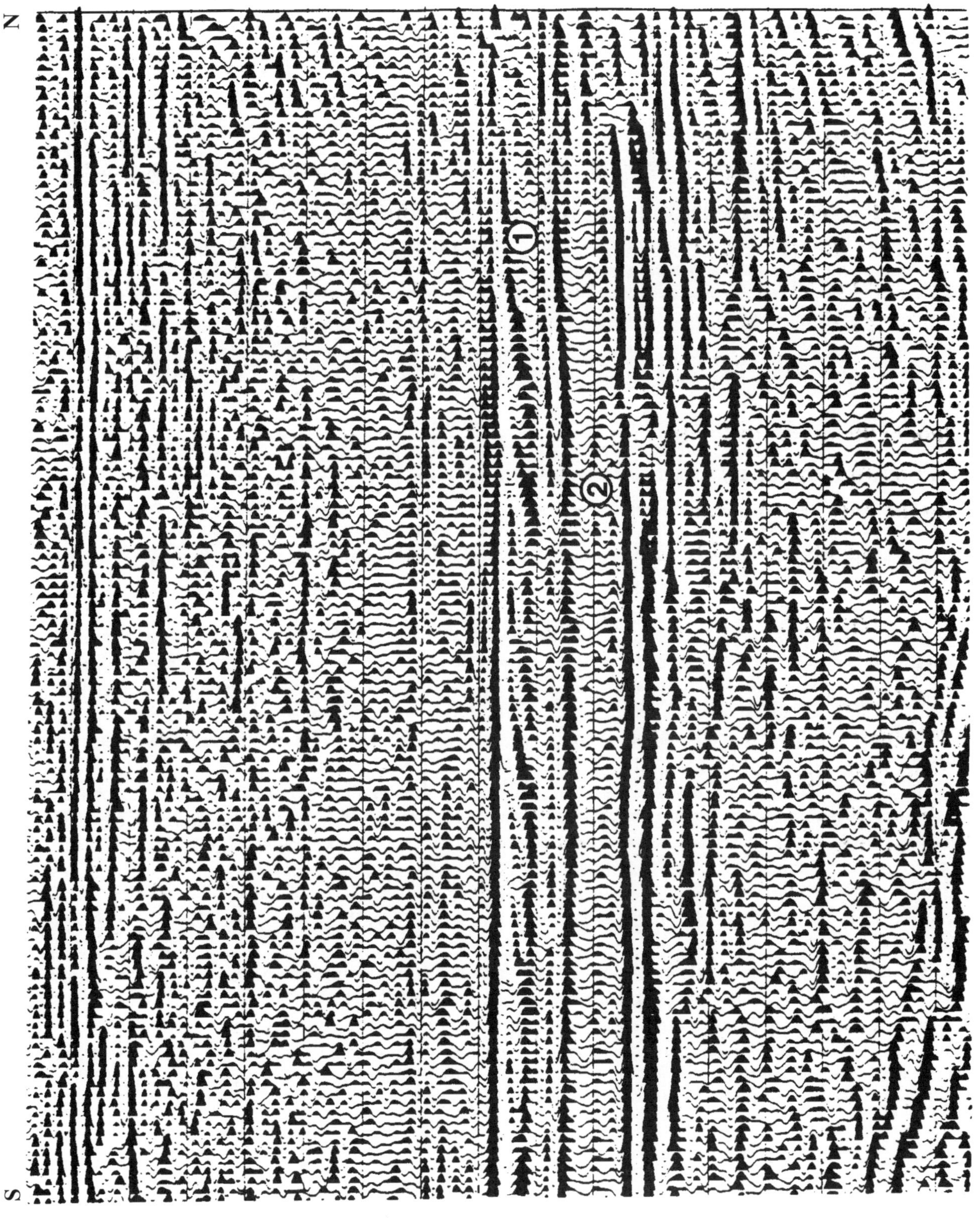

Fig. 6. Seismic profile from northern Louisiana or southern Arkansas. The shingled reflections (1) represent a prograding shallow shelf margin of the Jurassic Smackover Formation, in many places composed of oolites. The Louann Salt (2) underlies it. This is a good example of toplap. Note the interference phenomena as the shingled reflections approach the top boundary, which is on top of the continuous event.

nested convex appearance, but the convex reflections are discontinuous and offset. This pattern results from mass debris flows or collapse due to karstification (Figure 7). The upper surface of the unit is rough and bumpy. In contrast, *hummocky clinoforms* are convex reflections with a crisp convex image. They result from the supply of sediment from a point source (Figure 8). At any given moment, the prograding surface is convex in three dimensions, downlapping axially away from the sediment source but also spreading out radially to provide a convex pattern in cross section with downlap distally. The upper surface of the deposit is much smoother. Grain spilling from a tear in the side of a bag is a visual analog. Carbonate sediments being funneled downslope through lows in a bypass slope environment could produce this configuration. Because point sources are much rarer in carbonate environments than in siliciclastic ones, this pattern is rare in carbonates.

Mounded and Draped Reflections

Mounded and draped reflections are the most visible configurations to an observer. The usual assumption is that they represent reef or carbonate buildups, but care must be exercised. The most important feature is convexity of form, regardless of its origin. However, one must first determine if they are real geologic features. A number of phenomena can produce convex reflections, such as sideswipe, noise trains, overmigration, and diffractions; these are not geologic features but rather geophysical artifacts. Once these have been eliminated from consideration, then you must decide if the convex reflections are erosional, constructional, volcanic, or intrusional in nature. Next is to determine whether the mounded and draped reflections are composed of carbonates or siliciclastics. Remember that the reflection configuration suggests a depositional process, but it does not uniquely indicate lithofacies. Prediction of the lithofacies comes from regional analysis of facies and environments, calibration by cores and logs, velocity analysis, and reflectivity. Nothing should be assumed. For example, carbonate reefs occur in the siliciclastic tidally dominated Mahakham River Delta in Kalimantan, Indonesia, and the Devonian Nisku reefs of Alberta grew on underlying siliciclastic clinoforms.

Whether to call something a *reef* (a wave-resistant carbonate buildup) or not has consumed the energies of many individuals. Here I simply treat a convex feature produced by organic growth or entrapment as a buildup that shows up seismically as constructional in origin. The "cast" of builders has changed significantly over geologic time. Buildups are known from the Precambrian that were formed of microbial limestone with coarse flank deposits (James and Bourque, 1992). Stromatolites formed buildups as long ago as the Archean. In the Paleozoic, the biological components of reefs included archaeocyathids, sponges, stromatoporoids, bryozoa, phylloid algae, tubiphytes, and corals. New biota formed reefs in the Mesozoic, including scleractinian corals, calcisponges, rudists, and calcareous algae. The Cenozoic has been dominated by hermatypic scleractinian corals and calcareous algae (James and Bourque, 1992).

Carbonate buildups are often thought of as shallow water in origin, but this is not always the case. Mississippian "mud mounds" still remain enigmatic; these occur in the United States, Europe, and Africa. They are thought to have originated in outer shelf or slope environments. Work in the Straits of Florida by Neuman et al. (1977) resulted in the finding of carbonate buildups in more than 500 m of water on the floor of the Straits.

The great majority of organic buildups are found at shelf margins or the edge of carbonate ramps (see Wilson, Chapter 2, this volume). They are spectacular in outcrop as well as in seismic profiles (Figure 9). The classic facies change is from parallel, high-amplitude reflections characteristic of the shelf to convex, discontinuous, low-amplitude reflections representing the buildups to the prograding reflections of the slope environment. There are also spectacular examples of buildups in platform environments, including the late Paleozoic buildups north of Norway in Barents Sea and the shelf northwest of Sabah in southeastern Asia where more than 200 Miocene buildups have been recognized. Organic buildups can also occur on elevated features such as salt domes (Oligocene of Texas and present-day offshore Garden Bank), on elevated blocks (Miocene of Gulf of Suez; Burchette, 1988), on high points on volcanic ridges (Laccadive and Maldives, south of India), and on topographically higher areas (Glover's reef off Belize and Golden Lane in the Cretaceous of Mexico). Some can grow into pinnacle reefs, such as the Miocene of Papua New Guinea.

Reefs are the focus of exploration efforts in some basins, such as the Silurian of the Michigan Basin, and are one of the principal objectives in other efforts, such as the Devonian of Alberta and the Miocene of Indonesia. They range in visibility from slight velocity pullups (western Canada) or slight velocity continuity contrasts (Silurian of the Michigan Basin) to intermediate, seemingly endless mounds (Permian of the Barents Shelf) to stunning, bold mounds >1000 m thick (Cretaceous of the Gulf of Mexico).

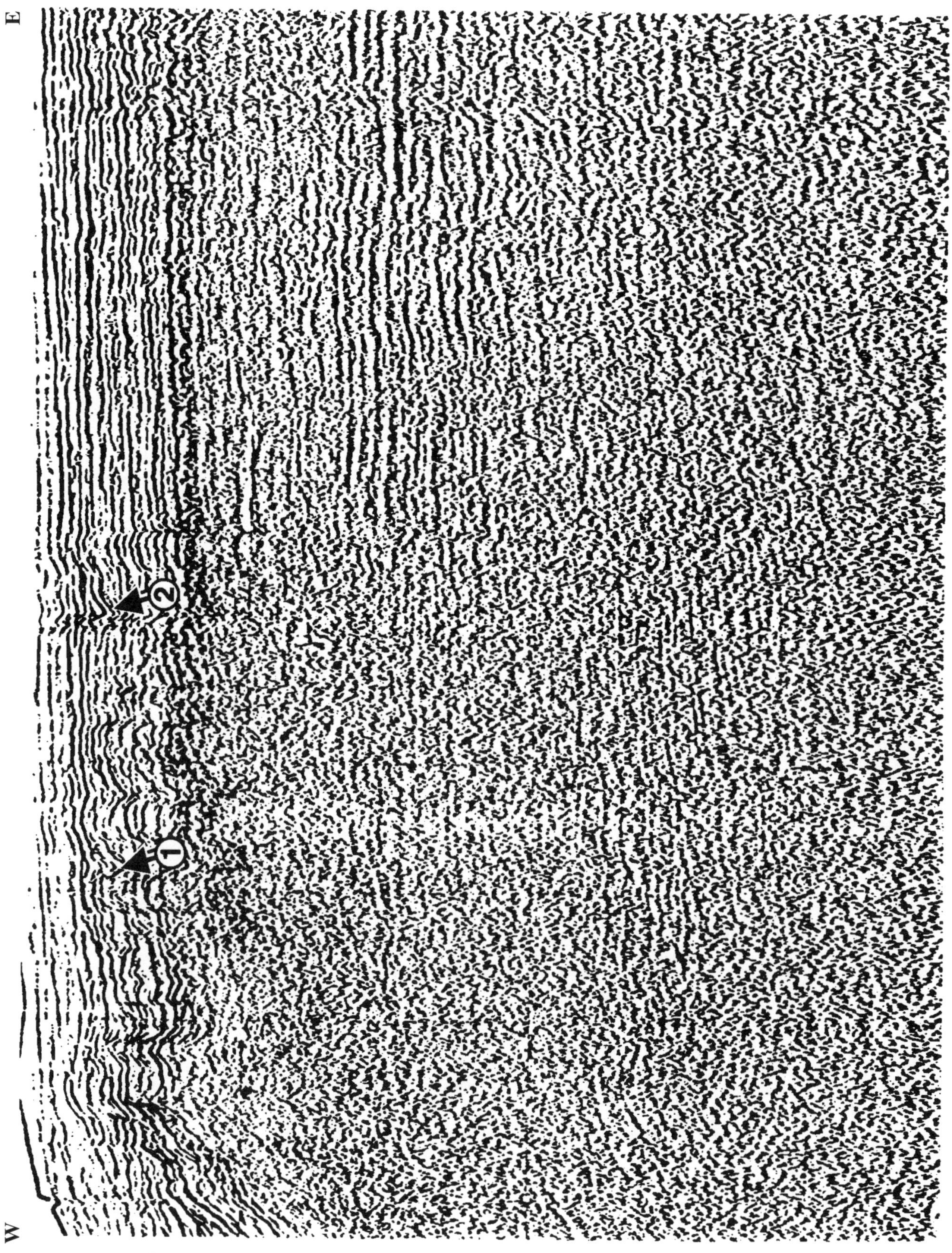

Fig. 7. Squeezed version of a seismic profile from offshore Florida. Visible solution phenomena (1, 2) have produced hummocky reflections.

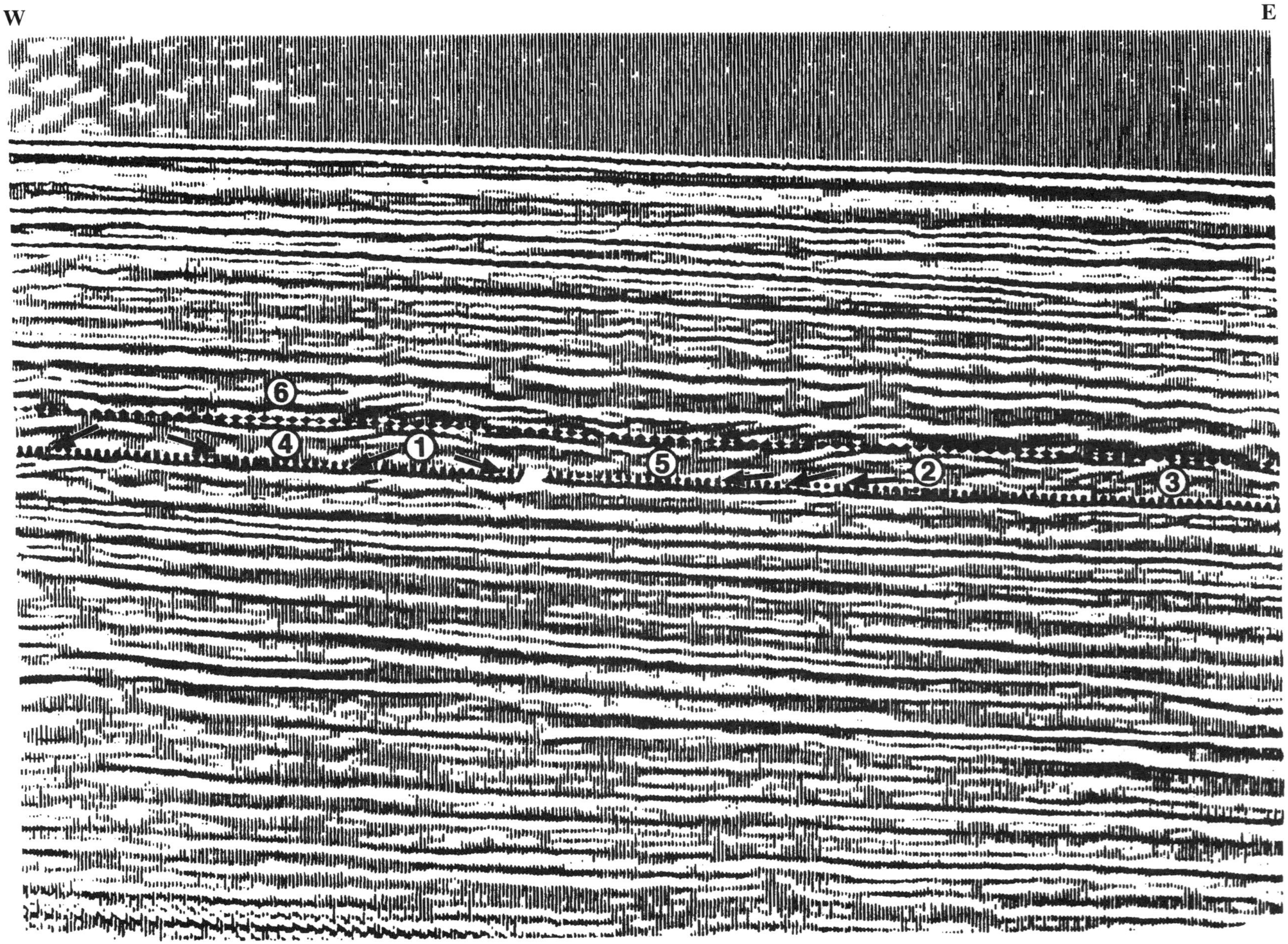

Fig. 8. Seismic profile of the continental slope in the Baltimore Canyon region on the central part of the western Atlantic margin. The mounded structures (1, 2, 3) are siliciclastic proximal turbidites (M1) formed by deposition from turbidity currents on the continental slope. They are hummocky clinoforms. The flatter reflections (4, 5) between the mounds represent interdistributary deposits. Another sequence (6) with similar mounded facies has prograded across the first one.

Regional examination of buildup facies are needed to be sure of the depositional setting. Some Waulsortian buildups are >300 m thick in the Republic of Ireland and Northern Ireland. These mud mounds had high topographic relief, as seen by the halo of bioclastic debris around them. Neuman's (1977) deep-water buildups in the Straits of Florida are only 20–30 m in height and thus would be hard to resolve seismically.

Some mounded carbonate facies are a consequence of lateral transport of sediments by deep-water currents, such as the carbonate contourites in the Straits of Florida where the peninsula of Florida and the Bahama Platform restrict the passage of water, causing an acceleration of currents. These are found in 7–800 m of water (Mullins et al., 1980). Other mounded carbonates are found in slope and basinal environments. Carbonate margins evolve from progradational to bypass to erosional. Ultimately, the margin can fail catastrophically. The western side of the Florida peninsula shows many erosional scars due to slope failure. A bathymetric map of the southeastern part of the Bahamas and the Turks and Caicos look like some great predator has taken great bites out of them. The imagery from the Gloria mapping program of the Exclusive Economic Zone (EEZ) shows spectacular photographic examples of these features north of Puerto Rico. Debris flows are well known on many old carbonate margins. Their internal patterns are hummocky to chaotic, and they have rough upper surfaces.

Karstification is one of the ongoing processes in carbonate terrains. The surface of the ground in parts of Croatia and Kentucky provide strong visual evidence of this solution process. It can happen both subaerially or in submarine settings, such as along the western side of the Florida Escarpment. The seismic effect is to produce a hummocky or chaotic reflection configuration. The ray paths are broken and refracted by the complex surface of the karstified surface, leading to poor data quality.

Until the end of the Jurassic, most biogenic carbonate production was confined to platform environments. The evolutionary development of planktonic foraminifera and coccoliths shifted carbonate sedimentation into the water column. While the bulk of carbonate production persisted in warm, shallow waters above the shelves, production occurred in the uppermost part of the water column over deep water areas as well. Upon settling, the carbonate tests and plates mantled the underlying topographic elements, producing a draped appearance. The visual clue of this feature in seismic data is concordance with underlying topography. The irony of the solubility of carbonates means that pelagic materials can be preserved in the shallower part of the basin, but can go into solution in the deeper part below the CCD.

In summary, mounded carbonate facies have many different origins and all possibilities should be considered. Care must be exercised in interpretation of depositional environments. If the rock that comes out of the core barrel on a "reef" you just drilled turns out to be black with hornblende phenocrysts, something went wrong. It has happened more than once.

Onlap and Onlap Fill Reflections

Seismic patterns of onlap and onlap fill imply a preexisting sloping surface that reflections or bedding planes can terminate against. The term *onlap* is used if terminations are against a single sloping surface; *onlap fill* is used for terminations against two opposing sides. A time gap exists between the underlying surface and the onlapping sediments. The onlap pattern is most commonly found in two settings: at the base of a slope (Figure 10) or on the shelf due to transgression. Older carbonate margins have become steep, and sediment bypasses the slope to accumulate at the base. This is common in many areas of the Bahamas today and in the Upper Cretaceous–lower Tertiary along the foot of the Florida Escarpment. Another example of where it can occur is in the talus at the base of many reefs found along platform margins. The gradient is too steep to allow more accumulation, so sedimentary bypassing becomes the predominant process.

The occurrence of onlap in shallow-water environments depends on a relative rise of base level to allow the carbonate "factory" to go into production. This is found in back-stepping reefal environments due to subsidence and shelf areas that undergo subsidence. A spectacular example from the Miocene of the Gulf of Suez was presented by Burchette (1988), who showed the onlap of Miocene platform carbonates onto Precambrian basement. This pattern is also common in volcanic terrains undergoing subsidence, such as in Hawaii, where organic reefs can take over as the land subsides below sea level. A. F. Embry (personal communication, 1984) found fragmental carbonates to be one of the three lithofacies deposited during the development of the transgressive-regressive cycles in the Triassic of the Sverdrup Basin in northern Canada.

Onlap fill implies that slopes occur on both sides of the feature with negative relief. Such conditions are met in carbonate platforms undergoing retrogradational slope failure where large embayments extend into the platform (Figure 11). Sediment accumulates at the base of the reentrant by slides, turbidites, and nepheloid sedi-

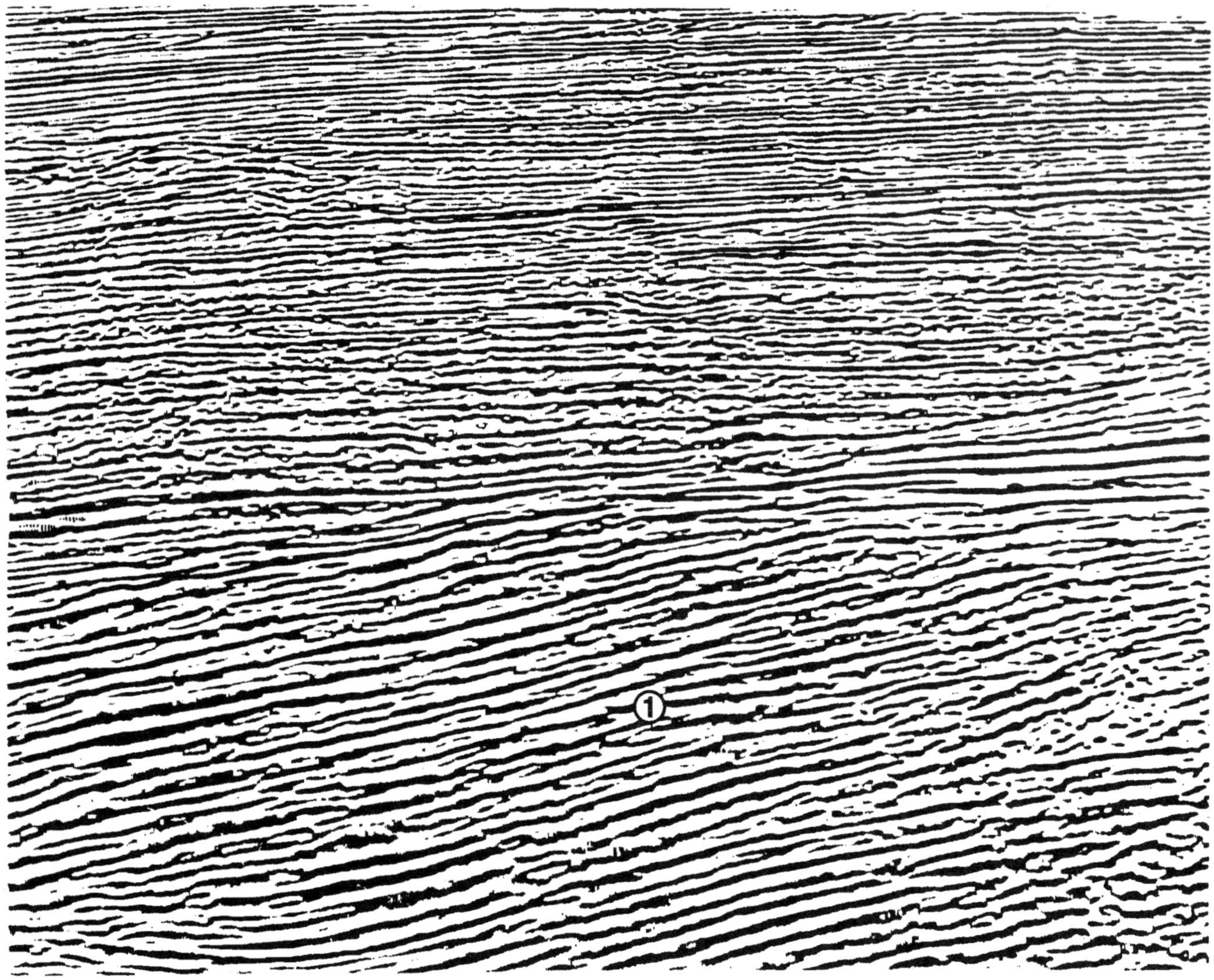

Fig. 9. Seismic profile (split into left and right panels) depicting Cretaceous Sligo buildups (1, 2 on right panel) on the shelf margin of the northeastern Gulf of Mexico. Rudists were the reef builders. The slope environment in front of the reef is extensive (1 on left panel). It is unclear whether the slope deposits are covered with the reef buildup or are separate in time (lowstand deposits). The reflections sloping landward (to the right) from the reef are characteristic of buildups that are carbonate source areas. Part of the slope may result from differential compaction. The flat-lying reflections below the reef are the Jurassic-Cretaceous Knowles Limestone, which progrades over an underlying siliciclastic delta seen here as hummocky reflections. (*Figure continues on next page.*)

mentation. An axial section of the embayment would show onlap, while a transverse section would show onlap fill. This pattern also occurs where there are extensive platform reefs; sediment fills the gap between adjacent mounds. Gullies are common on many carbonate slopes, and here again one can see onlap fill. If there are isolated topographic depressions or basins infilled by carbonate turbidites, a pattern of onlap fill can be found. Transportation by traction or suspension is implied.

Other Diagnostic Features of Carbonate Seismic Data

Amplitude

Well-layered carbonate sediments generally show high amplitudes (Figure 1). This can occur in a purely carbonate environment or one intermixed with evaporites (Saudi Arabia) or shales and marls (northeastern

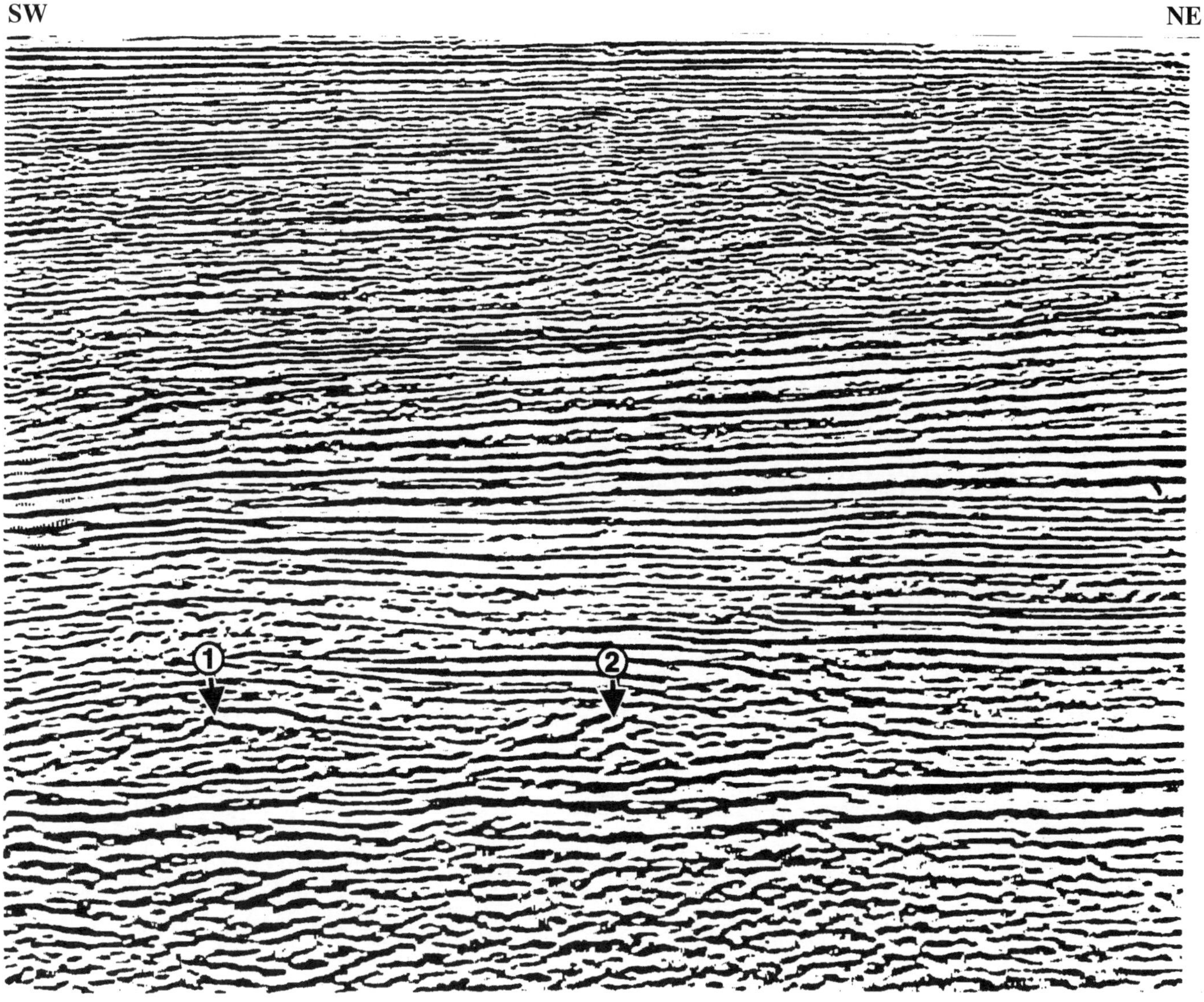

Gulf of Mexico). A high impedance contrast is implied. Amplitudes are usually high in platform settings. Spectacular examples occur in the Jurassic–Cretaceous carbonates rimming the Gulf of Mexico, the Tertiary of the Malta Platform, the top of the chalk in the Viking Graben, the lower Tertiary of northeastern South America, the Tertiary carbonates northwest of Australia, and the Cretaceous of Angola. Amplitudes are also high in basinal settings, as seen in the Gulf of Mexico, the Jurassic of the northeastern United States, the Straits of Florida, and the Jurassic of northwestern Africa. Prograding reflections are also usually characterized by high amplitudes, such as the Cretaceous of the Gulf of Mexico, the Permian of the Permian Basin of West Texas, the interior and west side of the Bahama Platform (Miocene–Recent), the lower Tertiary of northeastern South America, and the Jurassic of the eastern United States. Many of the clinoforms can loose amplitude toward their toes, making exact placement of the terminations difficult. High amplitudes are also associated with some organic buildups, such as the Miocene of northeastern South America and the Permian of the Barents Shelf.

In contrast, the interiors of carbonate buildups often show low-amplitude seismic character. Examples include the Cretaceous of the Porcupine Basin southwest of Ireland, the Jurassic of northwestern Africa, the Cretaceous of the Gulf of Mexico, the Lower Carboniferous of Khazakhstan, the Permian of West Texas, the Silurian of the Michigan Basin, the Pliocene of Indonesia, and the Miocene of the Gulf of Suez. This is usually attributed to the massive nature of the buildups or to the lack of any continuous horizons within it. Other convex reflections, such as mass debris flows, usually have high amplitudes, such as at the base of the west Florida Plat-

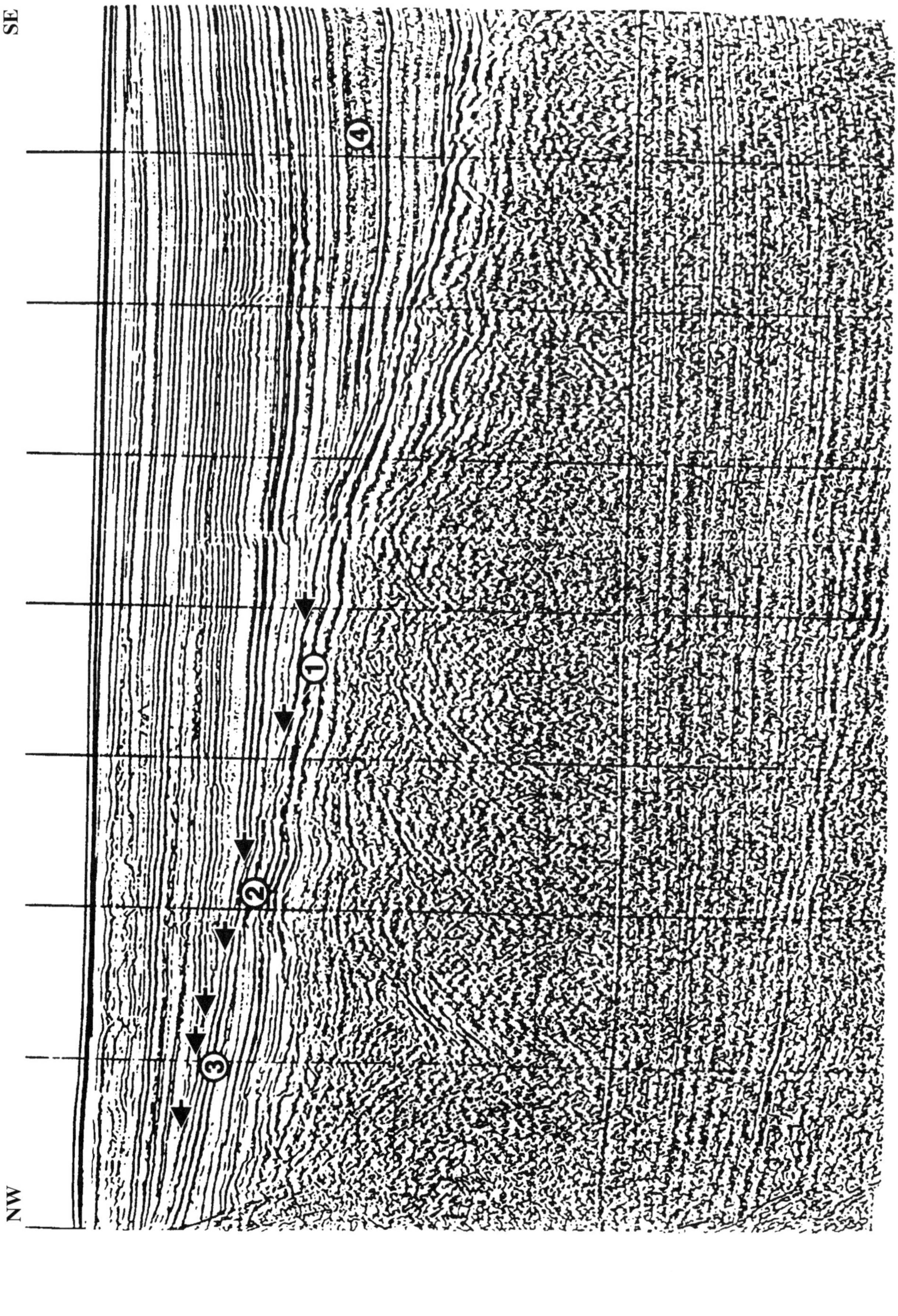

Fig. 10. Seismic profile of a deep-water setting in the southern part of the Bahamas. An island platform lies to the left just off the section. The onlap stratigraphy may appear simple at first, but careful analysis shows that there are several distinct intervals of onlap. Two arrows (to the left and right of 1) show onlap of a sloping reflector. Subsequent sediment accumulation ramped up against the escarpment on the left, in turn to be onlapped itself (arrows near 2). This pattern was then repeated (arrows near 3).

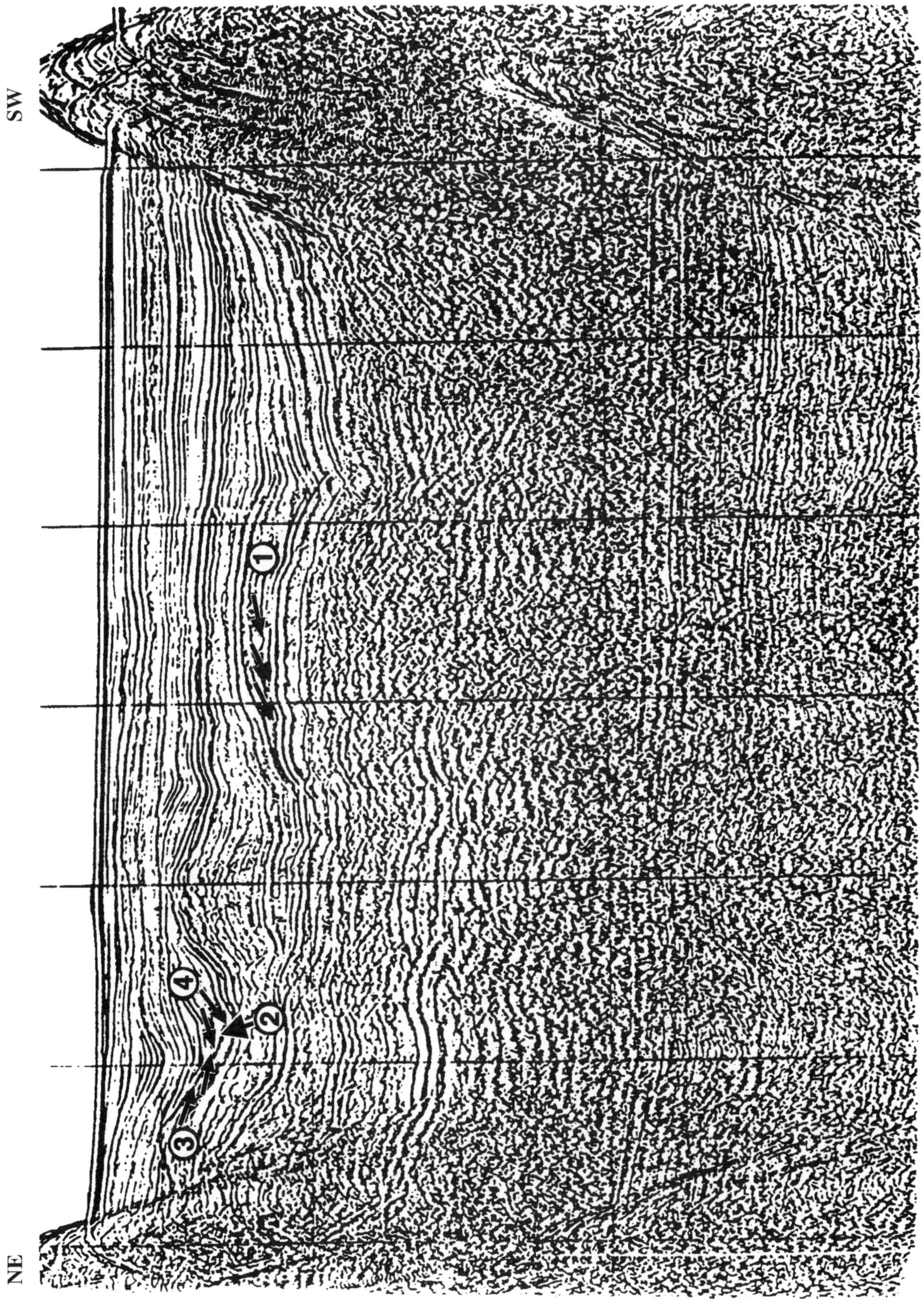

Fig. 11. Northeast-southwest seismic profile of the southern part of the Bahamas offshore from the carbonate banks. Proximal turbidites showing onlap fill have created the highlighted features. The localized nature of the currents caused mounds (1) to be constructed. The turbidity currents also formed a submarine canyon (2) that cut through older strata (3), which has been subsequently infilled (4). The sediments are primarily composed of carbonates.

form, the base of slope in the Bahamas, and the Cretaceous of the Blake Rise below the Blake Escarpment; these have convex to chaotic reflections. Karstified terrains also have high amplitudes, such as on the platforms east and west of Florida. Patterns of onlap and onlap fill are also usually characterized by high amplitude.

A change in the amplitude in a shelf environment proceeding shoreward usually means a major shift in environments. The most common occurrence is in mixed carbonate-siliciclastic environments where carbonate production is active on the outer shelf while siliciclastics are being deposited along the basin margin. Good examples of this are found at the Jurassic–Cretaceous boundary and in the Aptian of the northeastern Gulf of Mexico. The high amplitudes of the outer shelf fade away to almost nothing as the shelf passes into the siliciclastics of the basin margin.

It is illuminating to look at a relative amplitude section in which both siliciclastics and carbonates are present. Since such a section has no automatic gain control, it provides a closer approximation to the comparative amplitudes. The carbonates stand out in bold relief, particularly at carbonate-siliciclastic interfaces. Within one such series of profiles (from the northeastern Gulf of Mexico), the slope and basinal carbonates have the highest amplitudes, followed by those of the platform. The lack of amplitude in the buildups is most evident.

Another amplitude variation of predictive interest is in the transition from limestone to dolomite. Dolomites should be associated with higher amplitudes than adjacent limestones. Since the single most important carbonate reservoir environment globally is dolomitized shelf facies, this relationship should be carefully examined. If gas is present in buildups, it produces both a dimming effect and a velocity distortion.

Frequency

Frequency is measured by the number of seismic reflections per given vertical TWT. Since frequencies decline with depth and are often highly filtered by the processor, one must be aware of the TWT being used in the section and the filters applied to the data during acquisition and processing.

In general, carbonates are characterized by lower seismic frequencies than siliciclastics (Figure 1). Part of this is a function of their higher velocity, which affords less internal resolution. Basinal facies, perhaps because of their depth, usually have the lowest frequencies of all environmental settings and reflections are far apart. Prograding facies have variable frequencies, which seems to be a function of the particular trend and its age. Detailed work can be done on seismic data from some slope environments but not from others. The highest relative frequencies are found in platform environments. Good internal resolution is possible, and subtle buildups and prograding shoals can often be differentiated.

Continuity

No other environment can compete with platform carbonates for continuity of individual seismic reflections. The reflections within the Knowles Limestone, a series of carbonate ramps at the Jurassic–Cretaceous boundary in the northeastern Gulf of Mexico (Figure 9), and in some overlying Lower Cretaceous platform carbonates behind the Aptian–Albian buildups, show persistence for hundreds of kilometers southward from the Florida panhandle. These are the proverbial "railroad tracks." Such continuous reflections are present in other platform settings, such as the Tertiary northwest of Australia. Reflections with high continuity can also be present in basinal environments if they have not been interrupted by halokinesis. Slope environments, in contrast, usually have shorter reflections because of sedimentary processes (e.g., progradation, slumps, and slides), but these are still continuous in the sense that individual clinoforms are well imaged. There are some settings in platform carbonates where discontinuous reflections are present within the platform. Chalk deposits are also characterized by different styles of continuity. Major unit boundaries within the chalk in the Central Graben of the North Sea have excellent continuity, although internal reflection configurations are often discontinuous.

The carbonate deposits with the lowest continuity are buildups (Figure 12). Reflections may extend for 10–20 shotpoints, then fade out to be replaced by another equally short reflection adjacent to it. This is a function of the massive nature of the interiors of many buildups.

Analysis of Carbonate Environments

Platform and Supratidal Carbonates

Of all the carbonate environments, dolomitized shelf facies contain more hydrocarbons worldwide than any others. Therefore, proper analyses of these environments is important. Fine-scale seismic analysis of platform sediments shows upward-shoaling successions to be a repetitive theme from the Precambrian to the Recent. Are there larger scales of organization to these patterns that show up seismically? Modeling suggests that

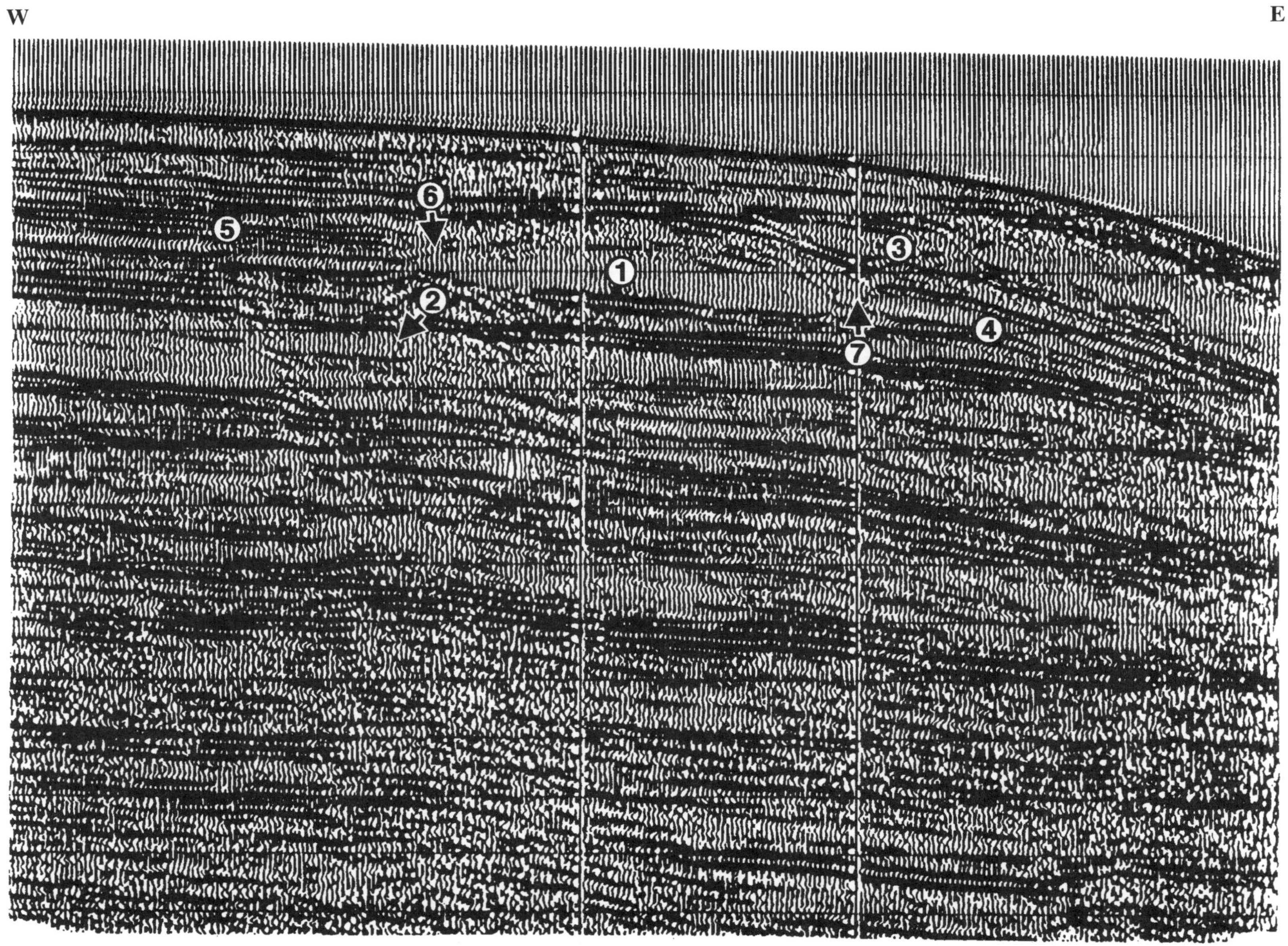

Fig. 12. East-west seismic profile of the edge of the Blake Plateau. The large mound (1) is a Cretaceous carbonate buildup, probably of rudistid clams. This large buildup was preceded by a smaller one (2). There was also a successive buildup seaward of the main one (3). The forereef (4) and backreef (5) facies are clearly visible. The termination of reflectors (6) marks an internal facies change within the sequence between the massive carbonates of the core and the bedded carbonates of the backreef; the same type of facies change occurs between the forereef and the core (7). The environment is that of a pinnacle or barrier reef.

the relative frequency of subtidal versus supratidal facies should show an increase of supratidal facies toward the basin margin and that individual parasequences should become thinner as a sequence boundary is approached (Goldhammer and Dunn, 1991). The high velocity of carbonates probably makes detailed application of these concepts difficult.

The outstanding facies patterns of platform carbonates is their parallel to subparallel reflection configurations and their high continuity. The Williston Basin in the United States has extensive platform and supratidal carbonates ranging in age from Ordovician to Mississippian. Principal unconformities can be determined by seismic sequence analysis alone (verified by tying back to wells), but the rate of convergence is low, and lines 25 km or longer are needed. An extreme example of low rates of convergence can be seen in the Lower Cretaceous platform carbonates of the Blake Plateau, where successive downlaps are 100 km apart. Individual reflections from the platform of the Upper Jurassic–Lower Cretaceous limestones of the northeastern Gulf of Mexico can be traced for hundreds of kilometers (Figure 1).

An excellent example of platform carbonates is in the northeastern Gulf of Mexico (Figure 9). Initial deposition occurred on top of an abandoned delta platform. Because of the lack of relief, this provided an ideal setting for the development of extensive platform facies. Three carbonate ramp environments occurred in succession. The reflections of this interval have the highest amplitude and continuity of any found in the entire Upper Jurassic–Lower Cretaceous interval. The overlying Lower Cretaceous carbonates extend to the uppermost Albian and comprise a thick accumulation exceeding 1 sec TWT, or >2500 m in thickness. These reflections are highly variable in amplitude (medium to high) and in continuity (poor to good), but their frequency is low. Regionally, these facies change drastically toward the basin margin as siliciclastics replace carbonates. The reflection configuration becomes discontinuous, amplitude drops sharply, and frequency increases.

The Permian Basin of West Texas is economically important because of its platform carbonates which are present on the Central Basin Platform and the Northwest Shelf. The seismic reflections from the carbonate platform are characteristically high in amplitude and continuity. Some of the highest amplitudes occur in the transition to the overlying evaporites. Last Chance Canyon in the Guadalupe Mountains has become a testing ground for many ideas in carbonate sequence stratigraphy (Sarg and Lehmann, 1986; Sonnenfeld, 1991).

The Williston Basin of North Dakota and Montana is another area with extensive platform and supratidal facies. Examples of high-frequency seismic data from this basin have been published by Andrews (1985). Although the parallel to subparallel nature of the reflections is apparent, different stratigraphic levels show differences in amplitude and continuity because the biota and depositional environments have changed through time. The Ordovician sequence is open marine, the Silurian is primarily supratidal, the Devonian forms part of an open platform interspersed with occasional evaporites, and the Mississippian has both open marine and supratidal environments. Other Paleozoic basins in North America with important hydrocarbon accumulations (Alberta Basin, Paradox Basin, and Michigan Basin) are similar in having high-amplitude, low-frequency reflections with variable continuity in their platform facies.

An excellent Mesozoic example can be found in the Jurassic of Morocco (Figure 13). This succession started as a ramp environment and evolved to a steep bypass or erosional carbonate margin. The platform environment is characterized by many individual sequences, each about 100 ms thick, attesting to many separate aggradational episodes. Amplitudes are medium to high, and the sequences have much internal character in contrast to some of the other cited examples. This may be a function of the data quality. No discussion of Mesozoic platform environments would be complete without mentioning the Jurassic of Saudi Arabia and the Ghawar field, which contains 80 billion bbl of hydrocarbons. The seismic signature is a single pair of parallel reflections 130 km long forming a gentle anticlinal closure. This lithofacies is a series of alternating carbonates and evaporites.

Platform interiors can provide space for spectacular buildups if there is regional subsidence or a relative rise of sea level slow enough to allow the reef builders to keep pace. Two of the most spectacular examples are in the Permian of the Barents Sea and the Miocene northwest of Sabah. Since there is positive expression of the reefs, these have kept up while the surrounding areas have not. Similar development occurred in the Devonian of the Alberta Basin, where reefs such as those of the Swan Hills form major hydrocarbon traps.

Platform Margins

The transition from shelf to slope provides some of the most rapid changes in the facies patterns of carbonates. The first pattern most interpreters look for are mounds representing reefs. Some of the best seismic examples are from the Cretaceous of the southeastern United States and Gulf of Mexico (Figures 9, 12), the Jurassic of Morocco, the Miocene of Papua New Guin-

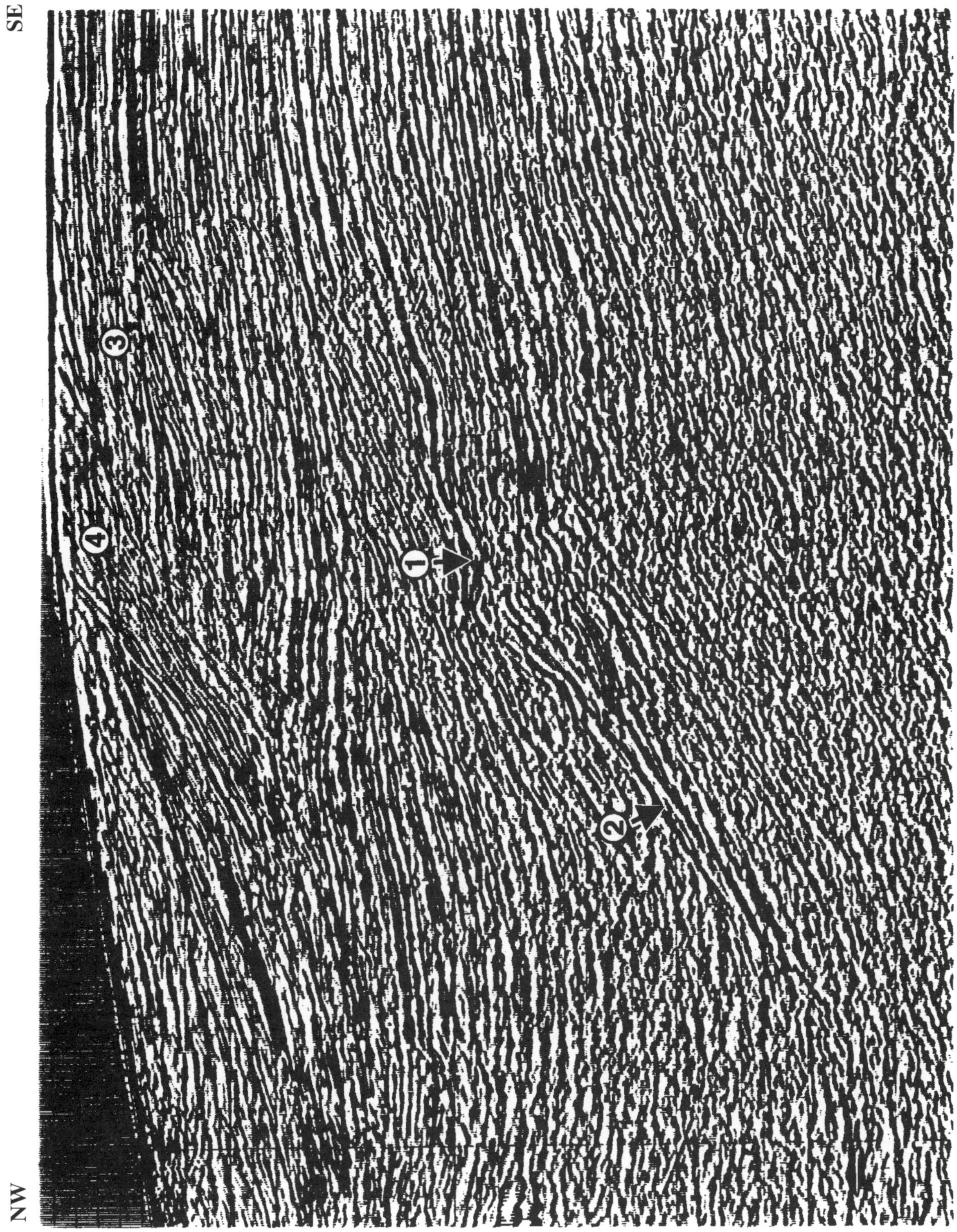

Fig. 13. Seismic profile from northwestern Morocco. The Jurassic carbonate margin is at (1) with the shelf facies to the right. The convex reflectors below (1) suggest a reefal buildup. The steep slope at (2) is a bypass margin. Extensive siliciclastics (3) were deposited in the Early Cretaceous. Subsequently, these were uplifted, eroded, and covered on their seaward edge by early Tertiary slope deposits (4). The Jurassic buildup contains nearly 1 billion bbl of dead oil. The shelf to basin relief is almost 3000 m.

ea, and the Permian of West Texas. Some buildups exceed 1000 m in height. One of the key signatures is to look for subtle shingled reflections that slope back into the shelf environment. These are a result of the transport of carbonate sediment by storms and currents from the reef crest into the platform interior. This may be enhanced by differential compaction. The internal signature of the buildups is usually a loss of amplitude and continuity, although this is not always true. Since primary dips on carbonate slopes can exceed 30û, the transition from a buildup into the upper slope can be abrupt.

Not all platform margins have buildups. This is particularly true of downwind margins such as the west side of the Bahama Banks. Here, the parallel reflections of the platform bend over into the upper slope without interruption. Sediment is carried off the platform by waves, currents, and storms. Platform margins such as in the Bahamas can prograde many tens of kilometers. Platform margins can be progradational (as already described), bypassing, or erosional. Seismic examples of erosional margins provide stunning images. The platform margin along the western edge of the Florida Escarpment bends abruptly, passing downward into slopes as high as 85° with relief up to 1500 m (Figure 14).

Slope Environments

Slope environments, by nature of the processes that operate in them, provide examples of some of the most variable facies to be found in carbonate environments. One of the key questions is whether slope deposits primarily represent relative lowstand deposits or are deposited primarily during highstands and thus represent highstand shedding. Analyses by Droxler and Schlager (1985) suggests that the offbank transport of carbonate sediments in the Bahamas is seven times higher during the transgressive and early highstand portions of a relative sea-level change than during a relative fall and lowstand. It is difficult to trace individual units any distance laterally in steeper slope environments.

Two different but complementary slope environments are presented as examples. The Lower Cretaceous (Aptian–Albian) carbonates of the northeastern Gulf of Mexico have a facies change from a progradational slope (Figures 9 and 14) to an erosional, nearly vertical slope as they are traced to the southeast. Following the drowning of this carbonate platform at the end of Albian time, the area inward from the old margin became a site of chalk and marl deposition. These deposits range from Late Cretaceous to Oligocene in age. The main site of carbonate production was set well back from the old margin, and these chalks and marls show distinct patterns of progradation toward the old shelf edge (Figures 1 and 12). Some material was also carried into the abyss beyond. Since there is such dramatic relief on the old shelf edge, the sediments thin toward the old edge with varying continuity in strike view.

Examination of the Lower Cretaceous slope deposits shows many prograding units that can be defined by successive downslopes and toplaps. Some of these are restricted to the slope (at least in the plane of the section), whereas others can be traced back into the shelf (Figure 9). Attempts to trace units (100 ms thick) laterally usually fail before 20 km has been reached. Internal fabric suggests some debris flows. Traced southward, the pattern of the clinoforms disappears to be replaced by chaotic reflections. South of De Soto Canyon, the margin has undergone collapse and erosion. The slope deposits resulting from collapse are hummocky and chaotic with great piles of debris.

The younger prograding slope, which is composed of chalks and marls, show long, continuous reflections in the in dip sections. As one approaches the old Lower Cretaceous margin, however, this continuity becomes disrupted due to thinning and erosion. Strike lines just inward of the Lower Cretaceous erosional edge are dramatic, showing submarine canyons and gullies. Individual units pinch and swell, in contrast to the apparent continuity in dip sections. The chalks and marls that went over the edge now form sheet-drape deposits over the rough topography of the Lower Cretaceous debris flows. Internally, the reflections parallel the upper surface. The contrast in rates of accumulation is stunning. The Gulf of Mexico formed ~160–170 Ma. The younger draped deposits, 100 ms thick, represent all the record we have for the past ~70 m.y.

Debris flows can also make great reservoirs. Along strike in Mexico, the Tuxpan Platform shed debris from the El Abra reef trend. The resultant deposits, the Poza Rica, contain several billion barrels of oil.

Other dramatic examples of carbonate slopes can be seen in both Mesozoic and Cenozoic sequences. The Jurassic of the Grand Banks has steep slopes with about 2000 m of relief (bypass-erosional). Its counterpart in Morocco has slopes with up to 3000 m of relief. The lower Tertiary of the Amazonas Basin in Brazil shows successive prograding wedges with varying continuity in reflections. These are variable in strike view. Submarine canyons notch the prograding margin at some time horizons.

Bosellini (1984) has studied and illustrated several different patterns of slope environments based on outcrop examples in the Mesozoic of northeastern Italy. These outcrops are on a scale comparable to many of

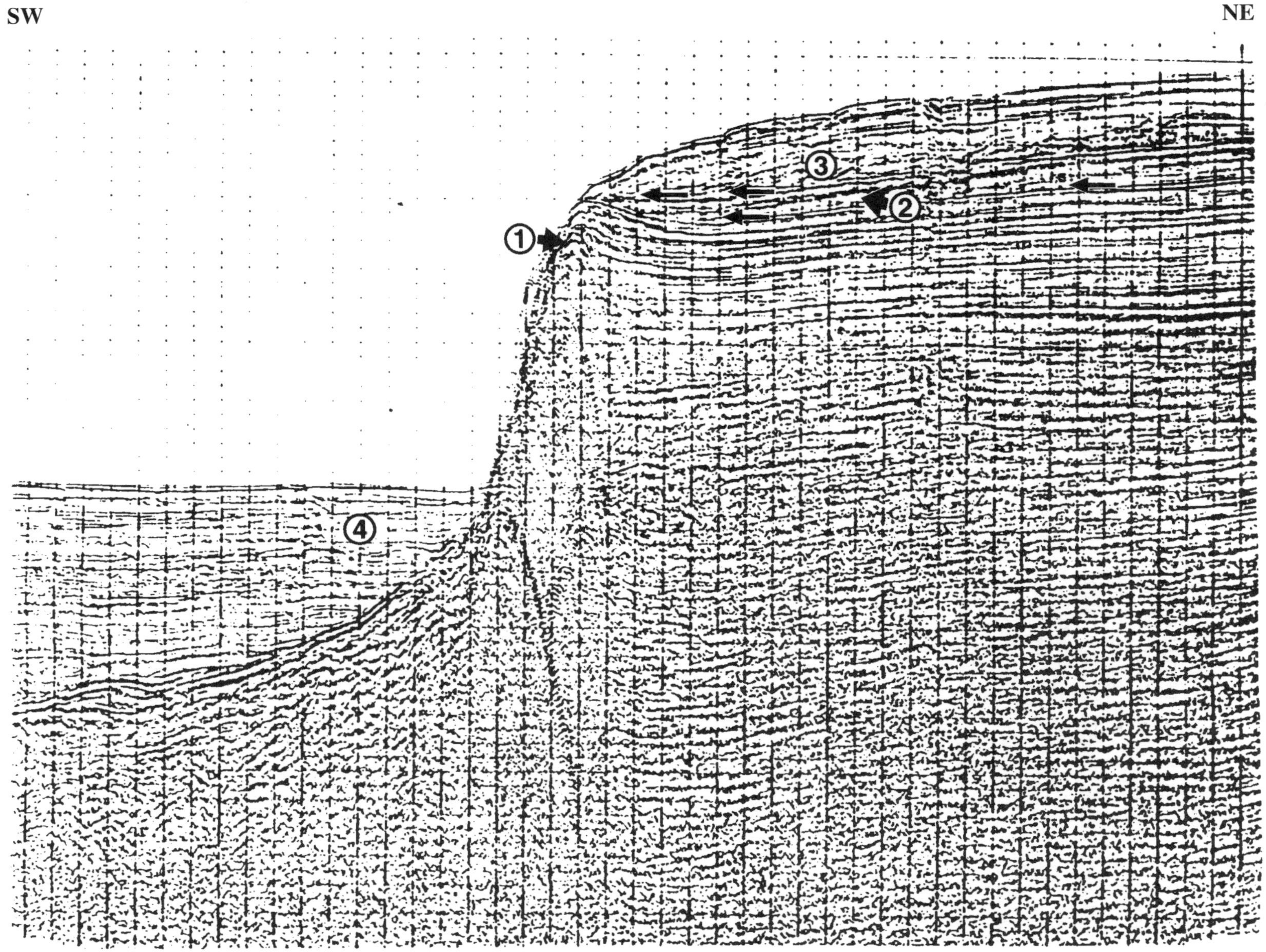

Fig. 14. Southwest-northeast seismic profile across the shelf margin in the northeastern Gulf of Mexico, west of Florida, showing a carbonate shelf platform. (Land is to the right.) Such margins are common in the Bahamas, Florida, and Yucatan Peninsula. The edge of the margin is steeper than 45°, so it appears as a series of diffractions (1) that obliterate any primary reflectors. The two sequences (2, 3) in the upper part of the section thin to a zero edge at the platform margin. The sediments at the base of the slope (4) contain a large amount of fine-grained clastics transported laterally along the base of the escarpment from the northwest. The sediments below sequence (2) are middle Cretaceous in age.

those seen seismically, providing an analog of seismic and outcrop examples.

Basinal Deposits

Basinal deposits mostly result from the gradual settling of pelagic particles onto the basin floor. Such deposits are well represented in many deep oceanic basins today where Cretaceous and Tertiary sediments have been drilled and cored by the DSDP and ODP programs. However, debris can travel many tens of kilometers as a result of the collapse of carbonate margins (volume of collapse debris ~200 km^3 in one Tertiary Pyrenean basin). Debris flows off the Bahama Platform rip up the older underlying sediments during their passage and have a rough upper surface due to their unsorted nature. Chalk debris flows, which moved onto the floor of the Viking Graben, provide the petroleum reservoir at Ekofisk field.

The deep ocean is not always a quiet, tranquil place. Contour currents formed by temperature or salinity currents transport sediment long distances along the base of slope and basin floor. The most dramatic examples of carbonate contourites come from the Straits of Florida, where currents constricted between the peninsula of Florida and the Bahama Bank pick up debris and redeposit it in long, prograding drifts. Fine carbonates also comprise an appreciable fraction of contourite drifts in the North Atlantic Ocean.

Concluding Remarks

The task seems clear: identify the seismic facies, predict the environments and attendant lithofacies, and use these as guides for where to drill. If only it were this simple in practice. More published case histories and detailed 3-D studies are needed. We have some excellent tools available, but we must use them more thoroughly and consistently to obtain the benefits they are capable of providing.

Acknowledgments

The author wishes to thank Schlumberger Geco-Prakla in Houston and Gatwick for permission to publish some of the figures in this chapter. In addition, thanks go to G. Eberli of the University of Miami for Figure 3. Appreciation is also extended to Norman Neidell for his many helpful suggestions which improved the quality of the manuscript.

References

Andrew, J. A., 1985, The art and the science of interpreting stratigraphy from seismic data, *in* Gries, R. R., and Dyer, R. C., Eds., Seismic exploration of the Rocky Mountain Region: Rocky Mountain Assoc. Geol. and Denver Geophy. Soc., 95-103.

Bosellini, A., 1984, Progradation geometrics of carbonate platforms: examples from the Triassic of the Dolomites, northern Italy: Sedimentology, **31**, 1-24.

Brown, L. P., Jr., and Fisher, W. L., 1980, Seismic stratigraphic exploration and petroleum exploration: Am. Assn. Petr. Geol. Educ. Course Notes, Ser. **16**.

Burchette, T. P., 1988, Tectonic control on carbonate platform facies distribution and sequence development, Miocene, Gulf of Suez: Sed. Geol., **59**, 179-204.

Droxler, A. W., and Schlager, W., 1985, Glacial versus interglacial sedimentation rates and turbidite frequency in the Bahamas: Geology, **13**, 799-802.

Eberli, G. P., and Ginsburg, R. N., 1987, Segmentation and coalescence of Cenozoic carbonate platforms in northwestern Great Bahama Bank: Geology, **15**, 75-79.

Fontaine, J. M., Cussey, R., Lacaze, J., Lanaud, R., and Yapaudjian, L., 1987, Seismic interpretation of carbonate depositional environments: Am. Assn. Petr. Geol. Bull., **71**, 281-297.

Goldhammer, R. K., and Dunn, P. A., 1991, Carbonate cyclic stratigraphy and sequence stratigraphy: a review, *in* Goldhammer, R. K., Lehmann, P. J., Todd, R. G., Wilson, J. L., Ward, W. C., and Johnson, C. R., Eds., Sequence stratigraphy and cyclostratigraphy of the Mesozoic of the Sierra Madre Oriental, northeast Mexico: a field guidebook: Soc. Econ. Paleont. Mineral., Gulf Coast Sec., 33-41.

Goldhammer, R. K., and Lehmann, P. J., 1991, Sequence stratigraphic summary, *in* Goldhammer, R. K., Lehmann, P. J., Todd, R. G., Wilson, J. L., Ward, W. C., and Johnson, C. R., Eds., Sequence stratigraphy and cyclostratigraphy of the Mesozoic of the Sierra Madre Oriental, northeast Mexico: a field guidebook: Soc. Econ. Paleont. Mineral., Gulf Coast Sec., 53-54.

James, N. P., and Bourque, P. A., 1992, Reefs and mounds, *in* Walker, R. G., and James, N. P., Eds., Facies models: response to sea level change: Geol. Assn. Can., 323-347.

Mitchum, R. M., Vail, P. R., and Sangree, J. B., 1977, Seismic stratigraphy and global changes in sea level, part 6: stratigraphic interpretation of seismic reflection patterns in depositional sequences, *in* Payton, C. E., Ed., Seismic stratigraphy—applications to hydrocarbon exploration: Am. Assn. Petr. Geol. Mem. **26**, 117-133.

Mullins, H. T., Neumann, A., Wilber, R. J., Hine, A. C., and Chinburg, S. J., 1980, Carbonate sediment drifts in northern Straits of Florida: Am. Assn. Petr. Geol. Bull., **64**, 1701-1717.

Neumann, A. C., Kofoed, J. W., and Keller, G. H., 1977, Lithoherms in the Straits of Florida: Geology, **5**, 4-11.

Sarg, J. T., 1988, Carbonate sequence stratigraphy, *in* Wilgus, C. K., Hastings, B. S., Ross, C. A., Posamentier, H., Van Wagoner, J., Kendall, G. G. St. C., Eds., Sea-level changes: an integrated approach: Soc. Econ. Paleont. Mineral. Spec. Pub. **42**, 155-181.

Sarg, J. T., and Lehmann, P. J., 1986, Lower–middle Guadalupian facies and stratigraphy, San Andros/Grayburg formations, Permian Basin, Guadalupe Mountains, New Mexico, *in* Moore, G. E., and Wilde, G. L., Eds., Field trip guidebook, San Andres/Grayburg formations, Guadalupe Mountains, New Mexico and Texas: Permian Basin Sec., Soc. Econ. Paleont. Mineral., **86-25**, 9-35.

Sonnenfeld, M. D., 1991, High cyclicity within shelf-margin and slope strata of the Upper San Andres Sequence, Last Chance Canyon, *in* Meader-Roberts, S., Candelaria, M. P., and Moore, G. E., Eds., Sequence stratigraphy, facies, and reservoir geometries of the San Andres, Grayburg, and Queen formations, Guadalupe Mountains, New Mexico and Texas: Permian Basin Sec., Soc. Econ. Paleont. Mineral., **91-32**, 11-52.

Chapter 7

Carbonate Sequence Stratigraphy and its Application to Hydrocarbon Exploration and Reservoir Development

W. M. Fitchen
Exxon Production Research Company
Houston, Texas, U.S.A.

Abstract

Carbonate strata differ fundamentally from siliciclastic strata and require their own set of sequence stratigraphic and facies models to aid in hydrocarbon exploration and reservoir development. Significant differences that impact sequence stratigraphic models include the following: (1) the largely autochthonous marine deposition of carbonates in the shallow-water photic zone; (2) secular changes in carbonate biota and mineralogy; (3) higher impedance of carbonates; (4) greater range of platform slope angles in carbonates due to early cementation, organic binding, and coarser grain size; (5) the virtual termination of platform growth and associated rapid freshwater cementation of carbonates during sea level lowstands; and (6) vigorous off-platform shedding of excess carbonate sediment during sea level highstands. Because carbonate deposition is linked to the photic zone, carbonates are regarded as better recorders of accommodation (sea level + subsidence) changes than siliciclastics.

Carbonate platforms are composed of a spatial and temporal hierarchy of stratal units, including parasequences, sequences, composite sequences, and supersequences, which records repetitive and cyclic changes in accommodation and sediment supply. Stratal units are recognized by their position in the stratal hierarchy and by bounding surface type and stacking patterns. Stacking patterns (landward-, vertical-, and seaward-stepping) describe changes in the geographic position and character of stratal units with respect to thickness, stratal geometry, facies types and distributions, and bounding surface type. The character of stratal units and hierarchies is a function of platform geometry, subsidence rate, amplitude and frequency of eustatic sea level fluctuations, siliciclastic supply, and environmental factors such as marine energy and climate. Furthermore, the character of smaller scale stratal units is a function of position within the larger scale stratal units. The recognition and analysis of stratal hierarchies is limited by the type, quality, location, and resolution of available data. Where possible, the analysis of stacking patterns within a stratal hierarchy is the key to improved correlation, better prediction of reservoir, source and seal facies distribution, geometry and continuity, and prediction of unconformities and related diagenesis that impact reservoir quality.

Introduction

Recent summary papers on sequence stratigraphy have provided a general framework for understanding the range of carbonate platform geometries and facies within a sequence stratigraphic and climatic context (Sarg, 1988; Handford and Loucks, 1993; Read, 1995). This paper focuses more on specific concepts and techniques of carbonate sequence stratigraphy and their application to hydrocarbon exploration and development. The reader is provided with a methodology and set of tools to construct sequence stratigraphic frameworks that will enhance prediction of lithofacies and porosity distributions. Fundamental lithostratigraphic and chronostratigraphic units used in carbonate sequence stratigraphy are defined. An emphasis is placed on describing the distribution of lithostratigraphic units within a nested hierarchy of chronostratigraphically significant stratal units. Next the basic sedimentologic and stratigraphic processes that form carbonate rocks, including the response of carbonate platforms to relative changes in sea level, are discussed. Third, the methods and tools of high-resolution sequence stratigraphic analysis are presented, along with their application to data sets of different types (seismic, well log, and outcrop). Finally, three case studies of carbonate sequence stratigraphic analyses in hydrocarbon exploration and development are presented that highlight the integration of different types of data in the construction of predictive sequence stratigraphic frameworks in carbonate settings.

Sequence stratigraphy is the study of genetically related sedimentary facies within a framework of chronostratigraphically significant surfaces (Van Wagoner et al., 1990). Sequence stratigraphy provides a conceptual basis and methodology for placing sedimentary strata (lithostratigraphic units) into a chronostratigraphic framework by integrating available seismic, well log, core, outcrop, biostratigraphic, petrographic, and geochemical data. The fundamental lithostratigraphic units used here are facies, facies tracts, and depositional systems. Lithostratigraphic units are packaged within chronostratigraphic units, which are bound by hiatal stratal surfaces separating older strata from younger strata (Figure 1). The fundamental chronostratigraphic units of sequence stratigraphy (as defined here) are parasequences, sequences, and composite sequences. Sequences and composite sequences are the stratal record of a cycle (rise and fall) of relative change in sea level. Parasequences are the record of relative changes in sea level (rise, or rise and fall) or changes in sediment supply. Climate can play an important indirect role in the development of stratal units, either by influencing sediment supply or global ice volumes (glacio-eustasy). Sequences and composite sequences are bound by subaerial unconformities that developed across subtidal platform lithofacies and by correlative conformities that developed in coeval marine settings. Parasequences are bound by marine flooding surfaces.

The fundamental stratal units of sequence stratigraphy comprise a spatial and temporal hierarchy from parasequences, which are relatively thin and short in time duration, to sequences, which are intermediate in thickness and time, to composite sequences, which are relatively thick and of longer duration. Within this hierarchy, parasequences stack into parasequence sets and sequences, which in turn stack into sequence sets and composite sequences. The origin of this stacking pattern is related to *composite* relative changes in sea level and to changes in sediment supply (Goldhammer et al., 1990; Schlager, 1993). Composite relative changes in sea level are the interference composite of sea level cycles of different periods and amplitudes. The recognition of stacking patterns in a hierarchy of stratal units is the key to (1) correlation; (2) better prediction of lithofacies distribution, geometry, and continuity (e.g., reservoir, source, and seal); and (3) better prediction of unconformities and associated diagenetic alteration zones (which can impact reservoir quality). Predictive, hierarchical stratigraphic frameworks are critical to the development of play concepts and prospects, reservoir characterization and development, and risk assessment.

The development of sequence stratigraphic frameworks depends on data resolution and coverage, depositional setting, carbonate platform geometry, and carbonate platform age. Data resolution dictates the scale of stratal units that can be recognized and mapped in a data set (Figure 2). Seismic sequences, delineated on seismic data, are typically composite sequences made up of sequences that are only poorly resolvable on seismic data. The recognition of sequences depends on the presence and correct interpretation of reflection (i.e., stratal) geometries and termination patterns in the data set.

Another factor in seismic data interpretation is the potential misinterpretation of reflection geometries and termination patterns in examples where thin bed effects and interference are important. Well log data provides 1-D resolution of beds down to fractions of a meter in thickness. However, because well logs do not resolve stratal geometries and termination patterns, sequence stratigraphic interpretation of well logs must rely on 1-D stacking pattern analysis of stratal units. Unconformities with significant truncation can sometimes be detected on well logs where the stratigraphic thickness and succession are well known. Well log cross-sections provide for 2-D interpretations of sequence stratigraphy, but require the interpreter to make assumptions about stratal geometries and stratal terminations. Conventional cores complement well logs by providing the needed resolution for lithofacies recognition, calibration to well logs, and delineation of parasequences. Outcrops can resolve seismic-scale stratal geometries and termination patterns and provide resolution of lithofacies and stratal units up to sequence or composite sequence scale. Outcrops can also provide critical data on interwell-scale lithofacies geometry and continuity.

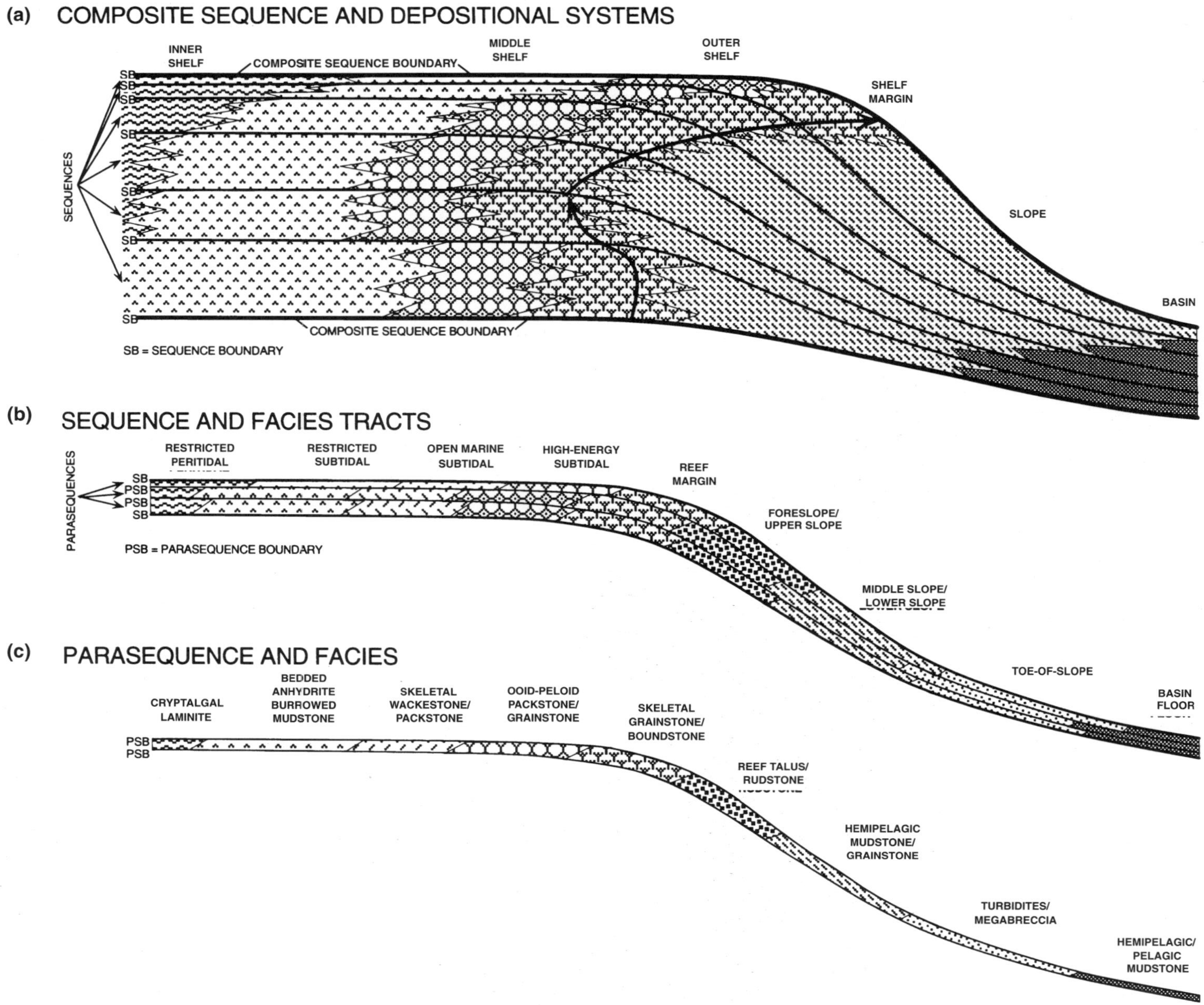

Fig. 1. Hierarchy of relative scales of stratal units and their component lithostratigraphic units. (a) Composite sequence and depositional systems. (b) Sequence and facies tracts. (c) Parasequence and facies. (No absolute scale implied.)

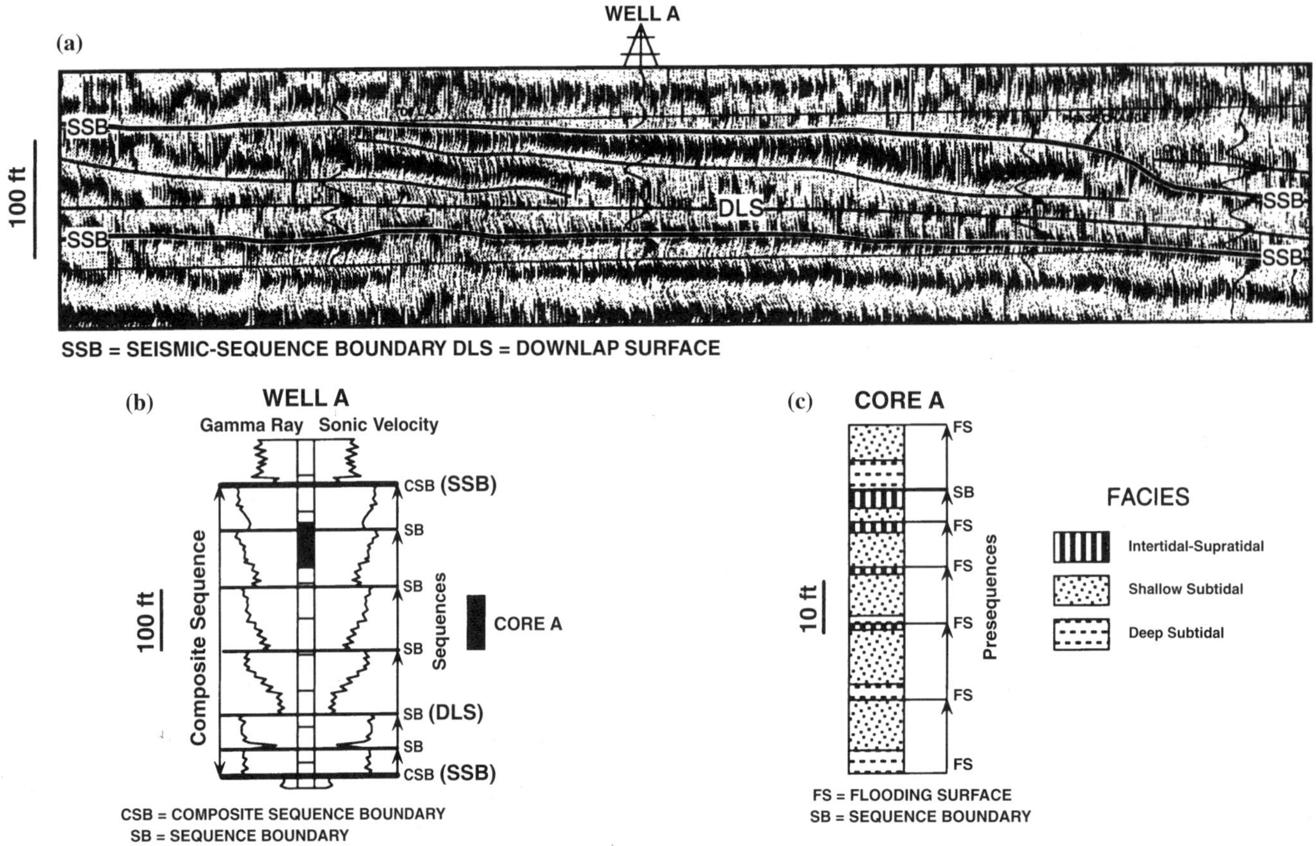

Fig. 2. Resolution of different scales of stratal units as a function of data type. (a) Seismic data resolving composite sequence. (b) Well log data resolving sequences within composite sequence. (c) Conventional core data resolving parasequences within sequence.

Depositional setting and platform geometry are important factors in carbonate sequence stratigraphic analysis. The depositional setting relative to data type and coverage is important because the types of sequence stratigraphic methods or tools used change according to the data set and the depositional setting. For example, seismic sequence resolution and recognition on seismic data are optimal around platform margins and slopes where thickness is greatest and stratal geometries and discordances are best developed. In inner platform settings where all strata are flat-lying, there may be no stratal discordance at sequence boundaries and thus sequence boundaries must be identified by parasequence stacking patterns or by subaerial exposure features seen in core. Sequence recognition on well log cross-sections is often optimal in inner platform settings where lithofacies heterogeneity and well log character are accentuated because the well log signature of platform margin carbonate successions is often monotonous.

Platform geometry encompasses factors such as platform and slope width and angle of inclination, which are reflected by stratal configuration and stratal termination patterns. Very low angles can be difficult to resolve depending on the data type and coverage and on structural deformation. High platform margin angles may be poorly imaged on seismic data or provide a conceptual obstacle to well log correlation. Biostratigraphy can provide critical time lines to constrain platform geometries in the absence of adequate seismic data or when the platform and basinal successions are decoupled by an unconformity with depositional bypass and nondeposition (e.g., Tyrrell, 1969). Cyclic and reciprocal sedimentation concepts (e.g., Wilson, 1967; Meissner, 1972; Dolan, 1989; Sonnenfeld and Cross, 1993), an understanding of depositional topography (e.g., Rich, 1951; Van Siclen, 1958), and stacking pattern analysis (e.g., Kerans et al., 1992) provide additional tools for correlating platform and basin strata in examples with inadequate data quality or coverage.

Carbonate platform age has major implications for carbonate sequence stratigraphy (Wilson, 1975). The evolution of marine ecosystems; secular variations in ocean water chemistry and carbonate mineralogy; first-order changes in plate configurations, sea level, and climate; and the presence or absence of continental glaciation (Vail et al., 1977; Crowell, 1978; Frakes, 1979; Fischer, 1982; Hallam, 1984) have all contributed greatly to the development and differentiation of carbonate platform styles through time. These factors determine the growth potential, subsidence history, and geometry of platforms, as well as the differential development of stratal hierarchies, diagenesis, and siliciclastic influx. For this reason, there is no single sequence stratigraphic model or template that can be used to characterize carbonate platforms of all ages. Rather, models must be specific to narrow ranges of (1) platform age (Wilson, 1975); (2) basin and margin type, configuration, and subsidence rate (e.g., Bally and Snelson, 1980; Read, 1985); (3) aridity (Handford and Loucks, 1993); (4) icehouse, greenhouse, and transitional climatic periods and their implications for amplitude and frequency of glacio-eustatic sea level changes and diagenesis (e.g., Read, 1995); and (5) siliciclastic influx (e.g., Corso et al., 1987; Meyer, 1989).

Stratigraphic Nomenclature and Definitions

This section describes nomenclature and definitions for the basic lithostratigraphic and chronostratigraphic units of carbonate platforms. My intent is to direct attention to the different scales of these units, the ways they can be defined based on the types of data used, and their practical use in exploration and production.

Lithostratigraphic Units

The basic lithostratigraphic units used in carbonate sequence stratigraphy are lithofacies, facies tracts, and depositional systems. These units record paleogeomorphic positions or depositional environments along a depositional profile. The analysis of geographic shifts in the position of lithostratigraphic units across a set of chronostratigraphic units (e.g., progradation of a shorezone lithofacies) provides a measure of changes in accommodation and sediment supply. In general, groups of lithofacies compose facies tracts and groups of facies tracts compose depositional systems. Delineation and mapping of different lithostratigraphic units depend on the scale of inquiry; depositional systems mapping is more applicable to regional stratigraphic analysis, whereas lithofacies mapping is more pertinent to reservoir description. Lithostratigraphic units are also important in terms of their petrophysical properties (reservoir, baffle, and seal) and organic content (source rock).

Lithofacies

Lithofacies are the basic lithostratigraphic building blocks of carbonate platforms. *Lithofacies* are defined as 3-D lithologic units characterized by a set of distinct attributes that may include rock type, rock fabric, grain type, physical and biogenic sedimentary structures, grain size, and sorting (Tucker and Wright, 1990).

The utility of lithofacies in stratigraphic analysis is that they provide an objective way to group observational sedimentologic data into genetic classes and to interpret these classes in terms of environments of deposition. The subdivision of strata into facies is often subjective and may be dictated by the larger objectives of a study. Studies that focus on regional paleogeography may group facies into facies tracts or depositional systems to enhance the bigger picture. Studies that delineate subenvironments within a depositional system, or petrographic studies of diagenesis or paleoecology, may subdivide facies into subfacies or microfacies (e.g., Flugel, 1982). Reservoir characterization studies may define rock fabric facies that are petrophysically significant (e.g., Lucia, 1983).

Facies can be defined on the basis of well logs (*well log facies*) or seismic reflection character (*seismic facies*). Where possible, such facies should be calibrated to conventional cores, sidewall cores, or cuttings.

Facies Tract

A *facies tract* is informally defined as a volume of strata composed of an assemblage of relatively contemporaneous lithofacies that reflect similar depositional processes, paleowater depths, and paleogeomorphic settings. The delineation of facies tracts within platforms of different ages and basin types and within different data sets is a matter of practicality. Facies tracts should be definable and mappable, have some paleogeographic significance, and contain facies representative of key play elements (reservoir, source, and seal). Delineation and mapping of facies tracts within chronostratigraphic units, such as parasequence sets or sequences, provides a basis for 2-D stacking pattern analysis and depositional modeling of carbonate platforms. The composition and architecture of facies tracts may exhibit significant secular variations that relate to variations in stratigraphic processes (accommodation, stratal preservation, and sediment supply) and their modulation of sedimentologic processes. These variations may be particularly evident as contrasts between transgressive and highstand systems tracts of sequences or between transgressive and highstand sequence sets of composite sequences.

Depositional Systems

Depositional systems, as defined by Fisher and McGowen (1967) and Brown and Fisher (1977), are 3-D assemblages of lithofacies, genetically linked by active (modern) or inferred (ancient) processes and environments that are the stratigraphic record or analog of deposition within major geomorphic systems (e.g., fluvial, deltaic, tidal flat, shoreface, or reef). Depositional systems are analogous to facies tracts as defined here, but in the past have had the connotation of being of larger scale, that is, incorporating a greater number of facies than would be incorporated into a facies tract.

Chronostratigraphic Units

The chronostratigraphic units used in sequence stratigraphy are parasequences, parasequence sets, systems tracts, sequences, sequence sets, composite sequences, and supersequences (Table 1). Figure 3 illustrates the stacking pattern or nesting of different orders of chronostratigraphically significant units within a hypothetical platform. These units may correspond to a parasequence/sequence/composite sequence hierarchy or to some other hierarchy. Examples of hierarchical stacking patterns in the stratigraphic record are provided by Grotzinger (1986), Goldhammer et al. (1990, 1991, 1993), Goldhammer and Lehmann (1991), Montanez and Osleger (1993), Pomar (1993), Sonnenfeld and Cross (1993), Barnaby and Ward (1995), Kerans (1995), and Kerans and Fitchen (1995), among others. The importance of a stratigraphic hierarchy is not so much the origin or exact duration of the stratal units but rather the recognition that a hierarchy with discrete scales of stratal units exists and that it can be used to predict stratigraphic correlations and facies distributions beyond control points.

Parasequences

In the stratigraphic record, lithofacies are most commonly packaged into repetitive vertical facies successions that record an upward-shallowing of inferred depositional environments. Repetitive vertical facies successions in carbonate strata have been referred to as upward-shallowing sequences (James, 1979), punctuated aggradational cycles (Goodwin and Anderson, 1985), high-frequency (fourth/fifth-order), meter-scale cycles (e.g., Goldhammer et al., 1990), and parasequences (Sarg, 1988). Parasequences are the fundamental chronostratigraphic building blocks of carbonate platforms. A *parasequence* is defined as a relatively conformable succession of genetically related beds or bedsets bound by marine flooding surfaces or their correlative surfaces (Van Wagoner et al., 1990). A *marine flooding surface* is defined as a surface that separates younger from older strata, across which there is evidence of an abrupt increase in water depth (Van Wagoner et al., 1990). This deepening, or landward facies offset, is commonly accompanied by minor submarine erosion and nondeposition, as inferred from the presence of marine cements, truncated, bored or burrowed fabrics (hardgrounds or firmgrounds), and lithoclasts overlying the flooding surface. Parasequence boundaries exhibit no evidence of subaerial exposure of subtidal facies (abnormal or allogenic subaerial exposure) or a basinward shift in facies tracts, although subaerial exposure of supratidal lithofacies (normal or autogenic subaerial exposure) may be present along the boundary. Exceptions to this definition occur at the base and top of sequences, where sequence boundaries exhibiting allogenic subaerial exposure coincide with parasequence boundaries.

No thickness scale or duration is implied or required by the term parasequence, although parasequences documented in the rock record have average durations of ~10–400 k.y. (Algeo and Wilkinson, 1987), which corresponds to the range of fourth- and fifth-order cycles (e.g., Goldhammer et al., 1991). This places their duration within the Milankovitch band, or within the temporal range of frequencies of earth's orbital parameters that are known to effect climate change and glacio-eustatic sea level oscillations (see Einsele et al., 1991; Schwarzacher, 1993; de Boer and Smith, 1994). A parasequence can be created during a relative rise in sea level, a relative rise and fall in sea level, or merely a change in sediment supply.

Figure 4 shows two basic models for parasequence development. In the first model, parasequence formation is driven by a self-regulating autocyclic process of tidal flat progradation in conjunction with uniform subsidence (after Ginsburg, 1971). In the second model, parasequences form in response to relative changes in sea level (e.g., Read et al., 1986). Parasequences in the stratigraphic record exhibit a range of lithofacies types, inferred environments, and lithofacies symmetry with some commonly recurrent types (Figure 5). Peritidal parasequences are capped by intertidal or supratidal lithofacies, whereas subtidal parasequences are entirely subtidal. Asymmetric parasequences are upward-shal-

Table 1. Definitions and characteristics of major chronostratigraphically significant units used in carbonate sequence stratigraphy.

Stratal Unit	Definition	Characteristics of Bounding Surfaces	Temporal Order	Synonymous Terms
Parasequence	A relatively comformable succession of genetically related beds or bedsets bounded by marine-flooding surfaces and their correlative surfaces; may be subtidal or peritidal symmetric or asymmetric	Subaerial exposure features developed on peritidal facies; marine firmground or hardground (burrowed/bored); submarine erosion at wavebase; landward shift in facies across surface.	4th-order (0.1-1.0 m.y. period) & 5th-order (0.01-0.1 m.y.)	Meter-scale, upward-shallowing sequence (James, 1979) or cycle (Goldhammer et al., 1987); upward-coarsening cycle (Aigner, 1985); punctuated aggradational cycle (PAC; Goodwin and Anderson, 1985); base-level transit cycle (Wheeler, 1964)
Parasequence Set	A succession of genetically related parasequences forming a distinctive stacking pattern (landward-, vertically-, or seaward-stepping) and commonly bounded by major marine-flooding surfaces and their correlative surfaces; may be coincident with systems tract (landward-stepping or retrogradational parasequences set = TST, seaward-stepping or progradational parasequence set = HST).	Subaerial exposure features developed on peritidal facies; marine firmground or hardground (burrowed/bored); submarine erosion at wavebase; intraclasts/lithoclasts above boundary; landward shift in facies across boundary. Parasequence set boundaries typically display greater facies offset and/or bounding surface alteration than parasequence boundaries within parasequence set. Separate distinctive parasequence-stacking patterns, may coincide with sequence boundaries or downlap surfaces.	4th-order (0.1-1.0 m.y.) & 5th-order (0.01-0.1 m.y.); may be 3rd-order (1-10 m.y.)	Cycle set (e.g., Kerans, 1995), bedding rhythm (Fischer, 1986), megacycle (Goldhammer et al., 1990); may be equivalent to systems tract in some cases
Systems Tract	A linkage of contemporaneous depositional systems (Brown and Fisher, 1977), defined objectively on the basis of types of bounding surfaces, their position within a sequence, and on parasequence and parasequence set stacking patterns; depositional systems are three-dimensional assemblages of lithofacies, genetically linked by active (modern) or inferred (ancient) processes and environments (Fisher and McGowen, 1967)	Highstand systems tract bounded below by maximum-flooding surface (downlap surface, and condensed interval in many cases) and bounded above by sequence boundary (subaerial exposure features); transgressive systems tract bounded below by sequence boundary and above by transgressive surface, which is restricted to basin in most carbonate settings	4th-order (0.1-1.0 m.y.) & 5th-order (0.01-0.1 m.y.); may be 3rd-order (1-10 m.y.)	Parasequence set, cycle set
Sequence, High Frequency Sequence	A relatively conformable succession of genetically related strata (parasequences and parasequence sets) bounded by unconformities and their correlative conformities	Subaerial exposure features developed on subtidal facies, possible submarine erosion of slope due to lowering of wave base and mass-wasting	3rd-order (1-10 m.y.) & 4th-order (0.1-1.0 m.y.); 5th-order (0.01-0.1 m.y. in areas of high subsidence/sediment supply. 4th-order more common during periods of glacial maxima and high-amplitude glacio-eustasy	Synthem (North American Stratigraphic Code); transgressive-regressive cycle (Johnson et al., 1985); genetic stratigraphic sequence (Galloway, 1989); depositional episode (Frazier, 1974)
Sequence Set, High Frequency Sequence Set	A succession of genetically related sequences arranged in a distinctive progradational (highstand and lowstand sequence sets), aggradational (highstand or transgressive sequence set) or retrogradational (transgressive sequence set) stacking pattern	Highstand sequence set bounded below by 3rd-order maximum-flooding surface (regional downlap surface and condensed interval in many (cases) and bounded above by composite sequence boundary (subaerial exposure features); transgressive sequence set bounded below by composite sequence boundary and above by maximum flooding surface; lowstand sequence set is bounded below by composite sequence boundary and above by 3rd-order transgressive surface, which is restricted to basin	3rd-order (1-10 m.y.) & 4th-order (0.1-1.0 m.y.)	Seismically-defined systems tracts in most cases (particularly 2D-seismic), except in basins with very high subsidence and sedimentation rates (e.g., Northern Gulf of Mexico Basin)
Composite Sequence	A succession of genetically related sequences in which the individual sequences stack into lowstand transgressive and highstand sequence sets, although not all sequence sets may be well-developed	Widespread and well-developed subaerial exposure features developed on subtidal facies; basinwide subaerial unconformities with stratal truncation, karst and/or caliche development, associated regional aquifer and meteoric phreatic diagenesis (dissolution, cementation)	2nd-order (10-100 m.y.) & 3rd-order (1-10 m.y.)	Seismically-defined depositional sequences in most cases (particularly 2D-seismic), except in basins with very high subsidence and sedimentation rates (e.g., Northern Gulf of Mexico Basin); may be equivalent to supersequence
Supersequence	A group of sequences or composite sequences that successively reach higher positions of encroachment onto the underlying unconformity surface, followed by one or more sequences with lower positions of encroachment (Mitchum, 1977)	Major interregional unconformities	2nd-order (10-100 m.y.)	Composite Sequence

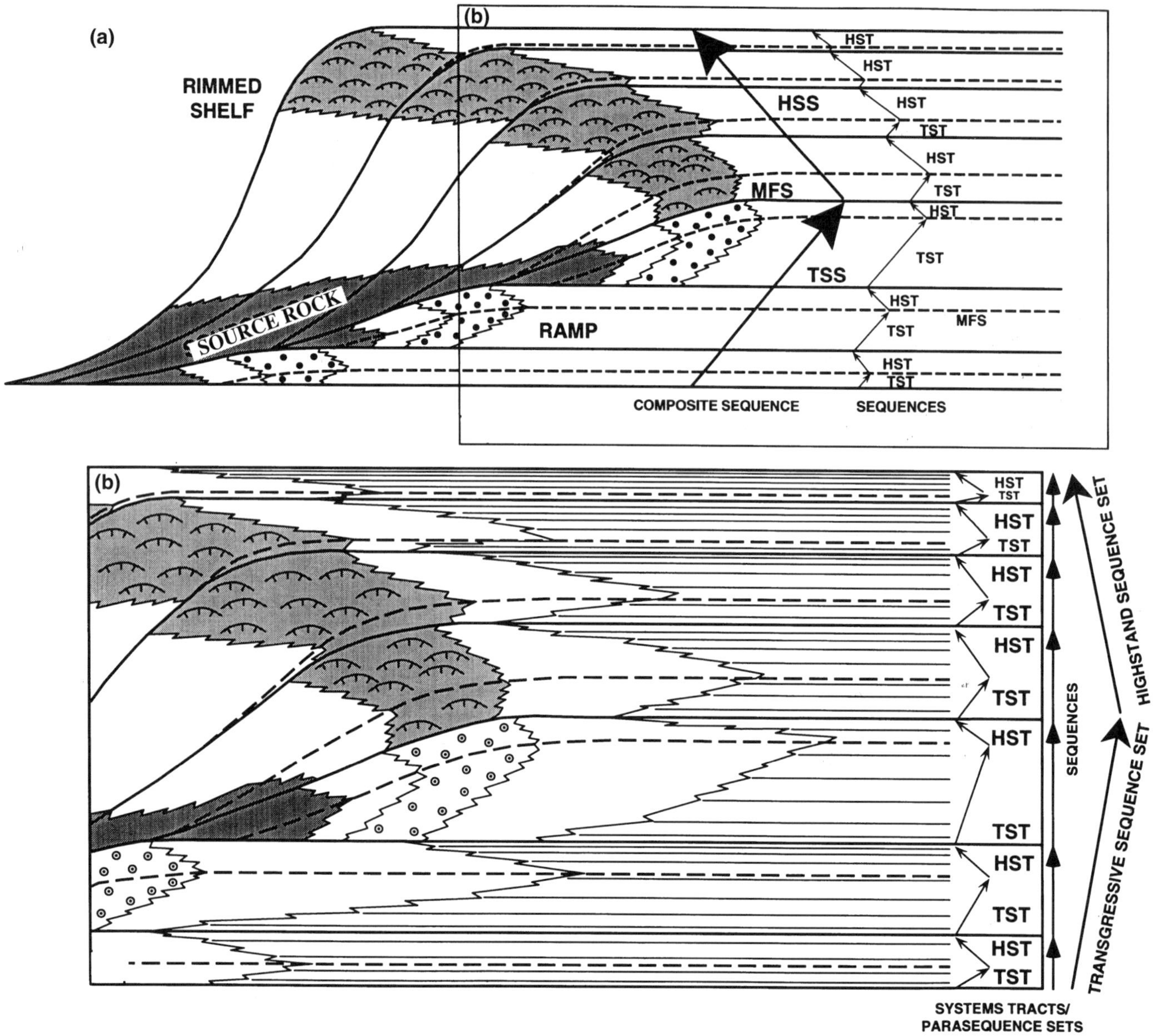

Fig. 3. Hierarchy of stratal units and stacking patterns in a hypothetical carbonate platform. (a) Composite sequence showing stacking pattern of component sequences. Key: TSS = transgressive sequence set, HSS = highstand sequence set, TST = transgressive systems tract, HST = highstand systems tract, MFS = maximum-flooding surface. (b) Detail showing stacking pattern of parasequences within sequences within composite sequence. Note stacking pattern of stratal units with respect to thickness, evolution of platform topography, and change in platform margin facies through large-scale stratal unit. (Courtesy of R. K. Goldhammer, personal communication, 1995.)

lowing lithofacies successions, whereas symmetric parasequences are upward-deepening to upward-shallowing successions. Symmetry generally increases in the direction of higher accommodation (generally basinward), and flooding surfaces may become more difficult to identify in such settings than in low-accommodation settings with asymmetric successions.

Parasequence Sets and Systems Tracts

Parasequences are packaged within parasequence sets and systems tracts. These units are defined on the basis of the stacking pattern of their component parasequences, their position within a sequence, and their bounding surfaces. A *parasequence set* is a succession

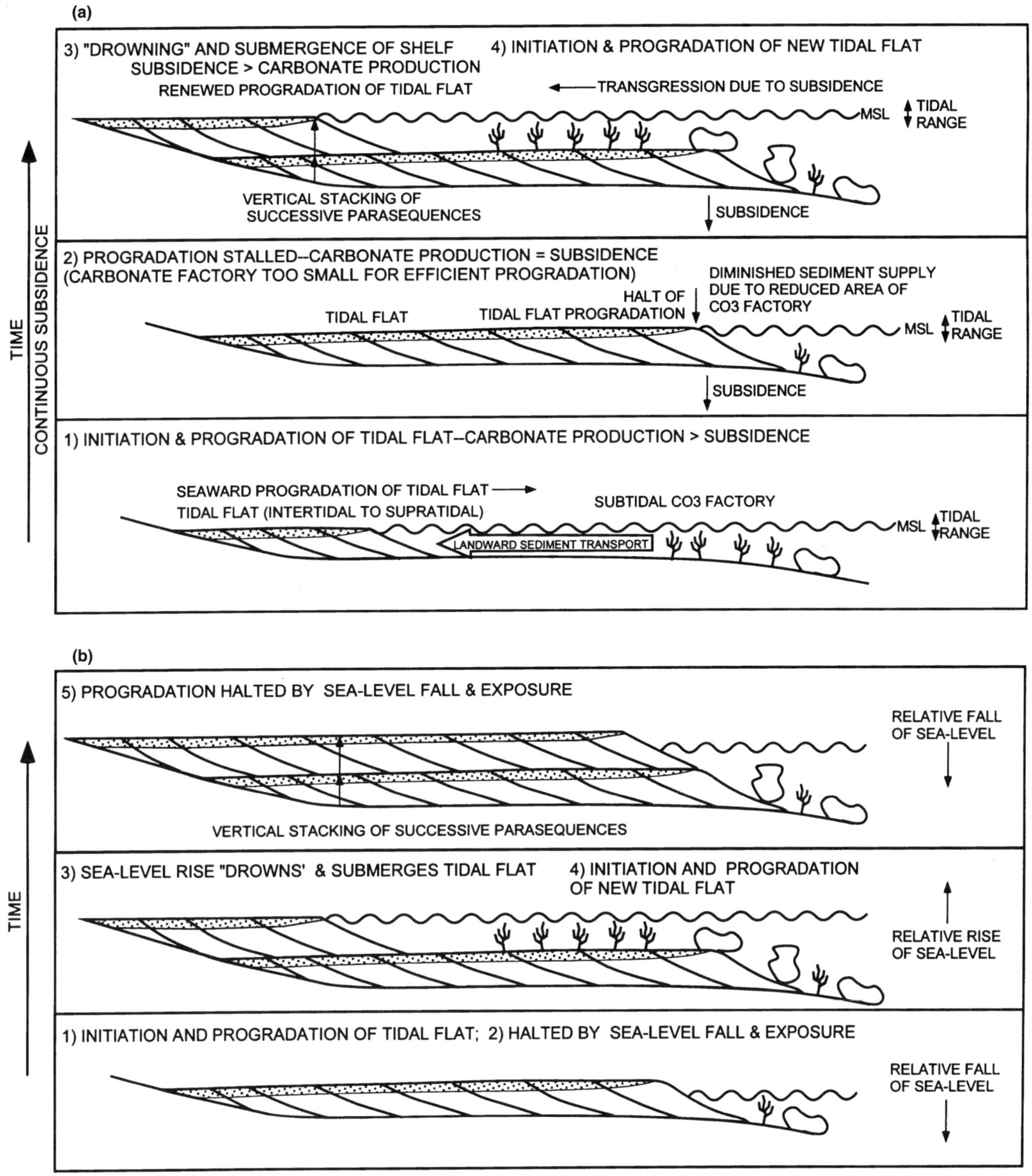

Fig. 4. Models of parasequence development. (a) Autocyclic mechanism. (b) Allocyclic mechanism.

of genetically related parasequences forming a landward, vertical or seaward stepping pattern, often bound by major marine-flooding surfaces (e.g., Van Wagoner et al., 1990). A *systems tract* is a linkage of contemporaneous depositional systems (Brown and Fisher, 1977) bound by relatively isochronous surfaces that separate older from younger strata.

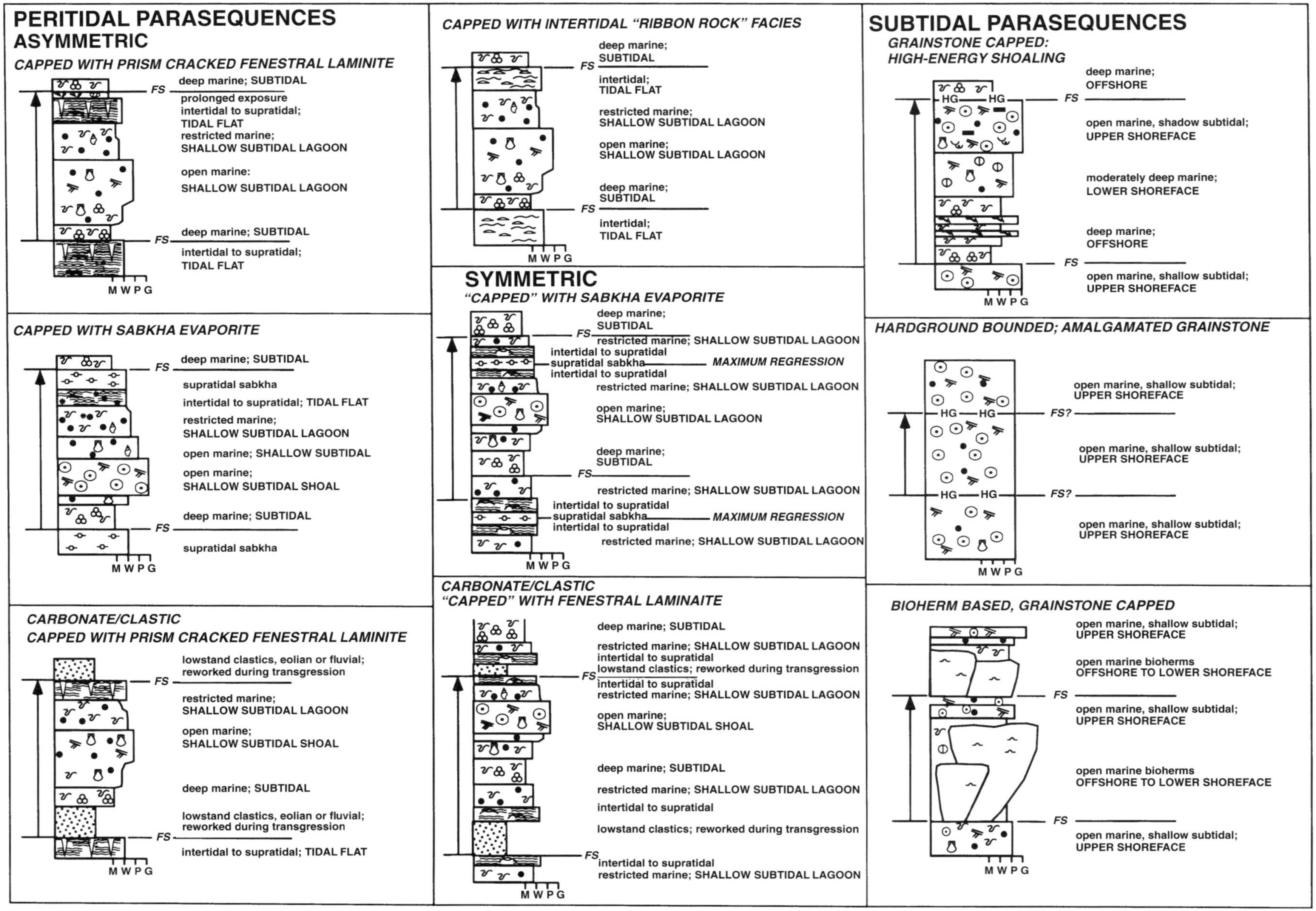

Fig. 5. Range of parasequence types common to platforms of various ages and settings, illustrating the spectrum of vertical facies successions and inferred environments. Basic end-member types are peritidal versus subtidal and asymmetric versus symmetric.

Sequences

Parasequence sets and systems tracts stack into sequences. *Sequences* are defined as relatively conformable successions of genetically related strata bound by unconformities and their correlative conformities. Sequences are defined by their bounding subaerial unconformities, which are necessarily products of a relative fall in sea level below the shelf top (Mitchum et al., 1977; Van Wagoner et al., 1990; Van Wagoner, 1995). However, subaerial exposure is a common *autogenic* process in peritidal settings. Supratidal deposits form by deposition of entrained sediments during storm surges and are exposed subaerially between storm events. Thus, the definition of sequence-bounding unconformities in carbonates should be restricted to cases of abnormal or demonstrably *allogenic* subaerial exposure, where subtidal carbonate facies are exposed. It is sometimes difficult to detect subaerial exposure surfaces in carbonates, thus parasequence stacking patterns should also be used to predict sequence boundaries.

Sequences stack into lowstand, transgressive, and highstand sequence sets, which are essentially the systems tracts of composite sequences. Composite sequences are bound by unconformities with widespread, well-developed subaerial exposure features developed on subtidal facies. Stratal truncation, karst, or caliche development and the development of regional meteoric aquifers and associated diagenesis (dissolution and cementation) are often associated with such boundaries. The maximum flooding surfaces and condensed sections of composite sequences are often widespread, rich in organics and fauna, and serve as regional biozone and mapping markers (Mitchum and Van Wagoner, 1991).

Composite sequence boundaries and maximum flooding surfaces represent important elements of regional hydrocarbon stratigraphic play analysis. Sequence boundaries can mark potential unconformity-related porosity development and stratigraphic traps (e.g., Schlee, 1984; Budd et al., 1995), while maximum flooding surfaces may bound intervals of potential source rock or seal facies (Loutit et al., 1988). Composite sequence boundaries and maximum flooding surfaces may also mark times of major paleogeographic reorganization and sediment supply variations within a basin (e.g., Galloway, 1989; Schlager, 1989).

Supersequences

Supersequences are groups of sequences or composite sequences that successively reach higher positions of encroachment onto the underlying unconformity surface, followed by one or more composite sequences with lower positions of encroachment (Mitchum, 1977). Supersequences are typically of second-order (10–100 m.y.) duration and are bound by major interregional unconformities. They are often the demonstrable record of pulses of tectonic subsidence, such as episodes of thrust sheet emplacement in foreland basins or rift, drift, and thermal cooling in passive margin basins (Vail et al., 1991). Alternatively, supersequences can be the record of long-term eustatic sea level changes driven by ridge spreading rates and ocean basin volume.

Sedimentologic and Stratigraphic Processes in Carbonate Systems

Sedimentology of Carbonate Platforms

Carbonate sediments are autochthonous, forming largely by organically mediated precipitation of calcium carbonate in marine environments. Once formed, carbonate grains and mud can be reworked and redistributed onshore, offshore, or along the strike of a carbonate platform. Many factors affect carbonate production, including light penetration, salinity, oxygenation and circulation, waves, tidal and storm currents, substrate type, and syndepositional cementation (Wilson, 1975; Chapter 2, this volume). All of these factors influence the growth, destruction, and redistribution of marine organisms and their precipitates. Carbonate production rates are a function of light intensity, which decreases exponentially with water depth; below 10–20 m, light intensity becomes progressively more growth limiting (Schlager, 1992) (Figure 6).

The lower limit of marine carbonate accumulation is controlled by the carbonate compensation depth (CCD), at ~4–5 km water depth, while the upper limit of carbonate sedimentation is the supratidal zone, tens of centimeters above sea level, which receives sediment during storm surges. The optimal zone of carbonate production is from sea level down to 10–20 m depth—the subtidal carbonate "factory." The sensitivity of carbonate-secreting organisms and carbonate production rates makes carbonate platforms sensitive recorders of climate and relative changes in sea level (Kendall and Schlager, 1981). Different carbonate sediment types are generally related to different marine environments of deposition. Because carbonate depositional environments are basically a product of water depth, they tend to be broadly parallel to strike, in contrast to point-sourced, dip-parallel siliciclastic depositional systems, such as fluvial-deltaic and submarine fan systems.

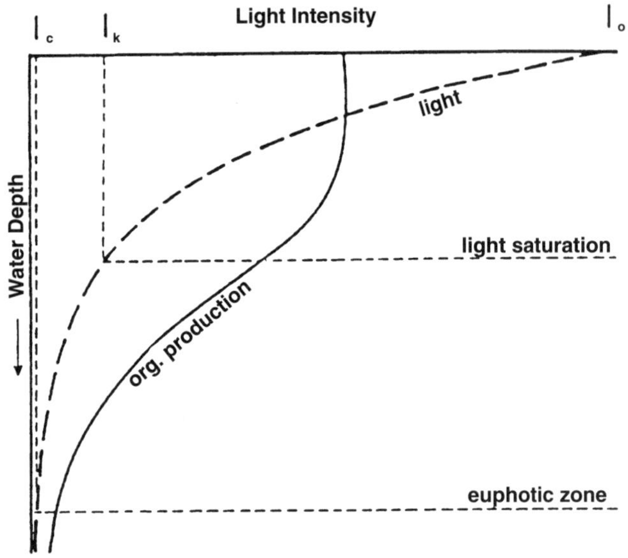

Fig. 6. Relationship between light intensity and organic production rates with increasing water depth. Key: Ic = light intensity at lower limit of euphotic zone, Ik = light intensity at lower limit of light saturation, and Io = light intensity at water surface. Depth at which organic production drops markedly (Ik) is 10–20 m, and the depth of the euphotic zone (Ic) is ~80 m. (From Schlager, 1992.)

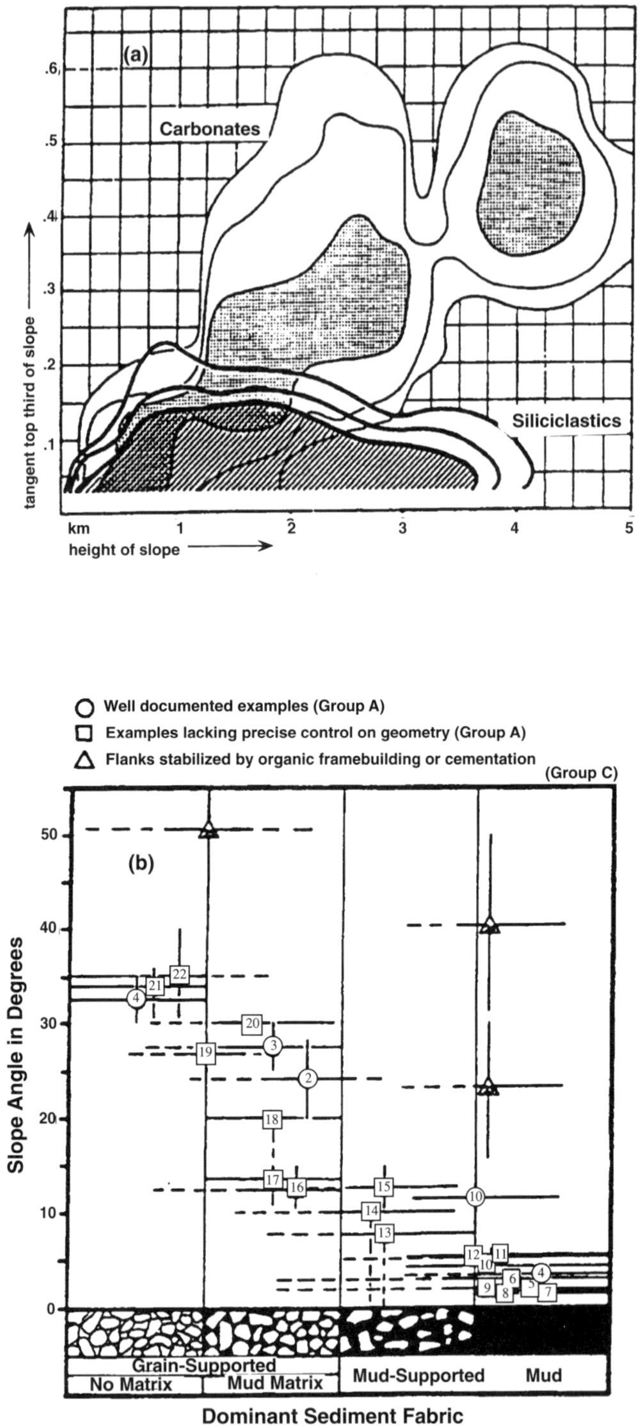

Fig. 7. (a) Slope angle versus platform height in carbonates versus siliciclastics. (From Schlager and Camber, 1986.) (b) Sediment fabric versus slope angle in carbonates. (From Kenter, 1990.)

In contrast to siliciclastic sediments, carbonate sediments have the potential to form rigid, wave-resistant edifices with significant depositional topography and steep slopes. This is due both to the construction potential of carbonate organisms and to early cementation. Carbonate sediments when formed contain both stable (low-Mg calcite) and metastable (high-Mg calcite and aragonite) carbonate minerals. Upon contact with meteoric water, metastable carbonate minerals either dissolve or are stabilized to low-Mg calcite. Dissolved carbonate is reprecipitated as cement, leading to lithification. In addition, carbonate sediments tend to lithify rapidly by cementation in the marine phreatic environment. Early lithification reduces the potential for carbonate sediments to be eroded during lowstands of sea level. Carbonate organisms form reef structures through organic binding. Organic binding and early cementation processes permit carbonate to build and maintain depositional topography, especially wave-resistant reef barriers with steep marginal slopes beyond the angle of repose. These processes, and the much greater range of grain size in carbonates, contribute to the greater range of slope angles seen in carbonates than in siliciclastics (Figure 7).

Carbonate platforms form in a variety of plate tectonic settings and basin types and exhibit a wide range of accumulation rates, platform morphologies, platform widths, terrigenous clastic content, and depositional systems (Wilson, 1975; Read, 1985; Crevello et al., 1989; Tucker et al., 1990; Loucks and Sarg, 1993). Carbonate

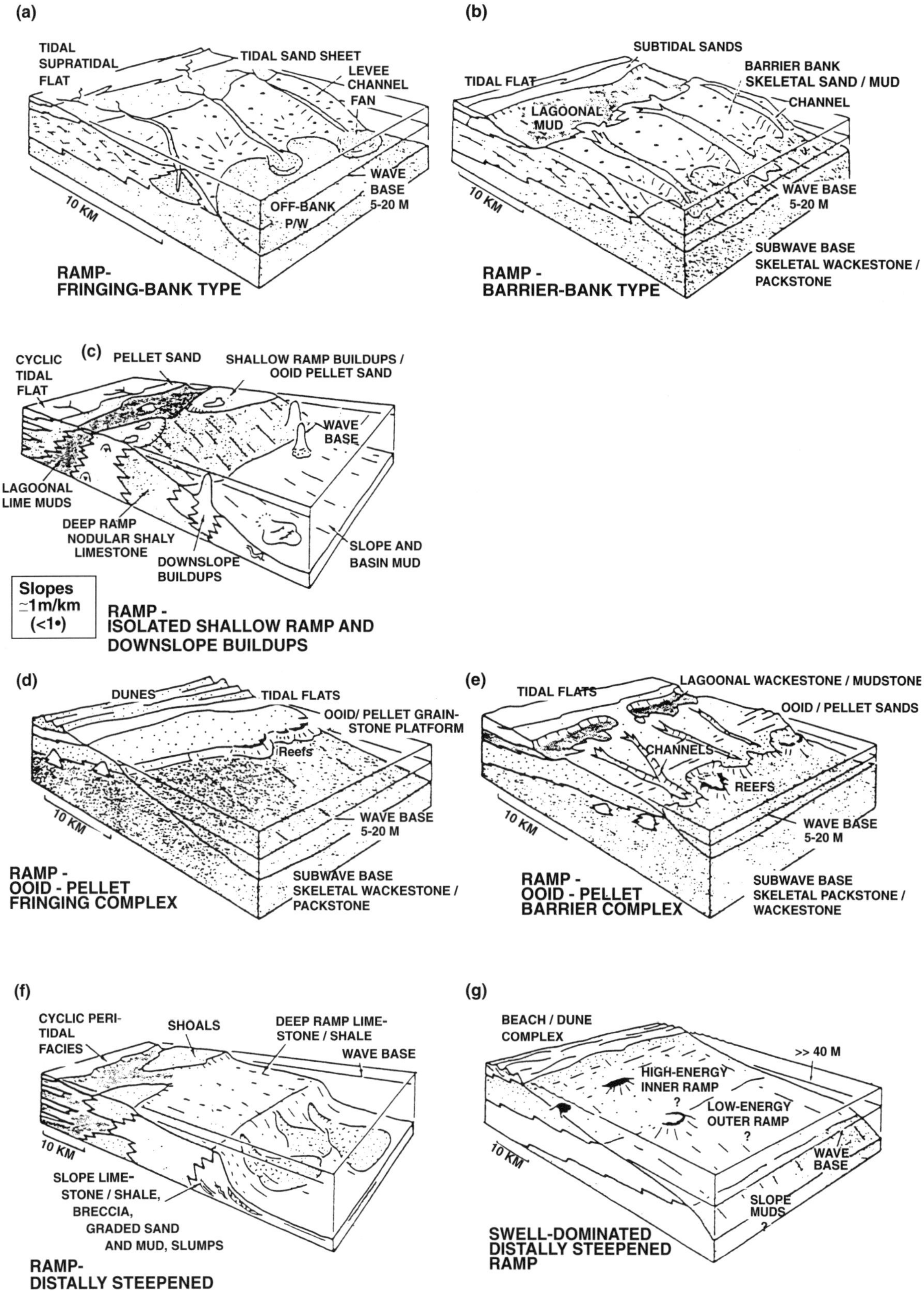

Fig. 8. Static depositional models of carbonate ramps. (a) ramp with fringing bank, (b) ramp with barrier bank, (c) ramp with isolated shallow ramp buildups and downslope buildups, (d) ramp with ooid-pellet fringing complex, (e) ramp with ooid-pellet barrier complex, (f) distally steepened ramp, and (g) swell-dominated distally steepened ramp. (From Read, 1985.)

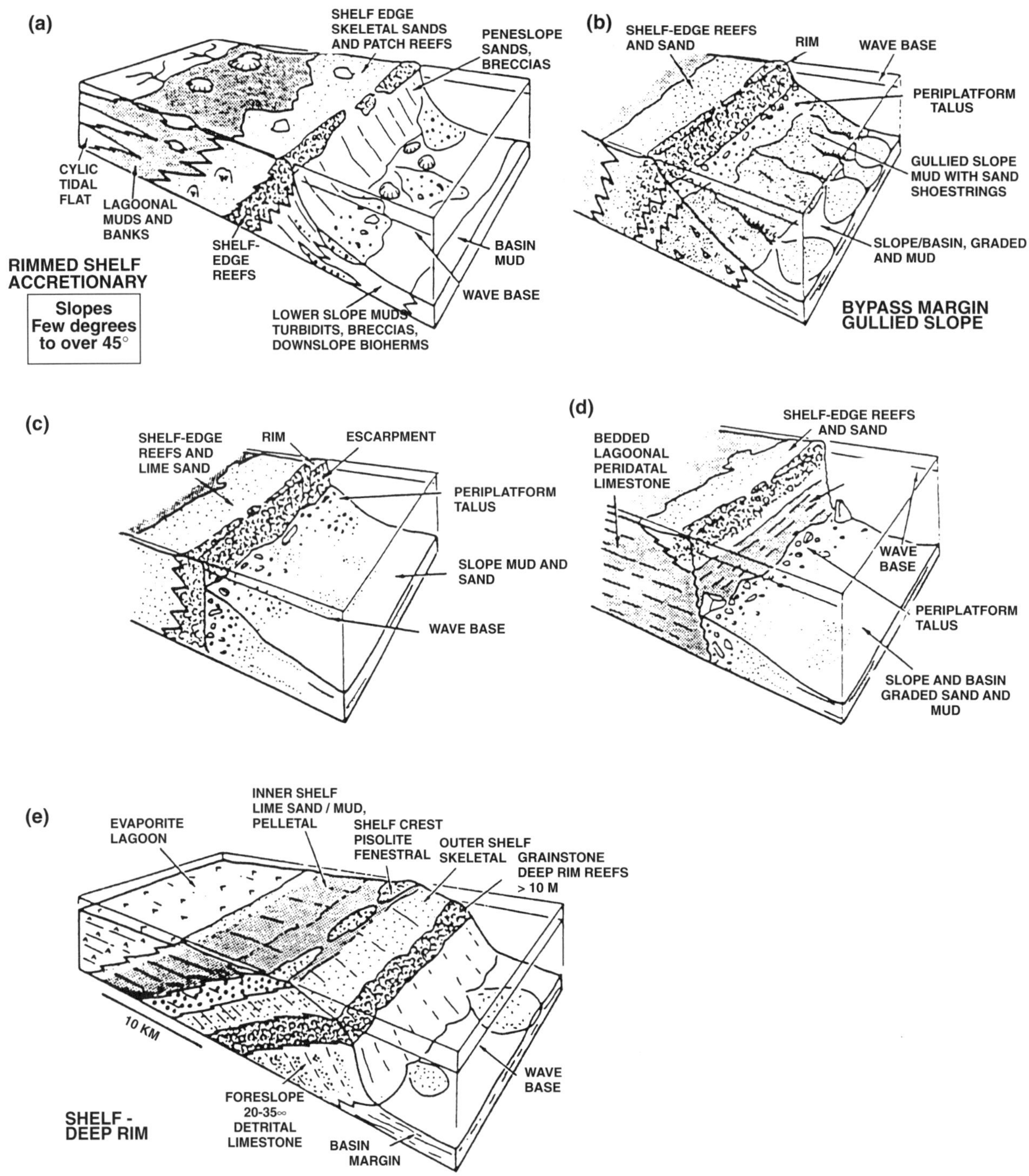

Fig. 9. Static depositional models of rimmed shelves. (a) accretionary rimmed shelf, (b) rimmed shelf with gullied bypass slope, (c) rimmed shelf with escarpment type bypass margin and onlapping toe-of-slope facies, (d) rimmed shelf with erosional margin and onlapping toe-of-slope facies, and (e) rimmed shelf with deep reefal rim and accretionary slope. (From Read, 1985.)

platforms can be differentiated into two major types—ramps and rimmed shelves—each with several subtypes.

Ramps (Figure 8) are broad platforms characterized by low-angle (typically <1°) depositional dips from the shallow water, high-energy zone to the deeper basin, thus they lack a well-defined break in slope that coincides with a high-energy shelf margin. Ramps can be differentiated into those with high-energy facies dominated by skeletal or ooid-pellet sediments and those in which the high-energy zone is either fringing (shore attached) or forms an offshore barrier complex with a back-barrier restricted lagoon. Ramps sometimes have isolated shallow-water buildups and conical downslope buildups. Some ramps are distally steepened, which may reflect basin margin structure or development above an antecedent drowned shelf. Some distally steepened ramps are dominated by high-energy oceanic conditions that result from the absence of a shelf margin barrier. These ramps are characterized by widespread, high-energy shelf and shorezone facies. Ramps have been common during geologic periods lacking in abundant reef-building faunas, and they are an expression of homoclinal antecedent topography and uniform subsidence patterns.

Rimmed shelves (Figure 9) are platforms characterized by a well-defined break in slope that coincides with the zone of highest energy facies on the platform. Shelf margins are relatively continuous along the strike of the platform. Slope angles seaward of the shelf margin break can be from several degrees to over 60∞° and usually decrease exponentially basinward (slopes are concave upward). Rimmed shelves can be differentiated into those with (1) accretionary margins with organic buildups forming either a shallow or slightly deeper rim, (2) bypass margins with either gullied or escarpment-type slopes, and (3) erosional margins. Erosional and escarpment-type bypass margins have associated onlapping, toe-of-slope aprons of allochthonous sediments, whereas accretionary and gullied-slope bypass margins are coupled to the toe-of-slope sediment apron. Isolated platforms are a common type of rimmed shelf with no attachment to a craton. They are relatively free of siliciclastics, typically contain a central low-energy lagoon, and often show marked differentiation of windward and leeward margin geometry and facies. Rimmed shelf development reflects the potential for shelf margin aggradation and is characteristic of geologic periods with luxuriant reef-building faunas and floras. Rimmed shelf development is a common response to relative rise in sea level, but it may also reflect antecedent topography (structural discontinuities and basin margin hinge zones) or differential subsidence patterns.

Figure 10 illustrates the dominant plate tectonic settings of carbonate platforms. On passive margins, carbonate platforms form thick successions that typically evolve from low-relief ramps to rimmed shelves; isolated platforms commonly nucleate above the margins of rotated extensional fault blocks. The demise of many Mesozoic passive margin platforms was caused by drowning and burial by siliciclastic sediments due to environmental degradation and terrigenous influx (Schlager, 1989), while many Paleozoic passive margin platforms were terminated by plate collision and siliciclastic burial. In convergent plate settings, carbonate platforms rim (1) accretionary prisms and oceanic plateaus on oceanic crust and (2) foreland margins on continental or transitional crust. Platform demise in convergent settings is due to plate collision, margin reversal, and burial by foreland siliciclastics. A third major setting of carbonate platforms is cratonic basins, including intracratonic sags, distal foreland basins, and pull-apart basins.

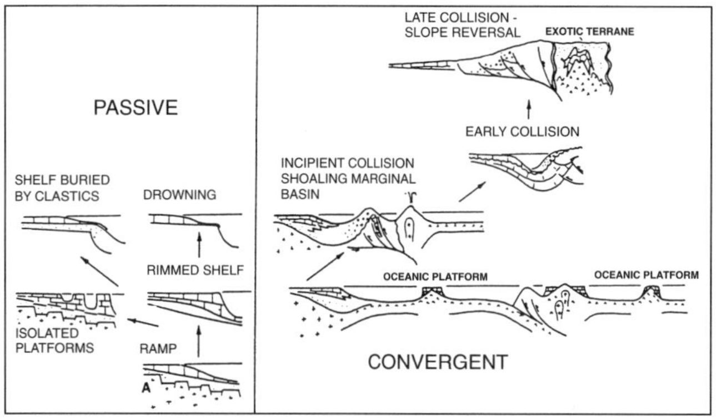

Fig. 10. Plate tectonic setting of carbonate platforms on both passive and convergent margins. (From Read, 1985.)

Controls on Carbonate Stratigraphy: Carbonate Platform Response to Sea Level Changes

Parasequences and sequences are stratal units that are the product of eustasy, subsidence, sediment supply, and climate. Eustasy and subsidence together control relative changes in sea level, or accommodation space, in the absence of sediment input. Sediment supply controls the rate at which accommodation space is filled, the rate at which underlying sediments compact, and the rate of isostatic adjustment due to sediment loading. Thus, there is a critical feedback between sediment supply and accommodation. Eustasy provides feedback by the process of water loading. As outlined earlier, sediment supply in carbonate systems is autochthonous and dependent on water depth and is thus fundamentally different from sediment supply in siliciclastic systems. The water depth-dependent nature of carbonate sedimentation makes carbonates particularly responsive to relative changes in sea level. This section discusses the basic response of carbonate platforms to relative changes in sea level and climatic changes.

Carbonate depositional systems respond to relative changes in sea level differently than do siliciclastic systems (Kendall and Schlager, 1981; Handford and Loucks, 1993). Relative changes in sea level in carbonate systems are inferred from stratal geometries (particularly on seismic data), parasequence stacking patterns, and biofacies. The response of carbonate platforms to relative changes in sea level is described with respect to the migration of facies or facies tracts among parasequences or sequences and with respect to changes in stratal geometries and termination patterns. Facies migrations in most platform settings are incremental rather than continuous and gradual, with facies shifts or offsets occurring across parasequence, parasequence set, and sequence-bounding surfaces.

Relative Rise in Sea Level

Carbonate systems generally have sufficient growth potential to keep pace with the most rapid relative rises in sea level (Schlager, 1981), including those induced by glacial melting (glacio-eustasy). However, this high growth potential may be specific to shallow-water reef-building faunas and associated facies, and thus other depositional environments may not keep pace with relative rises in sea level. During relative rises in sea level across an exposed platform top, platforms can respond in several basic ways (Figure 11). Initially during a rise, the sedimentation rate lags behind until sufficient subtidal depths are attained to activate the carbonate factory. Following this, sedimentation at the margins of platforms may either exceed, match, or lag behind rates of relative rise. If sediment accumulation at the margin matches and then exceeds the rise rate, then the margin initially aggrades and then progrades. This stratal response is generally reflective of slow rise rates and may be difficult to differentiate from the relative sea level highstand response.

If the rate of rise matches sediment accumulation, then the margin aggrades. In some cases, the margin may aggrade at a faster rate than the platform interior, forming a raised rim or bucket geometry (Schlager, 1992). This geometry may be progressively accentuated during successive cycles of relative rise of sea level or be smoothed out by inner platform or lagoon aggradation and platformward progradation of margin sediments during a relative stillstand or slower rise in sea level.

If the rate of sedimentation lags behind the rate of rise, the platform can be submerged below the photic zone and drowned. This process is common in isolated platforms that have no updip bathymetrically shallower areas. Many ancient isolated platforms exhibit a stepwise retreat in platform margins and corresponding decrease in the area of platform sedimentation during long-term relative rises in sea level (Rudolph and Lehmann, 1989; Wendte et al., 1992; Waite, 1993). This process reflects backstepping during short-term relative rises (with minor progradation during short-term relative falls), which is modulated by a longer term relative rise or by progressive environmental degradation (Schlager, 1989).

In shore-attached carbonate platforms, in cases where margin growth potential lags behind the rate of rise, the platform margin steps landward relative to the antecedent platform margin and reestablishes at a bathymetrically shallower position. The distance that the margin backsteps is a function of (1) the magnitude and duration of the relative rise, (2) shelf topography, and (3) environmental conditions. On mixed carbonate-siliciclastic platforms with updip fluvial-deltaic systems, there is a tendency toward greater landward encroachment of carbonates and landward migration of siliciclastic systems during relative rises in sea level, although siliciclastic supply may overwhelm carbonate production independent of relative changes of sea level.

Relative Highstand and Initial Fall in Sea Level

During relative highstand and early fall in sea level (sea level above the platform top), the growth potential of carbonate platforms typically exceeds rates of relative rise in sea level. Carbonate parasequences respond by aggradation of the platform to fill the available accommodation space and concurrent progradation of the platform margin (Figure 12). Progradation in carbonate systems is constrained by the platform margin slope angle, the water depth at which authochthonous sediment production decreases rapidly (particularly important for Paleozoic platforms with no pelagic faunal sediment input), the platform to basin relief, and the volume of off-platform sediment transport (windward versus leeward margin). Slope angles that exceed the angle of initial yield for unconsolidated sediments, or that are too steep to allow for accumulation of talus from reefs or platform margins, result in bypassing of sediments to the toe-of-slope sediment apron (bypass and erosional rimmed margins). In this case, the slope and shelf margin do not prograde or only do so at slow rates. The water depth at which sediment production drops off rapidly dictates the magnitude of platform to basin relief at which the platform margin can prograde readily. This relief dictates the volume of slope sediments necessary to prograde the platform margin.

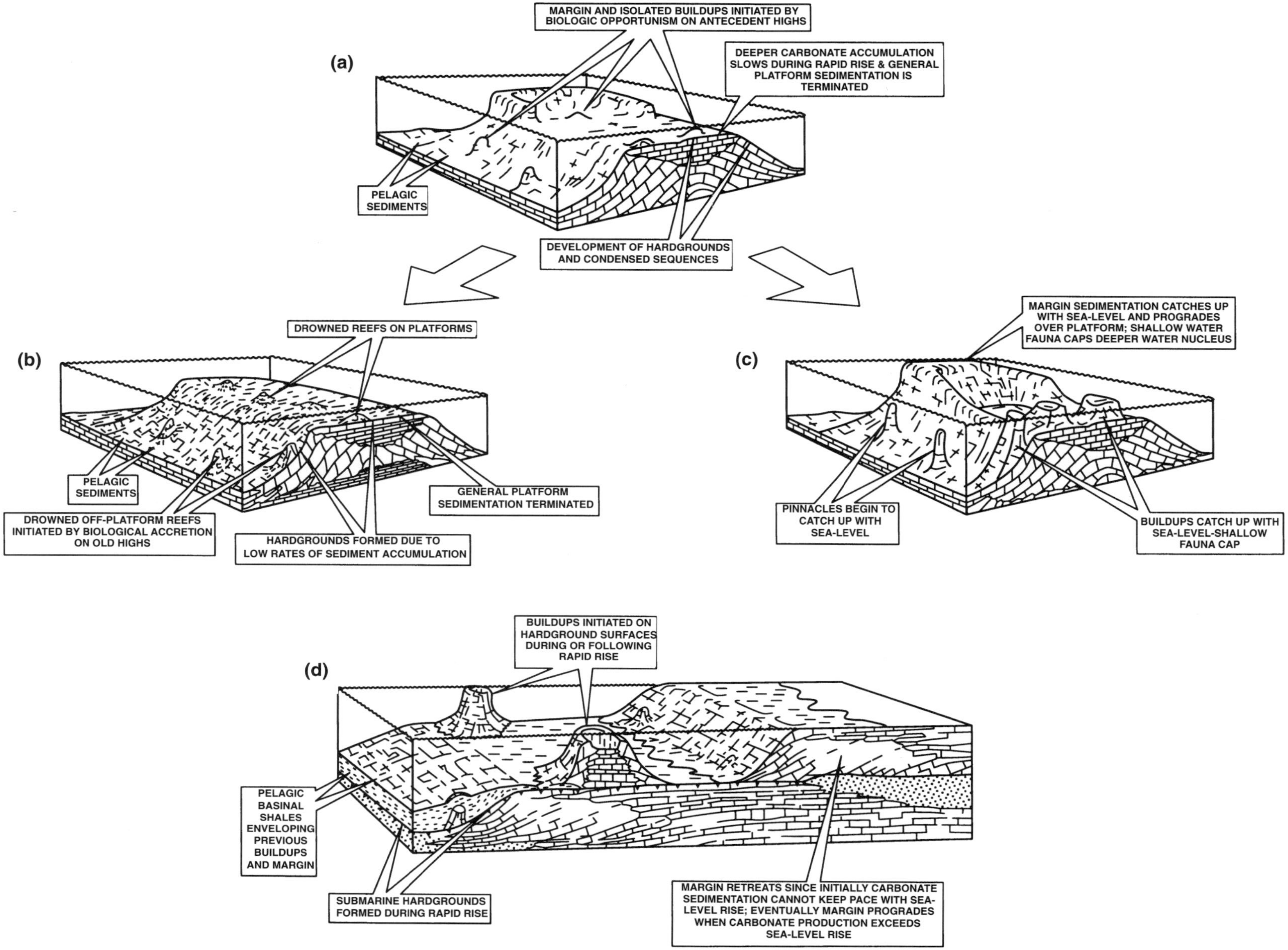

Fig. 11. Response of carbonate platforms to relative rise in sea level. (a) Relative rise in sea level on isolated platform leads to differential aggradation of platform rim and isolated buildups; (b) platform may respond by drowning or (c) by sedimentation catching up with the rate of rise. (d) On attached platforms (rimmed shelves and ramps), the common response to relative rise in sea level is platform margin retreat, followed by sedimentation catching up with rate of rise and subsequent platform aggradation and progradation. (Modified from Kendall and Schlager, 1981.)

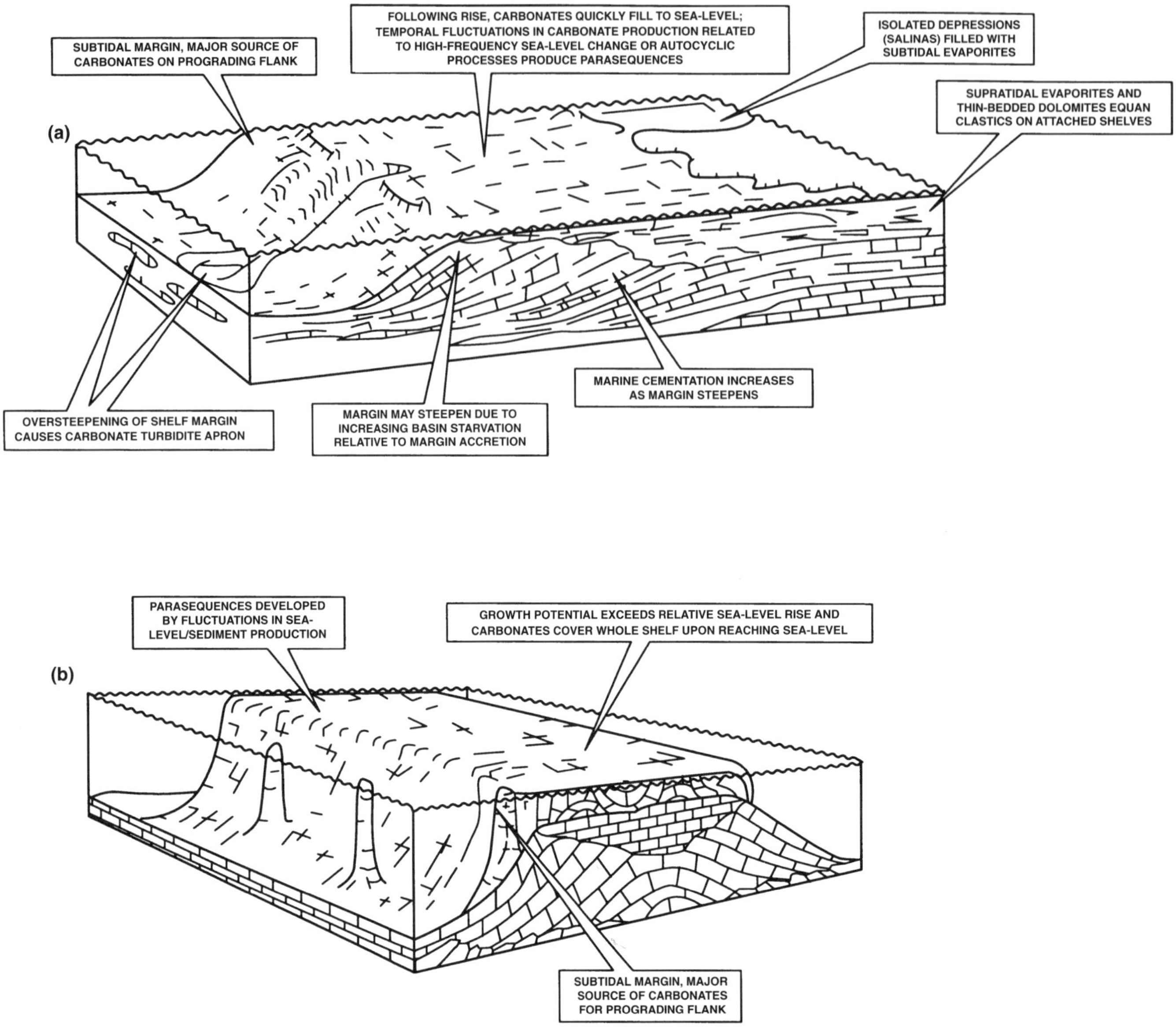

Fig. 12. Response of carbonate platforms to relative highstand and initial fall in sea level. (a) During late relative rise highstand and early fall in sea level, attached platforms (ramps and rimmed shelves) build to sea level and prograde (sediment accumulation exceeds accommodation potential). (b) On isolated platforms, carbonates shoal to sea level across entire platform and margin may prograde. (Modified from Kendall and Schlager, 1981.)

The volume of off-platform shed sediments, balanced by the platform to basin relief and slope angle, ultimately limits the rate of platform margin progradation. During initial relative fall in sea level, carbonate ramps and low-relief accretionary rimmed shelves may track sea level and develop geometries characteristic of a forced regressive wedge systems tract (Hunt and Tucker, 1993), with steepened ramp/shelf margin slopes, narrower inner ramp/shelf and ramp/shelf crest facies tracts, and a progression from sigmoid to oblique clinoform geometries. Progradation during relative fall in sea level on such platforms occurs coeval with subaerial exposure of the platform interior. Higher relief accretionary, bypass, and erosional rimmed shelves are effectively shutdown during relative falls in sea level, although a narrow reef fringe may track sea level down the slope. The inability of rimmed shelves to contribute significant sediment volume to the slope during relative lowstands of sea level inhibits progradation. In mixed carbonate-siliciclastic systems, fluvial-deltaic systems may encroach seaward across carbonate shelves during relative highstand and fall in sea level.

Relative Sea Level Lowstand

When sea level falls below the top of rimmed shelves (Figure 13), carbonate production largely ceases. Any continued production is limited to reef growth in a narrow band along the steep margin or slope. During this period of emergence, the platform top is subjected to meteoric diagenesis, the nature and magnitude of which are a function of climate (rainfall and temperature), duration of exposure, and topography (Budd et al., 1995). Also during this period of relative lowstand, prior to reflooding of the platform top, the exposed platform margin and upper slope may be subject to enhanced mass wasting processes that lead to the deposition of breccias and megabreccias at the toe-of-slope. Processes that may enhance the potential for mass wasting during lowstands include (1) fracturing and karstification of the exposed platform margin and upper slope, (2) bioerosion and undercutting of steep marginal slopes, and (3) translation of the zone of highest wave energy through the marginal slope (McNeill, 1990).

Carbonate ramps respond differently to relative falls in sea level. In pure carbonate ramp systems, lowstand strata are apparently poorly developed or are difficult to differentiate from late highstand strata (e.g., Everts et al., 1995; Kerans and Fitchen, 1995). During relative lowstands in sea level, the inner ramp, ramp crest, and ramp margin are subaerially exposed and subjected to meteoric diagenetic processes. In mixed carbonate-siliciclastic systems on both rimmed shelves and ramps, lowstands are characterized by siliciclastic bypass of eolian (e.g., Borer and Harris, 1991; Gardner, 1992) or fluvial (Brown et al., 1990) systems across the platform top and margin, and by gravity flow processes along the basin margin. In arid mixed carbonate-evaporite settings, lowstands may be characterized by basin-centered, shallow-water evaporites. In general, however, autochthonous carbonate sedimentation is negligible during relative lowstands in sea level. This is in marked contrast to siliciclastic systems, which are characterized by extensive lowstand strata (Van Wagoner et al., 1990).

The development of autochthonous carbonate sediments in lowstand wedge systems tracts has been a topic of some controversy (Mullins et al., 1983; Shanmugam and Moiola, 1984). Subsurface seismic studies (Rudolph and Lehmann, 1989; Eberli and Ginsburg, 1989; Sarg, 1989; Glaser et al., 1990) and outcrop studies (Jacquin et al., 1991) have observed reflection or stratal geometries that resemble autochthonous lowstand prograding wedge complexes, as depicted in the model diagrams of Jervey (1988), Posamentier and Vail (1988), and Sarg (1988).

It is now generally accepted that during relative lowstands in sea level on carbonate platforms, particularly on isolated platforms and attached fully aggraded rimmed shelves, platforms are emergent and subject to meteoric diagenesis and the carbonate factory is limited to a steep, narrow strip along the upper slope. In addition, sedimentation is dominantly skeletal in nature and derived from narrow, poorly developed fringing reefs, and the volume of lowstand-shed sediments is smaller than that shed from the platform during highstand (Droxler and Schlager, 1985; Handford and Loucks, 1993; Schlager et al., 1994). Meteoric diagenesis of the platform top results in rapid cementation and lithification of highstand platform carbonate sediments (Halley and Harris, 1979; Wanless and Dravis, 1989), which effectively limits the potential for erosion of such sediments and reworking into the marine environment (that is, it limits lowstand shedding) (Schlager et al., 1994).

The lowstand record of the lower slope and basin floor may contain some proportion of skeletal-rich turbidites, but in general, it is composed of carbonate lithoclasts—breccia and megabreccia. Breccia and megabreccia are products of mass wasting of the platform margin during sea level lowstand. Mass wasting may be favored by the following mechanisms (see Handford and Loucks, 1993): (1) loss of buoyancy effect on marine sediments during subaerial exposure (Schwarz, 1982); (2) wave loading and associated fluctuations in pore pressure in mud-rich platform margin sediments subject to lowered fairweather and stormweather wave base (McNeill, 1990); (3) depositional oversteepening and sediment loading associated with highstand shelf margin sediment reservoirs, (4) erosion and undercutting of shelf margin shore-zone sea cliffs by currents, waves, bioerosion, and mixing zone dissolution (Back et al., 1986); and (5) submarine erosion by persistent oceanographic currents that may shift position and accelerate during lowstands (e.g., Pinet and Popenoe, 1985; Popenoe, 1985). Karst processes and neptunian dikes can significantly weaken shelf margin strata and make them more susceptible to erosion during lowstands.

Carbonate Platform Response to Climate Change

Carbonate deposition is largely restricted to tropical and subtropical latitudes (between latitudes 30°N and 30°S today), where ambient marine water temperatures are generally above 15°C and salinities are in the range of 32–40‰. Temperate water carbonates deposited in waters of lower temperature exist beyond these limits

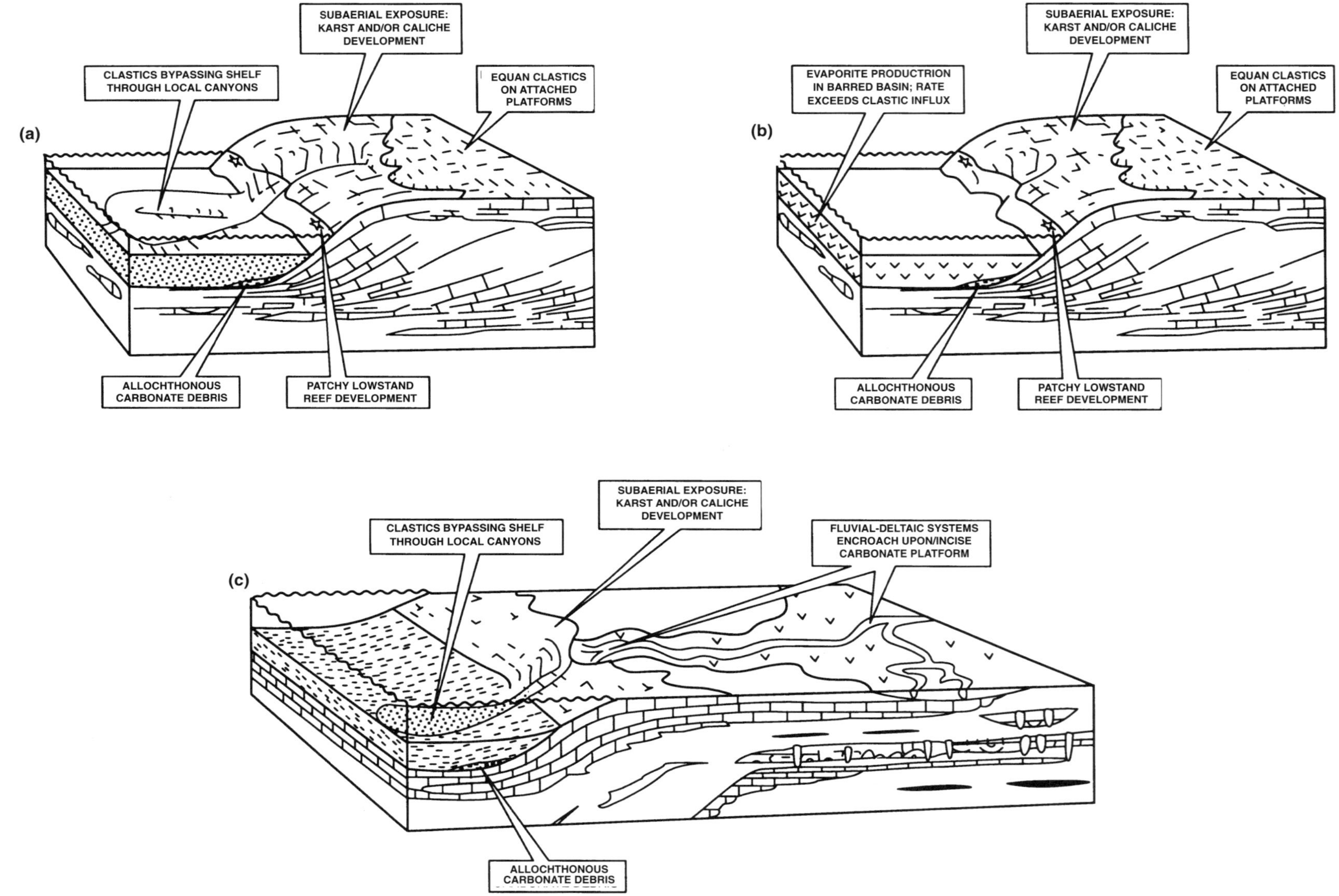

Fig. 13. Response of carbonate platforms to relative lowstand in sea level. (a) During relative lowstand in sea level on attached arid platforms, karst and caliche form on exposed platform, allochthonous debris may be deposited along slope, patchy lowstand reefs may form on slope, and eolian clastics may bypass to basin. (b) In arid silled basins, evaporites may form during lowstand. (c) On attached humid mixed carbonate-siliciclastic platforms, fluvial-deltaic systems encroach upon or incise carbonate platform and deposit siliciclastics in basin. (Modified from Kendall and Schlager, 1981.)

but are not volumetrically important today. Climate change on time scales ranging from 20 k.y. (precession of the equinoxes) to tens of millions of years (plate tectonic movements relative to paleolatitude) have a significant effect on changing styles of carbonate and siliciclastic sedimentation (Fischer, 1982, 1986). Climate change at temporal scales of 20–100 k.y. may affect changes in mean sea surface temperature, deep ocean circulation and oxygenation, ocean salinity, air temperature, humidity, rainfall, monsoonal effects, storm frequency and intensity, glacial ice buildup, and siliciclastic flux (e.g., Parrish and Barron, 1986).

In siliciclastic systems, changing climate has the greatest effect on changes in the amount, timing, and intensity of rainfall and seasonal snowmelt, which affects terrestrial runoff, sediment transport capacity, and seasonality of sediment transport. In carbonate systems, changing climate (colder and wetter) may cause a contraction of the tropical-subtropical latitudinal belt, such that the area of carbonate sedimentation is reduced. Increased terrestrial runoff during wet climatic cycles may be inimical to carbonate production in adjacent shallow-marine environments due to abrupt incursions of fresh water and increased turbidity from suspended terrigenous silt and clay. Warmer, dryer climatic phases may enhance carbonate production. Figure 14 provides a potential model for climatically driven mixed carbonate-siliciclastic parasequences.

Plate tectonic drift across the tropical-subtropical latitudes of carbonate production caused marked diachroneity of carbonate platform megasequences along certain plate margins. A well-documented example of this is the carbonate platform on the northeast Australia margin, which thins and becomes younger to the south due to northward motion of the Australia plate during the last 60 m.y. (Davies et al., 1989).

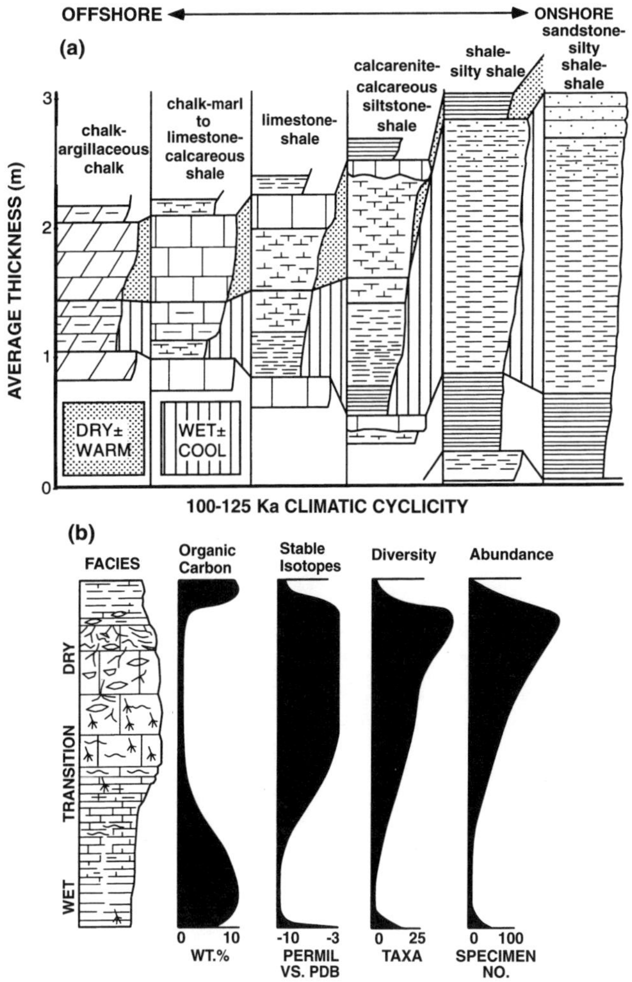

Fig. 14. Model for climatically driven parasequence development in mixed carbonate-siliciclastic settings, based on the middle Cretaceous of the Cretaceous Western Interior Basin. (a) Onshore-offshore oriented cross-section showing changes in facies with increasing water depth and distance from siliciclastic source. (b) Changes in organic carbon, $\delta^{18}O$, and diversity and abundance of taxa within climatically driven parasequence. (From Kauffman, 1988.)

Concepts, Methods, and Tools of Carbonate Sequence Stratigraphy

Introduction

Sequence stratigraphic analysis of carbonates can be performed using a variety of methods and tools whose applicability depends on the types of data available, the level of stratigraphic resolution desired, time constraints, and the types of problems one is trying to solve. The coarsest level of stratigraphic resolution is afforded by seismic reflection data. Seismic sequence analysis is based on large-scale stratal geometries and typically resolves composite sequences and their systems tracts (= sequence sets). Integration of well logs and conventional core data with seismic data, using synthetic seismograms, represents the next higher level of resolution in which sequences and their component parasequences are delineated. Interpretation of sequence stratigraphy at this scale of resolution involves primarily 1-D analysis of parasequence stacking patterns, although data sets with closely spaced conventional cores may permit 2-D analysis of stacking patterns. Finally, incorporation of 2-D and 3-D outcrop analogs, subsurface-based interpolative models, and geostatistical models provide leverage for 2-D stacking pattern analysis, which is used to address parasequence correlations and facies distribution at the interwell scale. The methods described here come from work by numerous authors, most notably Vail (1987), Sarg (1988), Goldhammer et al. (1990), Van Wagoner et al. (1990), Cross et al. (1993), and Kerans (1995).

Seismic Sequence Stratigraphy and Stratal Geometry

Seismic Sequence Interpretation

The seismic sequence stratigraphic approach is driven by the geologic interpretation of stratal geometries as inferred from seismic data (or in certain cases, seismic-scale outcrops). Seismic sequence interpretation involves the delineation and mapping of seismic sequences, systems tracts, and their contained seismic facies based on reflection configurations. These steps are followed by a geologic interpretation of seismic facies and an integration of geologic models into the interpretation (Sarg, 1988). The seven basic steps of seismic sequence stratigraphic interpretation are as follows: (1) seismic sequence analysis, (2) well log sequence analysis, (3) synthetic well-to-seismic ties and integration of biostratigraphic data, (4) seismic facies analysis, (5) interpretation of lithofacies and depositional environments, (6) forward or inverse seismic modeling, and (7) final interpretation and mapping (Vail, 1987).

Stratal Geometries and Seismic Sequence Analysis

Identification and tracing of stratal geometries and termination patterns (erosional truncation, onlap, and downlap) represents the basic tool for delineating sequences and systems tracts on seismic data. Such termination patterns can be mapped in outcrop in certain outstanding examples (e.g., Bosellini, 1984; Franseen et al., 1989; Hunt and Tucker, 1993; Pomar, 1993; Sonnenfeld and Cross, 1993; Kerans and Fitchen, 1995). Stratal geometries provide a basis for interpreting platform to basin depositional topography and the relative positions and strike-parallel orientation of depth-dependent facies tracts.

In seismic sequence analysis, reflection terminations (erosional truncation, toplap, onlap, and downlap) are marked and sequence boundaries and systems tracts interpreted on the basis of conceptual models (Vail, 1987) (Figure 15). Sequence boundaries are characterized by subaerial exposure features, erosional truncation or toplap below the boundary, and onlap above the boundary. In some cases, subaerial exposure in the form of karst can be recognized on seismic lines. Recognizable features include sinkhole depressions, faulted or chaotic reflection zones, high-reflection rugosity with abundant diffractions, regional erosional topography, and scattering and attenuation of energy at the karst surface resulting in poor subkarst imaging (del Olmo and Esteban, 1983; Popenoe et al., 1984; Lomando et al., 1993; Denny et al., 1994; Moldovanyi et al., 1995).

Sequence boundary recognition and differentiation of early and late highstand systems tracts can be inferred from progressive changes in offlap character and clinoform geometry. For example, progressive changes from sigmoid clinoform geometries to oblique clinoform geometries signal a decrease in the accommodation space to sediment supply ratio that commonly marks the early to late highstand transition and proximity to a sequence boundary (Sonnenfeld, 1991). Onlap above the sequence boundary typically occurs as reflections that lap out against the upper slope and outer shelf or that form toe-of-slope onlap wedges. Not all toe-of-slope onlap wedges mark sequence boundaries, however. Bypass and erosional rimmed shelf margins have onlapping base-of-slope carbonate aprons composed of allochthonous sediments (largely turbidites, breccias, and hemipelagic sediments). In these types of platforms, toe-of-slope marine onlap should not be confused with coastal onlap, which usually occurs above sequence boundaries on the platform top during a relative rise in sea level. Interpretation of baselap (onlap or downlap) patterns and their sequence stratigraphic significance can be complicated by bypass and erosional margins, other complex basin geometry, and out-of-plane sediment supply (e.g., a basin with axially fed siliciclastic turbidites and carbonates platforms prograding into the basin from either side) (Schlager, 1992; Handford, 1995). These complications highlight the importance of understanding the age relationships, lithologies, and sediment dispersal directions associated with marine baselap and potential sequence boundaries.

Systems tracts are defined by their internal reflection configuration and by reflection termination patterns that mark their boundaries (Figure 15). Highstand systems tracts are bound above by a sequence boundary and below by a downlap surface; internally, they consist of parallel to basinward divergent reflections representing stratal offlap. Highstand systems tracts are usually thinnest in the basin, thickest at the slope and shelf margin, and thin onto the platform. Transgressive systems tracts are bound above by downlap surfaces and below by sequence boundaries and are usually thickest at the platform margin and thin toward the basin. Uppermost reflections of the transgressive systems tract may exhibit

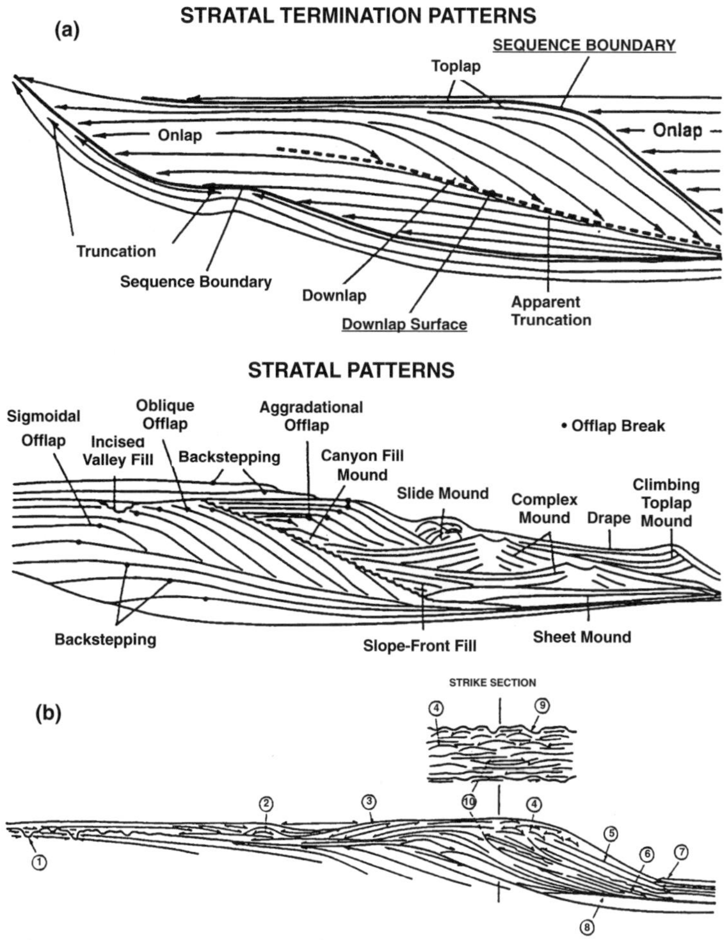

Fig. 15. (a) Generalized seismic sequence stratigraphic diagram showing reflection configurations and termination patterns associated with sequences and systems tracts. (From Vail et al., 1991.) (b) Seismic-scale stratal patterns particular to carbonate platforms. (From Handford and Loucks, 1993.) Key to circled numbers: (1) karst-related truncation and onlap fill; (2) shelf mounds that downlap at their bases and are onlapped/downlapped at their margins; (3) shelfward prograding carbonate clinoforms along leeward margins of rimmed shelf edges and associated onlap of flat-lying shelf strata; (4) steep-sided hummocky to lenticular mounds (bioherms) along rimmed shelf edges and in downslope areas; (5) depositional slopes up to and exceeding the angle of repose; (6) downlapping clinoforms at the toe-of-slope; (7) alternating downlap and onlap, or onlap; (8) clinoforms converging by thinning; (9) incisement of shelf edges at sequence boundaries or (10) within sequences.

apparent truncation beneath the downlap surface due to basinward thinning of strata, facies change, and tuning effects (Vail, 1987). Such apparent truncation must not be confused with actual truncation associated with sequence boundaries (Biddle et al., 1992; Stafleu et al., 1994). Lowstand systems tracts are bound above by transgressive surfaces and below by sequence boundaries.

Lowstand systems in pure carbonate settings typically consist of thin, mounded allochthonous toe-of-slope deposits that may be below seismic resolution. In mixed carbonate-siliciclastic or carbonate-evaporite settings, lowstands can be thick and form large basin-restricted onlapping wedges. Examples of basin-restricted siliciclastic lowstands include middle Permian platforms of the Delaware Basin (Meissner, 1972; Sarg, 1989; Kerans et al., 1992) (Figure 16), Aptian to middle Cenomanian platforms of the northern Gulf of Mexico Basin (e.g., Yurewicz et al., 1993), and Late Devonian platforms of the Canning Basin, Australia (Southgate et al., 1993). Examples of basin-restricted evaporite lowstand wedges include the Niagara Group of the Michigan Basin (Mesolella et al., 1974), the middle Pennsylvanian of the Paradox Basin (Peterson and Hite, 1969), the Upper Permian of the Zechstein Basin (Tucker, 1991), and the Middle Devonian Keg River of the Western Canada Sedimentary Basin (Schmidt et al., 1980; Campbell, 1992) (Figure 17).

Seismic Sequence Analysis and Resolution

The average resolution of conventional 2-D industry seismic data is typically adequate to delineate composite sequences in areas of shelf margin development. In the inner parts of platforms away from shelf or ramp margins, stratal geometries are planar and parallel and lacking in criteria to delineate sequences on the basis of seismic sequence stratigraphic methods alone. Certain geometries common to carbonates, particularly steeply inclined surfaces with heights below resolution, small buildups, and surfaces within isotropic strata, may not always be well imaged because of low bandwidth, poor spatial or temporal resolution, lack of data migration, or inappropriate acquisition parameters. Particular pitfalls

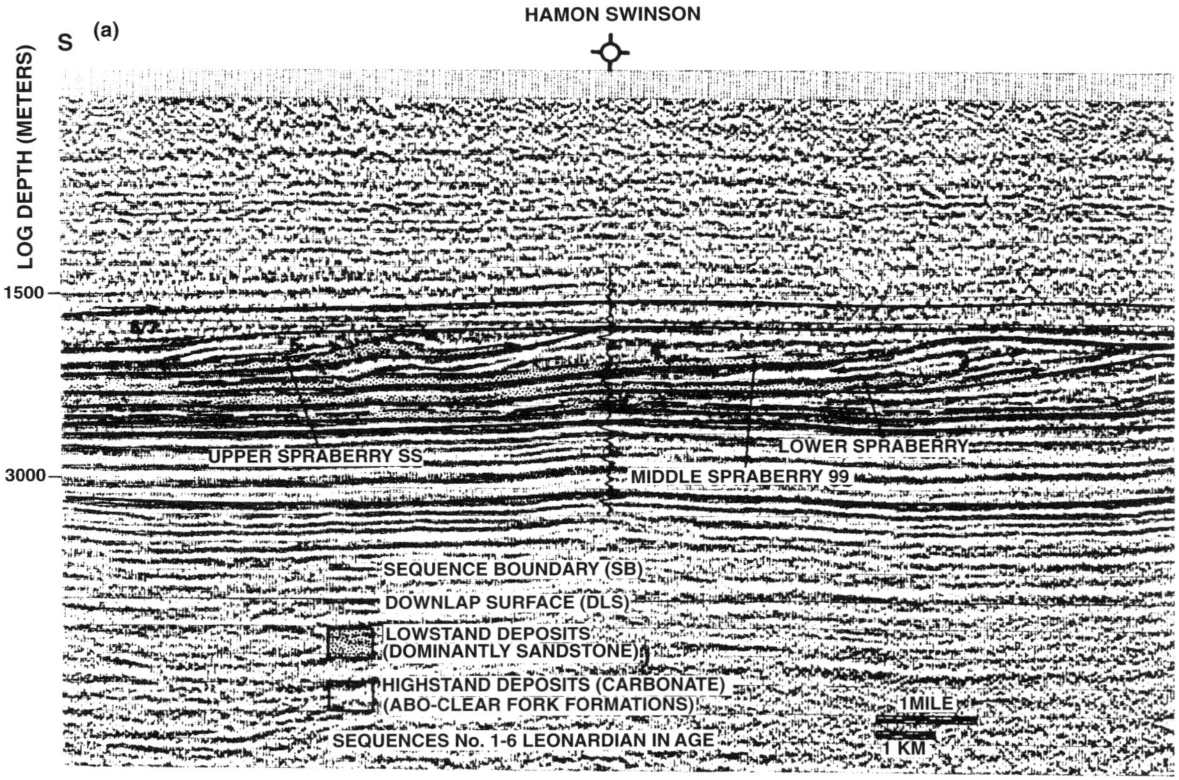

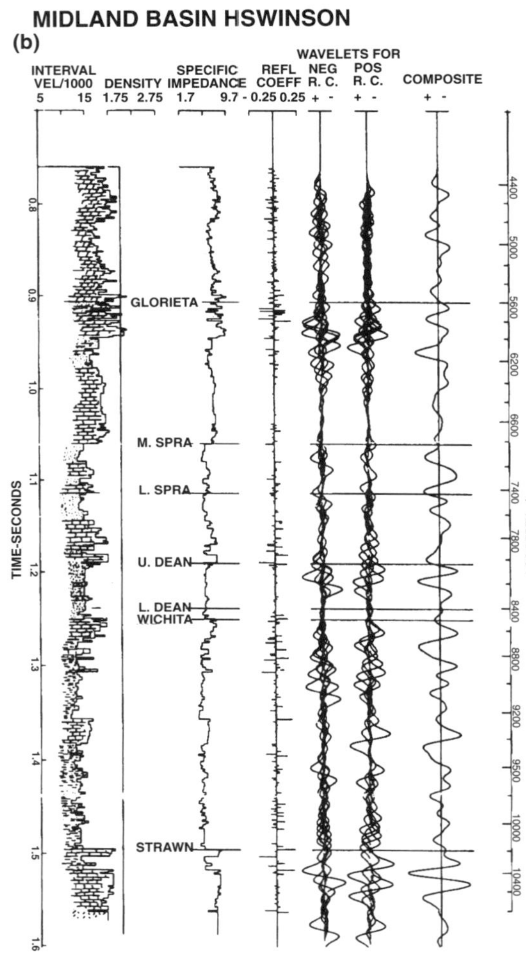

Fig. 16. (a) Seismic profile and (b) well log data characteristic of mixed carbonate-siliciclastic sequences with basin-restricted siliciclastic lowstand wedges. (Modified from Sarg, 1989.)

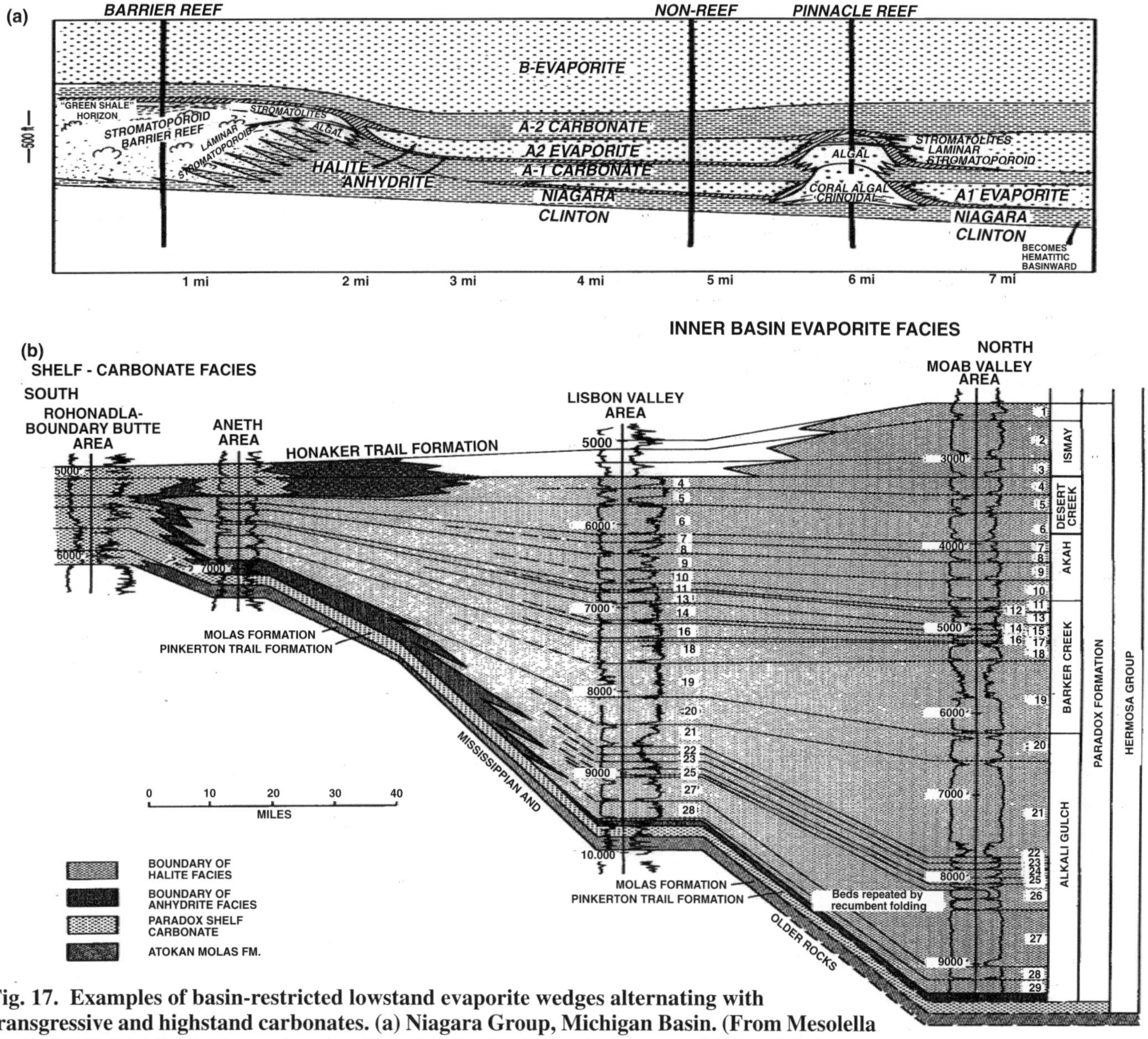

Fig. 17. Examples of basin-restricted lowstand evaporite wedges alternating with transgressive and highstand carbonates. (a) Niagara Group, Michigan Basin. (From Mesolella et al., 1974.) (b) Middle Pennsylvanian, Paradox Basin. (From Peterson and Hite, 1969.)

in seismic sequence analysis that relate to resolution limitations are (1) imaging of climbing toplap surfaces as oblique toplap (Stafleu and Schlager, 1993), (2) imaging of basin margin retrogradational facies stacking patterns as apparent truncation and onlap (Biddle et al., 1992; Stafleu et al., 1994), and (3) imaging of downlap as a surface despite significant basinal aggradation (Biddle et al., 1992).

Forward seismic models of well-constrained outcrop sections generally mimic conventional seismic response at input pulse frequencies of 25–30 Hz, with complex interference effects and poor imaging of known outcrop discontinuity surfaces. True outcrop geometries are generally well imaged in forward seismic models at input pulse frequencies of 50 Hz or more (Biddle et al., 1992; Campbell and Stafleu, 1992; Stafleu and Schlager, 1993; Stafleu et al., 1994; Stafleu and Sonnenfeld, 1994).

At the exploration scale, seismic sequence stratigraphic interpretation can yield key information on the stacking pattern of composite sequences with respect to seismic facies tracts, stratigraphic play elements, platform margin stratal geometries, and gross thickness (isochron) trends. When integrated with well logs, core, and biostratigraphic data, more refined sequence stratigraphic frameworks that resolve sequences may be erected. At the reservoir scale, the key benefit of seismic data in carbonate sequence stratigraphic analysis is to guide well log correlation with respect to stratal geometries of both structural and stratigraphic origin. In this type of analysis, synthetic seismograms are constructed from sonic logs and iteratively compared and adjusted to the actual seismic data to enhance its fit. Because individual seismic reflections are essentially interference composites of a number of bedding planes with varying reflection coefficients and impedance contrast, there is some ambiguity or stretch in resolution associated with seismic-guided well log correlation; nevertheless, the seismic data imposes an important constraint on well log correlations. Forward and inverse seismic modeling are additional techniques that can be applied in certain situations to better image reservoir development (e.g., Fischer et al., Chapter 16, this volume).

Seismic modeling can be used to better resolve stratigraphic and lithologic features whose scale is near or below tuning thickness. Seismic modeling is most often used to develop seismic criteria to assist in the development of plays, prospects, and fields. The interpretation and prediction of stratigraphy from seismic data are dependent on an understanding of the relationship between geology (bed thickness, bed geometry, density, sonic velocity, porosity, and fluid content) and seismic response (reflection geometry, frequency, amplitude, arrival time, and NMO velocity). The feasibility of seismic modeling depends on (1) impedance contrast, (2) bed resolution (dependent on frequency, S/N, and structure), and (3) the quality of the well-to-seismic tie in terms of peak-to-peak relative amplitude, time, uniqueness, and multiple well (same pulse) ties.

One-Dimensional Parasequence Stacking Patterns in Outcrop, Core, and Well Logs

One-dimensional lithofacies data sets (conventional core and outcrop) can be used to determine accommodation trends and thus define parasequences, parasequence sets, sequences, sequence sets, and composite sequences. The basic approach is first to define parasequences on the basis of models for Waltherian facies succession in the system of interest. Following this, parasequence stacking patterns are examined for their parasequence thickness trends, facies proportions, early diagenetic features, and parasequence symmetry. The term *Waltherian* refers to Walther's law of facies succession, which states that the sequence of facies found in a conformable vertical succession reflects the lateral migration (progradation or retrogradation) of contemporaneous, laterally adjacent depositional environments over one another (e.g., Tucker and Wright, 1990).

Delineation of Parasequences on Well Logs

Well logs represent a fundamental data type for analysis of stratigraphy at reservoir and flow unit scales. Well logs provide the link (through synthetic seismograms) from stratal architecture and reflection termination patterns of seismic sequences to the character, stacking patterns, and lithofacies distributions of parasequences. Modern well log suites can measure mineralogy, fluid properties, porosity, density, and sonic velocity. However, log response in carbonate rocks is difficult to interpret in terms of carbonate facies because of the low sensitivity of logging tools to carbonate mineralogic variations and subtle changes in rock fabric and grain types. Well log response must be calibrated to carbonate lithofacies as derived from examination of conventional cores. Then lithofacies and parasequence correlation from log response can be attempted in areas away from core control points. Examples of lithofacies and parasequence delineation based on logs are provided by Laferriere and Hattin (1989), Borer and Harris (1991), Ruppel (1992), and Saller et al. (1994). As these examples show, core-to-well log calibration is most reliable in successions characterized by parasequences with well-developed vertical variations in siliciclastic content (particularly clay or feldspathic sandstone), dolomite or evaporite content, organic matter content, or porosity.

Parasequence Thickness Trends

Systematic trends in parasequence thickness as determined from vertical stratigraphic sections are commonly used as an indicator of trends in longer term accommodation associated with sequences and composite sequences. The Fischer plot (Fischer, 1964; Read and Goldhammer, 1988) is a method of graphical analysis of parasequence thickness data that can be used to determine accommodation trends (Figure 18). In this method, the cumulative departure from mean parasequence thickness is plotted against parasequence number for a series of contiguous parasequences (Sadler et al., 1993). Trends characterized by thinning of successive parasequences are inferred to result from a decrease in accommodation, while trends of successively thickening parasequences are generally indicative of increasing accommodation. Thinning trends are a common expression of highstand systems tracts in fully aggraded, flat-topped platforms (e.g., Kerans and Lucia, 1989; Goldhammer et al., 1990, 1991; Montanez and Osleger, 1993), but may also occur in peritidal facies tracts in ramp settings, where they record progradation of parasequence sets (Elrick and Read, 1991). Thickening trends

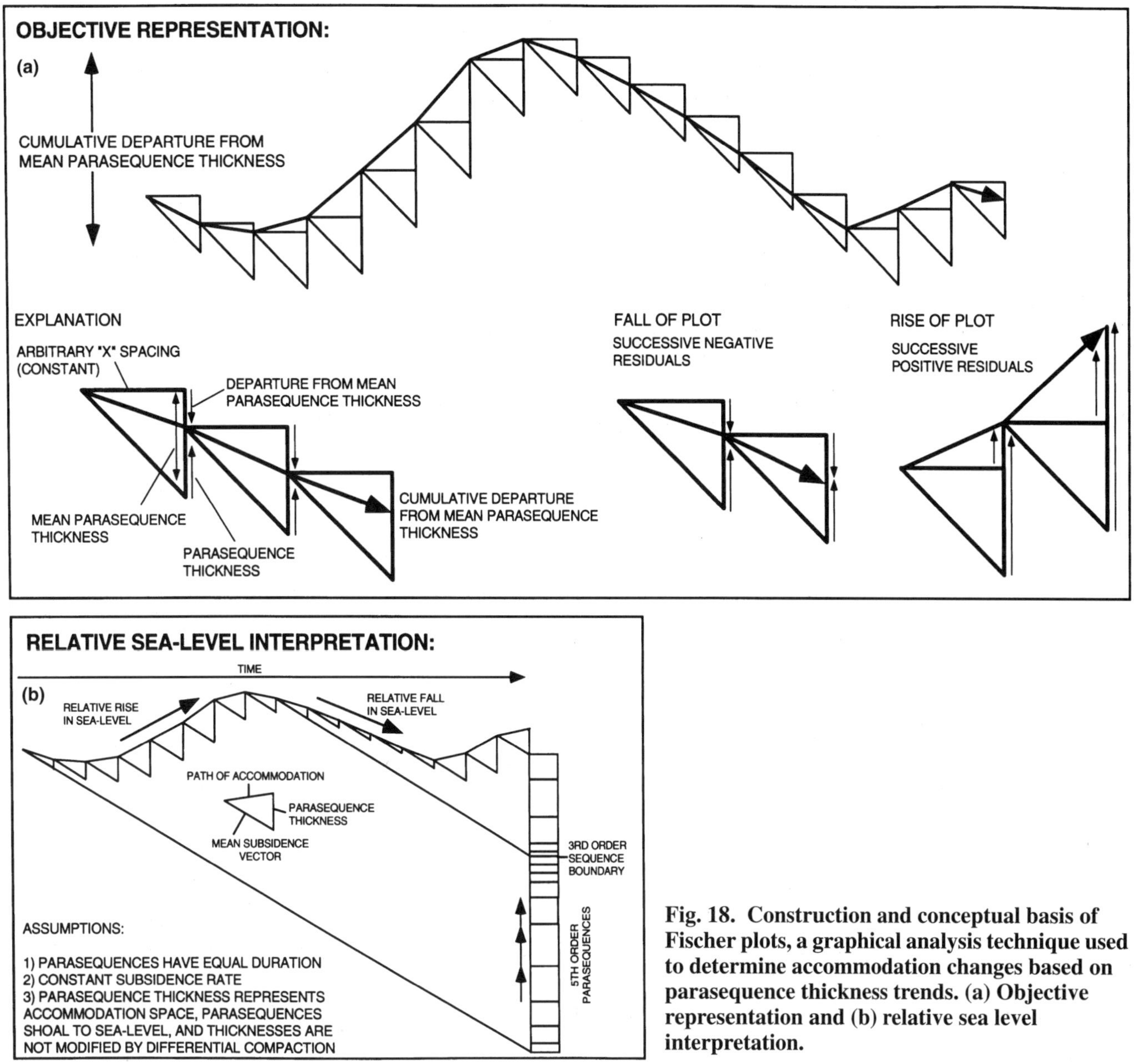

Fig. 18. Construction and conceptual basis of Fischer plots, a graphical analysis technique used to determine accommodation changes based on parasequence thickness trends. (a) Objective representation and (b) relative sea level interpretation.

are a common expression of transgressive systems tracts in fully aggraded flat-topped platforms, where rates of composite relative rise in sea level are low and sediment accumulation keeps pace with the rise (e.g., Goldhammer et al., 1991).

One pitfall in the interpretation of Fischer plots is that parasequence thinning may reflect decreasing sediment accumulation rates, attendant upon deepening below the photic zone during relative rise in sea level. Another problem is that parasequence thicknesses do not reflect filling of accommodation space but rather display a high degree of local variability (e.g., Gianniny and Simo, 1993). Major assumptions of the Fischer plot approach are that (1) all parasequences in the interval of interest fill the available accommodation space prior to generation of the next parasequence, (2) compactional modification of parasequence thicknesses can be neglected, (3) all parasequences have roughly the same duration, and (4) subsidence rates are relatively constant over the interval of interest. It should be noted that the upper boundary or base level of available accommodation space may be either the landward limit of storm surges in peritidal settings or an entirely subtidal base level such as fairweather or stormweather wavebase in subtidal settings (Collins, 1988; Enos, 1989; Osleger, 1991; Boreen and James, 1993).

Facies Proportions in Parasequences

The analysis of parasequence thickness patterns to determine accommodation trends can often yield ambiguous results. Parallel analysis of facies proportions within successive parasequences provides supporting evidence for accommodation trends based only on parasequence thickness stacking patterns. Computer modeling results and observation of parasequence stacking patterns in a variety of ancient platform settings suggest that systematic thinning of cycles in peritidal facies tracts due to decreasing accommodation (or increasing sediment supply) is accompanied by either a systematic increase in the proportion of intertidal and supratidal facies relative to subtidal facies (Goldhammer and Lehmann, 1991) or in a systematic increase in early diagenetic alteration of subtidal facies (Goldhammer et al., 1987, 1990).

The degree of early diagenetic alteration of intertidal and supratidal facies in parasequence caps may also increase systematically through a set of thinning parasequences. Types of early diagenetic alteration include dolomitization, meteoric leaching and cementation, formation of micritic crusts, and karstification. Conversely, systematic thickening of parasequences due to increasing accommodation is often accompanied by an increase in the proportion of subtidal facies relative to intertidal and supratidal facies and by a decrease in the magnitude and frequency of early diagenetic alteration. Figure 19 provides an example of integrated 1-D analysis of parasequence stacking patterns based on Fischer plots and histograms of facies proportions. The histograms provide a means to plot facies proportions in parasequence sets. As in analysis of parasequence thickness trends, caution must be exercised in certain platforms or

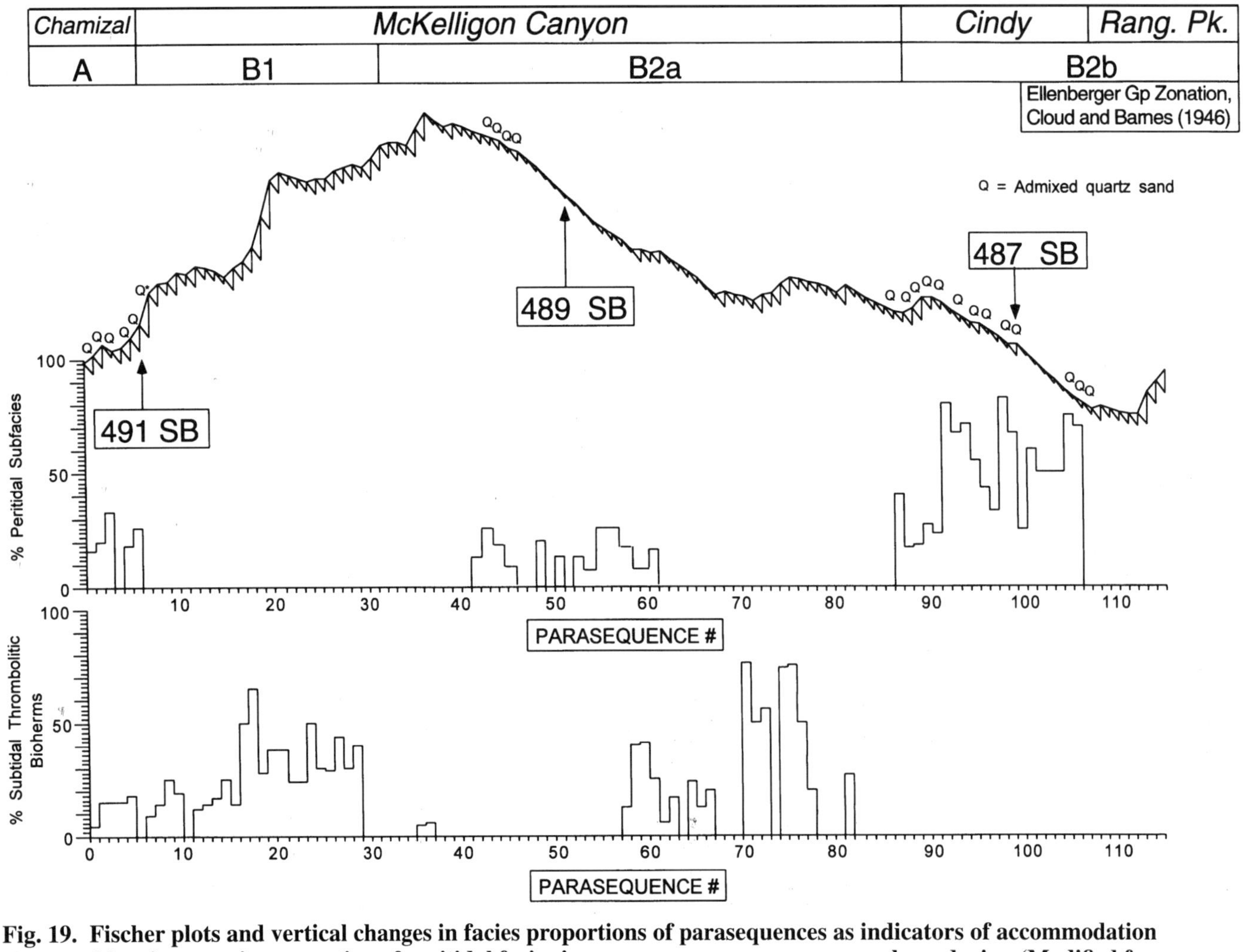

Fig. 19. Fischer plots and vertical changes in facies proportions of parasequences as indicators of accommodation changes. Note increase in proportion of peritidal facies in parasequences near sequence boundaries. (Modified from Goldhammer et al., 1993; formation stratigraphy by Lucia, 1968, 1971.)

particular facies tracts with low vertical and lateral facies continuity, in which facies proportions and thickness trends may be highly variable (e.g., Gianniny and Simo, 1993; Kerans and Fitchen, 1995).

Parasequence Symmetry

The symmetry of parasequences refers to the proportion of the parasequence composed of a facies succession reflecting a relative rise in sea level versus a relative fall in sea level. Asymmetric parasequences composed of facies successions indicative of a relative fall in sea level (e.g., CBA-CBA) are characteristic of parts of platforms (and periods of time) that have low accommodation. The majority of shallow-marine parasequences in the stratigraphic record are strongly asymmetric and upward-shallowing, which reflects an initial lag in sedimentation (lag time or lag depth) during the initial relative rise in sea level, followed by catch-up of the carbonate factory and shoaling during the later part of the rise and subsequent relative fall in sea level. The cause of the sedimentation lag during the relative rise, which is essential to producing asymmetric upward-shallowing sequences, has been attributed to the following: (1) poor circulation and water quality (schizohaline) during early transgression, (2) terrigenous clastic poisoning, (3) reworking and removal of transgressive sediments from underlying lithified substrates by high-energy tidal (Boss et al., 1992) and wave-generated currents (Enos, 1989), and (4) rates of relative rise in sea level that seemingly far exceed rates of sediment accumulation (Schlager, 1981).

Generally, asymmetric parasequences occur in inner ramp and ramp crest facies tracts, in the inner shelf, across the entire shelf in fully aggraded platforms, or associated with platform marginal shoals and islands in rimmed shelves. Symmetric parasequences (e.g., ABCBA-ABCBA) occur most often in deeper subtidal settings, where accommodation is greater over the entire cycle of relative change in sea level. Symmetric parasequences are characteristic of outer ramp, upper slope, and basinal settings in mixed carbonate-siliciclastic systems (Borer and Harris, 1991; Brown and Loucks, 1993; Sonnenfeld and Cross, 1993). The relative lowstand through rise in sea level is characterized by deposition and reworking of terrigenous clastics, whereas the relative highstand and fall in sea level is characterized by deposition of carbonates. Progradational parasequence sets may exhibit an upward increase in asymmetry, whereas retrogradational parasequence sets show an upward increase in symmetry.

One-Dimensional Stacking Patterns

Parasequence stacking patterns in carbonate platforms as observed in 1-D sections can be highly variable (e.g., Read, 1995). Carbonate platforms formed during periods of high global sea level and warm climates (greenhouse platforms) are characterized by fourth- and fifth-order parasequences and parasequence sets forming systems tracts of third-order sequences (e.g., Kerans and Lucia, 1989; Goldhammer et al., 1990, 1993; Goldhammer and Lehmann, 1991; Montanez and Osleger, 1993; Read, 1995) or by facies assemblages that form systems tracts of third-order sequences with poor development of parasequences (e.g., Calavan et al., 1992; Fitchen et al., 1995). Ice-house platforms, formed during periods of widespread continental glaciation, are characterized by fifth-order parasequences, which stack into fourth-order sequences (fifth- and fourth-order glacio-eustatic Milankovitch cycles), which in turn are grouped into third-order composite sequences (e.g., Goldhammer et al., 1991; Saller et al., 1994; Read, 1995). In any analysis of 1-D parasequence stacking patterns, published analogs of similar age and depositional setting can provide a reference framework for interpretation.

Concepts of Two-Dimensional Parasequence Stacking Patterns in Outcrop and Subsurface

Two-dimensional parasequence stacking pattern analysis applies to well log cross-sections with conventional core control (and seismic data) and high-continuity outcrop successions. Two-dimensional parasequence stacking pattern analysis builds upon 1-D stacking pattern analysis, attempting to provide the best or most stratigraphically reasonable image of changing facies distributions in successive parasequences as a function of position in a framework of parasequence sets, sequences, sequence sets, and composite sequences. Concepts or attributes that guide 2-D stacking pattern analysis to determine accommodation trends include stratal geometries, facies tract offset, volumetric partitioning of lithofacies, and lithofacies differentiation.

Analysis of 2-D stacking patterns can be enhanced by a thorough understanding of the sedimentologic processes, depositional setting, and geometry and continuity of facies in parasequences and how these characteristics change according to changing conditions of accommodation. The Holocene record provides actualistic analogs similar in temporal and spatial scale to ancient parasequences (e.g., Harris et al., 1993; Harris and Kowalik, 1994), which can be consulted to guide the stratigrapher to reasonable conclusions. Several well-constrained ancient examples, primarily from seismic-scale outcrops in the Guadalupe Mountains of New Mexico and Texas (Kerans et al., 1992; Sonnenfeld and Cross, 1993; Barnaby and Ward, 1995; Kerans, 1995; Kerans and Fitchen, 1995), provide 2-D images of facies distributions within parasequences and parasequence stacking patterns that can help guide the interpreter.

Facies Tract Offset

Two- and three-dimensional analysis of facies within genetic, chronostratigraphically significant increments of strata such as parasequences provides a model for normal Waltherian facies successions in the vertical dimension (measured sections, cores, and well logs). An understanding of the normal Waltherian facies succession provides the basis for grouping facies into facies associations and facies tracts. Parasequence stacking patterns in a sequence exhibit systematic seaward facies tract offsets in the case of a progradational parasequence set or highstand systems tract, and systematic landward facies tract offsets in the case of a retrogradational parasequence set or transgressive systems tract. The magnitude (in terms of distance in the dip direction) of facies tract offset, across parasequence boundaries, is to some degree a measure of the hiatus or lacuna (Wheeler, 1958) at the boundary, or at least a measure of the amplitude of relative change in sea level and water depth associated with the creation of the parasequence boundary.

Unusually large seaward facies tract offsets are often associated with sequence and composite sequence boundaries. These are the familiar downward shifts of coastal onlap (Mitchum, 1977) and consequently of facies tracts across a sequence boundary that have been cited previously (see also Van Wagoner et al., 1988). For example, a vertical succession from shallow subtidal carbonate facies to eolian carbonate facies would be indicative of such a seaward facies tract offset and would be a good candidate for a sequence boundary. Large landward facies tract offsets are commonly associated with maximum flooding surfaces, with platform drowning events (drowning unconformities; Schlager, 1989), or with sequence boundaries in and at the base of transgressive sequence sets of composite sequences (e.g., base of lower San Andres composite sequence; Kerans and Fitchen, 1995).

Volumetric Partitioning of Lithofacies

Parasequences, sequences, and composite sequences are often interpreted to contain packages of strata deposited during both a transgressive hemi-cycle (relative rise in sea level) and a regressive hemi-cycle (relative fall in sea level). The proportions and lithofacies composition of these hemi-cycles change according to geomorphic position along the depositional profile and stratigraphic position within a larger scale stacking pattern framework. In parasequences, regressive hemi-cycles typically predominate on the platform top (asymmetric parasequences). Both transgressive and regressive hemi-cycles may be developed at the platform margin and upper slope where there is greater accommodation space, particularly on ramps where slope angles are lower (symmetric parasequences). Either transgressive or regressive hemi-cycles may dominate in the lower slope and basin. In mixed carbonate-siliciclastic settings (e.g., Sonnenfeld and Cross, 1993) (Figure 20), the lower slope and basin may be dominated by transgressive hemi-cycles of siliciclastics. Lithofacies are volumetrically partitioned within parasequences according to the changing locus of accommodation space, across the depositional profile, during the relative rises and falls in sea level that create parasequences of a sequence. In other words, within a stack of parasequences, different volumes of strata are deposited in similar geomorphic settings of successive parasequences according to changes in longer term accommodation, sediment supply, and preservation potential (Sonnenfeld and Cross, 1993).

Lithofacies Differentiation

In the past, facies models have tended to present a generalized, static representation of lithofacies types and their distribution within depositional systems, with little regard to relative stand of sea level or position within a stratigraphic hierarchy of parasequences, sequences, and composite sequences (e.g., Walker, 1984). Although carbonate facies models have recognized that facies associations may differ according to age, climate, and platform type (Wilson, 1975; Read, 1985; Hardie and Shinn, 1986; Sarg, 1988), few past studies have recognized that platform facies models may change according to position within a cycle of relative change in sea level, of whatever temporal scale. Exceptions to this are studies that examined facies distributions during different time slices of the Holocene transgression (Enos, 1977; Harris, 1983).

More recently, attempts have been made to (1) understand the response of facies to sea level change

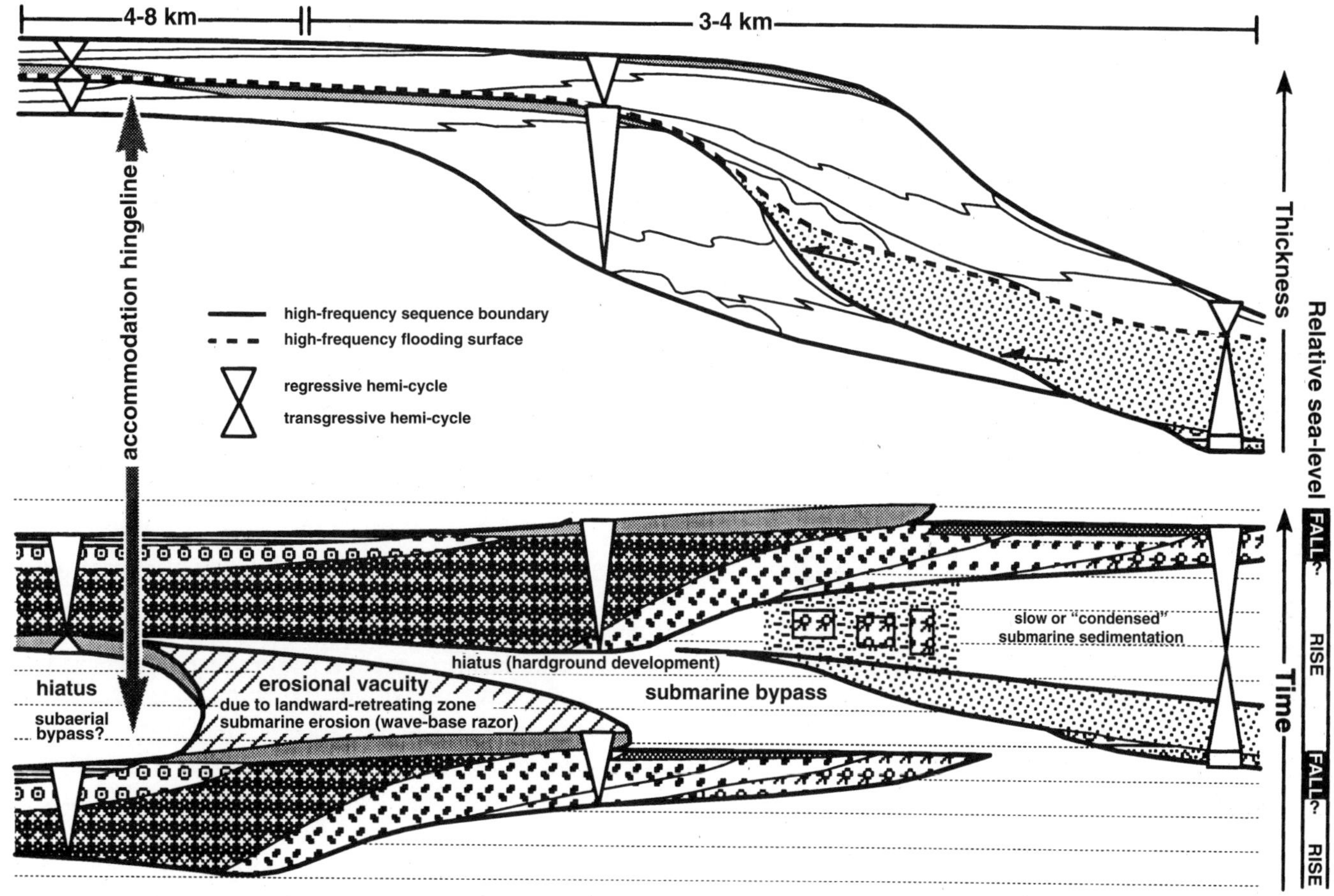

Fig. 20. Concept of volumetric partitioning as expressed in fifth-order sequences of upper San Andres Formation (Upper Permian), Last Chance Canyon, New Mexico. (From Sonnenfeld and Cross, 1993.)

(Walker and James, 1992), (2) tailor facies models to particular systems tracts of depositional sequences within different climatic and platform settings (Handford and Loucks, 1993), and (3) understand changing facies models as a function of ice-house versus greenhouse climatic phases, which are interpreted to have distinctly different frequencies and amplitudes of sea level change (Read, 1995; Read et al., 1991). Other recent studies have recognized that lithofacies differentiation is a common stratigraphic response to hierarchical sea level fluctuations (composite eustasy of Goldhammer et al., 1990), and thus facies models for particular facies tracts or paleogeomorphic positions change for each parasequence as a function of position within a stacking pattern of sequences and composite sequences (Sonnenfeld and Cross, 1993; Kerans and Fitchen, 1995).

Application of Carbonate Sequence Stratigraphic Concepts: Examples

Miocene Terumbu Formation, Natuna Field, Indonesia

The Sarawak Basin and Luconia Province in the South China Sea northwest of Borneo are areas of prolific middle–late Miocene carbonate buildup development (Figure 21). Buildups are isolated platforms or pinnacles composed of porous coral and red algal reefs with associated forereef, backreef, and lagoonal facies. Lithofacies distribution within buildups is mainly a function of relative changes in sea level driven by inferred glacio-eustasy and tectonic subsidence, as well as paleoenvironmental factors, including windward versus leeward margin orientation, siliciclastic poisoning, upwelling, and nutrient supply. Drilling activity in Miocene carbonate buildups since 1967 has resulted in the discovery of seven giant gas fields (reserves >1 tcf) and more than 20 smaller gas fields in central Luconia and a giant gas

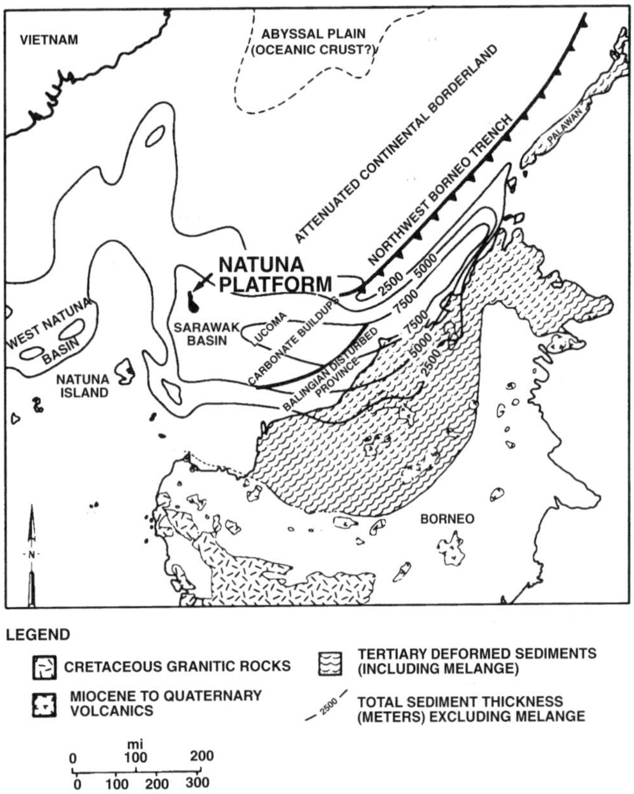

Fig. 21. Map of South China Sea showing location of Natuna L Structure and nearby Luconia buildup province. (From Rudolph and Lehmann, 1989.)

field, the Natuna field, in the East Natuna-Sarawak Basin (Epting, 1989; Rudolph and Lehmann, 1989). Miocene buildup plays in the South China Sea are stratigraphic traps that are instructive as examples of carbonate sequence stratigraphic development because they exhibit (1) a clear differentiation of seismic facies and good correlation of seismic facies to lithofacies where thickness is at seismic resolution, (2) differences in platform development on windward versus leeward margins, (3) interfingering with siliciclastics shed from neighboring landmasses, and (4) progressive decrease in platform area in younger sequences, followed by platform drowning and burial in a siliciclastic wedge. The Miocene serves as an analog for similar isolated platform reservoirs in the Middle Devonian of Canada (Wendte et al., 1992) and the Pennsylvanian–Lower Permian of West Texas (fields of the Horseshoe Atoll; Waite, 1993). This example focuses on integrated geologic and seismic sequence stratigraphic studies of the Natuna L-Structure buildup (Rudolph and Lehmann, 1989).

Regional Setting

The Natuna L-Structure is situated on the western edge of the Sarawak–East Natuna Basin about 360 km northwest of Borneo (Figure 21). Geohistory analysis and observed stratigraphy (Figure 22) subdivides the structural-stratigraphic record of Natuna into a late Oligocene–early Miocene rift phase, a middle–late Miocene passive sag phase, and a Pliocene–Recent phase of rapid downwarping related to subduction of the South China plate beneath Borneo. The oldest sedimentary rocks in the area are the synrift Oligocene–lower Miocene deltaic siliciclastics of the Arang and Gabus formations, which rest on a basement of extended granitic continental crust (Hamilton, 1979; Taylor and Hayes, 1983). The Arang Formation is inferred to be the source of gas present in the Natuna reservoir. The Terumbu Formation consists of carbonates deposited during the passive sag phase; this formation makes up the reservoir at Natuna. The Terumbu Formation is composed of platform and platform-margin facies that nucleated on a faulted, basement-involved structural high just east of a carbonate shelf (Figure 22). The Terumbu platform at Natuna was buried by a series of siliciclastic wedges of the Muda Formation, which were deposited during the downwarping phase. The lower part of the Muda at Natuna consists of prodelta to basinal shales deposited in about 900 m water depth, based on paleontologic and seismic interpretation.

Stratigraphic Framework

On the basis of seismic data analysis, ~290 m (960 ft) of conventional core and well logs from five wells, the Terumbu Formation at Natuna can be subdivided into eight seismic sequences (Figure 23), ranging in age from ~16.5 to 5.2 Ma (middle–late Miocene, third-order sequences) (Rudolph and Lehmann, 1989). Each seismic sequence contains lowstand, transgressive, and highstand systems tracts in variable proportions. Figure 24 (seismic line A) is representative of the basic sequence and systems tract geometries and seismic facies at Natuna. Sequence boundaries are characterized by truncation and toplap of underlying reflectors, onlap of overlying reflectors along the margins of the buildup, and petrographic and geochemical evidence for meteoric diagenesis below the boundary. Lowstands consist of thin toe-of-slope, basin-restricted wedges of siltstone and shale that were sourced from the adjacent landmasses (Sarawak, Natuna Arch, and Malaysia Peninsula). Transgressive systems tracts are characterized by thin,

(a)

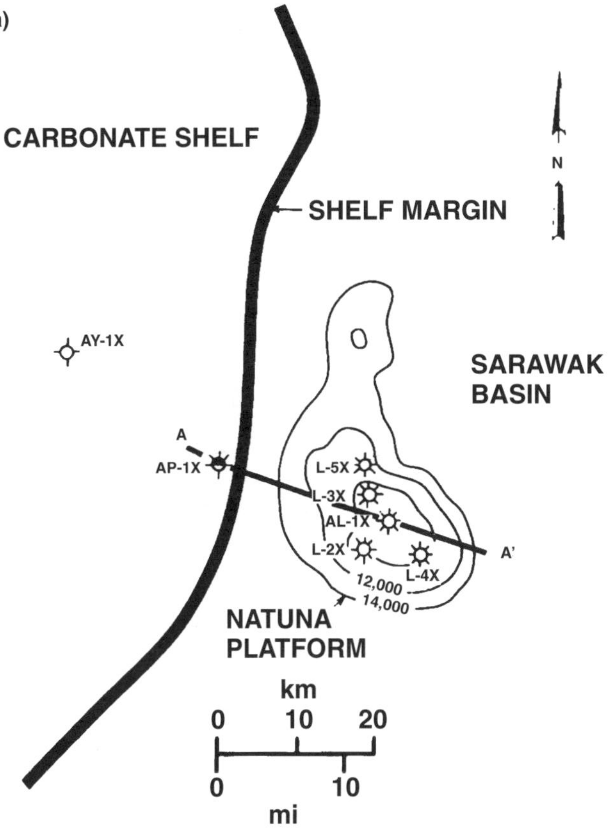

Fig. 22. (a) Depositional setting and regional stratigraphy of the Natuna platform. (b) Geohistory diagram of wells that penetrate the Natuna L Structure buildup showing major phases of basin subsidence. (From Rudolph and Lehmann, 1989.)

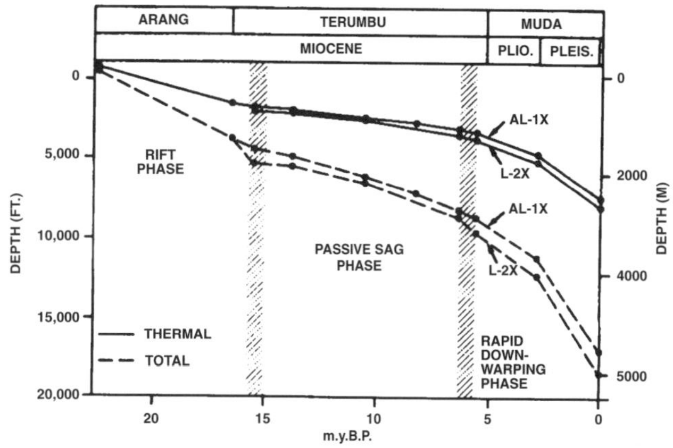

(b)

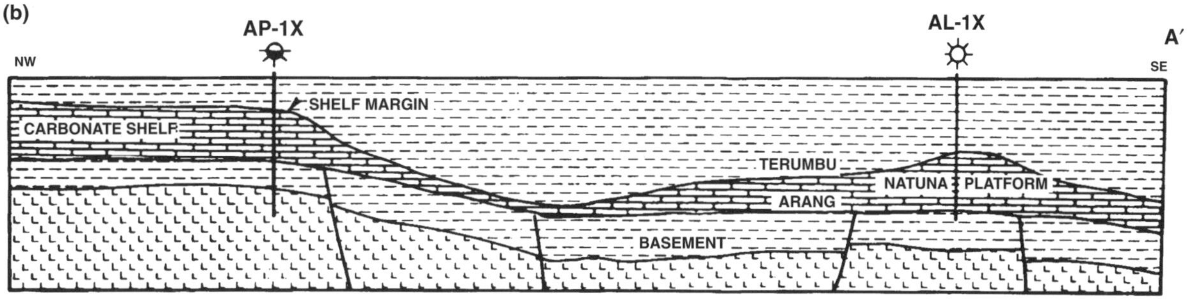

relatively isopachous onlapping and backstepping units above a subaerial unconformity. Highstand systems tracts consist of an early aggradational phase, during which broadly mounded reef and shoal facies reached their maximum relief, and a late phase when mounded margins prograded both seaward and bankward in response to a decrease in accommodation. Late highstand clinoforms are characterized by oblique toplap.

Figure 25 is a seismic facies map of combined middle–late Serravalian (11.5–13.8 Ma) highstand systems tracts. Low-amplitude obliquc scismic facies are related to high-energy platform lithofacies, which are dominantly coralgal reef, reef flat, and forereef facies with an average porosity of 15–25%. This facies is asymmetrically distributed along the eastern, windward margin of the platform. Low-amplitude, parallel to oblique seismic facies are moderate-energy platform facies composed of red algae, echinoderm, and mollusc packstones with an average porosity of 10–20%. High-amplitude parallel seismic facies are dominantly low-energy outer platform planktonic foraminifera and ostracod wackestones with 0–10% porosity. Medium-amplitude truncated facies are inferred to be eroded slope facies.

The major controls on reservoir continuity and quality at Natuna are the areal extent of sequences, the distribution of high-energy facies within sequences, and porosity enhancement by meteoric dissolution at sequence boundaries. The areal extent of sequences decreased throughout the evolution of the platform (Figure 26). The oldest carbonate highstands (15.5–13.8 Ma) were deposited on a widespread platform separated from a shelf by a shallow sag (Figure 22). These platforms exhibit only modest lateral variation in seismic facies and log porosity. Rudolph and Lehmann (1989) estimated that these platforms had ~300 m (~1000 ft) of relief.

In contrast, the youngest carbonate highstands (6.3–5.5 Ma) are areally restricted and exhibit much greater lateral variation in seismic facies and lithofacies. The greater variation in facies was due to the development of isolated reef pinnacles on different parts of the platform. Platform-to-basin relief increased to ~900 m (~3000 ft) in these sequences. The decrease in area in successive sequences at Natuna is due to periodic backstepping of the platform margin and an overall decrease in the area of healthy platform growth. This occurred primarily during the development of transgressive systems tracts, when the carbonate system is thought to have been unable to keep pace with the rate of relative rise in sea level. Progressive decrease in platform area can be related to increasing rates of long-term tectonic subsidence (Figure 22), which modulated shorter term third-order relative rises in sea level. Terrigenous clastic influx or nutrient excess related to terrigenous influx may have been an important factor as well (e.g., Ehrlich et al., 1993). The distribution of high-energy facies was probably a function of facing toward the windward margin. High-energy facies are best developed on the eastern side of the platform. This distribution may also be due to terrigenous clastic influx on the western side of platforms. High-energy reef-related facies have greater thickness and lateral continuity in highstands, when decreased accommodation favored progradation and the development of peri-reefal debris aprons. Trends of upward porosity increase within sequences are indicative of upward-shallowing highstand facies successions and meteoric dissolution at overlying sequence boundaries.

The Natuna Platform provides a good example of third-order sequence stacking patterns within the framework of a second-order tectonically driven cycle. This example is similar to many other productive buildups in terms of an isolated platform geometry, progressive areal restriction, deposition during a phase of long-term relative rise in sea level, and burial by prograding siliciclastic wedges (Greenlee and Lehmann, 1993). Figure 27 is a model of basin-filling and long-term accommodation based on hydrocarbon productive buildup development in the stratigraphic record. Note that in the progradational part of the model, carbonates can be replaced by (or be deposited alternately with) siliciclastics.

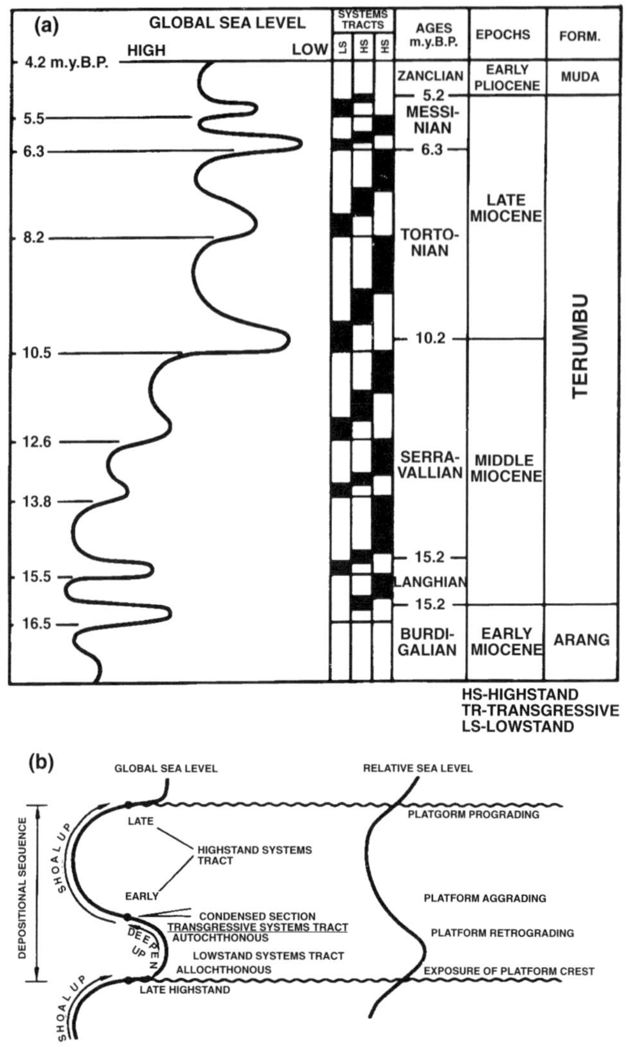

Fig. 23. Sequence stratigraphy of the Natuna Platform showing the age model for sequences and systems tracts and interpretation of systems tracts in terms of relative changes in sea level. (From Rudolph and Lehmann, 1989.)

Lower Cretaceous Shuaiba Formation, Bu Hasa Field, Abu Dhabi, U.A.E.

Bu Hasa field is located in eastern Abu Dhabi, ~55 km south of the Arabian Gulf (Figure 28). The discovery well was drilled in 1962 on a seismic anomaly. The field is a 35 km long, north-south trending, doubly-plunging anticline whose origin may be due to vertical basement-involved block faulting (Murris, 1980). Production at Bu Hasa is largely from porous platform carbonate facies of the Shuaiba Formation, with significant

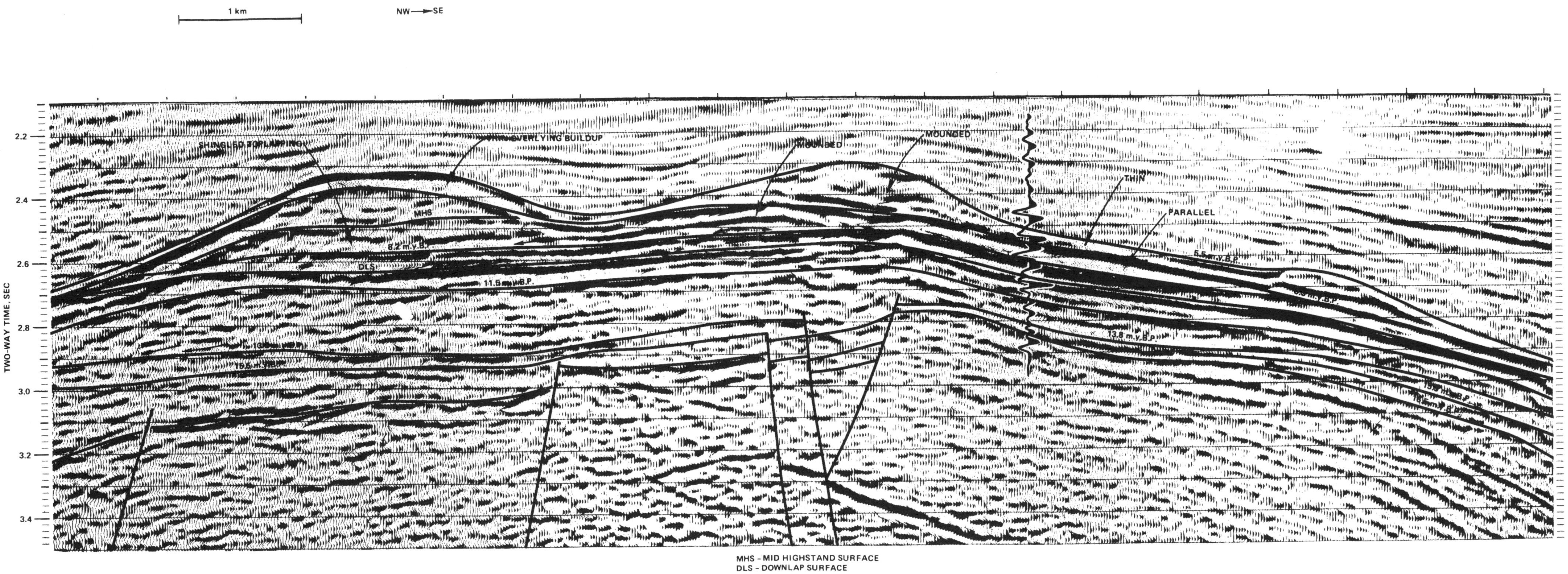

Fig. 24. Seismic line A showing development of Terumbu platform sequences above faulted basement structures. Note development of shingled toplapping reflectors at platform margins, laterally discontinuous mounds in youngest sequences, and progressive retreat of platform margins. See Figure 25 for location of seismic line.

facies-controlled permeability heterogeneity. A sequence stratigraphic framework for the Shuaiba reservoir at Bu Hasa was constructed to aid delineation of reservoir zones for interpolative porosity, permeability, and lithofacies modeling and reservoir simulation (Calavan et al., 1992). Stratigraphic correlation was conducted by iterative comparison and integration of geophysical, geologic, and pressure data.

Initially, a detailed reservoir zonation was established by wireline log correlation guided by reflection geometries mapped on seismic data, using synthetic seismogram well ties. The wireline log correlation framework was first determined for wells with synthetic seismograms, after which all other wells were incorporated into the framework based on vertical stacking patterns of lithofacies and log signatures. Later incorporation of lithofacies, petrographic, geochemical, biostratigraphic, and pressure data served to high-grade stratigraphic correlations of reservoir zones and provided a framework for 3-D interpolative computer models of lithofacies, porosity, and permeability. The high-resolution sequence stratigraphic and lithofacies framework for Bu Hasa has significant implications for hydrocarbon distribution and development strategies, and it provides a good example of Cretaceous ramp development (Calavan et al., 1992).

Regional Setting

The Shuaiba Formation of the Arabian Gulf is an Aptian aggradational to progradational carbonate ramp succession that is underlain conformably by platform carbonates of the Kharaib Formation (Hawar Member) and overlain unconformably by deeper water shales and siltstones of the Nahr Umr Formation (Figure 29). The Shuaiba is the uppermost formation of the Thamama Group and is a major hydrocarbon-producing unit in the United Arab Emirates and bordering countries. Most Shuaiba production occurs in giant structural and stratigraphic traps associated with anticlines and fractures. Shuaiba producing trends are centered around the lower Aptian Shuaiba intrashelf basin (Figure 28). Reservoirs around the periphery of this basin generally produce from matrix porosity associated with rudist and coral buildups, algal facies, and chalky skeletal facies of the middle Shuaiba. Within the basin proper, shallow-water, reservoir-prone facies occur in the lowest part of the Shuaiba and in the basin-restricted Bab Member, which represents the youngest part of the Shuaiba. Fractured deeper water facies of the middle Shuaiba may also produce in the basin area.

Sequence Stratigraphic Framework

At Bu Hasa, the Shuaiba is composed of parts of three third-order composite sequences and their component systems tracts and sequence sets (Figure 29) (Calavan et al., 1992). The lowest part of the Shuaiba, commonly referred to as Reservoir Unit A, is a widespread highstand deposit composed of chalky packstone to *Bacinella* boundstone; this unit overlies glauconitic deeper water facies of the Hawar interpreted as a condensed section. A sequence boundary (~112 Ma) is recognized at the top of the first Shuaiba highstand systems tract. This boundary is marked by a significant landward shift in facies distributions and deepening of the Shuaiba intrashelf basin, and it coincides with a readily mappable seismic reflection. The remainder of the Shuaiba is composed of two third-order composite sequences. The middle composite sequence contains an aggradational to retrogradational transgressive systems tract/sequence set, a sigmoid progradational early highstand systems tract/sequence set, and an oblique progradational late highstand systems tract/sequence set. The sequence boundary at the top of this composite sequence (~109.5 Ma) is marked by significant subaerial exposure and meteoric diagenesis. The upper Shuaiba composite sequence corresponds to the Bab Member. This is a basin-restricted composite sequence composed of transgressive and highstand systems tracts/sequence sets. The Nahr Umr transgressive and highstand systems tracts overstep the top of the Bab and main Shuaiba platforms.

Production at Bu Hasa is mostly from the middle Shuaiba composite sequence (~112–109.5 Ma), which exhibits strong differentiation of seismic facies from south to north that coincides with systems tracts/sequence sets (Figure 30). South Bu Hasa is characterized by flat-lying, weak, discontinuous reflectors of the transgressive and early highstand systems tracts/sequence sets. North Bu Hasa is characterized by shingled to oblique progradational clinoforms of the late highstand systems tract/sequence set, which toplap to the south and downlap to the north. There is an alternation of strong and weak reflection character along the toplap surface/sequence boundary in the late highstand. At the north side of Bu Hasa, the Bab is clearly imaged as a separate sequence that onlaps the middle Shuaiba basin margin.

Figure 31 is a seismic facies map of the upper part of the middle Shuaiba, showing differentiation of the early highstand platform in the south, the three zones of alternating strong and weak reflection character in the late highstand, and the distribution of the Bab in the intrashelf basin. The key feature of this map is the strike-

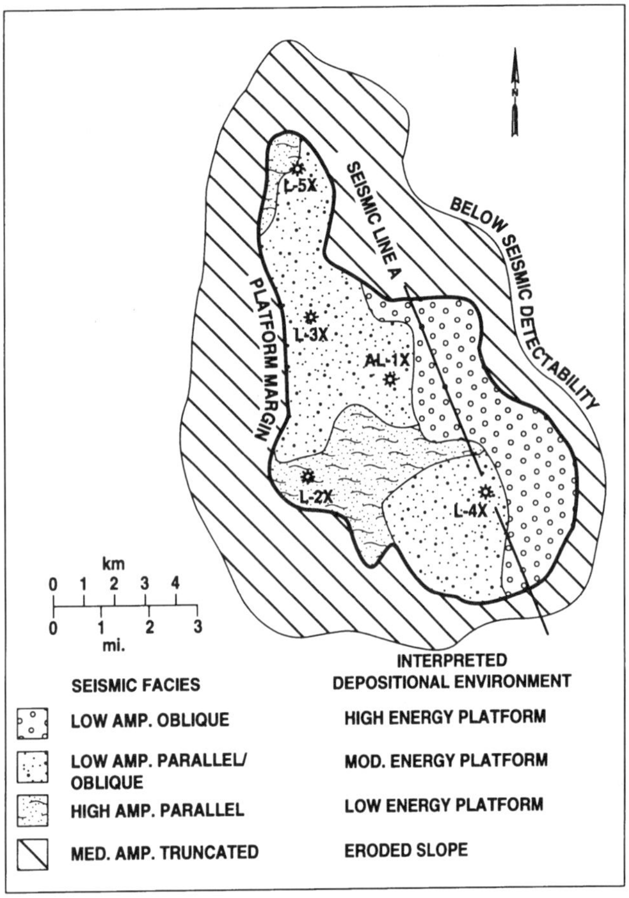

Fig. 25. Seismic facies map of combined middle–late Serravalian (13.8–11.5 Ma) highstand systems tracts. Note position of seismic line A, shown in Figure 24. (Modified from Rudolph and Lehmann, 1989.)

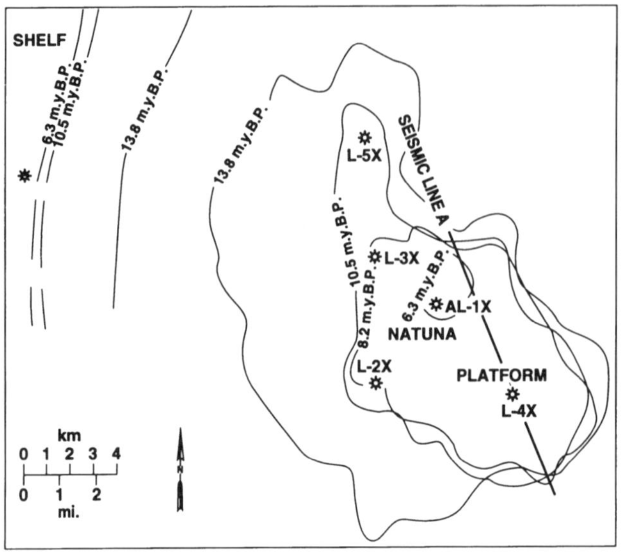

Fig. 26. Map of 13.8–6.3 Ma shelf margins of Natuna platform and adjacent shelf showing progressive backstepping. Note that backstepping is asymmetric, with greater shelf margin facies tract offset on the west and northwest edges of the platform and only minor offset on the east and northeast edges. (Modified from Rudolph and Lehmann, 1989.)

parallel distribution of low- and high-reflection amplitudes in the late highstand. Seismic modeling and core descriptions show that low-reflection amplitudes correspond to zones where the Nahr Umr overlies porous coral-rudist buildup facies, whereas high-reflection amplitudes occur where the Nahr Umr overlies dense, low-porosity (deeper water) carbonate.

Seismic-guided correlation of well logs and lithofacies from core descriptions provides a high-resolution sequence stratigraphic framework for the middle Shuaiba that better explains lithofacies and porosity-permeability distributions (Figure 32) (Calavan et al.,1992). The middle Shuaiba composite sequence appears to be composed of five (fourth-order) sequences. These consist of a transgressive sequence, an early highstand sequence, and three late highstand sequences. Sequences are recognized on the basis of the stacking patterns of lithofacies assemblages, commonly showing initially landward-stepping (retrogradational) facies followed by seaward-stepping (progradational) facies. This pattern can be interpreted in terms of systems tract development within a sequence. Parasequences are poorly developed in the Shuaiba at Bu Hasa, due both to the subtidal setting and to generally poor development in many Lower Cretaceous ramp successions.

Figure 32 is a south-north well log and lithofacies cross-section across Bu Hasa showing the distribution of sequences and porous versus nonporous zones. Sequences exhibit the following features: (1) a progressive increase in depositional slope, with a change from aggradation (transgressive systems tract, <1° slopes) to progradation (highstand systems tract, 5° slopes); (2) an increase in platform height from <30 to >120 m (<100 to >400 ft); (3) a trend from broad, ramp-margin, algal (*Lithocodium-Bacinella*) biostrome development (transgressive systems tract) to narrow, ramp-margin, coral-stromatoporoid-rudist buildup development (highstand systems tract); and (4) an overall increase in sequence-base, nonchalky mudstone-wackestone in the late highstand sequence set. The major reservoir facies at Bu Hasa include (1) *Lithocodium-Bacinella* wackestone-boundstone in the lower Shuaiba composite sequence and the transgressive systems tract of the middle composite sequence; (2) rudist bivalve-stromatoporoid boundstone-grainstone in the highstand systems tract of the middle composite sequence; and (3) chalky mudstone-wackestone that is common to both lower and

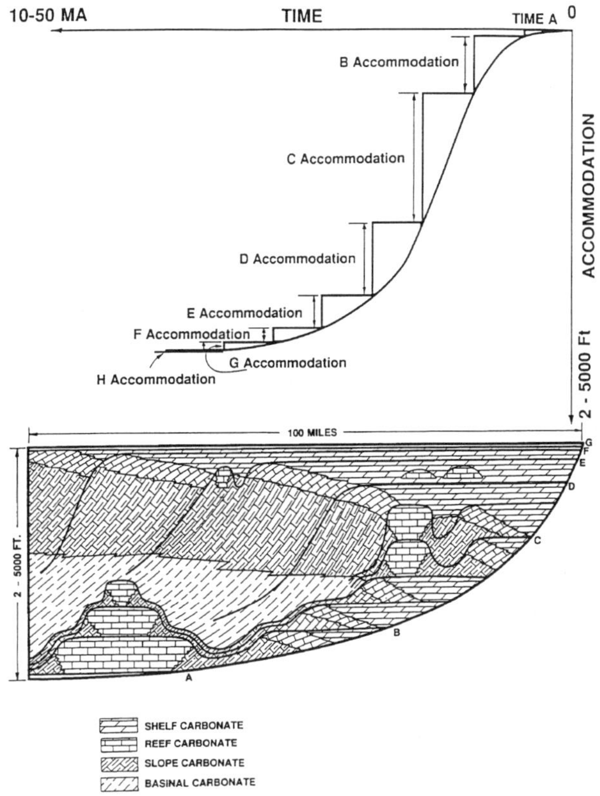

Fig. 27. Model for development of basin fill during period of increasing long-term accommodation followed by period of decreasing long-term accommodation. Many productive buildups form as isolated platforms or pinnacles during period of increasing long-term accommodation. (From Greenlee and Lehmann, 1993.

middle Shuaiba composite sequences. Reservoir porosity was enhanced by subaerial exposure and meteoric phreatic leaching at the ~112 Ma sequence boundary and at the 109.5 Ma sequence boundary. The Bab Member is interpreted to contain three sequences, with a prominent shale-rich unit in the basal sequence. Shuaiba reservoirs are sealed by the late Albian Nahr Umr Formation, which consists of deeper water shales and siltstones that are the downdip equivalent of the Burgan deltaic sandstones of Kuwait.

The sequence stratigraphy of the Shuaiba Formation shows the dominant control of two scales of stratal units on the stratal architecture, lithofacies distribution, and porosity distribution of a carbonate ramp. Sequence development is the cause of tongues of deeper water, low-porosity facies in the Shuaiba which serve as vertical barriers to flow. Facies tract offset at composite sequence boundaries influences regional reservoir compartmentalization. Examples of facies tract offset include landward offset across the ~112 Ma sequence boundary and seaward offset across the top middle Shuaiba sequence boundary associated with deposition of the Bab sequence in the basin. Facies tract offset associated with sequence boundaries in the middle Shuaiba late highstand systems tract/sequence set influences reservoir compartmentalization in North Bu Hasa. The change in ramp margin facies type from biostromal *Lithocodium-Bacinella* boundstone to biohermal rudist-stromatoporoid boundstone is an excellent example of lithofacies differentiation and a cause of variable lateral continuity of reservoir quality porosity in the middle Shuaiba highstand systems tract/sequence set.

Middle Permian San Andres Formation, Permian Basin, U.S.A.

The uppermost Leonardian–lower Guadalupian (Permian) San Andres Formation comprises a 390 m thick carbonate ramp system that rimmed the Delaware and Midland basins in West Texas and southeast New Mexico (King, 1942; Silver and Todd, 1969; Meissner, 1972; Kerans and Fitchen, 1995) (Figure 33). Basinal equivalents of the San Andres Formation consist of a 150 m thick basinal carbonate unit (Cutoff and upper Bone Spring formations) overlain by an ~660 m thick deep-water siliciclastic unit (Brushy Canyon and lower Cherry Canyon formations) (Figure 34). The San Andres Formation has long been of interest because it contains major oil reservoirs in the subsurface of the Permian Basin (cumulative production >9 billion bbl) (Kerans et al., 1994). The unit produces primarily from structural-stratigraphic traps below major intraformational unconformities or sequence boundaries that mark significant basinward shifts in sealing inner ramp evaporites. In this example, sequence stratigraphy provides a framework within which the distribution, thickness, and quality of reservoir- and seal-prone facies are highly predictable. A key element of this framework is the recognition that the composition and geometry of facies tracts, and in particular the ramp crest facies tract (which contains the most porous and permeable facies), changes as a function of position within sequences and composite sequences.

Regional Setting

The Delaware and Midland basins are part of a segmented foreland basin that formed in the Late Mississippian–middle Wolfcampian (Permian). The San Andres Formation represents a subtropical carbonate ramp to rimmed shelf system that formed in this foreland setting during a postdeformational period of relatively slow

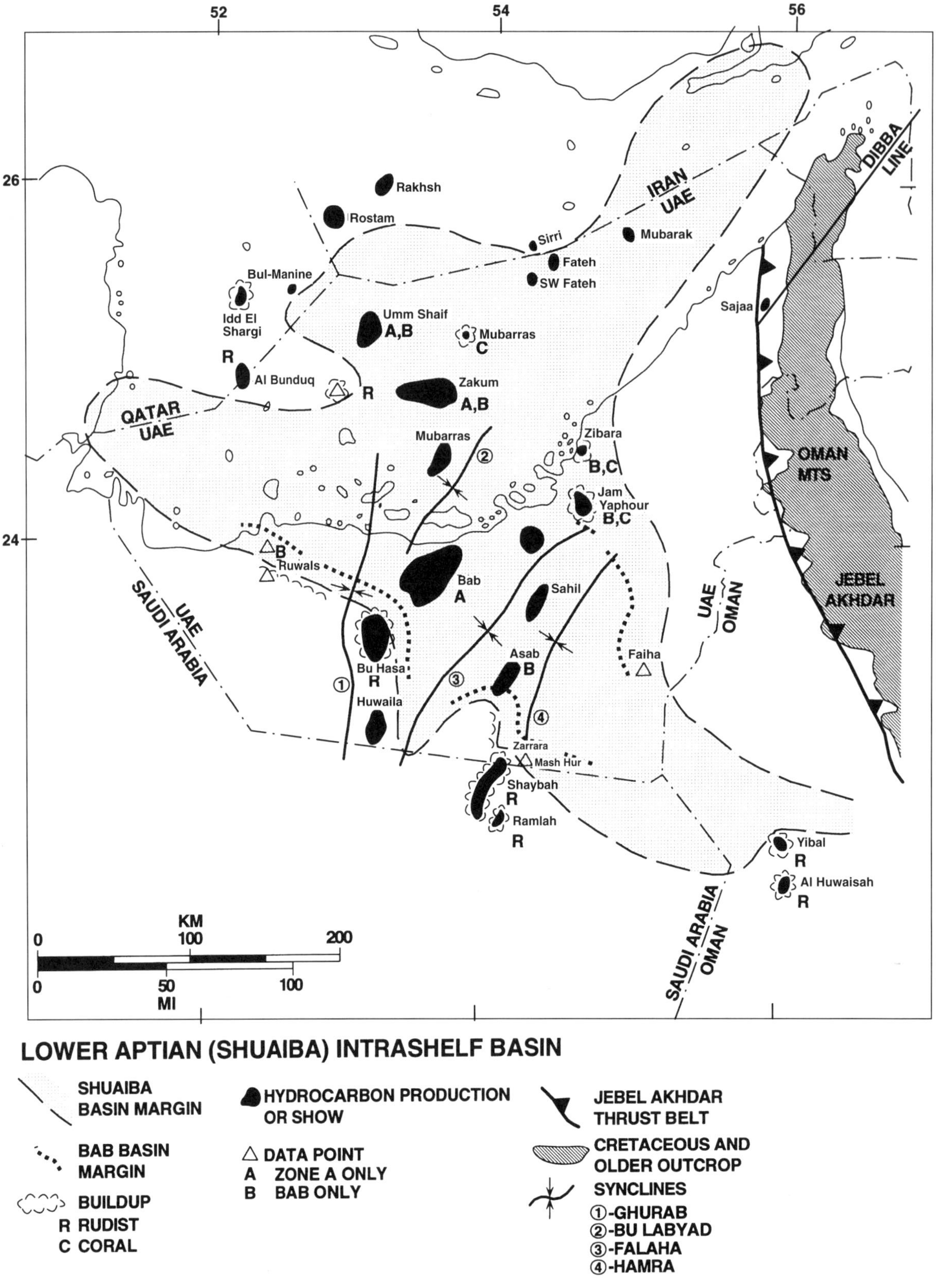

Fig. 28. Map of southern Arabian Gulf showing trend of lower Aptian ramp margin and intrashelf basin, distribution of fields that produce from the Shuaiba Formation, and associated structural trends. Note location of the Bu Hasa field on the southern margin of the intrashelf basin in the U.A.E.

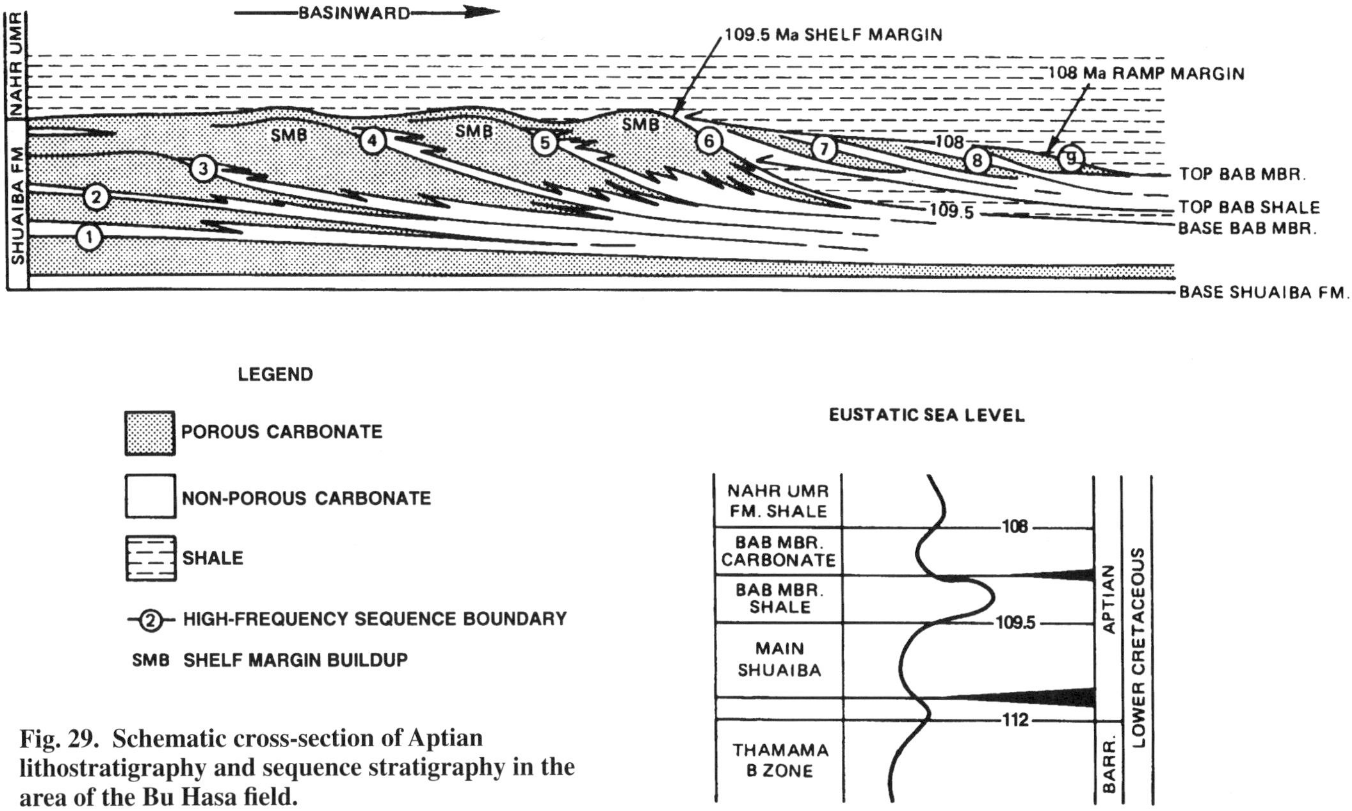

Fig. 29. Schematic cross-section of Aptian lithostratigraphy and sequence stratigraphy in the area of the Bu Hasa field.

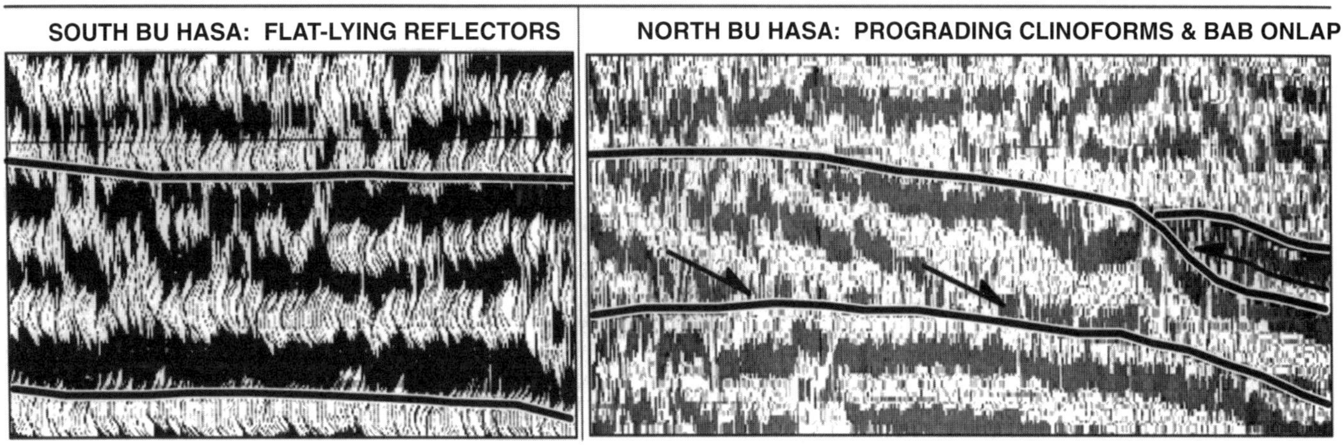

Fig. 30. Contrasting seismic character of the southern and northern parts of the Bu Hasa field.

subsidence. The majority of postdeformational sequences in the Permian Basin exhibit cyclic and reciprocal sedimentation patterns (Meissner, 1972; Wilson, 1967; Wilson, 1975) consisting of basin-centered siliciclastics bypassed across subaerially exposed platforms during lowstand and carbonates deposited on the platform during transgression and highstand. Siliciclastic delivery systems were dominantly eolian in origin and sourced from the Ancestral Rocky Mountains; basin-centered siliciclastics consist of eolian-marine turbidites (Fischer and Sarnthein, 1988). Sequences stack into aggradational and progradational sequence sets, which comprise composite sequences.

The San Andres Formation is bound below by the Clear Fork Group, which is part of a second-order composite sequence characterized in most parts of the Permian Basin by major ramp-margin progradation (Silver and Todd, 1969; Meissner, 1972; Handford, 1981; Glaser et al., 1990; Fitchen et al., 1995). The San Andres is overlain by the Grayburg Formation, which is part of a

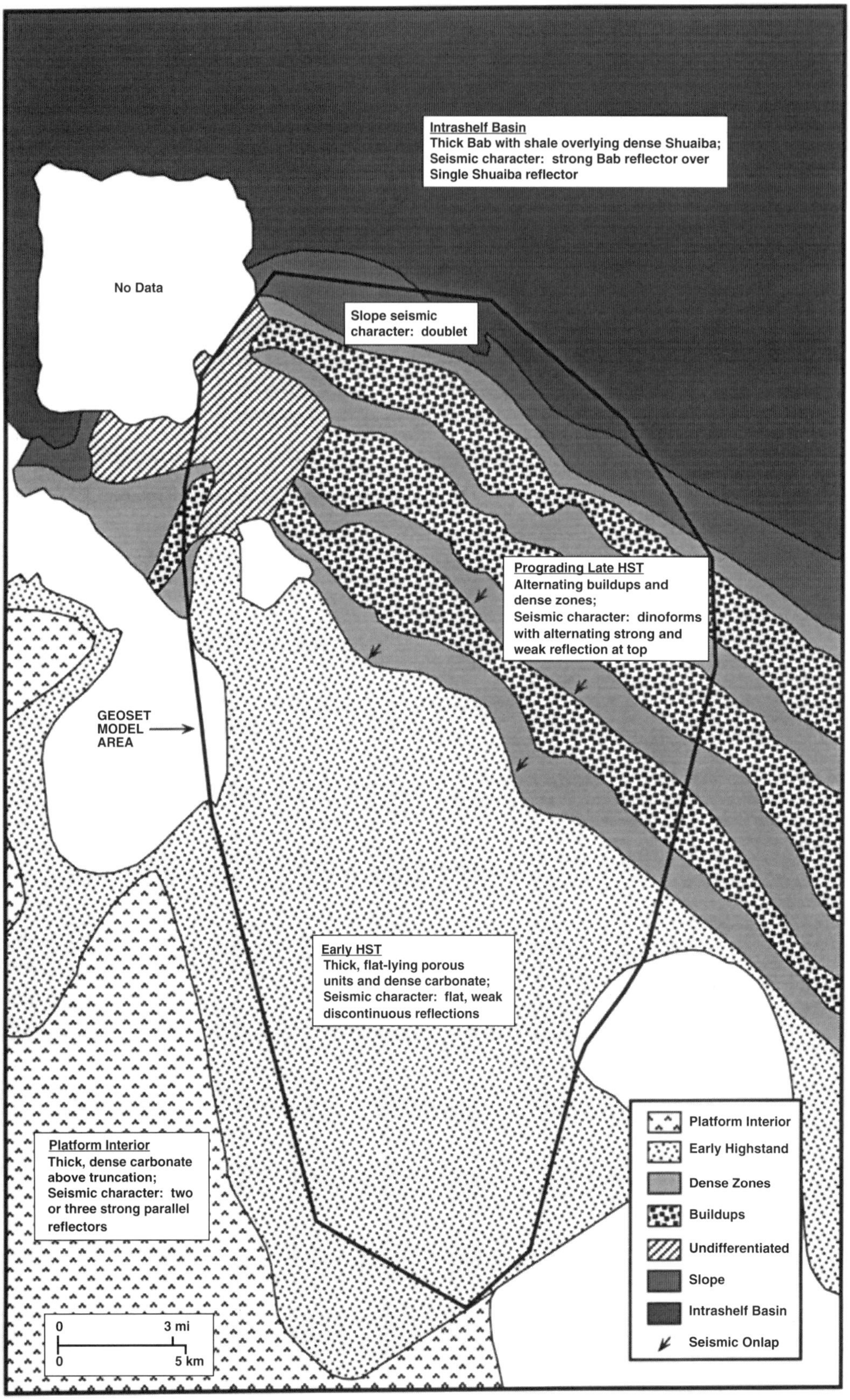

Fig. 31. Seismic facies map of the Bu Hasa field area, showing differentiation of early and late highstand systems tract and buildup trends of late highstand sequence set of the middle Shuaiba composite sequence.

younger third-order composite sequence. The San Andres has been subdivided into lower, middle, and upper lithostratigraphic units (Sarg and Lehmann, 1986). The lower San Andres Formation consists of aggradational carbonate ramp strata. The middle San Andres consists of progradational carbonate ramp strata that evolved from low-angle (<1°) to high-angle (3–5°) ramp margins. The top of the middle San Andres Formation marks a major period of sea level lowstand, unconformity development, and bypass of eolian siliciclastic systems across the ramp top and ramp margin (Kerans et al., 1993). This period is represented by the basin-centered Brushy Canyon Formation. Subsequent transgression of the ramp led to the deposition of the upper San Andres Formation, which consists of progradational, high-angle, ramp to grain-rich rimmed shelf strata. These strata evolved from pure carbonate systems to mixed carbonate-siliciclastic systems (Sonnenfeld, 1991; Kerans and Fitchen, 1995). The upper San Andres Formation is bisected by a subaerial unconformity that represents a second episode of eolian siliciclastic bypass to the basin (the Cherry Canyon Tongue). The youngest upper San Andres Formation consists of high-angle (5–15°), mixed carbonate-siliciclastic, outer shelf, and slope strata that prograded basinward over the Cherry Canyon Tongue.

Sequence Stratigraphic Framework

The San Andres Formation is subdivided into two third-order composite sequences that together represent a 2–4 m.y. duration (Figure 34). The lower composite sequence consists of six (fourth-order) sequences. The lower two are Leonardian and comprise a transgressive sequence set. The upper four are Guadalupian and comprise early and late highstand sequence sets. The upper composite sequence contains nine sequences. The lower seven are entirely basin-restricted, deep-water siliciclastic sequences equivalent to the Brushy Canyon Formation, which onlap a submarine erosional surface at the basin margin. These sequences compose a lowstand sequence set, which is equivalent to a regionally extensive karsted unconformity on the platform. The upper two sequences in this composite sequence represent transgressive and highstand sequence sets, respectively. Sequences are composed internally of parasequences, which can be mapped laterally through several different facies tracts (Kerans and Fitchen, 1995). The geometry of parasequences exerts an important control on the number of parasequences encountered in any one vertical section or well log. In steeply offlapping oblique progradational shelf margin and slope facies tracts of the youngest San Andres sequence, vertical sections may only intersect a fraction of the parasequences that make up the sequence. In low-angle aggradational ramp sequences such as the oldest San Andres sequence, all parasequences of the sequence may be intersected in a single vertical section.

In the lower San Andres composite sequence, there are systematic trends in (1) stratal geometry, (2) facies tract composition and width, (3) ramp margin progradation-aggradation ratio, (4) symmetry of systems tracts and parasequence sets, and (5) magnitudes of facies tract offset and subaerial exposure at sequence boundaries (Kerans and Fitchen, 1995) (Figure 35). These trends are interpreted to reflect a stratigraphic response to composite relative changes in sea level, as determined from a hierarchy of three scales of stratal units: third-order composite sequences, fourth-order sequences, and fifth-order parasequences.

The transgressive sequence set of the lower San Andres composite sequence is composed of two low-angle (<1°), ramp sequences (Leonardian 7 and 8, or L7 and L8). These sequences are characterized by >48 km (>30 mi) wide outer ramp facies tracts composed of skeletal wackestone and packstone. The ramp crest facies tract, consisting of grainstone and packstone that record shallow-water deposition within stormweather wave base, is a good basis for comparison among different sequences. The ramp crest facies tracts in the sequences are mostly greater than ~5 km (~3 mi) wide, are sheetlike at their base but narrow upward through aggradation, and are dominated by open-marine skeletal grains. Sequence boundaries in the L7 and L8 sequences are represented by seaward-stepping (progradational) to landward-stepping (retrogradational) turnarounds in parasequence stacking patterns, with no subaerial exposure present within ~48 km (30 mi) of the platform margin. Facies tract offset is greater than 40 km (25 mi) across the basal composite sequence boundary, is negligible to slightly landward for the boundary between the L7 and L8 sequences, and is greater than 13 km (8 mi) landward for the sequence boundary at the top of the L8 sequence.

The early highstand sequence set is composed of the Guadalupian 1 and 2 (G1 and G2) sequences. These are low-angle (0.5–1°) sigmoid progradational sequences with a significant component of aggradation. The G1 sequence downlaps the top of the L8 sequence at a low angle and contains the maximum flooding surface of the lower San Andres composite sequence. The ramp crest facies tract of the G1 sequence is not developed in the study area, but presumably is present somewhere landward of the area. The ramp crest facies tract of the G2

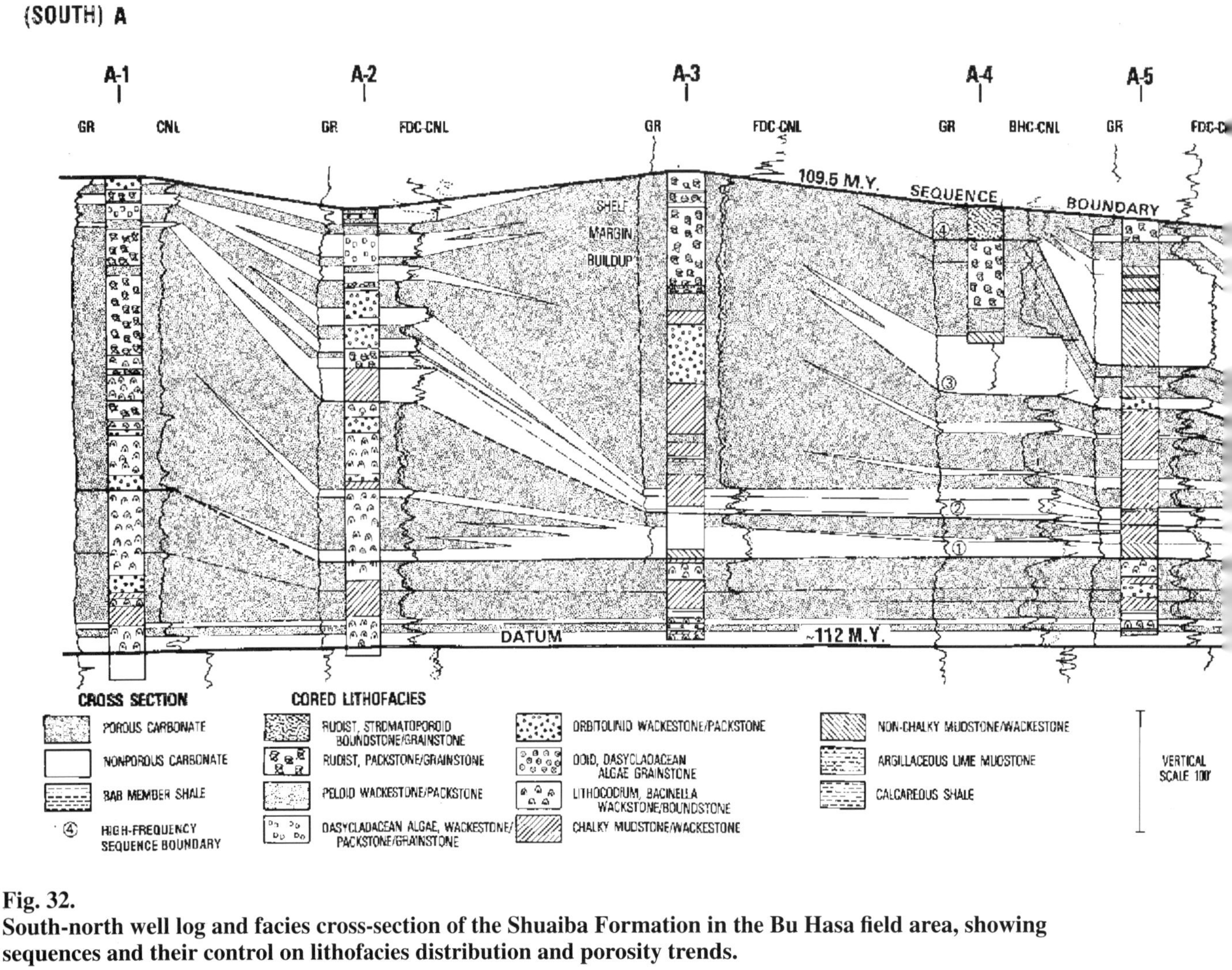

Fig. 32.
South-north well log and facies cross-section of the Shuaiba Formation in the Bu Hasa field area, showing sequences and their control on lithofacies distribution and porosity trends.

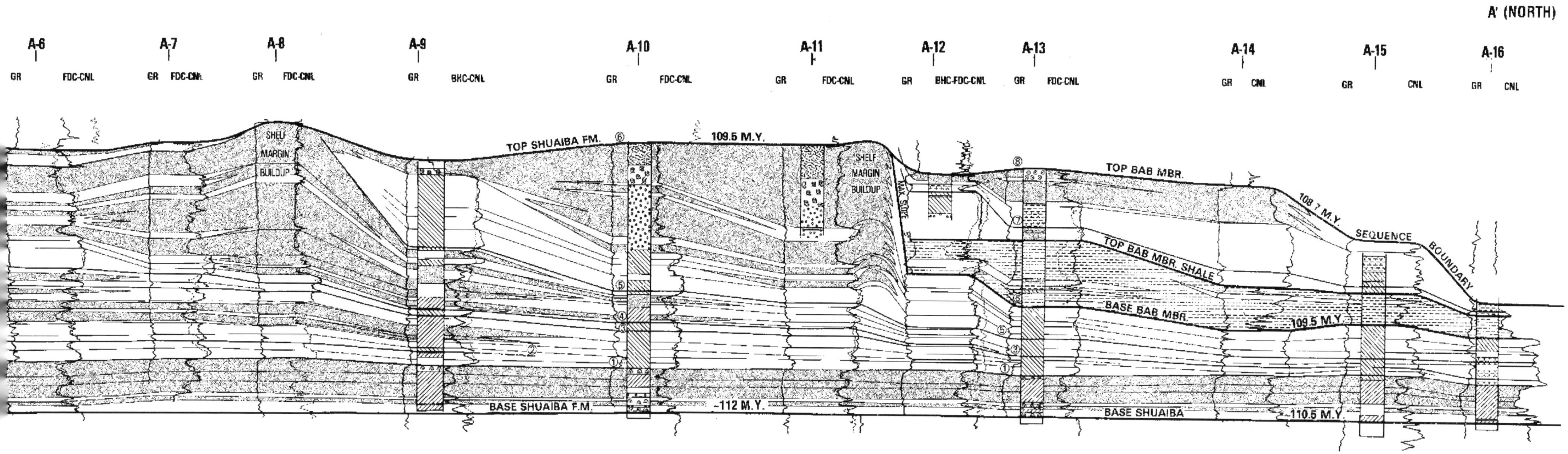

A' (NORTH)
A-6
A-7
A-8
A-9
A-10
A-11
A-12
A-13
A-14
A-15
A-16
GR
FDC-CNL
BHC-CNL
BHC-FDC-CNL
CNL
SHELF MARGIN BUILDUP
TOP SHUAIBA FM.
109.5 M.Y.
TOP BAB MBR.
108.7 M.Y.
SEQUENCE BOUNDARY
TOP BAB MBR. SHALE
BASE BAB MBR.
109.5 M.Y.
BASE SHUAIBA F.M.
~112 M.Y.
BASE SHUAIBA
~110.5 M.Y.

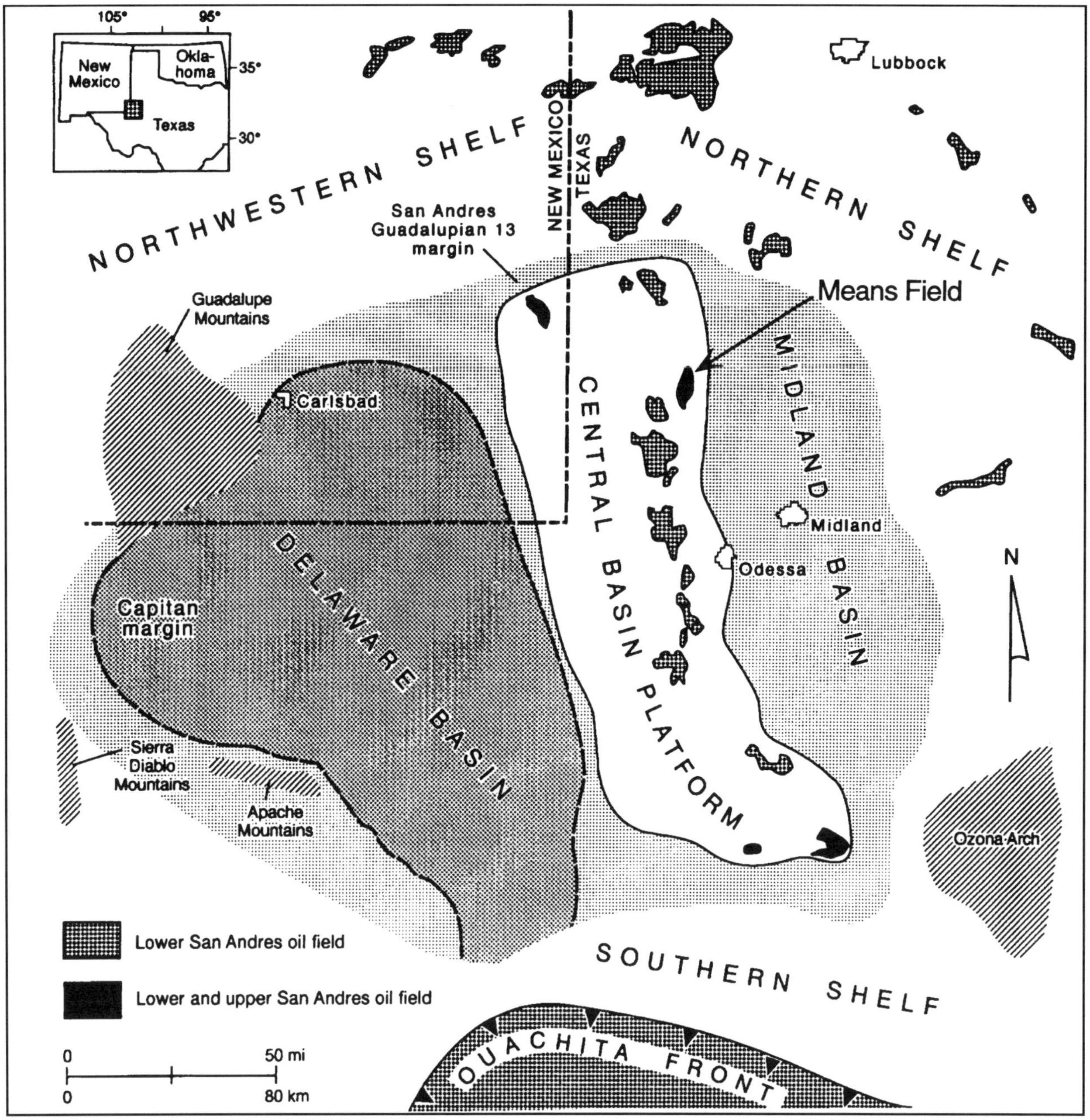

Fig. 33. Map of the Permian Basin region of West Texas and New Mexico, showing San Andres paleogeography, the location of major (>10 million bbls of oil) reservoirs, and the locations of important San Andres outcrops. Note the location of Means field. (From Kerans and Fitchen, 1995.)

sequence extends for up to 8 km (5 mi) in a dip direction and consists of an aggradational succession of discontinuous ooid-peloid packstone and grainstone facies that pass shelfward into deeper water middle ramp facies. The landward offset of the ramp crest facies tract in the G1 sequence, followed by the seaward offset in the G2 sequence, represents the landward- to seaward-stepping turnaround in stacking pattern in the composite sequence. No subaerial exposure is developed in either of the two early highstand sequences. The G2 sequence appears to represent relatively unique conditions of accommodation, topography, and sediment production in that it contains the majority of outer ramp to ramp crest bioherms documented from the lower San Andres composite sequence on the outcrop.

The late highstand sequence set consists of the Guadalupian 3 and 4 (G3 and G4) sequences (Figure 35). These contain relatively higher angle (1–2°, locally up to 15°), sigmoid progradational (G3) to oblique progradational (G4) stratal geometries, with higher

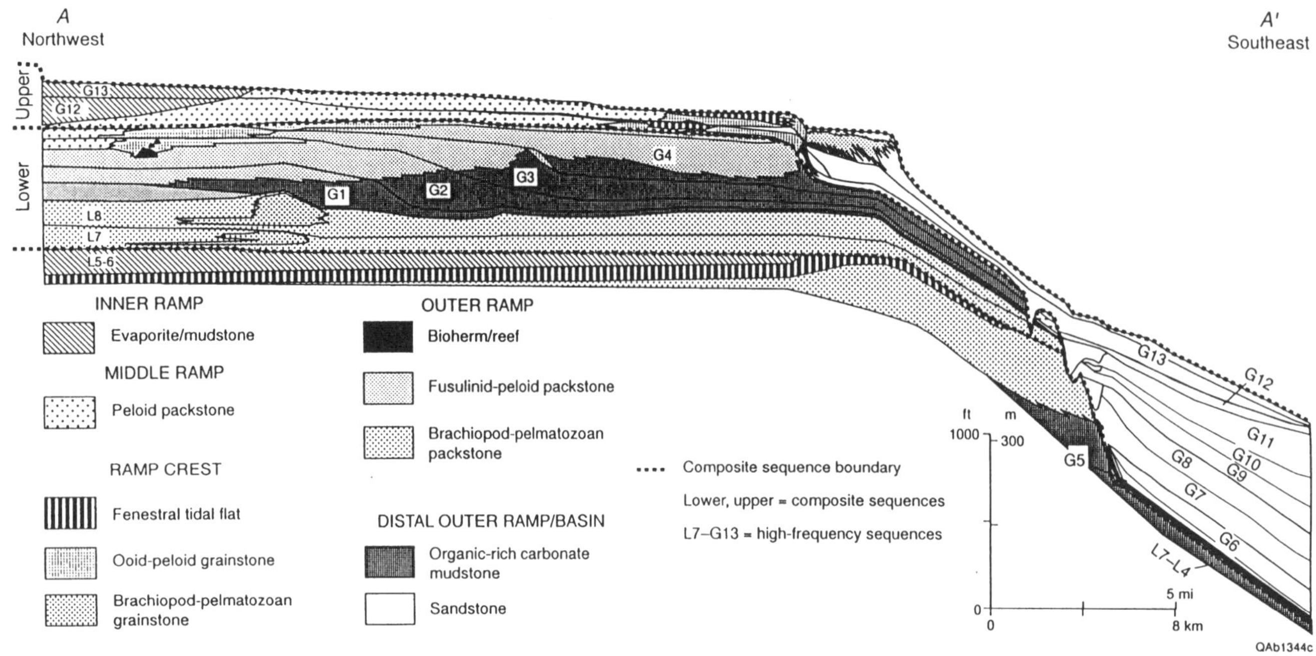

Fig. 34. Stratigraphic cross-section showing lithofacies distribution, lithostratigraphy, and sequence stratigraphy for the San Andres Formation and basinal equivalents in the Guadalupe and Brokeoff mountains, New Mexico. (From Kerans and Fitchen, 1995.)

progradation-aggradation ratios and thinner and more dip-elongate ramp crest geometries. The dip width of the G4 sequence ramp crest is >16 km (10 mi). However, because this sequence is oblique progradational, this width reflects ramp crest progradation rather than the static width of the ramp crest in individual parasequences. The facies composition of late highstand, ramp crest facies tracts varies from (1) offshore (shore-detached), ooid-peloid grainstone and packstone deposited in tidal bar belts in the G3 sequence to (2) ooid-peloid grainstone and packstone deposited in shore-attached shoreface complexes in the G4 sequence. There is a pronounced 1.6 km (1 mi) seaward offset in the ramp crest facies tract within the late highstand sequence set, from the G3 to G4 sequence. Subaerial exposure at sequence boundaries is first developed at the top of the G3 sequence, across the ramp crest, and extends for ~1.6 km (~1 mi) in a dip direction. Karst features extend across the ramp crest at the top of the G4 sequence for a dip distance of >11 km (>7 mi) and penetrate up to 12 m (40 ft) into the sequence. This sequence boundary coincides with the composite sequence boundary and is considered to be a major eolian siliciclastic bypass surface, equivalent to >450 m (>1500 ft) of deep-water sandstone in the Delaware Basin.

The upper San Andres composite sequence reflects lower third-order accommodation than the lower San Andres composite sequence, with the majority of sequences volumetrically partitioned into the lowstand sequence set, and abundant eolian-marine siliciclastics in slope and basinal facies tracts of all sequences of the composite sequence (Figure 34). The lowstand sequence set (Guadalupian 5 to 11) consists of seven basinally restricted sequences that comprise a marine onlap wedge. This wedge onlaps a submarine erosional surface at the basin margin and is equivalent to a subaerial unconformity at the top of the lower San Andres composite sequence on the platform. Sequences consist of sets of parasequences, which are composed of sandstone units bound by siltstones. Sequence boundaries are marked by turnarounds from siltstone-dominated parasequences, reflecting platform margin flooding and landward migration of an eolian erg, to sandstone-dominated parasequences, reflecting migration of an erg to the platform margin (Fischer and Sarnthein, 1988; Gardner, 1992).

The transgressive sequence of the composite sequence (Guadalupian 12) consists of a siliciclastic lowstand systems tract and carbonate transgressive and highstand systems tracts. The lowstand consists of a thick siltstone section at the basin margin, reflecting a landward migration of eolian systems, and minor sandstone nearer to the ramp margin. Platform carbonates of the transgressive and highstand are relatively thin and exhibit a low progradation-aggradation ratio. The G12 sequence is capped by an unconformity with minor karst development near the ramp margin. The highstand se-

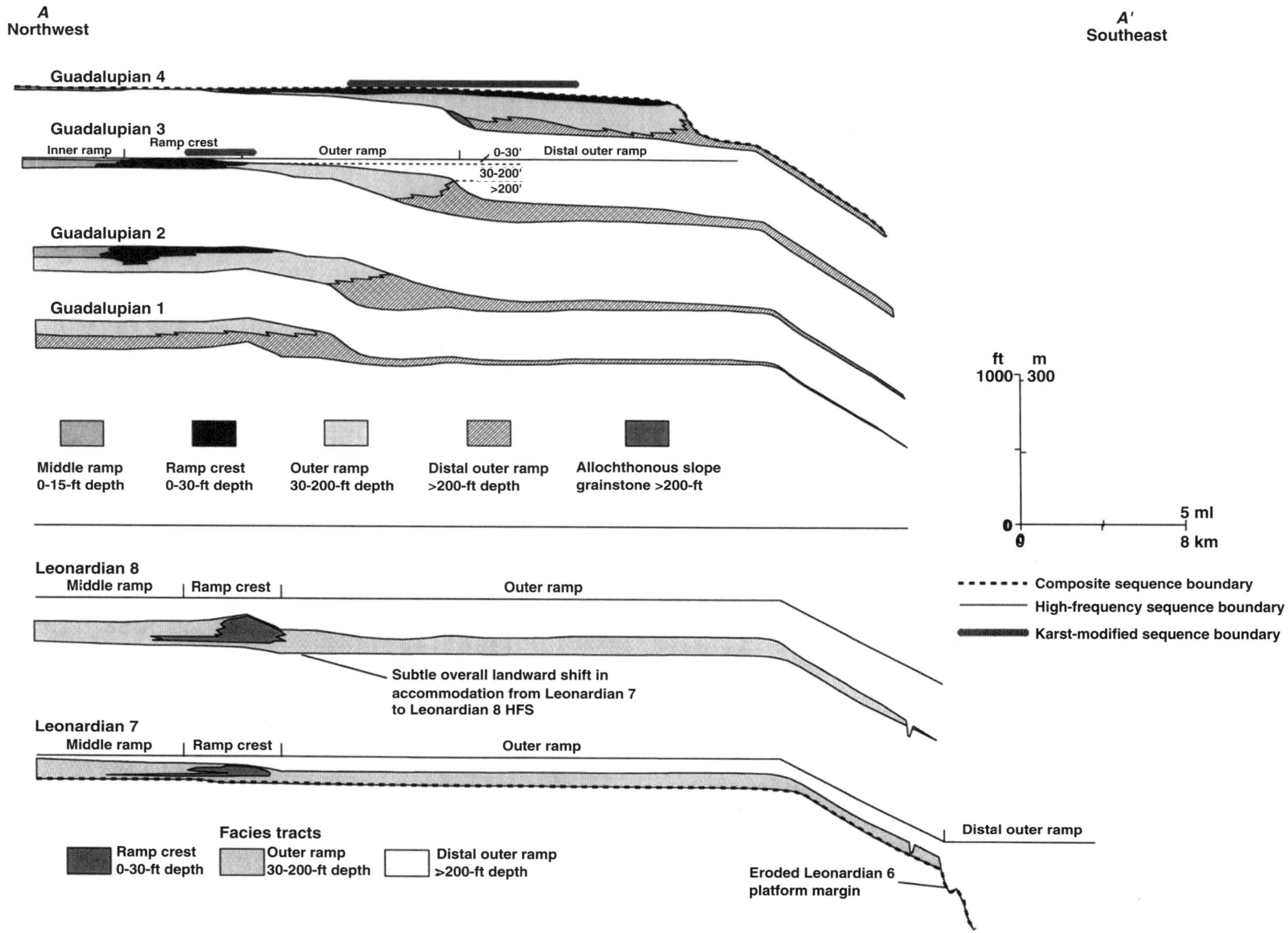

Fig. 35. Stacking pattern diagram of L7–G4 sequences of lower San Andres composite sequence, illustrating changes in sequence attributes as a function of the third-order accommodation cycle.

quence of the composite sequence (Guadalupian 13) contains (1) a thick, onlapping basin and slope lowstand sandstone, (2) a thin, mixed carbonate-siliciclastic, transgressive systems tract, and (3) an offlapping, mixed carbonate-siliciclastic, highstand systems tract. The G13 highstand has a high progradation-aggradation ratio and exhibits sigmoid (early highstand) to oblique (late highstand) clinoform geometries (Sonnenfeld and Cross, 1993; Fitchen, 1993). The G13 sequence is capped by a regionally extensive subaerial unconformity and is onlapped by overlying sequences equivalent to the Grayburg Formation.

Integrated Sequence Analysis: Means Field Example

The Means field provides a case study for the application of the outcrop-derived sequence stratigraphic framework of the San Andres to the subsurface. Means field is located on the eastern side of the Central Basin Platform (Figure 33) in a position that straddles the Leonardian Clear Fork basin margin. The Means field data set includes 2-D seismic, well logs, and core. Production at Means is from the lower and upper San Andres composite sequences. A seismic profile across Means field (Figure 36) shows geometries of the San Andres that are similar to those seen in the Guadalupe Mountains, except that at Means, these geometries occur across a much narrower dip transect.

The top of the Clear Fork platform is marked by toplap and truncation at the platform margin and by onlap at the toe-of-slope. Above this is an 80 ms seismic sequence, which is composed of shingled clinoforms that downlap the top Clear Fork sequence boundary and toplap a major seismic sequence boundary. At the basin margin, this upper sequence boundary is onlapped by a basin-restricted seismic sequence consisting of two or three onlapping reflections succeeded by offlapping, toplapping clinoforms. The major downward-basinward shift associated with this sequence boundary suggests that the lower seismic sequence is equivalent to the lower San Andres composite sequence, while the upper seismic sequence is equivalent to the lowstand sequence set (Brushy Canyon Formation) of the upper San Andres composite sequence. Above the second seismic sequence at Means, there are two additional seismically resolvable sequences that appear to offlap one another. The lower sequence may downlap the top of the basin-restricted seismic sequence and has a geometry similar to that of the G12 sequence in outcrop. The upper sequence exhibits onlap at the basin margin. On the platform it is one cycle thick and thus does not exhibit internal reflection geometries. The upper sequence boundary of this sequence is defined by onlap of reflections in the overlying sequence (Grayburg equivalent). Its gross geometry and position is similar to that of the G13 sequence, and as on the outcrop, the overlying Grayburg contains onlapping shelf sandstones.

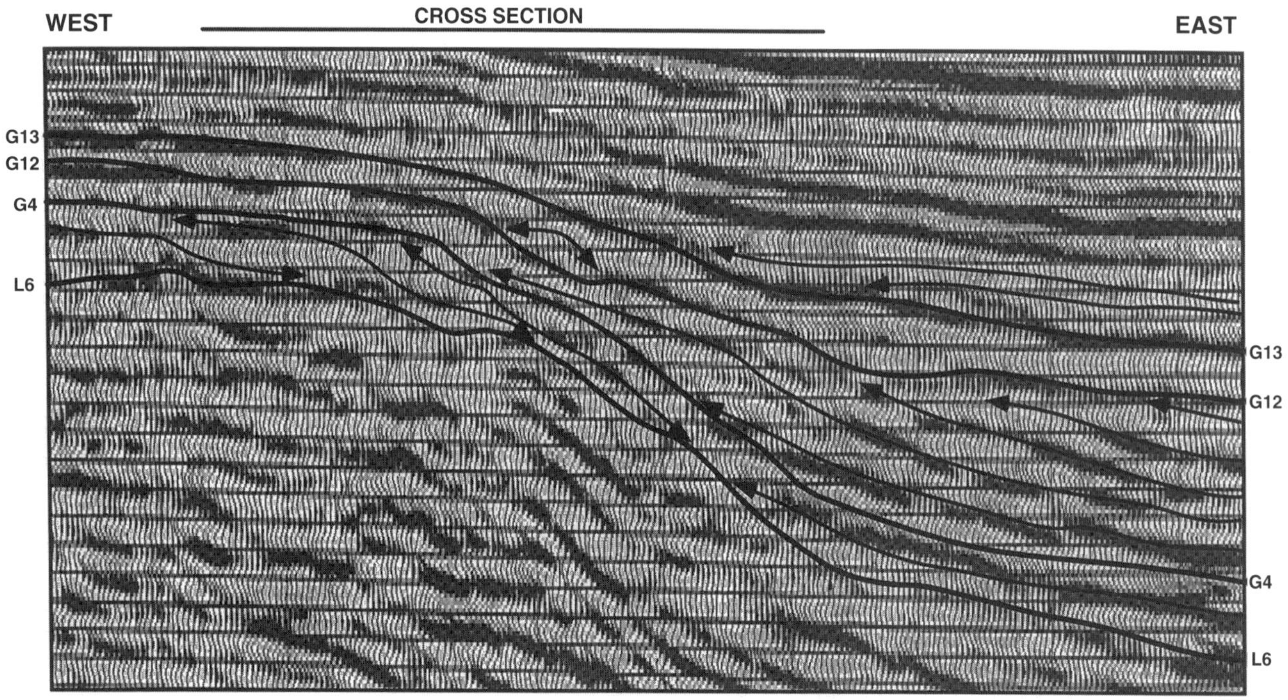

Fig. 36. Interpreted dip-oriented seismic line across Means field, Central Basin platform, Texas, showing seismic sequences of the San Andres Formation. Cross-section line indicates location of parallel cross-section shown in Figure 37.

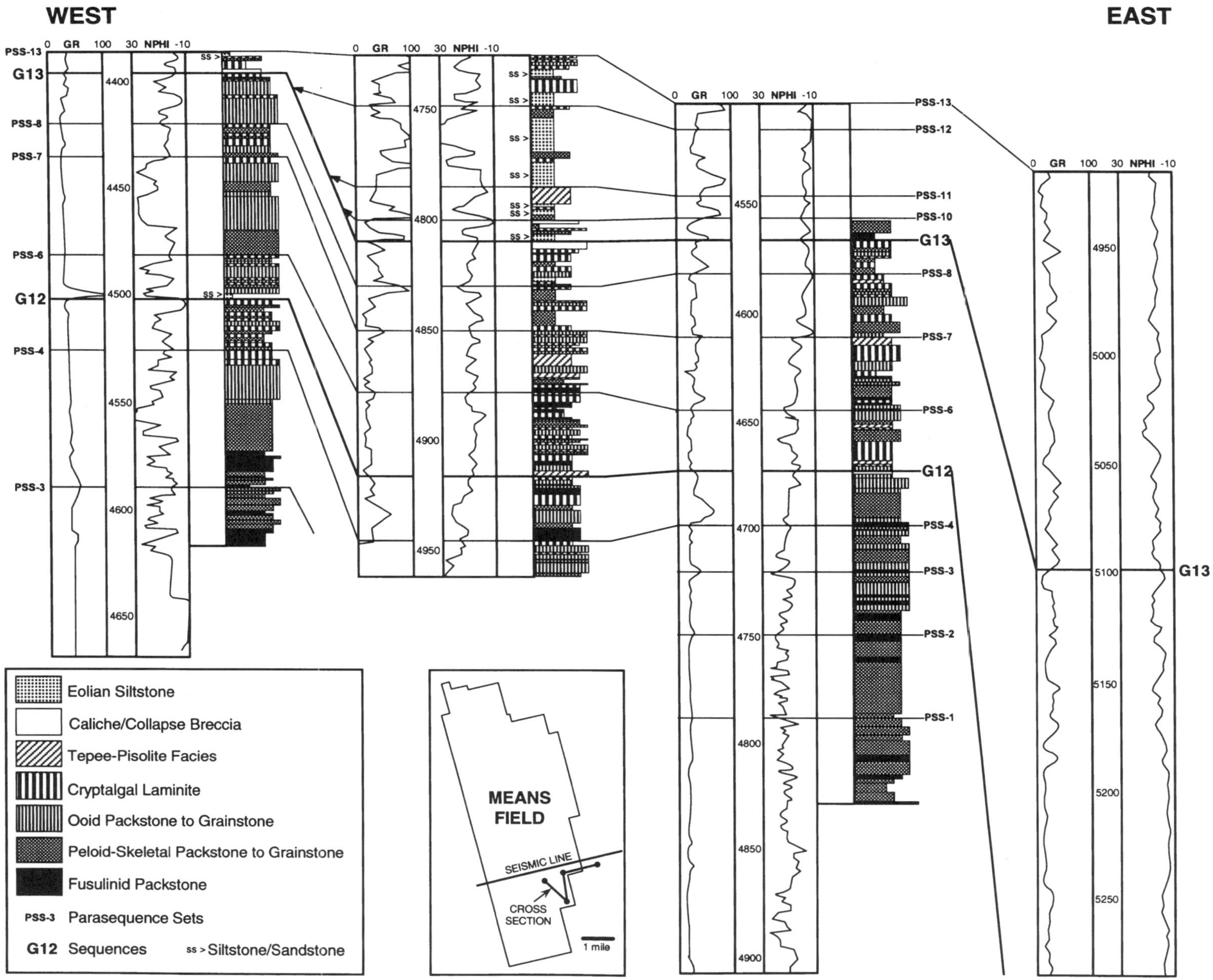

Fig. 37. Dip-oriented well log and core cross-section across Means field located parallel to the seismic line of Figure 36. Note the development and correlation of parasequences and parasequence sets within seismically defined sequences.

The gross seismic geometries present at Means field suggest a good correspondence to the outcrop model in terms of composite sequence resolution. In general, it is more difficult to resolve sequences in the lower San Andres composite sequence due to their low-angle ramp margin geometries, pure carbonate lithology, and poorly developed (pure carbonate) lowstand systems tracts. In the upper San Andres composite sequence, all three sequence sets are well imaged, and general accommodation trends are similar to the outcrop.

Figure 37 illustrates the stacking patterns and lateral correlation of parasequences, parasequence sets, and systems tracts within seismically defined sequences for a part of Means field. Parasequences were defined from core descriptions on the basis of upward-shallowing lithofacies successions. Parasequence sets were defined as groups of parasequences that form an overall thinning- and shallowing-upward succession. Core-defined parasequence sets were calibrated to well log response and correlated across the reservoir. Particularly evident at Means is the overall shallowing-upward character of the upper San Andres with a distinct seaward facies tract offset at the top G12 boundary, coincident with subaerial exposure and siliciclastic bypass to the basin. Also note the seaward thinning of peritidal (cryptalgal laminite) facies in the upper part of the G12 sequence. Exposure, siliciclastic bypass, and seaward facies tract offset are also well developed at the top of the San Andres and in the basal Grayburg. These relationships further corroborate the validity of the outcrop framework as an analog to Means field.

Conclusions

Carbonate sequence stratigraphy provides a set of concepts and methodologies to increase our understanding of the distribution, geometry, and chronostratigraphic relationships of lithostratigraphic units in carbonate platforms, by integrating available and pertinent data. Important to the real-world analysis of carbonate sequence stratigraphy is the idea that stratigraphic resolution is largely dependent on the data set (regional coverage, density, and resolution), platform geometry and age, and depositional setting.

The major chronostratigraphically significant units in sequence stratigraphy are parasequences, parasequence sets, systems tracts, sequences, sequence sets, composite sequences, and supersequences. All of these stratal units are products of cyclical changes in accommodation and/or sediment supply (specifically, the accommodation to sediment supply ratio). Stratal units are characterized by systematic vertical and lateral distributions of lithofacies, facies tracts, and depositional systems.

The construction of predictive sequence stratigraphic frameworks in carbonate platform successions depends on the recognition of a hierarchy of stratal units. In stratal unit hierarchies, the stratigraphic processes that form smaller scale units (e.g., parasequences) are modulated by stratigraphic processes that form intermediate- and large-scale units (e.g., sequences and composite sequences). An understanding of this modulation leads to enhanced prediction of lithofacies distributions at all scales. Stacking pattern analysis is a means of delineating hierarchies of stratal units.

Stratal patterns (geometries and terminations) are a stratigraphic response to changes in accommodation and sediment supply as well as to antecedent topography. Analysis of stratal patterns (seismic sequence analysis) provides an important technique for delineating large-scale stratal units. Stratal patterns should be calibrated by lithofacies distributions whenever possible. Stratal patterns combined with 1-D or 2-D stacking pattern analysis represents the most integrated approach to carbonate sequence stratigraphy. Stacking pattern analysis (1-D and 2-D) is the analysis of incremental shifts or changes in the sedimentologic character, volume, geometry, and geographic distribution of lithostratigraphic units among contiguous chronostratigraphic units. Analysis of rock volumes is complemented by the analysis of changes in the geometry, character, and diagenesis of hiatal stratal surfaces that bound chronostratigraphic units. Stacking pattern analysis provides the means to calculate and interpret facies offset, volumetric partitioning, and lithofacies differentiation.

Acknowledgments

The author wishes to thank Exxon Production Research Company for permission to publish this chapter, and acknowledges the major contributions of current and former Exxon stratigraphers P. A. Dunn, R. K. Goldhammer, S. M. Greenlee, P. J. Lehmann, D. J. Lehrmann, J. C. Mitchell, J. F. Sarg, and K. W. Rudolph to the development of concepts, case studies, and text figures included here. I credit L. F. Brown, Jr., W. Galloway, M. Gardner, R. Goldhammer, C. Kerans, M. Kominz, M. Sonnenfeld, and J. C. Van Wagoner for shaping my understanding of stratigraphy. An early draft of this manuscript benefitted from the reviews of W. Burgis, J. Kaufman, D. J. Lehrmann, R. G. Loucks, and J. C. Van Wagoner; however, any mistatements or mistakes that remain are the sole responsibility of the author.

References

Aigner, T., 1985, Storm depositional systems: dynamic stratigraphy in modern and ancient shallow-marine sequences, *in* Lecture Notes in Earth Sciences: Springer-Verlag.

Algeo, T. J., and Wilkinson, B. H., 1987, Periodicity of mesoscale Phanerozoic sedimentary cycles and the role of Milankovitch orbital modulation: J. Geol., **96**, 313-322.

Back, W., Hanshaw, B. B., Herman, J. S., and van Driel, J. N., 1986, Differential dissolution of a Pleistocene reef in the ground-water mixing zone of coastal Yucatan, Mexico: Geology, **14**, 137-140.

Bally, A. W., and Snelson, S., 1980, Realms of subsidence, *in* Miall, A. D., Ed., Facts and principles of world petroleum occurrence: Can. Soc. Petr. Geol. Mem., **6**, 9-75.

Barnaby, R. J., and Ward, W. B., 1995, Sequence stratigraphic framework, high-frequency cyclicity and 3-D heterogeneity: Grayburg Formation, Brokeoff Mountains, New Mexico, *in* Pause, P. H., and Candelaria, M. P., Eds., Carbonate facies and sequence stratigraphy: practical applications of carbonate models: PBS-SEPM Publ., **95-36**, 37-49.

Biddle, K. T., Schlager, W., Rudolph, K. W., and Bush, T. L., 1992, Seismic model of a progradational carbonate platform, Picco di Vallandro, the Dolomites, northern Italy: AAPG Bull., **76**, 14-30.

Boreen, T. D., and James, N. P., 1993, Holocene sediment dynamics on a cool-water carbonate shelf: Otway, southeastern Australia: J. Sed. Pet., **63**, 574-588.

Borer, J. M., and Harris, P. M., 1991, Lithofacies and cyclicity of the Yates Formation, Permian Basin: implications for reservoir heterogeneity: AAPG Bull., **75**, 726-779.

Bosellini, A., 1984, Progradation geometries of carbonate platforms: examples from the Triassic of the Dolomites, northern Italy: Sedimentology, **31**, 1-24.

Boss, S. K., Rasmussen, K. A., and Neumann, A. C., 1992, Early sequence evolution during carbonate platform flooding: a case study on Great Bahama Bank (abs.): IAS/SEPM Research Conference, Abstracts with Prog., 21-22.

Brown, Jr., L. F., and Fisher, W. L., 1977, Seismic-stratigraphic interpretation of depositional systems: examples from Brazilian rift and pull-apart basins, *in* Payton, C. E., Ed., Seismic stratigraphy—applications to hydrocarbon exploration: Am. Assn. Petr. Geol. Mem. **26**, 213-248.

Brown, Jr., L. F., Solis-Iriarte, R. F., and Johns, D. A., 1990, Regional depositional systems tracts, paleogeography, and sequence stratigraphy, upper Pennsylvanian and Lower Permian strata, north- and west-central Texas: Univ. Texas at Austin, Bur. Econ. Geol., Rept. of Invest. **197**.

Brown, A. A., and Loucks, R. G., 1993, Influence of sediment type and depositional processes on stratal patterns in the Permian basin-margin Lamar Limestone, McKittrick Canyon, Texas, *in* Loucks, R. G., and Sarg, J. F., Eds., Recent advances and applications of carbonate sequence stratigraphy: Am. Assn. Petr. Geol. Mem. **57**, 133-156.

Budd, D. A., Saller, A. H., and Harris, P. M., Eds., 1995, Unconformities and porosity in carbonate strata: Am. Assn. Petr. Geol. Mem. **63**.

Calavan, C. W., Hagerty, R. M., Mitchell, J. C., and Schutter, S. R., 1992, Integrated reservoir study and modeling of the Shuaiba reservoir, Bu Hasa field, Abu Dhabi, UAE: Proceedings, Soc. Petr. Eng., fifth Petr. Cong.

Campbell, A. E., and Stafleu, J., 1992, Seismic modeling of an Early Jurassic drowned carbonate platform: the Djebel Bou Dahar, High Atlas, Morocco: AAPG Bull., **76**, 1760-1777.

Campbell, C. V., 1992, Upper Elk Point megasequence, *in* Wendte, J., Stoakes, F. A., and Campbell, C. V., Eds., Devonian–Early Mississippian carbonates of the Western Canada Sedimentary Basin: a sequence stratigraphic framework: SEPM Short Course **28**, 145-162.

Collins, L. B., 1988, Sediments and history of the Rottnest Shelf, southwest Australia: a swell-dominated non-tropical carbonate margin: Sedimentary Geol., **60**, 15-49.

Corso, W., Austin, Jr., J. A., and Buffler, R. T., 1989, The Early Cretaceous platform off northwest Florida: controls on morphologic development of carbonate margins: Mar. Geol., 86, 1-14.

Crevello, P. D., Wilson, J. L., Sarg, J. F., and Read, J. F., Eds., 1989, Controls on carbonate platform and basin development: Soc. Econ. Paleont. Mineral. Special Publ. **44**.

Cross, T. A., Baker, M. R., Chapin, M. A., Clark, M. S., Gardner, M. H., Hanson, M. S., Lessenger, M. A., Little, L. D., McDonough, K.-J., Sonnenfeld, M. D., et al., 1993, Application of high-resolution sequence stratigraphy to reservoir analysis, *in* Eschard, R., and Doligez, B., Eds., Subsurface reservoir characterization from outcrop observations: Editions Technip, Paris, 11-33.

Crowell, J. C., 1978, Gondwanan glaciation, cyclothems, continental positioning, and climate change: Am. J. Sci., **278**, 1345-1372.

Davies, P. J., Symonds, P. A., Feary, D. A., and Pigram, C. J., 1989, The evolution of the carbonate platforms of northeast Australia, *in* Crevello, P. D., Wilson, J. L., Sarg, J. F., and Read, J. F., Eds., Controls on carbonate platform and basin development: Soc. Econ. Paleont. Mineral. Spec. Publ. **44**, 233-258.

del Olmo, W. M., and Esteban, M., 1983, Paleokarst development, *in* Scholle, P. A., Bebout, D. G., and Moore, C. H., Eds., Carbonate depositional environments: Am. Assn. Petr. Geol. Mem. **33**, 93-95.

Denny III, W. M., Austin, Jr., J. A., and Buffler, R. T., 1994, Seismic stratigraphy and geologic history of middle Cretaceous through Cenozoic rocks, southern Straits of Florida: AAPG Bull., **78,** 461-487.

de Boer, P. L., and Smith, D. G., Eds., 1994, Orbital Forcing and Cyclic Sequences: Internat. Assn. Sediment., Spec. Publ. **19**.

Dolan, J. F., 1989, Eustatic and tectonic controls on deposition of hybrid siliciclastic/carbonate basinal cycles: discussion with examples: AAPG Bull., **73**, 1233-1246.

Droxler, A. W., and Schlager, W., 1985, Glacial versus interglacial sedimentation rates and turbidite frequency in the Bahamas: Geology, **13**, 799-802.

Eberli, G. P., and Ginsburg, R. N., 1989, Cenozoic progradation of northwestern Great Bahama Bank, a record of lateral platform growth and sea level fluctuations, *in* Crevello, P. D., Wilson, J. L., Sarg, J. F., and Read, J. F., Eds., Controls on carbonate platform and basin development: Soc. Econ. Paleont. Mineral. Spec. Publ. **44**, 339-351.

Ehrlich, R. N., Longo, A. P., Jr., and Hyare, S., 1993, Response of carbonate platform margins to drowning: evidence of environmental collapse, *in* Loucks, R. G., and Sarg, J. F., Eds., Recent advances and applications of carbonate sequence stratigraphy: Am. Assn. Petr. Geol. Mem. **57**, 241-266.

Einsele, G., Ricken, W., and Seilacher, A., Eds., 1991, Cycles and events in stratigraphy: Springer-Verlag.

Elrick, M., and Read, J. F., 1991, Cyclic ramp-to-basin carbonate deposits, lower Mississippian, Wyoming and Montana: a combined field and computer modeling study: J. Sed. Pet., **61**, 1194-1224.

Enos, P., 1977, Holocene sediment accumulations of the south Florida shelf margin, *in* Enos, P., and Perkins, R. D., Eds., Quaternary sedimentation in south Florida: Geol. Soc. Am. Mem. **147**, 1-130.

Enos, P., 1989, Lag time—is it simply storm wave base? (abs.), *in* Franseen, E. K., and Watney, W. L., Eds., Sedimentary modeling—computer simulation of depositional sequences: Kansas Geol. Surv., Subsurface Geol. Ser. **12**, 41-42.

Epting, M., 1989, Miocene carbonate buildups of Central Luconia, offshore Sarawak, *in* Bally, A. W., Ed., Atlas of seismic stratigraphy: Am. Assn. Petr. Geol. Studies in Geology **27**, 168-173.

Everts, A. J. W., Stafleu, J., Schlager, W., Fouke, B. W., and Zwart, E. W., 1995, Stratal patterns, sediment composition, and sequence stratigraphy at the margin of the Vercors carbonate platform (Lower Cretaceous, SE France): J. Sed. Res., **B65**, 119-131.

Fischer, A. G., 1964, The Lofer cyclothems of the Alpine Triassic: Kansas Geol. Surv. Bull., **169**, 107-150.

Fischer, A. G., 1982, Long-term climatic oscillations recorded in stratigraphy, *in* Climate in earth history: National Academy Press.

Fischer, A. G., 1986, Climatic rhythms recorded in strata: Ann. Rev. Earth & Planet. Sci., **14**, 351-376.

Fischer, A. G., and Sarnthein, M., 1988, Airborne silts and dune-derived sands in the Permian of the Delaware Basin: J. Sed. Pet., **58**, 637-643.

Fisher, W. L., and McGowen, J. H., 1967, Depositional systems in the Wilcox Group of Texas and their relationship to occurrence of oil and gas: Gulf Coast Assn. Geol. Soc. Trans., **17**, 105-125.

Fitchen, W. M., 1993, Sequence stratigraphic framework of the upper San Andres Formation and equivalent basinal strata in the Brokeoff Mountains, Otero County, New Mexico: 4th Field Conf., New Mexico Geol. Soc. Guidebook, 185-193.

Fitchen, W. M., Starcher, M. A., Buffler, R. T., and Wilde, G. L., 1995, Sequence stratigraphic framework of Lower Permian carbonate platform margins, Sierra Diablo, West Texas, *in* Garber, R., Ed., Wolfcampian–Leonardian shelf margin facies of the Sierra Diablo: seismic models for subsurface exploration: West Texas Geol. Soc. Publ. **95-97**, 23-66.

Flügel, E., 1982, Microfacies analysis of limestones: Springer-Verlag.

Frakes, L. A., 1979, Climates throughout geologic time: Elsevier.

Franseen, E. K., Fekete, T. E., and Pray, L. C., 1989, Evolution and destruction of a carbonate bank at the shelf margin: Grayburg Formation (Permian), western escarpment, Guadalupe Mountains, Texas, *in* Crevello, P. D., Wilson, J. L., Sarg, J. F., and Read, J. F., Eds., Controls on carbonate platform and basin development: Soc. Econ. Paleont. Mineral. Spec. Publ. **44**, 289-304.

Frazier, D. E., 1974, Depositional episodes: their relationship to the Tertiary stratigraphic framework in the northwest portion of the Gulf Basin: Univ. Texas at Austin, Bur. Econ. Geol., Geol. Circ. **74-1**, 28.

Galloway, W. E., 1989, Genetic stratigraphic sequences in basin analysis I: architecture and genesis of flooding-surface bounded depositional units: AAPG Bull., **73**, 125-142.

Gardner, M. H., 1992, Sequence stratigraphy of eolian-derived turbidites: deep-water sedimentation patterns along an arid carbonate platform and their impact on hydrocarbon recovery in Delaware Mountain Group reservoirs, west Texas: West Texas Geol. Soc. Publ. **92-91**, 7-11.

Gianniny, G. L., and Simo, J. A., 1993, Kilometer-scale facies variability on a low angle carbonate/siliciclastic ramp, lower Desmoinesian of the Paradox Basin, SE Utah (abs.): Am. Assn. Petr. Geol. Ann. Meet., Program with Abs., 108.

Ginsburg, R. N., 1971, Landward movement of carbonate mud—new model for regressive cycles in carbonates (abs.): AAPG Bull., **55**, 340.

Glaser, K. S., Jordan, J. E., and Vail, P. R., 1990, Tectonic and eustatic controls on the carbonate stratigraphy of the Leonardian–Guadalupian (Permian) section, northwestern Delaware Basin, New Mexico and Texas (abs.): AAPG Bull., **74**, 662.

Goldhammer, R. K., Dunn, P. A., and Hardie, L.A., 1987, High frequency glacio-eustatic sea level oscillations with Milankovitch characteristics recorded in Middle Triassic platform carbonates in Northern Italy: Am. J. Sci., **287**, 853-892.

Goldhammer, R. K., Dunn, P. A., and Hardie, L.A., 1990, Depositional cycles, composite sea level changes, cycle stacking patterns, and the hierarchy of stratigraphic forcing: examples from Alpine Triassic platform carbonates: Geol. Soc. Am. Bull., **102,** 535-562.

Goldhammer, R. K., and Lehmann, P. J., 1991, The origin of fifth-order cycles in the Lower Cretaceous Cupido Fm. and one-dimensional forward stratigraphic modeling, *in* Goldhammer, R. K., Lehmann, P. J., Todd, R. G., Wilson, J. L., Ward, W. C., and Johnson, C. R., Eds., Sequence stratigraphy and cyclostratigraphy of the Mesozoic of the

Sierra Madre Oriental, northeast Mexico: Gulf Coast Sect., Soc. Econ. Paleont. Mineral. Field Guidebook, **67-81**.

Goldhammer, R. K., Oswald, E. J., and Dunn, P. A., 1991, Hierarchy of stratigraphic forcing: example from middle Pennsylvanian shelf carbonates of the Paradox Basin, *in* Franseen, E. K., Watney, L. W., Kendall, C. G. St. C., and Ross, W., Eds., Sedimentary modeling: computer simulations and methods for improved parameter definition: Kansas Geol. Surv. Bull., **233**, 361-413.

Goldhammer, R. K., Lehmann, P. J., and Dunn, P. A., 1993, The origin of high-frequency platform carbonate cycles and third-order sequences (Lower Ordovician El Paso Gp., West Texas): constraints from outcrop data and stratigraphic modeling: J. Sed. Pet., **63**, 318-359.

Goodwin, P. W., and Anderson, E. J., 1985, Punctuated aggradational cycles: a general hypothesis of episodic stratigraphic accumulation: J. Geol., **93**, 515-533.

Greenlee, S. M., and Lehmann, P. J., 1993, Stratigraphic framework of productive carbonate buildups, *in* Loucks, R. G., and Sarg, J. F., Eds., Carbonate sequence stratigraphy: recent developments and applications: Am. Assn. Petr. Geol. Mem. **57**, 43-62.

Grotzinger, J. P., 1986, Cyclicity and paleoenvironmental dynamics, Rocknest platform, northwest Canada: Geol. Soc. Am. Bull., **97**, 1208-1231.

Hallam, A., 1984, Pre-Quaternary sea level changes: Ann. Rev. Earth & Planet. Sci., **12**, 205-244.

Halley, R. B., and Harris, P. M., 1979, Fresh-water cementation of a one thousand year old oolite: J. Sed. Pet., **49**, 969-988.

Hamilton, W., 1979, Tectonics of the Indonesian region: U.S. Geol. Surv. Prof. Paper **1078**, 345 p.

Handford, C. R., 1981, Sedimentology and genetic stratigraphy of Dean and Sprayberry formations (Permian), Midland Basin, Texas: AAPG Bull., **65**, 1602-1616.

Handford, C. R., 1995, Baselap patterns and the recognition of lowstand exposure and drowning—a Mississippian ramp example and its seismic signature: J. Sed. Res., **B65**, 323-337.

Handford, C. R., and Loucks, R. G., 1993, Carbonate depositional sequences and systems tracts—responses of carbonate platforms to relative sea level changes, *in* Loucks, R. G., and Sarg, J. F., Eds., Carbonate sequence stratigraphy: recent developments and applications: Am. Assn. Petr. Geol. Mem. **57**, 3-41.

Hardie, L. A., and Shinn, E. A., 1986, Carbonate depositional environments, modern and ancient, part 3: tidal flats: Colorado School of Mines Quarterly, **81**, 74 p.

Harris, P. M., 1983, The Joulters ooid shoal, Great Bahama Bank, *in* Peryt, T. M., Ed., Coated grains: Springer-Verlag, 132-141.

Harris, P. M., Kerans, C., and Bebout, D. G., 1993, Ancient outcrop and modern examples of platform carbonate cycles—implications for subsurface correlation and understanding reservoir heterogeneity, *in* Loucks, R. G., and Sarg, J. F., Eds., Carbonate sequence stratigraphy: recent developments and applications: Am. Assn. Petr. Geol. Mem. **57,** 475-492.

Harris, P. M., and Kowalik, W. S., 1994, Satellite images of carbonate depositional settings—examples of reservoir- and exploration-scale geologic facies variation: Am. Assn. Petr. Geol. Methods in Exploration Series **11**, 147 p.

Hunt, D., and Tucker, M. E., 1993, Sequence stratigraphy of carbonate shelves with an example from the mid-Cretaceous (Urgonian) of southeast France: Internat. Assn. Sedimentol. Spec. Publ. **18**, 307-341.

Jacquin, Th., Arnaud-Vanneau, A., Arnaud, H., Ravenne, C., and Vail, P. R., 1991, Systems tracts and depositional sequences in a carbonate setting: a study of continuous outcrops from platform to basin at the scale of seismic lines: Mar. & Petr. Geol., **8**, 122-139.

James, N. P., 1979, Shallowing-upward sequences in carbonates, *in* Walker, R. G., Ed., Facies models: Geoscience Can. Repr. Ser. **1**, 109-119.

Jervey, M. T., 1988, Quantitative geological modeling of siliciclastic rock sequences and their seismic expression, *in* Wilgus, C. K., Hastings, B. S., Kendall, C. G. St. C., Posamentier, H. W., Ross, C. A., and Van Wagoner, J. C., Eds., Sea-level changes: an integrated approach: Soc. Econ. Paleont. Mineral. Spec. Publ. **42**, 47-69.

Johnson, J. G., Klapper, G., and Sandberg, C. A., 1985, Devonian eustatic fluctuations in Euramerica: Geol. Soc. Am. Bull., **99**, 567-587.

Kauffman, E. G., 1988, Concepts and methods of high-resolution event stratigraphy: Ann. Rev. Earth & Planet. Sci., **16**, 605-654.

Kendall, C. G. St. C., and Schlager, W., 1981, Carbonates and relative changes in sea level: Mar. Geol., **44**, 181-212.

Kenter, J. A. M., 1990, Carbonate platform flanks: slope angle and sediment fabric: Sedimentology, **37**, 777-794.

Kerans, C., 1995, Use of one- and two-dimensional cycle analysis in establishing high-frequency sequence frameworks, *in* Read, J. F., Kerans, C., and Weber, L. J., Eds., Milankovitch sea level changes, cycles and reservoirs on carbonate platforms in greenhouse and ice-house worlds: Soc. Econ. Paleont. Mineral., Short Course Notes **35**, pt. 2, 1-20.

Kerans, C., and Fitchen, W. M., 1995, Sequence hierarchy and facies architecture of a carbonate-ramp system: San Andres Formation of Algerita Escarpment and western Guadalupe Mountains, West Texas and New Mexico: Univ. Texas at Austin, Bur. Econ. Geol., Rept. Invest. **235**.

Kerans, C., and Lucia, F. J., 1989, Recognition of second, third, and fourth/fifth order scales of cyclicity in the El Paso Group and their relation to genesis and architecture of Ellenburger reservoirs, *in* Cunningham, B. K., and Cromwell, D. W., Eds., The lower Paleozoic of West Texas and southern New Mexico—modern exploration concepts: Permian Basin Sect., Soc. Econ. Paleont. Mineral. Publ. **89-31**, 105-110.

Kerans, C., Fitchen, W. M., Gardner, M. H., Sonnenfeld, M. D., Tinker, S. W., and Wardlaw, B. R., 1992, Styles of sequence development within uppermost Leonardian through Guadalupian strata of the Guadalupe Mountains,

Texas and New Mexico: West Texas Geol. Soc. Publ. **92-91**, 1-6.
Kerans, C., Fitchen, W. M., Gardner, M. S., and Wardlaw, B. R., 1993, A contribution to the evolving stratigraphic framework of middle Permian strata of the Delaware Basin, Texas and New Mexico, *in* Love, D. W., et al., Eds., New Mexico Geol. Soc. Guidebook, 4th Field Conf., 175-184.
Kerans, C., and Ruppel, S. C., 1994, San Andres sequence framework, Guadalupe Mountains: implications for San Andres type section and subsurface reservoirs, *in* Garber, R. A., and Keller, D. R., Eds., Field guide to the Paleozoic section of the San Andres Mountains: Permian Basin Sect., Soc. Econ. Paleont. Mineral. Publ. **94-35**, 105-116.
Kerans, C., Lucia, F. J., and Senger, R. K., 1994, Integrated characterization of carbonate ramp reservoirs using Permian San Andres Formation outcrop analogs: AAPG Bull., **78**, 181-216.
King, P. B., 1942, Permian of west Texas and southeastern New Mexico: AAPG Bull., **26**, 535-763.
Laferriere, A. P., and Hattin, D. E., 1989, Use of rhythmic bedding patterns for locating structural features, Niobrara Formation, United States Western Interior: AAPG Bull., **73**, 630-640.
Lomando, A. J., Harris, P. M., and Orlopp, D. E., 1993, Casablanca field, Tarragon Basin, offshore Spain, a karsted carbonate reservoir, *in* Fritz, R. D., Wilson, J. L., and Yurewicz, D. A., Eds., Paleokarst related hydrocarbon reservoirs: Soc. Econ. Paleont. Mineral. Core Workshop **18**, 201-225.
Loucks, R. G., and Sarg, J. F., Eds., 1993, Carbonate sequence stratigraphy: recent developments and applications: Am. Assn. Petr. Geol. Mem. **57**.
Loutit, T. S., Hardenbol, J., Vail, P. R., and Baum, G. R., 1988, Condensed sections: the key to age determination and correlation of continental margin sequences, *in* Wilgus, C. K., Hastings, B. S., Kendall, C. G. St. C., Posamentier, H. W., Ross, C. A., and Van Wagoner, J. V., Eds., Sea-level changes: an integrated approach: Soc. Econ. Paleont. Mineral. Spec. Publ. **42**, 183-213.
Lucia, F. J., 1983, Petrophysical parameters estimated from visual description of carbonate rocks: a field classification of carbonate pore space: J. Petr. Tech., 626-637.
McNeill, D. F., 1990, Contributions of sea level-induced erosional events in platform/periplatform deposits to lowstand sedimentation in the Bahamas (abs.): AAPG Bull., **74**, 717.
Meissner, F., 1972, Cyclic sedimentation in middle Permian strata of the Permian Basin, *in* Elam, J. G., and Chuber, S., Eds., Cyclic sedimentation in the Permian Basin (second edition): West Texas Geol. Soc. Publ. **72-60**, 203-232.
Mesolella, K. J., Robinson, J. D., McCormick, L. M., and Ormiston, A. R., 1974, Cyclic deposition of Silurian carbonates and evaporites in Michigan Basin: AAPG Bull., **58**, 34-62.
Meyer, F. O., 1989, Siliciclastic influence on Mesozoic platform development: Baltimore Canyon trough, western Atlantic, *in* Crevello, P. D., Wilson, J. L., Sarg, J. F., and Read, J. F., Eds., Controls on carbonate platform and basin development: Soc. Econ. Paleont. Mineral. Spec. Publ. **44**, 213-232.
Mitchum, Jr., R. M., 1977, Seismic stratigraphy and global changes of sea level, part II: glossary of terms used in seismic stratigraphy, *in* Payton, Charles E., Ed., Seismic stratigraphy—applications to hydrocarbon exploration: Am. Assn. Petr. Geol. Mem. **26**, 205-212.
Mitchum, Jr., R. M., Vail, P. R., and Thompson III, S., 1977, The depositional sequence as a basic unit for stratigraphic analysis, *in* Payton, C. E., Ed., Seismic stratigraphy—applications to hydrocarbon exploration: Am. Assn. Petr. Geol. Mem. **26**, 53-62.
Mitchum, Jr., R. M., and Van Wagoner, J. C., 1991, High-frequency sequences and their stacking patterns: sequence-stratigraphic evidence of high-frequency eustatic cycles, *in* Biddle, K. T., and Schlager, W., Eds., The record of sea level fluctuations: Sed. Geol., **70**, 131-160.
Moldovanyi, E. P., Wall, F. M., and Yan, Z. J., 1995, Regional exposure events and platform evolution of Zhujiang Formation carbonates, Pearl River Mouth Basin: evidence from primary and diagenetic seismic facies, *in* Budd, D. A., Saller, A. H., and Harris, P. M., Eds., Unconformities and porosity in carbonate strata: Am. Assn. Petr. Geol. Mem. **63,** 125-140.
Montanez, I. P., and Osleger, D. A., 1993, Parasequence stacking patterns, third-order accommodation events, and sequence stratigraphy of Middle to Upper Cambrian platform carbonates, Bonanza King Formation, southern Great Basin, *in* Loucks, R. G., and Sarg, J. F., Eds., Carbonate sequence stratigraphy: recent developments and applications: Am. Assn. Petr. Geol. Mem. **57**, 305-326.
Mullins, H. T., Shanmugam, G., and Moiola, R. J., 1983, Comment and reply: eustatic control of turbidites and winnowed turbidites: Geology, **11**, 57-58.
Murris, R. J., 1980, Middle East: stratigraphic evolution and oil habitat: AAPG Bull., **64**, 597-618.
Osleger, D., 1991, Subtidal carbonate cycles: implications for allocyclic vs. autocyclic controls: Geology, **19**, 917-920.
Parrish, J. T., and Barron, E. J., 1986, Paleoclimates and economic geology: Soc. Econ. Paleont. Mineral. Short Course **18**.
Peterson, J. A., and Hite, R. J., 1969, Pennsylvanian evaporite-carbonate cycles and their relation to petroleum occurrence, southern Rocky Mountains: AAPG Bull., **53**, 884-908.
Pinet, P. R., and Popenoe, P. 1985, A scenario of Mesozoic-Cenozoic ocean circulation over the Blake Plateau and its environs: Geol. Soc. Am. Bull., **96**, 618-626.
Pomar, L., 1993, High-resolution sequence stratigraphy in prograding Miocene carbonates: application to seismic interpretation, *in* Loucks, R. G., and Sarg, J. F., Eds., Carbonate sequence stratigraphy: recent developments and applications: Am. Assn. Petr. Geol. Mem. **57**, 389-407.
Popenoe, P., 1985, Cenozoic, depositional and structural history of the North Carolina margin from seismic strati-

graphic analyses, *in* Poag, C. W., Ed., Geologic evolution of the United States Atlantic margin: Van Nostrand Reinhold, 125-187.

Popenoe, P., Kohout, F. A., and Manheim, F. T., 1984, Seismic-reflection studies of sinkholes and limestone dissolution features on the northeastern Florida shelf, *in* Beck, B. F., Ed., Sinkholes: their geology, engineering and environmental impact: Proc. First Multidisciplinary Conf. on Sinkholes, 43-57.

Posamentier, H. W., and Vail, P. R., 1988, Eustatic controls on clastic deposition II—sequence and systems tract models, *in* Wilgus, C. K., Hastings, B. S., Kendall, C. G. St. C., Posamentier, H. W., Ross, C. A., and Van Wagoner, J. C., Eds., Sea-level changes: an integrated approach: Soc. Econ. Paleont. Mineral. Spec. Publ. **42**, 125-154.

Read, J. F., 1985, Carbonate platform facies models: AAPG Bull., **69**, 1-21.

Read, J. F., 1995, Overview of carbonate platform sequences, cycle stratigraphy, and reservoirs in greenhouse and icehouse worlds, *in* Read, J. F., Kerans, C., and Weber, L. J., Eds., Milankovitch sea level changes, cycles and reservoirs on carbonate platforms in greenhouse and icehouse worlds: Soc. Econ. Paleont. Mineral. Short Course Notes **35,** pt. 1, 1-102.

Read, J. F., Grotzinger, J. P., Bova, J. A., and Koerschner, W. F., 1986, Models for generation of carbonate cycles: Geology, **14**, 107-110.

Read, J. F., and Goldhammer, R. K., 1988, Use of Fischer plots to define third-order sea level curves in peritidal cyclic carbonates, Ordovician, Appalachians: Geology, **16**, 895-899.

Read, J. F., Osleger, D., and Elrick, M., 1991, Two-dimensional modeling of carbonate ramp sequences and component cycles, *in* Franseen, E. K., Watney, L. W., Kendall, C. G. St. C., and Ross, W., Eds., Sedimentary modeling: computer simulations and methods for improved parameter definition: Kansas Geol. Surv. Bull., **233**, 473-488.

Rich, J. L., 1951, Three critical environments of deposition and criteria for recognition of rocks deposited in each of them: Geol. Soc. Am. Bull., **62**, 1-20.

Rudolph, K. W., and Lehmann, P. J., 1989, Platform evolution and sequence stratigraphy of the Natuna Platform, South China Sea, *in* Crevello, P. D., Wilson, J. L., Sarg, J. F., and Read, J. F., Eds., Controls on carbonate platform and basin development: Soc. Econ. Paleont. Mineral. Spec. Publ. **44**, 353-361.

Ruppel, S. C., 1992, Styles of deposition and diagenesis in Leonardian carbonate reservoirs in West Texas: implications for improved reservoir characterization: Soc. Petr. Engineers, 67th Ann. Tech. Conf. Proc., 313-320.

Sadler, P. M., Osleger, D. A., and Montanez, I. P., 1993, On the labeling, length, and objective basis of Fischer plots: J. Sed. Pet., **63**, 360-368.

Saller, A. H., Dickson, J. A. D., and Boyd, S. A., 1994, Cycle stratigraphy and porosity in Pennsylvanian and Lower Permian shelf limestones, eastern Central Basin Platform, Texas: AAPG Bull., **78,** 1820-1842.

Sarg, J. F., 1988, Carbonate sequence stratigraphy, *in* Wilgus, C. K., Hastings, B. S., Kendall, C. G. St. C., Posamentier, H. W., Ross, C. A., and Van Wagoner, J. C., Eds., Sea-level changes: an integrated approach: Soc. Econ. Paleont. Mineral. Spec. Publ. **42**, 155-181.

Sarg, J. F., 1989, Middle–Late Permian depositional sequences, Permian Basin, west Texas–New Mexico, *in* Bally, A. W., Ed., Atlas of seismic stratigraphy: Am. Assn. Petr. Geol. Studies in Geol. **27**, 140-157.

Sarg, J. F., and Lehmann, P. J., 1986, Lower–Middle Guadalupian facies and stratigraphy, San Andres/Grayburg Formations, Permian Basin, Guadalupe Mountains, New Mexico, *in* Moore, G. E., and Wilde, G. L., Eds., San Andres/Grayburg formations, Guadalupe Mountains, New Mexico and Texas: Permian Basin Section, Soc. Econ. Paleont. Mineral. Publ. **86-25**, 1-35.

Schlager, W., 1981, The paradox of drowned reefs and carbonate platforms: Geol. Soc. Am. Bull., **92,** 197-211.

Schlager, W., 1989, Drowning unconformities on carbonate platforms, *in* Controls on carbonate platform and basin development: Soc. Econ. Paleont. Mineral. Spec. Publ. **44**, 15-25.

Schlager, W., 1992; Sedimentology and sequence stratigraphy of reefs and carbonate platforms: Am. Assn. Petr. Geol. Cont. Ed. Course Notes Ser. **34**.

Schlager, W., 1993, Accommodation and supply—a dual control on stratigraphic sequences: Sed. Geol., **86**, 111-136.

Schlager, W., and Camber, O., 1986, Submarine slope angles, drowning unconformities, and self-erosion on limestone escarpments: Geology, **14,** 762-765.

Schlager, W., Reijmer, J. J. G., and Droxler, A., 1994, Highstand shedding of carbonate platforms: J. Sed. Res., **B64,** 270-281.

Schlee, J. S., Ed., 1984, Interregional unconformities and hydrocarbon accumulation: Am. Assn. Petr. Geol. Mem. **36**.

Schmidt, V., McDonald, D. A., and McIlreath, I. A., 1980, Growth and diagenesis of Middle Devonian Keg River cementation reefs, Rainbow field, Alberta, *in* Halley, R. B., and Loucks, R. G., Eds., Carbonate reservoir rocks: Soc. Econ. Paleont. Mineral. Core Workshop **1**, 43-63.

Schwarz, H. U., 1982, Subaqueous slope failures—experiments and modern occurrences, *in* Fuchtbauer, H., Lisitzyn, A. P., Milliman, J. D., and Seibold, E., Eds., Contributions to sedimentology: E. Schweizerbartsche Verlagsbuchhandlung.

Schwarzacher, W., 1993, Cyclostratigraphy and Milankovitch theory: developments in sedimentology 52: Elsevier.

Shanmugam, G., and Moiola, R. J., 1984, Eustatic control of calciclastic turbidites: Mar. Geol., **56**, 273-278.

Silver, B. A., and Todd, R. G., 1969, Permian cyclic strata, northern Midland and Delaware basins, west Texas and southeastern New Mexico: AAPG Bull., **53,** 2223-2251.

Sonnenfeld, M. D., 1991, High-frequency cyclicity within shelf-margin and slope strata of the upper San Andres sequence, Last Chance Canyon, *in* Meader-Roberts, S., Candelaria, M. P., and Moore, G. E., Eds., Sequence stratigraphy, facies, and reservoir geometries of the San Andres, Grayburg, and Queen formations, Guadalupe Mountains, New Mexico and Texas: Permian Basin

Section, Soc. Econ. Paleont. Mineral. Publ. **91-32**, 11-51.

Sonnenfeld, M. D., and Cross, T. A., 1993, Volumetric partitioning and facies differentiation within the Permian upper San Andres Formation of Last Chance Canyon, Guadalupe Mountains, New Mexico, *in* Loucks, R. G., and Sarg, J. F., Eds., Recent advances and applications of carbonate sequence stratigraphy: Am. Assn. Petr. Geol. Mem. **57,** 435-474.

Southgate, P. N., Kennard, J. M., Jackson, M. J., OBrien, P. E., and Sexton, M. J., 1993, Reciprocal lowstand clastic and highstand carbonate sedimentation, subsurface Devonian reef complex, Canning Basin, Western Australia, *in* Loucks, R. G., and Sarg, J. F., Eds., Recent advances and applications of carbonate sequence stratigraphy: Am. Assn. Petr. Geol. Mem. **57**, 157-179.

Stafleu, J., and Schlager, W., 1993, Pseudo-toplap in seismic models of the Schlern-Raibl contact (Sella platform, northern Italy): Basin Res., **5**, 55-65.

Stafleu, J., and Sonnenfeld, M. D., 1994, Seismic models of a Permian shelf margin: J. Sed. Res., **B64**, 481-499.

Stafleu, J., Everts, A. J. W., and Kenter, J. A. M., 1994, Seismic models of a prograding carbonate platform: Vercors, SE France: Mar. & Petr. Geol., **11**, 514-527.

Taylor, B., and Hayes, D. E., 1983, Origin and history of the South China Sea Basin, *in* Hayes, D. E., Ed., The tectonic and geologic evolution of southeast Asian seas and islands, part 2: Lamont-Doherty Geol. Observ., Geophys. Monogr. **27**, 23-57.

Tyrrell, Jr., W. W., 1969, Criteria useful in interpreting environments of unlike but time-equivalent carbonate units (Tansill-Capitan-Lamar), Capitan reef complex, West Texas and New Mexico, *in* Friedman, G. M., Ed., Depositional environments in carbonate rocks: Soc. Econ. Paleont. Mineral. Spec. Publ. **14**, 80-97.

Tucker, M. E., 1991, Sequence stratigraphy of carbonate-evaporite basins: models and application to the Upper Permian (Zechstein) of northeast England and adjoining North Sea: J. Geol. Soc. London, **148**, 1019-1036.

Tucker, M. E., and Wright, V. P., 1990, Carbonate sedimentology: Blackwell Scientific Publ.

Tucker, M. E., Wilson, J. L., Crevello, P. D., Sarg, J. F., and Read, J. F., Eds., 1990, Carbonate platforms—facies, sequences and evolution: Internat. Assn. Sed. Spec. Publ. **9**.

Vail, P. R., Mitchum, Jr., R. M., Todd, R. G., Widmier, J. M., Thompson III, S., Sangree, J. B., Bubb, J. N., and Hatlelid, W. G., 1977, Seismic stratigraphy and global changes in sea level, *in* Payton, C. E., Ed., Seismic stratigraphy—applications to hydrocarbon exploration: Am. Assn. Petr. Geol. Mem. **26**, 49-212.

Vail, P. R., 1987, Seismic stratigraphy interpretation using sequence stratigraphy, *in* Bally, A. W., Ed., Atlas of seismic stratigraphy, vol. 1: Am. Assn. Petr. Geol. Studies in Geol. **27,** 1-10.

Vail, P. R., Audemard, F., Bowman, S. A., Eisner, P. N., and Perez-Cruz, C., 1991, The stratigraphic signatures of tectonics, eustasy, and sedimentology: an overview, *in* Einsele, G., Ricken, W., and Seilacher, A., Eds., Cycles and events in stratigraphy: Springer-Verlag, 617-659.

Van Siclen, D. C., 1958, Depositional topography: examples and theory: AAPG Bull., **42**, 1897-1913.

Van Wagoner, J. C., 1995, Overview of sequence stratigraphy of foreland basin deposits: terminology, summary of papers, and glossary of sequence stratigraphy, *in* Van Wagoner, J. C., and Bertram, G. T., Eds., Sequence stratigraphy of foreland basin deposits: Am. Assn. Petr. Geol. Mem. **64**, ix-xxi.

Van Wagoner, J. C., Posamentier, H. W., Mitchum, R. M. Jr., Vail, P. R., Sarg, J. F., Loutit, T. S., and Hardenbol, J., 1988, An overview of the fundamentals of sequence stratigraphy and key definitions, *in* Wilgus, C. K., Hastings, B. S., Ross, C. A., Posamentier, H. Van Wagoner, J. C., and Kendall, C. G. St. C., Eds., Sea-level changes: an integrated approach: Soc. Econ. Paleont. Mineral. Spec. Publ. **42**, 39-45.

Van Wagoner, J. C., Mitchum, R. M., Campion, K. M., and Rahmanian, V. D., 1990, Siliciclastic sequence stratigraphy in well logs, cores, and outcrops: concepts for high-resolution of time and facies: Am. Assn. Petr. Geol. Methods Expl. Ser. **7**.

Waite, L. E., 1993, Upper Pennsylvanian seismic sequences and facies of the eastern and southern Horseshoe Atoll, Midland Basin, West Texas, *in* Loucks, R. G., and Sarg, J. F., Eds., Recent advances and applications of carbonate sequence stratigraphy: Am. Assn. Petr. Geol. Mem. **57**, 213-240

Walker, R. G., Ed., 1984, Facies models, second edition: Geoscience Can. Rep. Ser. **1**.

Walker, R. G., and James, N. P., Eds., 1992, Facies models: response to sea-level change: Geol. Assn. Can.

Wanless, H. R., and Dravis, J., 1989, Carbonate environments and sequences of Caicos Platform: Am. Geophys. Union, Field Trip Guidebook **T374**, 1-75.

Wendte, J., Stoakes, F. A., and Campbell, C. V., Devonian–Early Mississippian carbonates of the Western Canada Sedimentary Basin: a sequence stratigraphic framework: Soc. Econ. Paleont. Mineral. Short Course **28**.

Wheeler, H. E., 1958, Time-stratigraphy: AAPG Bull., **42**, 1047-1063.

Wheeler, H. E., 1964, Base-level, lithosphere surface, and time-stratigraphy: Geol. Soc. Am. Bull., **75**, 599-610.

Wilson, J. L., 1967, Cyclic and reciprocal sedimentation in Virgilian strata of southern New Mexico: Geol. Soc. Am. Bull., **78**, 805-818.

Wilson, J. L., 1975, Carbonate facies in geologic history: Springer-Verlag.

Yurewicz, D. A., Marler, T. B., Meyerholtz, K. A., and Siroky, F. X., 1993, Early Cretaceous carbonate platform, north rim of the Gulf of Mexico, Mississippi and Louisiana, *in* Simo, J. A., Scott, R. W., and Masse, J.-P., Eds., Cretaceous carbonate platforms: Am. Assn. Petr. Geol. Mem. **56**, 81-96.

Chapter 8

Characteristics of Natural Fractures in Carbonate Strata

John C. Lorenz
Sandia National Laboratories
Albuquerque, New Mexico, U.S.A.

Helen E. Farrell
Phillips Petroleum Company
Bartlesville, Oklahoma, U.S.A.

Catherine L. Hanks
University of Alaska
Fairbanks, Alaska, U.S.A.

William D. Rizer
Conoco Inc.
Ponca City, Oklahoma, U.S.A.

Mark D. Sonnenfeld
Colorado School of Mines
Golden, Colorado, U.S.A.

Abstract

Natural fractures are present in carbonates in settings ranging from undeformed, near-horizontal beds to fold and thrust belts. Most fractures in the carbonate examples described here show evidence of an origin in extension. Lithology is one of the primary controls on fracturing; dolomites and dolomitic mudstones are typically more intensely fractured than limestones, and fractures are commonly limited vertically by shale breaks. Structural position is another important control on fracture character, with more fractures occurring at the hinges of folds than on fold limbs. Fractures caused by faulting in moderately deformed strata are commonly present only within 100 m or less of the fault. Older fractures may be overprinted by younger fracture sets in the more highly deformed strata. Structure, petrofacies, and lithofacies combine to control fracture characteristics in the thicker mechanical units under these conditions. Different types of karsted zones may also be superimposed on fractured carbonate strata.

Introduction

Carbonates, especially dolomites, are usually considered to be more commonly and more highly fractured than clastic strata. Although no definitive data base has been published to support this, carbonate reservoirs are usually recognized to be fracture-enhanced plays, whereas many clastic reservoirs have only recently been widely recognized as such.

Fracture characteristics in carbonates are as variable as the range of carbonate lithologies, fabrics, and structural settings. Structure and tectonics cause the stresses that initiate fracturing within a package of rock, while lithology is the primary control on fracture susceptibility. Of these two controls, lithology is commonly more important than structural position in determining the degree of fracturing in carbonate strata (e.g., Nelson and Serra, 1993), although structure becomes more important as deformation increases (see the Madison Limestone example later).

The relatively high solubility of carbonates is an additional factor in carbonate fracture characterization. Formation fluids within the fractures may have dissolved much of the adjacent rock, leaving mismatched fracture walls and significant volumes of vuggy, fracture-related porosity. Mobile carbonate may then have been reprecipitated within the open voids, partially or completely filling them with crystalline mineralization. If the carbonate layer was exposed to surface weathering conditions at any time during its history, solution-enhanced fractures may have developed into cavernous porosity, or even into irregular karst topographies along subsequently reburied unconformities (e.g., Cater and Gillcrist, 1994; MacDonald et al., 1994). Where little or no solution has occurred, however, fractures in carbonates may be regular in strike and uniform in character and similar to fractures in sandstones.

The primary objective of this paper is to present fracture characterizations that would be useful in interpreting seismic data. To this end, we present fracture descriptions from a number of different settings, starting with relatively simple fracture patterns and finishing with more complex ones. Our focus is on the meter-scale fractures that are most easily observable in outcrop, although all scales of fracturing affect seismic data. We have not addressed microscopic fractures, and we discuss faults only in their relationship to fractures. The important effects of fractures on reservoir permeability and hydrocarbon production (e.g., Teufel and Rhett, 1992) are not discussed here.

There are a number of caveats in making subsurface fracture interpretations from indirect data such as seismic signatures. These are not discussed in this paper since our purpose is descriptive, but we list them here to assist readers in the use of this material:

- The scale of seismic measurement is typically larger than the scale of individual fractures. It is the cumulative effect of fractures on the rock mass that leaves a seismic signature.
- In-situ stresses can maintain fracture apertures or can partially or completely close them, depending on the relative orientation of stress components and fractures and on the magnitude of the stress differential. Since aperture affects seismic characteristics, the in-situ stresses should be taken into account when interpreting fractures from seismic data.
- Interpretations of fractures from seismic signals can become difficult where the fractures are filled with minerals having properties similar to those of the host rock, such as calcite filling fractures in limestone.

Fractures in Mildly Deformed Strata

Introduction

Fractures are common even in carbonate beds that have not been folded or otherwise tectonized. Fractures in such relatively flat-lying strata can usually be attributed to combinations of basin maturation and far-field stresses derived from basin subsidence, uplift, or distant tectonic events (e.g., Lorenz et al., 1991). Such fractures are called *regional fractures* and form parallel or subparallel to the regional maximum horizontal compressive stress. They are extensive in area and commonly oriented normal to bedding. They probably form under differential stresses significantly less than those required to fold and fault most strata because they often appear to have formed before the strata were significantly bent or broken.

The vertical extent of regional fractures depends on the nature of the bedding discontinuities and the lithologic heterogeneity of the formation. Thicker shale partings and marly intervals typically arrest the vertical propagation of fractures, and fractures in some settings may even terminate at subtle carbonate-on-carbonate bedding contacts or at horizontal stylolites (Nelson, 1981). Regional fracture lengths occur at a number of scales and, allowing for en echelon offsets, are commonly limited only by the horizontal extent of the beds in which they occur.

Different lithologies in a given area of regional fracturing typically display different fracture intensities and spacing. Regional fracture spacing is usually not related directly to mechanical bed thickness as it can be when strata are folded. Regional fracture spacing is also a function of (1) the intensity and duration of stresses that caused fracturing and (2) the mechanical properties of the strata; thus, the spacing of regional fractures is difficult to predict.

Younger fracture sets may be superimposed at high angles onto regional fractures during later tectonic events, during stress release on uplift and erosion, or during later development of structural folding or faulting of the strata. The younger fractures commonly terminate at high-angle intersections with preexisting regional fractures. Generalizations about fracture spacing become even more difficult to make where such a mixture of regional fractures, release fractures, and local structurally controlled fractures is present.

The following examples are characteristic of regional fractures in little-deformed strata. The Fort Riley Limestone is located on the gentle flanks of a broad, low-amplitude arch, while the Austin Chalk covers an area dominated by small, widely spaced, normal faults. Low structural dip angles are typical of both settings.

Fort Riley Limestone, Kay County, North-Central Oklahoma[1]

Introduction

This study was part of a larger research effort (Queen and Rizer, 1990) to investigate seismic techniques for characterizing natural fracture systems in the subsurface. The approach was to examine relationships among natural fractures, microfractures, seismic anisotropy, and stress. The study area in Kay County, north-central Oklahoma (Figure 1), was chosen because of the availability of (1) good outcrop at quarry exposures for mapping surface fracture systems and (2) core and well-bore data (at Conoco Borehole Test Facility) suitable for characterizing subsurface fractures.

The area is located on the Nemaha Ridge, well removed from the structural complexities of the Ouachita fold and thrust belt and the Ozark uplift. In Kay County, the Nemaha Ridge consists of a ~50 km wide zone (Figure 2) of differentially uplifted blocks bordered by northeast- and northwest-trending faults that were active in Paleozoic time (Luza and Lawson, 1981). Shallow structure in this area is characterized by a regional dip of <1° to the west-southwest (Gasteiger, 1980), except where interrupted by low-amplitude folds overlying some of the block-bounding faults. The low-amplitude folds include the Mervine anticline, a broad, asymmetric (steeper on the east) drape over a northeast-trending fault (Smith, 1954). Dips on the west limb vary from 1.5° near the hinge to a regional dip of 0.5° (Gasteiger, 1980). Key localities in this area are the outcrop at Vap's Pass, located on the plunging nose of the Mervine anticline, and Conoco Borehole Test Facility, just off the west limb of the structure.

Stratigraphically, the study area is located in what is called the Northern Shelf (Luza and Lawson, 1981). The strata consist of interbedded limestone and shale in the Permian section, with increasing amounts of sandstone in the Pennsylvanian section. Surface rocks are primarily limestone and shale of the Lower Permian Chase Group, described by Chaplin (1988). The carbonate unit

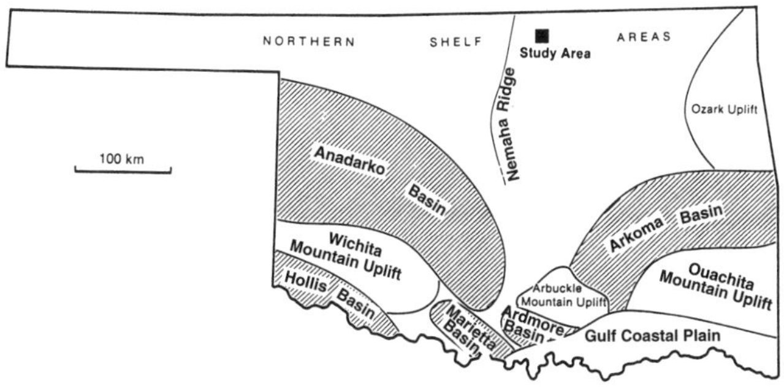

Fig. 1. Location map of the Fort Riley Limestone study area in north-central Oklahoma.

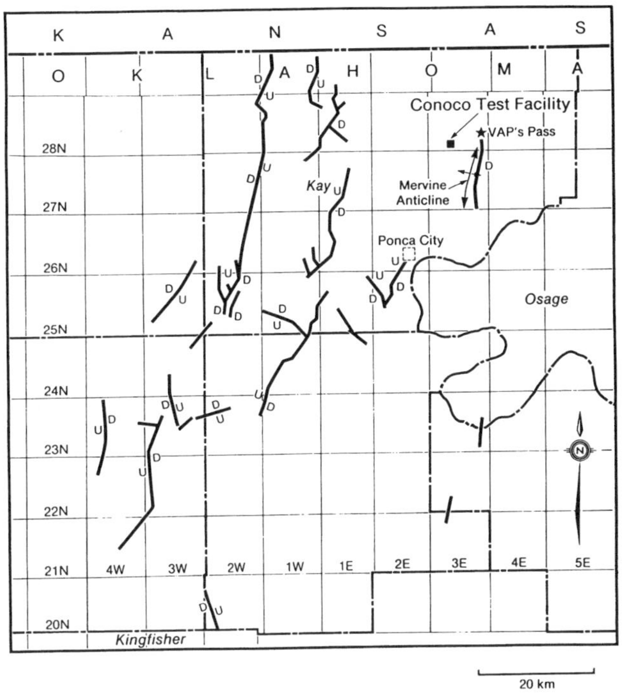

Fig. 2. Map of the Mervine anticline, Vap's Pass, and the Conoco Borehole Test Facility in Oklahoma, showing local fault patterns.

containing the Fort Riley Limestone is about 20 m thick, with the Fort Riley Limestone itself about 15 m thick (Rizer and Queen, 1986).

Hagan (1972) mapped two orthogonal sets of vertical fractures in these strata in easternmost Kay and western Osage counties: a systematic set striking N 70° E and a nonsystematic set striking N 25° W. Following the terminology of Hodgson (1961), *nonsystematic fractures* generally terminate against *systematic fractures* which are generally more throughgoing in plan view. In a surface study of Mervine anticline, Gasteiger (1980) mapped vertical orthogonal fracture sets striking east-northeast and north-northwest and an additional vertical

[1]William D. Rizer

fracture set striking N 30° E, parallel to the fold axis of the Mervine structure. That set, best developed on the steep limb, is a local feature likely related to folding. The northeasterly and northwesterly fracture sets appear to be part of a regional system of orthogonal natural fractures mapped in eastern Oklahoma by Melton (1929).

North-central Oklahoma is part of the Mid-continent stress province, characterized by northeast-southward trending maximum horizontal compression (Zoback and Zoback, 1980). Wellbore breakouts indicate that the maximum horizontal compressive stress is oblique to this regional trend in the Anadarko Basin, southwest of the study area, which strikes N 80° E (±19°) (Dart, 1987). Impression packers of induced hydraulic fractures in a well in Kingfisher County indicate an orientation of N 63° E (±15°) for the maximum horizontal stress (von Schonfeldt et al., 1973), while analysis of acoustic borehole televiewer images of hydraulic fractures and Anelastic Strain Recovery tests at Mounds (northeastern Oklahoma) indicate a strike of N 86° E (±15°) (Smith et al., 1986).

Methods

The surface fracture pattern was mapped from an excellent exposure of Lower Permian Fort Riley Limestone at an abandoned quarry and roadcut (locally known as Vap's Pass), ~5 km east-northeast of the Borehole Test Facility (Figure 2). Two orthogonal fracture sets are clearly visible on the horizontal pavement (bedding) surface. In plan view, the fractures are visible because of the vegetation growing in the weathered fracture zones. In cross section, along the roadcut, the weathered fractures form complex zones up to 1 m wide (Figure 3). All mappable segments of fractures were surveyed using an HP 3810B Medium Range Total (Laser) Station. Small-scale changes in fracture orientation, that is, changes of a few degrees over a few feet of fracture length, were not mapped. Aerial photographs were used to determine whether the fracture pattern observed at Vap's Pass could be traced on surface limestone outcrops to the Borehole Test Facility.

About 690 m of Schlumberger Borehole Televiewer (BHTV) images (Zemanek et al., 1970), along with oriented and unoriented core from those wells, were analyzed to determine the large-scale subsurface fracture pattern. Following Plumb and Hickman (1985), BHTV images were analyzed to characterize the orientation of the maximum horizontal compressive stress.

Microfabric was inferred from examination of thin sections cut parallel to bedding and from the induced

Fig. 3. Field photograph of complex fracture zone in outcrops of the Fort Riley Limestone, north-central Oklahoma.

fracture anisotropy determined by Point Load testing of surface and subsurface samples (Queen and Rizer, 1990). Point load tests were run on 35 samples from an oriented block of limestone excavated from the roadcut at Vap's Pass and on 9 samples obtained from a section of oriented shale core. Thin sections obtained from 10 failed limestone samples were examined microscopically to determine whether the induced fracture geometries are related to oriented microfractures or to other microfabric elements.

Results: Natural Fracturing, Stress, and Microfabric

The Conoco Borehole Test Facility and surroundings were chosen as the study area because of the relative simplicity of the geologic structure and because of the availability of good outcrop and extensive well log data to characterize the surface and subsurface natural fracture orientations. The surface fracture pattern (Figure 4) is orthogonal, consisting of an east-northeast

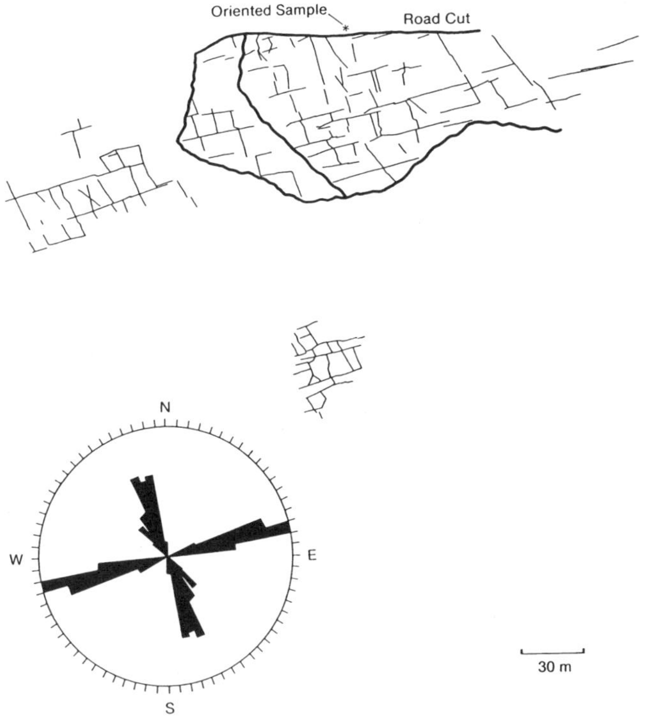

Fig. 4. Outcrop bedding-plane map of plan-view fracture patterns in the Fort Riley Limestone. The rose diagram in lower left shows orientations of surface fractions.

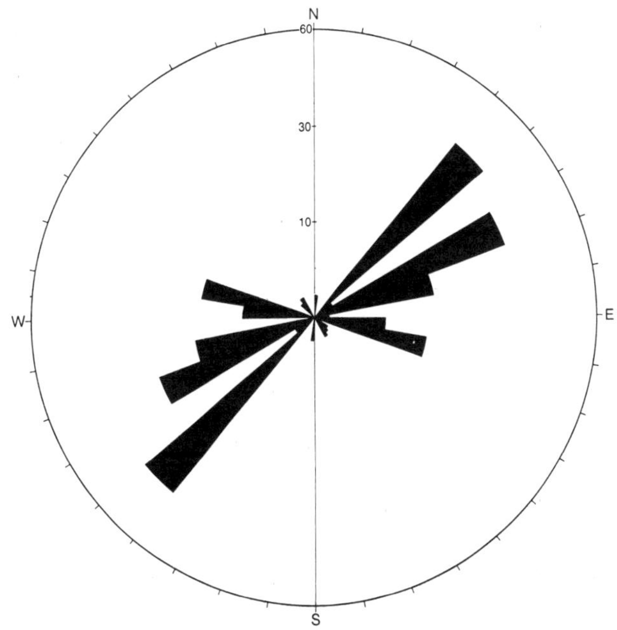

Fig. 5. Rose diagram of subsurface fracture orientations at the Conoco Borehole Test Facility. Compare with the rose diagram of surface fracture orientations in Figure 4.

systematic set and a north-northwest nonsystematic set, as mapped in detail on the limestone pavement at Vap's Pass and observed on aerial photographs throughout the study area. Both fracture sets are essentially normal to bedding. The outcrop fractures are commonly weathered and open, whereas the subsurface core fractures are partially mineralized. Thinner limestone beds are typically more highly fractured than thicker beds, but the relationship does not appear to be simple and has not been studied in detail.

The subsurface fracture system mapped from BHTV images is more complex. The rose diagram in Figure 5 suggests that three sets of vertical to near-vertical fractures are present in the subsurface. Evidence that these sets are distinct and do not represent poor sampling of a single widely distributed set comes from an oriented section of indurated shale core (from one of the Borehole Test Facility wells) which contains all three (calcite-filled) fracture sets.

Analyses of breakouts on BHTV images indicate that the maximum horizontal compressive stress is oriented N 55°–75° E, parallel to the regional maximum horizontal compressive stress in the Mid-continent and in good agreement with local measurements. Results of the Point Load tests on both surface and subsurface samples indicate small-scale mechanical (microcrack) fabric anisotropy subparallel to the maximum horizontal compressive stress and to the dominant (systematic) direction of fracturing. A secondary northwest-striking fabric direction is approximately normal to the interpreted in-situ maximum compression. Analysis of thin sections indicates that this anisotropy is not related to a depositional fabric, such as preferred alignment of grains.

Despite the structural simplicity of the study area, the stress-microfabric-fracture system is complex, with four distinct directional trends present. The east-northeast direction is likely the most significant because it is common (within probable error) to the stress, microfabric, and one of the dominant fracture directions mapped on the surface and in the subsurface. Differences are observed between the natural fracture system mapped on outcrop and from subsurface data. Similar differences between surface and subsurface data from the Piceance Basin have been reported by Lorenz et al. (1989).

Austin Chalk, Texas[2]

General Setting

The Austin Chalk varies from about 130 to 180 m thick. It is composed of three subunits of about equal

[2]John C. Lorenz

thickness: an upper and lower chalk with average porosities of less than 10% and a central marly unit that is richer in clay. This formation extends across much of southeastern Texas with dips averaging <1°. Dips increase slightly where the strata are cut by zones of normal faults, such as the Balcones fault zone. Offsets of about 30 m are common on the larger faults within these zones (Collins et al., 1992).

Fracture Characteristics

The degree of fracturing in the Austin Chalk is governed in part by lithology. Collins et al. (1992, their Figure 2) report that fracture intensity in the upper and lower chalks is greater than the fracture intensity in the intervening marl unit by a factor of two to four. Corbett et al. (1991) note that most fractures in the chalk beds terminate vertically at those marl beds exceeding 10 cm in thickness.

Fracture spacing is commonly on the order of bed thickness in areas where repetitive marl and shale partings break the massive chalks into thinner, regular horizontal units (Friedman and McKiernan, 1993). Spacing has a negative exponential distribution in thicker, more massive beds, where small fracture spacing is common, (i.e., most fractures are within 20 cm of another fracture) (Corbett et al., 1991). Rarer, larger spacing separates "swarms" of closely spaced fractures. Friedman and McKiernan (1993) suggest that the spacing of the dominant, oldest set of fractures in outcrop correlates well with fracture spacing measured in boreholes, although Belfield and Sovich (1993) suggest that a more homogeneous distribution may exist in much of the subsurface.

Fracture occurrence and character are also governed by local faults and folds. Friedman and McKiernan (1993) and Collins et al. (1992) note "halos" of more intense fault-parallel fracturing within 60 m of faults. More widely spaced regional fractures are present between these structure-associated fractures. Locally, the regional fractures occur in swarms that die out along strike (Collins et al., 1992).

Local monoclines and grabens are also associated with increased numbers of natural fractures. The fractures trend northeast, parallel to these structures (Corbett et al., 1987, 1991), and are present in both surface outcrops and in the subsurface. Several younger sets of orthogonal, oblique, and random fractures that are present in outcrops are rare to absent in the subsurface (Belfield and Sovich, 1993; Friedman and McKeirnan, 1993).

Horizontal cores cutting across the dominant fracture trend show parallel fractures that are both mineralized and unmineralized (Figure 6), suggesting that fractures with the same orientation formed during several different stress events. Calcite is the dominant mineral present in the fractures.

Fig. 6. Photographs of natural fracture surfaces in a horizontal core from the Austin Chalk showing plumose surface ornamentation. (a) Unmineralized fracture with a gently radiating plume; dark line indicates horizontal. (b) Fracture striking parallel to core in part (a), but tightly mineralized, unornamented, and with small offsets. (Cores courtesy of Oryx Energy.)

Fractures in Moderately Deformed Strata

Introduction

Carbonate strata need not become excessively deformed to develop multiple sets of fractures. The following case studies present fracture characteristics from two areas: (1) the Madera Limestone, located in a generally block-faulted extensional tectonic setting with a probable strike-slip overprint, and (2) the chalks of the Ekofisk oil field, a large offshore domal structure located over a salt plug. Successive generations of essentially unrelated fractures have been superimposed on these strata.

It is interesting to compare the irregularly distributed fractures in the Austin Chalk (porosity 10%) just described with the intensely fractured Ekofisk chalks (porosity 30%). "Chalk" is a generalized term that covers rocks with a wide range of properties and fracture characteristics.

Although carbonate rocks are relatively ductile when deformed in the laboratory under confining stress, most carbonate strata in nature fracture in a brittle fashion, suggesting that pore pressures largely negate confining stresses. The higher temperatures and pressures that lead to ductile deformation obviate any interest by the hydrocarbon industry in most strata of metamorphic grade.

Fractures tend to cross more lithologic boundaries and more lithologies as strata are increasingly deformed, yet the smaller fractures of a population still commonly terminate against the more plastic lithologies such as shale. Stylolites, created by pressure solution of the relatively soluble carbonates, become more common, along with the small local fractures that often form adjacent to stylolites. Short, wide, en echelon tension gashes can also occur. Fracture surfaces can become smoothed as asperities are sheared during minor slip, creating slickensided textures. Highly irregular fracture aperture, however, can also be created if the normal stress across the slipping fracture is insufficient to plane off the asperities. Later mineralization may or may not fill all or part of the resulting irregular void space.

Madera Limestone, Central New Mexico[3]

Introduction

The Pennsylvanian Madera Limestone is exposed in road cuts southeast of Albuquerque, New Mexico, along the dip slope of a tilted fault block that was uplifted during Cenozoic normal faulting. The unit is primarily a bioclastic limestone, with infrequent layers of marl and siliciclastic shale up to 1 m thick. Limestone bedding ranges from a few centimeters to 4–5 m thick.

Located at the southwestern corner of the extensional Basin and Range Province, the formation is cut into kilometer-scale blocks by faults that show evidence for several senses and phases of motion. The latest motion is commonly indicated to be strike-slip, possibly related to tens of kilometers of strike-slip offset across the Tijeras fault immediately to the north. However, the exposures occur in an overall structural unit that consists of a 50 km long block fault, upthrown on the west and dipping east. Smaller normal faults and small thrust faults are also present in this mixed tectonic setting. Local mild flexures of the strata add to the structural diversity.

Fracture Characteristics

Despite the numerous faults, fracturing in most of the Madera beds is relatively simple. Most fractures are normal to bedding regardless of dip and are spaced one to several meters apart. Fractures commonly terminate or are offset at interbedded marl and shale layers. Fracture spacing is most regular where 0.5 to 1 m thick limestone beds are separated by slightly thinner shale beds (Figure 7). Fracture propagation was not influenced significantly by inhomogeneities such as limestone-on-

Fig. 7. Field photograph of the Madera Formation, central New Mexico, showing regularly distributed fractures in thinner limestones (middle of photo) and more widely and irregularly spaced fractures in the thicker beds. The thin interbeds are shales. Outcrop is ~5 m high.

[3]John C. Lorenz

limestone bedding contacts, thus many fractures cross internal bedding planes.

Road cuts show that most of the fractures have been completely mineralized with calcite. The associated faults have a few tens of meters of offset, and irregular void space has been created by offset asperities. This void space is partially filled with coarsely crystalline (centimeter-scale) calcite, and significant fracture porosity remains.

Fracture swarms are not common. Fractures occur in a variety of heights and lengths as pervasively distributed bed-normal planes in the massive limestones over 1 m thick (Figure 8). In one exposure, however, several fracture swarms comprising seven to ten fractures each over a lateral interval of 1–2 m are spaced about 10 m apart. Other fractures at this outcrop are more uniformly distributed, and the swarms are inferred to be younger and to represent localized incipient faulting similar to the nearby strike-slip fault. The fracture swarms are subvertical, are more continuous vertically, and have larger (although completely mineralized) fracture widths than do the older, vertical, and more evenly distributed garden-variety fractures.

Although faults do not seem to be the direct cause of most fracturing, fracturing and faulting may be related in this unit. The dominant fracture set in the Madera Limestone trends parallel to many of the faults, and some of the larger fractures have surface textures (incipient slickensides) that suggest minor strike-slip motion across the fracture plane. However, fracture intensity does not increase with proximity to normal or strike-slip faults as it does in the Austin Chalk or in the Lisburne carbonates (described later). Only within the sharply flexed limbs of the local, small thrust faults does fracture intensity increase significantly.

Fig. 8. Photograph of pervasive fracture distribution in massive limestones of the Madera Formation, central New Mexico (see quarter dollar coin for scale). Note that several fractures terminate or are offset at stylolitic bedding planes and that both calcite-mineralized (white) and unmineralized fractures (adjacent to coin) occur.

A feature in a road-cut outcrop of the Madera Limestone illustrates how different carbonate lithologies may have different responses to the same strain. In this example, extension in a grainstone is accommodated by a small-displacement (<0.5 m) normal fault that terminates upward at a thin shale layer. The same extension is accommodated by an array of en echelon veins in the micritic limestone above the shale layer.

Stress orientations during fracturing were not necessarily uniform throughout the strata, even where faults do not disrupt the pattern. Although most outcrops contain fractures that strike within about 5° of one another, fractures in some limestone beds have strikes as much as 20° oblique to the fractures in the overlying or underlying beds, yet all other characteristics indicate that they are part of the same set of fractures.

Ekofisk Field Fractures[4]

Introduction

Ekofisk field is the largest of nine chalk reservoirs in the southern part of the Norwegian sector of the North Sea. It is an elliptical dome, elongate in a north-south direction, that lies within the Central Graben. The reservoir consists of two fractured chalk intervals separated by a relatively impermeable layer of more siliceous and argillaceous chalk. The chalks are of Maastrichtian and Danian age and are named the Tor and Ekofisk formations, respectively. The chalks are draped over what is assumed to be a salt diapir, although the salt has not yet been penetrated by wells.

The chalk is primarily comprised of whole or fragmented clay-sized unicellular planktonic algae (coccolithophorids). The chalks are composed predominantly of low-magnesium calcite, which forms over 80% of the Ekofisk Formation and over 95% of the Tor Formation. The bulk of the remainder is quartz, with subordinate clay minerals. Porosities average over 30% in both formations, but range from 5 to 50%.

The Shetland Group chalk reservoirs (of which the Tor and Ekofisk formations are members) in the Central Graben all contain natural fractures. Reservoir permeability is controlled by fracturing in the fields of the greater Ekofisk area; permeabilities inferred from the analysis of well tests are as high as 150 md (Brown, 1987), which is at least two orders of magnitude greater

[4]Helen E. Farrell

than the matrix permeabilities of 1–6 md measured in cores (Thomas et al., 1987). Natural fractures recognized in cores from the Shetland Group chalks have been classified into five major types. Tectonic, stylolite-associated, slump, and irregular fractures all enhance reservoir permeability. Healed fractures can reduce matrix permeability, at least locally. Compaction as a result of production has created numerous artificial fractures over much of the field (Teufel and Farrell, 1991; Teufel and Rhett, 1992), but this class of fractures is not described here.

Tectonic Fractures

Tectonic fractures are the most common fracture type in the Ekofisk Formation. They are throughgoing planar features that dip from 65° to 80° and form well-developed parallel sets (Figure 9a). Where the fracture intensity is high, an associated conjugate set is also developed. Fracture spacing is variable. In the most highly fractured zones, the fracture spacing is as small as 5 cm. Elsewhere, spacing of 15–100 cm is more common, and fracture spacing rarely exceeds 100 cm.

Offsets across the fractures are rarely seen, largely because of the scarcity of markers within the cores, but offsets of a few millimeters in a normal sense are typical where they can be documented. Slickensides and surface striations also indicate shear in a dip-slip sense across these fractures. The length of individual tectonic fractures cannot be determined from core because fracture terminations are rarely captured within these cores from deviated holes. By analogy with similar features observed in onshore chalk outcrops, individual fractures may range in length from a few meters to tens of meters. Tectonic fractures form an interconnected network throughout much of the productive formation.

Stylolite-Associated Fractures

Stylolite-associated fractures (Figure 9b) are common in the Tor Formation but are relatively rare in the Ekofisk Formation. Most moderate- to high-amplitude stylolites in the cores have fractures associated with them. The fractures are opening-mode tension cracks that initiated at the stylolite seam. The stylolite-associated fractures usually dip at >85° and are parallel to the stylolite columns. They form subparallel, anastomosing networks within which the fractures are interconnected. The fractures usually extend less than 10 cm from the stylolite columns, but may be as much as 20 cm in length. The lengths of the stylolite-associated fractures are usually considerably shorter than the distance between adjacent stylolites. As a result, where stylolite-associated fractures occur alone, they do not form an interconnected network throughout the formation.

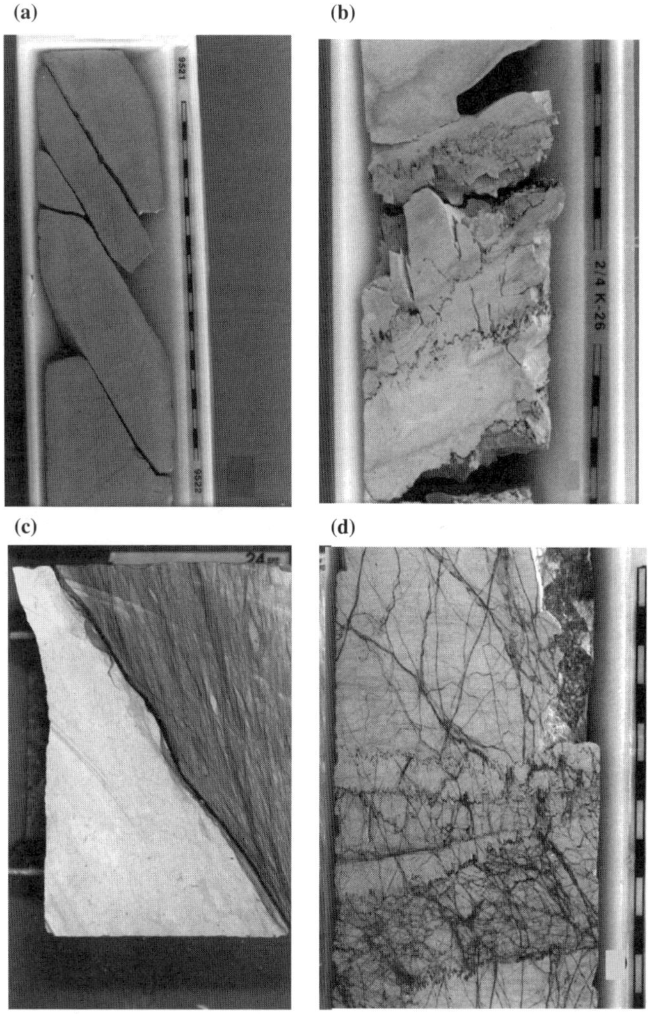

Fig. 9. Photographs of types of fractures found in cores from the Ekofisk field: (a) tectonic fractures, (b) stylolite-associated fractures, (c) slump fractures, and (d) healed fractures. Scales are in 5-cm increments.

Slump Fractures

Slump fractures (Figure 9c) have characteristics that are common to tectonic fractures and also to slump and glide surfaces. They usually cut the entire core but are rarely smooth and planar. Slickensides are present on some of the fracture surfaces. These fractures are parallel or subparallel to local bedding dips and occur exclusively where bedding dips are steep (>15°). The frequent parallelism of slump fractures and deformed laminations indicates that the fractures originated as glide surfaces and were reactivated as shear fractures during later deformation.

Irregular Fractures

Irregular fractures are defined as discontinuous, commonly nonplanar fractures that usually terminate at both ends within the core. The lengths of these fractures ("length" as used here is the maximum dimension measurable in the deviated core) are typically less than 20 cm. They often appear to be genetically associated with the tectonic fractures. These fractures may locally enhance permeability, but they are poorly interconnected on a large scale.

Healed Fractures

Healed fractures (Figure 9d) are narrow zones of comminuted chalk matrix that have been cemented and healed by diagenetic processes. They appear on the core as networks of fine, dark brown lines that are less porous, and as a result harder, than the adjacent chalk matrix. The healed fracture networks occur in allochthonous chalk facies and are thought to be the remnants of syndepositional deformation along microfaults that occurred before diagenetic processes had fully lithified the chalk matrix. These fractures are no longer permeable and do not form channels for fluid flow in the reservoir. Scanning electron microscopic (SEM) analysis suggests that there is a reduction in pore size as well as porosity within the healed fractures. The resulting increase in capillary forces in the pores of the fractures is thought to have resulted in more oil staining of the fractures than of the adjacent chalk matrix because the fractures are less easily flushed by the drilling fluids.

Fracture Mineralization

Any of the permeability-enhancing fractures may be completely or partially filled by secondary mineralization. The most common fracture-filling mineral is calcite, but quartz, kaolinite, chalcedony, pyrite, marcasite, and barite have also been observed. When the fractures are completely filled with a secondary mineral, their permeability is significantly reduced, but partial mineralization of fractures can be beneficial during pressure drawdown. Fracture mineralization is rare within the porous zones that form the best quality reservoir rock, especially in the Ekofisk field. In contrast, fracture mineralization is common where diagenetic processes have resulted in matrix cementation and hence lower porosity.

Fracture Intensity

The fracture characteristics of the Ekofisk and Tor formations differ (Figure 10). Stylolites and stylolite-associated fractures dominate the Tor Formation (95% chalk), whereas tectonic fractures are the most common type in the Ekofisk Formation (80% chalk). The presence or absence of stylolites appears to have controlled the style of fracturing: tectonic fractures occur where stylolites are absent, and stylolite-associated fractures dominate where stylolites are common. Bulk mineralogy also probably influenced the development of stylolites.

Within each formation, the intensities of tectonic and stylolite-associated fractures vary systematically in association with lithologic characteristics. Three chalk categories are recognized, based primarily on bedding characteristics: (1) laminated, (2) bedded, and (3) massive or contorted. In addition to changes in bedding style, the average matrix porosity increases from category 1 to category 3. Moreover, category 1 chalks tend to be slightly more argillaceous than the other two groups.

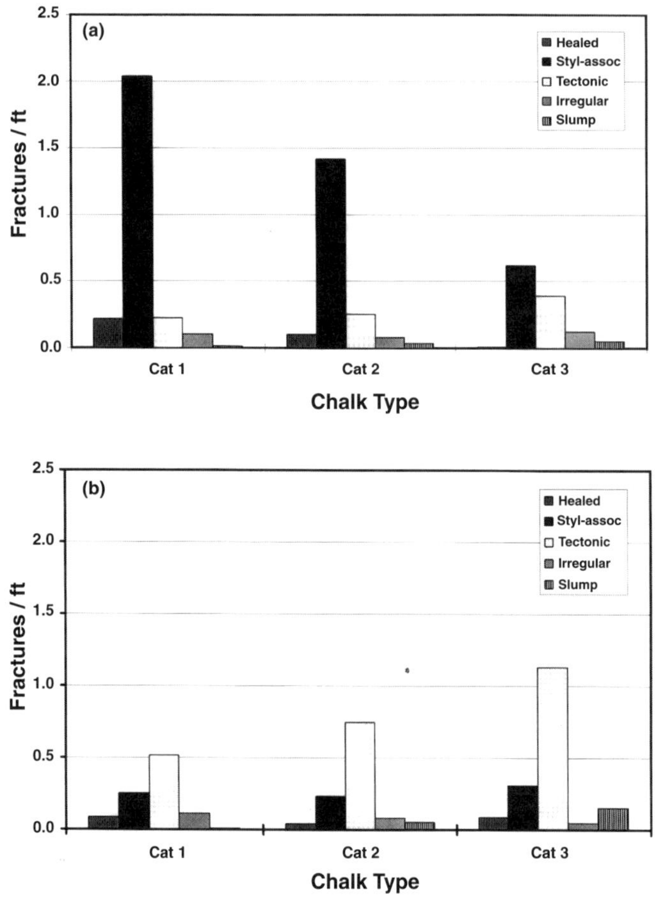

Fig. 10. Plots of lithology versus fracture intensity for different chalk types in the two reservoir formations at the Ekofisk field: (a) Tor Formation; (b) Ekofisk Formation.

Also, the intensity of the tectonic fractures in the Ekofisk Formation (and to a lesser degree in the Tor formation) increases from category 1 to category 3 chalks (Figure 10). The intensity of the stylolite-associated fractures that are more common in the Tor Formation follows a reversed pattern. The intensities of the other fracture types are relatively constant in all chalk types.

Fracture Orientations

Fracture orientations were measured from oriented core and are shown in Figure 11. A strong trend of NNE-SSW striking tectonic fractures is present in most of the wells. Fractures with this trend form a conjugate set dipping steeply to both the east and west. This NNE-SSW trend is parallel to major normal faults that have been mapped from seismic data in the northwest quadrant of the field. These fractures are inferred to have developed at the same time and in response to the same stresses as the faults, during the last pulses of the Mesozoic extension that formed the Central Graben.

Elsewhere, the tectonic fractures are oriented radially around the dome. On the east flank of the field, the fractures trend approximately east-west, and on the southeast flank they are oriented northwest-southeast. Again, these fractures form conjugate pairs with fractures dipping to the north and south. These radially oriented fractures are interpreted to be genetically related to radial and tangential stresses developed during doming of the field (Withjack and Scheiner, 1982; Teufel and Farrell, 1991).

Fractures in Highly Deformed Strata

Introduction

Strata within fold and thrust belts have typically been subjected to both regional and local structural stress events and may thus display superimposed fracture sets, especially if an early set has healed and the rock was effectively intact at the time of later fracturing. Preexisting *un*healed fractures may, however, accommodate much of the strain of a later event, thus a fracture record of the later event may not have been created. As shown by the Lisburne example that follows, the horizontal compression inherent in thrust belts commonly creates regional fractures normal to the thrust front (and to subsequent fold axes) prior to initial thrusting. The strata may then be flexed and fractured accordingly as they conform to ramping thrust planes or as they are folded into anticlines and synclines.

Strata that have undergone significant flexure also typically display hinge-parallel fractures in the region of maximum curvature. The spacing of such fractures is commonly related to the mechanical bed thickness. Mechanical thickness may be defined by formation boundaries, bedding plane boundaries, or (as shown by the Madison example that follows) boundaries defining packages of strata that are deformed as a unit, the fracture characteristics of which are related despite differing internal lithologies.

As the strata become more tightly folded, more bedding-parallel planes fail in shear (flexural slip) to accommodate folding, thus the mechanical thicknesses that control fracture spacing in flexed strata are controlled by lithologic heterogeneity, the tightness of the flexure, and the flexural history. More bedding planes slip as the folds tighten, and therefore mechanical thicknesses decrease and fractures become more closely spaced. In severe cases, bedding-parallel extension fractures and shear planes can be created within a sedimentologically homogeneous bed, and the bed reworked tectonically (R. A. Nelson, personal communication, 1995). With increased flexure, the hinge ultimately fails, accommodating a thrust fault.

Nelson and Serra (1993) have described the relative influences of lithology and structure on fractures in a thrusted carbonate sequence. They found that dolomites contain more than three times as many fractures as limestones in the relatively planar backlimb areas of a fold. The fracture intensity of both lithologies increases toward the fold hinge, but significantly more in limestones (6–8 times) than in the already highly fractured dolomites (2.2 times).

Below a certain degree of deformation, fracture intensity is typically less within the well-cemented, clean grainstones and packstones than in the carbonate mudstones. This is probably due to the twinning capabilities of calcite crystals. In an example from a thrust block in the Appalachians, Wiltschko et al. (1985) noted that similar strain magnitudes were accommodated by fracturing and stylolites in the finer, poorly cemented lithologies and primarily by calcite twinning in the coarser grained, cleaner, well-cemented, and less intensely fractured grainstones. However, as shown by the Madison Limestone example, "shattered" fracture facies may occur in any and all lithologies and in a wide range of bed thicknesses if significant bedding-parallel shear (to the point where the term *detachment zone* may be applicable) has occurred along interbedded evaporite facies.

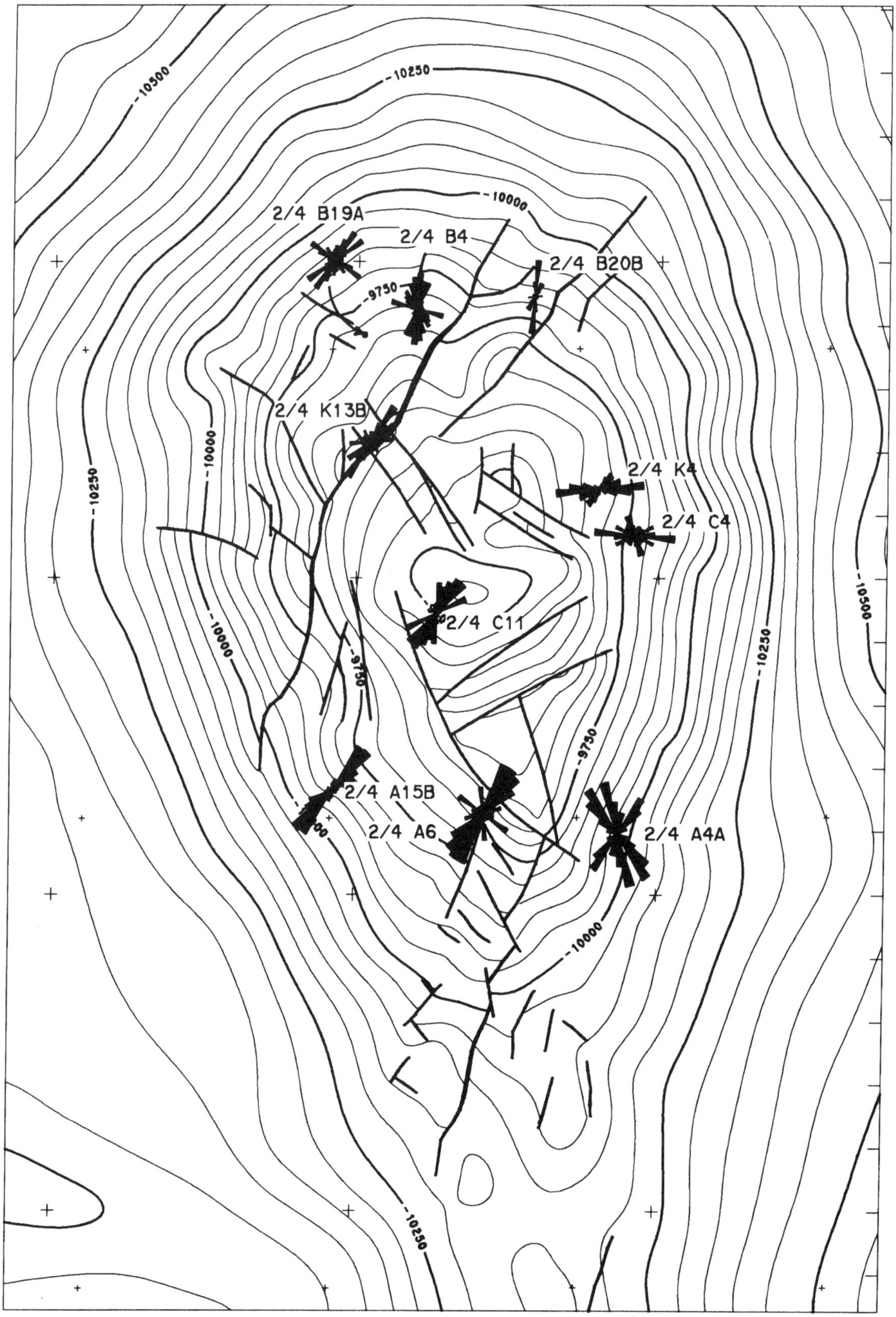

Fig. 11. Map of the Ekofisk field, North Sea, showing fracture orientations and structure contours (in meters subsea). Dome is ~4 km across.

Lisburne Limestone, Northeastern Brooks Range, Alaska[5]

Introduction

The Lisburne Group is exposed throughout the Brooks Range fold and thrust belt of northern Alaska (Figure 12). The Lisburne is also an important fractured carbonate reservoir in the subsurface of northern Alaska. The Lisburne Group is a thick (hundreds of meters) sequence of platform carbonates that was deposited on a south-facing continental margin during Carboniferous time.

The northernmost exposures of the Lisburne Group are in the Sadlerochit Mountains of the northeastern Brooks Range, ~120 km southeast of the Lisburne field on the North Slope of Alaska. The Lisburne Group at this locality was deformed during Cenozoic thrusting and related folding. The degree of folding of the Lisburne Group throughout the Brooks Range varies according to the amount of shortening and the presence or absence of an underlying decollement horizon (Wallace and Hanks, 1990). In contrast, the producing Lisburne field is in an extensional structural regime related to Cretaceous formation of the northern margin of Alaska and is now influenced by far-field compressive stresses related to Cenozoic compression (Jamison et al., 1980; Edrich, 1987). However, it has not experienced folding or faulting.

Fracture Patterns and Their Distribution

Two sets of fractures in the Lisburne Group are exposed in the northeastern Brooks Range. In the Sadlerochit Mountains where the Lisburne is only gently folded and is not underlain by a decollement horizon, an early fracture set strikes north-northwest and is normal to bedding (Figure 12). These fractures are in general evenly and widely spaced and continue vertically for several meters. Fracture spacing is a function of lithology rather than bedding thickness (Figure 13). Most of these early fractures were probably filled with calcite, although they are now locally unfilled because of weathering.

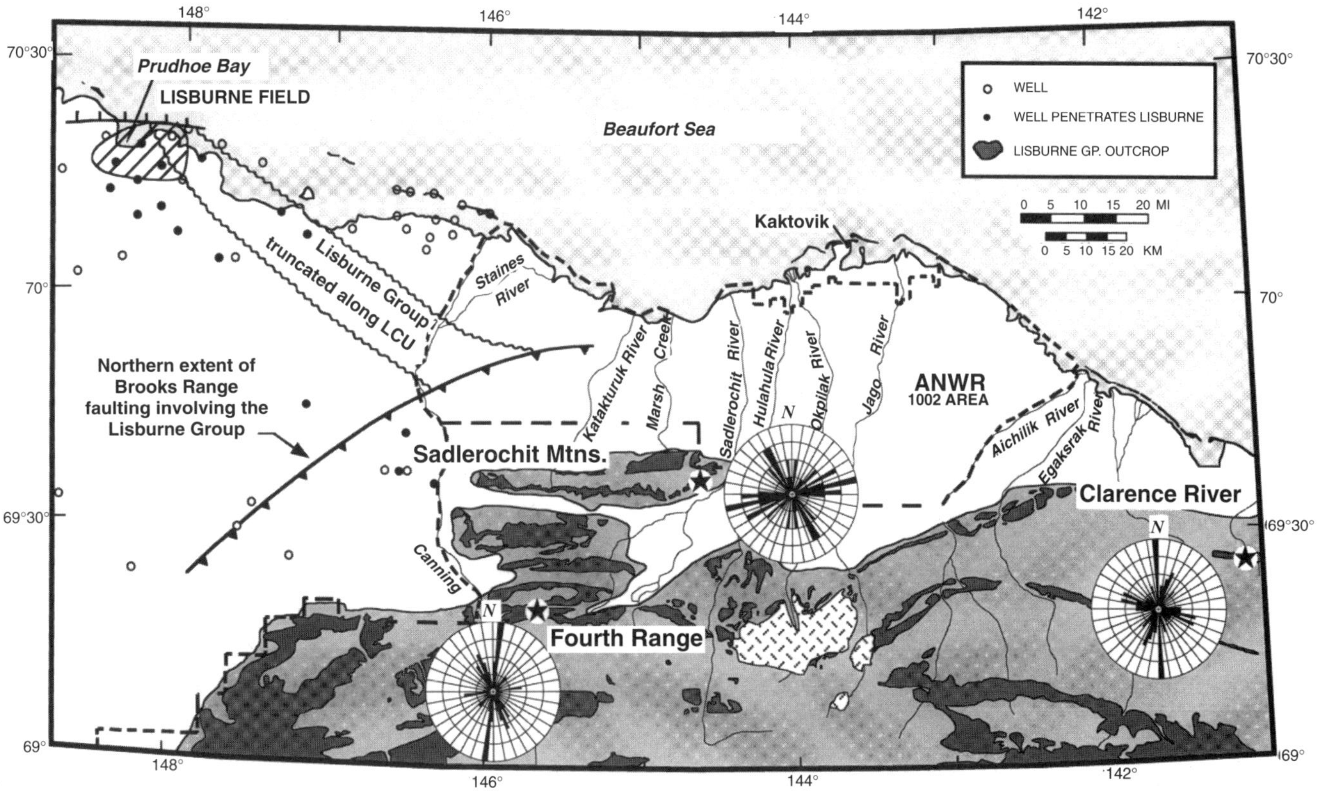

Fig. 12. Map showing the distribution of the Lisburne Group in the northeastern Brooks Range of northeastern Alaska. Also shown are the fracture patterns for the eastern Sadlerochit Mountains, the Fourth Range, and the Clarence River areas. (Modified after Bader and Bird, 1986; Bird et al., 1987; courtesy of K. Watts.)

[5]Catherine L. Hanks and John C. Lorenz

Although no surface ornamentation is preserved on these fractures, fractures with the same orientation in overlying clastic rocks have plumose structures indicating that the fractures are extensional in origin. The orientations of the plume axes are consistent with north-northwest-directed maximum horizontal compressive stresses. This early set of fractures is interpreted to be the result of regional stresses that developed in flat-lying strata in the foreland in front of the growing fold and thrust belt.

The Lisburne Group strata containing early fractures were subsequently incorporated into structures related to the thrust belt, and younger structure-related fractures were superimposed on the older fractures. Fractures of this later set dip steeply and strike east-northeast (Figure 12). These fractures are often filled with calcite, are often more closely spaced than the earlier fractures (Figure 13), and often terminate vertically at bedding planes. In the Sadlerochit mountains, these east-northeast fractures parallel a nearby thrust fault and the axial planes of local map-scale folds. The fractures are not purely extensional, and they often have a component of shear on them consistent with bedding-parallel slip during folding of the carbonates. These observations suggest that this second set of fractures has a genetic relationship with the associated folds and faults. However, the fractures commonly maintain a bedding-normal orientation where beds have been folded, indicating that they may have formed during the earliest stages of folding.

These same two orientations of fractures occur in more intensely folded Lisburne Group rocks found elsewhere in the northeastern Brooks Range (e.g., Fourth Range and Clarence River areas; see Figure 12). However, younger fractures that are subparallel to both previously described fracture sets have developed in the more intensely folded rocks, particularly near fold hinges. These secondary fracture sets appear to be related to the folding event and are oriented both perpendicular and parallel to the fold axes, similar in orientation to those predicted by Stearns and Friedman (1972). The early, north-northwest striking, prefold regional fracture set is largely obscured by the later development of north-south striking fractures in the most intensely folded strata.

A strong relationship exists among fracture density, character, and lithology in the Lisburne Group carbonates, in both surface exposures and subsurface reservoirs (Figure 13). The density of fractures generally increases with increasing carbonate mud content. Grainstones are the least densely fractured, and carbonate mudstones, especially if dolomitized, are the most densely fractured. This may be due to the increase in carbonate mud, an increase in dolomitization in the muddier lithologies, or more complete cementation of the grainstones and packstones.

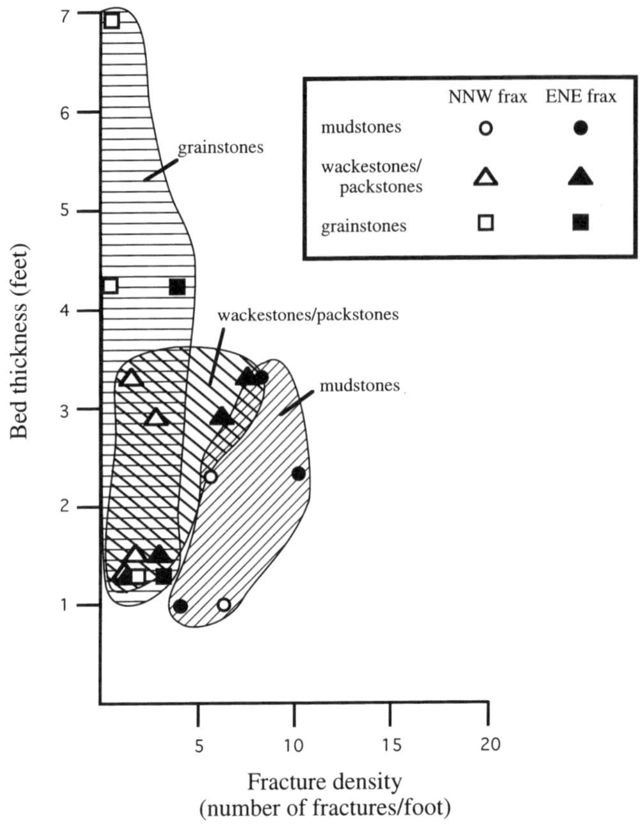

Fig. 13. Graph of fracture density versus bed thickness and lithology in carbonates of the Lisburne Group exposed in the eastern Sadlerochit Mountains. Note that fracture density is poorly related to bed thickness, especially for the reservoir grainstones.

The character of the fractures also changes with lithology and, to a lesser extent, with clast composition. While fractures are abundant in the dolomitic mudstones, most terminate at the over- and underlying boundaries with cleaner limestones. Fractures are commonly narrow, anastomosing, and multistranded in mudstones that are not highly dolomitized, suggesting that multiple episodes of fracturing opened new fractures and did not reactivate old ones. In contrast, fractures in the grainstones are typically more widely spaced, have wider apertures, and are planar and throughgoing. This is especially true of the early north-northwest striking fractures. Clast composition in the grainstones appears to affect the fracture morphology to some degree, with oolitic grainstones having more planar and more evenly spaced fractures than the skeletal grainstones.

No obvious correlation exists between stratigraphic bed thickness and fracture density. Increased bed thickness within the grainstone horizons does seem to increase the possibility of fractures terminating against stylolitic horizons or thin lime mudstone interbeds, if only because these bedding-parallel features are more common in thicker beds.

Siliciclastic shale horizons are interbedded within the Lisburne Group carbonates both in the Clarence River area and in the subsurface. In the Clarence River area, both the north-northwest regional fractures and local east-northeast fractures commonly terminate against shale beds as thin as 10–20 cm. In addition, shale horizons locally act as surfaces of bedding-parallel slip in those sections involved in folding. Such shales are highly deformed and contain calcite-filled fractures that have been folded and sheared.

Comparison with the Subsurface

Two fracture sets also occur in the relatively undeformed subsurface strata of the Lisburne field 120 km northwest of the northeastern Brooks Range. An early northeast-striking set parallels fault trends in the reservoir. A later north-northwest striking set is similar in character to the north-northwest striking set exposed in outcrops, and these fractures, both surface and subsurface, are inferred to be related to horizontal compression during Cenozoic thrusting from the south. These fractures are often open or only partially filled (Belfield, 1988; Sampson and Marcou, 1988; Missman and Jameson, 1991), whereas the east-northeast striking fractures are more commonly mineralized. The north-northwest set is inferred to be analogous in origin and character to the north-northwest striking regional fractures in outcrops. The early relative age of the east-northeast set in the subsurface suggests that this set does not have the same origin as the late-formed fractures with similar strikes found in outcrops.

Cores from deviated Lisburne wells allow for the measurement of fracture spacing in the subsurface. These data show, first, that fracture spacing in the same beds is consistently greater for the north-northwest set than for the east-northeast set, implying that the nature of the stress event controlled spacing rather than bed thickness. Second, the fractures of both sets are more widely spaced, by a significant margin, in the cleaner grainstones and packstones than in the muddier wackestones and mudstones, showing the strong effect of lithology on fracture intensity (Figure 14).

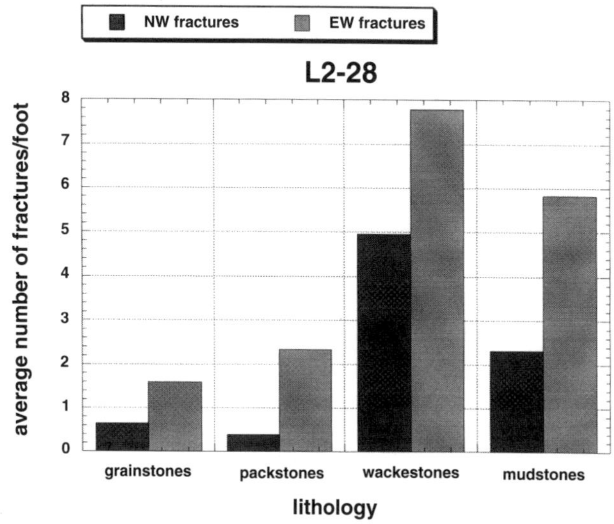

Fig. 14. Graph of fracture spacing in cores from well L2-28 in the Lisburne field. Note that lithology is a strong control on fracture spacing, but that the spacing of the east-west oriented fractures is consistently smaller that of the northwest-trending fractures in the same strata, implying that the nature of the stress event (e.g., magnitude and differential) also effects fracture spacing.

Moreover, the measured horizontal permeability anisotropy across the field is greatest in the north-northwest direction, parallel to the least numerous north-northwest fractures. This suggests that preferential fracture mineralization and the north-northwest maximum compressive horizontal in-situ stress are playing roles in closing the east-northeast fractures, diminishing their effectiveness in enhancing permeability despite their numbers. Seismic waves might be similarly effected.

An unconformity that developed pre-Permian exposure and erosion of the Lisburne Group is present at the top of the formation in the subsurface reservoir, below the overlying Kavik Shale. The 1–5 m thick zone of alteration associated with the unconformity is characterized by fractures with more extensive vertical development, dissolution of the fracture-filling calcite, and up to 50 m of vertical relief across the field (Belfield, 1988; Missman and Jameson, 1991). Spinner logs show that the enhanced permeability through fractures at this unconformity makes it both a thief zone during drilling and the primary production zone, although minimal core and the absence of a similar zone in outcrop make it difficult to characterize further.

Mechanical Stratigraphy of the Madison Limestone, Bighorn Basin, Wyoming and Montana[6]

Definition of Mechanical Stratigraphy

The term *mechanical stratigraphy* has been used to describe layering that reflects variations in deformation type and intensity caused by subtle petrologic and petrophysical changes (Stearns, 1978; Nelson, 1985; W. G. Brown, 1987, 1993). This perspective emphasizes the stratigraphic (usually lithologic) control of structural style and fracture distribution and tries to separate the influence of intrinsic petrologic rock properties from extrinsic or geometric attributes, such as regional stress analysis, depth of burial, radius of curvature, and fault proximity (Nelson, 1985).

Efforts at fracture prediction based on intrinsic rock properties tend to focus on the following attributes, usually only one at a time: (1) mineralogic composition (as it relates to rock strength), (2) prevailing grain size, (3) porosity, (4) lithologic contrasts, (5) bed thickness, and (6) preexisting fractures or other fabric anisotropy. Mechanical stratigraphic units are practical for reservoir analysis because they represent an integration of intrinsic variables. Fracture generalizations correlated with individual intrinsic attributes are apt to be site-specific, whereas characterizing fracture styles and distributions by mechanical stratigraphic units is a "portable" methodology dependent on a detailed level of stratigraphic analysis.

Geologic Setting

The study area is located in the northern Bighorn Basin of Wyoming and Montana. During the Laramide orogeny (latest Cretaceous–Eocene), numerous structural basins such as the Bighorn Basin were superimposed across the vast area of the former Mississippian Madison carbonate platform. The Bighorn Basin has numerous, mostly northwest-southeast oriented folds concentrated around its periphery. Most of these Laramide compressional structures are asymmetric, doubly plunging anticlines of varying axial vergences (Hoppin and Jennings, 1971; Stone, 1967, 1993). The following generalizations on fracture style and frequency reflect direct observations from core descriptions of the Elk Basin, Garland, and Torchlight Madison fields, coupled with study of Sheep Mountain anticline, a surface structure analogous to those producing from the Madison Limestone in the subsurface.

Stratigraphic Framework

To understand mechanical layering within the Madison Limestone, one must first appreciate the complex hierarchy of depositional sequences embedded in it, as summarized in Figure 15. Not all stratigraphic bedding or cycle boundaries are mechanical breaks. Therefore, sequence stratigraphic analyses should not rely on 1-D summaries, such as Fischer plots, which focus on cycle thickness only, nor should lithologic variations (or any other intrinsic rock attribute) be graphically emphasized to the exclusion of all others. The graphic summary in Figure 15 portrays interpreted cyclic hierarchies, depositional facies (and by inference, grain size and fabric), lithology, and stratigraphic bed thicknesses, making the interpretive transition to mechanical stratigraphic units a more data integrative process.

The entire Madison forms a second-order sequence of about 15 m.y. duration, bounded above and below by regional subaerial unconformities of tectonic origin, each representing a hiatus of more than 10 m.y. (Rose, 1976; Sando, 1976, 1978). Embedded within the Madison are six third-order depositional sequences (I–VI). Numerous small-scale, high-frequency cycles embedded in each third-order sequence are defined by facies successions indicative of subtidal storm-proximity trends, prograding grainstone shorefaces, and shallowing-upward peritidal conditions (Figure 15).

Third-order sequence boundaries are placed at culminations of thinning-upward sets of high-frequency cycles (Figure 15). The presence of multiple, thinly spaced, evaporite solution-collapse breccias representing sabkha-capped cycles below third-order sequence boundaries contrasts with thickening-upward cycle sets that occur above sequence boundaries. Although these intervals involve deepening-upward facies successions, they commence with a lowstand interval coinciding with interpreted slow rates of relative sea level rise.

During this time, the deposition of evaporitic supratidal facies, or perhaps shallow subaqueous evaporites, kept pace with the increasing space available for deposition caused by subsidence. These intervals are critical to the Madison's large-scale mechanical stratigraphy because interbedded among carbonates are formerly evaporitic, argillaceous intervals that are much thicker than the thin-bedded sabkha deposits found below sequence boundaries. The 3–10 m thick argillaceous evaporite collapse breccias at the base of sequences III, IV, and V form the most fundamental mechanical breaks

[6]Mark D. Sonnenfeld

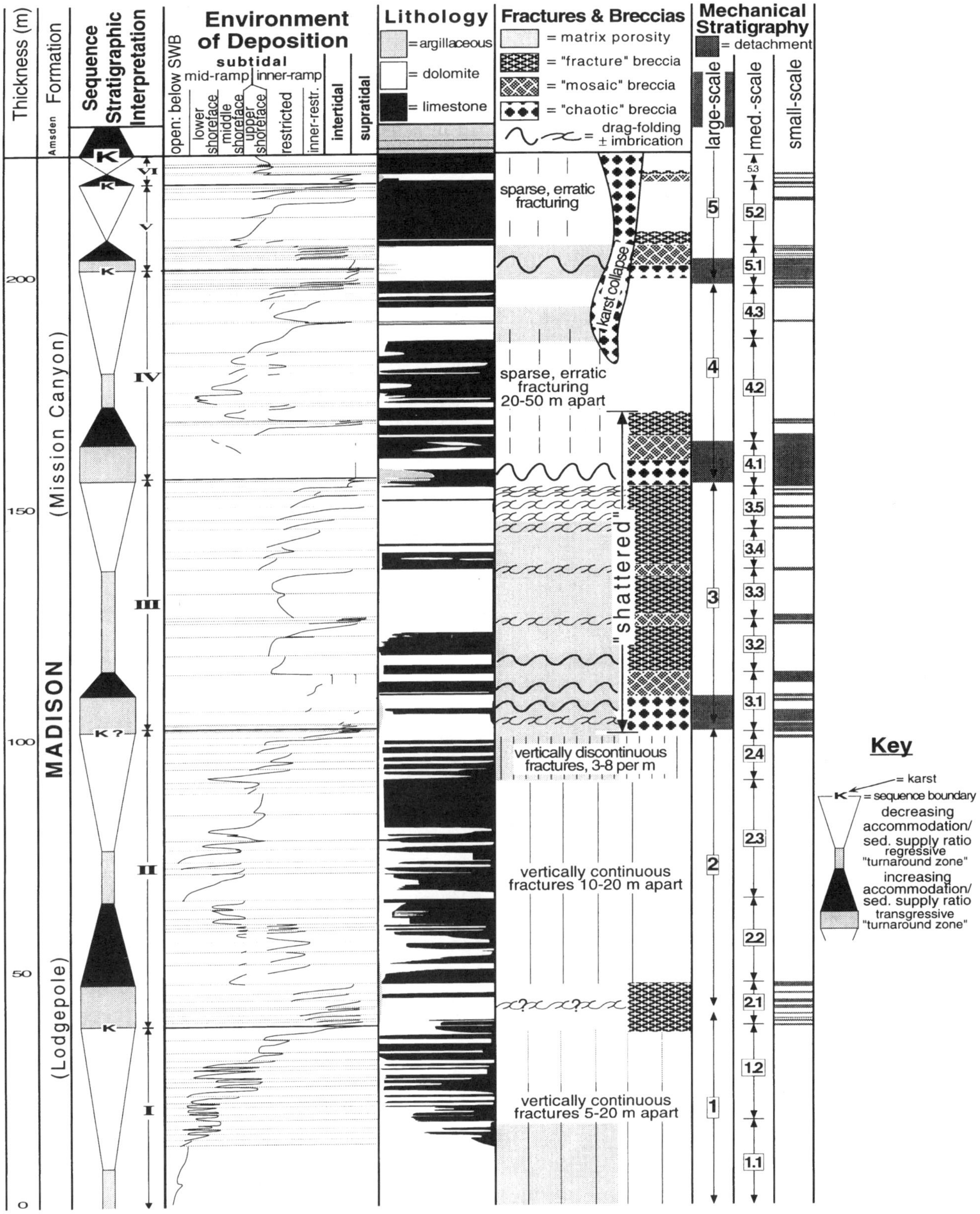

Fig. 15. Physical stratigraphy of the Madison Limestone at Sheep Mountain emphasizing the relationships between sequence stratigraphic and mechanical stratigraphic units. Chaotic breccias reflect intrastratal karst collapse, often of formerly evaporitic intervals. "Mosaic" (rotated, but nearly in-situ clasts) and "fracture" (fitted, in-situ, shattered fabric) breccias may be cave roof facies, purely tectonic in origin, or reflect pretectonic collapse breccias that were tectonically rebrecciated due to roles as intraformational detachments.

in the Madison Limestone (Figures 15, 16). Although most if not all evaporites within the Madison were leached during protracted post-Madison karsting, concentration of argillaceous residuum, coupled with the "ball-bearing" effect of clasts, focused flexural slip along these collapse breccia horizons.

Sequence III in Figure 16 shows an imbricated mini thrust belt resulting from very thin bedded mechanical stratigraphy due in large part to cyclic peritidal deposits capped by thin evaporite collapse breccias. Sequence IV is characterized by tighter, grainier, deeper water limestones than the fine-grained, restricted lagoonal to supratidal dolomites of sequence III. Note the collapsed caves within sequence IV, which have stoped upward above the stratigraphically controlled solution-collapse breccia level. Stoping and karst processes appear to have been enhanced along Mississippian normal faults or fractures that were reactivated and rotated during Laramide folding.

Mechanical Stratigraphy

Common slickensides along bed contacts at Sheep Mountain (Hennier and Spang, 1983) indicate bedding-plane slip. At this smallest scale of investigation one or two intrinsic rock parameters such as bed thickness and lithology seem to effectively explain contrasts in fracture distribution across bed boundaries. Fracture styles of progressively larger scale mechanical units depend increasingly on the integrated effects of multiple intrinsic parameters. For example, the hierarchy of stratigraphic cycles controls the distribution of small-scale evaporite collapse breccias (Figure 15), which in turn create a fracture-prone mechanical stratigraphy due to a combination of the following intrinsic factors: (1) thin bedding, (2) predominantly dolomitic lithologies, (3) fine grain sizes, and (4) accentuated bedding-plane slip due to mechanical contrasts between relatively argillaceous supratidal evaporite collapse breccias and intervening competent peritidal dolomite beds.

Large-scale mechanical units 1 and 2 (Figure 16) are approximately equivalent to sequences I and II, respectively. Although bed-to-bed variations in the closer fracture spacing (one to three per meter) probably reflect variations in lithology, bed thickness, porosity, and grain size, much more vertically continuous fracture sets spaced 5–20 m apart crosscut small- and medium-scale mechanical boundaries but are contained within mechanical units 1 and 2, respectively.

Spacing of strike-oriented extension fractures (type II; see Stearns, 1968) in these sets shows a clear correspondence to fold curvature. In Madison fields such as Elk Basin and Garland, the fracture system within the dolomitic reservoir at the base of sequence I contributes to high flow rates from wells that have historically been subject to coning, leaving significantly less depleted matrix blocks between the throughgoing, widely spaced (5–20 m) fractures (Figures 15, 16). Similar widely spaced fractures span the height of mechanical unit 2 (35 m), cutting the predominantly tight ooid-skeletal lime packstones and grainstones of sequence II and fostering vertical fluid communication through otherwise impermeable intervals (Figures 15, 16) (Demiralin et al., 1994).

Although fractures in the zone of thin-bedded evaporite collapse breccias and fine-grained peritidal dolomites between sequences I and II (mechanical unit 2.1) are more closely spaced (5 to 15 per meter) than those in the over- and underlying predominantly limestone mechanical units 1.2, 2.2, and 2.3, the former units lack the shattered appearance of analogous dolomitic deposits within overlying sequence III (mechanical unit 3). This is probably because unit 2.1 lacks major detachment zones and small-scale bed imbrication due to its relatively thin collapse breccias, which contain less argillaceous residuum than the collapse breccias within and bounding mechanical unit 3.

Depositional sequence III corresponds to mechanical unit 3, a 60 m thick zone with a generally shattered nature that tends to have random-looking fractures creating small centimeter- to decimeter-scale matrix blocks. Throughout the Bighorn Basin, this interval is the most pervasively fractured and productive of the entire Madison Limestone. Permeability enhancement 5 to 15 times that measured in core plugs has been observed based on injectivity tests at Elk Basin field (D. L. Cox, personal communication, 1994). The fractured and porous nature of this sequence consistently make it the most recessive slope-forming interval of the entire Madison. Stafleu et al. (in press) have proposed a semi-quantitative correspondence between weathering profiles or "erosive hardness" and impedance or "acoustic hardness," suggesting that the microfracture contribution to porosity of mechanical unit 3 might ultimately prove seismically resolvable.

Figure 15 shows this interval to be composed of comparable to slightly thicker than average beds, yet its pervasively fractured to shattered appearance probably reflects thinner than average mechanical bedding resulting from numerous interspersed evaporite solution collapse breccias. In addition, the muddy precursor fabrics of predominantly lagoonal to peritidal depositional sequence III yield fine-grained, hence more fracture-prone, dolomite.

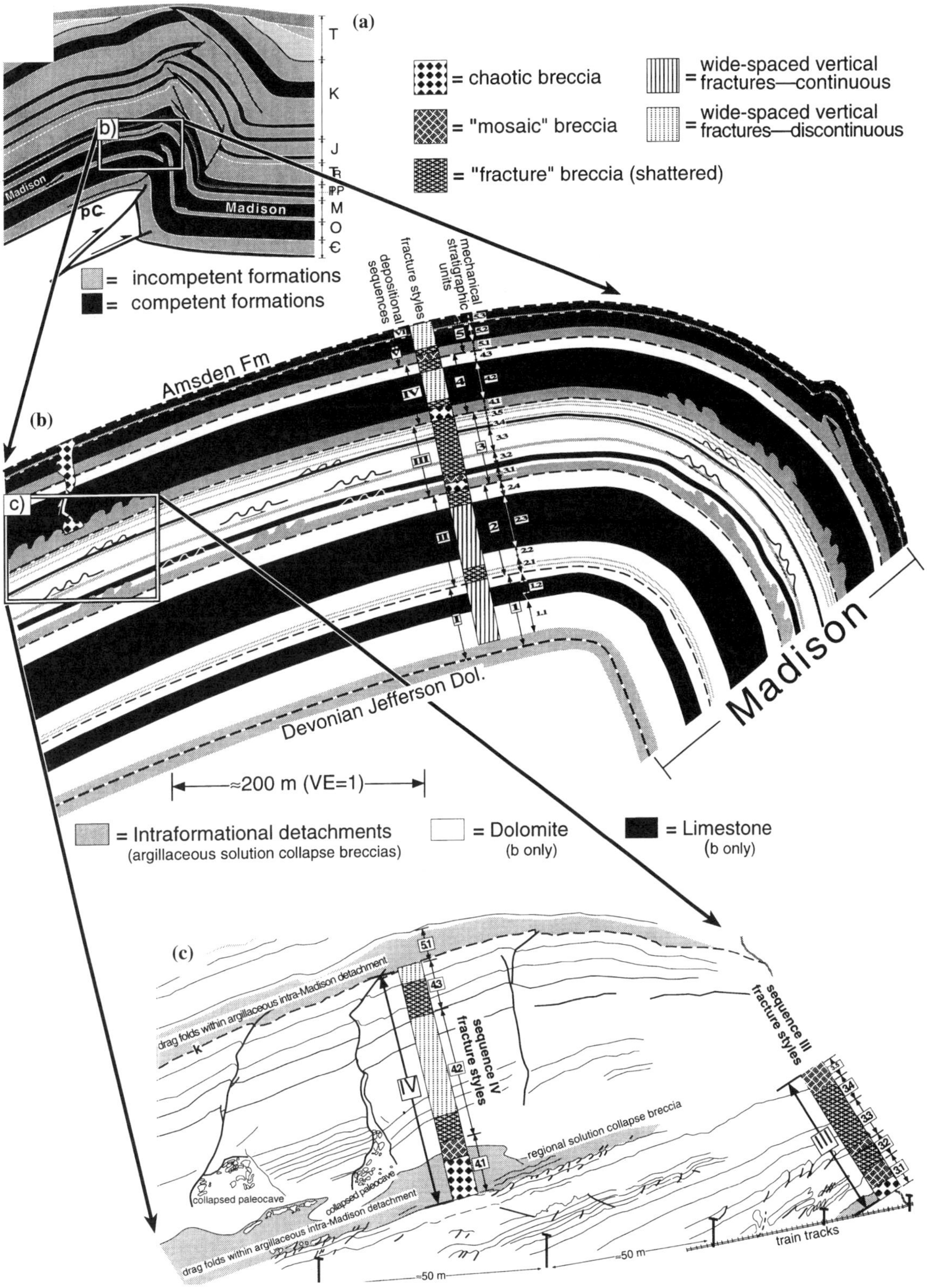

Fig. 16. **(a) Composite foreland basin flank anticline showing faulting and folding styles in the Paleozoic and Mesozoic section. (From Brown, 1987.) (b) Sketch of Madison at Sheep Mountain anticline showing simplified lithologic distribution, internal depositional sequences (I–VI), mechanical units (1–5), and gross fracture styles. (c) Tracing of photomosaic of west limb of Sheep Mountain anticline emphasizing the thick mechanical stratigraphy of the sequence IV relative to shattered sequence III. See text for more detail.**

Perhaps the most significant contributors to the fractured nature of sequence III (mechanical unit 3) are the thick, 5–10 m thick argillaceous evaporite solution collapse zones that bound it above (unit 4.1) and below (unit 3.1). These two zones are pervasive around the Bighorn Basin area and behaved as incompetent decollement zones during Laramide basin formation and smaller scale folding. Seismically unresolvable small-scale drag folds (<1–5 m amplitude) and reverse faults (<1–5 m throw) that originate and terminate within unit 3 detachments are common throughout the entire interval bounded by these two larger scale decollement zones (Figure 16). This structural style has an attendant mixture of shear and extensional fractures that show far less organization (a shattered, multiply brecciated look) than the throughgoing strike-parallel (type II) extensional fractures of mechanical units 1 and 2, whose spacing tends to be relatively predictable through analysis of extrinsic parameters such as large-scale fold curvature.

Like depositional sequence III, predominantly limestone depositional sequence IV (mechanical unit 4) is bounded by major argillaceous decollement zones (unit 4.1 and unit 5.1). However, the net fracture style of mechanical unit 4 is completely different because of variable overprinting of the original depositional layering by post-Madison karst collapse, brecciation, and calcite cementation. The throughgoing detachment zone at the base of this mechanical unit (unit 4.1) was a regional stratiform cave collapse system resulting from post-Madison evaporite dissolution (Sando, 1988). These collapse systems have a characteristic chaotic matrix- to clast-supported breccia fabric. Associated cave roof breccias were rebrecciated when underlying argillaceous cave fill and evaporite residuum were further contorted by drag folding during Laramide deformation.

In outcrop, the remainder of unit 4 is usually composed of tight limestones with nonexistent to highly erratic throughgoing fracture patterns coupled with calcite-filled fractures related to karst collapse. Karst brecciation within mechanical unit 4 resulted in less laterally continuous, less stratigraphically confined, and more vertically oriented collapse breccias (Figure 16). Although Demiralin et al. (1994) observed better development of fractures in tight limestones of sequences I and II than in porous dolomites, throughgoing fractures are suppressed in the comparably tight limestones of sequence IV, possibly because karst overprinting of the depositional fabric rendered the effective mechanical bed thicknesses much greater or perhaps simply more variable than original stratigraphic bedding.

Summary of Madison Fracture Characteristics

Single, or perhaps limited combinations of, intrinsic rock parameters cannot confidently predict variations in fracture styles and intensity within thick, vertically heterogeneous intervals such as the Madison Limestone. Consequently, stratigraphic cycles provide an effective scale of analysis in the quest for true mechanical stratigraphic units that are defined by common fracture styles. This approach integrates numerous intrinsic factors governing rock fracturing, such as (1) mineralogic composition (rock strength), (2) prevailing grain size, (3) porosity, (4) lithology contrasts, and (5) bedding thickness, because these attributes are likely to show a greater degree of organization and predictability when considered in a sequence stratigraphic context.

The mechanical stratigraphy approach should be viewed as a framework for fracture prediction distinct from, yet complementary to, evaluations of extrinsic controls on fracturing such as regional stress analysis (past and present), depth of burial and confining pressure, radius of curvature, and fault proximity. Mechanical stratigraphic units derived from sequence stratigraphic building blocks tend to emphasize the vertical distribution of flexural slip in concentric folding. The vertical distribution of fractures and small-scale faulting spatially associated with intraformational detachment zones and preferred horizons of drag folding are more predictable when stratigraphically controlled variations in flexural slip are understood. Finally, close attention to the hierarchy of stratigraphic cycles should enhance our appreciation for comparable mechanical stratigraphic hierarchies that may help explain distinct yet spatially coexisting fracture styles and lengths.

Conclusions

Natural fractures are as much a part of carbonate strata as calcium. They occur not only in beds that have been extensively deformed but also in essentially flat-lying beds. Many fractures in carbonate strata originated in extension (mode I fractures), although evidence for fracture origins in shear is not uncommon in tectonized regions. Later shear on extensional fractures can also occur. Regional fractures are widely distributed in most beds, and more localized sets of structure-related fractures occur in association with faults and fold hinges. Fracture distributions vary, with some fractures occurring in swarms interspersed with relatively unfractured rock, and other fractures being pervasively distributed within the strata. Multiple unrelated fracture sets may be superimposed on one another in some places, whereas in other areas, later stresses may only reactivate an existing set.

Interbedded evaporites and solution breccias may allow the formation of detachment zones within a folded formation, bounding packages of strata that deform as a mechanical unit. Within such units, the different beds may show relatively uniform fracture characteristics that are controlled by a combination of structure and lithology. Dolomites, in particular, can be found as shattered fracture facies in these settings.

Fractures are typically filled by calcite, but other mineral precipitates are common. Fractures in carbonates may also have been widened and lengthened by postfracturing dissolution. Most fractures, except in highly deformed strata, are oriented normal to bedding regardless of the current bedding orientation. Alternately, "conjugate" sets of high-angle fractures may have a common strike but opposite senses of dip.

Although more fractures occur at the hinges of folds than on fold limbs, lithology is the primary control on the degree of fracturing in most beds until the later stages of deformation are reached. Thus, dolomites and dolomitic mudstones are commonly more intensely fractured than limestones in the same structural settings. Fractures caused by or directly associated with faulting are commonly present only near the fault.

Acknowledgments

Amoco Production Company's Northwestern Business Unit is thanked for supporting Mark Sonnenfeld's Ph.D. research on the Madison Limestone at Elk Basin. Dennis Cox, Randy Kobbe, Don Sieger, and Paul Wehrle of Amoco are thanked for their input and support of the Elk Basin Madison study over the past several years. Discussions with Neil Hurley and Tom Oesleby of Marathon Oil Co. and with Dick Inden further broadened the foundation of the Madison study.

Helen Farrell would like to acknowledge Phillips Petroleum Company Norway and coverturers, including Fina Exploration Norway, Norsk Agip, Elf Petroleum Norge, Den Norske Stats Oljeselskap, Total Norge, Elf Rex Norge, and Norminol for their financial support and permission to publish this paper. The opinions expressed are those of the authors and do not necessarily represent those of Phillips Petroleum Company and coverturers.

John Lorenz's work was supported by the U.S. Department of Energy under contract DE-AC04-76DP00789.

We would like to thank Ron Nelson for his review of the manuscript and suggestions for its improvement.

References

Bader, J. W., and Bird, K. J., 1986, Geologic map of the Demarcation Point, Mt. Michelson, Flaxman Island and Barter Island quadrangles, northeastern Alaska: U.S. Geol. Surv. Misc. Invest. Ser. Map **I-1791**, scale 1:250:000, 1 sheet.

Belfield, W. C., 1988, Characterization of a naturally fractured carbonate reservoir: Lisburne field, Prudhoe Bay, Alaska: 63rd Annual Tech. Conf. & Exhib., Soc. Petr. Eng. Paper **18174**, 677-685.

Belfield, W. C., and Sovich, J. P., 1993, Fracture statistics from horizontal wellbores: Can. SPE/CIM/CANMET Internat. Conf. on Recent Advances in Horizontal Well Applications, Can. Soc. Petr. Eng. Paper **HWC94-37**.

Bird, K. J., Griscom, S. B., Bartsch-Winkler, S., and Giovanetti, D. M., 1987, Petroleum reservoir rocks, *in* Bird, K. J., and Magoon, L. B., Eds. Petroleum geology of the northern part of the Arctic National Wildlife Refuge, northeastern Alaska: U.S. Geol. Surv. Bull. **1778**, 79-100.

Brown, D. A., 1987, The flow of water and displacement of hydrocarbons in fractured chalk reservoirs, *in* Goff, J. C., and Williams, B. P., Eds., Fluid flow in sedimentary basins and aquifers: Geol. Soc. London Spec. Publ. **34**, 201-218.

Brown, W. G., 1987, Structural style of the Laramide Orogeny, Wyoming foreland: Ph.D. thesis, Univ. of Alaska–Fairbanks.

Brown, W. G., 1993, Structural style of Laramide basement-cored uplifts and associated folds: *in* Snoke, A. W., Steidtmann, J. R., and S. M. Roberts., Eds., Geology of Wyoming: Geol. Surv. of Wyoming Mem. **5**, 313-371.

Cater, J. M. L., and Gillcrist, J. R., 1994, Karstic reservoirs of the mid-Cretaceous Mardin Group, SE Turkey: tectonic and eustatic controls on their genesis, distribution and preservation: J. Petr. Geol., **17**, 253-278.

Chaplin, J. R., 1988, Lithostratigraphy of Lower Permian rocks in Kay County, north-central Oklahoma, and their stratigraphic relationships to lithic correlatives in Kansas and Nebraska: Soc. Econ. Paleont. Mineral. Spec. Publ. **1**, 79-111.

Collins, E. W., Hvorka, S. D., and Laubach, S. E., 1992, Fracture systems of the Austin Chalk, north-central Texas, *in* Schmoker, J. W., Coalson, E. B., and Brown, C. A., Geological studies relevant to horizontal drilling: examples from western North America: Rocky Mtn. Assn. Geol., 129-142.

Corbett, K. P., Friedman, M., and Spang, J., 1987, Fracture development and mechanical stratigraphy of Austin Chalk, Texas: AAPG Bull., **71**, 17-28.

Corbett, K. P., Friedman, M., Wiltschko, D. V., and Hung, J., 1991, Characteristics of natural fractures in the Austin Chalk outcrop trend: Austin Chalk Explor. Symp., Abstracts and Short Papers of South Texas Geol. Soc., 3-11.

Dart, R. L., 1987, South-central United States wellbore breakout-data catalogue: U.S. Geol. Surv. Open-File Rept. **87-405**.

Demiralin, A. S., Hurley, N. F., and T. W. Oesleby, 1994, Influence of karst fabrics on reservoir heterogeneity, Madison Formation, Garland field, Wyoming, *in* Dolson, J. C., Hendricks, M. L., and W. A. Wescott, Eds., Unconformity-related hydrocarbons in sedimentary sequences: Rocky Mtn. Assn. Geol., Unconformity Controls 1994 Symp. Vol., 219-230.

Edrich, S. P., 1987, The geological setting of Prudhoe Bay/ North Slope oil fields (abs.), *in* Tailleur, I., and Wimer, P., Eds., Alaska North Slope geology: Pacific Section, Soc. Econ. Paleont. Mineral. and Alaska Geol. Soc., Book **50**, 41.

Friedman, M., and McKiernan, D. E., 1993, Extrapolation of fracture data from outcrops of the Austin Chalk in Texas to corresponding petroleum reservoirs at depth: Can. SPE/CIM/CANMET Internat. Conf. on Recent Advances in Horizontal Well Applications, Can. Soc. Petr. Eng. Paper **HWC94-36**.

Gasteiger, C. M., 1980, Strain analysis of a low amplitude fold in north-central Oklahoma using calcite twin lamellae: M.Sc. thesis, Univ. of Oklahoma.

Hagen, K. B., 1972, Mapping of surface joints on air photos can help understand waterflood performance problems at North Burbank unit, Kay and Osage counties, Oklahoma: M.Sc. thesis, Univ. of Oklahoma.

Hennier, J., and Spang, J. H., 1983, Mechanisms for deformation of sedimentary strata at Sheep Mountain anticline, Bighorn Basin, Wyoming, *in* Boberg, W. W., Ed., Geology of the Bighorn Basin: 34th Ann. Field Conf., Wyoming Geol. Assn. Guidebook, 97-111.

Hodgson, R. A., 1961, Regional study of jointing in Comb Ridge—Navajo Mountain area, Arizona and Utah: Am. Assn. Petr. Geol. Bull., **45**, 1-38.

Hoppin, R. A., and Jennings, T. V., 1971, Cenozoic tectonic elements, Bighorn Mountain region, Wyoming-Montana, *in* A. R. Renfro, Ed., Symposium on Wyoming tectonics and their economic significance: Wyoming Geol. Assn. 23rd Field Conf. Guidebook, 39-47.

Jamison, H. C., Brocket, L. D., and McIntosh, R. A., 1980, Prudhoe Bay—a 10-year perspective, *in* Halbouty, M. T., Ed., Giant oil and gas fields of the decade: 1968-1978: Am. Assn. Petr. Geol. Mem., **30**, 289-314.

Lorenz, J. C., Warpinski, N. R., Branagan, P. T., and Sattler, A. R., 1989, Fracture characteristics and reservoir behavior of stress-sensitive fracture systems in flat-lying formations: J. Petr. Tech., **41**, 615-622.

Lorenz, J. C., Teufel, L. W., and Warpinski, N. R., 1991, Regional fractures, I: a mechanism for the formation of regional fractures at depth in flat-lying reservoirs: AAPG Bull., **75**, 1714-1737.

Luza, K. V., and Lawson, Jr., J. E., 1981, Seismicity and tectonic relationships of the Nemaha uplift, part III: Oklahoma Geol. Surv., Spec. Publ. **81-3**.

Macdonald, R. W., Jones., B., and Martindale, W., 1994, Cavity and fracture systems in Winnipegosis buildups (Givetian), Tableland area, southeastern Saskatchewan: Bull. Can. Petr. Geol., **42**, 232-244.

Melton, R., A., 1929, A reconnaissance of the joint systems in the Ouachita Mountains and central plains of central Oklahoma: J. Geol., **37**, 729-746.

Missman, R. A., and Jameson, J., 1991, An evolving description of a fractured carbonate reservoir: the Lisburne field, Prudhoe Bay, Alaska: Internat. Arctic Tech. Conf., Soc. Petr. Eng. Paper **22161**, 699-718.

Nelson, R. A., 1981, Significance of fracture sets associated with stylolite zones: AAPG Bull., **65**, 2417-2425.

Nelson, R. A., 1985, Geologic analysis of naturally fractured reservoirs: Gulf Publ. Co.

Nelson, R. A., and Serra, S., 1993, Vertical and lateral variations in fracture spacing in folded carbonate sections and its relation to locating horizontal wells: Can. SPE/CIM/ CANMET Internat. Conf. on Recent Advances in Horizontal Well Applications, Can. Soc. Petr. Eng. Paper **HWC94-41**.

Plumb, R. A., and Hickman, S. H., 1985, Stress-induced borehole elongation: a comparison between the four-arm caliper and borehole televiewer in the Auburn geothermal well: J. Geophys. Res., **90**, 5513-5521.

Queen, J. H., and Rizer, W. D., 1990, An integrated study of seismic anisotropy and the natural fracture system at the Conoco Borehole Test Facility, Day County, Oklahoma: J. Geophys. Res., **95**, 11,255-11,273.

Rizer, W. D., and Queen, J. H., 1986, Fracturing in the Fort Riley Limestone at Vap's Pass, Kay County, northern Oklahoma: 4th Ann. Mtg., Soc. Econ. Paleont. Mineral. Midcontinent Sect., Field Guidebook, 159-173.

Rose, P. R., 1976, Mississippian carbonate shelf margins, western United States, *in* Hill, J. G., Ed., Geology of the Cordilleran hingeline: Rocky Mtn. Assn. Geol. Symp. Guidebook, Denver, 135-151.

Sampson, L. E., and Marcou, J. A., 1988, Interference test in a fractured carbonate: a Lisburne case history: 63rd Annual SPE Tech. Conf. and Exhib., Soc. Petr. Eng. Paper **18138**, 303-317.

Sando, W. J., 1976, Mississippian history of the northern Rocky Mountains region: U.S. Geol. Surv. J. Res., **4**, 317-338.

Sando, W. J., 1988, Mississippian limestone (Madison) paleokarst: a geologic synthesis, *in* N. P. James and P. W. Choquette, Eds., Paleokarst: Springer-Verlag.

Smith, E., W., 1954, Subsurface geology of eastern Kay County, Oklahoma, and southern Cowley County, Kansas: M.Sc. thesis, Univ. of Oklahoma.

Smith, M. B., Ren, N.-K., Sorrells, G.G., and Teufel, L.W., 1986, A comprehensive fracture diagnostics experiment, part 2, comparison of fracture azimuth procedures: SPE Prod. Eng., **1**, 423-431.

Stearns, D. W., 1968, Certain aspects of fractures in naturally deformed rocks, in Riecker, R. E., Ed., Rock mechanics seminar; Air Force Cambridge Res. Lab., 97-118.

Stearns, D. W., 1978, Fault and forced folding in the Rocky Mountains foreland, *in* Matthews III, V., Ed., Laramide folding associated with basement block faulting in the western United States: Geol. Soc. Am. Mem. **151**, 1-37.

Stearns, D. W., and Friedman, M., 1972, Reservoirs in fractured rocks, *in* King, R. E., Ed., Stratigraphic oil and gas fields—classification, exploration methods, and case histories: Am. Assn. Petr. Geol. Mem. **16**, 82-106.

Stafleu, J., Schlager, W., Everts., A. J. W., Kenter, J. A. M., Blommers, G., and Van Voorden, A., in press, Topography as a proxy of acoustic impedance: Geophysics.

Stone, D. S., 1967, Theory of Paleozoic oil and gas accumulation in Bighorn Basin, Wyoming: AAPG Bull., **51**, 2056-2114.

Stone, D. S., 1993, Basement-involved thrust-generated folds as seismically imaged in the subsurface of the central Rocky Mountain foreland, *in* Schmidt, C. J., Chase, R. B., and Erslev, E. A., Eds., Laramide basement deformation in the Rocky Mountain foreland of the western United States: Geol. Soc. Am. Spec. Paper **280**, 271-318.

Teufel, L. W., and Farrell, H. E., 1991, Interrelationship between in situ stress, natural fractures and reservoir permeability, a case study of the Ekofisk field: Fractures and Jointed Rock Mass Conf., Intl. Soc. Rock Mechanics.

Teufel, L. S., and Rhett, D. W., 1992, Effect of reservoir stress path on permeability of fractures in chalk: Proc. 4th North Sea Chalk Symp., Deauville, France.

Thomas, L. K., Dixon, T., Evans, C., and Vienot, M., 1987, Ekofisk pilot waterflood: J. Petr. Tech., **39**, 221-232.

von Schonfeldt, H. A., Kehle, R. O., and Gray, K. E., 1973, Mapping of the stress field in the upper earth's crust of the U.S., U.S. Geol. Surv. Final Tech. Rep., Grant 14-08-0001-1222.

Wallace, W. K., and Hanks, C. L., 1990, Structural provinces of the northeastern Brooks Range, Arctic National Wildlife Refuge, Alaska: AAPG Bull., **74**, 1100-1118.

Wiltschko, D. V., Medwedeff, D. A., and Millson, H. E., 1985, Distribution and mechanisms of strain within rocks on the northwest ramp of Pine Mountain block, southern Appalachian foreland: a field test of theory: Geol. Soc. Am. Bull., **96**, 426-435.

Withjack, M., and Scheiner, C., 1982, Fault patterns associated with domes—an experimental and analytical study: AAPG Bull., **66**, 302-316.

Zemanek, J. E., Glenn, E., Norton, L. J., and Caldwell, R. L., 1970, Formation evaluation by inspection with the borehole televiewer: Geophysics, **35**, 254-269.

Zoback, M. L., and Zoback, M., 1980, State of stress in the conterminous United States: J. Geophys. Res., **85**, 6113-6156.

Chapter 9

Remarks on Exploration Tools: Integrated Exploration Strategy Applied to Carbonate Environments[1]

K. Fischer
M. Holling
R. Marschall
J. Mau
Geco-Prakla (ECM)
Hannover, Germany

Abstract

Applying an integrated exploration strategy requires (1) establishment of a data base containing the given observations and (2) development of a model data base. The model here is defined as a tool to aid in organizing the given observations and contains interrelated subgroups of detailed submodels, which include facies trends, velocity models, or maps resulting from global log correlations. Development of the submodels involves these steps:

- Global basin analysis (uplift or subsidence, basin history, hydrocarbon potential, migration paths, and trapping)
- Structural interpretation and sequence analysis of the seismic data (2-D or 3-D) and integration of nonseismic data (gravity and magnetics)
- Classification of intervals and facies trends for selected intervals (A-B/C code), mapping of additional tops, and application of geostatistics
- Refinement and extension of mapped interfaces
- Final depth conversion using improved velocity functions
- Application of modeling tools and special processing techniques to improve resolution and verify interpretation

Aspects of carbonate settings that can be investigated by following these guidelines include identification of buildups, mapping of shelf trends, and verification of shelf trends in terms of lithofacies units (individual impedance values) by finite-difference forward modeling (e.g., zero offset application). Further investigations include application of special processing tools, such as L_1-Deconvolution and AVO-analysis, and derivation of a final velocity model for time-depth conversion using velocity checking wavefronts.

The final result of seismostratigraphic work improves the effectiveness of hydrocarbon evaluation and production due to detailed knowledge of lateral and vertical lithofacies trends, permeability distribution, local permeability barriers, and sealing potential. The chronostratigraphic information derived from seismic sequence analysis allows for improved well-to-well correlation. This is especially useful for 3-D data and the resulting 3-D stratigraphic model, which forms the basis of a reservoir model.

Introduction

Integrated exploration strategy is defined here as the integration of structural and stratigraphic interpretation with interactive feedback using the same data base. Figure 1 shows a flow chart summary of this strategy. This concept can be applied to carbonate environments.

A schematic cross-section of marine carbonate depositional environments is shown in Figure 2 (Pettijohn, 1975). In general, there are five types of modern carbonate deposits:

- Shallow-water marine deposits
- Deep-water marine deposits
- Carbonates of evaporitic basins
- Carbonates of freshwater lakes and springs
- Eolian carbonates

The shallow-water environment is complex and includes many subenvironments within it, such as tidal and supratidal flats, extensive shelf and bank areas, marginal reefs, and backreef lagoons (see Wilson, Chapter 2, this volume).

Using a data example from Africa, we discuss the integrated exploration strategy in detail. This data example involves a prograding carbonate platform (Mayouf et al., 1993), to which a subset of steps as described in Figure 1 is applied to demonstrate the concept.

Statement of the Problem

The common procedure in hydrocarbon exploration using seismic data is to construct a "structural" interpretation of the data. Prominent seismic markers, normally representing regional unconformities, are picked on the seismic sections (Figure 3) and mapped. Markers are generally picked as close as possible to the assumed target level. Also, the markers should reflect vertical variations of the velocity field, that is, they should separate intervals of different lithologies (e.g., carbonates, siliciclastics, and anhydrite) from one another. The output is a collection of time-contour maps of various levels. A depth conversion, normally based on well data using an apparent average velocity field or regional velocity laws, results in a detailed picture of the subsurface. The use of irregularly distributed well data leads to many uncer-

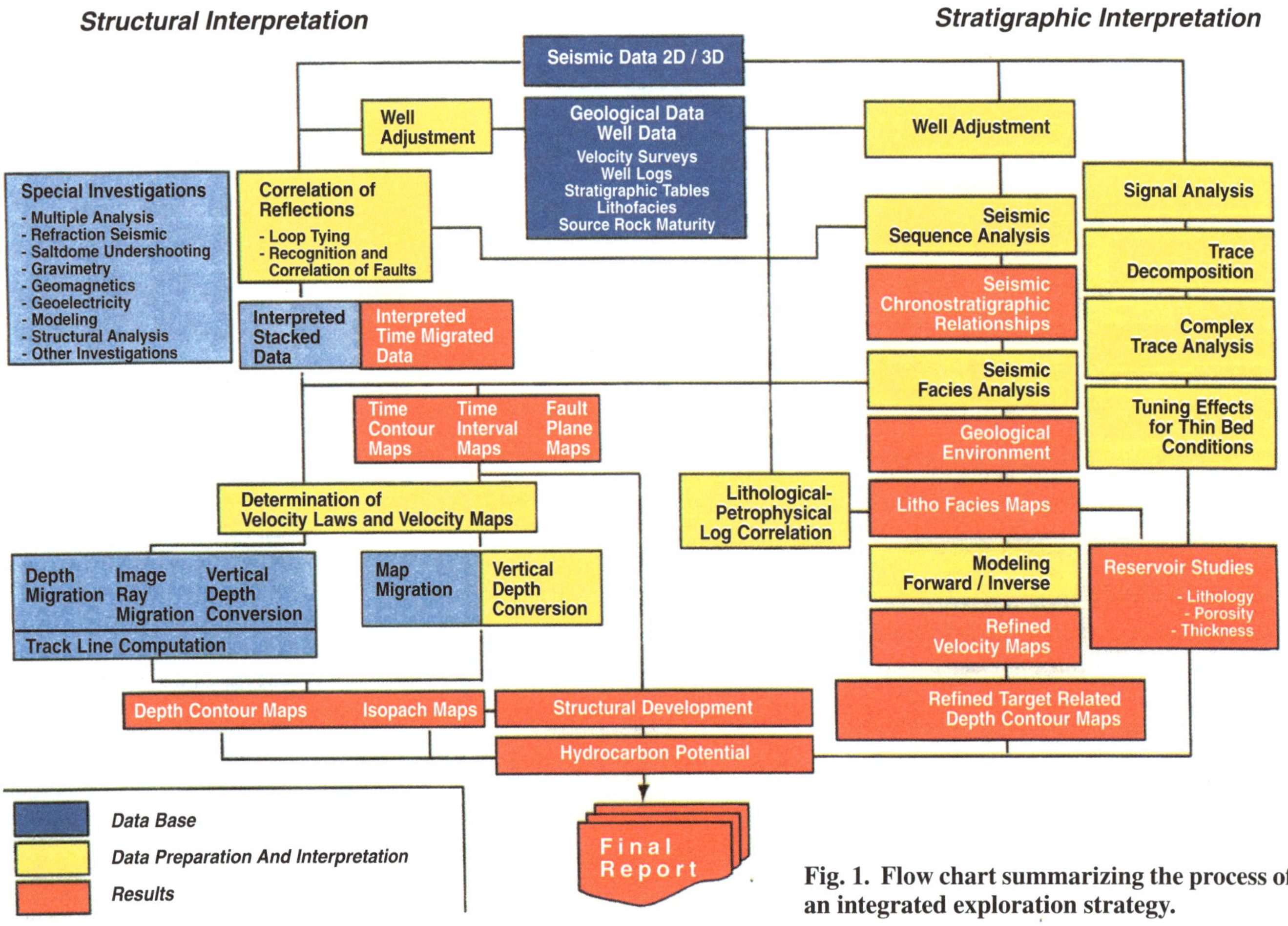

Fig. 1. Flow chart summarizing the process of an integrated exploration strategy.

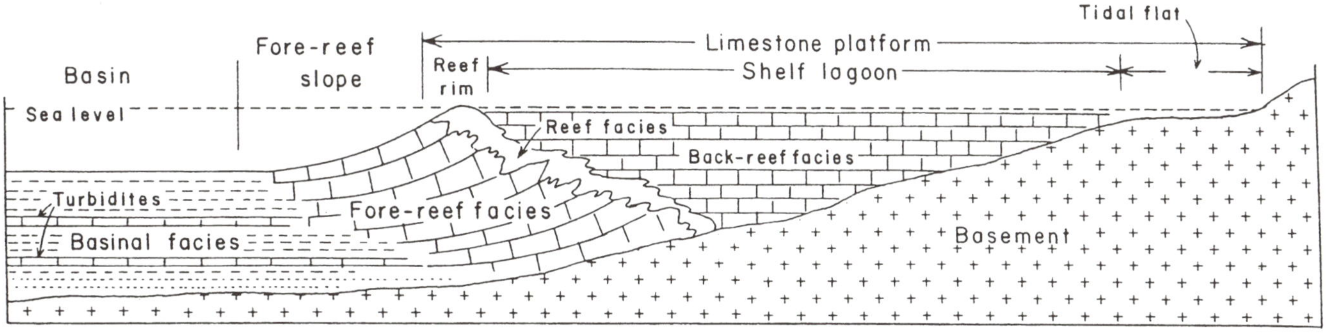

Fig. 2. Schematic cross-section showing principal marine carbonate depositional environments. From Pettijohn (1975).

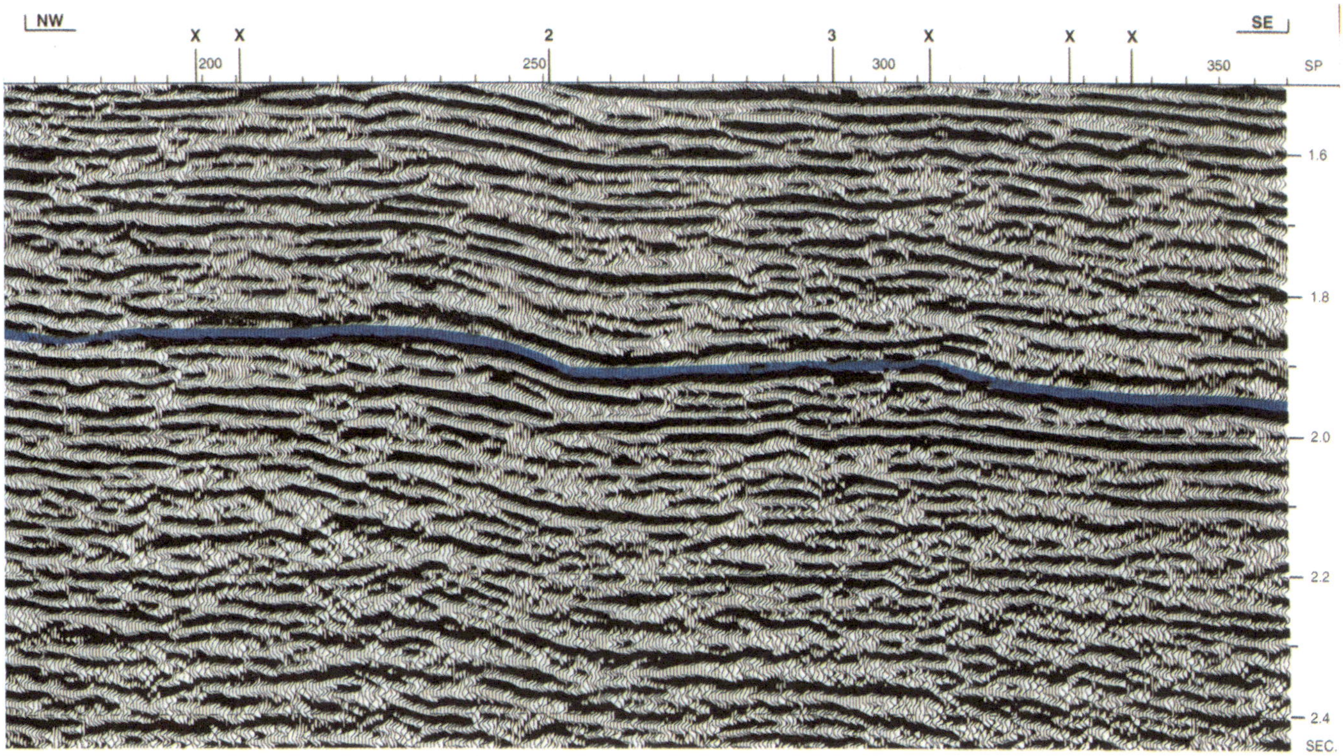

Fig. 3. Example of structural interpretation of seismic data (line 1). The seismic marker (blue) indicates a regional unconformity.

tainties, such as the velocity field between the wells, which must be interpolated.

This structural interpretation of seismic data results in knowledge of the structural setting of an area, including the location and trends of faults and anticlinal and synclinal structures. By using isopach maps, the structural development of the area can be outlined, such that the timing of deformation and the areas of prevailing relative uplift and subsidence can be identified. Based on these and geochemical investigations, a drilling location on a four-way dip closure is normally defined.

In the example shown in Figure 3, two minor closures were identified (Figure 4). They were located within the flank of a major anticline. From structural investigations, the northeast area was shown to have undergone strong subsidence, whereas the southwest area showed relative uplift. However, both structures appeared to be too small to justify drilling a well. Thus, on the basis of just this structural interpretation, the area would be regarded as not very prospective. Application of seismic stratigraphic concepts to the data can increase the prospect potential of the area because it provides additional information on the reservoir.

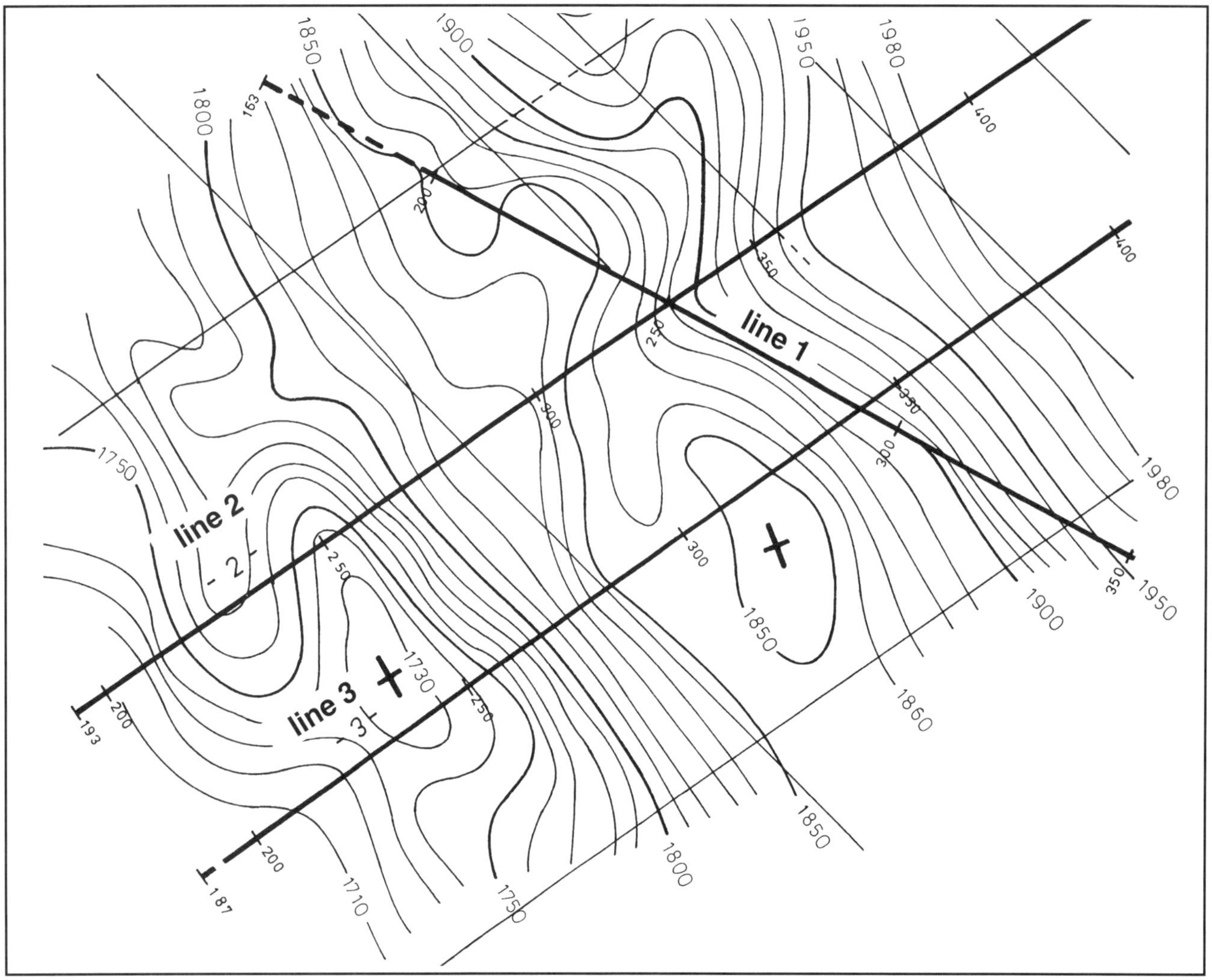

Fig. 4. Example of a time structure map representing the structural interpretation of seismic data presented in Figure 3. Two minor closures (crosses) were identified in the flank of an anticline.

Stratigraphic Interpretation

The normal progression of steps in constructing a seismic stratigraphic interpretation, based on carefully processed and zero-phased seismic data, is as follows:

- Seismic sequence analysis
- Well tie
- Sea level analysis
- Seismic facies analysis
- Derivation of a geologic model
- Derivation of a lithologic model
- Forward modeling and verification loop

A *seismic sequence* is defined as a relatively conformable succession of reflections on a seismic section interpreted as genetically related strata. This succession is bounded at its top and base by surfaces of discontinuity marked by reflection terminations and interpreted as unconformities or their correlative conformities (Mitchum et al., 1977). Applying this to the seismic section shown in Figure 3, six seismic sequences were distinguished according to toplap, downlap, and onlap reflection terminations (Figure 5). The same sequence boundaries were also identified on the crossline sections shown in Figures 6 and 7. When these unconformities were traced to a well located about 5 km to the north, the sequence boundaries were seen to coincide well with

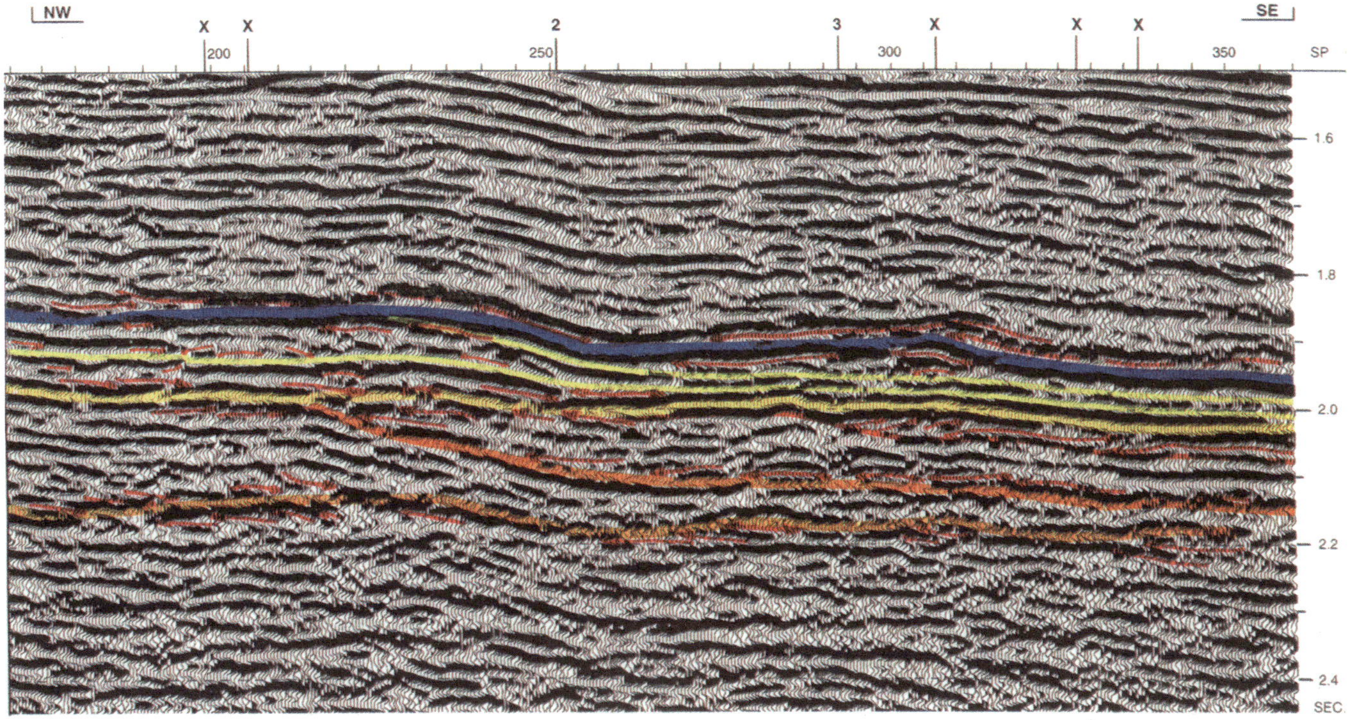

Fig. 5. Seismic section line 1 (from Figure 3) showing seismic sequence analysis. Five seismic sequences (yellow and orange) are distinguished on the basis of toplap, downlap, and onlap. Red arrows indicate reflection terminations.

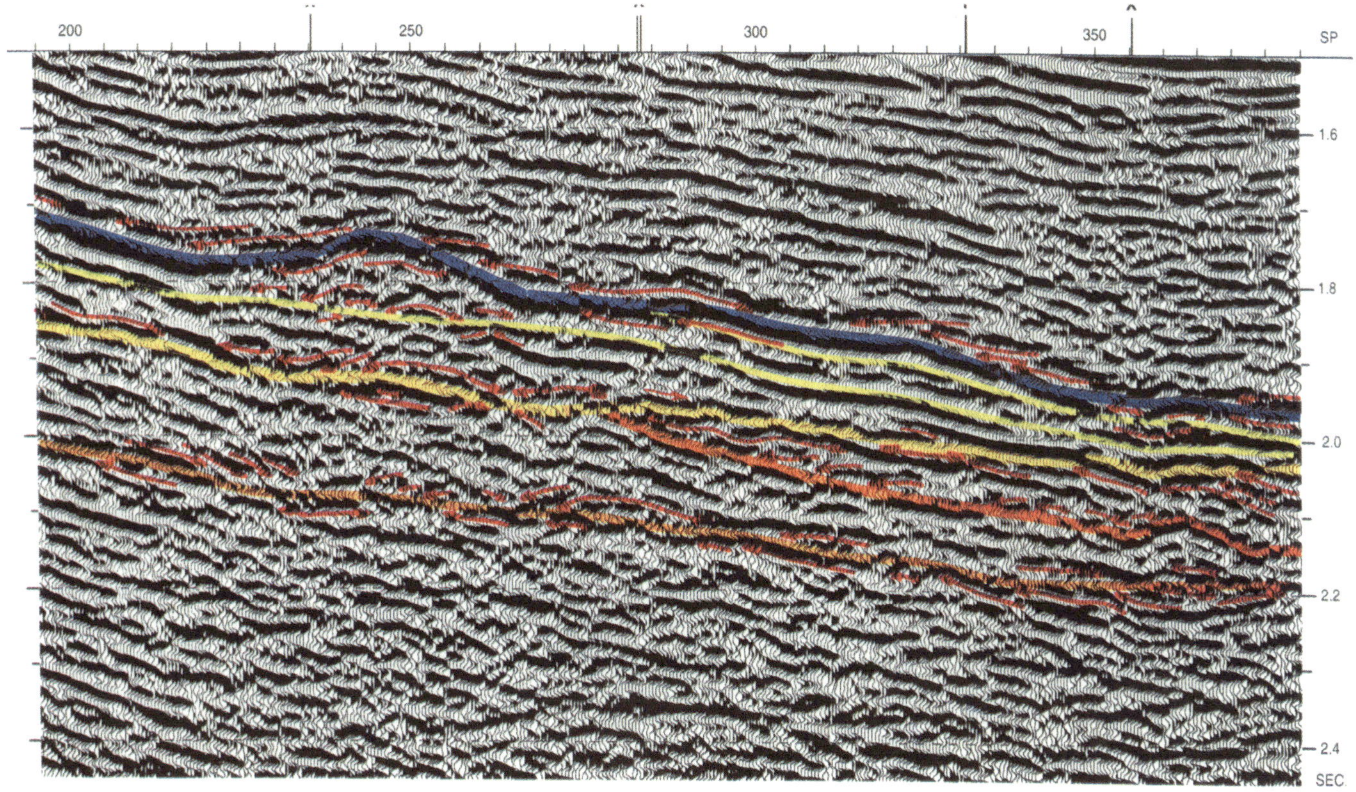

Fig. 6. Seismic section line 2 (first crossline of Figure 5) showing five seismic sequences (yellow and orange), as in Figure 5.

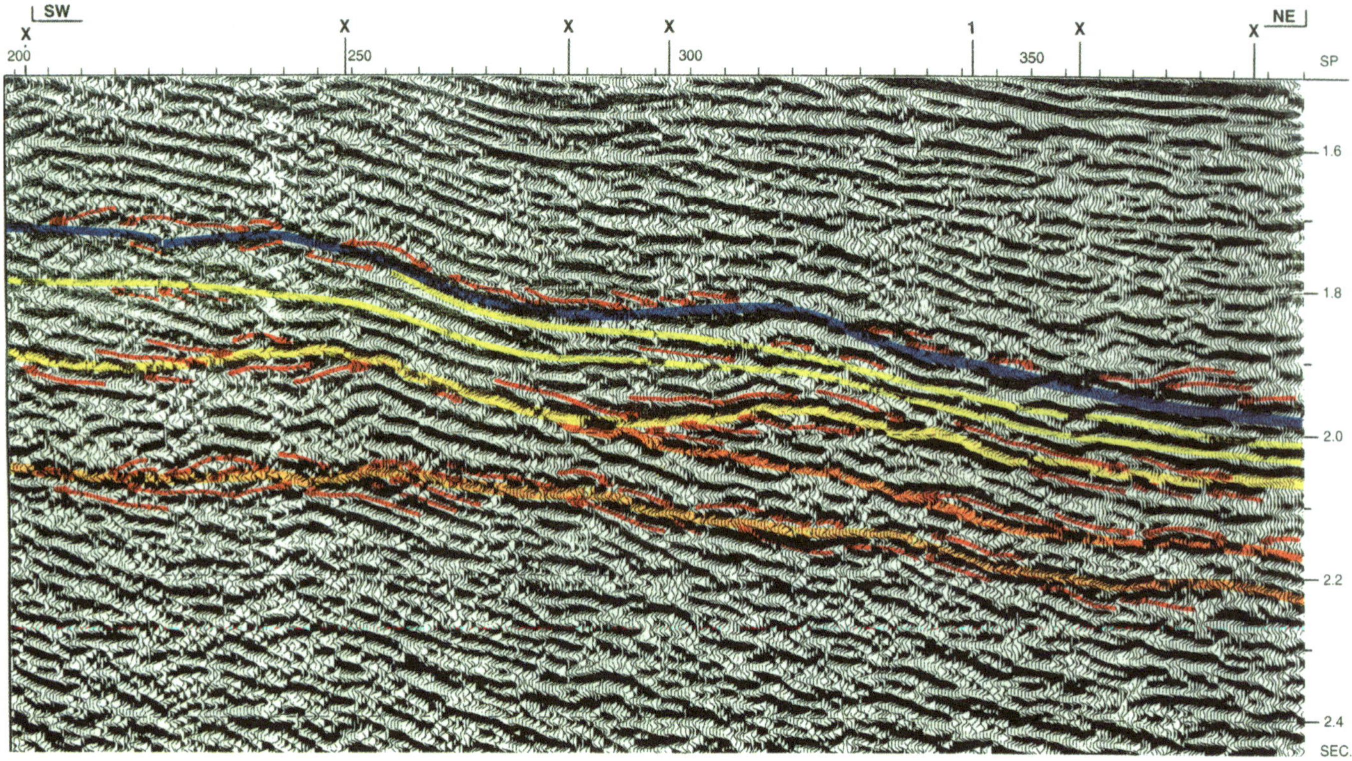

Fig. 7. Seismic section line 3 (second crossline of Figure 5) showing five seismic sequences (yellow and orange), as in Figures 5 and 6.

prominent log markers within a carbonate sequence. If the relevant information for this well is available, this would allow for a stratigraphic classification of the sequences identified.

Generally, a seismic sequence should be regarded as part of a depositional sequence, which is interpreted to be deposited during a cycle of eustatic change of sea level, starting and ending in the vicinity of the inflection points on the falling limbs of the sea level curve. The top and base boundaries are unconformities or their correlating conformities (after Vail, 1987). A schematic sketch of a depositional sequence shows the subdivision into lowstand, transgressive, and highstand systems tracts, each distinguished by specific boundaries and internal geometries (Figure 8). When this model was transferred to the example under investigation, it resulted in a classification of the six sequences defined as a succession of highstand, lowstand, and transgressive systems tracts (Figure 9). The transgressive systems tract represents a relatively thin interval, which might be partially beyond the resolution of the seismic data. The uppermost three sequences represent three stages of the highstand systems tract, probably separated by maximum flooding surfaces.

The next step comprises the seismic facies analysis. The A-B/C code (Ramsayer, 1979) was used to describe the character of the reflection terminations at the upper (Above) and lower (Below) sequence boundaries and the internal reflection Configuration. By using this code, three different seismic facies types were distinguished within each of the upper sequences in line 1 (Figure 10):

- C-C/P in the northeastern and southwestern areas, which represent a succession of parallel (P) reflections that are concordant (C) at the top and the base of the sequence,
- TP-DN/SO between the concordant patterns, representing toplap (TP) and downlap (DN) terminations at the top and the base of the sequence and an internal sigmoid-oblique (SO) pattern
- C/TP-DN/M mainly found to the northeast of the preceding pattern and representing mounded (M) configurations

These same seismic facies were also found in the crosslines to line 1 (Figures 11 and 12). From this delineation of the seismic facies in section, these facies can now be mapped in plane view, taking into account all

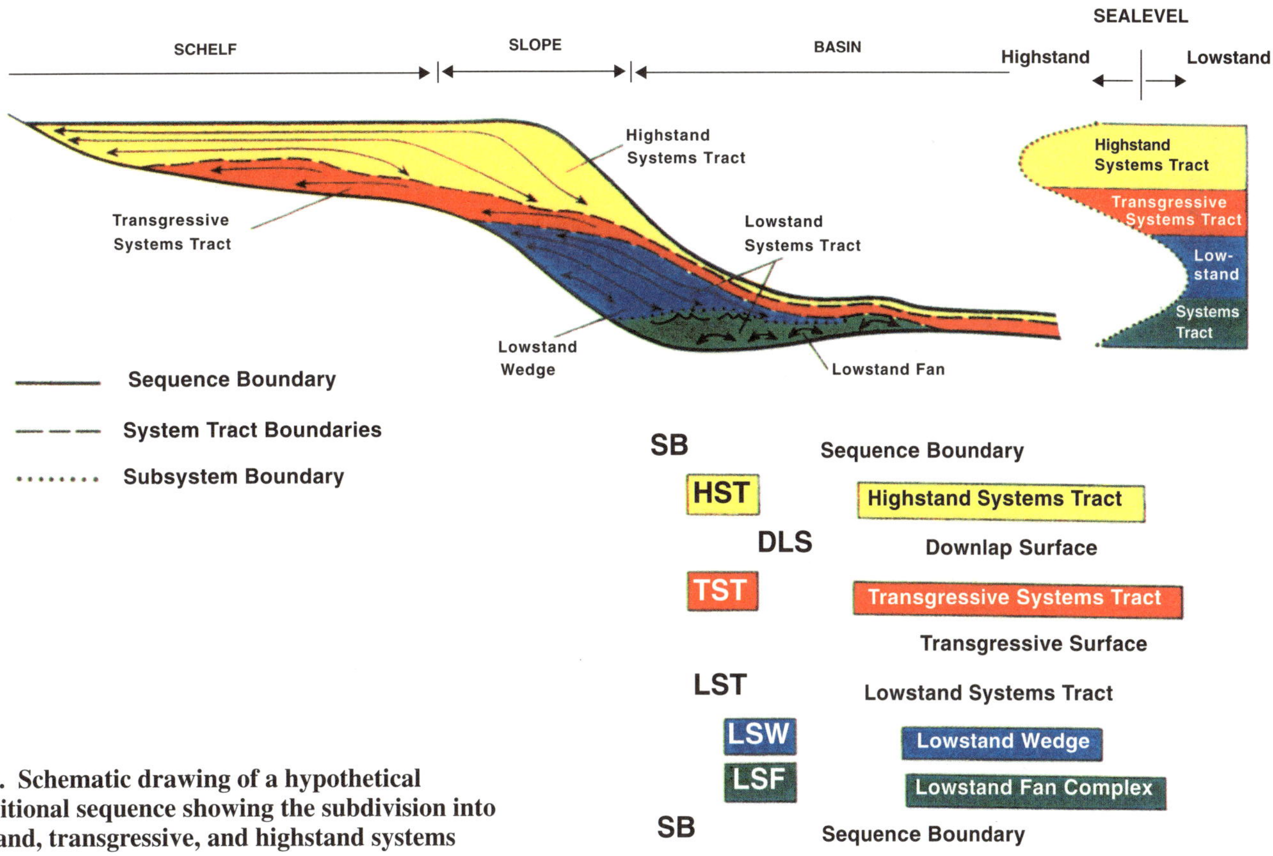

Fig. 8. Schematic drawing of a hypothetical depositional sequence showing the subdivision into lowstand, transgressive, and highstand systems tracts.

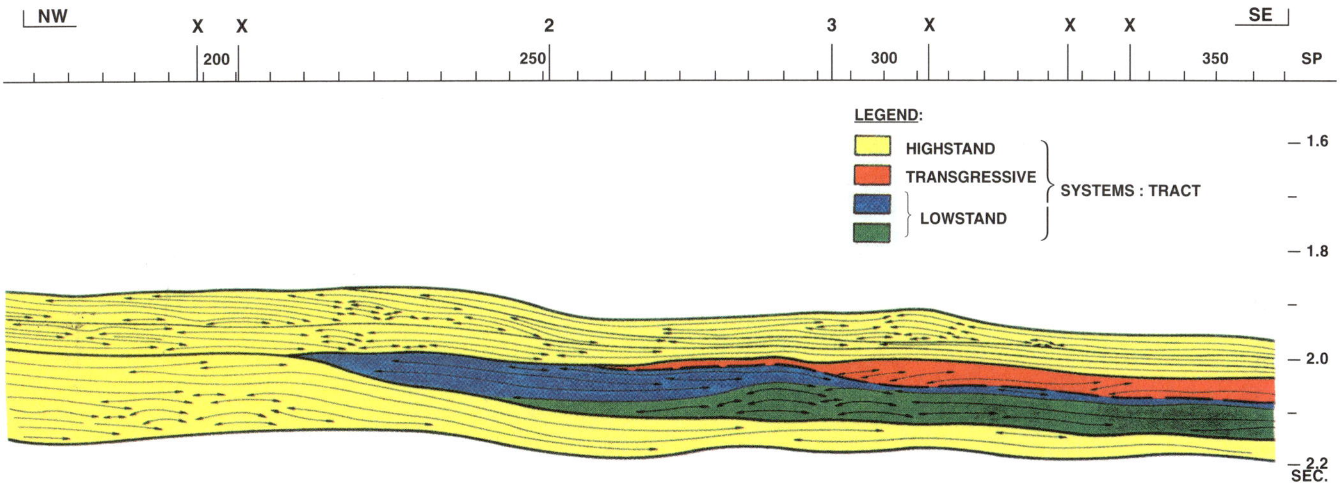

Fig. 9. Interpreted section of line 1 (Figure 5) showing depositional sequences and systems tracts. Five sequences are defined as a succession of highstand and lowstand (transgressive) systems tracts.

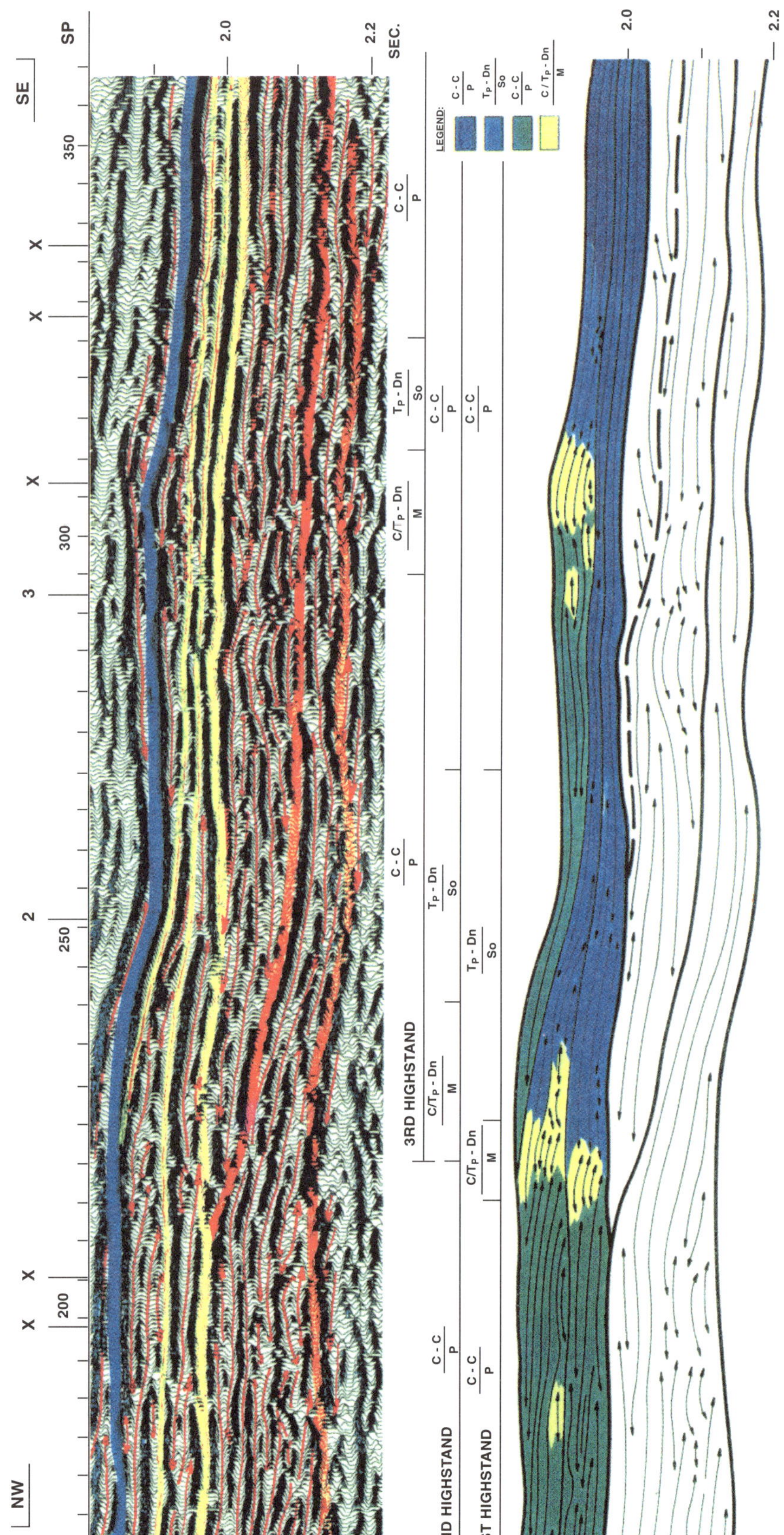

Fig. 10. Interpreted seismic section (top) and schematic drawing (bottom) of line 1 showing the seismic facies analysis. Colors on seismic section are as in Figure 5. The seismic facies are A-B/C coded (following Ramsayer, 1979).

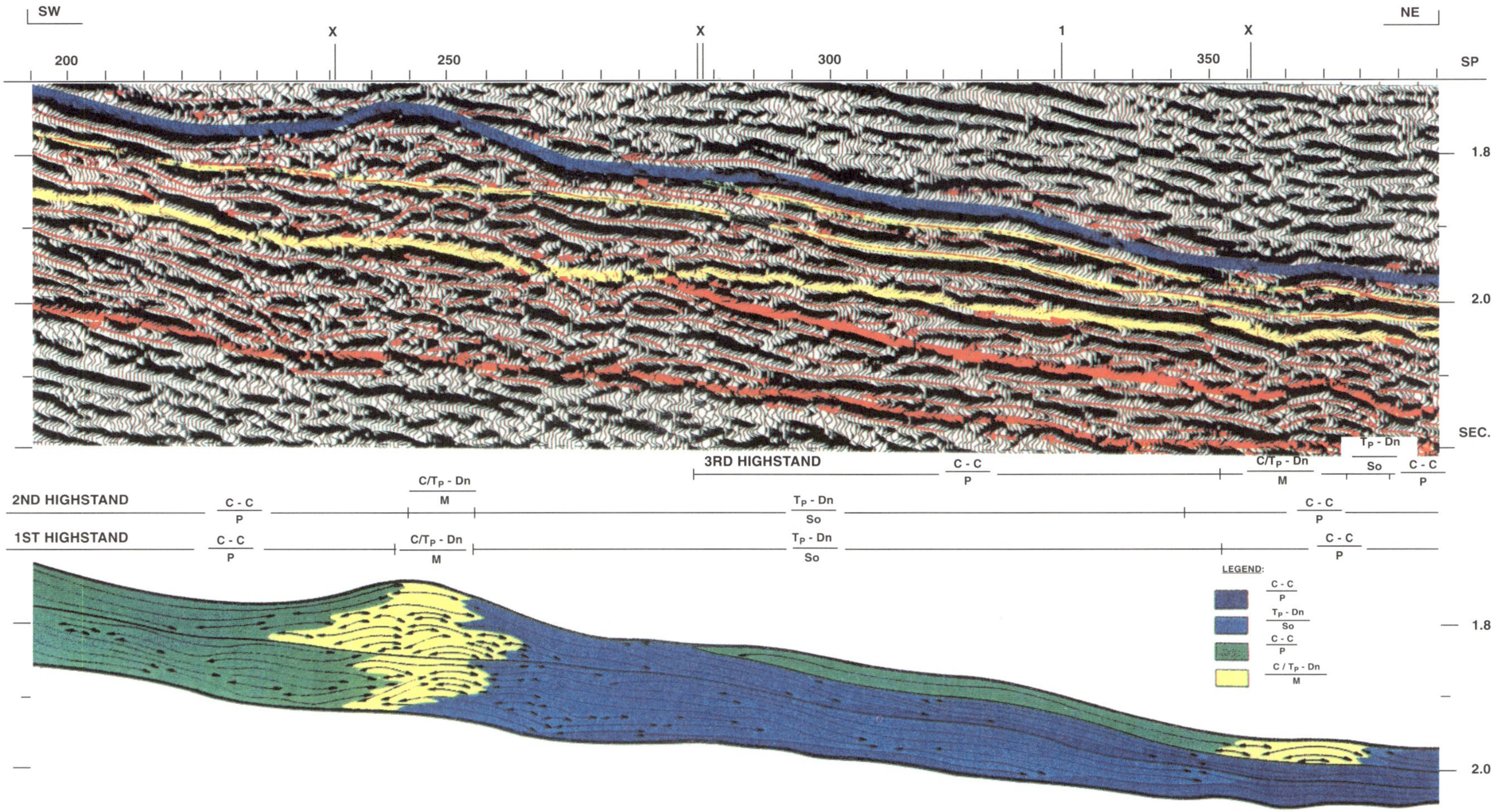

Fig. 11. Interpreted seismic section (top) and schematic drawing (bottom) of line 2 showing the seismic facies analysis. Colors on seismic section are as in Figure 5. The seismic facies are A-B/C coded (following Ramsayer, 1979).

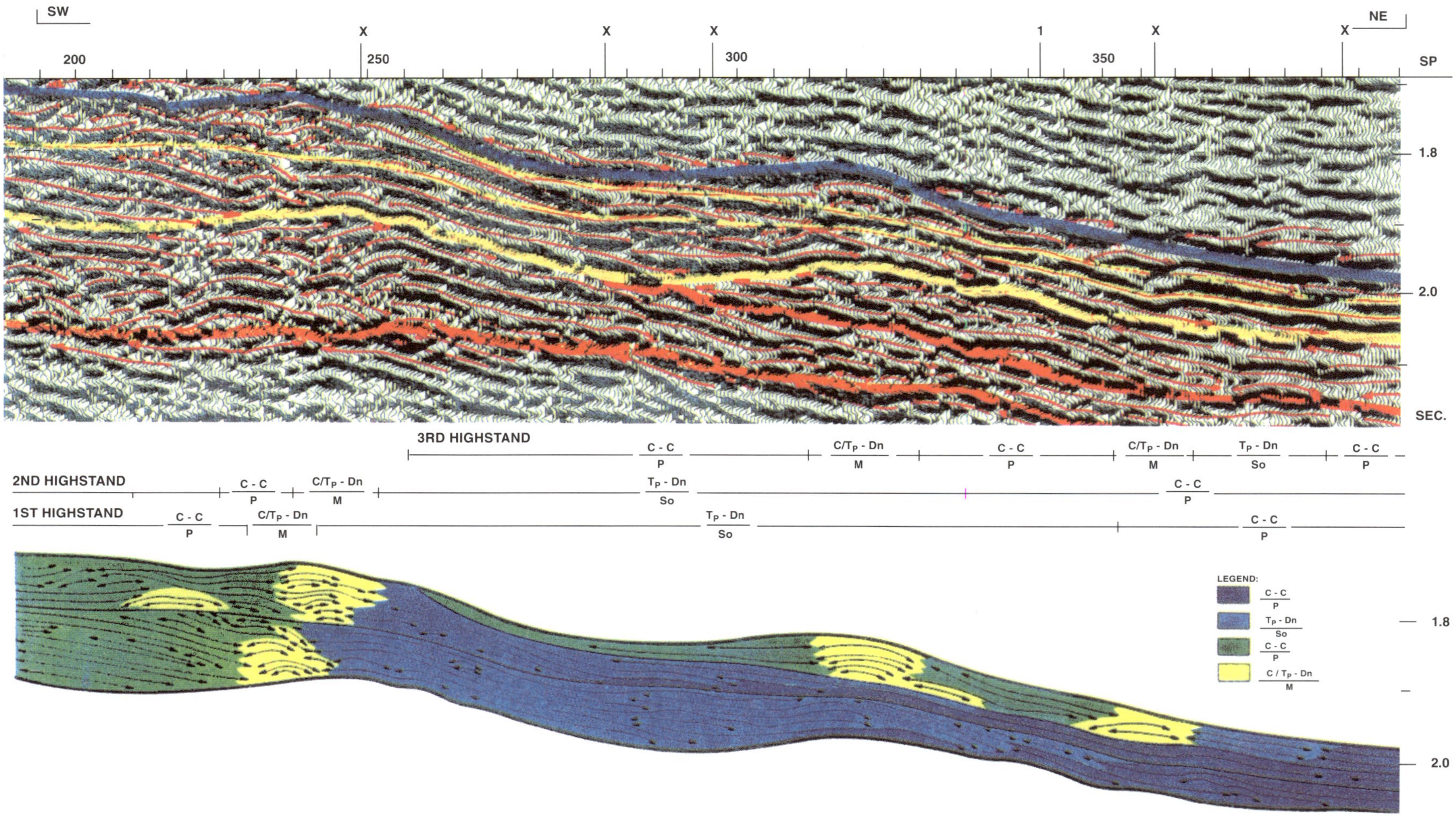

Fig. 12. Interpreted seismic section (top) and schematic drawing (bottom) of line 3 showing the seismic facies analysis. Colors on seismic section are as in Figure 5. The seismic facies are A-B/C coded (following Ramsayer, 1979).

available seismic lines to obtain their lateral distribution. Figures 13, 14, and 15 are maps of the seismic facies at the times of three separate highstands.

As a result of the preceding investigations, a geologic model of a prograding carbonate platform can be derived. During a relative sea level highstand, a carbonate platform developed and prograded basinward. A relative fall of sea level caused the termination of carbonate production on the platform, and sedimentation became restricted to the previous basinal area. Then, the gradual rise of sea level caused onlapping of these lowstand sediments onto the former shelf slope. After strong transgression, a carbonate platform again developed in a similar position as during the previous highstand. Three highstands can be distinguished by internal reflection configurations and terminations, each showing a lateral differentiation into platform-slope-basin. The shelf edge is marked by mounded features interpreted as shelf margin shoals or buildups. The first and second highstands show aggradation (Figures 13 and 14), while the third shows strong basinward progradation (Figure 15) caused by a relative fall of sea level, implying subareal exposure of the shelfal part of the second highstand.

This geologic model allows for derivation of a lithologic model using available data from wells or the literature (Figure 16) (e.g., Wilson, 1975). For the basinal area, a low-energy depositional environment (mudstone, wackestone, and shale) is assumed, with the energy level gradually increasing in the slope area (packstone-grainstone to grainstone-packstone). High-energy deposits would be expected at the shelf edge with its mounds and buildups (grainstone–packstone, grainstone, and boundstone). On the platform, medium- to

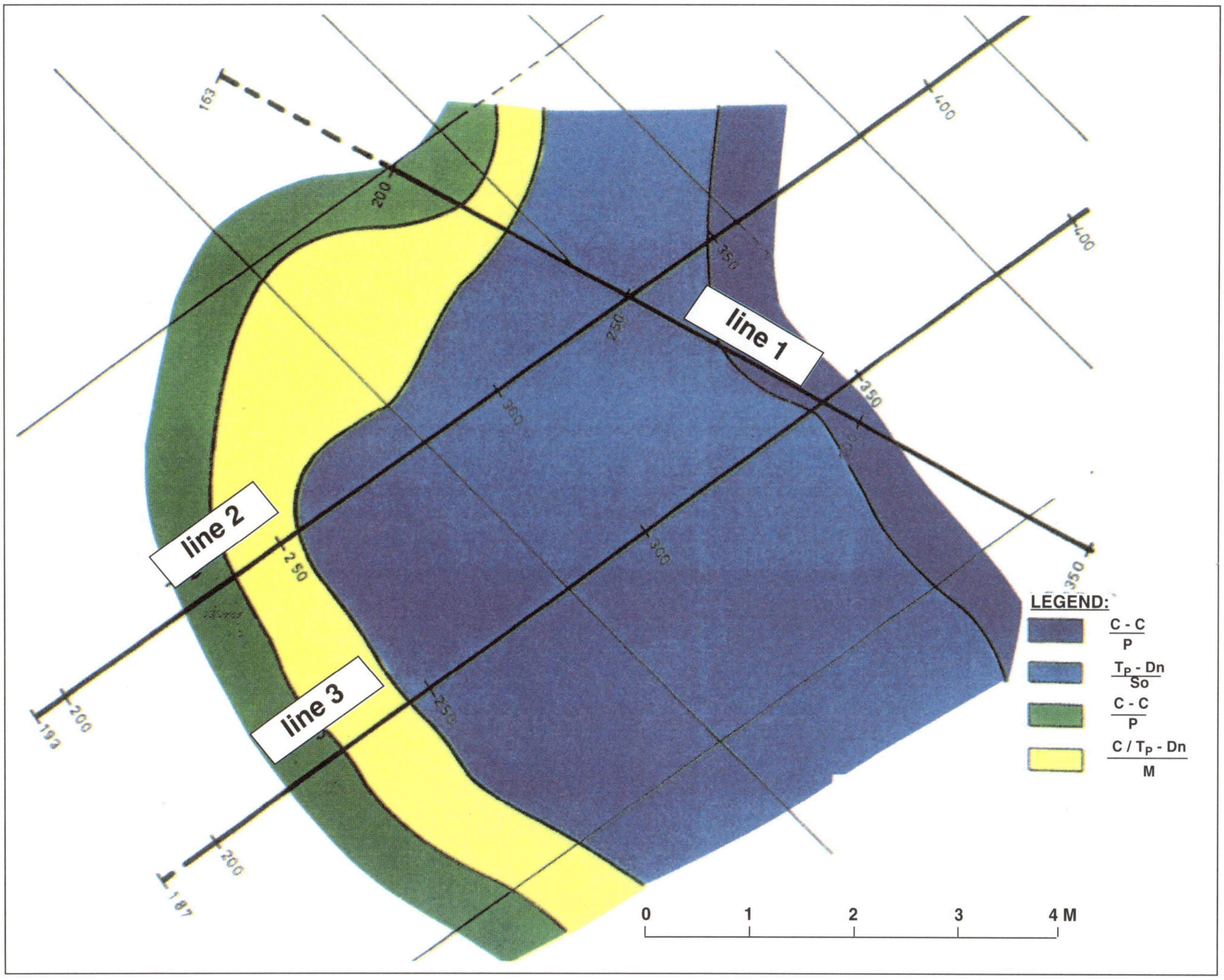

Fig. 13. Map showing seismic facies distribution at the time of the first highstand (based on seismic facies sections in Figures 10, 11, and 12).

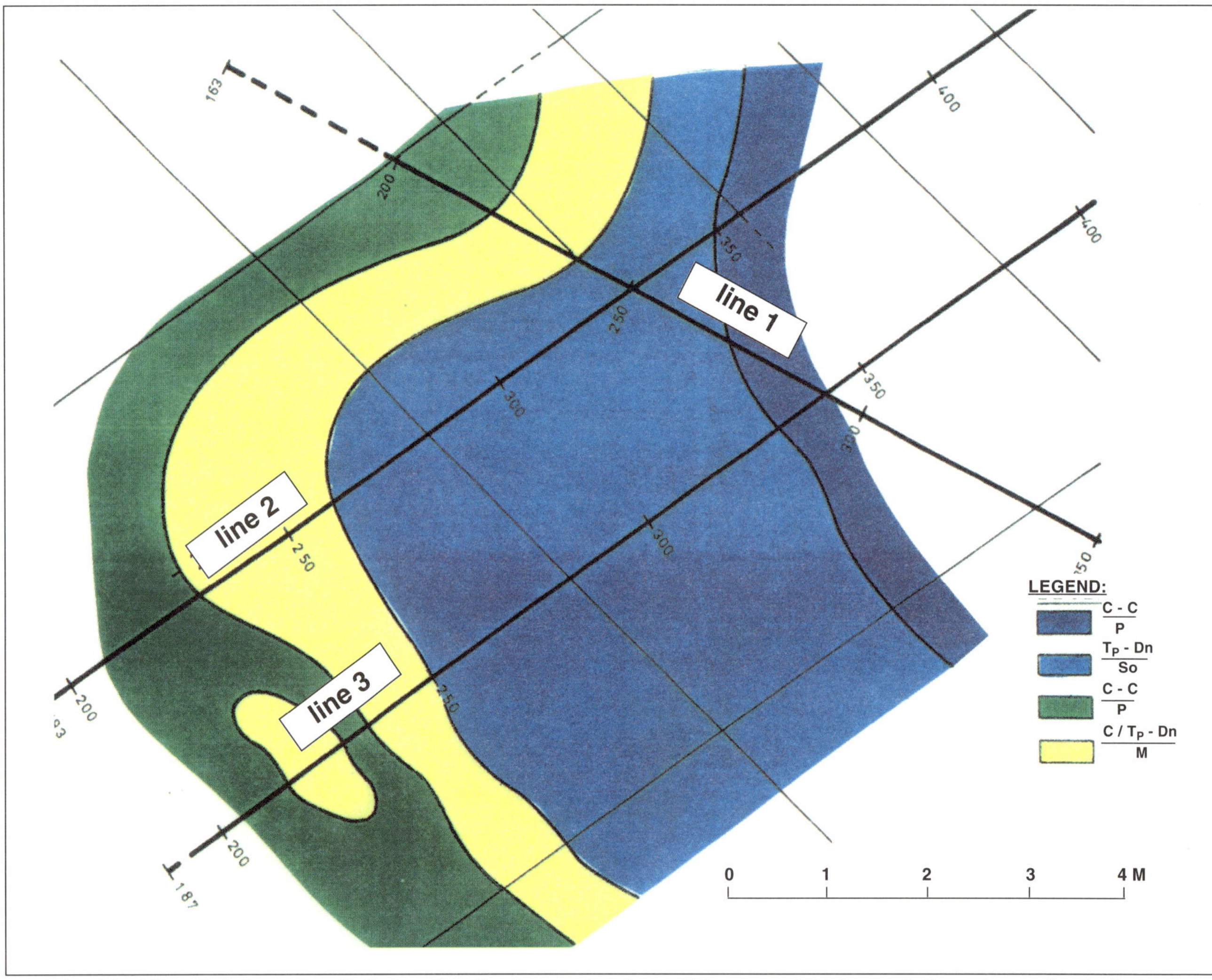

Fig. 14. Map showing seismic facies distribution at the time of the second highstand (based on seismic facies sections in Figures 10, 11, and 12).

low-energy sediments are deposited (packstone–grainstone), except for platform interior shoals, which are medium- to high-energy (grainstone) (Figure 17). For these lithologies, impedance values (density and velocity) are derived from well data by statistical investigations, or if no well data are available, impedance values can be used from comparable areas or from the literature. Impedance values were derived for part of the interpreted section (Figure 18) and were used as a first approach to a lithologic model. This leads to the next important step in the integrated approach—verification.

Verification Loop

The essential step in the verification loop is to forward model the established model in terms of its acoustic and elastic response in either prestack or poststack mode to validate or correct the interpretation. For this particular example, the poststack or exploding reflector option (i.e., modeling of stacked section) was applied, resulting in the model response shown in Figure 19. To achieve a comparison with the real seismic data, *f-k* migration was applied as a final step. Comparison of the synthetic section with the original section showed an overall good match of the configuration of reflections, but there were major discrepancies in amplitude in the areas of assumed high-energy deposits. Therefore, the initially assumed highly porous grainstones were replaced in the first iteration by less porous boundstones (higher velocity and density), and the adjacent packstones–grainstones were replaced by more porous grainstones (lower velocity and density). This results in an overall better match of synthetic and original data (Figure 20). After comparison, further iterative corrections are implemented to achieve the optimum model within the given bandwidth of the seismic data.

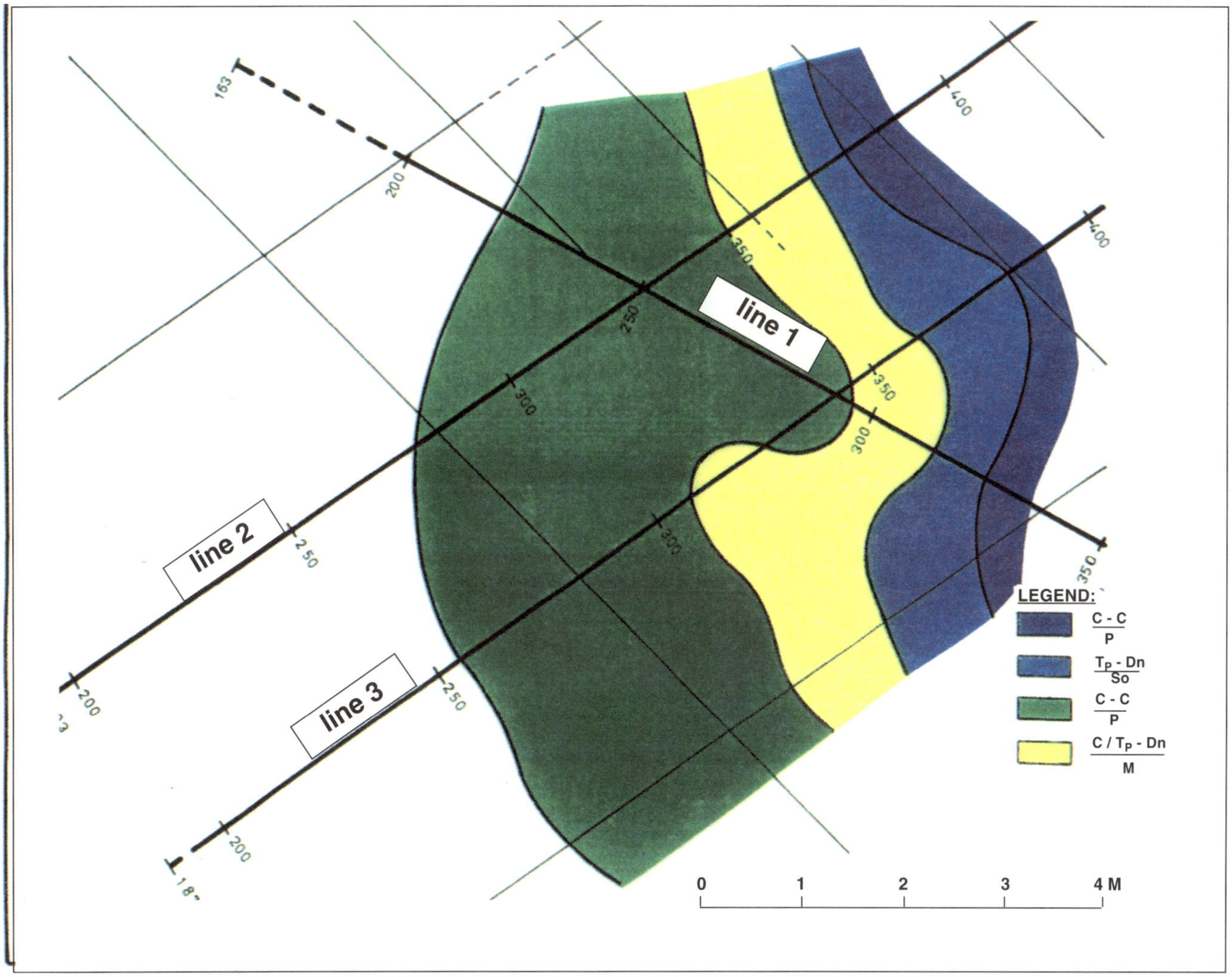

Fig. 15. Map showing seismic facies distribution at the time of the third highstand (based on seismic facies sections shown in Figures 10, 11, and 12).

The stratigraphic interpretation combined with the verification loop results in a new evaluation and play concept for the area (Figure 21). Due to a relative drop in sea level after deposition of the second highstand sequence and subsequent subareal exposure, the shelfal area is interpreted to show higher porosity caused by leaching effects from meteoric waters. Within the third highstand, the high-energy deposits of the shelf margin are not expected to be the optimum reservoir. Better porosities are related to adjacent grainstone deposits on the shelf. Thus, an updip stratigraphic trap can be delineated within the pinchout of shelfal deposits against the slope sediments of the second highstand (permeability barrier).

The underlying velocity model can be checked prior to its use in forward modeling by the "velocity checking by wavefronts" method (Marschall, 1991), which uses prestack traveltimes from constant offset sections to verify and correct the assumed velocity model by depth migration of these traveltimes. At least two offsets with corresponding traveltimes per common midpoint are needed. For the final risking and ranking, the geophysical risk factors are based on the results of the verification loop and the velocity checking.

Additional Steps

When geochemical logging is applied, the resulting logs can provide accurate compositional and petrophysical interpretations. This is useful because a strong correlation exists in some complex carbonate formations between mineralogy and permeability (Herron et al., 1992). This must be evaluated in the light of dolomitization (dolomitic replacement), which occurs volume for volume rather than molecule for molecule. This volume-for-volume replacement results in an overall volume re-

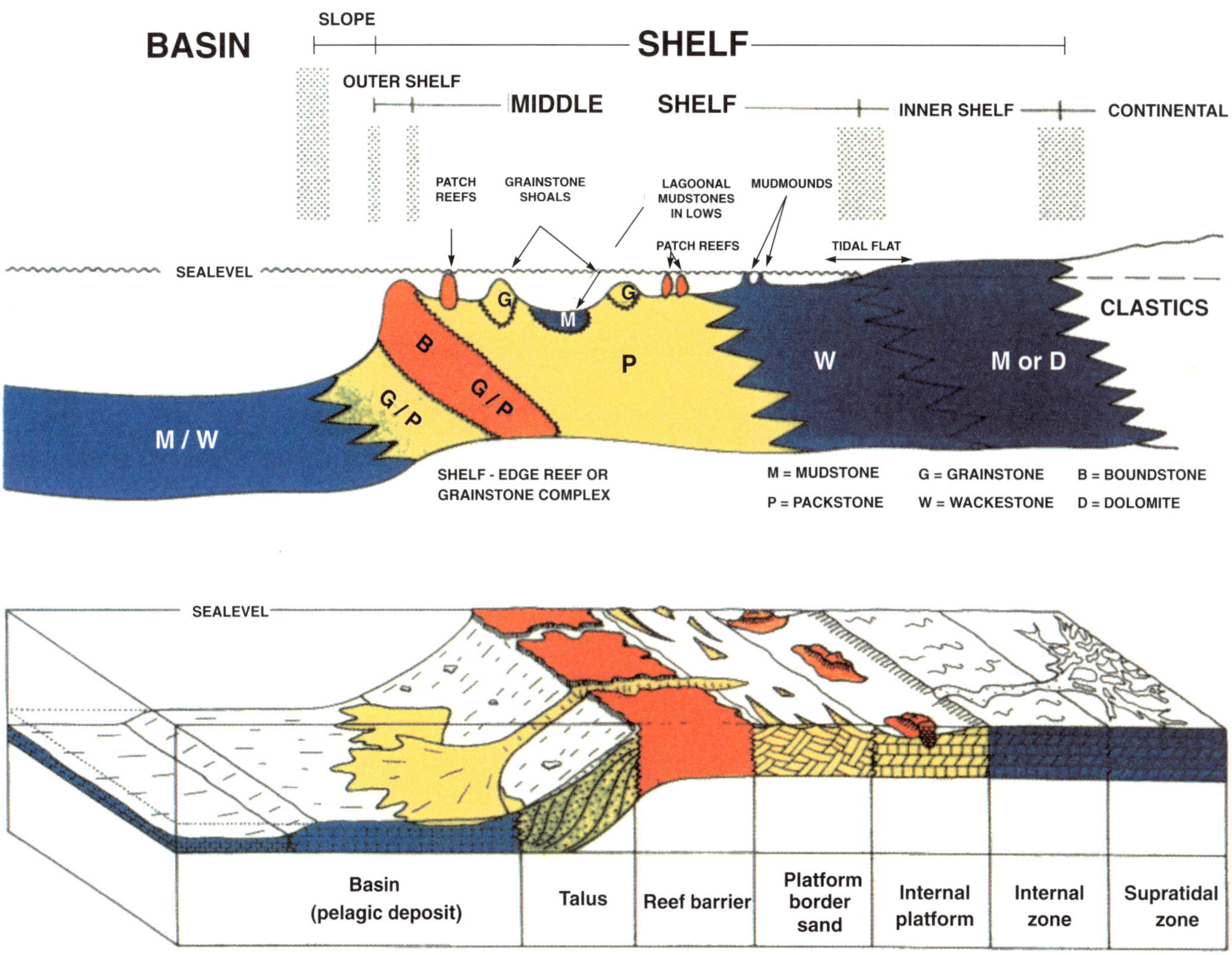

Fig. 16. Schematic sketch of carbonate depositional environments (after Wilson, 1975, 1983).

duction at a ratio of 100:88, with a consequent increase in porosity (Pettijohn, 1975).

Furthermore, compaction also has a considerable effect on rock velocities. Pettijohn (1975) states that carbonate sands might be expected to show little or no compaction effects, whereas carbonate mud might be compactible. In our model, the oversteepened dip of the reef-flanking beds indicated a greater compaction of these beds compared to the reef core. For decompaction, we use Athy's (1930) relationship, after making adjustments for carbonates:

$$P = P_{\mathrm{o}}^{-cz} \tag{1}$$

where P is the porosity, P_{o} is the average porosity of the surface layer, z is the depth of burial, and c is a constant (km^{-1}). We recommend using the value for P_{o} given by Schmoker and Halley (1982) of $P_{\mathrm{o}} = 0.42$ rather than the value published by Sclater and Christie (1980) of $P_{\mathrm{o}} = 0.70$. With $P_{\mathrm{o}} = 0.42$, the constant c can be derived from a given well as

$$c = -\{\ln[P(z_{\mathrm{w}})/P_{\mathrm{o}}]\}/z_w \tag{2}$$

where $P(z_{\mathrm{w}})$ is the porosity at depth z_{w} (given) (Marschall, 1992). This adjusted equation can then be applied in a straightforward manner.

Conclusions

The objective of this paper was to provide a comprehensive and rational approach to the process of integrated interpretation. The following steps should taken to generate a well-defined result:

- Optimum processing of the given seismic data, including inversion

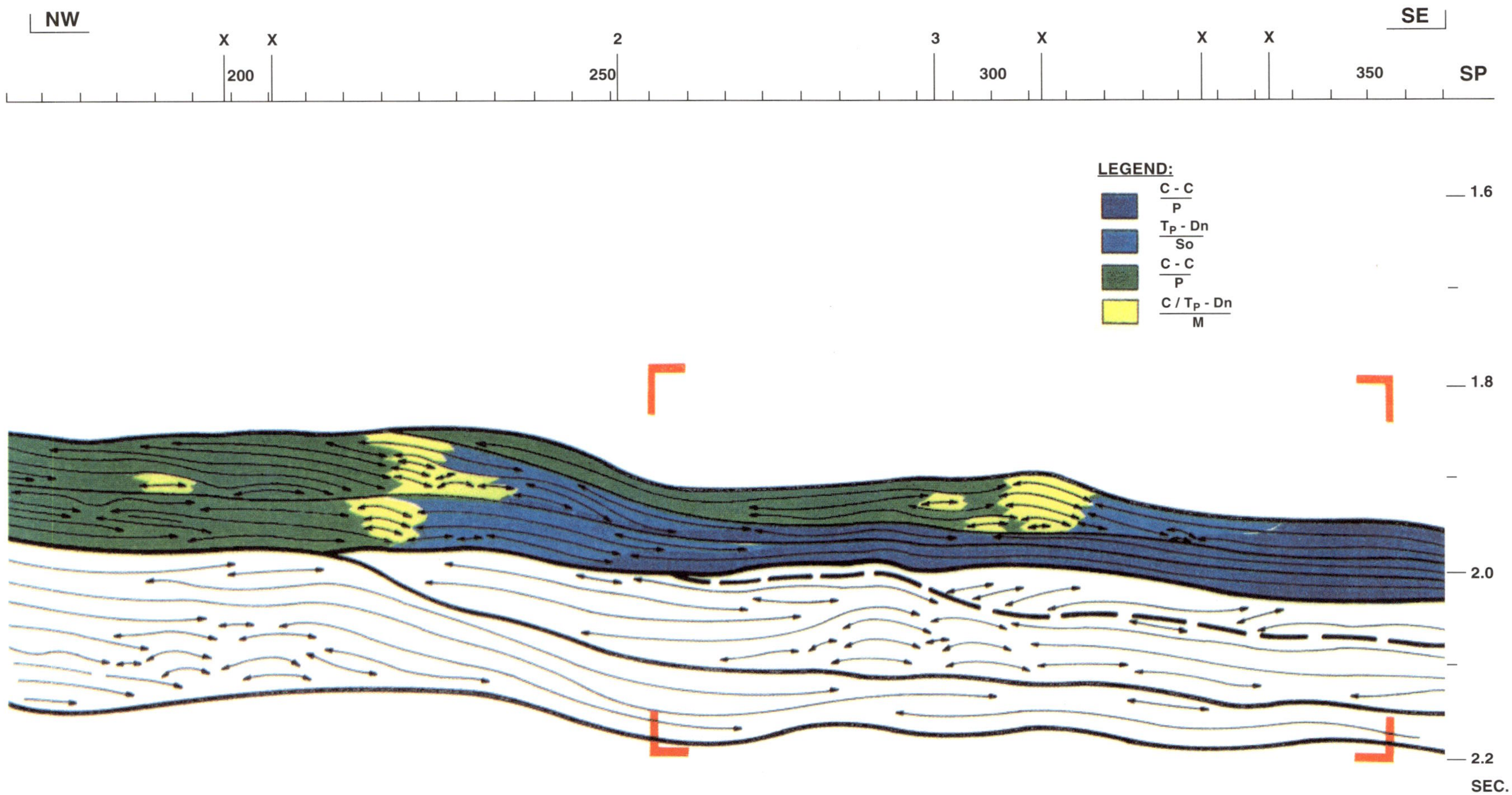

Fig. 17. Interpreted seismic section of line 1 (as in Figure 10) showing interpretation of depositional environments. Red brackets show portion of section enlarged in Figure 18.

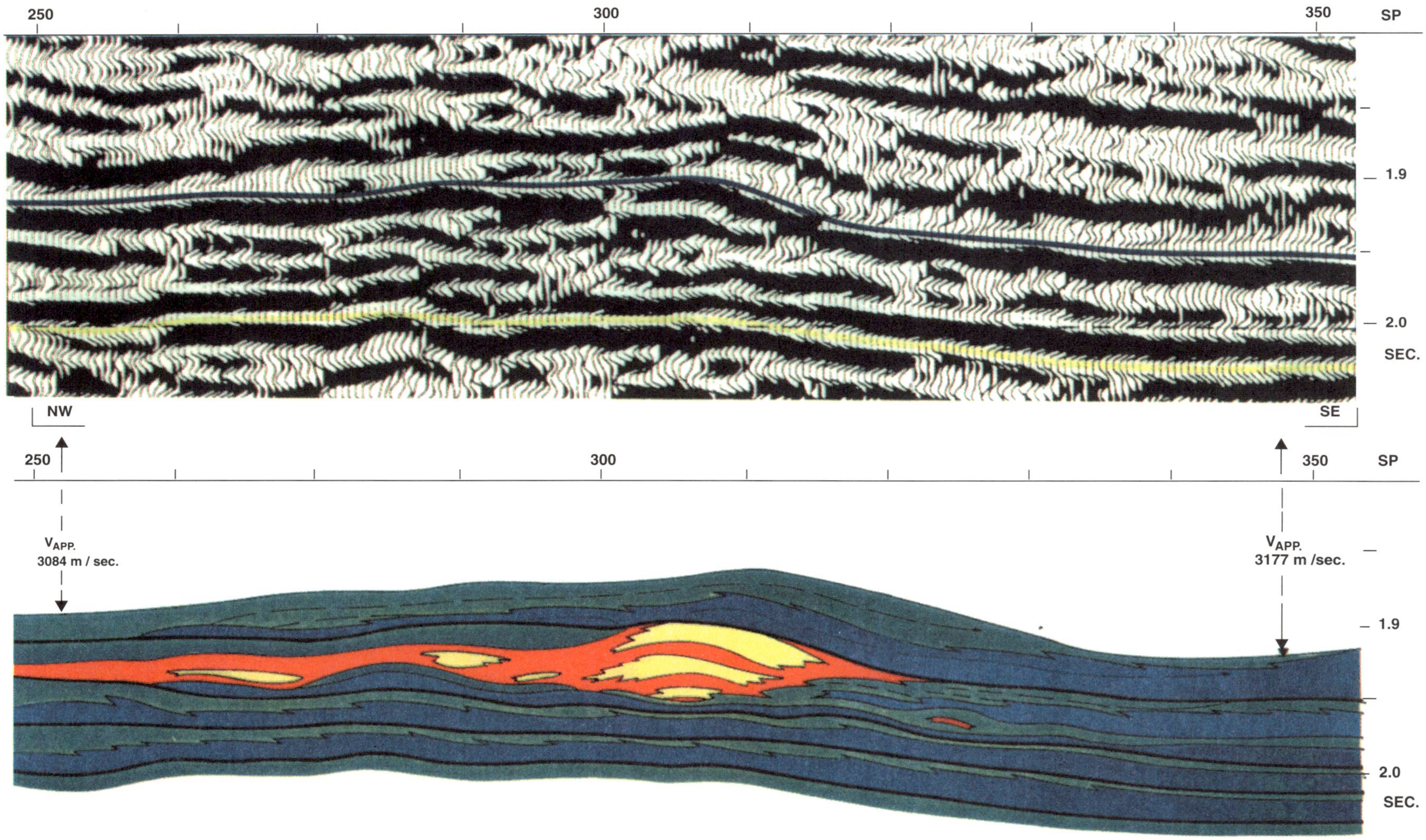

Legend	Impedance [105 g/cm2 s]	Velocity [m/s]	Density [g/cm3]
Biocl. Grainstone, Packstone	9.67	4159	2.325
Packstone, Grainstone	9.16	3940	2.325
Lime-Mudstone, Biocl. Packstone, Wackestone	13.07	5042	2.59
Shale	8.83	3560	2.48

Fig. 18. Interpreted seismic section (top) and schematic drawing (bottom) of a portion of line 1 showing the lithologic model.

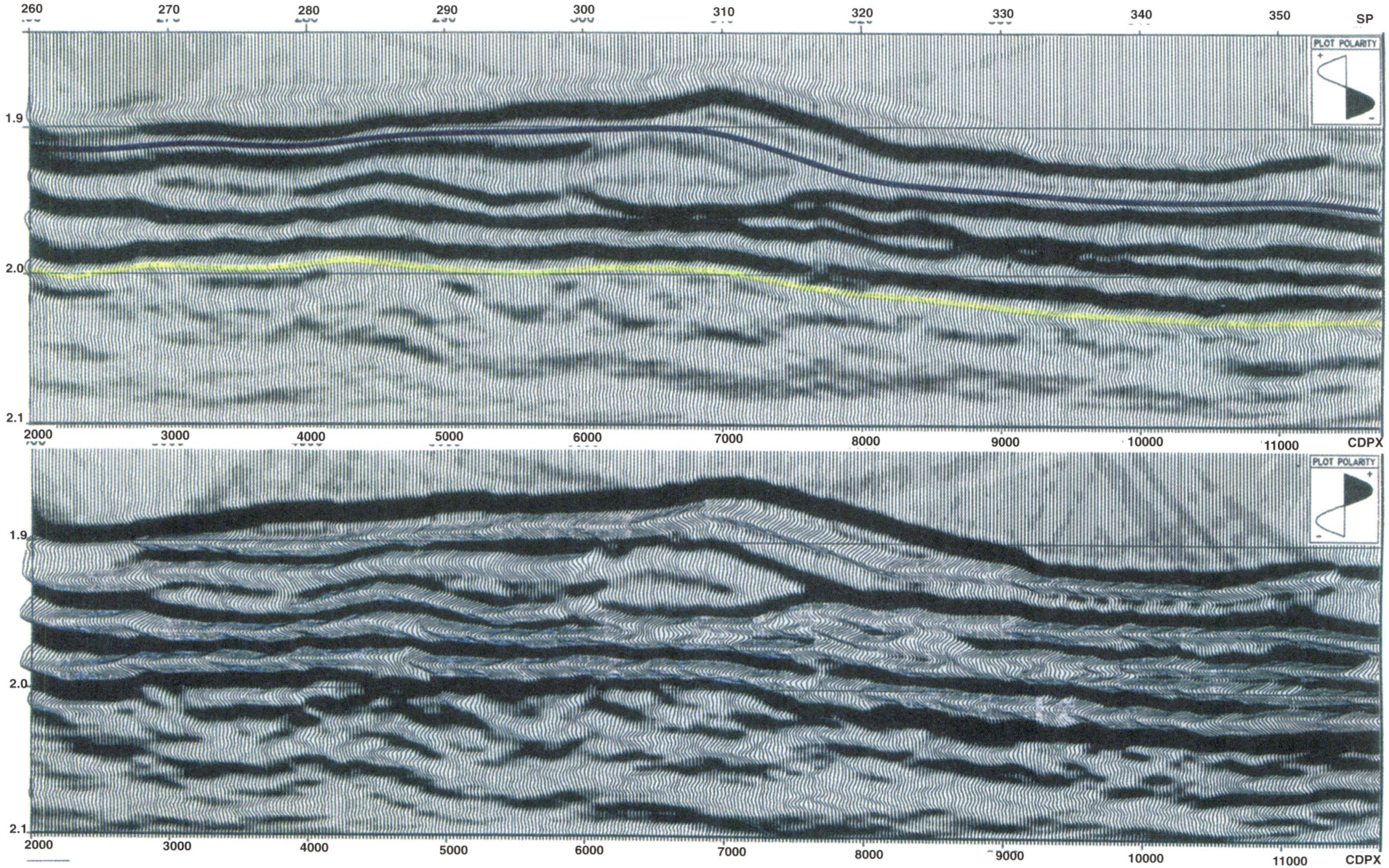

Legend	Impedance [105 g/cm2 s]	Velocity [m/s]	Density [g/cm3]
Biocl. Grainstone, Packstone	9.67	4159	2.325
Packstone, Grainstone	9.16	3940	2.325
Lime-Mudstone, Biocl. Packstone, Wackestone	13.07	5042	2.59
Shale	8.83	3560	2.48

Fig. 19. Synthetic seismic sections showing the first approach to the lithologic model for line 1. Reverse (top) and normal (bottom) polarity according to SEG norm.

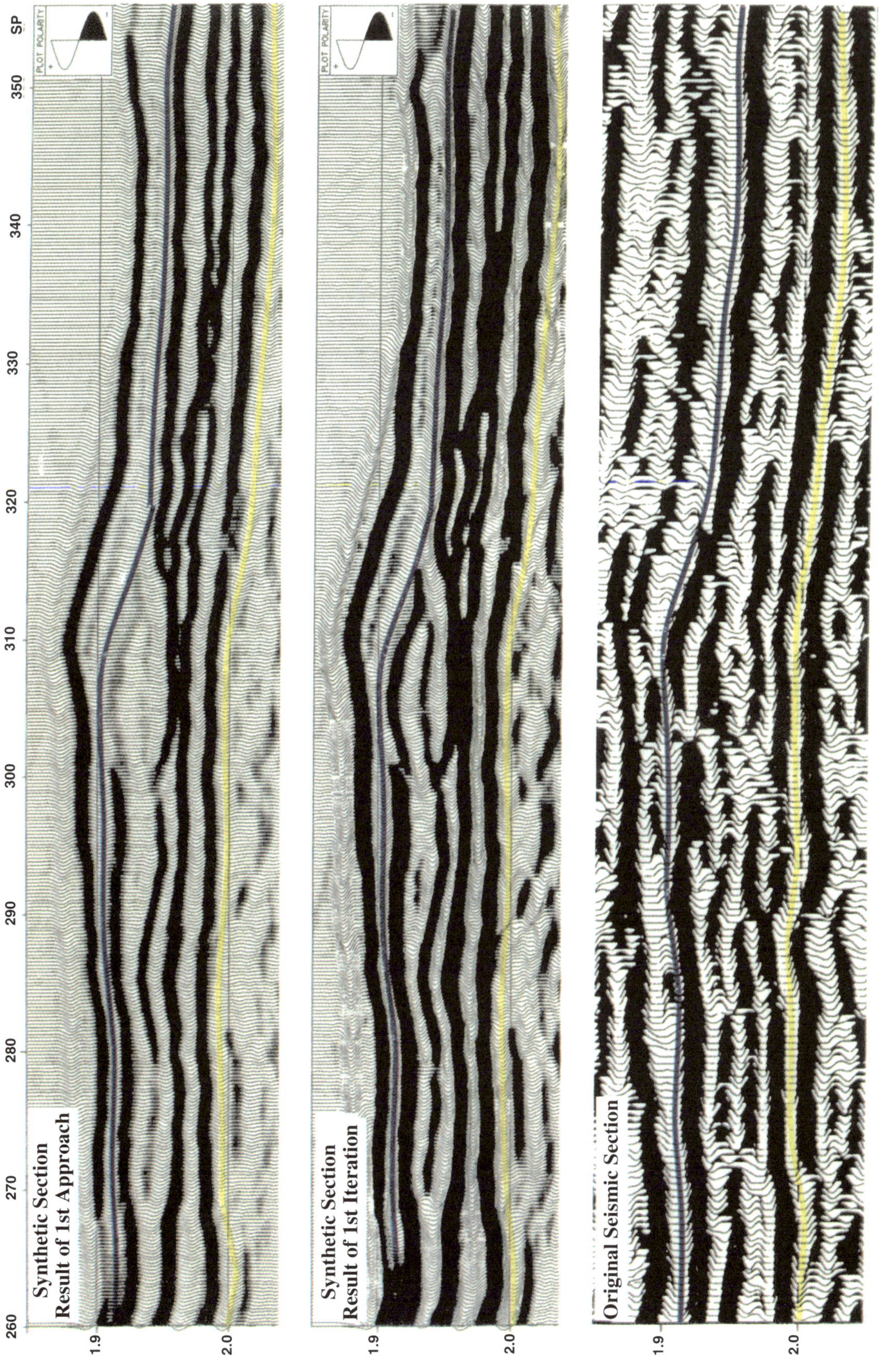

Fig. 20. Comparison of the synthetic seismic sections (first approach and first iteration) with the original seismic section for line 1.

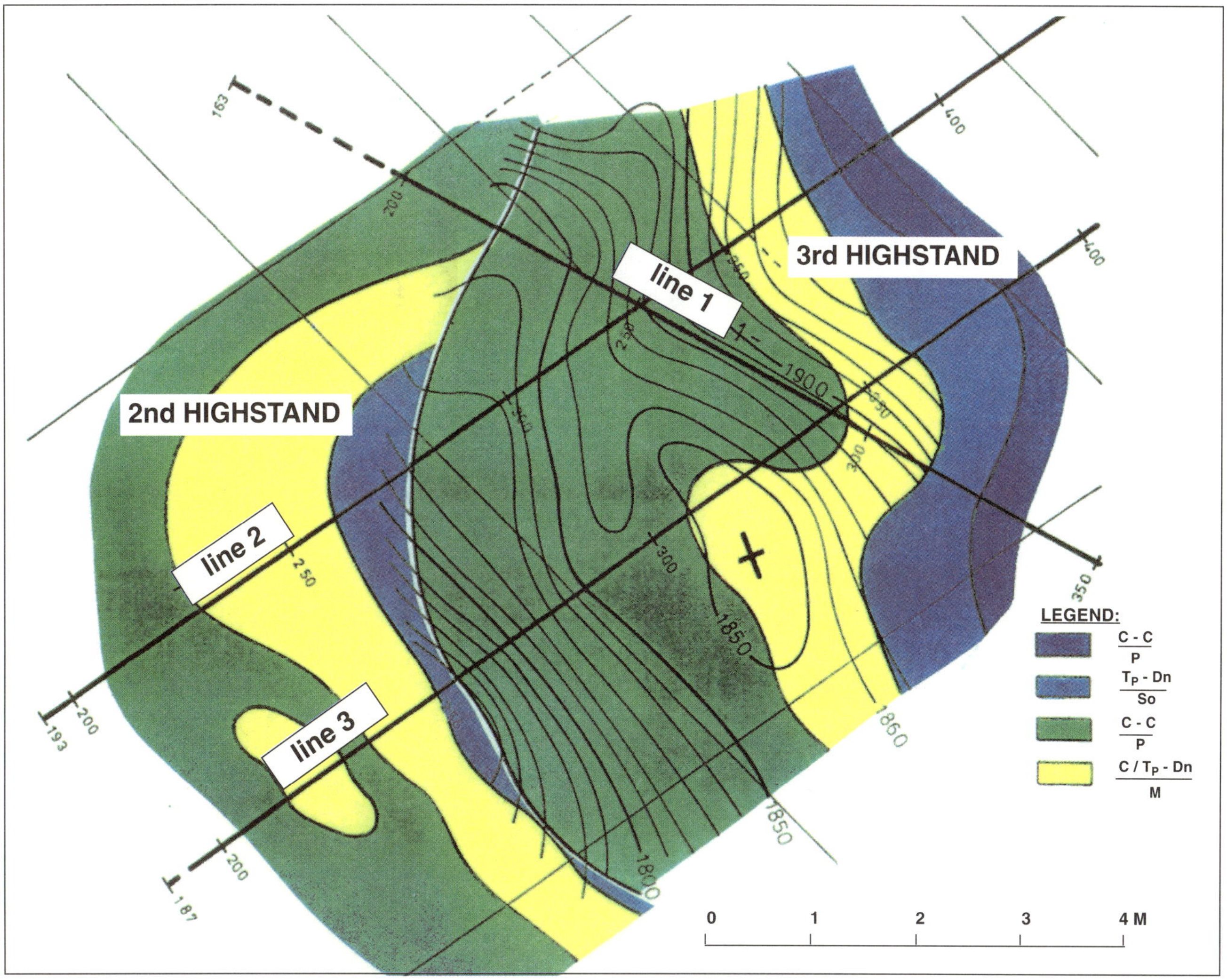

Fig. 21. Seismic facies map showing new play concept. The possible stratigraphic updip closure shown is based on the results of forward modeling.

- Structural interpretation, followed by stratigraphic interpretation.
- Verification of the resulting lithologic model by forward modeling
- Verification of the resulting velocity model by velocity checking by wavefronts
- Appropriate decompaction
- Definition of geophysical risk factors that properly account for the results of verification

References

Athy, L. F., 1930, Density, porosity, and compaction of sedimentary rocks: AAPG Bull., **14**, 25-35.

Herron, S. L., Petricola, M. J. C., and Dove, R. E., 1992, Geochemical logging of a Middle East carbonate reservoir: J. Petr. Tech., **44**, 1176-1183.

Marschall, R.,1991, Macro model optimization and depth migration: Proc. 2nd Intl. Congr. Braz. Geoph. Soc., Paper WOR **1.6**.

Marschall, R., 1992, Decompaction of carbonates in UAE: Unpublished report, Geco-Prakla (ECM).

Mayouf, A., Fischer, K., Holling, M., Marschall, R., 1993, Ein integriertes Explorationskonzept fr Karbonat-abfolgen: Mintrop-Seminar, Arnsberg, **13**.

Mitchum, R. M., Vail, P. R., and Thompson, S., 1977, Seismic stratigraphy and global changes of sea level, part 2, the depositional sequence as a basic unit for stratigraphic analysis, *in* Payton, C. E., Ed., Seismic stratigraphy—applications to hydrocarbon exploration: Am. Assn. Petr. Geol. Mem. **26**, 53-62.

Pettijohn, E. J., 1975, Sedimentary rocks: Harper & Row.

Ramsayer, G. R., 1979, Seismic stratigraphy, a fundamental exploration tool: Offshore Tech. Conf. Proc., **3**, 1859-1867.

Schmoker, J. W., and Halley, R. B., 1982, Carbonate porosity versus depth: a predictable relation for south Florida: AAPG Bull., **66**, 2561-2570.

Sclater, J. G., and Christie, P. A. F., 1980, Continental stretching, an explanation of the post-mid-Cretaceous subsidence of the Central North Sea Basin: J. Geoph. Res., **85**, 3711-3739.

Vail, P., 1987, Seismic stratigraphy interpretation procedure, *in* Bally, A. W., Ed., Atlas of seismic stratigraphy, vol. 1: Am. Assn. Petr. Geol. Studies in Geol. **27**, 1-10.

Wilson, J., 1975, Carbonate facies in geologic history: Springer-Verlag.

Chapter 10

Seismic Data Processing Flow in Areas of Shallow Carbonates

M. Turhan Taner
Seismic Research Corporation
Houston, Texas. U.S.A.

Abstract

Developing a specific data processing flow in areas of shallow carbonates, as in other seismically difficult areas, requires a clear understanding of the problems associated with such environments. Once the initial problems are defined, a specific processing procedure is designed to achieve the objectives of seismic data processing. This design properly starts from the end, at the final required results, and proceeds backward to identify the processing tools needed to achieve these results. A number of severe data problems are associated with the shallow carbonate environment, the most obvious one being the strong multiples generated by the surface and carbonate interfaces. Water-trapped multiples propagate with water velocity, hence they are more easily discernible (due to their NMO differences) from the deeper, higher velocity primary reflections. The surface-connected and interbed multiples generated by carbonate interfaces, however, propagate with higher velocities, hence they are not as easily identified as the low-velocity multiples. As a result, a variety of multiple suppressing methods must be included in the processing stream.

Carbonate layers also cause the generation of horizontally traveling waves, such as the refracted events on marine data and large-amplitude ground roll on land data. These must be suppressed before proceeding with processes such as deconvolutions and wavelet processing. On land data, carbonates are normally covered by weathered low-velocity layers, and in some areas, carbonate layers are cut by old riverbeds. These conditions create short-wavelength disturbances in traveltimes and thus adversely affect the imaging process. The processing flow must therefore include computation and removal of these time disturbances as well. Large-velocity differences between carbonates and clastic rocks cause most of the pressure wave energy to be reflected, with little transmitted to the layers below. This results in low signal-to-noise ratios in the primary reflections below carbonate layers. The velocity difference also causes the transversely isotropic layer velocity anisotropy, which in turn affects the apparent NMO velocities of longer offset data.

Introduction

Designing a seismic data processing sequence depends on the final objectives, on the problems that may be encountered, and on the experience and ability to recognize such problems and to select appropriate corrective processes. After several decades of digital seismic processing, many types of problems, signal-to-noise (S/N) ratios, geologic settings, and adverse conditions have been encountered. This paper presents a processing sequence that can be used in areas of shallow carbonates. I present a logical explanation of each step based on the identified problem and its possible solution available at this time. With experience and new developments, a completely different sequence may be developed in the future. If this happens, it is a good sign that this profession has progressed. The procedure presented here can be used in other recording environments and with both 2-D and 3-D acquisition geometries.

Problems Associated with Carbonate Layers

I first discuss some of the problems with carbonate strata, or with any high-velocity environment embedded in low-velocity layers. Later, I cover unsolved problems, some possible solutions, and related processing sequences.

One of the main characteristics of carbonate layers is that they have much higher interval velocities than their neighboring clastic layers. This creates a multitude of problems. First, if the carbonate layer is located at or near the surface on land or close to the seafloor in marine settings, only a small portion of the pressure wave front penetrates the carbonate layer. Most of the source wavefront is reflected or forced to propagate in a horizontal direction. The portion that penetrates the carbonate layer is severely refracted and spread over a large area by the time it reaches the base of the carbonate (Figure 1). The reflected and transmitted pressure wave energy is relatively low. This produces poor S/N for base carbonate and all lower reflectors.

A second problem results from both carbonate base and top interfaces having high acoustic impedance contrasts. This causes both interfaces to be strong contributors to the multiple-generating process. In the marine environment, they generate strong surface and interbed multiples (Figure 2).

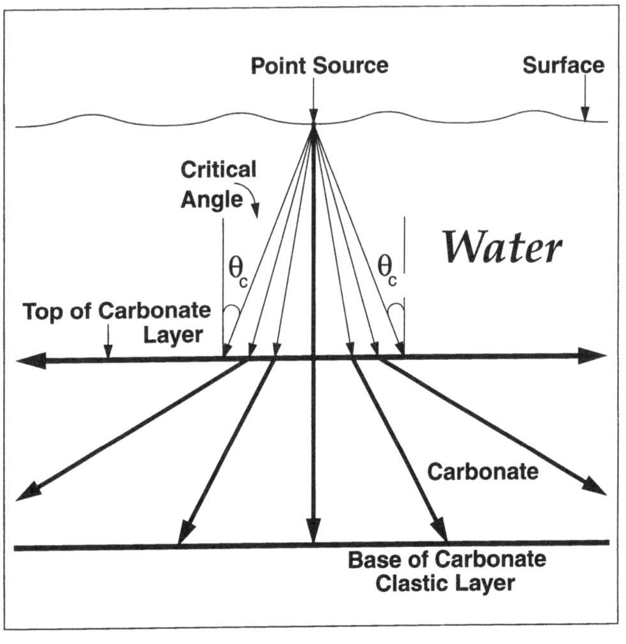

Fig. 1. Propagation angle relationship at the water–carbonate interface.

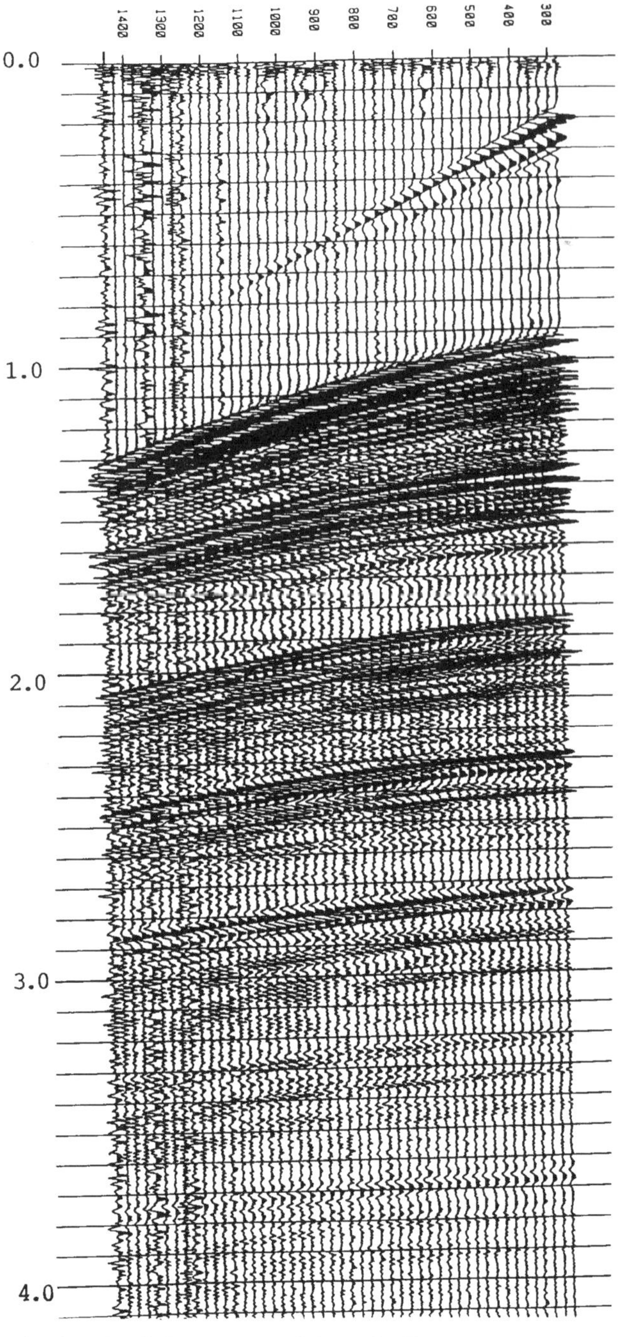

Fig. 2. Multiples generated by a shallow carbonate layer at the bottom of a deep-marine seismic record.

A third problem occurs when a carbonate layer is close to the surface on land. The low-velocity surface layer generates strong horizontal ground roll that masks low-amplitude reflected pressure wave events. Figure 3 shows a typical land seismic record with a severe ground roll problem. In marine records, multiply-refracted events occur instead of ground roll (Figure 4).

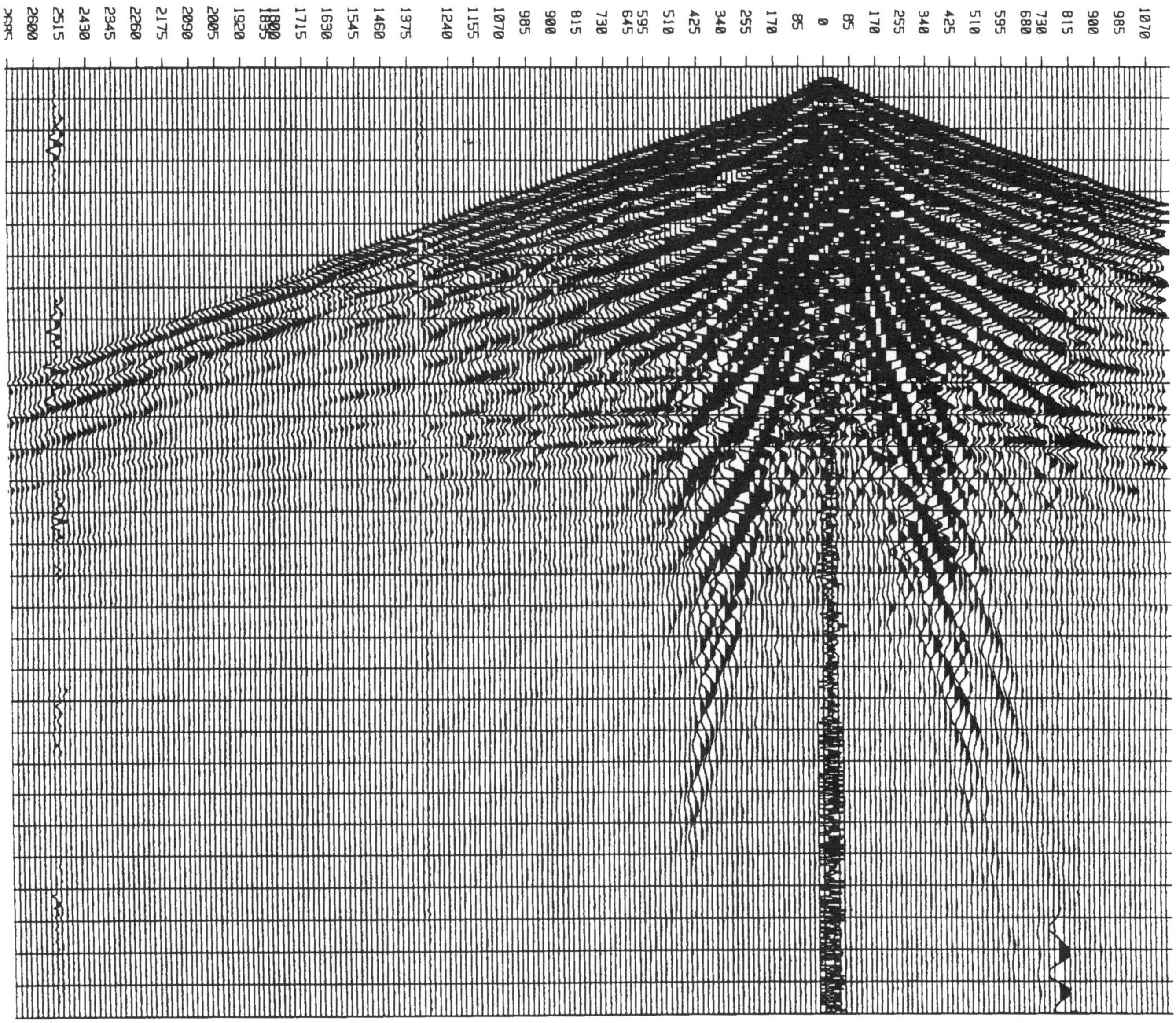

Fig. 3. A land seismic record showing ground roll contamination.

These refracted waves have a multiple structure similar to the reflected waves since the refracted wavefront becomes a plane wavefront after reflection from the surface. Figure 5 depicts a characteristic *P*-wave reflection and transmission response. A carbonate interface reflects almost 100% of the energy at the critical angle. This portion of the reflected wavefront travels within the water layer with minimum loss. The travel path of the critical reflection can be seen along a radial line, shown schematically in Figure 6. Figure 7 shows an example of trapped critical angle reflection. Since postcritical reflection coefficients become complex, the reflected waves exhibit phase distortion with respect to the source wavelet.

A fourth problem with carbonate layers is that high-velocity contrast creates problems with the NMO application and seismic image generation. The velocity contrast creates high-velocity heterogeneity. It has been recognized that transversely isotropic (TI), horizontally layered media cause velocity anisotropy (see velocity anisotropy bibliography). To demonstrate this phenomenon, we can look at a two-layered model (Figure 8a). The velocity computed for the vertical path on this model is simply the average velocity, $\bar{V}$, which is equal to the total travel distance, *d*, divided by total one-way traveltime, *t*:

$$\bar{V} = \frac{(d_1 + d_2)}{(t_1 + t_2)}. \quad (1)$$

Fig. 4. A marine seismic record showing multiply reflected refractions (linear events) and repetition of high-amplitude critical angle reflections on the radial direction.

This average velocity, $\bar{V}$, is shown in Figure 8b for the zero incidence angle. If we increase the incidence angle to an arbitrary value, θ_i, less than the critical angle, θ_c, the average velocity on the new path is given by

$$\bar{V}(\theta) = \frac{(v_1 \cdot t_1(\theta)) + v_2 \cdot t_2(\theta)}{(t_1(\theta)) + t_2(\theta)}. \qquad (2)$$

Since the ray bends according to Snell's law (consistent with Fermat's principle) such that the observable seismic energy propagates along minimum travel paths, the ray has a longer path in the high-velocity layer and a

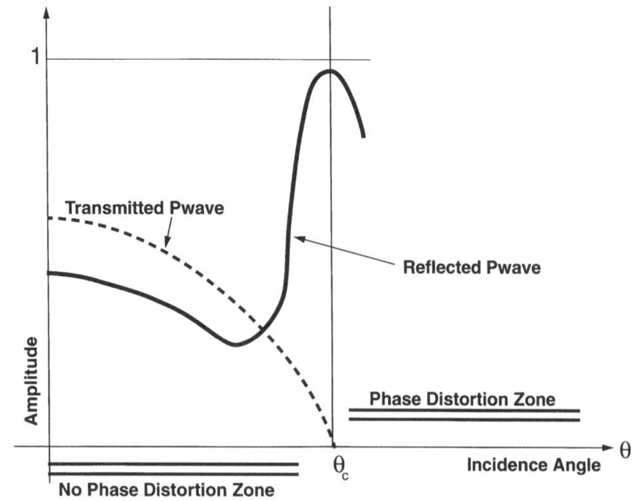

Fig. 5. Transmission and reflection response of a high-contrast interface.

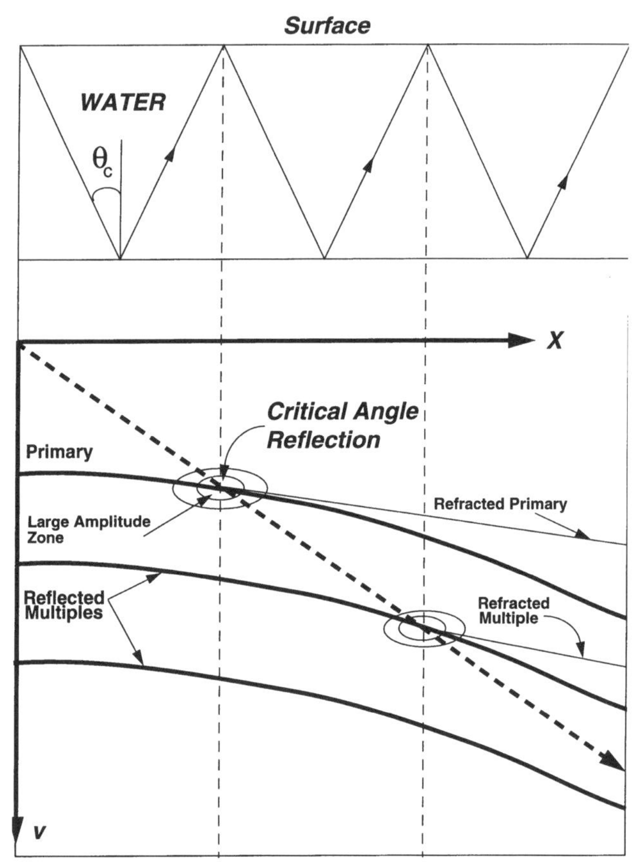

Fig. 6. Geometry of multiple and critical angle reflected event arrivals.

shorter path in the low-velocity layer. Equations (1) and (2) indicate that, as the initiation angle increases, the corresponding average travel velocity increases. It approaches the higher velocity, v_2, asymptotically. This simple example shows that the propagation velocity in-

Fig. 7. Common-source record with trapped critical angle reflections and multiply-reflected refractions.

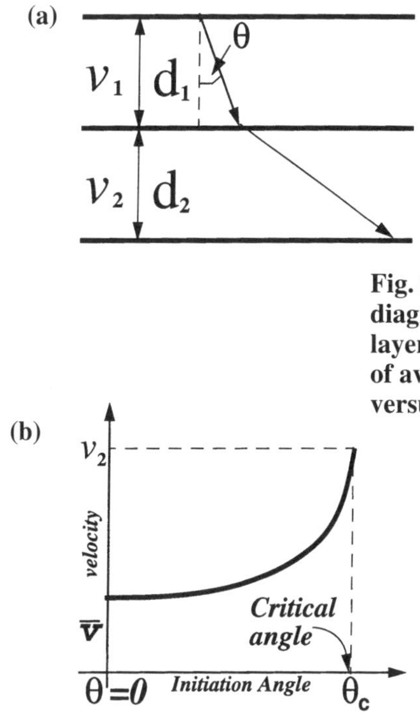

Fig. 8. (a) Schematic diagram of the two-layer model. (b) Graph of average velocity versus initiation angle.

creases with initiation angle and, consequently, with increasing offset. This is most visible on shallow reflectors because they record a wider range of propagation angles. This effect gradually diminishes on deeper events. As seen from equation (2), the greater the velocity difference between v_1 and v_2, the larger the velocity variation between the near and far offsets.

Let's look at the velocities in a different way. Consider a stack of layers as shown in Figure 9, where *v(n), d(n),* and $\Delta t(n)$ represent the interval velocity, layer thickness, and layer transit time of the *n*th layer. If we accumulate all transit times for layers with the same interval velocity and display them, we get the velocity density plot shown in Figure 10. Any amplitude on the graph represents the accumulated traveltime for all layers with the same velocity. By definition, the average velocity is given as

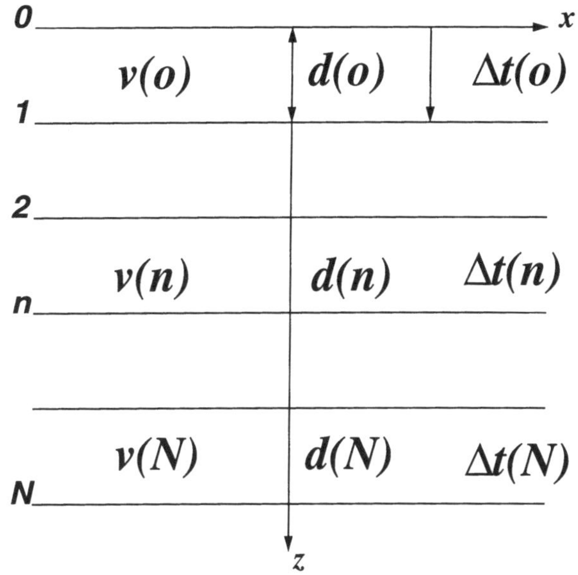

Fig. 9. Schematic representation of the multilayer model with horizontal interfaces.

$$\bar{V} = \frac{\sum d(n)}{\sum \Delta t(n)},$$

or

$$\bar{V} = \frac{\sum v(n)\,\Delta t(n)}{\sum \Delta t(n)}. \tag{3}$$

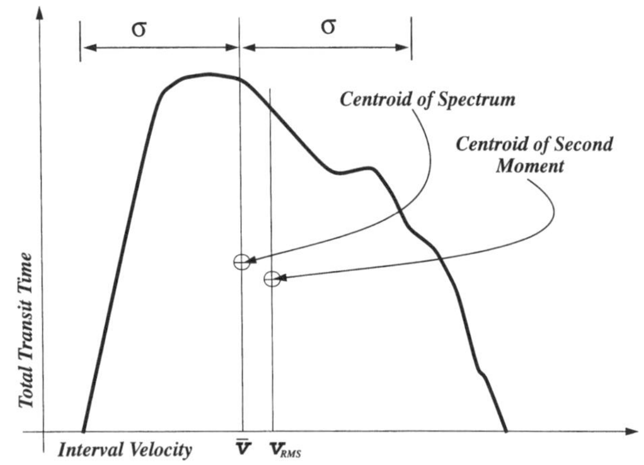

Fig. 10. Total transit time density spectrum with respect to interval velocity.

From mechanics, we know that average velocity corresponds to the center of gravity of the velocity–transit time density spectrum. This time average velocity is essentially the propagation velocity at the zero-offset. It has been shown that NMO velocities determined by conventional velocity analysis are closer to rms velocities, as given by

$$V_{\text{rms}}^2 = \frac{\sum v^2(n)\,\Delta t(n)}{\sum \Delta t(n)},$$

or

$$V_{\text{rms}}^2 = \frac{\sum v(n)\,d(n)}{\sum \Delta t(n)}. \tag{4}$$

Again, from mechanics, we recognize that this expression represents the calculation of the centroid of the second moment of the velocity–transit time density spectrum. From statistics, we can form the relationship

$$\alpha^2 = V_{\text{rms}}^2 - V^2 \tag{5}$$

which represents the variance (standard deviation squared). This shows that the velocity heterogeneity is half the velocity bandwidth of the stack of horizontal layers. We can conclude from this and earlier observations that the larger the heterogeneity, the larger the velocity increase with respect to offset.

Now consider the density spectrum with respect to total layer thickness, as sketched in Figure 11. Similar to the average velocity, the centroid is defined by

$$\tilde{V} = \frac{\sum v(n)\,d(n)}{\sum d(n)}, \tag{6}$$

which represents the depth-averaged velocity. This is the velocity observed from the well logs and differs from the time-averaged velocities. If we substitute time-averaged and rms velocity functions into equation (6), we get

$$\tilde{V} = \frac{V_{\text{rms}}^2}{\bar{V}},$$

or

$$V_{\text{rms}}^2 = \bar{V} \times \tilde{V}. \tag{7}$$

The seismically computed rms velocity is thus shown to be the harmonic average of the time- and depth-averaged velocities. It can also be deduced that the time-averaged velocity is less than both the depth-averaged and rms velocities. They are the same if and only if there is no velocity variation in the total depth under consideration. This corresponds to zero velocity heterogeneity. Another concern is the error in depth or layer thickness estimation.

A final problem with carbonate layers is that, due to the similar high-velocity contrasts at the top and bottom of a layer, the critical angle for incident waves is small on both sides. As shown in Figure 12, waves penetrating through the carbonate layer are strongly vertical (or more concentrated in the direction normal to the carbonate interfaces). Waves reflecting from beds parallel to the base carbonate are transmitted through the carbonate layers back to the surface. However, waves reflected

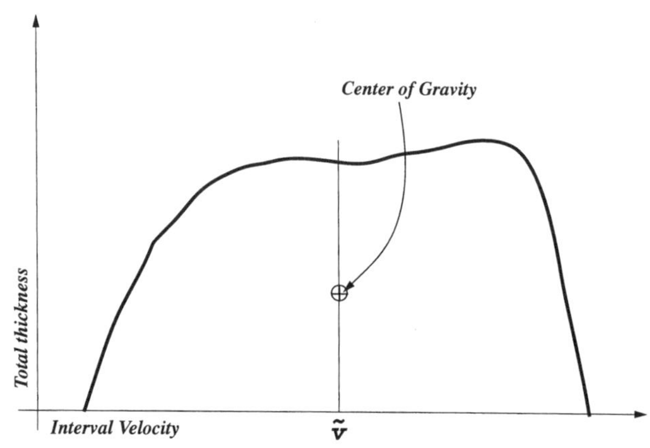

Fig. 11. Total thickness spectrum with respect to interval velocity.

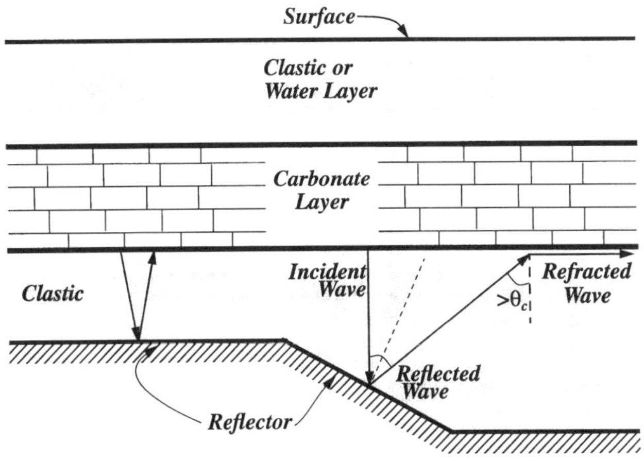

Fig. 12. Schematic cross-section showing the geometric effects on seismic waves of a carbonate layer between two lower velocity layers.

from dipping beds or beds not parallel to the base carbonate have incidence angles larger than the critical angle, hence they are totally refracted by the base carbonate reflector and do not reach the surface. Thus, only a limited angle of reflected waves are recorded at the surface, such that only the beds roughly parallel (in the range of the critical angle) to the base carbonate can be seen; the others are lost. In addition, because only a limited angle is recorded, the spatial wave number bandwidth is narrow, such that images appear smeared spatially. This causes the loss of definition of events in both the depth and spatial directions.

Low S/N, excessive horizontally traveling waves, and multiple reflected waves are the principal concerns in designing the processing sequence. The question is, "Is there any useful reflected energy left after all the multiples and noise are suppressed?" Use of a large source energy is recommended, provided there is sufficient recording dynamic range. Also, higher recording multiplicity increases redundancy and thus helps with the poor S/N. The processing objective, aside from multiple suppression and signal-to-noise improvement, is to generate a seismic section and corresponding time or depth images that are as close as possible to the true image of the subsurface, with the highest possible resolution. If there are any wells, it is desirable for the seismic section to match the synthetic seismograms, without some questionable filtering. Since the objectives are well stated, it is best to start from the end product and set the process sequences to achieve those objectives.

Resolution, Sampling, and Bandwidth

It is the constant goal of the geophysicist to achieve the greatest possible resolution, constrained by the condition that a high coherency level be maintained. This objective translates, in the time domain, to a wavelet possessing the following qualities:

- compactness (short duration)
- symmetry (zero-phase)
- a limited number of low-amplitude side lobes ("non-ringy")
- a smooth, rapidly converging envelope
- uniformity of the wavelet among the traces in the data set

In the frequency domain, this wavelet is characterized by a broad, smooth signal amplitude spectrum and zero-phase.

A useful model of the construction of a wavelet is given by the product of a cosine wave and smooth function, which plays the role of an envelope. Figure 13 shows such a construction. The corresponding amplitude spectrum is given in Figure 14. Note that the peak frequency of the spectrum is 50 Hz, which is the frequency of the modulated cosine wave. Also note that the spectral width measured at the half-amplitude points is approximately 25 Hz, the reciprocal of the envelope width used in constructing the wavelet.

This model provides us with a simple link to the frequency domain. The frequency of the sinusoid becomes the dominant frequency of the spectrum, and the shape of the spectrum is given by the transform of the envelope. If the shape of the envelope is approximately bell-shaped or Gaussian, then its spectrum is similarly bell-shaped, with the widths in time and frequency related reciprocally. If one approximates the width of the envelope (pulse width, ΔE) as the half-amplitude width, the spectral width (bandwidth, ΔF), approximated in the same way, relates as

$$\Delta F \approx 1/\Delta E. \quad (8)$$

A more formal definition of temporal or spectral width is given by the width of a rectangle having the same height and area as the envelope or spectrum, respectively. With this definition, there is an exact reciprocal relationship, $\Delta F = 1/\Delta E$. For practical purposes, however, the half-amplitude width is adequate and facilitates an easy connection between resolution and bandwidth.

If resolution is expressed quantitatively as the minimum perceivable time separation, ΔT, between adjacent major reflections (such as the top and bottom of a carbonate bed), then ΔT can be represented in terms of pulse width and bandwidth as

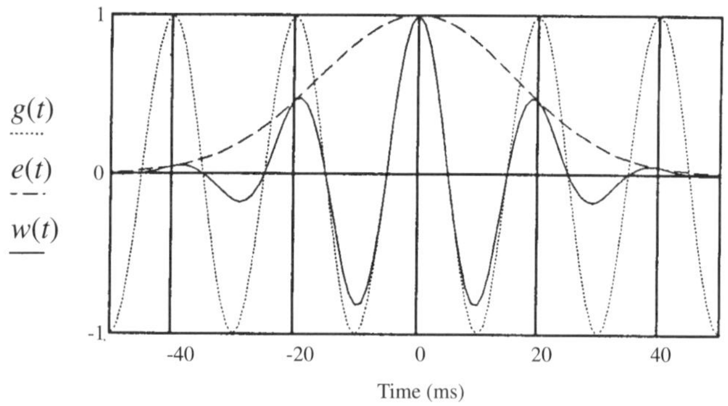

Fig. 13. Wavelet model in the time domain showing the construction of a wavelet, *w(t)*, as the product of a cosinusoid function, *g(t)*, and an envelope function, *e(t)*.

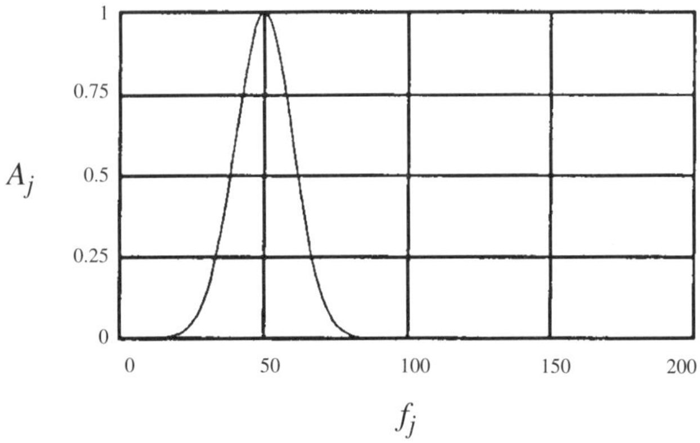

Fig. 14. Amplitude spectrum of the wavelet shown in Figure 13.

$$\Delta T > \Delta E = 1/\Delta F. \tag{9}$$

The logic of equation (9) is that, if the separation of the reflection events is less than ΔE, then the two reflection wavelets could potentially coalesce into a wavelet with a single envelope with one peak—no discernible event separation. However, if the separation is greater than the half-width of the envelope, there will always be distinguishable peaks on the envelope pair. It should be noted that the envelope function, as its name implies, represents the outer limit of the wavelet amplitude excursion. Consequently, the sum of the envelopes of two closely spaced events expresses the outer limit of their sum and their combined envelopes.

Maximum resolution is thus a matter of minimizing the envelope width by maintaining as broad a spectrum as possible, with the vital constraint that the phase spectrum be linear, preferably with zero slope and intercept. Since resolution and, ultimately, interpretation are enhanced by reduction in size and number of the side lobes of the basic waveform, it may also be necessary that the envelope width be less than the period of the dominant frequency. A wavelet that approaches the ideal is depicted in Figure 15.

Figure 16 shows an example of resolution ranging from poor to complete. Pairs of closely spaced events, with $\Delta T = 10$, 20, and 40 ms, respectively, are convolved with the wavelet of Figure 15. Once the ΔT separation exceeds the envelope width, ΔE, the reflections are essentially resolved. Note that the amplitudes of unresolved reflections are also distorted. The amplitudes are no longer proportional to reflectivity and may be either enhanced or reduced by the constructive or destructive interference patterns that result from the combination. In theory, this "tuning" phenomenon can be used to estimate bedding thickness, but it is preferable to obtain time resolution and use tuning as a last resort measure, especially in light of the many other factors that affect amplitude. These influences include porosity, pore fluid, mineralogy, fracturing, and other rock properties of importance to the explorationist. Whenever possible, it is desirable to estimate the bed thickness by

$$\Delta Z = \frac{1}{2}[V\Delta T], \tag{10}$$

where ΔZ is the bed thickness and V is the velocity in the interval.

The discussion up to this point has not dealt with the practical realities of seismic recording and processing. These adverse factors, which limit our ability to achieve the ideal wavelet by restricting usable bandwidth, include sampling, noise, data length, and conventional filter design.

The proper sampling of seismic data in time and space is an important and often abused concept in acquisition. Spatially, for example, we must sample the total wavefield finely enough and with proper alias protection so that (1) the coherent noise is safely reduced and (2) the entire signal spectrum can be correctly migrated without dispersion and other aliasing artifacts. Unfortunately, the guidelines for sampling spatially are rarely followed, and as a result, both signal and noise are aliased in the field. While it is desirable to image interfaces as sharply as possible (the ultimate being a spiked waveform), this is achievable only in theory and by infinite bandwidth recording and processing. In time- and space-sampled recording and processing as presently practiced, one can represent the seismic data only within the bandwidth governed by the temporal and spatial sampling interval. All data beyond the limits becomes subject to aliasing.

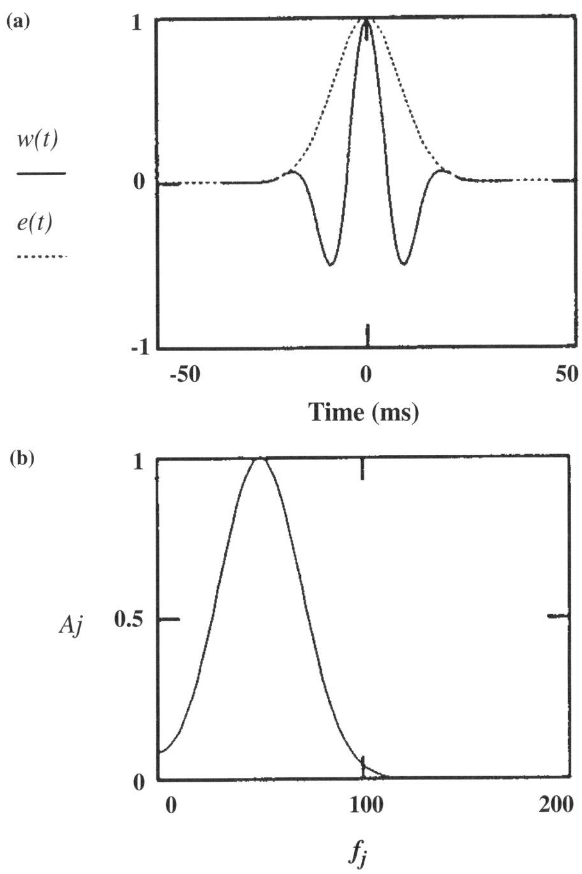

Fig. 15. (a) The ideal wavelet is approached as the envelope width narrows relative to the dominant frequency, (b) corresponding to a broader spectrum.

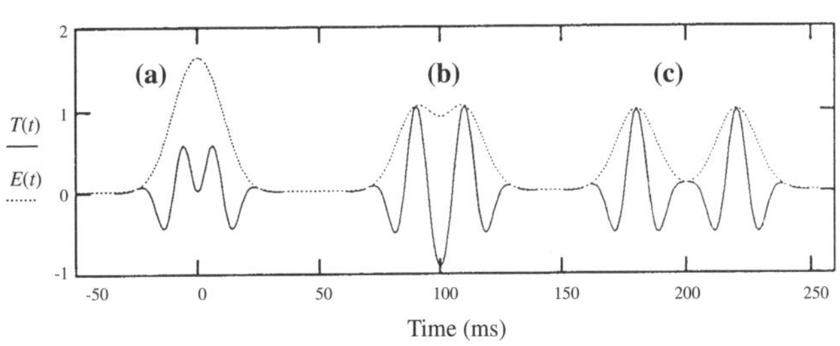

Fig. 16. Resolution of closely spaced events as a function of envelope width relative to event spacing, ΔT. (a) $\Delta T = 0.5\,\Delta E = 10$ ms; (b) $\Delta T = \Delta E = 20$ ms; (c) $\Delta T = 2\,\Delta E = 40$ ms.

Temporal aliasing results in a frequency ambiguity, with higher frequencies masquerading as lower frequencies within the band of interest. An illustration of this is shown in Figure 17. A 100-Hz sinusoid, which can be regarded as noise, is sampled at 8 ms, or at a sampling frequency of 125 Hz. Sampling theory predicts that if any frequency is sampled at a rate equal to or less than twice that frequency, then aliasing occurs. In this figure,

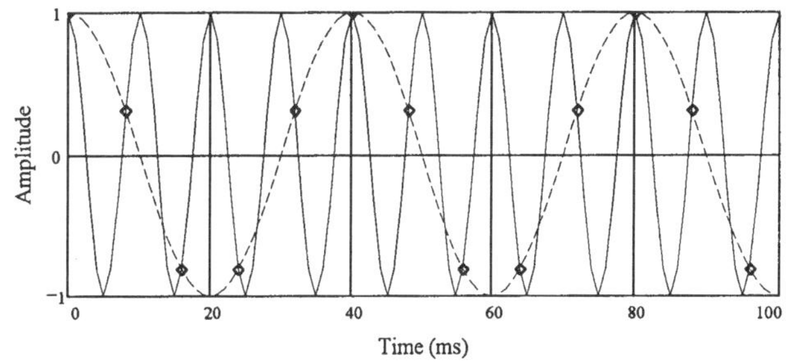

Fig. 17. Example of aliasing—the consequence of undersampling. A 100-Hz sinusoid (solid) sampled at 8 ms (sampling frequency 125 Hz) has disguised itself (aliased) to 25 Hz. Any multiple of 125 ±100 Hz (e.g., 150, 225, 275, and so on) will also fit the original samples (diamonds).

we see that the samples (denoted by diamonds) represent 25 Hz as well as the actual frequency of 100 Hz. In actuality, these samples fit an infinity of frequencies that are replicas of the original spectrum but now centered at integer multiples of 125 Hz. All these frequencies (other than ±100 Hz), are called *alias spectra*. The one that adversely effects the data quality is the lowest alias, in this case, 25 Hz. If all the aliases are greater than the sampled frequencies, there is no problem since the original spectrum remains intact and is fully recoverable. To ensure this condition, we filter out any frequencies higher than half the sampling frequency before sampling (or resampling) takes place. In the time domain, these are temporal frequency filters that attempt to reduce to negligible any frequencies above half of the sampling frequency, also known as the *Nyquist frequency*, or f_N. For example, if the 100-Hz input had been filtered before sampling with the required 62.5-Hz highcut, there would have been no significant residual 100 Hz to alias to 25 Hz and therefore no distortion of the real (perhaps signal) frequency residing at that location. After sampling, the original data are recovered by applying the same filter.

The necessity of an alias filter limits the meaningful bandwidth to a maximum of the Nyquist frequency, that is, half of the sampling frequency, f_s. In practice, the useful maximum is considerably less. The geophysicist must therefore choose a sampling frequency significantly greater than the highest frequency anticipated to be in the reflection signal, or the highest frequency required to satisfy the resolution objectives. The currently favored technique is to choose a sampling frequency according to the following principle:

$$f_s > 4 f_{s\max}, \tag{11}$$

where f_{smax} represents the maximum usable signal frequency. With this technique, the use of the appropriate alias filters do not harm the signal.

Sampling in space, where aliasing manifests itself as a dip ambiguity of either signal or noise or both, should follow guidelines similar to the temporal rule. The basic sampling interval in space is the group interval (or trace spacing), ΔG. This corresponds to a spatial sampling frequency of $k_s = 1/\Delta G$. The required alias filtering is the function of the receiver and source arrays. Their sole purpose is to significantly reduce any frequencies higher than K_N, the spatial Nyquist. Extension of the principles learned in temporal sampling suggest that the proper choice of k_s should be based on $k_s > 4k_{smax}$, where k_{smax} represents the highest spatial frequency in the signal spectrum. This frequency can be shown as

$$k_{smax} = \frac{f_{smax}}{V_{min}}, \tag{12}$$

where V_{min} is the lowest realistic horizontal velocity of the signal. This corresponds to the steepest slope of the curve of normal moveout (NMO). Combining these concepts, we find that

$$\Delta G < \frac{V_{rms}}{2f_{smax}} \tag{13}$$

where V_{rms} represents the V_{min} or reciprocal slope of the NMO curve at the time of interest at the maximum conceivable offset. Although equation (13) is the equivalent of the temporal sampling rule, the spatial implementation is almost never done in practice. This presents a serious problem for the processor and, ultimately, the interpreter. The length of the data sampled imposes limitations on the accuracy with which the low-frequency components can be measured. This uncertainty, coupled with a dominance of residual noise at the low end, forces the choice of usable spectrum to the practical limits (Figure 18).

The presence of noise spread generously across the spectrum requires that the final spectrum be shaped to emphasize those spectral components relatively rich in signal. Wiener filtering principles can be applied here by "optimum" specification of the filter, $W(f)$, with the response at each frequency as

$$V(f) = \frac{[s(f)]^2}{[s(f)]^2 + [n(f)]^2} \tag{14}$$

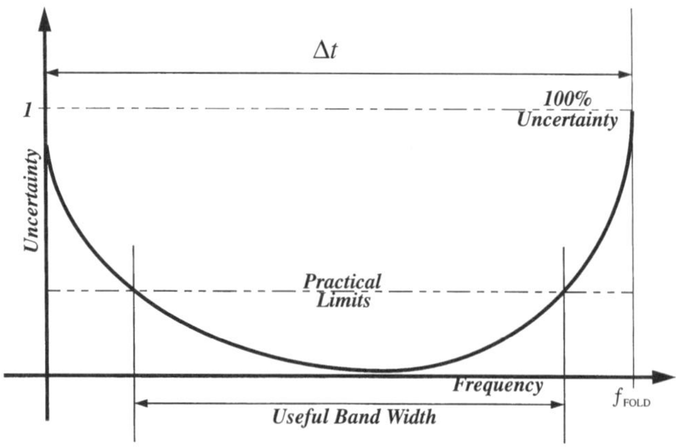

Fig. 18. Schematic diagram showing the uncertainty of discrete sampling of continuous data.

where $[s(f)]^2$ is the signal power spectrum, $[n(f)]^2$ is the noise power spectrum, and $[s(f)]^2 + [n(f)]^2$ is the total power spectrum.

While this filter is zero-phase, as required, and is optimum in terms of maximum signal power on output, it generally lacks the necessary spectral smoothness that produces a desirable, high-resolution wavelet. Normally, the Wiener filter design is based on auto- and cross-correlation functions among the traces in question. While this may prove beneficial in a general way, spectral shaping requires a compromise to achieve smoothness. The sharp corners and steep cut-offs of the industry standard "Ormsby" filter illustrate good intentions gone astray. The trapezoidally specified filter shape typically results in a long oscillatory wavelet, which is unsuitable for high-resolution work. It also produces zeroes on the unit circle in the reject zones. Figures 19 through 23 show the spectral band-pass and resulting wavelets of the Butterworth filter design. As can be seen, the wavelets benefit from gentle roll-off slopes at both ends of the frequency spectrum, as opposed to the severe slopes and attempted low-response regions of the Ormsby filter. As demonstrated later, this concept of gentle and smooth attenuation, outside the band of usable frequencies, is critical to the resolution and coherency of carbonate reflectors.

Content of Stacked Sections

The objective in generating seismic images is to accurately reflect the subsurface. An interface between low acoustic impedance above and high acoustic impedance below has a positive reflection coefficient and is represented by a spike with a magnitude proportional to the reflection coefficient.

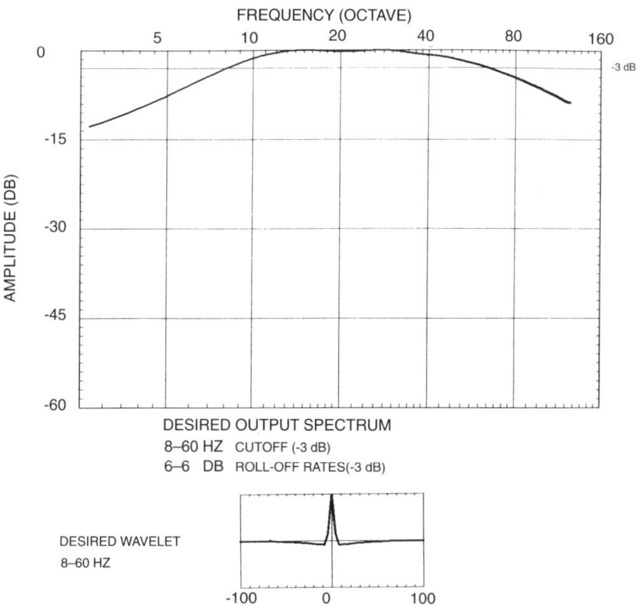

Fig. 19. Band-pass filter response (8–60 Hz; 6 dB/octave slopes), with associated wavelet (bottom).

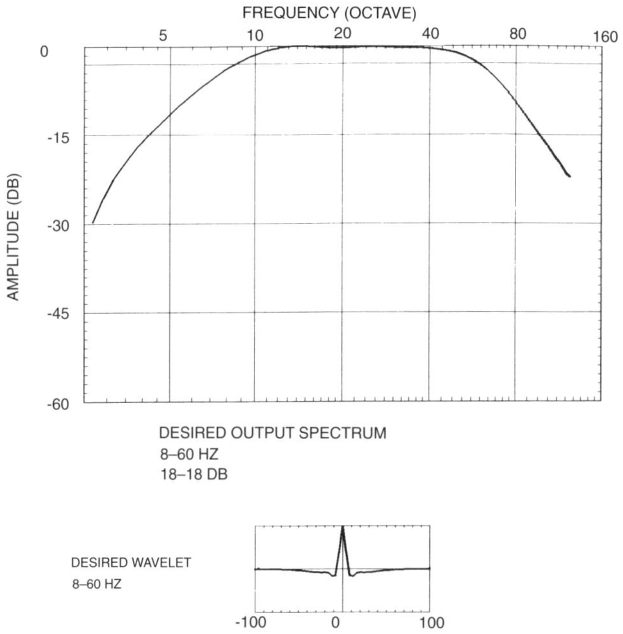

Fig. 20. Band-pass filter response (8–60 Hz; 18 dB/octave slopes), with associated wavelet (bottom).

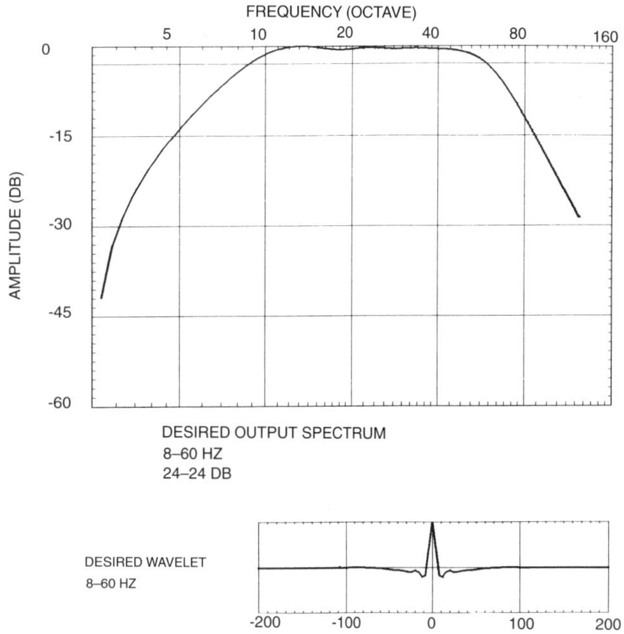

Fig. 21. Band-pass filter response (8–60 Hz; 24 dB/ octave slopes), with associated wavelet (bottom).

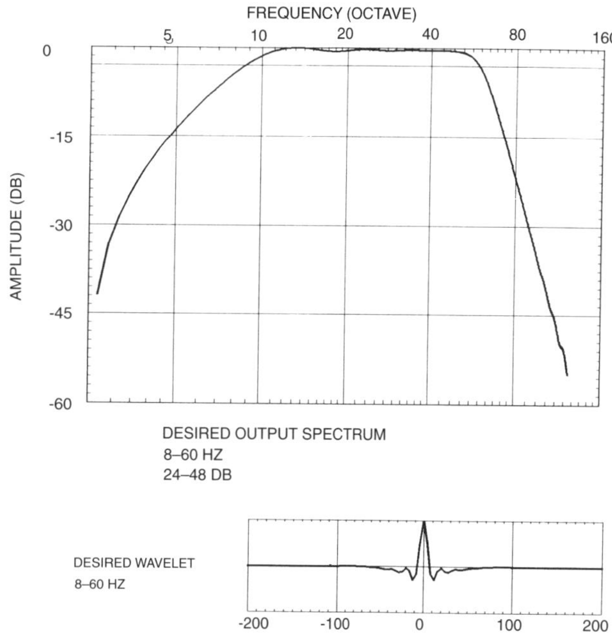

Fig. 22. Band-pass filter response (8–60 Hz; 48 dB/octave slopes), with associated wavelet (bottom).

This is the infinite bandwidth representation. In the sampled and band-limited world, the reflection takes the form of a zero-phase wavelet, in which the peak of the wavelet corresponds to the maximum of the envelope and zero instantaneous phase (Figure 24). Conversely, high acoustic impedance to low would be represented by a negative spike in the infinite bandwidth and a wavelet with a negative extreme (trough) at reflection time. This corresponds to an instantaneous phase of 180° (Figure 25).

Subcritical angle reflection coefficients are scalar quantities ranging between +1 and –1. In limited bandwidth, this corresponds to the zero and 180° phase wavelets. However, in sampled data stacked sections, there is more than just wavelet polarity. For a high acoustic impedance slab embedded in low acoustic impedance, there are positive and negative spikes at a distance equal to the two-way traveltime within the high-impedance layer. If this time is less than the wavelet resolution, then a 90° phase-shifted wavelet occurs, with a single peaked envelope (Figure 26). If the case is reversed, then a –90° wavelet occurs (Figure 27). These responses occur if the wavelets are stacked perfectly, that is, if they are balanced in amplitude, time, and phase aligned. Wavelets, such as multiple reflections, that are stacked with residual moveout have an additional phase of about –45° imposed (Figure 28). Thus, the stacked sections have events with eight different instantaneous phase characteristics.

In theory, we expect to have reflection coefficients attain only scalar values within the critical angle. This represents positive and negative polarities (zero and 180° phases). The band-limited nature of the sampled data, however, is the cause of a composite ±90° phase wavelet. A few investigators have noted that thin beds are abundant in clastic environments. Events for a single interface (zero or 180° phase wavelets) occur only rare-

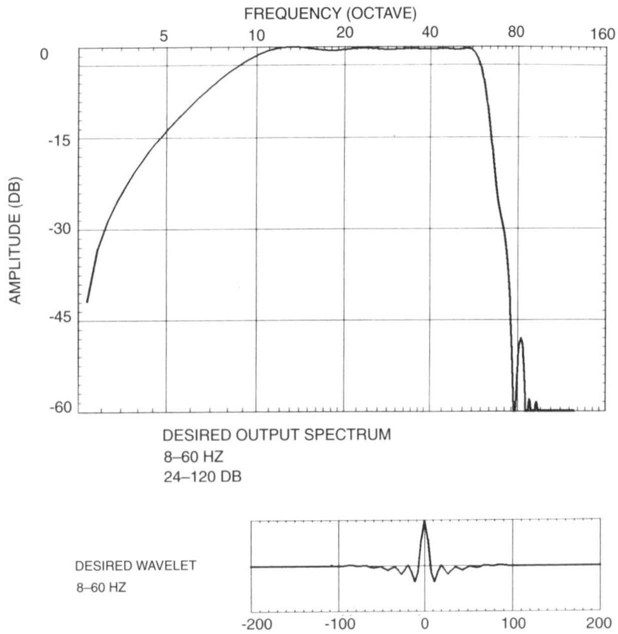

Fig. 23. Band-pass filter response (8–60 Hz; 120 dB/octave slopes), with associated wavelet (bottom).

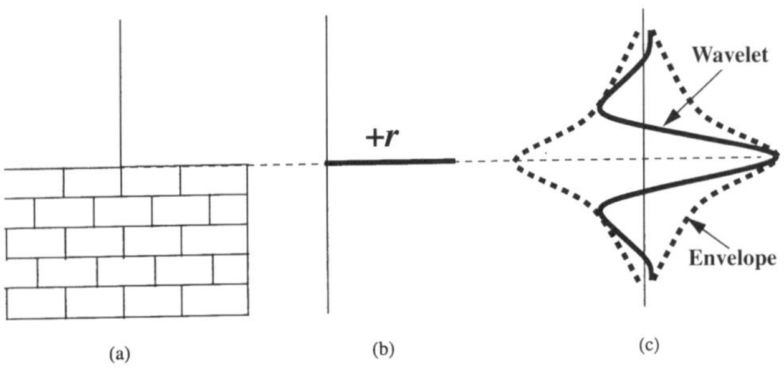

Fig. 24. Low to high acoustic impedance contrasting interface: (a) schematic diagram of interface, (b) theoretical reflection response, and (c) recorded band-limited zero-phase wavelet.

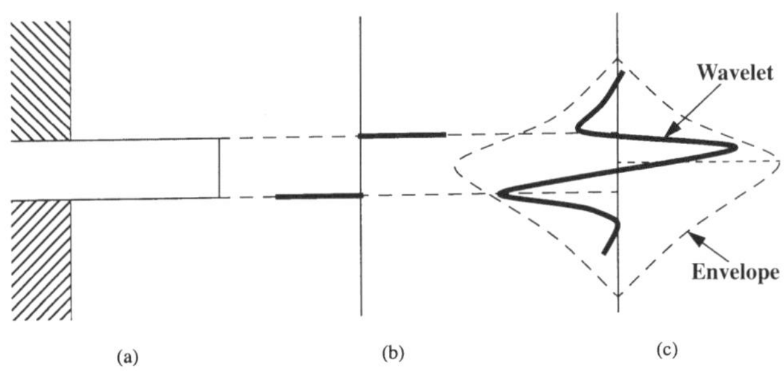

Fig. 26. High acoustic impedance layer between two low acoustic impedance layers: (a) schematic diagram of interface, (b) theoretical reflection response, and (c) recorded band-limited wavelet (+90° phase).

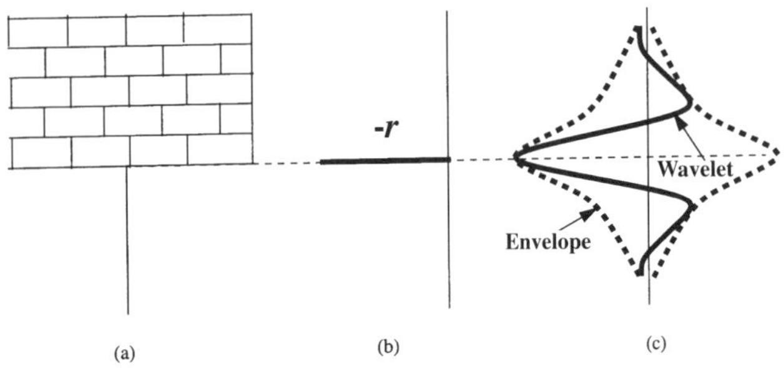

Fig. 25. High to low acoustic impedance contrasting interface: (a) schematic diagram of interface, (b) theoretical reflection response, and (c) recorded band-limited wavelet (180° phase).

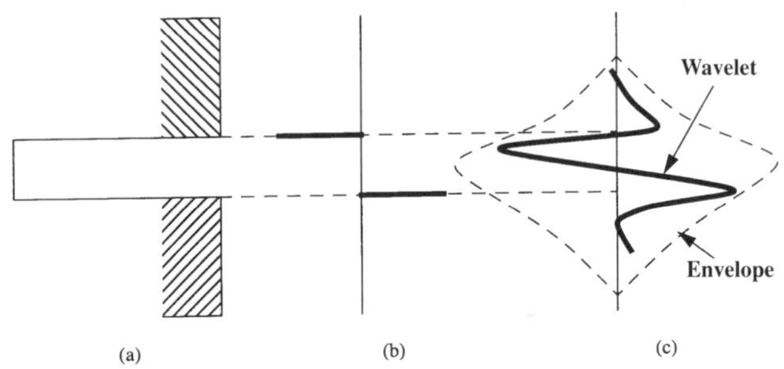

Fig. 27. Low acoustic impedance layer between two high acoustic impedance layers (a) schematic diagram of interface, (b) theoretical reflection response, and (c) recorded band-limited wavelet (–90° phase).

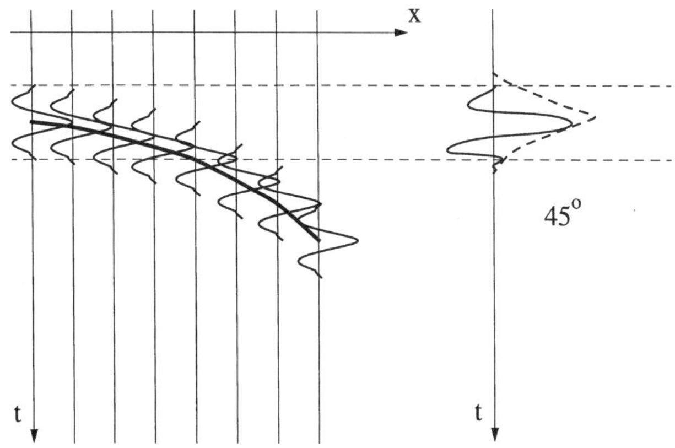

Fig. 28. Rotated wavelet (45° phase) resulting from the sum of undercorrected traces. Multiples are shown on the stacked traces.

ly, such as at gas-water contacts and perhaps at the strong acoustic impedance contrast formed on unconformity surfaces. Therefore, in clastic sequences, there are more seismic reflections from the thin beds than from single interfaces. Thus, in a zero-phase seismic section, most of the events appear to be ±90° phase rather than zero or 180° phase wavelets. This helps to determine the phase correction needed before the computation of band-limited acoustic impedance.

The wavelets with their variety of phase components also show that stacking is not a simple scalar number addition but rather a vector summation. Thus, before stacking, all of the traces must be corrected for misalignment of amplitudes, time, and phase content of the signal. Since these misalignments should not be corrected arbitrarily (which might result in artificial stratigraphic and structural configurations), they must be investigated and computed in a surface-consistent manner.

Surface-Consistent Processing

The surface-consistency condition assumes that all traces belonging to a common source have similar effects caused by the source and near-surface conditions. Thus, all traces recorded with this source have essentially the same source and near-surface effects, including time, amplitude, and wave shape. The same thing can be said for the common-receiver effects, that is, all traces recorded at the same receiver position have the same effects in common. The common-source effects are generally those induced on the downward traveling wavefront, and conversely, the common-receiver effects are induced on the upward traveling wavefront. The common-offset effects include the initiation and reception angle and raypath dependent effects, such as NMO and spreading. If these signal misalignments and distortions are to be properly corrected, then all the processes that affect them must be applied in a surface-consistent manner. For example, spike deconvolution, wavelet shaping, amplitude correction, and static corrections for long- and short-wavelength statics must be computed and applied in a surface-consistent manner.

Long-period multiples are not surface-consistent, hence they cannot be suppressed by surface-consistent deconvolution. They are part of a source-generated propagating wavefield, thus their prediction and removal must be based on wave equations. Unfortunately, no single satisfactory multiple suppression method presently exists. However, reverberations and short path multiples generated by and near the source can be suppressed by surface-consistent deconvolution. The other main surface-consistent process is the estimation of the source signature. There have been many conflicting discussions on this subject. Some prefer to record the source signature in the field. Conceptually, these deterministic methods are preferable to purely statistical methods. However, field recording gives only one facet of the source signature, the one near the source where large pressure changes occur. The source wavefront is also variable with respect to the direction of propagation. Most of the time we need to estimate the far-field signature, which is close to the shape of the source signature as it exists on the seismic records. This is estimated from near-field recording by modeling the near surface. In some marine cases, the seismic signature is recorded by geophones placed below the boat in deep water. This is used as the initial source wavelet estimate, and statistical methods are used to correct the residual changes to the final desired form.

With land data, geophones similar to the uphole phones are used to record the source signature. When there is no field-recorded source signature, it must be determined by statistical means. Statistical methods are also used for determining the receiver-consistent and offset-consistent corrections because there are no deterministic means to record them in the field. These wavelet determinations are necessary so they can be shaped to the forms that give the best possible resolution within the available bandwidth. The corrections are performed by shot-, receiver-, and offset-consistent modes, so any distortion introduced by these processes remains surface and raypath consistent. For example, the wavelet-shaping processes might leave an average amplitude and linear component of phase in the wavelet spectrum uncorrected. A filtering process such as deconvolution does not correct the time statics. Note that the residual static corrections are not only time shifts but also include surface-consistent phase and amplitude components. Surface consistency throughout the processing flow allows for further refinement at later stages, retaining the surface consistency and data integrity.

Design of Seismic Data Processing Flow

Based on the available tools and the final objectives, it is now possible to design a processing stream for seismic data having shallow carbonate problems. A processing flow diagram is shown in Figure 29. It begins by combining the recording geometry and near-surface information with the corresponding seismic data. Near-surface data include cable geometry, depth of the cable, source depth, and water depth for marine recordings. Data for land recordings include surface elevation, shot depth, uphole time, and coordinates of all shots and receivers. The data are edited to eliminate bad traces, spikes, and other data problems.

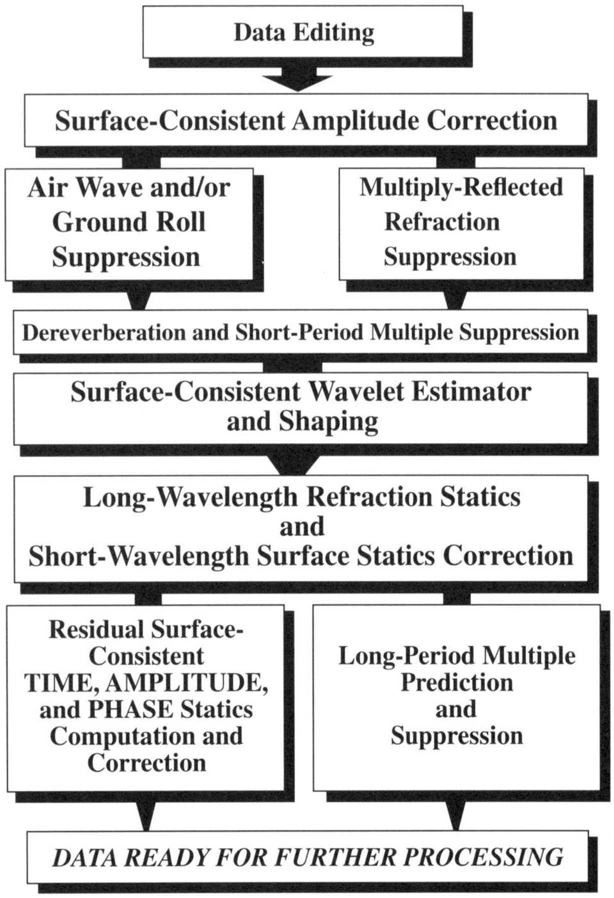

Fig. 29. Flow diagram of seismic data processing steps.

The next step is to make the surface-consistent amplitude correction. This process computes shot and receiver amplitude corrections that suppress the shot-to-shot and receiver-dependent amplitude variations. This prepares the data amplitudes for ground roll or multiple suppression procedures since they require the wavefield to be recorded in a balanced manner. This procedure also computes the offset and time-dependent amplitude decay, which can be used as an overall correction to balance the amplitudes in a controlled manner rather than the much-used automatic gain control (AGC).

The next step in land data cases is to suppress the air wave or ground roll contamination. For marine data, we can go directly to the deconvolution stage since there is no air wave contamination. Some ground roll may occur in shallow-marine cases, but the usual multiply-reflected refracted arrivals can be suppressed by *f-k* filtering.

Reverberation and short-path multiples are generated by source coupling and shallow near-surface layers. These reverberations are naturally minimum-phase and can be suppressed by a short minimum-phase inverse (spiking deconvolution) or a short prediction lag deconvolution process. I have found that in poor S/N data sets, it is better to use prediction filters with 24–60 ms prediction lags. They produce an output section with better S/N than the spiking deconvolution filters. Because these deconvolution filters are generated in the vicinity of sources and receivers, they can be used in a surface-consistent manner.

The next step is to estimate and shape the source and receiver signatures and produce a set of records that would be generated by a constant source wavelet of our design and recorded by a set of receivers with constant recording characteristics. Therefore, any variations in the reflected wavelets represent subsurface changes rather than arbitrary surface anomalies. To produce this, we determine one operator for each shot gather and another one for each common-receiver gather. In some cases, we want to suppress the offset-related near-surface effects, such as the geophone array directionality. Although the array response depends on the exit angle of the reflectors, this can be approximately corrected by an additional set of operators, one for each common-offset trace gather. The surface-consistent process produces traces with all near-surface distortions suppressed, except for the time constant amplitude, intercept, and slope of linear phase distortion factors. These are considered in the surface-consistent statics computation section.

The statics step, as for air wave suppression, is different for land and marine recordings. On land, long-wavelength refraction statics and short-wavelength near-

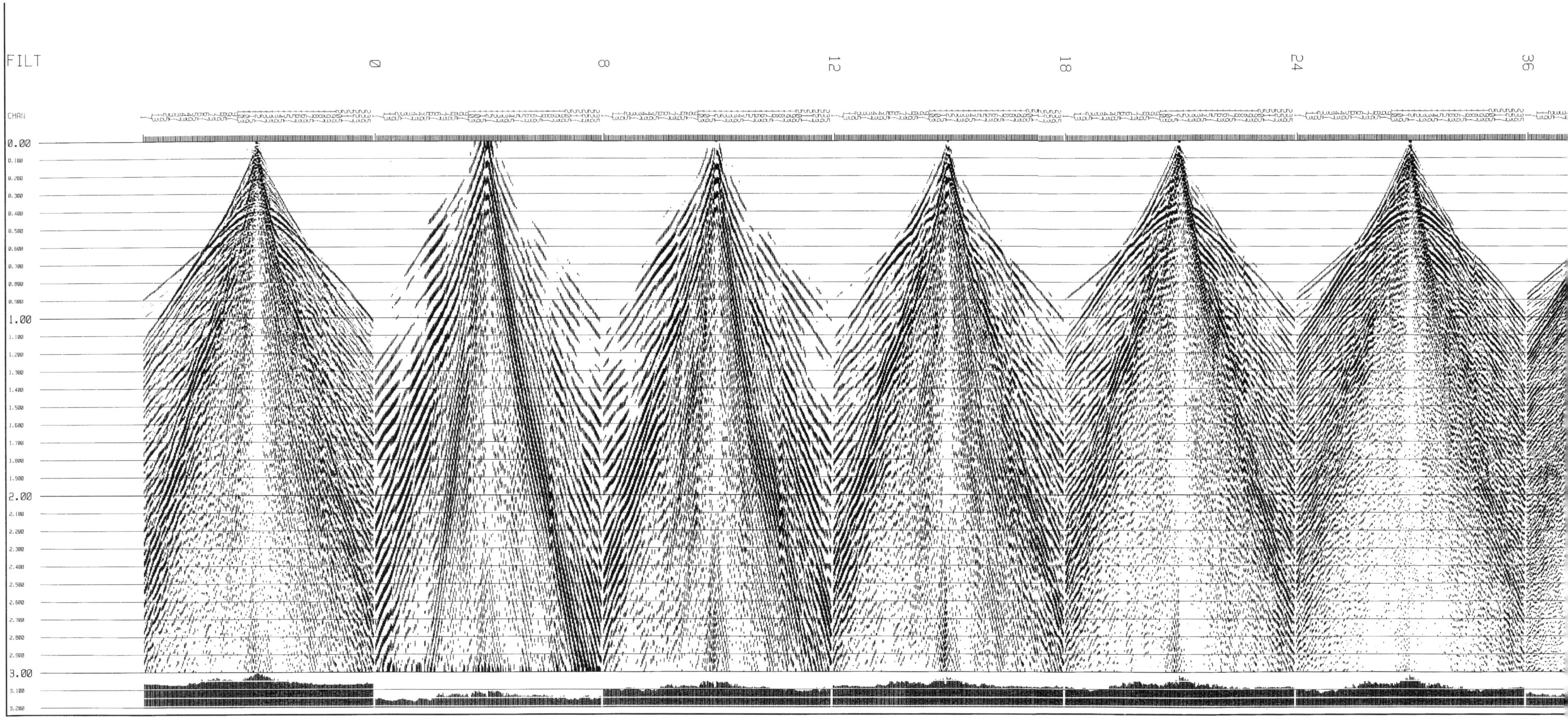

Fig. 30. Filter panel. A series of narrow-pass band filters have been applied to the input record at the far left.

Unfold to view Fig. 30

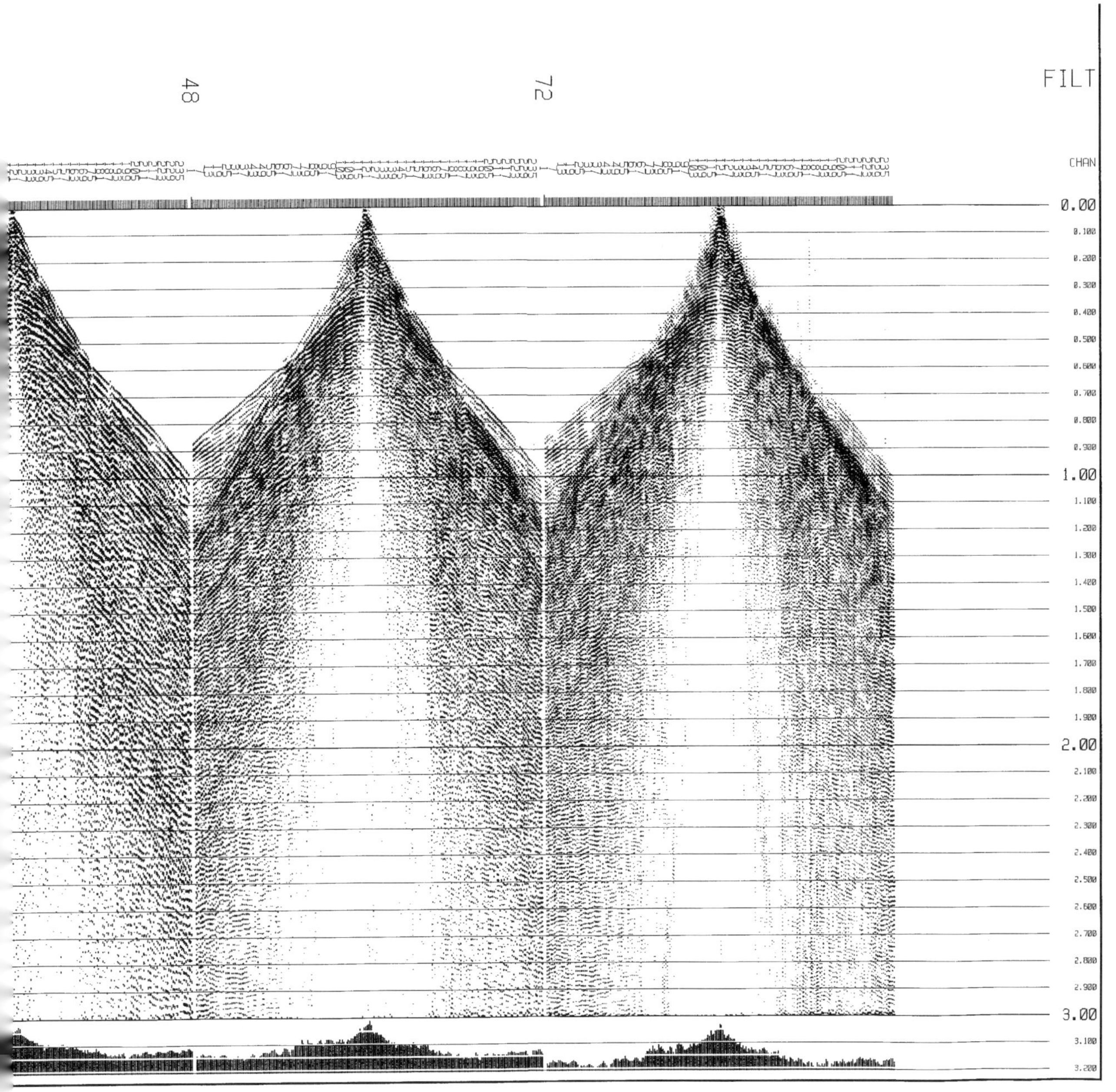
FILT
CHAN
48
72
0.00
1.00
2.00
3.00

surface statics are computed from first-break arrival times. Application of these statics removes laterally varying near-surface depth and time effects, and the time statics are reduced to residual levels (less than one-half dominant period). Surface-consistent residual statics computation and corrections for time, phase, and amplitude statics are performed next. Application of static corrections produces traces within CDP gathers that are amplitude balanced and time and phase aligned. Since CDP stacking is a vectorial addition, summing these traces does not disturb the wavelet shapes and amplitudes adversely.

In the marine case, the next step is long-period multiple suppression. There are a number of proposed methods to do this, but even the most effective methods are not completely successful. The general method used to tackle the problem of all surface-connected multiples is to remove the water surface and compute the reflection response of the water bottom only. Unfortunately, when all surface-connected multiples are removed, all interbed multiples connected to the top and bottom of the carbonate layer are left. This multiple suppression process is the weakest link in the processing chain. Better multiple suppression can help to develop better velocities below the carbonate layer, hence better images of the beds.

The next step is velocity analysis and stack section generation. The stack section can be in different forms, such as conventional CDP stack, CDP stack with dip moveout (DMO) applied, and pre- or poststack time- or depth-migrated sections. CDP stack or time-migrated sections can later be used in complex seismic trace attribute analysis.

Carbonate Processing: Theory, Practice, and Examples

This section presents some processing examples to illustrate their problems and possible solutions. In some instances, a number of other viable solutions exist, while others have no acceptable solutions. An extensive bibliography at the end of this paper directs the interested reader to more detailed articles. The following processing steps are kept essentially parallel to those suggested earlier.

Frequency Bandwidth Analysis

Figure 30 shows the filter panel generated from one shot record along the seismic line. For convenience, the display on the left shows the original data after amplitude and exponential decay corrections. Each panel on the right of the original data represents the output of narrow band-pass filtering with 120 dB/octave roll-off rates. The filter panel is generated for 0–8, 8–12, 12–18, 18–24, 24–36, 36–72, 72–125 Hz frequency bands. One of the principal uses of the filter panel is to point out the frequency bands where the signal is laterally continuous. This is one of the most convenient ways to specify the signal. It is assumed that all laterally discontinuous data represent ambient noise. Other types of source-generated events such as air waves, ground roll, and multiply-reflected refractions are considered to be coherent noise since they are usually not used in the normal flow of seismic processing. The refracted first arrivals are used for statics computations.

Ground roll is clearly visible on lower frequency bands in Figure 30. A low-cut filter with 16–18 Hz low-cut frequency suppresses this ground roll considerably. However, a 2-D Fourier transform study shows that there may be useful data at these frequencies, which are suppressed along with the ground roll when the low-cut filter is used. Since ground roll usually has a much lower propagation velocity, it can be filtered out with a 2-D filter.

Surface-Consistent Amplitude Analysis and Correction

Figures 31a and b represent the quality control output from a surface-consistent amplitude analysis program. Figure 31a shows the surface diagram, which is used to confirm the acquisition geometry and locate traces with anomalous amplitudes. The horizontal axis indicates the common-receiver positions, and the vertical axis shows the common-source location of each trace. The CDP and common-offset positions are in slant directions. Trace amplitudes are measured as rms values over selected windows and compared against the surface-consistent solutions. Traces with errors more than 10 times the correct value need to be investigated more closely. Since this type of anomaly represents a local condition rather than a surface-consistent condition, it is a good indication of a recording problem.

Figure 31b shows the performance factors of each shot, receiver, and offset group. The display indicates the performance as the relative coupling performance of each shot and receiver group. For example, all shots between stations 100 and 200 have larger (positive) amplitudes and thus better coupling performance than the shots after station 200. It also indicates that there is poor coupling between stations 200 and 300. Receiver performance shows gradual decay as the station numbers increase. This is partially related to shot performance and

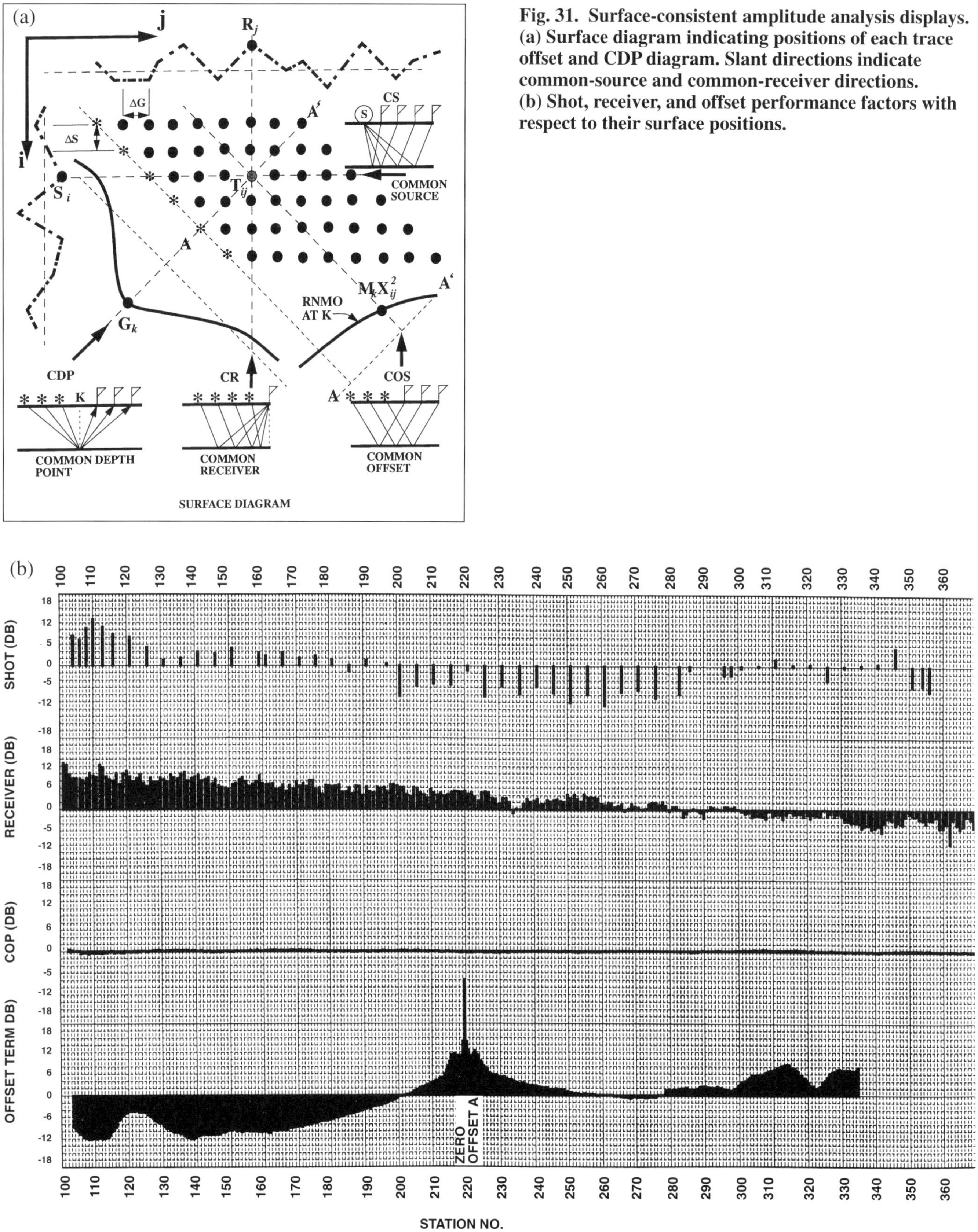

Fig. 31. Surface-consistent amplitude analysis displays. (a) Surface diagram indicating positions of each trace offset and CDP diagram. Slant directions indicate common-source and common-receiver directions. (b) Shot, receiver, and offset performance factors with respect to their surface positions.

may indicate that varying surface conditions are causing gradually poorer coupling toward the larger station numbers. This can be confirmed by reviewing the seismic records. Amplitude variation with respect to offset is, as expected, a decaying function (nearly logarithmic) in the increasing offset direction. Based on these observations, the user can correct the traces with surface-consistent solutions and can eliminate some of the traces based on their anomalous local amplitudes. These corrections provide a seismic recording with balanced shot and receiver performances that prepares the data for later application of multichannel processes.

Ground Roll Analysis and Suppression

Ground roll is identified by its slow propagation velocity and low-frequency band. The most common way to suppress it is by using 2-D or *f-k* filters. However, these filters are effective only if the data amplitudes are reasonably balanced with offset and time and do not have large time statics. The surface-consistent amplitude statics correction takes care of the first problem. The time-varying amplitude imbalance can be temporarily corrected with storable (removable) exponential gain correction. The ground roll is an indication of a low-velocity surface layer over a high-velocity consolidated zone. This gives rise to both short- and long-wavelength time statics problems, thus it is better to remove the ground roll after both the time and amplitude statics are corrected. Unfortunately, the ground roll statics correspond to the shear wave statics, which are not the same as the pressure wave statics computed from first-break arrival times or from reflected pressure wave traveltime differences. As a compromise, we can assume that the statics are small with respect to the long wavelength of the low-frequency ground roll and therefore do not adversely affect the data continuity of the ground roll events.

The efficacy of 2-D filter suppression of ground roll can be confirmed by actual trials. Figure 32 shows records with ground roll contamination. The zone is easily identifiable by the low frequency and the steep slope (low propagation velocity). The 2-D Fourier transform of one of these records is shown in Figure 33a (the area below 10 Hz has been suppressed for visual clarity). The area of ground roll (arrow) shows a zone of low frequency and lower velocity separated from similar frequency but higher propagating velocities which can be seen

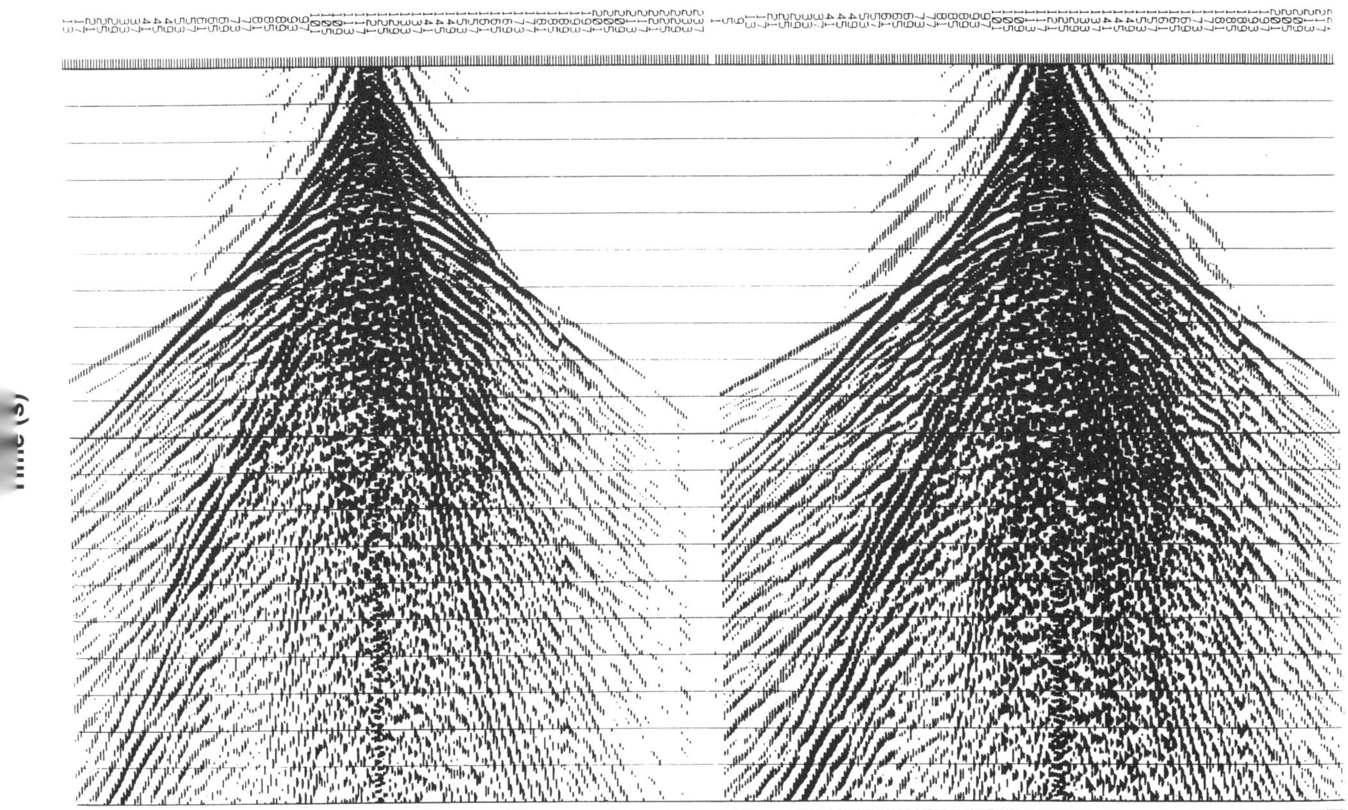

Fig. 32. Common-source land seismic data. Note the refracted events generated by the high-velocity carbonate layer and the horizontally traveling ground roll.

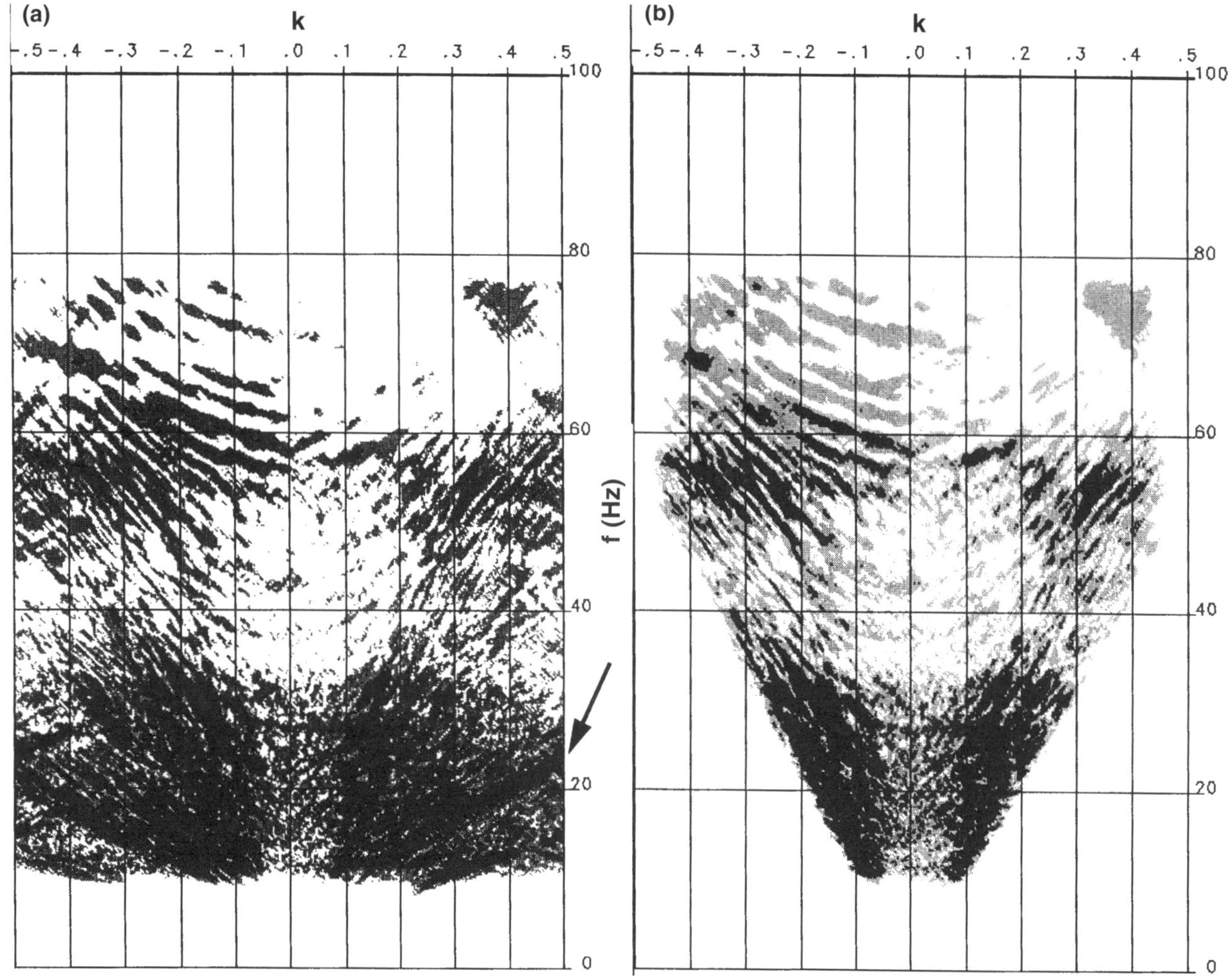

Fig. 33. 2-D Fourier transform (a) of original data (Figure 32) and (b) of ground roll suppressed data (Figure 34).

closer to the vertical frequency axis. The *f-k* filter process suppresses the ground roll zone by excluding it before the data are transformed back to the time-space domain. Figures 33a, 33b, and 34 show the *f-k* transform of Figure 32, with the ground roll suppressed and the output record in *x-t* space.

A comparison of Figures 32 and 34 shows that a considerable amount of ground roll contamination has been removed. Apparently, the time statics were not detrimental in this case, and the reflection signal has been preserved. Thus, this process can be used safely in the processing flow.

Over the years, several other processes have been suggested for ground roll suppression, but none of them are universally effective. Ground roll is a function of the surface conditions (its velocity structure, low-velocity layer thicknesses, surface topography, and underlying high-velocity topography) and is therefore time and space variable. Thus, time- and space-varying adaptive filters may prove to be more effective in the future than the time- and space-constant 2-D Fourier processes of today.

Exponential Decay Analysis and Correction

The amplitudes of seismic traces decay with time and offset due to the spherical expansion and transmission losses occurring across the interfaces. The losses can be on the order of tens or hundreds of times the earlier amplitudes. This creates computational and visual display difficulties. One of the most commonly used methods of correction is spherical correction. However, this depends on the propagation velocities, which are not known in advance, and the amount of correction is the ratio of time-velocity squares with respect to some shallow zone. Therefore, corrections are not uniform along the line. I recommend application of a space constant exponential correction in processes prior to the velocity computation. As for the amplitude decay correction, exponential expansion is the least harmful correc-

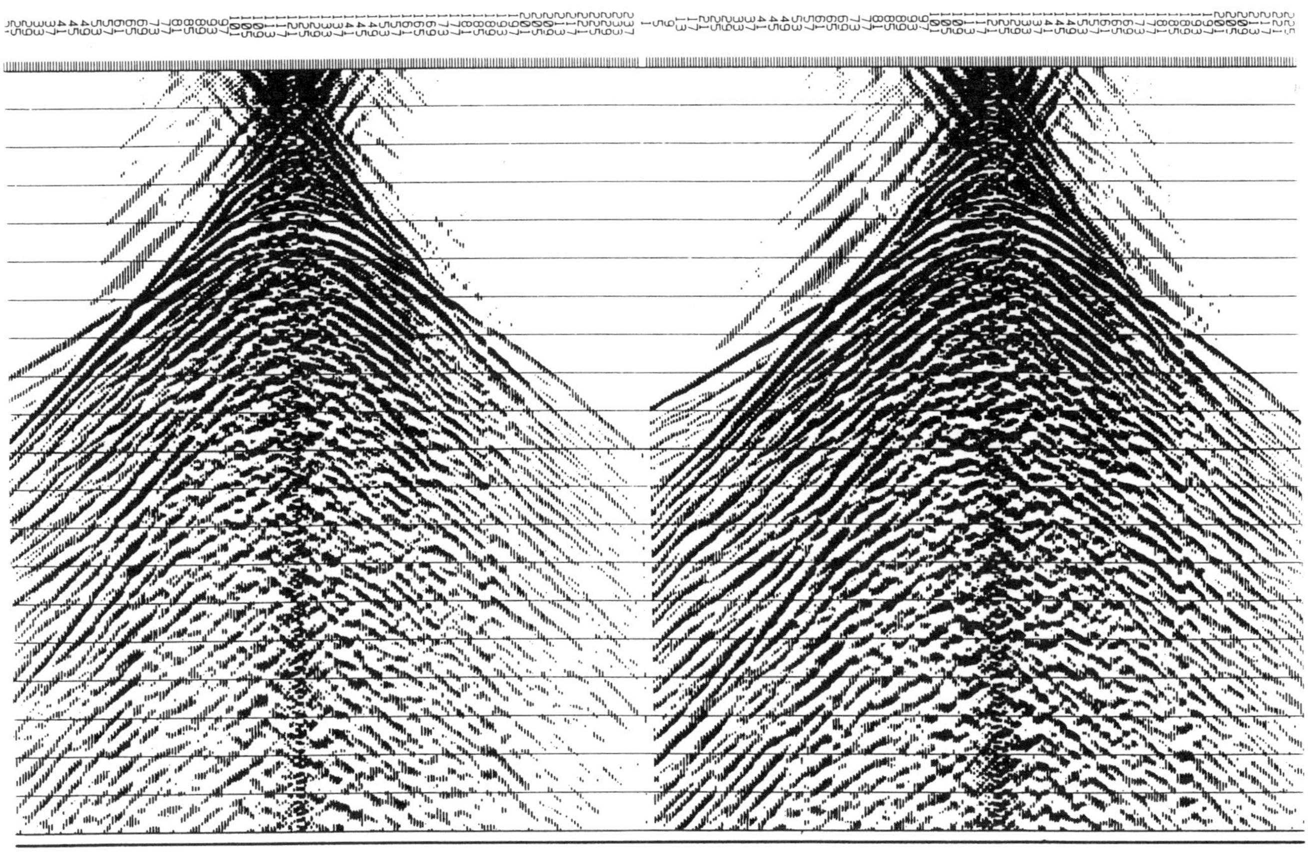

Fig. 34. Common-source land data (Figure 32) after ground roll suppression.

tion for deconvolution and wavelet estimation, which have long time windows. Exponential expansion retains amplitude variations, and it changes only their ratios over similar time windows. Furthermore, this expansion can be removed exactly and replaced by the spherical divergence correction at a later stage of processing. To illustrate the exponential decay analysis, Figure 35 shows the measured amplitudes of a selected shot record plotted in two-way traveltime versus logarithmic decay. A decay rate of 8 dB/s has been chosen from the plot and applied to the data. Figures 36a and 36b show the seismic data before and after the exponential correction.

Deconvolution Analysis and Application

All filtering processes, as well as single and multichannel deconvolutions (predictive or inverse) and wavelet-shaping filtering, are passive operators or filters. That is, they take away information and pass what is left over. Band-pass filters are described by limits of pass zones where suppressions are less than one-half power. Predictive deconvolutions subtract predicted portions from the data and output the rest, which is called the prediction error. The signal injected into the subsurface is an undesirably shaped pulse that must be rehabilitated. The shaping process, like the work of a sculptor, depends on the raw material available. For example, if a vibratory signal is generated within 10–40 Hz, then the shaped signal cannot contain information less than 10 Hz or more than 40 Hz. The final shaped signal must be within the original bandwidth. Since each filtering operation shaves part of the signal, each step of the processing must be planned ahead based on the content of the original seismic signal.

The generally accepted model of a seismic trace is the convolution model, represented here in both t- (time) and z-transform domains:

$$t\text{-domain:} \qquad f(t) = s(t) \times r(t) + n(t), \qquad (15)$$

$$z\text{-domain:} \qquad f(z) = s(z) \times r(z) + n(z), \qquad (16)$$

where the seismic trace f is the result of convolving the reflectivity series, r, with the seismic wavelet, s, and

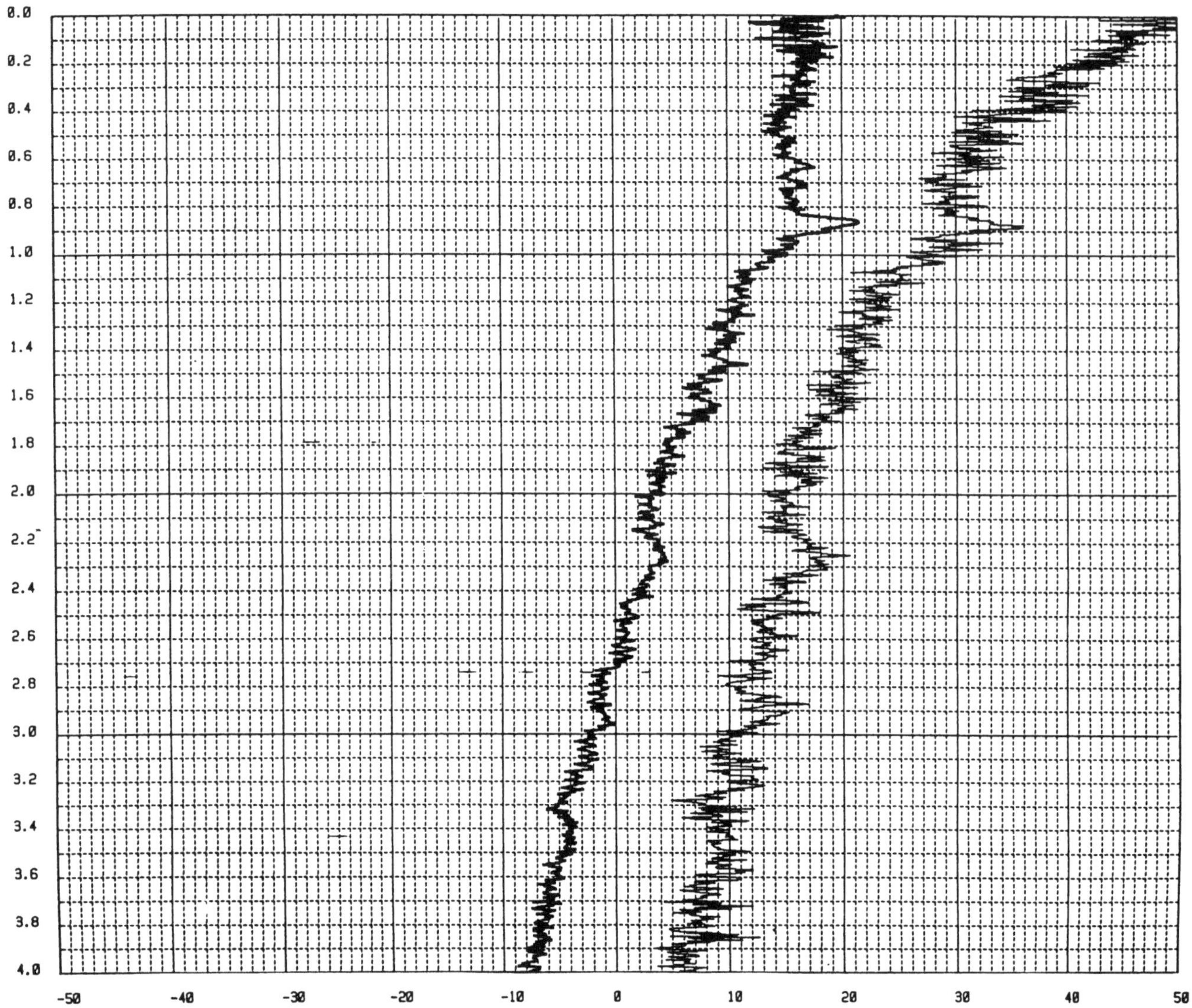

Fig. 35. Amplitude decay analysis showing amplitude (in decibels) versus time (in seconds). The fine line represents maximum values, and the heavy line represents the log-average.

adding noise, n. In practice, the seismic wavelet includes both the influence of the coupling of the seismic source and receiver with the earth and the multiples that are generated by the highly reflective near-surface layers. There are several types of multiples: surface reverberations and the longer lag multiples generated by deeper reflectors. Reverberations exist in both onshore and offshore seismic data sets. They are a form of multiple reflection, generated by the shallow layers with a lag close to the dominant frequency period of the seismic wavelet. These types of multiples elongate the seismic wavelet, but they are usually suppressed by conventional single-channel deconvolution. The second type of reverberation is generated during the coupling of the seismic energy source with the earth. The coupling reverberations are not clearly separable, but they naturally exist on all seismic records. They are not well understood and are usually neglected in wavelet processing, leading to erroneous results.

Theoretically, reverberations and multiples are infinitely long and are minimum-phase processes, hence they are best represented by their inverse, as an autoregressive process. Therefore, a more practical representation of the seismic trace is in the form of an autoregressive moving average (ARMA) model:

$$f(z) = \frac{w(z)\,r(z)}{d(z)} + n(z), \tag{17}$$

and identifying the rational filter as

$$s(z) = \frac{w(z)}{d(z)}, \qquad (18)$$

where $d(z)$ represents the reverberation generating process, $r(z)$ is the reflectivity series with long-period multiples, and $w(z)$ is the finite-length wavelet. Since the actual seismic data reverberations are stable and decay with time, they represent minimum-phase processes, hence their inverses are also minimum-phase processes. Consequently, reverberations can be removed by a minimum phase inverse filter such as the popular spiking deconvolution. Unfortunately, spiking deconvolution tries to remove not only the reverberations but considers the wavelet, $w(z)$, as part of the minimum-phase process and tries to shape the seismic wavelet into a spike. Since a spike cannot exist in sampled data, the resulting wavelet is arbitrarily shaped, although time limited. This wavelet contains phase rotations under the influence of the S/N of the data set and other unknown factors. These effects are thus time and space varying and cannot be undone later by any other process, even by surface-consistent analyses. Such results are detrimental to the stratigraphic interpretation because the wavelet phase carries stratigraphic information.

In mature areas with higher S/N, a short-spiking (80–120 ms long) deconvolution can be used without appreciable data deterioration. A short-spiking deconvolution mainly works on reverberation, reducing the main peaks of the amplitude spectrum by about 15–20 dB. This leaves the wavelet portion reasonably unaffected. However, in severe reverberation areas, this is not sufficient, requiring a second pass to be performed. For example, the shallow-marine record in Figure 37 shows severe reverberation contamination, which can also be seen on the autocorrelation display in Figure 38. A single pass spiking deconvolution produces the record and its autocorrelation given in Figures 39 and 40. Both figures indicate that the deconvolution has not suppressed the reverberation effectively, since they are still apparent on the output record. This record was specifically selected for its low S/N, which demonstrates that multiple-pass deconvolution is sometimes necessary.

A second pass of spiking deconvolution produces the record shown in Figure 41 and the autocorrelation in Figure 42. Here the effectiveness of the reverberation suppression is visible on both the seismic record and the autocorrelation. However, there is an obvious deterioration of the S/N. In theory, a second pass should not be needed, but in practice, this is necessary for a number of reasons. First, the operator is not exact from a computational point of view (e.g., the linear mean square, or lms, solution is simplified to a Toeplitz matrix, the autocorrelation is not correct, and white noise is added to make it stable). Second, the data are assumed to be stationary, and third, computer words have a limited number of digits, which controls the dynamic range of the processes. Therefore, each application reduces the reverberation to some degree, and the results and the autocorrelation functions show us how successful the suppression has been.

The same reasoning can be used for predictive deconvolution. However, predictive deconvolution does not shape the portion of the wavelet within the prediction lag. Figures 43 and 44 show the deconvolved record and its autocorrelation after applying an operator with a prediction distance of 80 ms and a length of 240 ms. Since operators are basically designed for reverberation suppression, the results are more effective than spiking deconvolution (compare with Figures 39 and 40). The results of the second pass of predictive deconvolution are given in Figures 45 and 46. Both the seismic data and corresponding autocorrelation show an appreciable degree of reverberation suppression. We also see that the S/N has not deteriorated as much as with spiking deconvolution (Figures 41 and 42).

The use of spiking deconvolution (a commonly used process in the industry) versus predictive deconvolution is a matter of S/N, and to decide which to use is a matter of personal choice. In the previous example, both types of deconvolution were applied, and the reasons for preserving the signature portion of the reverberation were given. In practice, whether spiking or predictive, deconvolution is run with parameters left over from previous generations. The use of autocorrelation is practically forgotten. This may be acceptable in some quick production processing, but in difficult areas such as shallow carbonates and volcanics, a study of all processes is important.

Wavelet Analysis and Shaping

The objective of wavelet processing is to remove the space-varying effects caused by the source, the source coupling with the earth, near-surface filtering on the down- and upgoing wavefront, the receiver array and coupling, and the recording instruments. If this can be achieved, then the time- and space-varying effects are due solely to the geology and lithology of the subsurface. The principal part of the process is capturing the seismic wavelet. Ideally, the best way to capture the wavelet is to record it in the field. It is definitely desirable to record the source signature in the field, particularly for complex signatures such as the vaporchoc and water gun. The practical recording is of the near-field

(a)

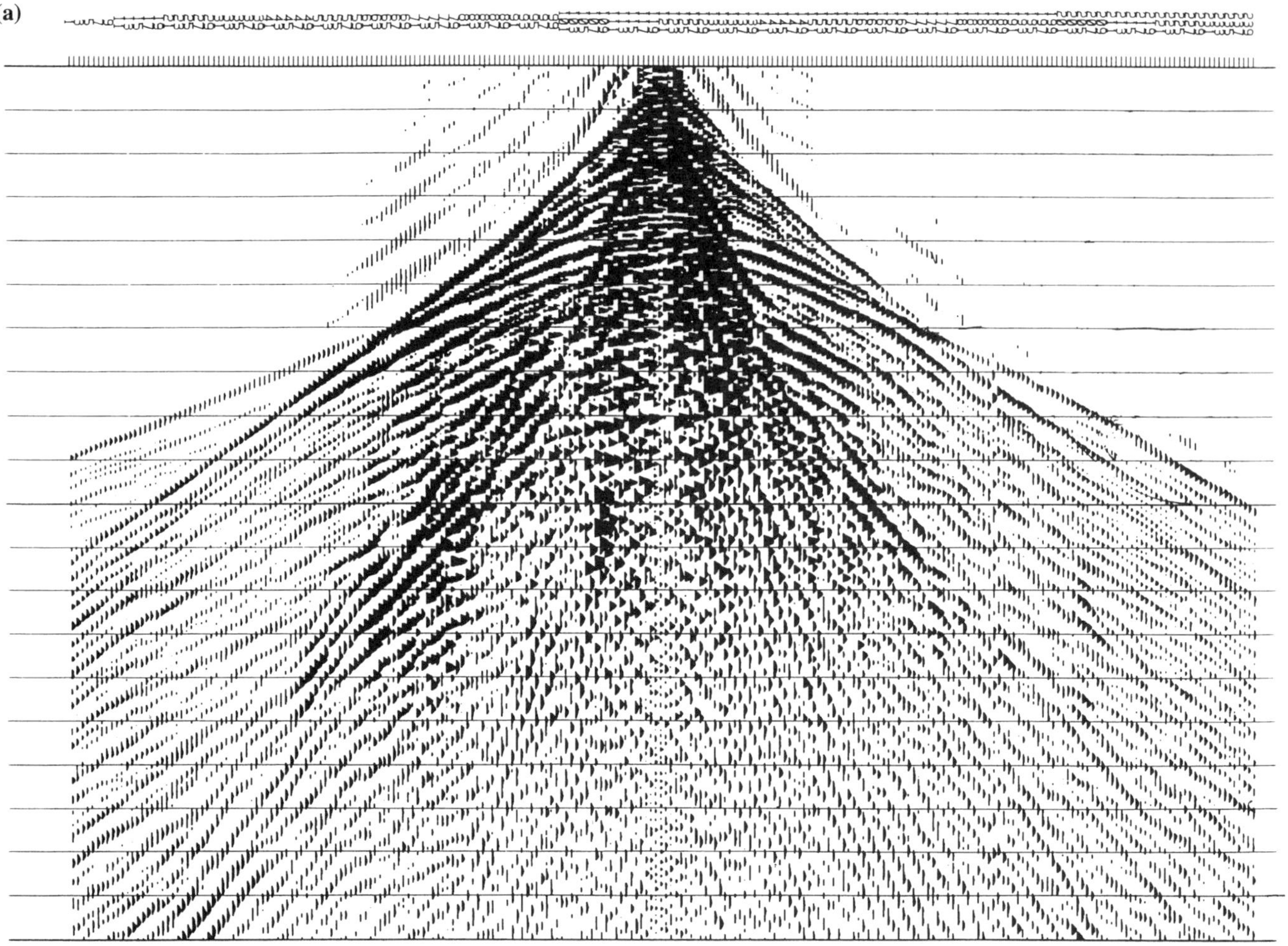

Fig. 36. Exponential amplitude correction example. (a) Common-source data after surface-consistent amplitude correction but before exponential amplitude correction. (b) Common-source data after exponential amplitude correction. ***(see part b on facing page)***

signature, from which the far-field signature is estimated and used for wavelet shaping. In some instances, the far-field signature for a particular gun combination is recorded in deep water, which is used as the reference signature for the boat. Recording the seismic signature is more complicated on land. When a vibratory source is used, the base plate monitor can give an estimate of the phase of the outgoing sweep, which is helpful during correlation. However, the receiver array response must also be recorded to suppress or shape their effects. At present, there is no viable method to do this, thus statistical methods are needed to fill the missing parts.

Wiener filters are designed to shape field-recorded signatures into desirable wavelet shapes in a deterministic manner. Since they contain only the outgoing wavelet estimate, the output is chosen as a minimum-phase wavelet, which retains the causality of the resulting wavelet. This produces a wavelet shape that is suitable for use in statistical methods.

Four separate factors generally affect the shape of the seismic wavelet. The first is the source signature, which is the source waveform itself plus the coupling with the near surface. The second factor is on the receiver side and includes individual geophone responses, array response, the near surface around the receiver array, transmission lines, and recording instrument responses, which are summarized in the receiver response. Offset-related effects comprise the third factor, which includes all offset-related initiation and reception angle dependent factors and marine hydrophone array responses. The fourth factor is the subsurface response, which is assumed to be the CMP constant response.

(b)

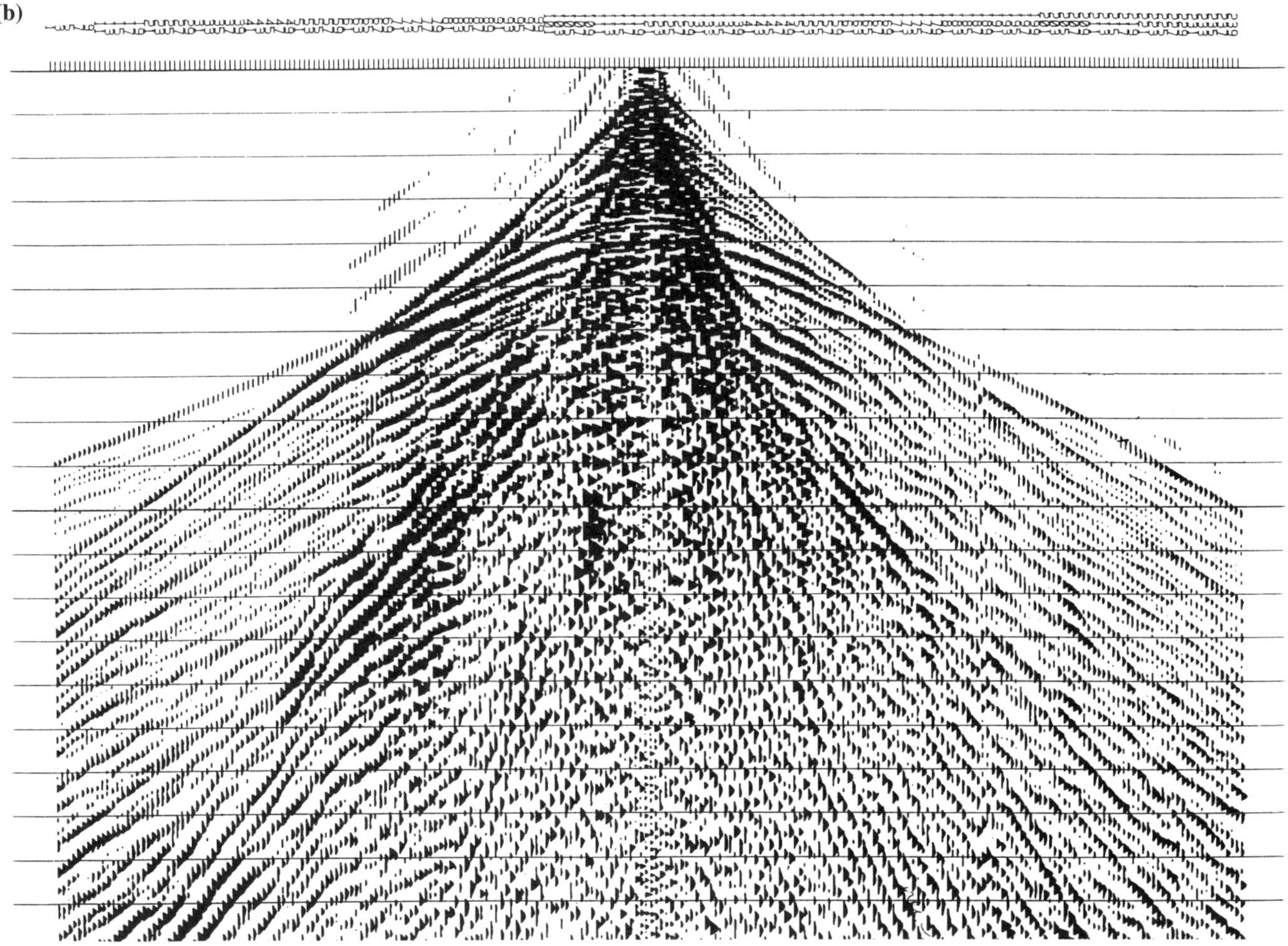

Surface-consistent processing attempts to eliminate all surface initiation and reception responses and output seismic sections containing only the subsurface-related responses from which geologic and lithologic parameters are computed. The assumption of surface consistency considers that all common shot data have the same shot-related responses; similar assumptions are made for common-receiver and common-offset traces. In the processing sequence, the source signature is first estimated and shaped into a minimum-phase wavelet since the source signature contains the most effective source pulse. Receiver and offset-related effects are secondary to the source wavelet. The second pass is on the common-receiver traces to balance the spatial variation of receiver array responses. A shot was selected to demonstrate the wavelet analysis. A zero-phase 8–80 Hz bandwidth was chosen with an 18 dB/octave roll-off rate on both sides. The time and frequency responses of the selected wavelet are shown in Figure 47a. The low cut of 8 Hz with 18 dB/octave is usually used in the field as a standard filter, which is why it was chosen here. We must finalize the wavelet selection after the analysis when the bandwidth of the actual signature is determined.

The wavelet is estimated using a special iterative technique that is at the core of the statistical program used for wavelet shaping. The method forces the actual unknown wavelet into a minimum-phase condition by successive testing of exponential decay curves. Once the wavelet is judged to be minimum-phase, the original mixed (nonminimum) phase wavelet is reliably determined with maximum stability. Wavelets estimated with respect to decay rates of 1.0–0.98 are shown in Figure 47b. The figure shows that the correlation coefficient between decay rates of 0.994 and 0.992 is equal to 1.0, that is, they are essentially identical. This indicates that the decay rate of 0.994 induces the wavelet into a minimum-phase wavelet. Once the wavelet becomes minimum-phase, then it can be estimated by several conventional means.

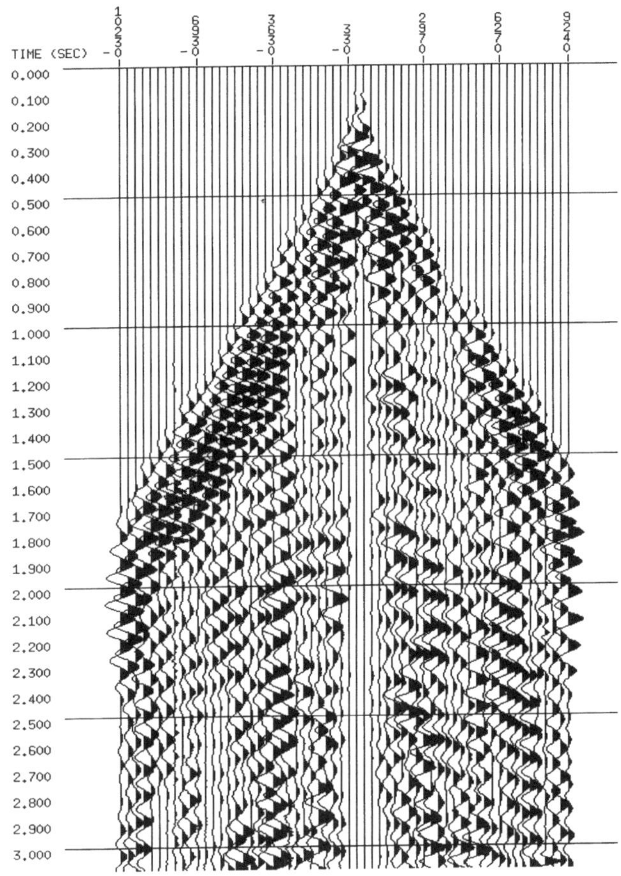

Fig. 37. Shallow-marine common-source seismic data showing heavy ringing.

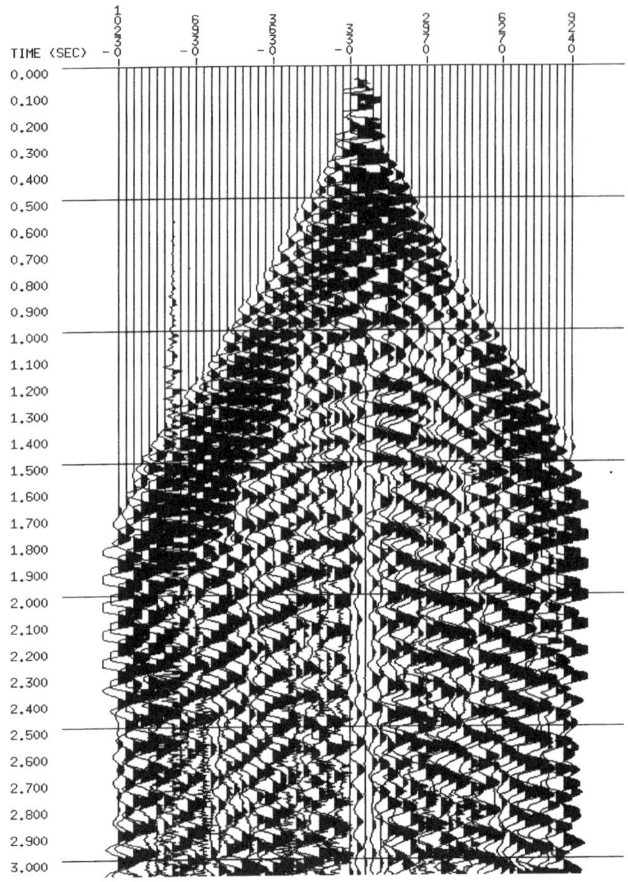

Fig. 39. Output of first-pass spiking deconvolution of Figure 37. Deconvolution operator length is 240 ms.

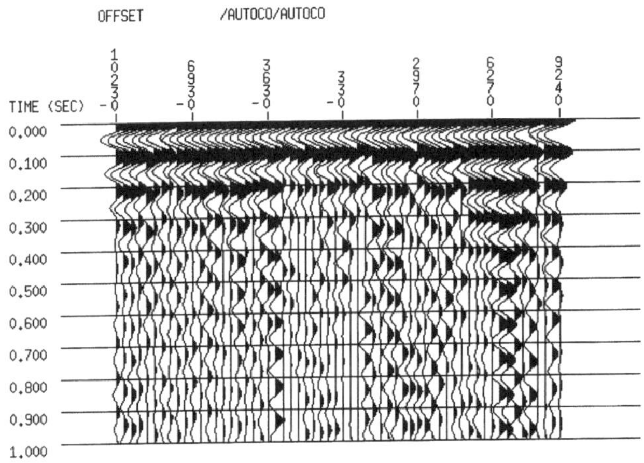

Fig. 38. Autocorrelation function of seismic data shown on Figure 37.

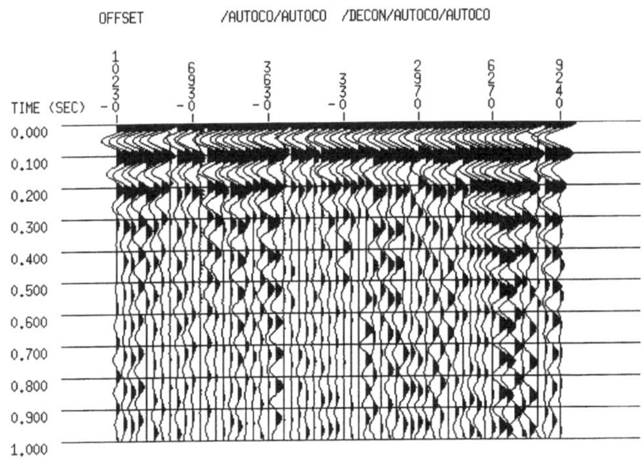

Fig. 40. Autocorrelation of deconvolved output shown in Figure 39.

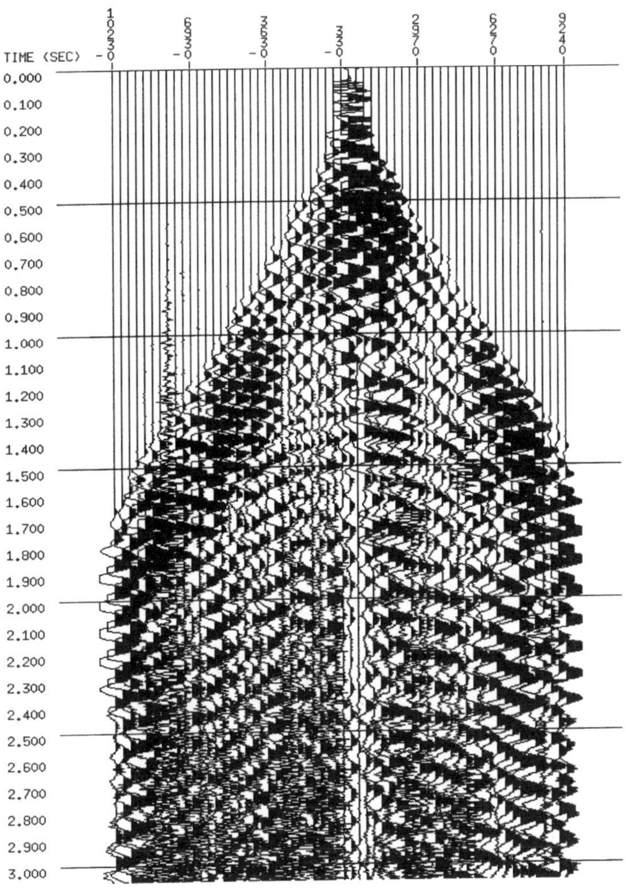

Fig. 41. Second-pass spiking deconvolution with same parameters as in the first pass.

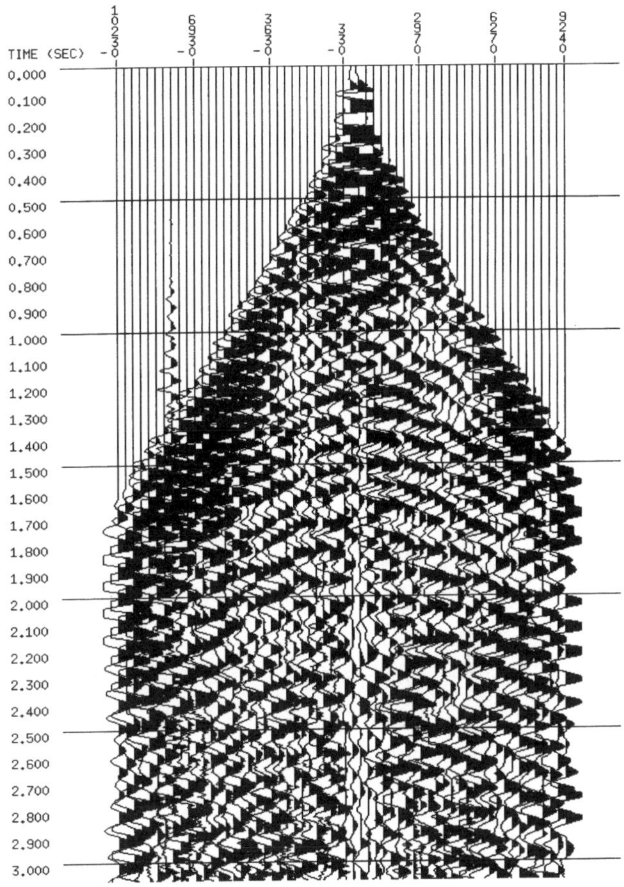

Fig. 43. First-pass predictive deconvolved output of original input data shown in Figure 37. Deconvolution operator has 80 ms of prediction lag and 240 ms of total lag.

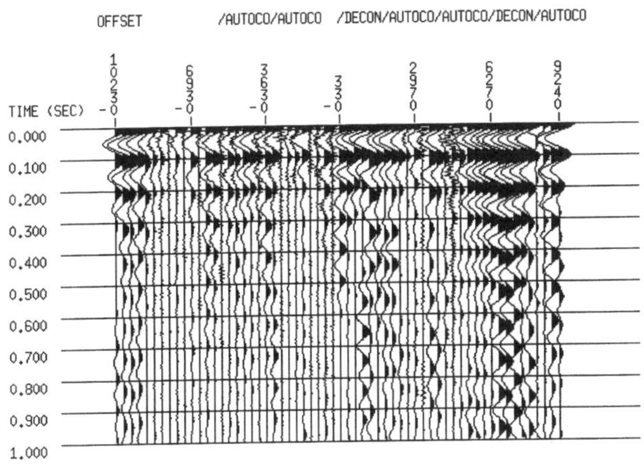

Fig. 42. Autocorrelation of data after second-pass deconvolution shown in Figure 41.

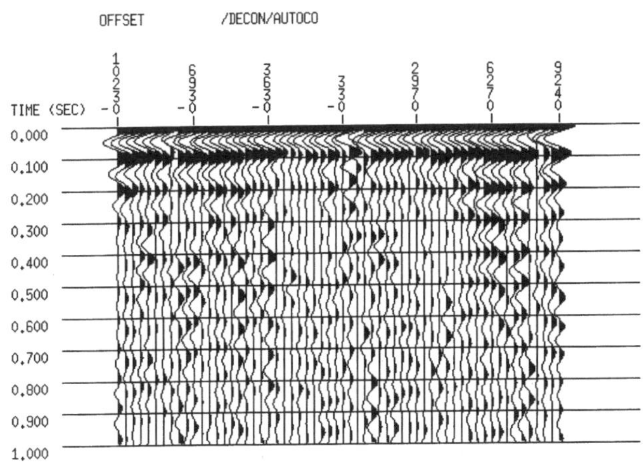

Fig. 44. Autocorrelation function of first-pass predictive deconvolved data shown in Figure 43.

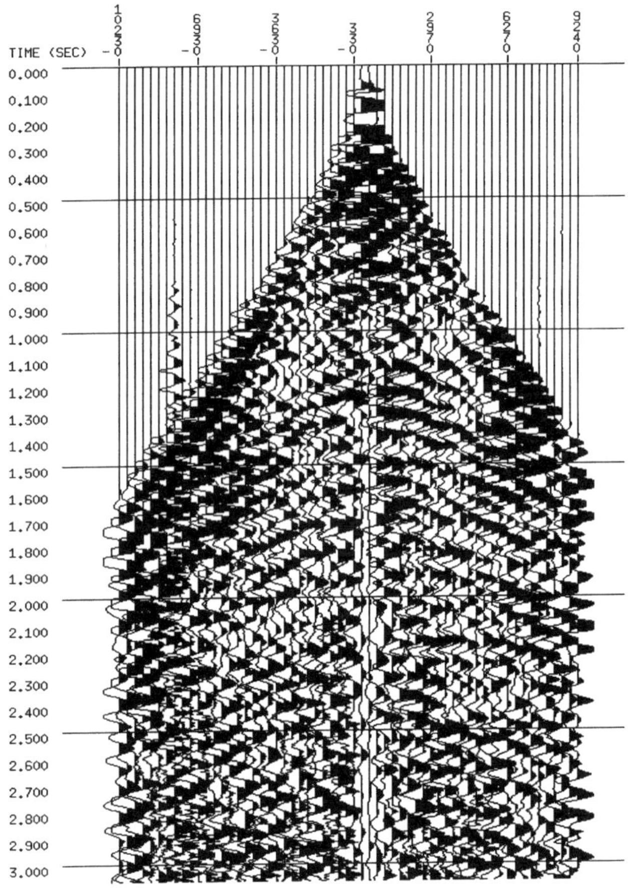

Fig. 45. Second-pass predictive deconvolved data with same parameters as in the first pass.

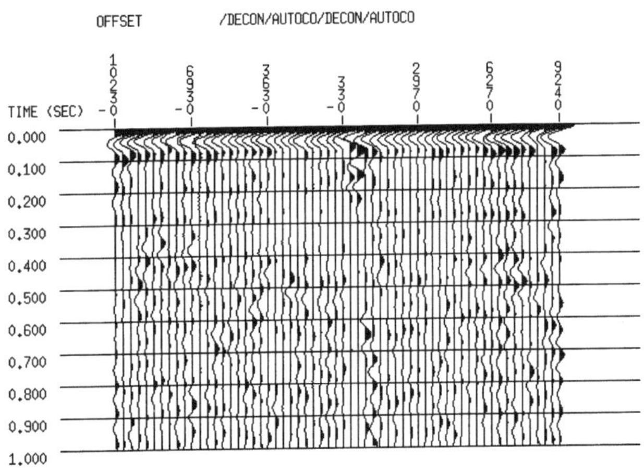

Fig. 46. Autocorrelation function of second-pass predictive deconvolved output shown in Figure 45.

The statistically estimated wavelet is then used to design a Wiener-shaping filter that can induce the computed mixed-phase seismic wavelet into the desired shape. Figure 47c shows the chosen wavelet, the corresponding Wiener-shaping operator, and the result of its convolution with the input wavelet (the actual output). The computed wavelet shows low coherency on the high side. Similar ripples on the high side of the actual output indicate possible noise contamination of the final output data. All the filters, including deconvolution, are passive filters, that is, they do not contribute anything new; they can only suppress the data in varying amounts. In other words, a filtered output will be equal to or less than the input. Thus, if the data have a poor S/N over some frequency bands, it cannot be improved within that band by filtering. Therefore, balancing the spectrum must be done with respect to the lowest level of the spectrum; all other parts are brought down to that level. If we want the output to have a 8–80 Hz bandwidth, then we must reduce all higher energy level bands and shape them down to the 80-Hz level of the original data. Consequently, the output will be noisier.

Figure 45 shows the input data (after predictive deconvolution), and Figure 48 shows the wavelet-shaped output data. These figures indicate the dominant high-frequency noise levels on the output data. If one is striving for a wide band output, these results may be accepted and processing can proceed. Redundancy of the data is the only help in estimating and suppressing the noise. Therefore, high-fold stacking improves the chances to obtain a better S/N on the resulting section. However, in this case, more accurate estimates of time statics and velocities must be made because any small error destroys the high-frequency side of the wavelets and the wide band advantage with it. If the attempt to use the high-fold coverage is unsuccessful, then the desired output must be designed to use the coherent energy contained in the source wavelet. This is particularly a problem for vibratory sources.

For example, if the source sweep is 10–60 Hz, the desired output must fit within these input limits. For a 10–60 Hz sweep, the vibrator starts at 10 Hz and gradually builds its amplitude and frequency up to 60 Hz. As it approaches the end of the sweep, the amplitude is gradually tapered down. The half power points cannot be at 10 and 60 Hz because these are the starting and ending frequencies; thus, frequencies between these limits must be chosen. For example, we can experiment with a wavelet with a 16–48 Hz bandwidth and look at the results. If the output in Figure 48 is not acceptable, then a wavelet more suitable to the input wavelet spectrum shown in Figure 47c can be used. Figure 49 shows

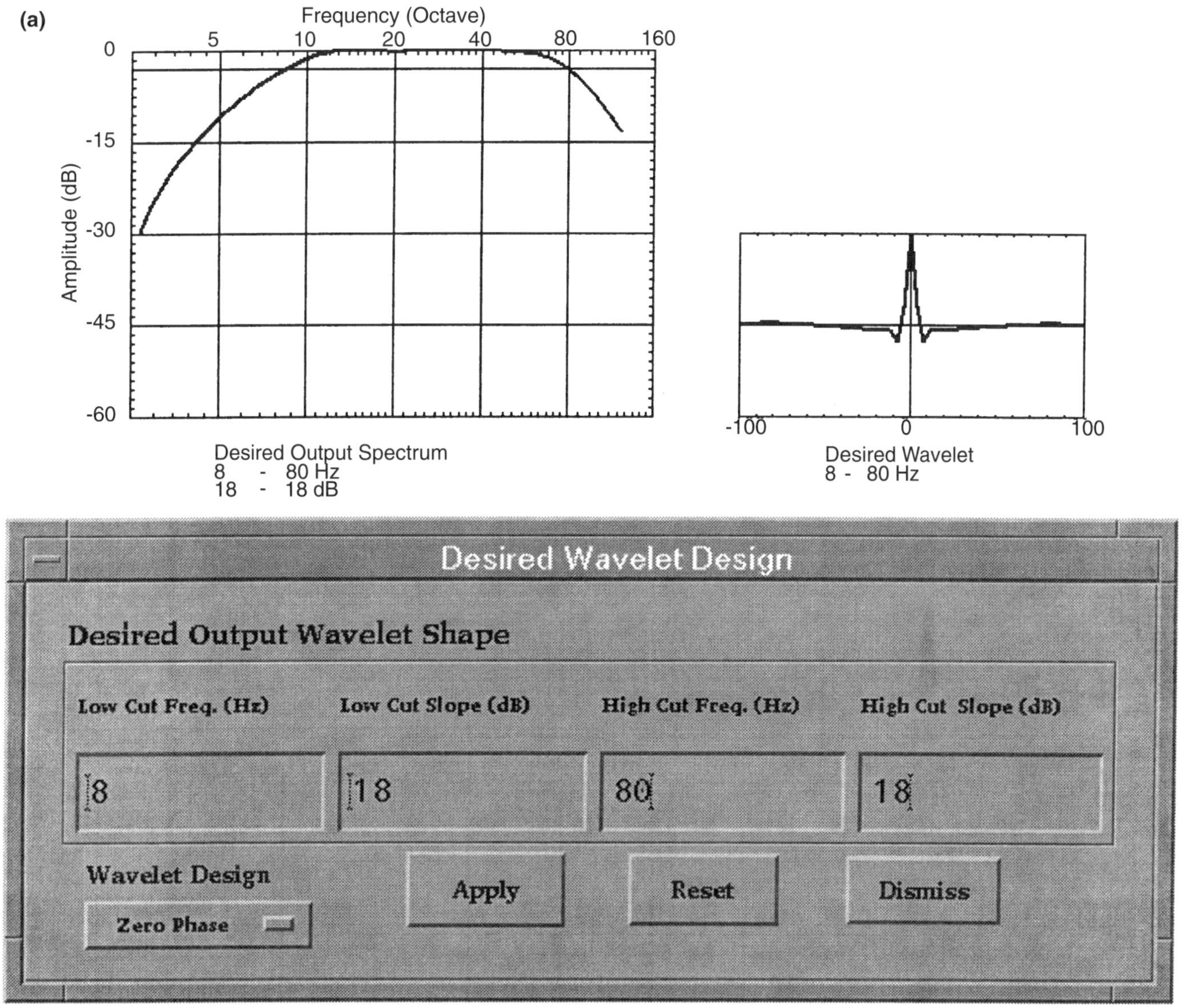

Fig. 47. Wavelet analysis plots. (a) Desired wavelet parameters and plot in time and frequency domain responses. (b) Computed wavelets with respect to different exponential decay parameters. Note that the correlation coefficient between decay rate 0.994 and 0.992 is 1.0 and remains 1.0 thereafter. (c) Selected wavelet, shaping filter, and frequency domain responses. Computed wavelet shows noise undulations above 40 Hz. At the top are the amplitude spectra of the computed mixed-phase wavelet and the final shaped wavelet. Spectra graph scales are decibels on the amplitude axis and octaves on the frequency axis. *(see parts b and c on next two pages)*

the experiment with the desired output wavelet of 8–36 Hz with 18 dB/octave roll-off rates. In this case, the actual output is a much closer fit to the input spectrum, and the processed output (Figure 50) shows more continuity and less noise contamination.

Long-Period Multiples Analysis and Suppression

Because carbonate layers constitute high-velocity zones that contrast with weathered layers or with lower velocity sandstone and shale sequences, they are bounded by highly reflective interfaces. These interfaces are the natural source of large-amplitude multiples, which overpower the weak primary reflections arriving from deeper interfaces. In marine cases, the surface of the water joins the highly reflective carbonate interfaces to create surface-connected multiple and complex interbed multiples.

A number of approaches can suppress multiples, but none of them work universally. The method of predictive deconvolution is used to suppress short-period

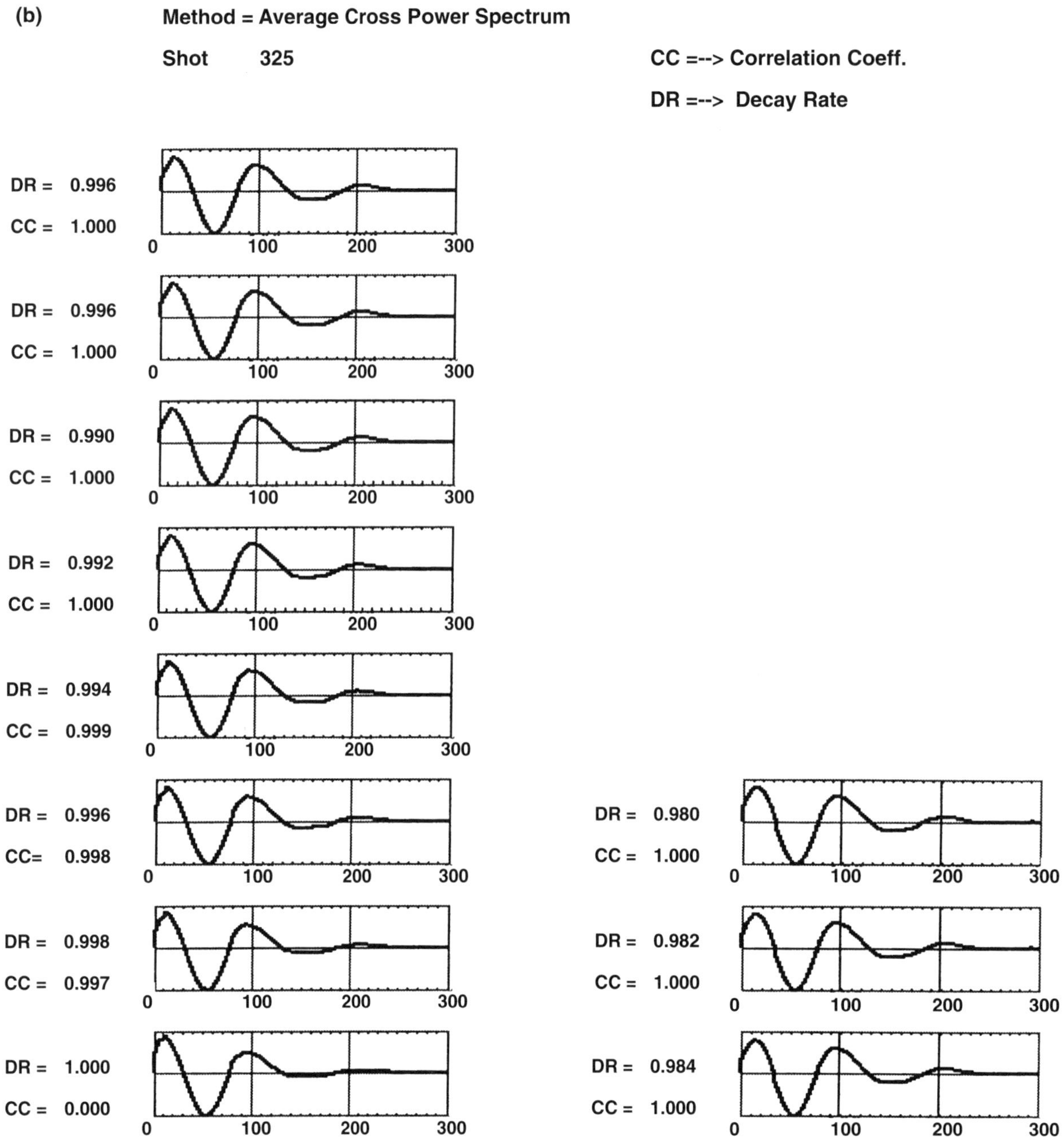

(<150 ms) multiples and is a single-channel process. Multiples with periods of >200 ms cannot be predicted by a single-channel process. Multiples are recordings of expanding spherical wavefronts by stationary geophone arrays, hence they do not arrive at equal time increments. As indicated in Figure 51, multiples can be modeled as if they were generated by point sources located at their reflected image positions with respect to the multiple-generating layer. Therefore, a point source at the surface can appear at depth in multiples of twice the water depth, as shown in the figure. A receiver located at location G records arrivals from sources located at points P, M1, M2, etc. As apparent from the geometry, the distance from M1 to G is not twice the distance from P to G. None of the multiple-generating locations are multiples of the distance from P to G, which is the primary arrival time. Only the vertical line (equivalent to the zero-offset trace in this model) has multiples arriving at equal time increments. All others have gradually decreasing time intervals.

For short-period multiples, these differences remain sufficiently small to be handled by ordinary deconvolu-

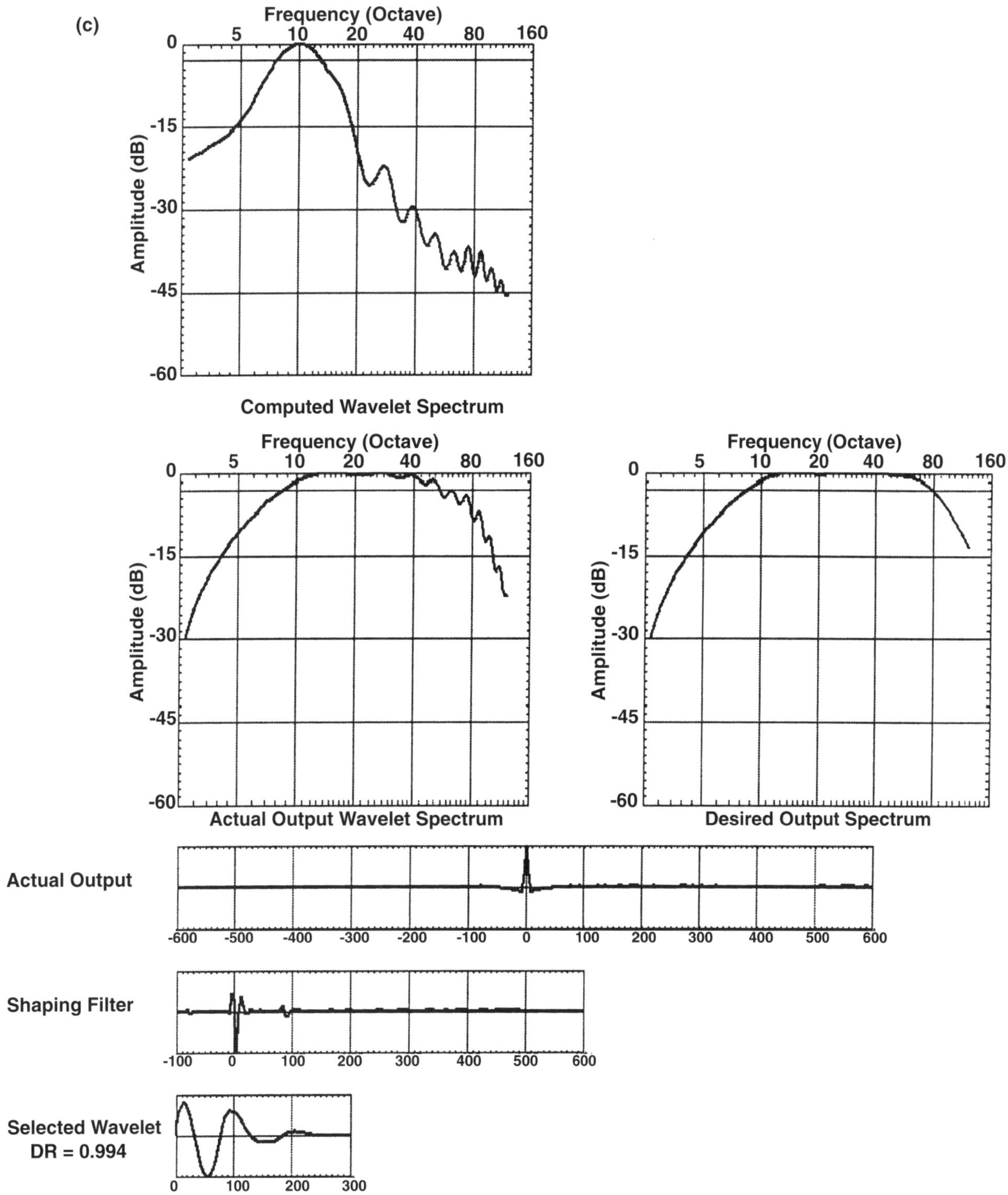

tion methods. However, for multiples with longer periodicity, this approximation of single-channel periodicity cannot be used. I demonstrate this by the use of autocorrelation functions. Multiples must be identified to be suppressed, which is done by autocorrelation functions. All events appearing on the autocorrelation functions represent events appearing periodically on the original seismic traces. Figure 52 shows the single-channel autocorrelation function of Figure 54a. Near offset is to the right of the figure, where several multiples are present

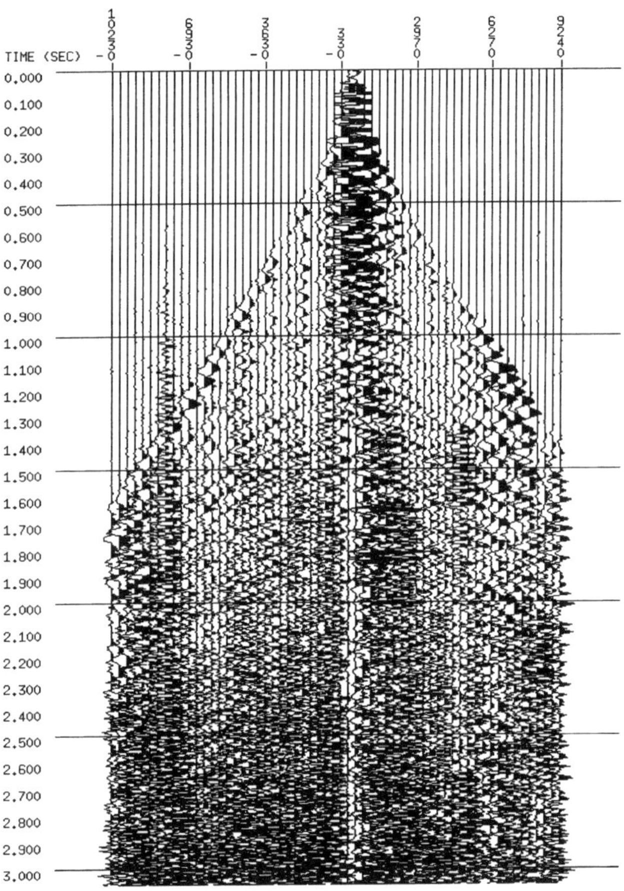

Fig. 48. Wavelet-shaped output data. Desired output wavelet is 8–80 Hz with 18 dB/octave roll-off rates. High-frequency noise can be predicted from the computed wavelet spectrum (Figure 47c).

and repeat at a periodicity of ~450-ms. The periodicity and number of multiples decreases toward the longer offsets (to the left). This indicates a lack of periodicity with offset.

If we assume that the beds causing the multiples are essentially horizontal and cylindrical with respect to the recording line, then the wave propagation can be simplified to a 2-D environment. Hence, we can investigate the multiples by a 2-D autocorrelation function (Figure 53). The vertical axis Figure 53 represents the time lag (as in 1-D autocorrelation function), and the horizontal axis is the spatial lag (as in time lag). It shows the spatial shift during the computation, and the resulting autocorrelation shows 2-D periodicity. A comparison of Figures 52 and 53 reveals that multiples prominently appear on the 2-D autocorrelation function. 2-D predictive deconvolution is an extension of 1-D deconvolution into the 2-D domain, where a set of multichannel operators are determined that minimizes the difference between the predicted and the actual traces in a least squared error manner. Note that the difference is the prediction error, thus the minimization process produces output traces with least-squared amplitudes or the least total power. The absolute minimum could be the zero output trace. This is not our objective and can occur in multichannel cases where the number of operators are equal to the number of samples in the minimization window. It can also happen if the input data are completely predictable.

Figure 54a shows the original input data and Figure 54b the results of 2-D predictive deconvolution. Figure 54c is the autocorrelation function of the output, which shows considerable multiple suppression (as in Figure 53). It also shows the possible interbed multiple corresponding to the water bottom and the base of the high-velocity layer. Figures 55 and 56 show the stack sections produced before and after long-period multiple suppression by 2-D predictive deconvolution.

One of the most promising ways to suppress multiples is to change the point source records into synthetic line source records, thus simulating multiples due to expanding cylindrical wavefronts (plane waves in 2-D) instead of expanding spherical wavefronts. Plane wave sections have less spatial distortions, so the multiples can be more easily predictable by single-channel procedures.

Time, Amplitude, and Phase Statics: Analysis and Correction

As outlined earlier, statics computations and corrections are carried out in three stages. The first stage involves refraction and short-wavelength surface statics computation and correction. In the second stage, reflection statics are computed in a surface-consistent manner, and in the third stage, local trim statics are computed and applied to improve the coherency of the final section.

In the second stage, reflection statics are usually applied twice, each followed by a new velocity analysis. This is required because of the interaction of velocities and statics. The use of the term *statics* (in contrast to *dynamics*) implies that all of the processes applied are time constant. The time statics are time shifts applied to the data to correct the time delays caused by the varying thicknesses and velocities of the shallow layers. Time shifts are justifiable if the layer being corrected is low velocity. In this case, the travel path through the low-velocity layers is nearly vertical and thus constant for all reflections. If the offending near-surface layer is high velocity (e.g., carbonate), then the travel path is slanted, such that a time shift corresponding to a vertical path

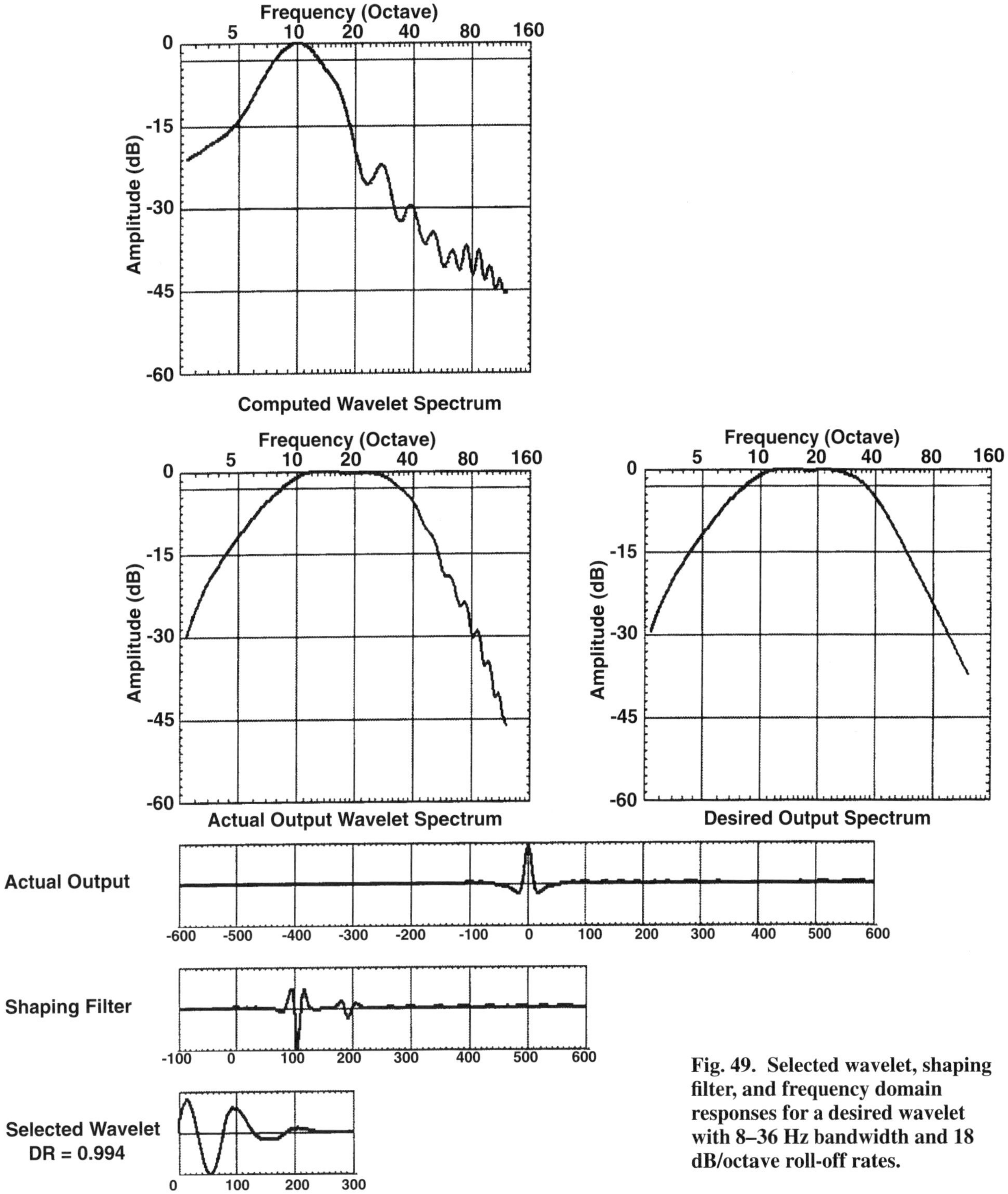

Fig. 49. Selected wavelet, shaping filter, and frequency domain responses for a desired wavelet with 8–36 Hz bandwidth and 18 dB/octave roll-off rates.

will not be correct. This situation also occurs in permafrost areas where the velocity in the permafrost is much higher than in the sandstone and shale layers below. In both of these cases, simple time shifts do not produce the necessary correction.

Data having residual reflection statics produce erroneous velocities. The amount of error depends on the magnitude of misalignments due to time statics. The velocities computed initially from the data serve as the first estimate and are used in the computation of the residual statics, considering the effect of the residual NMO. Resulting statics are then applied to the original data after which a second set of velocities are computed. Velocities computed from data with excessive statics show

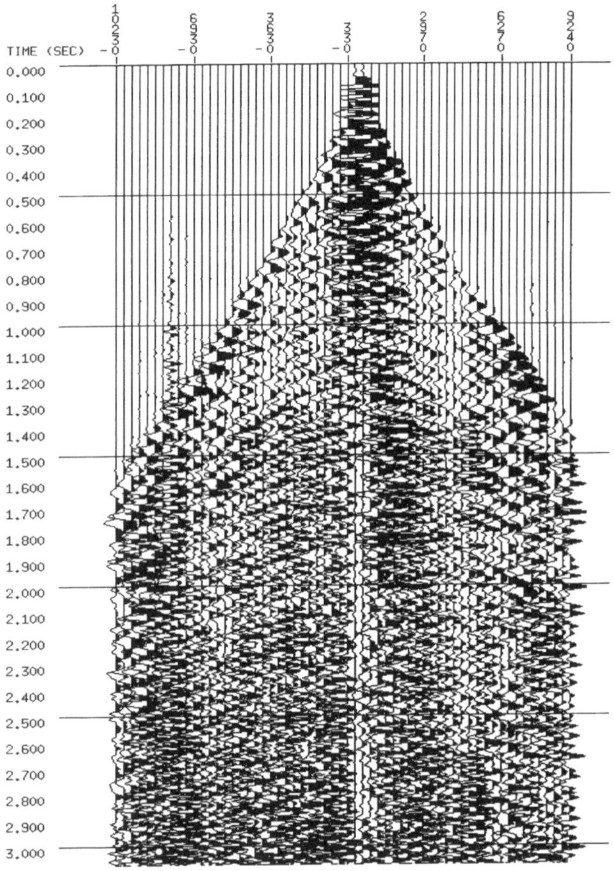

Fig. 50. Wavelet-shaped output with 8–36 Hz bandwidth desired wavelet.

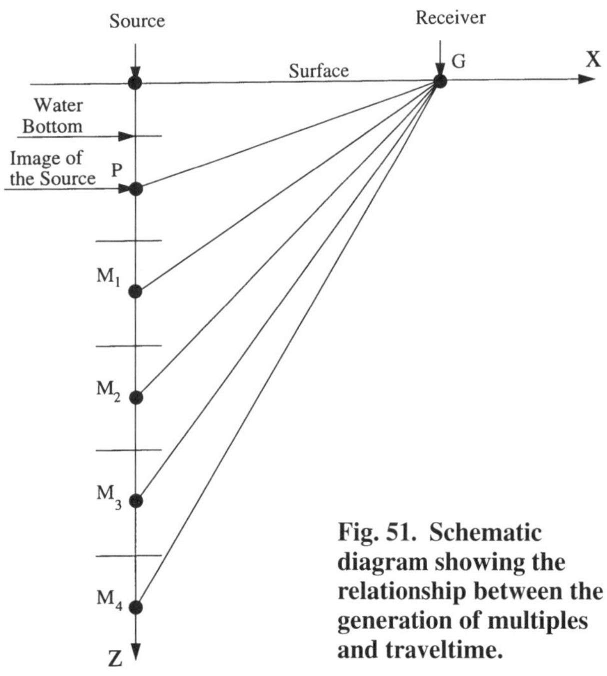

Fig. 51. Schematic diagram showing the relationship between the generation of multiples and traveltime.

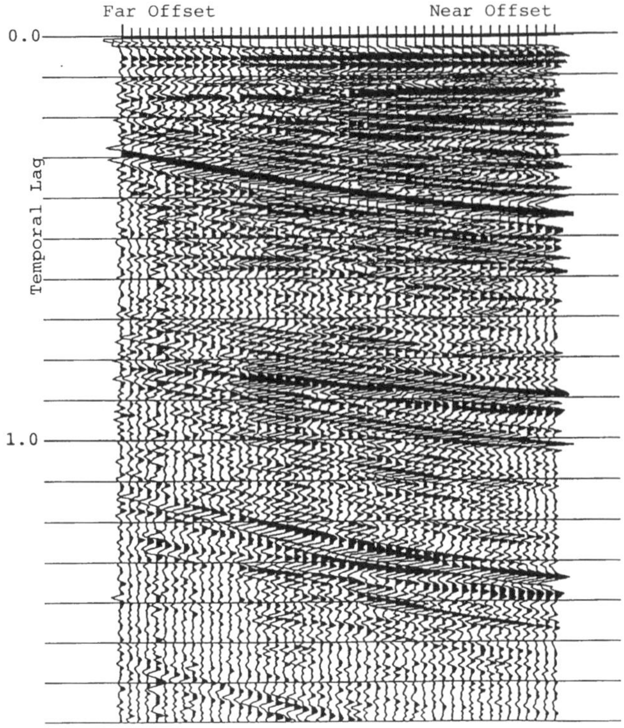

Fig. 52. Single-channel or 1-D autocorrelation function of the seismic record shown in Figure 54a.

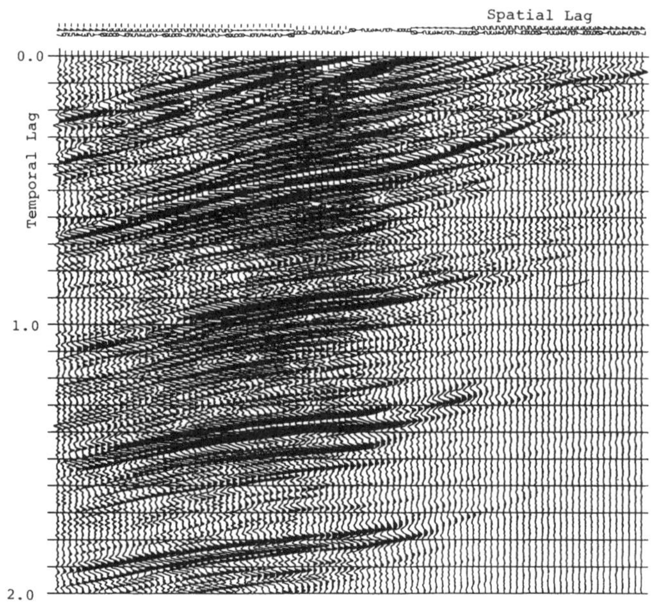

Fig. 53. 2-D autocorrelation function of data shown in Figure 54a.

large spatial variations in velocity. Ultimately, this recursive procedure converges to a minimum residual on successive passes and the iteration can be terminated with these final statics and velocities used in the final stacking. Another problem is zones of large topographic changes. Since corrections to a horizontal datum represent large time shifts, they cannot accommodate the proper transfer of source and receiver positions (from the wave propagation point of view) from the corrugated surface to the datum. In such cases, the best choice is to use a smooth floating datum that follows the surface and correct the data to it. This results in smaller corrections that correspond to the short wavelength undulation of the surface. Final correction to a horizontal datum can be applied to the resulting final stack section. If a time- or depth-migrated section is needed, then the imaging process can start from the smoothed floating datum, thus avoiding the approximate correction to the datum by simple time shifts.

The most important static corrections on land data are refraction and surface static corrections, which are computed from the picked first arrivals. In shallow carbonate areas, the carbonate layer itself represents the top of the refracting zone. It is covered by the weathered layer of low-velocity material that causes a considerable amount of time statics contamination. Fortunately, the first-break times contain both the long-wavelength refraction and the short-wavelength surface statics. The only problem is the quality of the first breaks themselves. Noise obscures the actual first-break times. I have found the filter panel analysis to be a good tool to determine the frequency band that best shows the first breaks.

Normally, correlated vibratory data first breaks have interfering precursors. Theoretically, since the data were correlated to a zero-phase Klauder wavelet, they have a precursor and tail. The coupling reverberation makes this precursor extend a considerable distance in front of and following the first-break area, which makes the first-break picking difficult. Long tails and precursors are an indication of peaks on the amplitude spectrum of the correlated wavelet. Spectral balancing of the first-break area reduces the peaks and, consequently, the precursor contamination. If good sweep monitors are available, the data can be correlated into a minimum-phase wavelet, reducing the precursor contamination to a minimum.

Figure 57a shows the first-break area of raw vibrator source data. Spectral balancing has helped to reduce the precursors (Figure 57b). Figure 58a shows shot records with noise and surface statics contamination, and static corrected data in Figure 58b show increased lateral continuity.

Most of the methods assume a simplified model of the near surface; a high-velocity refractor overlain by a low-velocity zone continues up to the surface. This model does not consider a surface-weathered layer, which is the main cause of short-wavelength statics. A more accurate surface model divides the near surface into two zones. The first zone, the weathered layer at the surface, consists of low-velocity material. It is assumed that its surface and base have varying undulations (Figure 59) and that its velocity varies spatially. Field crews measure elevations and the coordinates of all source and receiver locations, thus the elevation of the top of the weathered layer is known. The thickness and traveltimes through the layer are computed from the short-wavelength portion of the first-break pick times. The layer below the weathered zone is more consolidated and continues uniformly down to the refracting layer. In the computational stage, it is often assumed that this layer and the top of the refracting layer vary uniformly in the spatial direction, both in thickness and velocity.

Picked first breaks contain both long- and short-wavelength undulations. Due to the expanding Fresnel zone of waves arriving from the deeper refracting surface, only the longer wavelength undulations of the first-break times can be attributed to the refracted first arrivals. Thus, we assign the long-wavelength portion of the first-break picks to the refractor statics computation and the short-wavelength portion to the near-surface statics computation. The general model (Figure 59) is defined as follows:

$$\begin{aligned} t(n,m) &= S(n) + R(m) \\ &\quad + \sum [\sigma(k)\,\delta(k)] + s(n) + r(m) \end{aligned} \tag{19}$$

where $t(n,m)$ is the picked first-break traveltime for shot and receiver locations, n and m, respectively; $S(n)$ is the refractor delay time at the source location, n; $R(m)$ is the refractor delay time at the receiver location, m; $\sigma(k)$ is the slowness at the cell, k, located along the path between the source and receiver locations; $\delta(k)$ is the subpath length in the k-th cell of the path between source and receiver locations; $s(n)$ is the surface short-wavelength statics at the n-th source locations; and $r(m)$ is the surface short-wavelength statics at the m-th receiver location.

This model depends only on the source and receiver positions, thus it can be used for all shooting geometries (2-D, 3-D, and crooked line). Figures 60a and 60b show

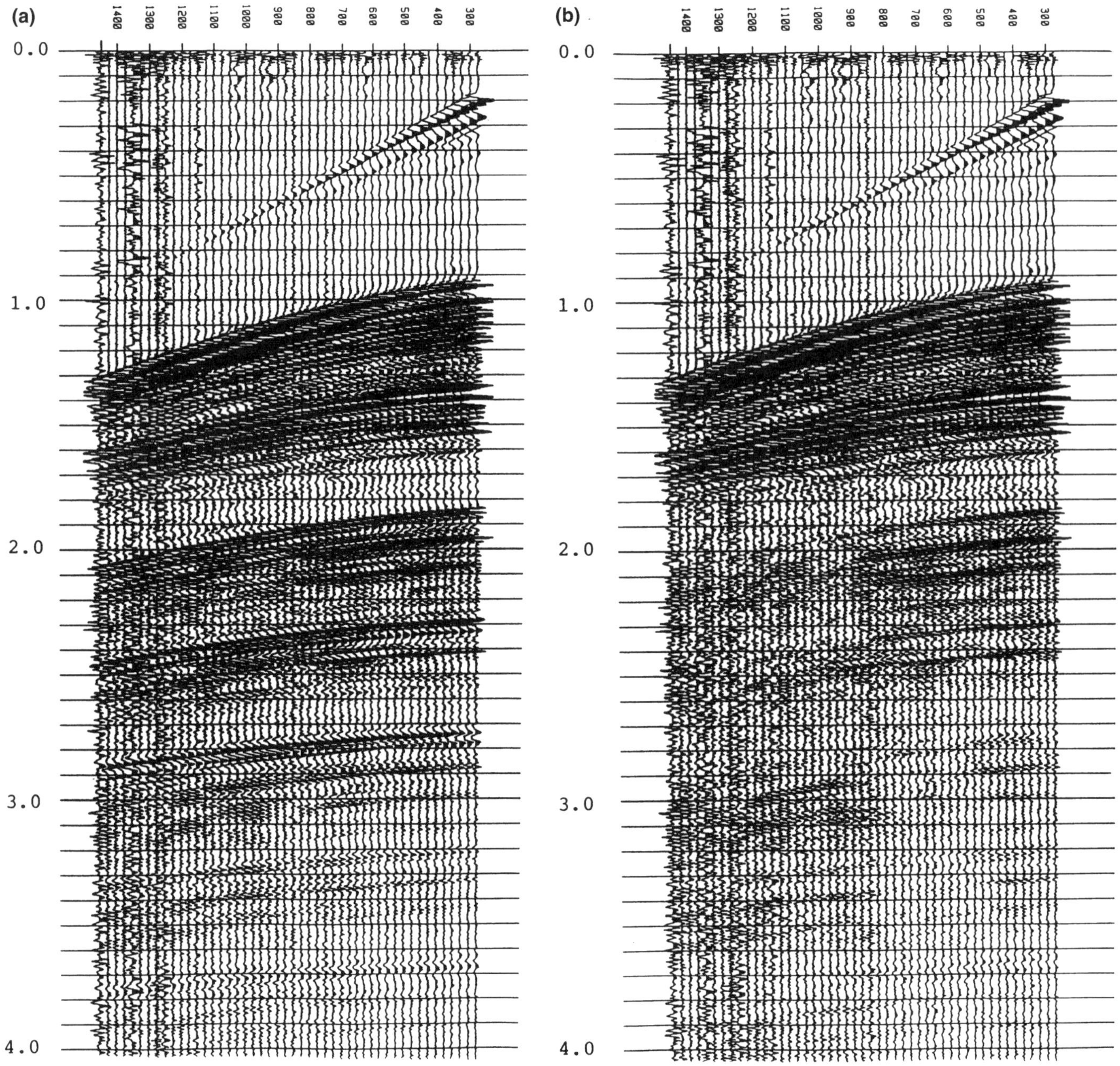

Fig. 54. Long-period multiple suppression example. (a) Shot record before water layer multiple suppression. (b) Shot record after water layer multiple suppression. (c) 2-D autocorrelation function of shot record after multiple suppression. Note the high-amplitude interbed multiple around 600 ms temporal lag and between 0 and –40 trace intervals (TI) spatial lag. *(see part c on facing page)*

stacked sections before and after surface and refraction statics were applied. Enhanced lateral continuity is clearly observable on Figure 60b. Unfortunately, refracted arrival times do not give sufficient information to allow separate computation of the overburden velocity (interval velocity of the layer above the refractor) and depth of the refractor. Therefore, this velocity must be determined from reflected arrivals. In most cases, shallow refractors require recording near-offset data with closer group intervals.

Another problem to consider is that, since the refracted arrivals are generated at the interface between the low and high acoustic impedance contrast, only the top of a high-velocity layer can be seen. The base of the

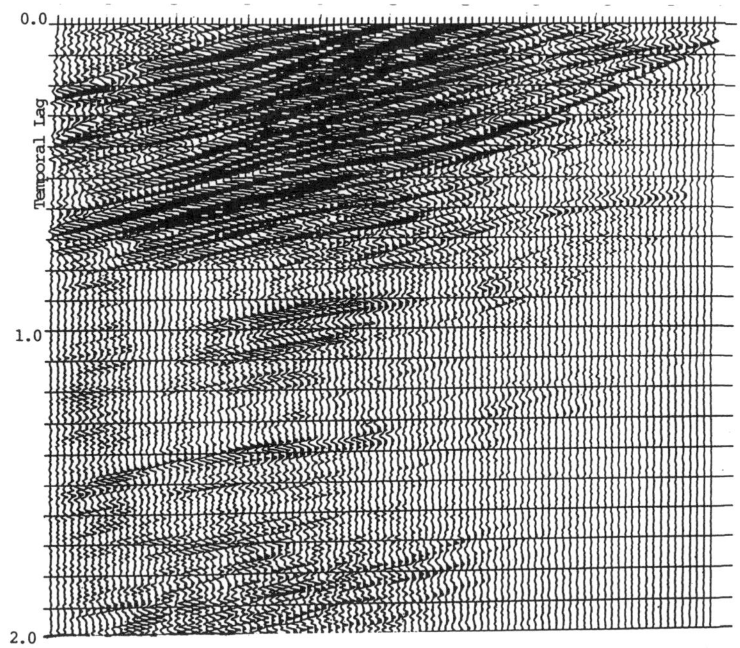

high-velocity layers do not generate refracted waves and are thus invisible. They can be detected only by reflected waves. This condition prevents us from detecting the thickness of the refracting layer by refracted arrivals. The use of refracted arrivals requires an understanding of the nature and limitations of the refracted waves, so some of its shortcomings can be anticipated and addressed by reflected waves.

Properly computed refraction and surface statics cover the major portion of the time statics, leaving only residual statics. In this computation, however, we now consider not only the time statics, but also the amplitude and phase statics. Wavelet shaping has corrected all of the spatially varying amplitude and phase spectra of the source and receiver responses, leaving only the mean amplitude, phase rotation (zero-frequency intercept of the phase function), and the time shift (slope of the phase function) as unknowns. These are determined by surface-consistent computations. Figure 61 shows seismic wave propagation from source to receiver, indicating the factors affecting wave shape. Let $F(\omega)$ be the recorded data in the frequency domain, $S(\omega)$ the source response, $R(\omega)$ the receiver response, $C(\omega)$ the subsurface response, and $D(\omega)$ the offset response. The recorded seismic trace is the result of cascaded convolution (multiplication in the frequency domain) of all of these responses:

$$F_{n,m}(\omega) = S_n(\omega) R_m(\omega) C_k(\omega) D(\omega), \qquad (20)$$

where n and m are the source and receiver locations, k is the mid-point, and l is the distance between n and m. The objective is to extract $C(\omega)$ from $F(\omega)$. Wavelet estimation and shaping remove spatial variations of $S(\omega)$, $R(\omega)$, and $D(\omega)$, with the exception of mean amplitude levels and the linear component of the phase spectra (intercept phase and slope of the phase spectrum). The natural log of the surface-consistency equation produces linear equations since

$$F(\omega) = |F(\omega)| e^{i\theta(\omega)},$$

then

$$\ln [F(\omega)] = \ln|F(\omega)| + i\theta(\omega). \qquad (21)$$

Complex values in the frequency domain separate into real and imaginary parts. The real part is the log amplitude spectrum and the imaginary part is the phase spectrum. These can now be separately computed. The real part represents the surface-consistent amplitude relationship. Since the wavelet processing has removed variations in the amplitude spectra except the mean amplitudes, then after wavelet shaping, we are left with the following relationship:

$$F(n,m) = S(n) + R(m) + C(k) + D(l) \qquad (22)$$

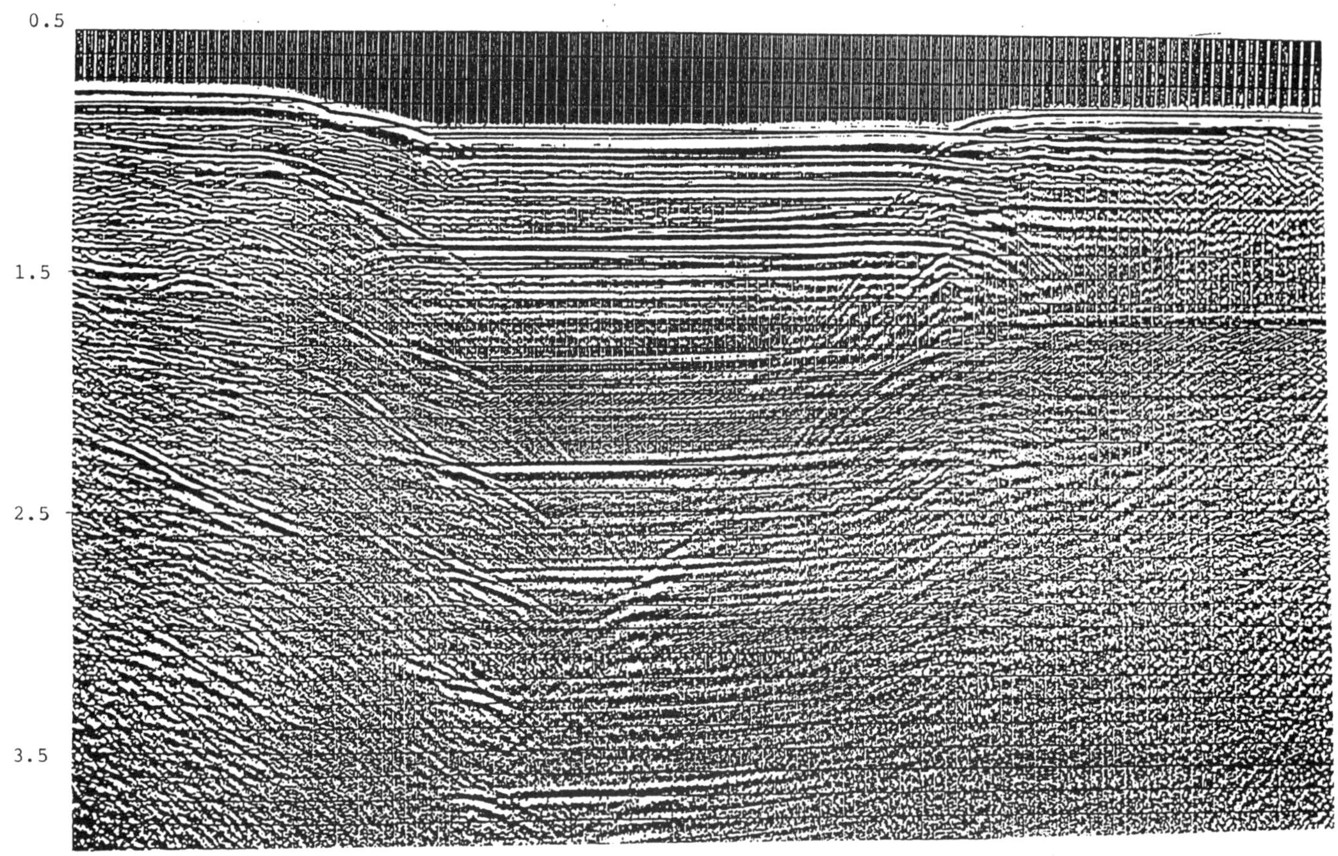

Fig. 55. Stacked section before long-period multiple suppression.

where all parameters represent mean amplitude levels. In practice, $F(n,m)$ is the natural log of the rms value of the trace. Computed values of $S(n)$, $R(m)$, and $D(l)$ represent performance factors of the source, receiver, and offset components. Since there will be one equation of this type for each trace, the unknowns can be estimated by standard least squared error techniques. However, this set of equations has one indeterminate value, which is the overall mean amplitude, a value that cannot be determined independently.

The imaginary part of the wavelet-corrected equations contains two sets of unknowns: the intercept phase (phase rotation of the wavelet) and the slope (time shift of the wavelet). These can be written separately in a form similar to the amplitude relationship:

$$f(n,m) = s(n) + r(m) + c(k) + d(l) \tag{23}$$

where $s(n)$ and $r(m)$ are the phase rotations to be applied to the common-source and common-receiver traces, respectively, and $d(l)$ is the phase rotation for the common-offset traces. The time shift (or statics) equations are slightly altered to fit the real data problems more realistically. The offset-related time statics are usually due to erroneous NMO functions computed from the statics contaminated data set. Therefore, instead of the original relationship as in the general surface-consistency equations, it is in the form

$$t(n,m) = S(n) + R(m) + C(k) + M(k)(m-n)^2 \tag{24}$$

where $M(k)$ is the average influence of the residual NMO on the k-th midpoint location and $t(n,m)$ represents the total static. Phase and time differences are determined by the complex cross-correlation function computed between individual and reference traces (Figure 62). The location of the envelope peak with respect to the zero-lag gives the time difference Δt between the two traces. This indicates the group delay. The instantaneous phase computed at the envelop peak corresponds to the (rotational) phase difference between the two traces.

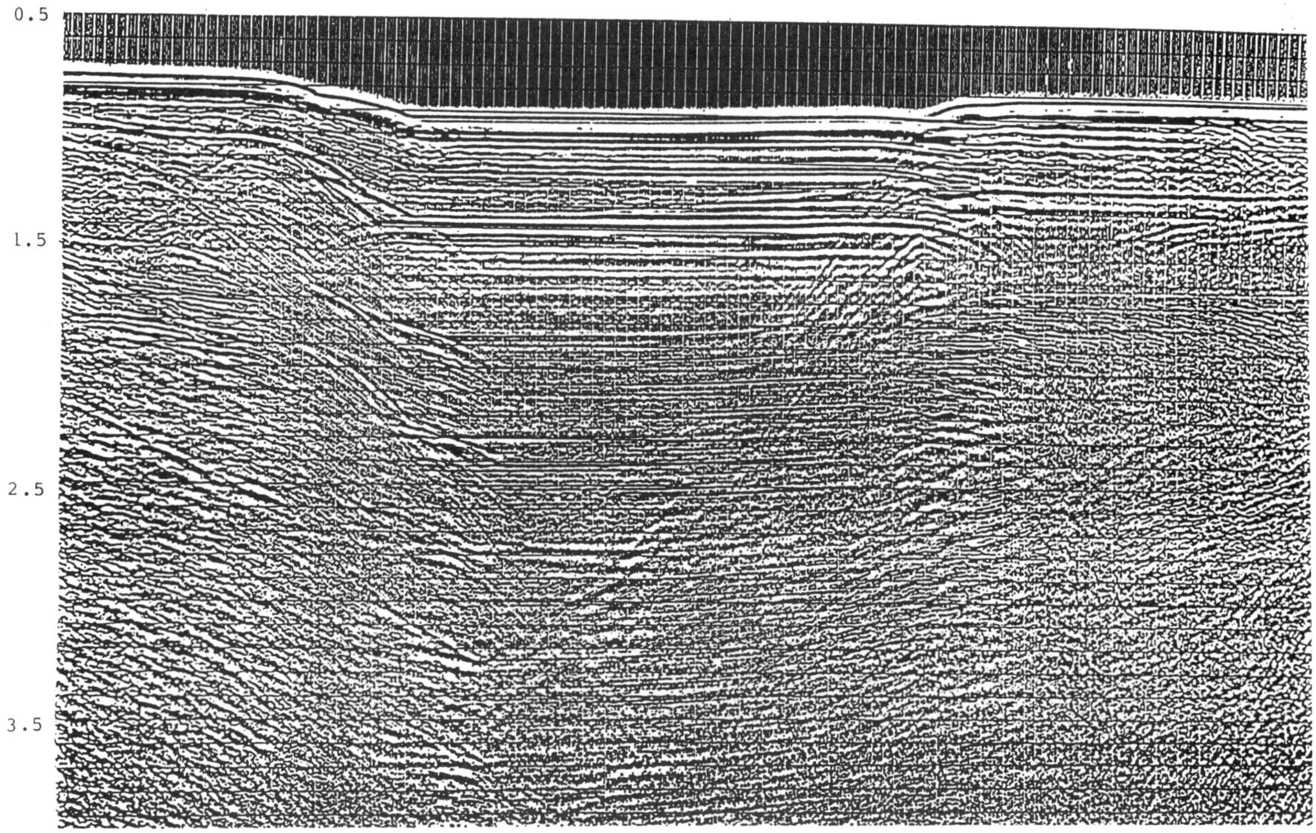

Fig. 56. Stacked section after long-period multiple suppression.

Velocity Analysis

Velocity of propagation of sound waves in the subsurface is a useful physical parameter for helping to determine lithology and porosity. Four other velocities are also useful to determine in the path of the seismic processing stream. Unfortunately, the relationships of these velocities are not clearly understood, and hence, they are incorrectly used interchangeably. To clarify these relationships, I have considered a simple model (Figure 63a) with a single interface comprising horizontal and dipping parts. The corresponding CDP time response is sketched in Figure 63b, where t_1 and t_2 represent the two-way normal incidence times over normal incidence distances d_1 and d_2. The first utility function is the NMO velocity, which is computed from CMP gathers to produce alignment of interpreted primary events prior to stacking. In horizontal layer areas, NMO velocities obtained from near-offset traces are good approximations of rms velocities computed with respect to time. They provide some estimates of the interval velocities. These interval velocities are the rms velocities computed between two events where NMO velocities are picked. They give an estimate of the propagation velocities.

As discussed earlier, transversely isotropic (TI) effects are noticeable on the long offsets with large initiation angles of the wavefront. In areas of dipping beds, derived velocities are under the influence of dips and anisotropic effects of TI layers. Both of these make the NMO velocities higher than the flat-lying beds, which gives erroneous values for interval velocity and depth estimates. In Figure 63c, the dipping bed with shorter two-way normal incidence time has a higher velocity and earlier time than the horizontal layer. Note that the normal incidence positions are at different positions, but they appear on the same CDP trace and on the same velocity spectra (for NMO analysis). Hence, any interval velocity estimation between these two events becomes meaningless since they both belong to the same reflector. Also, as discussed in the statics section, any time delay causes erroneous values for the computed NMO velocities. Therefore, in complicated areas, NMO velocities are not reliable measurements for interval velocity and depth estimates.

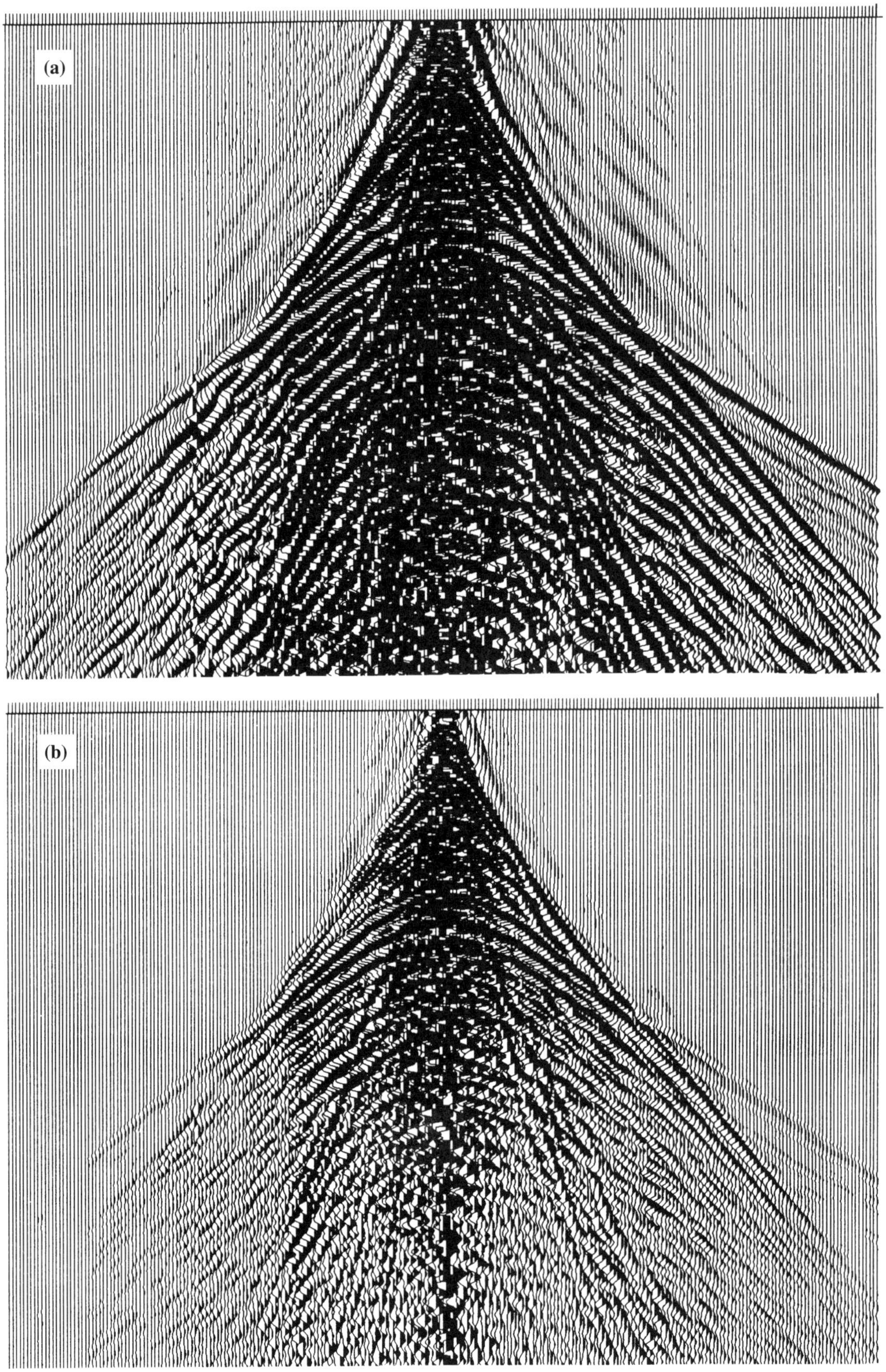

Fig. 57. First-break area spectral balancing. (a) Original data first-break area. (b) Spectral balanced first-break area.

The second velocity type is the NMO velocity computation after the dip moveout (DMO) application, which when applied correctly, produces data with reduced effects of dips on arrival times. Hence, the NMO velocities are relatively independent of dip (Figure 63d). Note that DMO application is an approximate procedure, hence the results should be viewed by an experienced eye. Also, TI effects are still present. If the structures are not complicated, then we can obtain more reliable interval velocities. Unfortunately, one of the most widely used practices in the industry is to use the NMO or DMO-corrected NMO velocities as the time-migration velocity field. NMO velocities are computed along the two-way normal incidence (to the reflector) paths, which are all separate paths if reflectors have varying dips.

The migration velocity field must be given according to the migrated positions of the reflectors, which is generally neglected by processors. In fact, some papers claim that the results of time-migrated sections generated by velocity fields equal to some percentage of NMO velocities are erroneous. They fail to consider that these velocities are computed at unmigrated positions of the reflectors.

The third velocity is the time migration function. Figures 63e and 63f show two migration velocity analyses run on locations A and D. One velocity is at each location. In time migration, the image is formed along the image raypath, which starts vertically downward from the surface and refracts across beds to the image point. Due to refraction across dipping beds, there are some spatial and temporal distortions. However, all diffractions are compressed and reflectors are imaged in a less complicated manner. Time migration velocities are computed in a manner similar to NMO velocities. Image traces are generated from common-source or common-offset trace gathers. The Kirchhoff integral method allows generation of one image trace at any desired surface location without generating many other traces, as in other methods. These traces are generated for a sweep of velocities and displayed as common-image gathers. Image traces generated from different source or common-offset gathers for the same migration velocities are displayed side by side to show the alignment of images. The velocities that produce events at the same time (as in NMO analysis) are picked as the time migration velocities. This guarantees that, if we use these velocities to migrate the prestack data, all primary reflections (or the reflections picked as primaries) appear at the same temporal and spatial position, thus they align properly for an optimal stack. Since they produce similar images for different gathers, they are also proper to use as the poststack migration velocity field.

Time migration velocities are computed for all times in one step, hence overburden effects are not considered. In the fourth computation, which is the depth migration velocity analysis, these overburden effects are taken into consideration, hence computations are done in a layer-stripping fashion. By this method, velocities are determined one layer at a time, hence they do not need any iterative corrections. After one layer is determined, it is added to the previously computed layers. Once the new velocity field is computed, they are kept constant and a new layer's velocities are scanned. This is a recursive form of velocity-depth model building. In other methods, an initial velocity model is used and image coincidence is checked for validity. If any misalignments of events are found, then the small corrections are applied to the velocity field. The iteration of checking and correction continues until satisfactory results are reached. Complete prestack migrations may be needed before a satisfactory condition is reached. I consider the layer-stripping methods to be more viable. Figures 63g and h show the velocity analyses run at locations A and D.

Time and Depth Migration

Seismic imaging is a developing art in the mid-1990s. A number of papers propose different methods to produce time and depth images from pre- or poststack seismic data. Seismic data can be imaged many different ways. It all depends on the chosen velocity-depth model, which is like an optical lens that can produce images of the incoming wavefield. The most important characteristic of this lens is that its image must be the lens' own shape. In other words, we essentially know the image before we produce it. Any arbitrary velocity field produces an image. But only the correct one produces its own image. Some approaches use an approximate velocity-depth model developed from previous knowledge of the area. For example, in the Gulf of Mexico, initial velocity is a linear function that increases with depth. Trial and error procedure is used to improve the estimate. The final image is the result of knowledgeable interpretation of the seismic data and of all the analyses.

Hence, instead of attempting to produce a final depth image, it is better to proceed with the imaging process step by step. The original seismic data provide the most complicated view of the subsurface. Both point diffractors and reflecting surfaces appear as reflected point sources in the form of hyperbolas. NMO is applied to these data to simulate the zero-offset traces and sum them to improve the S/N. This produces the CDP stack

(a)

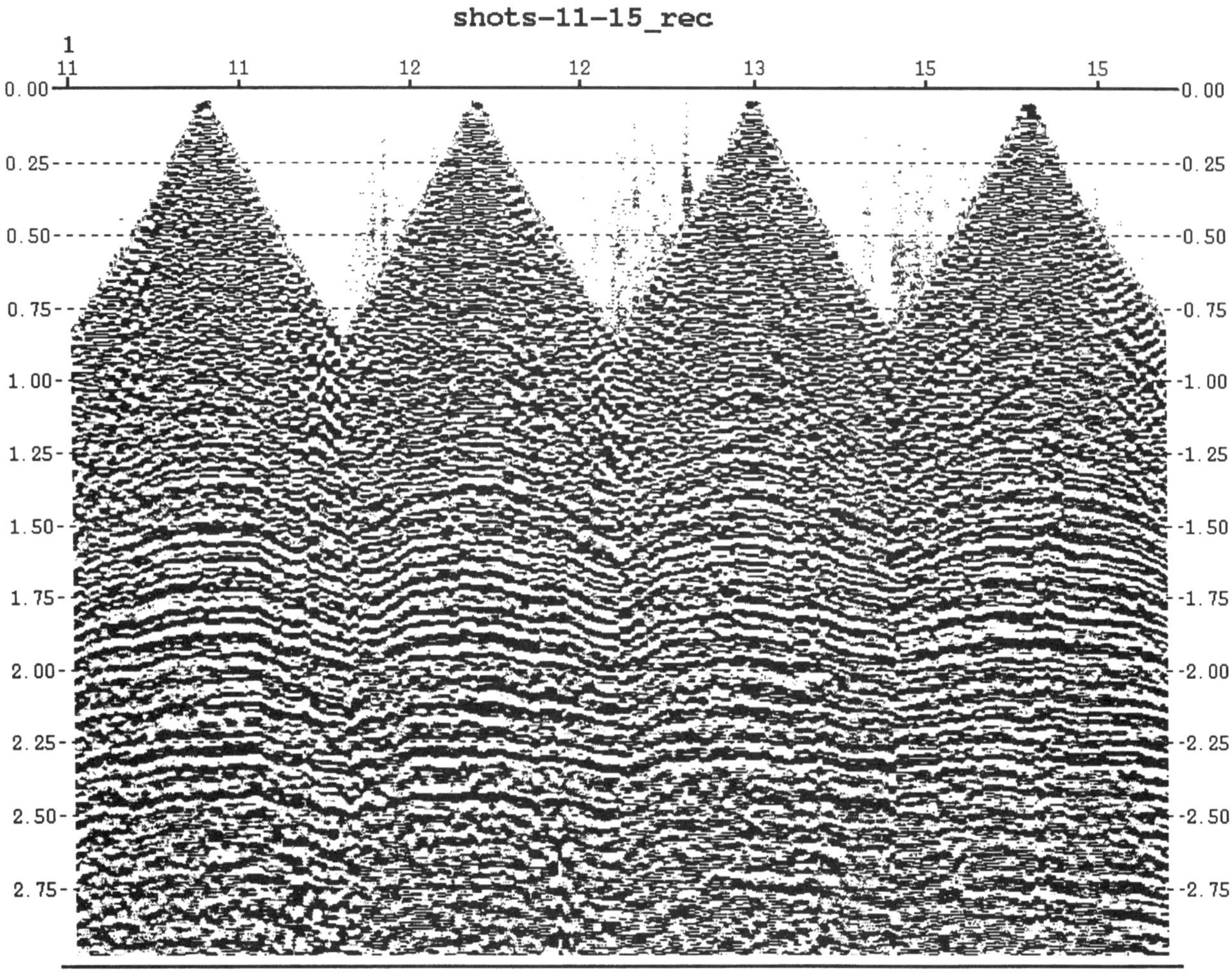

Fig. 22. Example of static correction. (a) Original shot record. (b) Shot record after application of statics correction. ***(see part b on facing page)***

section. CDP stack sections are actually time images of the subsurface, except that the image is with respect to two-way traveltime, which originates perpendicular to the reflector and travels to the surface. On these sections, point diffractors appear as diffraction hyperbolas, and reflectors appear distorted, but as continuous events. To obtain a properly stacked CDP section, NMO velocities are developed from velocity spectra or from velocity panels, based on generating stack sections over a velocity sweep and computing corresponding semblance, which measures the optimality of event alignment. Time migration velocity analysis, as with NMO velocity analysis, measures average velocities from the surface to the reflector, since they are determined without considering any lateral velocity inhomogeneity in the shallower layers.

In NMO velocity analysis, NMO-applied traces are used. In migration velocity analysis, migrated traces generated by trial velocities are similarly used; these are called *image traces*. Image traces generated at the same surface coordinates are called *common-image traces*. Each trace of the common-image gather is generated from a different shot gather or common-offset gather. Only the correct imaging velocity produces events appearing at the same time or depth (horizontal alignment) on the common-image gathers. Figure 64 shows a stack section, and Figure 65 shows the common-image gathers computed for velocities from 2000 to 3200 m/s. Note the change of "smile" patterns to "frown" patterns as the velocity increases. Horizontally aligned events show the correct imaging velocities. If one computes the semblance of each of the constant-velocity image panels, a conventional-looking velocity spectra is generated (Figure 66). The semblance peaks on the velocity spectra correspond to the horizontally aligned velocity-time pairs of the common-image gathers. Thus, the time mi-

(b)

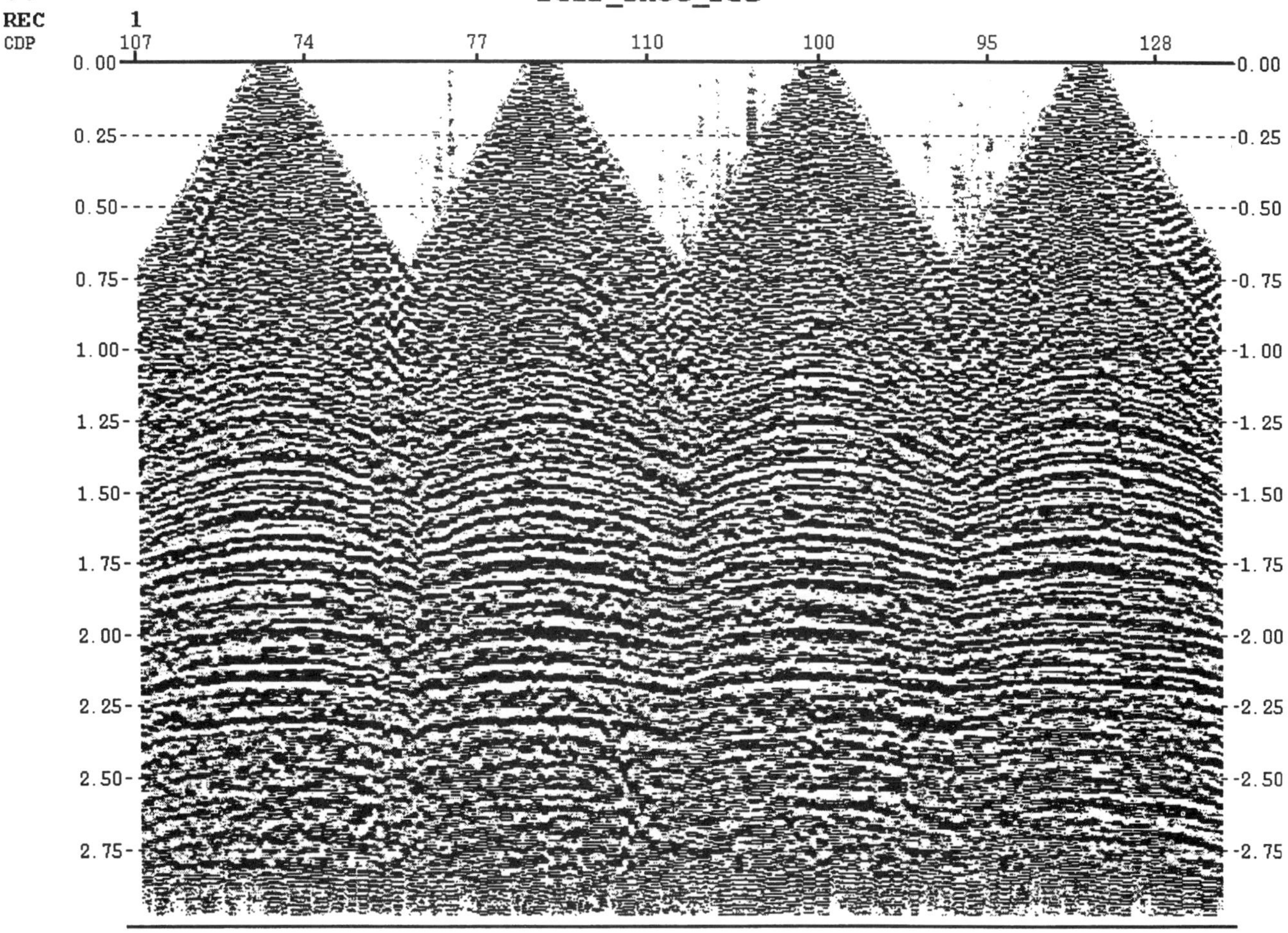

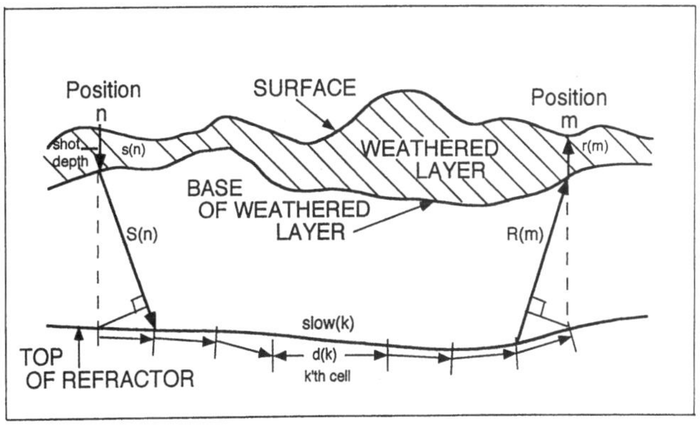

Fig. 23. Simple surface model for short- and long-wavelength statics computations.

gration velocities can be determined in a manner analogous to the NMO velocities. Note that the time migration velocities are determined from common-image gathers, which is a prestack process. Hence, these velocities can be used for pre- and poststack migration with equal validity.

A comparison of the time-migrated section in Figure 67 with the stack section in Figure 64 shows the absence of all diffraction patterns. The vertical scale is still time, but here it represents two-way traveltime starting vertically from the surface down to the reflector. This path is called the *image raypath*. However, the image raypath contains refraction across dipping interfaces, hence images appear spatially shifted from their correct position. Furthermore, since this is the traveltime, it does not reflect the travel distance; it has time pull-ups below high-velocity layers or pull-downs below low-velocity layers. Nevertheless, time-migrated sections are much closer to the final objective of the depth image and are easily obtained. These sections can be used for seismic attribute analysis and are reliable for determining macrolayers for depth migration velocity analysis. Figure 67 clearly shows the fault planes and top of the salt strubcture.

Depth migration velocities are determined one macrolayer at a time. Once one layer is determined, then this layer is kept constant and new imaging velocities are searched in the unknown layer below. In the depth-mi-

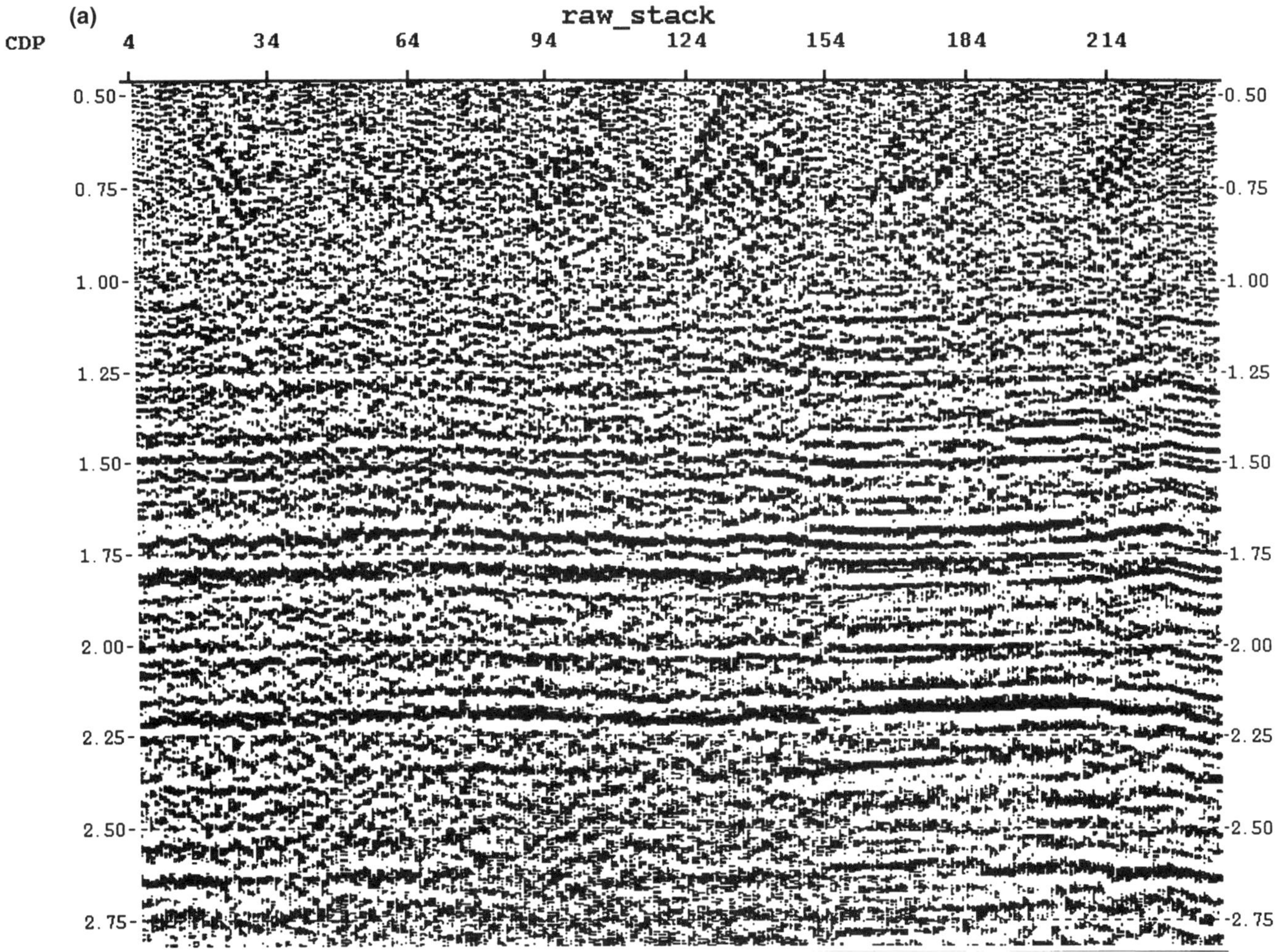

Fig. 24. Example of static correction. (a) Stacked section without any statics correction. (b) Stacked section after application of short- and long-wavelength statics computed from first arrival times. ***(see part b on facing page)***

grated section of Figure 68, velocity changes are clearly observable. Here, the higher velocity zone events appear to be lower frequency because they are stretched to show the increased wavelength. The lower velocity zones have higher frequency, shorter wavelength events. This shows that resolution is lost at increasing velocity (and depth) due to an increase of corresponding wavelengths. Thus, by using both time- and depth-migrated sections, we should be able to minimize the disadvantages of both while improving data interpretability.

Concluding Remarks

I have covered seismic data processing with specific attention to shallow carbonate areas. A number of other favorable approaches are equally applicable. For this reason, I have tried to explain some of the background reasons for the approach recommended here. I have also avoided the use of references in the text to not interrupt the train of thought. However, to help the interested reader and to give proper credit, an extensive bibliography has been included.

Acknowledgments

I would like to express my appreciation to Graham Jago and Jeff Taner and especially to Mike Graul for their tireless review and editing of the text, to Jeff and Orhan Yilmaz for generating actual data examples, and to John Taner for generating all figures and text editing. I also want to thank Cogniseis Development for the use of their Focus and Disco systems to generate the actual data examples. This papers reflects my experience and what I gained from the experiences of many other researchers and processors over more than 30 years. The views expressed are mine, hence I am responsible for them. During the editing, Mike Graul and I had many constructive discussions and expressed differing points

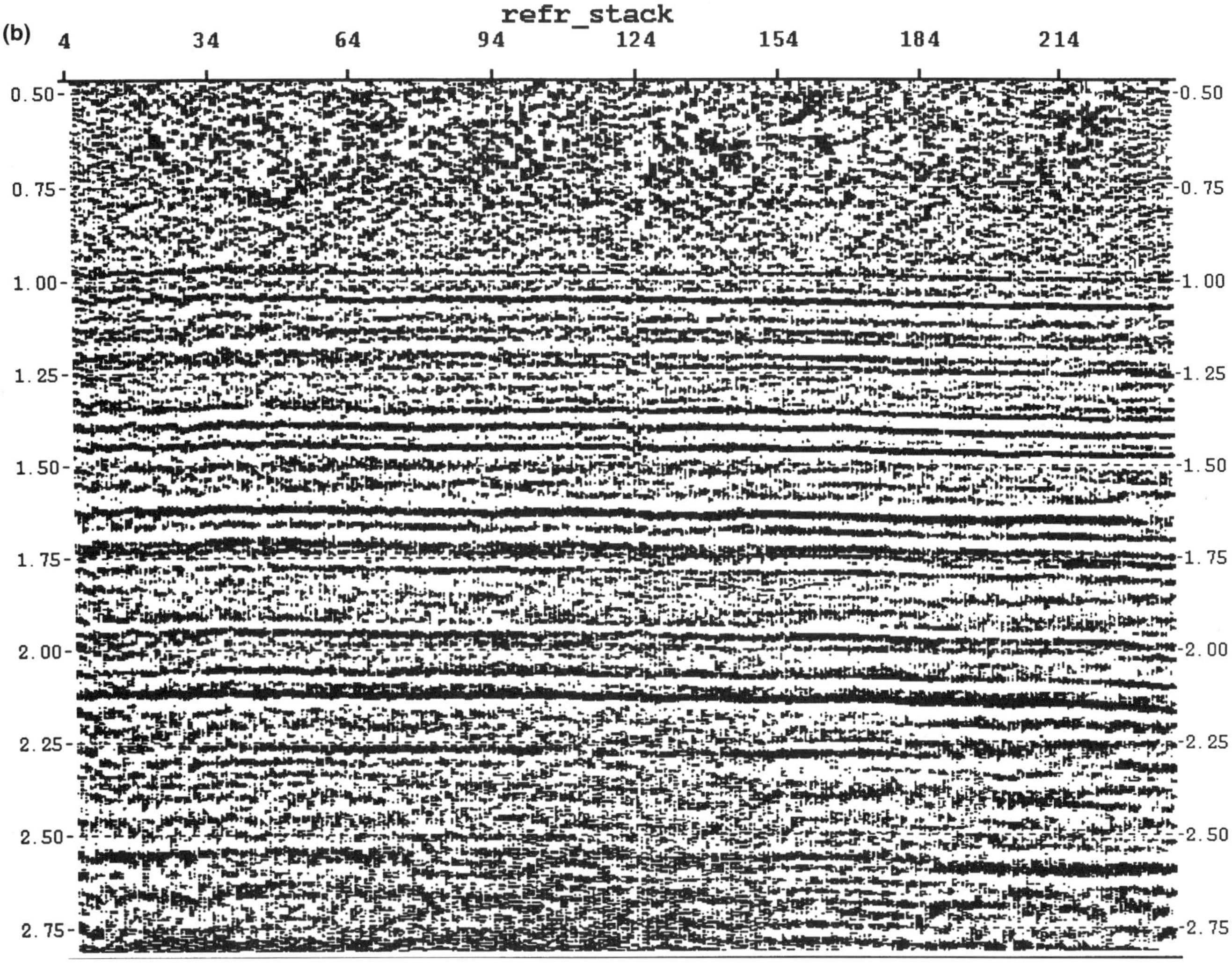

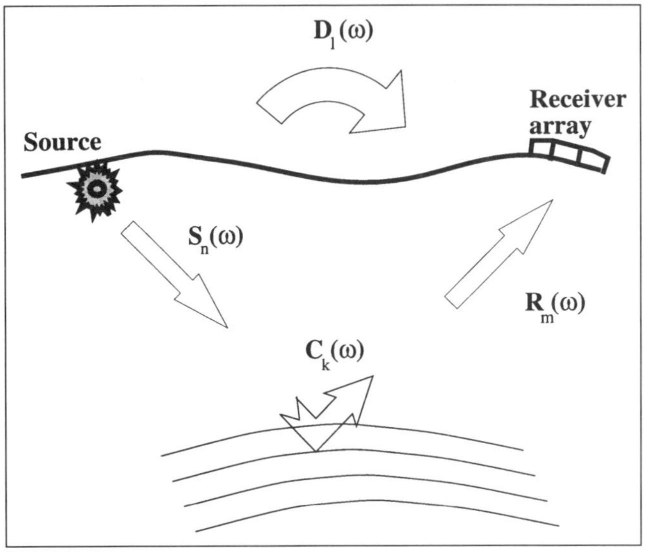

Fig. 25. Simple surface model for surface-consistent computations.

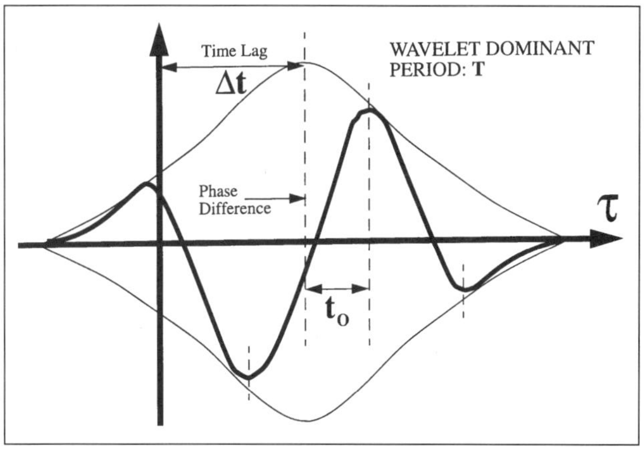

Fig. 26. Complex correlation function. The distance from zero-lag to the envelope peak represents the time lag, and instantaneous phase at envelope maximum represents the phase difference.

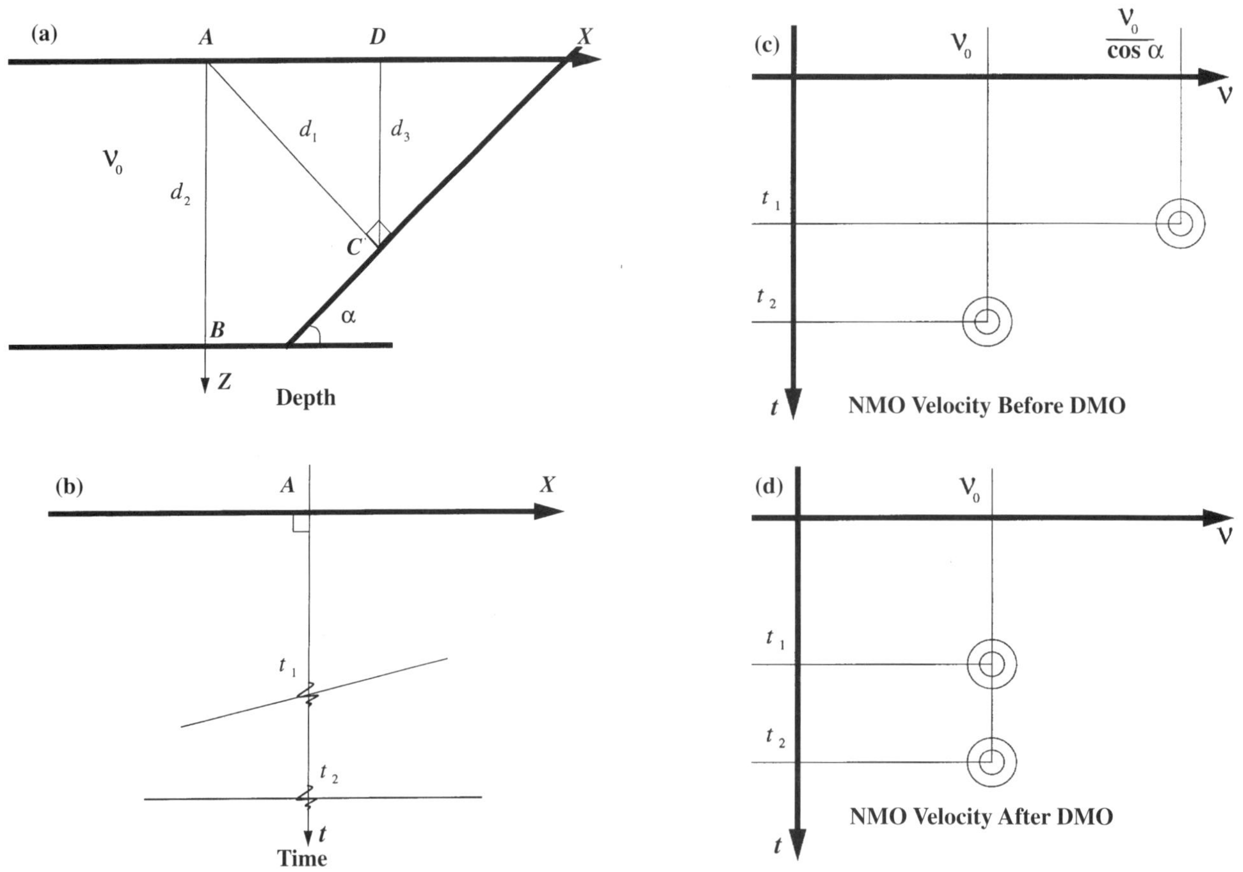

Fig. 27. Relationships of different velocities. (a) Depth section showing a layer with horizontal and dipping components. (b) CDP stacked (zero-offset) time section. (c) NMO velocities before DMO application. (d) NMO velocities after DMO application. ***(next page)*** **(e) Time migration velocity at location A. (f) Time migration velocity at location D. (g) Depth migration velocity at location A. (h) Depth migration velocity at location D. The vertical axis is in depth, and events appear at corresponding depths of beds under points A and D.** ***(see parts e, f, g, and h on facing page)***

of views on some issues. I gratefully acknowledge that Mike Graul is the author of the section on resolution because of the clarity of his discussion on the subject.

Bibliography

The following references are grouped by subject according to specific problems encountered in shallow carbonates. The lists were compiled and condensed from the Society of Exploration Geophysicists' World Wide Web Site.

Anisotropy and Related Effects

Anderson, J. E., Alkhalifah, T. A. and Tsvankin, I., 1994, Fowler DMO and time migration in transversely isotropic media: 64th Ann. Internat. Mtg., Soc. Expl. Geophys., Expanded Abstracts, 1217-1220.

Clark, D., 1987, Near-surface effects workshop: The Leading Edge, **6**, No. 1, 8-9.

Crampin, S., 1984, Anisotropy in exploration seismics: First Break, **2**, No. 3, 19-21.

Crampin, S., 1989, Suggestions for a consistent terminology for seismic anisotropy: Geophys. Prosp., **37**, 753-770.

Crampin, S., Chesnokov, E. M. and Hipkin, R. A., 1984, Seismic anisotropy—the state of the art: First Break, **2**, No. 3, 9-18.

Hake, H., 1986, Slant stacking and its significance for anisotropy: Geophys. Prosp., **34**, 595-608.

Helbig, K., 1984, Anisotropy and dispersion in periodically layered media: Geophysics, **49**, 364-373.

Helbig, K., 1993, Longitudinal directions in media of arbitrary anisotropy: Geophysics, **58**, 680-691.

Krey, T., and Helbig, K., 1956, A theorem concerning anisotropy of stratified media and its significance for reflection seismics: Geophys. Prosp., **4**, 294-302.

Levin, F. K., 1984, Anisotropy due to bedding—a computer study: Geophys. Prosp., **32**, 187-197.

Macbeth, C., and Crampin, S., 1991a, Automatic processing of seismic data in the presence of anisotropy: Geophysics, **56**, 1320-1330.

Macbeth, C., and Crampin, S., 1991b, Comparison of signal processing techniques for estimating the effects of anisotropy: Geophys. Prosp., **39**, 357-386.

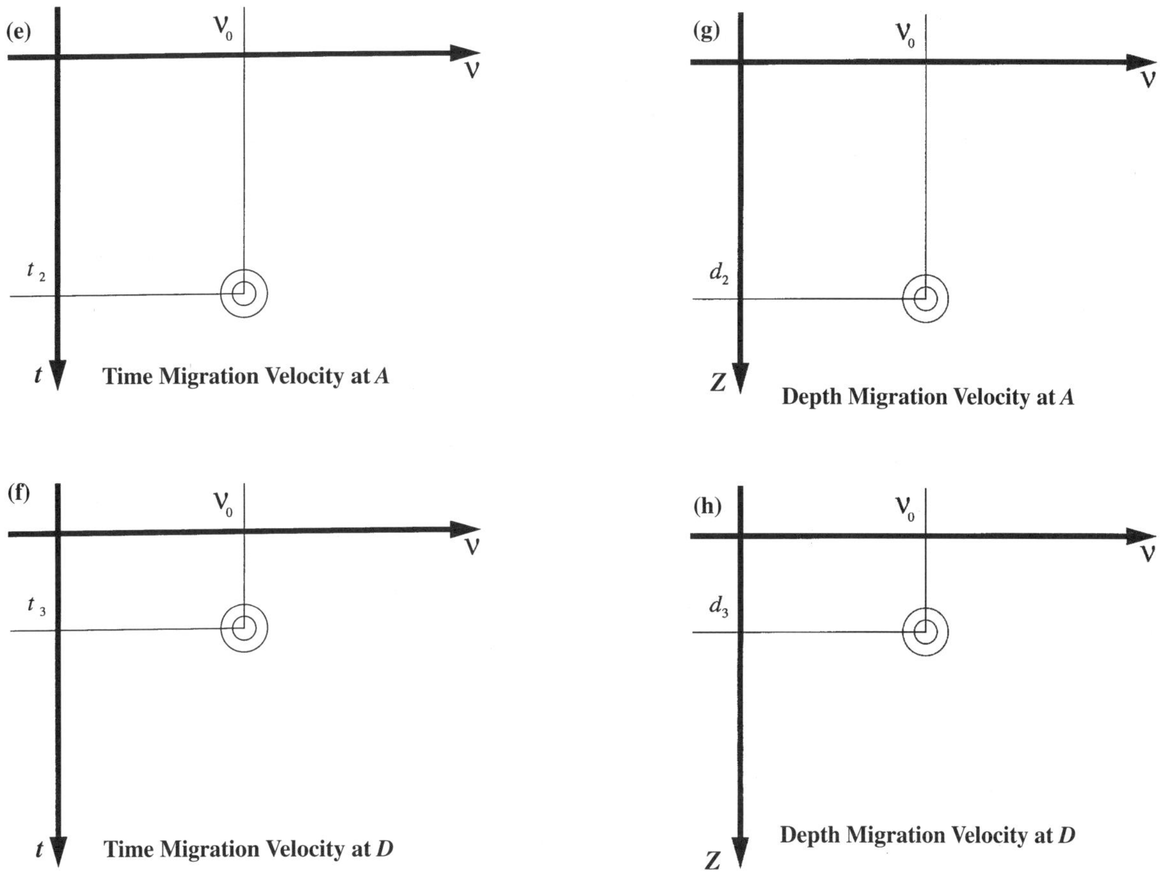

McCollum, B., and Snell, F. A., 1947, Asymmetry of sound velocity in stratified formations: Early Geophysical Papers, **1**, 216-227.

Melia, P. J., and Carlson, R. L., 1994, An experimental test of *P*-wave anisotropy in stratified media, *in* Ebrom, D. A., and McDonald, J. A., Eds., Seismic physical modeling: Soc. Expl. Geophys., 312-316.

Mueller, M. C., 1992, Using shear waves to predict lateral variability in vertical fracture intensity: The Leading Edge, **11**, No. 2, 29-35.

Postma, G. W., 1955, Wave propagation in a stratified medium: Geophysics, **20**, 780-806.

Samec, P., and Blangy, J. P., 1992, Viscoelastic attenuation, anisotropy, and AVO: Geophysics, **57**, 441-450.

Spencer, T. W., and Chi, H. C., 1991, Thin-layer fracture density: Geophysics, **56**, 833-843.

Stockli, R. F., 1986, Errors in depth computation (short note): Geophysics, **51**, 414-416.

Thomsen, L., 1992, Weak elastic anisotropy, *in* Wang, Z., and Nur, A., Eds., Seismic and acoustic velocities in reservoir rocks, 2—theoretical and model studies: Soc. Expl. Geophys., 395-407.

Uhrig, L. F., and van Melle, F. A., 1955, Velocity anisotropy in stratified media: Geophysics, **20**, 774-779.

Winterstein, D. F., 1986, Anisotropy effects in *P*-wave and *SH*-wave stacking velocities contain information on lithology: Geophysics, **51**, 661-672.

Winterstein, D. F., 1990a, Anisotropy effects in *P*-wave and *SH*-wave stacking velocities contain information on lithology, *in* Byun, B. S., Ed., Velocity analysis on multichannel seismic data: Soc. Expl. Geophys., 498-509.

Winterstein, D. F., 1990b, Velocity anisotropy terminology for geophysicists: Geophysics, **55**, 1070-1088.

Winterstein, D. F., 1992, How shear-wave properties relate to rock fractures: simple cases: The Leading Edge, **11**, No. 9, 21-28.

Carbonate Rocks

Angeleri, G. P., Bricchi, G., Marfurt, K. J., and Singer, P. J., 1993, 3-D noise characterization and suppression in a deep-water carbonate terrane using wideline profiles: 63rd Ann. Internat. Mtg., Soc. Expl. Geophys., Expanded Abstracts, 1293-1297.

Ball, M. M., 1972, Exploration methods for stratigraphic traps in carbonate rocks, *in* King, R. E., Ed., Stratigraphic oil and gas fields—classification, exploration methods, and case histories: Am. Assn. Petr. Geol. and Soc. Expl. Geophys., 64-81.

Carlson, R. L., Schaftenaar, C. H., and Moore, R. P., 1984, Causes of compressional wave anisotropy in carbonate-bearing deep-sea sediments: Geophysics, **49**, 525-532.

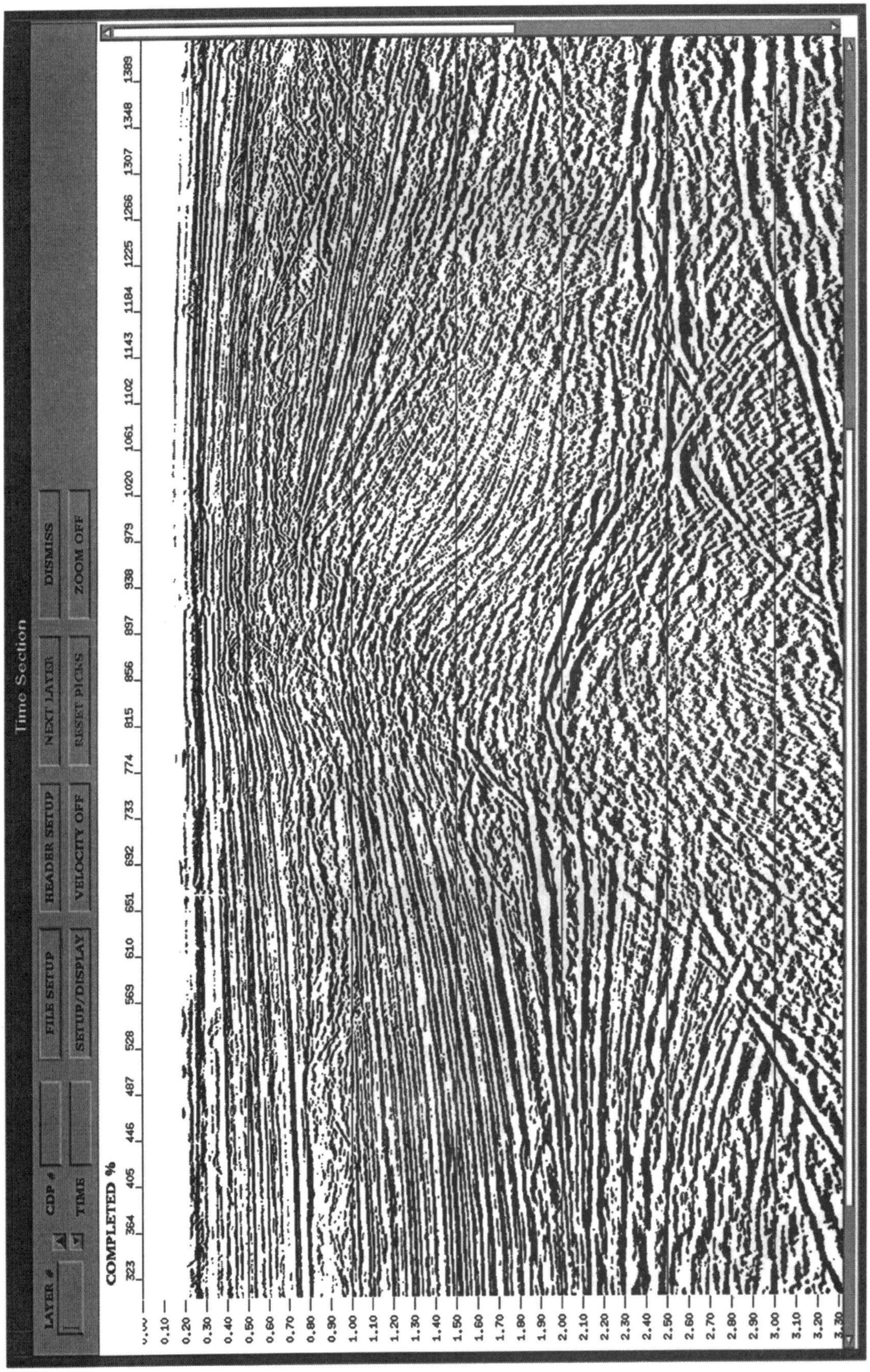

Fig. 28. CDP stacked seismic section showing a salt structure.

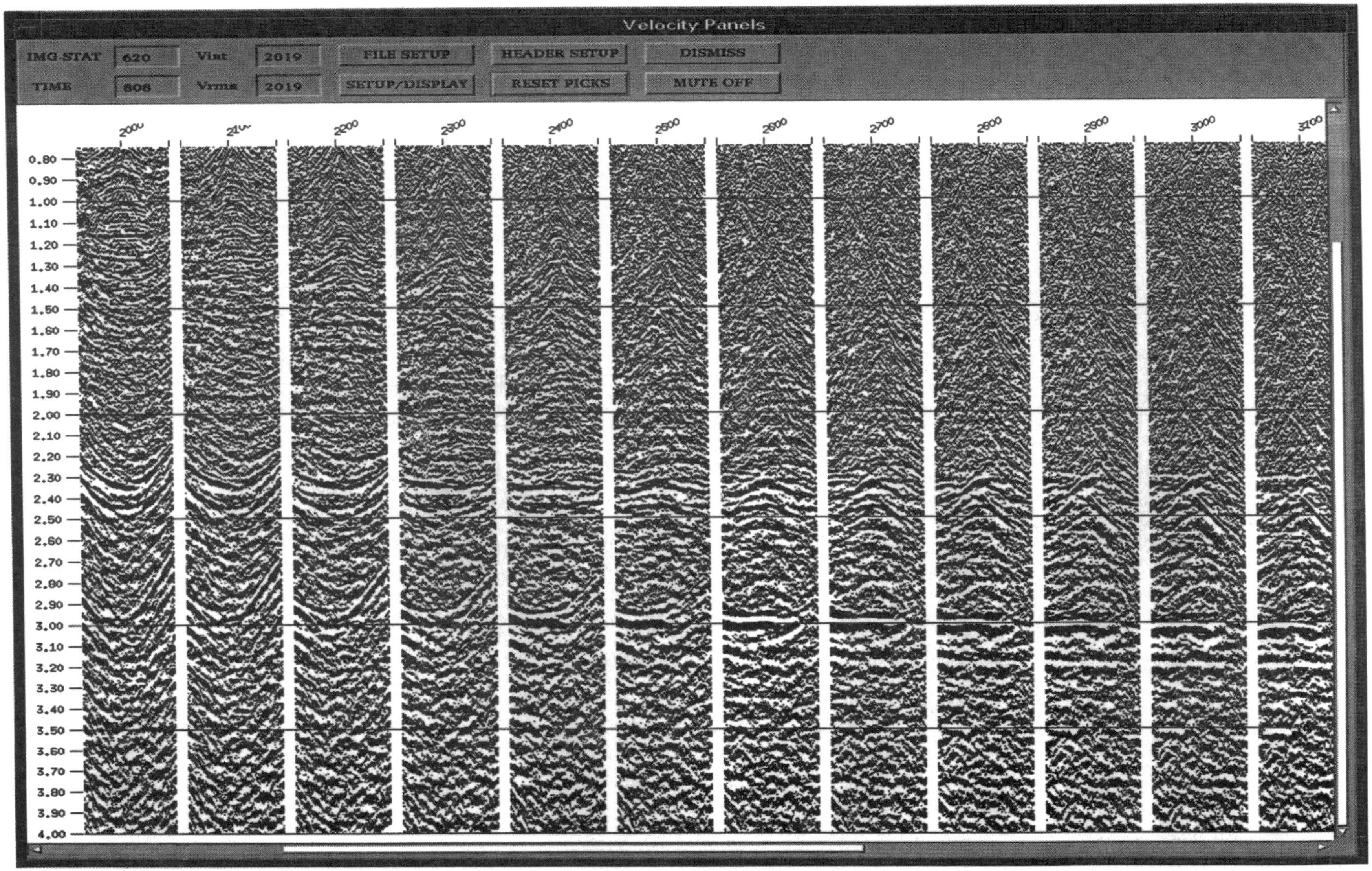

Fig. 29. Time migration (common-image gathers) velocity panel. Each trace is an image trace generated from different shots from the vicinity of the analysis location with a sweep of velocities. The section shows a velocity range from 2000 to 3100 m/s. Horizontal alignments indicate correct imaging velocity and time.

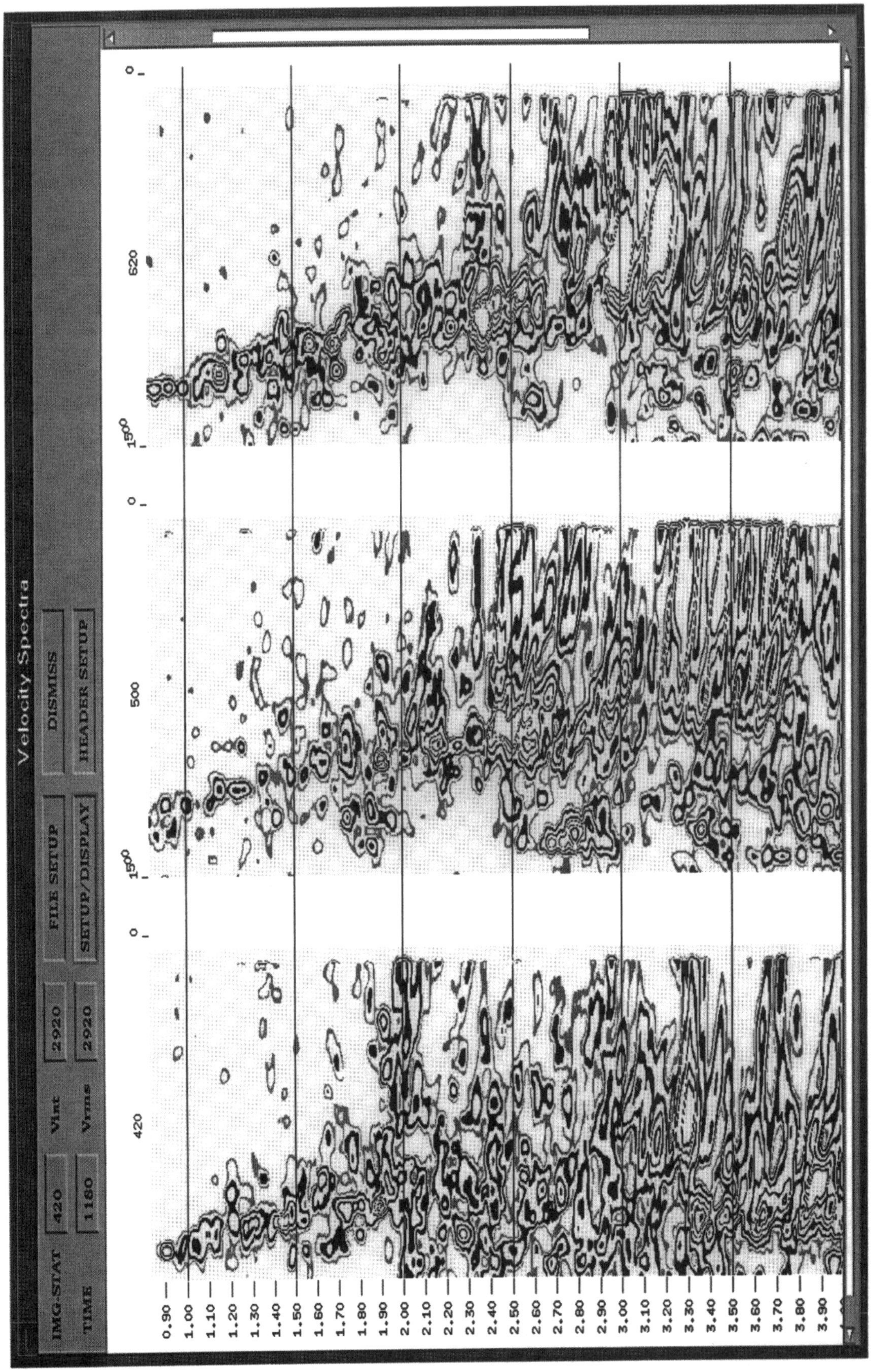

Fig. 30. Velocity spectra of time migration velocities computed from common-image gathers shown in Figure 65.

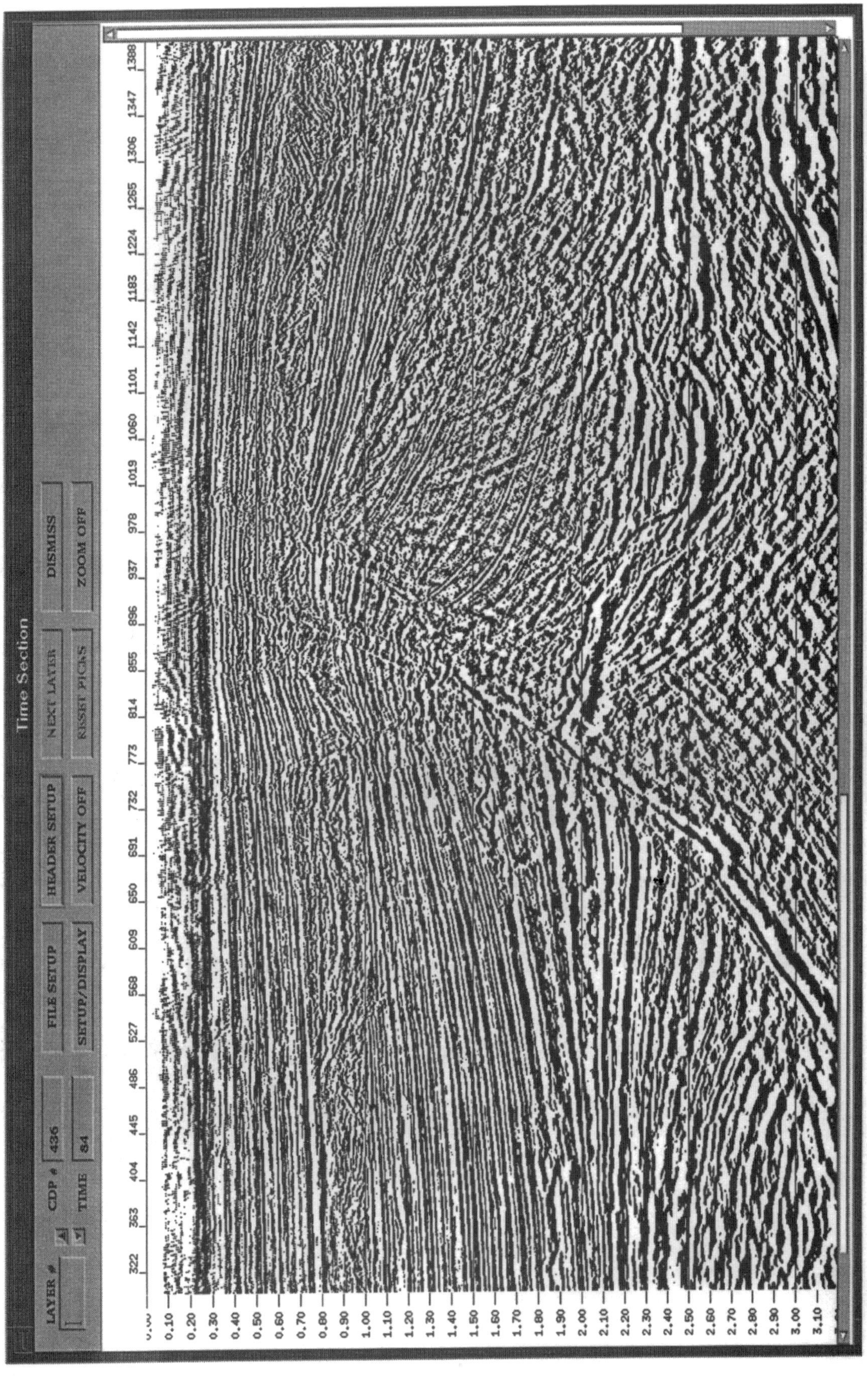

Fig. 31. Poststack time-migrated section of the CDP stacked section shown in Figure 64.

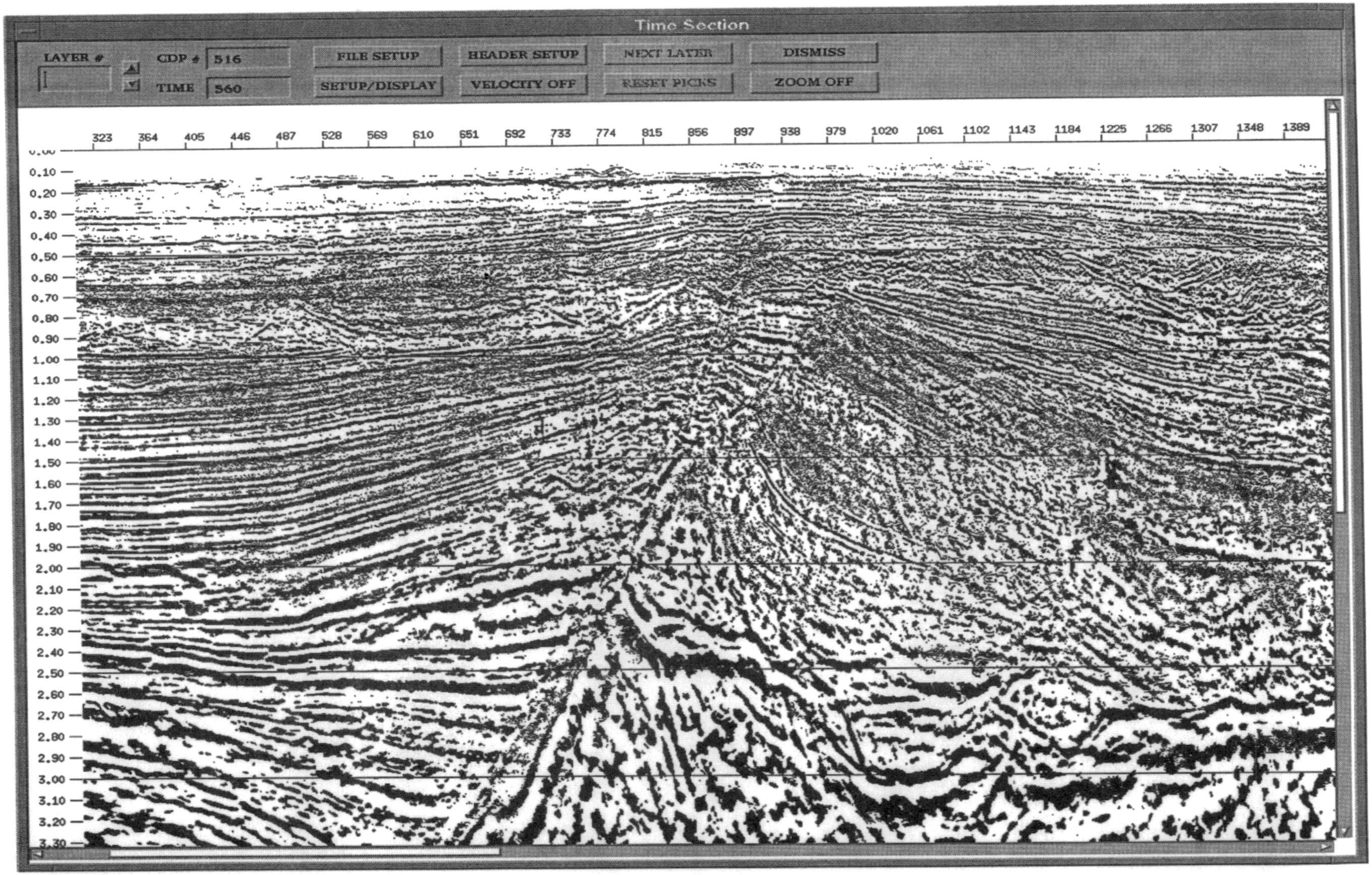

Fig. 32. Poststack depth-migrated section of CDP stacked section shown in Figure 64. Note lower frequency events associated with the salt body due to long wavelengths in the high-velocity zones.

Davis, T. L., 1980, Application of reflection seismology to Mississippian carbonate porosity and direct hydrocarbon detection: J. Can. Soc. Expl. Geophys., **16**, 19-25.

Davis, T. L., 1992, SPG/SEG 1990 carbonate exploration symposium: The Leading Edge, **11**, No. 1, 38-41.

Harvey, P. J., 1993, Porosity identification using amplitude variations with offset in Jurassic carbonate, offshore Nova Scotia: The Leading Edge, **12**, 180-184.

Harvey, P. J., and MacDonald, D. J., 1990, Seismic modelling of porosity within the Jurassic aged carbonate bank, offshore Nova Scotia: J. Can. Soc. Expl. Geophys., **26**, 54-71.

Jankowsky, W., 1970, Empirical investigation of some factors affecting elastic wave velocities in carbonate rocks: Geophys. Prosp., **18**, 103-118.

Jifon, F., and Goulty, N. R., 1985, Seismic reflection surveying offshore Durham England—a carbonate problem area: First Break, **3**, No. 10, 17-24.

Kotcher, J. S., Gardner, G. H. F., and McDonald, J. A., 1984, Modeling the effect of static errors in areal seismic data caused by glacial erosion over carbonate reefs: Geophysics, **49**, 1-16.

Lucia, F. J., Bebout, D. G., and Kerans, C., 1991, The Archie approach to characterizing carbonate reservoirs: Proc., First Archie Conf., Am. Assn. Petr. Geol., 79.

Maureau, G. T. F. R., and van Wijhe, D. H., 1979, The prediction of porosity in the Permian (Zechstein 2) carbonate of eastern Netherlands using seismic data: Geophysics, **44**, 1502-1517.

Missman, R. A., and Jameson, J., 1991, An evolving description of a fractured carbonate reservoir: the Lisburne field, Prudhoe Bay, Alaska: Proc., First Archie Conf., Am. Assn. Petr. Geol., 204-223.

Nardon, S., Marzorati, D., Bernasconi, A., Cornini, S., Gonfalini, M., Mosconi, S., Romano, A., and Terdich, P., 1991, Fractured carbonate reservoir characterization and modelling: a multidisciplinary case study from the Cavone oil field, Italy: First Break, **9**, 553-565.

Neff, D. B., 1993, Amplitude map analysis using forward modeling in sandstone and carbonate reservoirs: Geophysics, **58**, 1428-1441.

Ozkanli, M., and Standen, E. J. W., 1993, Fracture morphology in Cretaceous carbonate reservoirs from south-east Turkey: First Break, **11**, 323-333.

Petta, T. J., and Bryant, J. G., 1991, Resolution of carbonate reservoirs at depth—an integrated approach to lower development risk, Smackover trend, Alabama: Proc. of first Archie Conf., Am. Assn. Petr. Geol., 196.

Qing-Xian, O., 1987, Reconnaissance and exploration for oil gas in marine carbonate areas of south China: The Leading Edge, **6**, No. 2, 39.

Robertson, J. D., 1987, Carbonate porosity from *S/P* traveltime ratios: Geophysics, **52**, 1346-1354.

Stanulonis, S. F., and Tran, H. V., 1992, Method to determine porosity/thickness directly from 3-D seismic amplitude within the Lisburne carbonate pool, Prudhoe Bay: The Leading Edge, **11**, No. 1, 14-21.

Multiples

Berkhout, A. J., 1988, Seismic inversion in terms of prestack migration and multiple elimination, *in* Lines, L. R., Ed., Inversion of geophysical data: Soc. Expl. Geophys., 404-416.

Booker, A., and Ong, C.-Y., 1971, Multiple-constraint adaptive filtering: Geophysics, **36**, 498-509.

Carrion, P. M., 1986, A layer-stripping technique for the suppression of water-bottom multiple reflections: Geophys. Prosp., **34**, 330-342.

Cassano, E., and Rocca, F., 1973, Multichannel linear filters for optimal rejection of multiple reflections: Geophysics, **38**, 1053-1061.

Dillon, P. B., and Evans, R. H., 1994, VSP-based surface-seismic multiple suppression: First Break, **12**, 409-417.

Dix, C. H., 1948, The existence of multiple reflections: Geophysics, **13**, 49-50.

Domenico, S. N., 1979, Multiple reflections and head-waves in the Gulf of Suez: Geophys. Prosp., **27**, 539-563.

Ellsworth, T. P., 1948, Multiple reflections: Geophysics, **13**, 1-18.

Estevez, R., and Claerbout, J. F., 1982, Wide-angle diffracted multiple reflections: Geophysics, **47**, 1255-1272.

Ferber, R.-G., 1994, Migration to multiple offset and velocity analysis: Geophys. Prosp., **42**, 99-112.

Foster, D. J., and Mosher, C. C., 1992, Suppression of multiple reflections using the Radon transform: Geophysics, **57**, 386-395.

Griffiths, L. J., Smolka, F. R., and Trembly, L. D., 1977, Adaptive deconvolution—a new technique for processing time-varying seismic data: Geophysics, **42**, 742-759.

Haase, A. B., 1992, Slope data water-bottom multiple attenuation: Geophys. Prosp., **40**, 403-428.

Hampson, D., 1986, Inverse velocity stacking for multiple elimination: J. Can. Soc. Expl. Geophys., **22**, 44-55.

Hampson, D., 1991, Inverse velocity stacking for multiple elimination, *in* Gardner, G. H. F., and Lu, L., Eds., Slant-stack processing: Soc. Expl. Geophys., 332-343.

Hardy, R. J. J., and Hobbs, R. W., 1991, A strategy for multiple suppression: First Break, **9**, 139-144.

Hardy, R. J. J., Warner, M. R., and Hobbs, R. W., 1989, Labeling long-period multiple reflections (short note): Geophysics, **54**, 122-126.

Hardy, R. J. J., Hobbs, R. W., Warner, M. R., and Jones, I. F., 1990, The reliability of multiple suppression: First Break, **8**, 297-304.

Higgins, G. E., 1949, A note on multiple reflections: Geophysics, **14**, 357-360.

Hryhor, W., 1969, Spread design for maximum multiple attenuation: J. Can. Soc. Expl. Geophys., **5**, 7-23.

Johnson, C. H., 1948, Remarks regarding multiple reflections: Geophysics, **13**, 19-26.

Kelamis, P. G., and Chiburis, E. F., 1992, Land data examples of radon multiple suppression: First Break, **10**, 275-280.

Koehler, K., Krey, Th., and Wachholz, H., 1974, Cancelling multiple reflections by feedback: Geophys. Prosp., **22**, 652-666.

Krey, T., 1978, Seismic stripping helps unravel deep reflections: Geophysics, **43**, 899-911.

Lokshtanov, D. E., 1993, Adaptive multiple suppression in tau–p domain: 63rd Ann. Internat. Mtg., Soc. Expl. Geophys., Expanded Abstracts, 1086-1089.

Lokshtanov, D. E., and Helle, H. B., 1992, An efficient *f-k* modeling approach in multiple elimination: 62nd Ann. Internat. Mtg., Soc. Expl. Geophys., Expanded Abstracts, 1077-1080.

Mayne, W. H., 1967, Practical considerations in the use of common-reflection point techniques: Geophysics, **32**, 225-229.

Meissner, R., 1965, Multiple events in refraction shooting: Geophys. Prosp., **13**, 617-658.

Michon, D., Wlodarczak, R., and Merland, J., 1971, A new method of cancelling multiple reflections—Souston: Geophys. Prosp., **19**, 615-625.

Monk, D. J., 1993, Wave-equation multiple suppression using constrained gross-equalization: Geophys. Prosp., **41**, 725-736.

Mouritsen, S. A., 1963, Multiple reflection problems in Western Canada: Geophysics, **28**, 882-893.

Pflueger, J. C., 1982, Multiple reflection analysis in *t-t* space: J. Can. Soc. Expl. Geophys., **18**, 34-42.

Reiter, E. C., and Toksöz, M. N., 1990, Joint imaging with primary reflections and deep water multiples: 60th Ann. Internat. Mtg., Soc. Expl. Geophys., Expanded Abstracts, 1294-1296.

Riley, D. C., and Claerbout, J. F., 1976, 2-D multiple reflections: Geophysics, **41**, 592-620.

Schneider, W. A., Prince, Jr., E. R., and Giles, B. F., 1978, A new data-processing technique for multiple attenuation exploiting differential normal moveout, *in* Webster, G. M., Ed., Deconvolution: Soc. Expl. Geophys., 307-321.

Taner, M. T., O'Doherty, R. F., and Baysal, E., 1991, Multiple suppression by multi-channel predictive deconvolution: 53rd Mtg. Eur. Assn. Expl Geophys., Abstracts, 22-23.

Verschuur, D. J., and Berkhout, A. J., 1993, Integrated multiple elimination and source wavefield estimation: 63rd Ann. Internat. Mtg., Soc. Expl. Geophys., Expanded Abstracts, 1095-1098.

Verschuur, D. J., Berkhout, A. J., and Wapenaar, C. P. A., 1992, Adaptive surface-related multiple elimination: Geophysics, **57**, 1166-1177.

Watson, R. J., 1965, Decomposition and suppression of multiple reflections: Geophysics, **30**, 54-71.

Williams, R. G., Haugen, A., and Cooper, N., 1991, Multiple suppression in the Barents Sea using the generalised radon transform as a notch velocity filter: 53rd Mtg. Eur. Assn. Expl Geophys., Abstracts, 58-59.

Static Computation and Correction

Blackburn, G. J., 1981, Seismic static corrections in irregular or steeply dipping water-bottom environments: Bull. Austr. Soc. Expl. Geophys., **12**, 93-100.

Booker, A. H., Linville, A. F., and Wason, C. B., 1976, Long wavelength static estimation: Geophysics, **41**, 939-959.

Coppens, F., 1985, First-arrival picking on common-offset trace collections for automatic estimation of static corrections: Geophys. Prosp., **33**, 1212-1231.

Cunningham, A. B., 1974, Refraction data from single-ended refraction profiles: Geophysics, **39**, 292-301.

deGroot-Hedlin, C., 1991, Removal of static shift in two dimensions by regularized inversion (short note): Geophysics, **56**, 2102-2106.

Hanot, F., and Renoux, P., 1991, Petrophysical variation in the Senonian chalk of the Paris Basin and their influence on static corrections: First Break, **9**, 515-526.

Hileman, J. A., Embree, P., and Pflueger, J. C., 1968, Automated static corrections: Geophys. Prosp., **16**, 326-358.

Kotcher, J. S., Gardner, G. H. F., and McDonald, J. A., 1984, Modeling the effect of static errors in areal seismic data caused by glacial erosion over carbonate reefs: Geophysics, **49**, 1-16.

Laski, J. D., 1970, Simultaneous estimation of parameters of reflection events (depth dip velocity) and relative static corrections: Geophys. Prosp., **18**, 269-276.

Lawton, D. C., 1989, Computation of refraction static corrections using first-break traveltime differences: Geophysics, **54**, 1289-1296.

Marsden, D., 1993a, Static corrections—a review, part 1: The Leading Edge, **12**, 43-49.

Marsden, D., 1993b, Static corrections—a review, part 2: The Leading Edge, **12**, 115-120.

Marsden, D., 1993c, Static corrections—a review, part 3: The Leading Edge, **12**, 210-216.

Saghy, G., and Zelei, A., 1975, Advanced method for self-adaptive estimation of residual static corrections: Geophys. Prosp., **23**, 259-274.

Steeples, D. W., Miller, R. D., and Black, R. A., 1990, Static corrections from shallow-reflection surveys: Geophysics, **55**, 769-775.

Taner, M. T., and Koehler, F., 1981, Surface consistent corrections: Geophysics, **46**, 17-22.

Taner, M. T., Koehler, F., and Alhilali, K. A., 1974, Estimation and correction of near-surface time anomalies: Geophysics, **39**, 441-463.

Taner, M. T., Lu, L., and Baysal, E., 1991, Static corrections: time, amplitude and phase: 61st Ann. Internat. Mtg., Soc. Expl. Geophys., Expanded Abstracts, 1399-1401.

Wiest, B., and Edelmann, H. A. K., 1984, Static corrections for shear-wave sections: Geophys. Prosp., **32**, 1091-1102.

Wiggins, R. A., Larner, K. L., and Wisecup, R. D., 1976, Residual static analysis as a general linear inverse problem: Geophysics, **41**, 922-938.

Zelei, A., and Saghy, G., 1980, Residual static corrections—iterative solution analysis: Geophys. Prosp., **28**, 175-184.

Seismic Velocity Analysis

Al-Chalabi, M., 1973, Series approximation in velocity and traveltime computations: Geophys. Prosp., **21**, 783-795.

Al-Chalabi, M., 1994, Seismic velocity—a critique: First Break, **12**, 589-596.

Al-Yahya, K. M., 1989, Velocity analysis by iterative profile migration: Geophysics, **54**, 718-729.

Ameely, L., Krey, Th., Muhtadie, F., Rau, H. F., and Rist, H., 1983, Migration in the presence of a rugged interface with high-velocity contrast: Geophys. Prosp., **31**, 561-573.

Anstey, N. A., 1991, Velocity in thin section: First Break, **9**, 449-457.

Artley, C., and Hale, D., 1994, Dip-moveout processing for depth-variable velocity: Geophysics, **59**, 610-622.

Beasley, C. J., and Lynn, W., 1992, The zero-velocity layer: migration from irregular surfaces: Geophysics, **57**, 1435-1443.

Becquey, M., Lavergne, M., and Willm, C., 1979, Acoustic impedance logs computed from seismic traces: Geophysics, **44**, 1485-1501.

Berryhill, J. A., 1986, Submarine canyons—velocity replacement by wave equation datuming before stack: Geophysics, **51**, 1572-1579.

Bickel, S. H., 1990, Velocity-depth ambiguity of reflection traveltimes: Geophysics, **55**, 266-276.

Bleistein, N., Cohen, J. K., and Hagin, F. G., 1985, Computational and asymptotic aspects of velocity inversion: Geophysics, **50**, 1253-1265.

Brzostowski, M. A., and McMechan, G. A., 1992, 3-D tomographic imaging of near-surface seismic velocity and attenuation: Geophysics, **57**, 396-403.

Byun, B. S., Corrigan, D., and Gaiser, J. E., 1989, Anisotropic velocity analysis for lithology discrimination: Geophysics, **54**, 1564-1574.

Cao, D., Beydoun, W. B., Singh, S. C., and Tarantola, A., 1990, A simultaneous inversion for background velocity and impedance maps: Geophysics, **55**, 458-469.

Cook, E. E., and Taner, M. T., 1969, Velocity spectra and their use in stratigraphic and lithologic differentiation: Geophys. Prosp., **17**, 433-448.

Crampin, S., 1986, Anisotropy and transverse isotropy: Geophys. Prosp., **34**, 94-99.

Cross, G. M., and Knoll, M. D., 1991, Diffraction-based velocity estimates from optimum offset seismic data: Geophysics, **56**, 2070-2079.

Cunningham, A. B., and Heffring, H. H., 1980, Interpretation of velocity spectra: Geophysics, **45**, 1741-1752.

Deregowski, S. M., 1990, Common-offset migrations and velocity analysis: First Break, **8**, 224-234.

Domenico, S. N., 1984, Rock lithology and porosity determination from shear and compressional wave velocity: Geophysics, **49**, 1188-1195.

Douze, E. J., and Laster, S. J., 1979, Statistics of semblance (short note): Geophysics, **44**, 1999-2003.

Eckhardt, W., 1994, Velocity model updating using image gathers: Geophys. Prosp., **42**, 975-986.

Ferber, R.-G., 1994, Migration to multiple offset and velocity analysis: Geophys. Prosp., **42**, 99-112.

Gardner, G. H. F., French, W. S., and Matzuk, T., 1974, Elements of migration and velocity analysis: Geophysics, **39**, 811-825.

Gardner, G. H. F., Gardner, L. W., and Gregory, A. R., 1974, Formation velocity and density—the diagnostic basics for stratigraphic traps: Geophysics, **39**, 770-780.

Gardner, G. H. F., Gardner, L. W. and Gregory, A. R., 1985, Formation velocity and density—the diagnostic basics for stratigraphic traps: Geophysics, **50**, 2085-2095.

Garotta, R., and Michon, D., 1967, Continuous analysis of the velocity function and of the moveout corrections: Geophys. Prosp., **15**, 584-597.

Gazdag, J., and Sguazzero, P., 1984, Interval velocity analysis by wave extrapolation: Geophys. Prosp., **32**, 454-479.

Green, C. H., 1938, Velocity determinations by means of reflection profiles: Geophysics, **3**, 295-305.

Gulunay, N., 1991, High resolution CVS: generalized covariance measure: 61st Ann. Internat. Mtg., Soc. Expl. Geophys., Expanded Abstracts, 1264-1267.

Hagedoorn, J. G., 1954, A practical example of an anisotropic velocity-layer: Geophys. Prosp., **2**, 52-60.

Hale, D., and Artley, C., 1993, Squeezing dip moveout for depth-variable velocity: Geophysics, **58**, 257-264.

Harlan, W. S., Claerbout, J. F., and Rocca, F., 1984, Signal/noise separation and velocity estimation: Geophysics, **49**, 1869-1880.

Helbig, K., 1994, Geometric reasoning: different kinds of average velocity: The Leading Edge, **13**, 117-118.

Hughes, D., 1985, Velocity without tears: The Leading Edge, **4**, No. 2, 50-52.

Ivansson, S., 1985, A study of methods for tomographic velocity estimation in the presence of low-velocity zones: Geophysics, **50**, 969-988.

Justice, J. H., 1986, Interval velocity analysis from VSP surveys: J. Can. Soc. Expl. Geophys., **22**, 33-43.

Landa, E., Kosloff, D., Keydar, S., Koren, Z., and Reshef, M., 1988, A method for determination of velocity and depth from seismic reflection data: Geophys. Prosp., **36**, 223-243.

Landa, E., Thore, P., Sorin, V., and Koren, Z., 1991, Interpretation of velocity estimates from coherency inversion: Geophysics, **56**, 1377-1383.

Larner, K., and Cohen, J. K., 1993, Migration error in transversely isotropic media with linear velocity variation in depth: Geophysics, **58**, 1454-1467.

Lavergne, M., and Willm, C., 1977, Inversion of seismograms and pseudo velocity logs: Geophys. Prosp., **25**, 231-250.

Levin, F. K., 1971, Apparent velocity from dipping interface reflections: Geophysics, **36**, 510-516.

Lines, L. R., Rahimian, F., and Kelly, K. R., 1993, A model-based comparison of modern velocity analysis methods: The Leading Edge, **12**, 750-754.

Lines, L., 1993, Ambiguity in analysis of velocity and depth (short note): Geophysics, **58**, 596-597.

Lynn, W., Gonzalez, A., and Mackay, S., 1991, Where are the fault-plane reflections?: 61st Ann. Internat. Mtg., Soc. Expl. Geophys., Expanded Abstracts, 1151-1154.

Naess, O. E., and Bruland, L., 1981, Velocity analysis using iterative stacking: Geophys. Prosp., **29**, 1-20.

Neidell, N. S., and Taner, M. T., 1971, Semblance and other coherency measures for multichannel data: Geophysics, **36**, 482-497.

Purnell, G. W., 1992, Imaging beneath a high-velocity layer using converted waves: Geophysics, **57**, 1444-1452.
Resnick, J. R., Shuey, R. T., Spindler, D. A., and Wyld, H. W., 1985, Tomographic determination of velocity and depth in laterally varying media: Geophysics, **50**, 903-923.
Robertson, J. D., 1987, Carbonate porosity from *S/P* traveltime ratios: Geophysics, **52**, 1346-1354.
Ross, W. S., 1994, Velocity-depth ambiguity in seismic traveltime data: Geophysics, **59**, 830-843.
Sattlegger, J. W., 1975, Migration velocity determination—part I, philosophy: Geophysics, **40**, 1-5.
Schneider, W. A., and Backus, M. M., 1968, Dynamic correlation analysis: Geophysics, **33**, 105-126.
Sena, A. G., and Toksöz, M. N., 1993, Kirchoff migration and velocity analysis for converted and nonconverted waves in anisotropic media: Geophysics, **58**, 265-276.
Shah, P. M., 1973, Use of wavefront curvature to relate seismic data with subsurface parameters: Geophysics, **38**, 812-825.
Sherwood, J. W. C., and Poe, P. H., 1972, Continuous-velocity estimation and seismic wavelet processing: Geophysics, **37**, 769-787.
Squires, L. J., Blakeslee, S. M., and Stoffa, P. L., 1992, The effects of statics on tomographic velocity reconstructions: Geophysics, **57**, 353-362.
Stickland, P., Beresford, G., and Mair, D., 1991, Improving velocity analysis beneath a submarine canyon using velocity re-datuming: Expl. Geophys., **22**, 583-589.
Taner, M. T., 1991, Migration velocity computation using plane wave decomposed data: 53rd Mtg. Eur. Assn. Expl. Geophys., Abstracts, 50-51.
Taner, M. T., and Koehler, F., 1969, Velocity spectra—digital computer derivation and applications of velocity functions: Geophysics, **34**, 859-881.
Taner, M. T., Cook, E. E., and Neidell, N. S., 1970, Limitations of the reflection seismic method—lessons from computer simulations: Geophysics, **35**, 551-573.
Taner, M. T., Lu, L., Postma, R. W., and Baysal, E., 1991, Depth migration velocity analysis: 61st Ann. Internat. Mtg., Soc. Expl. Geophys., Expanded Abstracts, 1218-1221.
Tieman, H. J., 1994, Investigating the velocity-depth ambiguity of reflection traveltimes: Geophysics, **59**, 1763-1773.
Uhrig, L. F., and van Melle, F. A., 1955, Velocity anisotropy in stratified media: Geophysics, **20**, 774-779.
Wyatt, K. D., Wyatt, S. B., Towe, S. K., Layton, J. E., von Seggern, D. H., and Brockmeier, C. A., 1994, Building velocity-depth models for 3-D depth migration: The Leading Edge, **13**, 862-866.
Yilmaz, O., 1989, Velocity stack processing: Geophys. Prosp., **37**, 357-382.
Yilmaz, O., and Chambers, R. E., 1984, Migration velocity analysis by wavefield extrapolation: Geophysics, **49**, 1664-1674.

Wavelet Estimation

Angeleri, G. P., 1983, A statistical approach to the extraction of the seismic propagating wavelet: Geophys. Prosp., **31**, 726-747.
Barnes, A. E., 1991, Instantaneous frequency and amplitude at the envelope peak of a constant-phase wavelet (short note): Geophysics, **56**, 1058-1060.
Berkhout, A. J., 1977, Least-squares inverse filtering and wavelet deconvolution: Geophysics, **42**, 1369-1383.
Bois, P., 1968, Determination de l'impulsion sismique: Geophys. Prosp., **16**, 4-20.
Brown, R. L., McElhattan, W., and Santiago, D. J., 1988, Seismic interpretation 12—wavelet estimation, an interpretive approach: The Leading Edge, **7**, 16-19.
Buttkus, B., 1975, Homomorphic filtering—theory and practice: Geophys. Prosp., **23**, 712-748.
Canales, L. L., 1989, Stable wavelet estimation: 59th Ann. Internat. Mtg., Soc. Expl. Geophys., Expanded Abstracts, 1217.
Carrion, P. M., dos S. Sacramento, S. and da C. Pestana, R., 1990, Source wavelet and its angular spectrum from plane-wave seismograms: Geophysics, **55**, 1026-1035.
de Voogd, N., 1974, Wavelet shaping and noise reduction: Geophys. Prosp., **22**, 354-369.
Kelamis, P. G., and Chiburis, E. F., 1988, Post critical wavelet estimation and deconvolution: Geophys. Prosp., **36**, 504-522.
Lange, J. N., and Almoghrabi, H. A., 1988, Lithology discrimination for thin layers using wavelet signal parameters: Geophysics, **53**, 1512-1519.
Lazear, G. D., 1984, An examination of the exponential decay method of mixed-phase wavelet estimation: Geophysics, **49**, 2094-2099.
Lazear, G. D., 1993, Mixed-phase wavelet estimation using fourth-order cumulants: Geophysics, **58**, 1042-1051.
Levy, S., and Oldenburg, D. W., 1987, Automatic phase correction of common-midpoint stacked data: Geophysics, **52**, 51-59.
Lindsey, J. P., 1988, Measuring wavelet phase from seismic data: The Leading Edge, **7**, No. 7, 10-16.
Loewenthal, D., Lee, S. S., and Gardner, G. H. F., 1985, Deterministic estimation of a wavelet using impedance type technique: Geophys. Prosp., **33**, 956-969.
Morgan, N. A., 1970, Wavelet maps—a new analysis tool for reflection seismograms: Geophysics, **35**, 447-460.
Mosegaard, K., 1985, Insufficiently sampled Q-values—a cause of wavelet distortion in the generation of synthetic seismograms: Geophys. Prosp., **33**, 817-827.
Newman, B. J., 1986, Deconvolution of noisy seismic data—a method for prestack wavelet extraction: Geophysics, **51**, 34-44.
Oldenburg, D. W., Levy, S., and Whittall, K. P., 1981, Wavelet estimation and deconvolution: Geophysics, **46**, 1528-1542.
Poggiagliolmi, E., and Allred, R. D., 1994, Detailed reservoir definition by integration of well and 3-D seismic data using space adaptive wavelet processing: The Leading Edge, **13**, 749-754.
Ricker, N., 1944, Wavelet functions and their polynomials: Geophysics, **9**, 314-323.

Ricker, N., 1953, Wavelet contraction wavelet expansion and the control of seismic resolution: Geophysics, **18**, 769-792.

Thomas, J. E., 1986, Multichannel estimate of the seismic wavelet (short note): Geophysics, **51**, 838-843.

Tygel, M., Huck, H., and Hubral, P., 1991, Mixed-delay wavelet deconvolution of the point-source seismogram: Geophysics, **56**, 1405-1411.

Walden, A. T., 1991, Wavelet estimation using the multitaper method: Geophys. Prosp., **39**, 625-642.

Wang, L.-X. and Mendel, J. M., 1992, Adaptive minimum prediction-error deconvolution and source wavelet estimation, using Hopfield neural networks: Geophysics, **57**, 670-679.

Ziolkowski, A. M., Lerwill, W. E., March, D. W., and Peardon, L. G., 1980, Wavelet deconvolution using a source scaling law: Geophys. Prosp., **28**, 872-901.

Chapter 11

Measurement and Identification of 3-D Coherent Noise Generated from Irregular Surface Carbonates

Carl J. Regone
Amoco EPTG
Tulsa, Oklahoma, U.S.A.

Abstract

One of the primary causes of poor seismic data quality is the masking of desired reflection signals by seismic waves scattered from topography or near-surface heterogeneities. For successful seismic acquisition and processing, one must be able to identify this scattered noise and quantify its strength relative to the desired signal so that appropriate measures can be taken to suppress it. An economical procedure has been developed that allows analysis of scattered noise. A field test consisting of a series of shots into a 3-D grid of receivers, whose dimensions depend on the range of noise velocities to be analyzed and the choice of temporal frequencies over which to make the analysis, makes it possible to identify events by measuring the direction and magnitude of their horizontal velocity. This 3-D grid also allows the formation of a series of nested areal arrays with graduated attenuation levels that can be used to quantify the signal-to-noise ratio by attenuating the scattered noise by known amounts and observing the level at which desired signal appears. For any given shot, the seismic traces recorded by the grid can be analyzed and a graphic workstation can be used to produce radar-like displays of seismic events showing their arrival time, horizontal velocity, azimuth, temporal frequency bandwidth, and amplitude. With this information, one can design field arrays of the appropriate size and attenuation level to suppress the scattered noise. Often the required arrays are so large that they must be recorded in segments to prevent loss of desired signal due to necessary static and cross-line dip corrections. This array approach is most suitable for 2-D and wide-line profile acquisition schemes. Noise suppression levels of 20–30 dB have often been achieved, which has successfully revealed signal in many areas, such as Edwards Plateau in West Texas, karstic limestone regions of the former Yugoslavia, and Black Warrior Basin of Mississippi.

Introduction

Poor seismic data quality is often due to topography or near-surface heterogeneities that cause seismic waves propagating away from the source and traveling horizontally on or near the surface to be scattered back into the recording spread. These scattered surface or near-surface events carry little information about the subsurface and create problems when they encounter the recording spread at levels higher than the desired signal. These problems have long been known and extensively studied. For a thorough review of work done on scattering phenomena in the near surface, I refer the reader to Levander (1990).

Two important phenomena that apply to the problem of seismic acquisition on carbonates are the scattering of surface waves from topography (Hudson and Knopoff, 1967) and the conversion of body waves to surface waves by topography (Aki and Richards, 1980). The rough topography often associated with carbonate terrains is the major source of scattered surface wave energy in these areas, and the reflection coefficients can be very large. The conversion of body waves to surface waves caused by topography means that surface wave energy can be present throughout the entire seismic shot record and is not confined to the surface wave's direct-arrival noise cone. Thus, one cannot simply record long source-to-receiver offsets and successfully "out run" the surface wave problem.

Numerous efforts in the area of seismic exploration have focused on attenuating surface waves on carbonates through the use of field arrays. Parr and Mayne (1955) showed improvements in the quality of seismic data recorded on surface carbonates by using large areal Chebychev-weighted source arrays designed to suppress the generation of surface wave energy. Husseini et al. (1981) attempted to optimize field arrays to attenuate surface waves on carbonates by predicting surface wave characteristics from *P*-wave velocity measurements in shallow uphole surveys.

The effect of scattered noise events on 2-D marine multifold seismic lines was studied by Larner et al. (1983), who showed that the common midpoint (CMP) stacking process can actually enhance the scattered energy. They showed noise reduction benefits of *f-k* filtering in both the common shot and common receiver domains. The performance of this method on problematic land data recorded with typical 2-D geometries is often hindered because the scattered noise events are often badly aliased in at least one of these domains. I have observed the failure of this method to adequately suppress scattered noise on 2-D land data even when both the common shot and common receiver domains were properly sampled for the noise. An example of such a failure is shown later.

As Parr and Mayne (1955) showed, adding an areal dimension to the acquisition procedure is often effective in improving the seismic data quality when scattered noise is present, but it is difficult to determine when these areal schemes must be used and what form they should take to ensure that the desired signal level is achieved. Therefore, a test is needed that facilitates both the identification of any scattered noise events that are present and the measurement of their strength relative to the desired signal.

Most field tests used to measure source-generated coherent noise, such as the walkaway noise test, are designed primarily for direct arrival events and therefore use a recording spread that has little or no areal extent. Seismic noise events that encounter such a recording spread from the cross-line direction are not easily recognized because they have hyperbolic moveout and often create a complex interference pattern that appears as a cone of seemingly random noise. Often, this noise cone prevents observation of even the in-line backscattered events, which normally would be easily recognized by their linear moveout. In many areas of poor signal quality, the desired reflection events are buried in this noise. For seismic data recorded with typical 2-D geometries, it is usually impossible to tell how deeply they are buried.

This paper expands on earlier work by Regone and Rethford (1990). It presents a field procedure and processing method that takes advantage of the large number of channels available with modern recording systems and the graphic capabilities of modern workstations to identify and quantify scattered noise problems.

Test Configuration and Design Parameters

In many problem areas, 2-D seismic records exhibit only noise events, and without discernible signal, no direct measure of the signal-to-noise ratio (S/N) is possible. My approach in solving this problem is to infer the strength of the noise (relative to the desired signal) by forming a series of nested areal arrays with graduated attenuation levels. By passing the data through this series of arrays and observing the attenuation level at which signal appears, an indication of the S/N is obtained.

As indicated in Figure 1, an array with element spacing *d* attenuates spatial wavelengths from some maximum L_{max} to some minimum L_{min} and imposes a us-

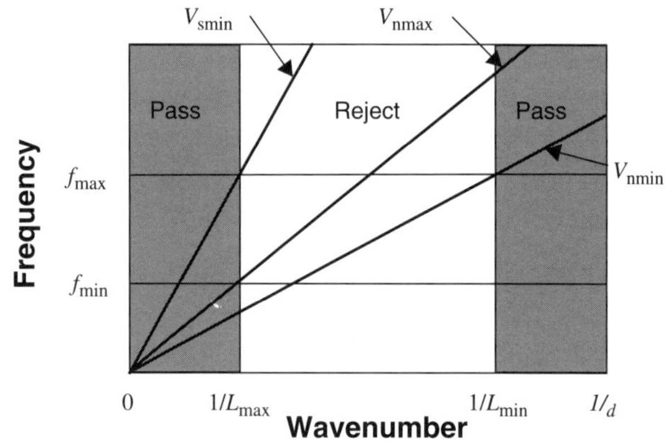

$$L_{max} = \frac{V_{nmax}}{f_{min}} \qquad L_{min} = \frac{V_{nmin}}{f_{max}}$$

$$d = \frac{L_{max}L_{min}}{L_{max}+L_{min}} = \frac{L_{max}}{\beta}, \text{ where } \beta \equiv \frac{L_{max}}{L_{min}} + 1 = \frac{V_{nmax}}{V_{nmin}} \times \frac{f_{nmax}}{f_{nmin}} + 1$$

Fig. 1. An array imposes a usable temporal frequency bandwidth on seismic data.

able temporal frequency bandwidth on the data. For frequencies above f_{max}, the signal is lost and noise is passed, while for frequencies below f_{min}, noise is passed. The noise below f_{min} and above f_{max} can be removed by applying a temporal bandpass filter to the recorded data. The noise between f_{min} and f_{max} must be attenuated by the array.

For the test procedure described here, it is necessary to be able to vary the amount of attenuation provided by the array while keeping the reject-zone bandwidth constant. Attenuation levels far above that provided by unit-weighted arrays are also needed. Such high-performance arrays can be constructed by combining weighted elements in the field before recording or by recording each element individually and synthesizing the array in processing. A problem common to all high-performance arrays is the sensitivity of these arrays to variations in element spacing and weighting. Because earth-coupling effects may cause geophone-to-geophone amplitude variations within the array and because physical objects such as trees and bushes often get in the way, the desired weights and element spacing, and therefore, the desired attenuation levels are usually not achieved for field-implemented high-performance arrays (Newman and Mahoney, 1973).

If subarrays of unit-weighted geophones are placed closely enough in the field so that the recorded output of each subarray can be treated as one element of a larger synthesized array and if the number of geophones in the subarray is large enough, variations in the location of individual geophones in the subarray will average statistically so that the effective center of the subarray will be very close to the desired location. Because the subarray output is a recorded seismic trace, amplitude variations due to earth coupling can be eliminated during data processing. Synthesized high-performance arrays thus have a much better chance of succeeding than their field-implemented counterparts. Therefore, synthesized arrays are used for this test procedure.

Designing a high-performance array for the desired test procedure requires a selection of the range of noise velocities to be analyzed, V_{nmax}/V_{nmin}, the desired temporal frequency band over which to view the data, f_{max}/f_{min}, and the maximum attenuation level to be attained during the analysis. These choices determine the width of the reject zone, L_{max}/L_{min}, the maximum number of elements n to use in the array synthesis, and the element spacing d. Once n and d are determined, we can make an areal array by placing an $n \times n$ grid on the ground. Figure 2 shows a typical field test arrangement for a 15×15 grid. The element spacing d is the same in both the in-line and cross-line directions, where *in-line* is defined as being parallel to the source line and *cross-line* is defined as perpendicular to it. Each of the 225 elements is an individually recorded areal subarray with an effective array length in both the in-line and cross-line directions equal to the element spacing of the grid. Because the grid is usually square, the appearance of the receiver spread is that of a large box. For this reason, we refer to the entire test procedure as a *box wave test.*

In addition to the channels required to record the grid, some extra recording channels are sometimes used to record a series of additional arrays all centered within the grid. These extra arrays might be various trial field arrays for a production-line scheme or arrays of three-component geophones for particle motion analysis or polarization filtering tests.

Since the means by which we hope to measure the strength of the noise relative to desired signal yields, for each source point, one output trace per array type, it is important that the source interval, *SI*, be small enough that the slowest signal event is not spatially aliased in the common-receiver domain. This implies that

$$SI \leq \frac{V_{smin}}{2f_{max}}, \tag{1}$$

where V_{smin} is the minimum horizontal velocity for the desired signal. Then, if these source points are spaced along a line that extends about a normal production spread length away from the grid, a series of common-receiver records can be constructed and compared for signal quality.

Although high-performance arrays can be constructed in a number of ways, the use of Chebychev weights is particularly attractive. Because the width of the reject zone and the level of attenuation are easily controlled by changing the element weights, Chebychev arrays allow great design flexibility (Parr and Mayne, 1955; Holzman, 1963). As shown by Holzman (1963), the minimum attenuation attained in the reject zone of a Chebychev array is a function of the number of elements in the array and the effective width of the reject zone. If we characterize the width of the reject zone in terms of a dimensionless bandwidth parameter β,

$$\beta \equiv \frac{L_{max}}{L_{min}} + 1 = \frac{V_{nmax}}{V_{nmin}} \times \frac{f_{max}}{f_{min}} + 1 = \frac{L_{max}}{d}, \tag{2}$$

then the attenuation provided by a Chebychev array with n elements is

$$A = 20 \log \left\{ \cosh\left[(n-1)\cosh^{-1}\left(1/\cos\frac{\pi}{\beta} \right) \right] \right\}, \tag{3}$$

Fig. 2. Typical grid geometry for a box wave test. Each element of the grid is individually recorded for each source point. The source interval is usually determined by spatial aliasing criteria for the desired signal.

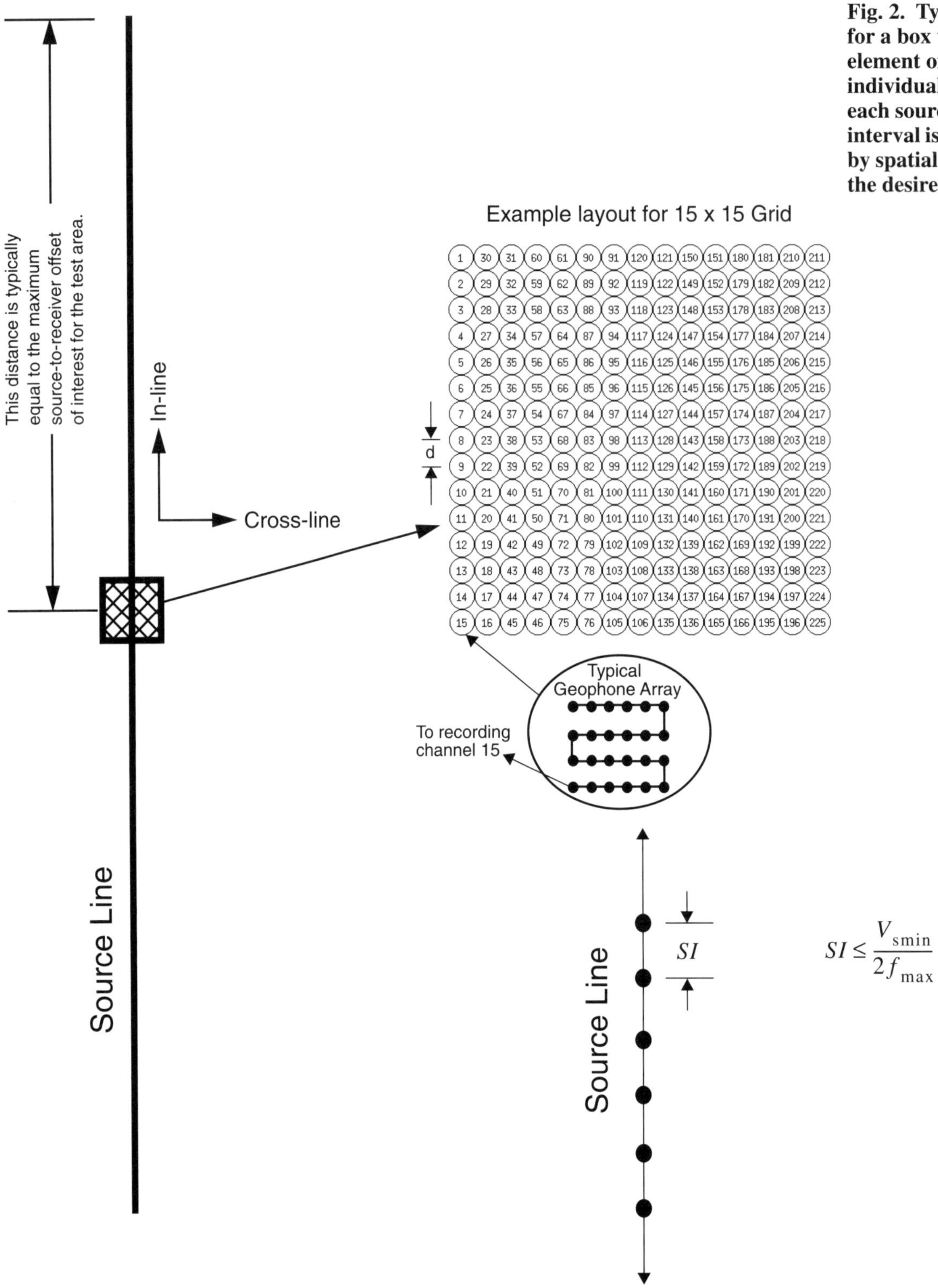

If we hold the width of the reject zone constant (β = constant), the usable frequency bandwidth for the array remains constant, and changes in the number of elements in the array merely raise or lower the attenuation level achieved. This can be seen graphically in Figure 3, where the attenuation levels achieved by a Chebychev array are plotted for the range of values for n and β that have found to be practical for a box wave test.

It is important to remember that there are two goals for this test procedure. The first is to identify any scattered coherent noise events and to characterize them by their horizontal velocity. The second is to quantify their

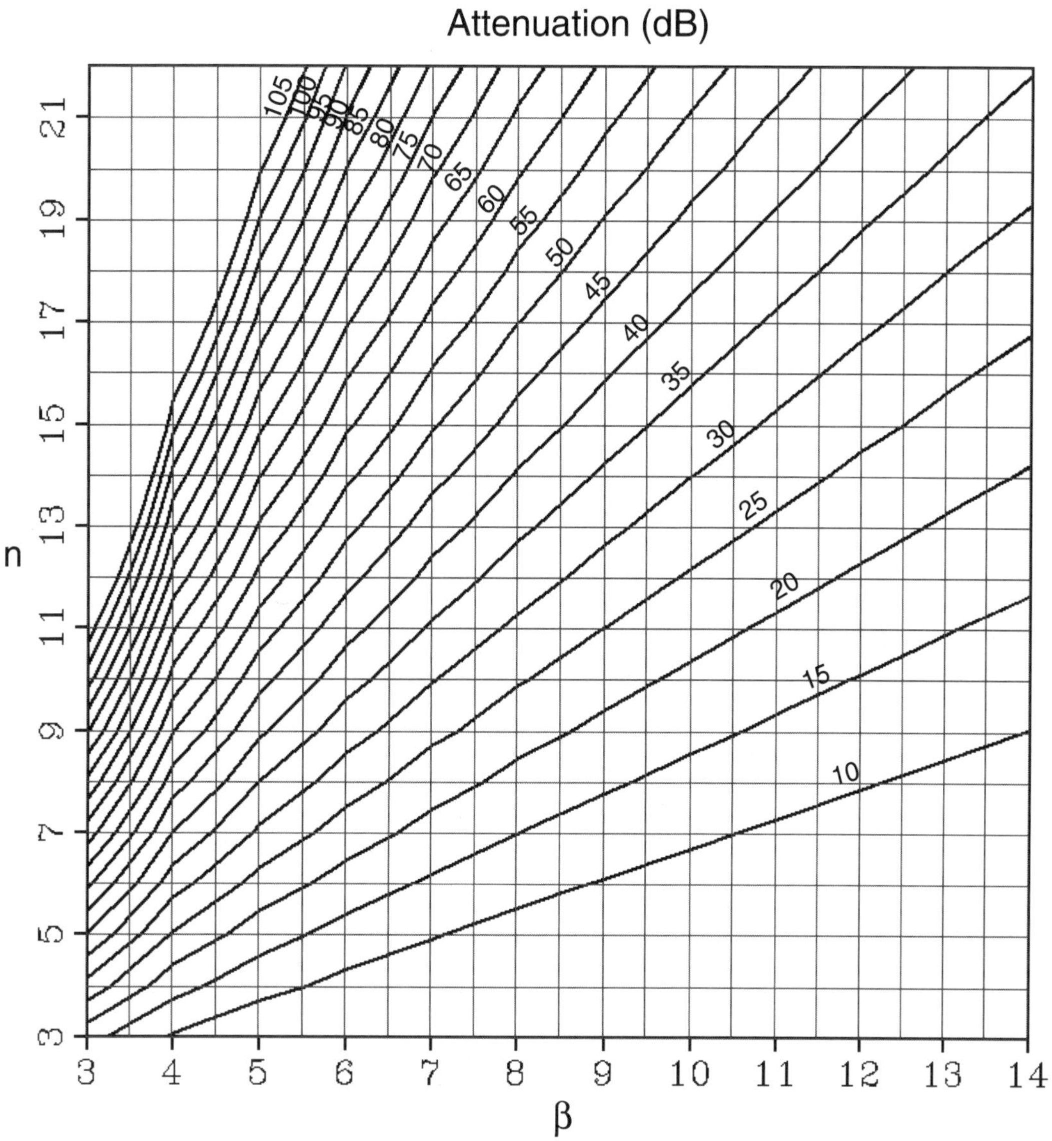

Fig. 3. The attenuation level achieved by a Chebychev array is a function of *n* and β only. These are the attenuation levels achieved for the values of *n* and β that are most practical for box wave tests. See equation (3) in the text.

strength relative to the desired signal. To see how this can be accomplished, consider three examples of field tests of this analysis method. In the first example, the field experiment was conducted on a massive carbonate, and the coherent noise consisted primarily of scattered Rayleigh waves. In the second and third examples, the coherent noise consisted of both scattered ground roll and head wave events. Comparisons are made in the second example between a 2-D line with in-line only arrays and a 2-D line with areal arrays of sufficient size to attenuate the ground roll events but not the head waves. Comparisons are made in the third example among a 2-D line with in-line only arrays, a 2-D line with areal arrays of sufficient size to attenuate the ground roll events but not the head waves, and a wide-line profile of sufficient width to attenuate both types of scattered noise events.

West Texas Example

The test location shown in this example is in the Val Verde basin northeast of Sanderson, Texas. This is mesa and canyon country and is part of the Edwards Plateau. The surface rocks are primarily massive carbonates. In some of the primary drainage valleys, the in-situ carbonates have been replaced by carbonate gravel. Thus, the mechanical properties of the rocks occasionally change dramatically from mesa to valley floor. The grid for the Sanderson area box wave test was located on top of a massive carbonate mesa. The source line used for the test extended off the mesa and onto a feature called Big Canyon, which is one of the major drainage pathways in which the in-situ carbonates have been replaced by gravel. The source used for this test was vibroseis, and there were a few small skips due to access problems. These source skips manifest themselves as dead traces in some of the following figures.

The field geometry for the Sanderson box wave test is shown in Figure 4. With this 13 × 13 grid, a large number of arrays can be synthesized. I illustrate the process by choosing $\beta = 9$. With $d = 75$ ft (23 m) and $V_n =$ 6750 ft/s (2060 m/s), this choice of β corresponds to *f*min = 10 Hz. From Figure 3, $\beta = 9$ and $n = 13$ indicate an attenuation level of 31 dB. Other choices of β alter *f*min and the ultimate attenuation level achievable by a 13 × 13 grid. For example, raising *f*min to 15 Hz reduces β to 6, and from Figure 3 the attenuation level for $n = 13$ and $\beta = 6$ is 51 dB.

For each source record, the elements that are shaded in Figure 5c should be summed in the cross-line direction after being filtered, scaled, and given the appropriate Chebychev weights corresponding to a β of 9 and

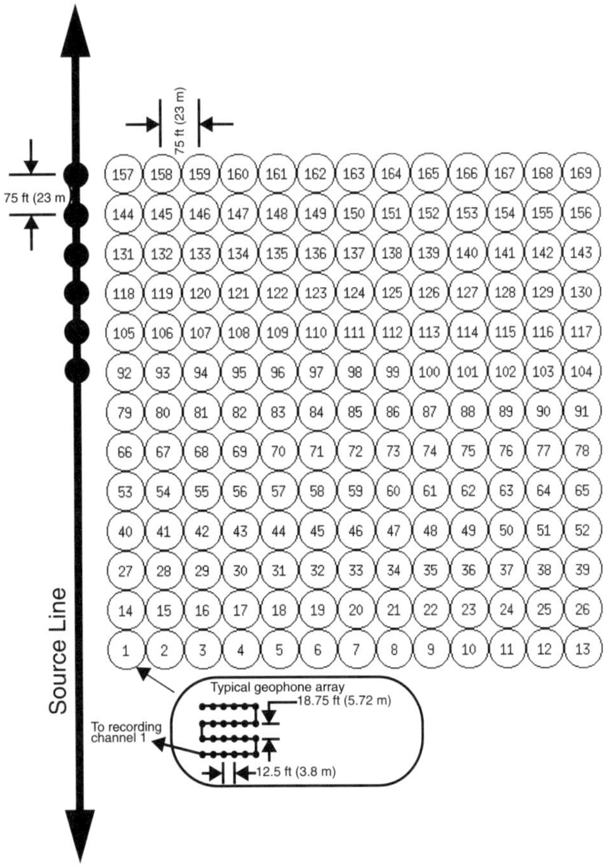

Fig. 4. Geometry description for the Sanderson, Texas, area box wave test. Each of the 169 elements of the grid were recorded for every source point.

$n = 11$. Then, the 121 shaded elements can be reduced to an 11-element column oriented in the in-line direction. If these 11 elements are given the same weights relative to their position in the array as before and summed, a single output trace is produced that has the attenuation level shown in Figure 6c in both the in-line and cross-line directions.

A similar process produces the areal arrays for the remaining parts of Figure 5 (a, b, d). Figure 6 shows the resulting attenuation levels achieved. Note that the width of the reject zone is the same in all of the array response plots in Figure 6. The effective length of each of the four arrays in both the in-line and cross-line directions is $L_{max} = \beta d = 9 \times 75 = 675$ ft (206 m).

Identifying Scattered Noise

The first goal of a box wave test is to detect the presence of scattered noise and to identify it based on its horizontal velocity. There are numerous ways of accomplishing this once a 3-D grid of receivers has been recorded. A few of them are shown here. I begin by showing the seismic record (Figure 7) obtained by extracting only the center trace from each source record. To the right of the skipped shots around shot 632, the direct arrival surface wave is nondispersive, which is consistent with a massive carbonate. To the left of shot 632, however, the surface waves have become highly dispersive, the head wave is stronger, and the general character of the seismic traces is markedly different than those to the right. What has happened is that the mesa ends here, and the shots to the left of 632 are located on the gravel-filled Big Canyon feature mentioned earlier. The mechanical properties of the rocks have changed radically, and the appearance of the seismic data reflects this.

For the shots located on the mesa (to the right of shot 632), many events resulting from scattered surface waves are seen. Some of these have hyperbolic shape, and none of the events seen on this record represent primary reflection data that we would call desired signal. The nature of these events is indicated in Figure 8. In Figure 8a, a Chebychev array has been formed from the elements of the grid that are shaded. For this in-line only array, β is 7, n is 13, and the attenuation achieved is about 42 dB. The in-line backscattered events have been eliminated, and the direct-arrival surface wave has been greatly reduced in strength. In Figure 8b, the same type of array has been formed, but this time in the cross-line direction. Now the in-line scattered events dominate the record, and the direct-arrival surface wave is undiminished. The disappearance of the hyperbolic events that

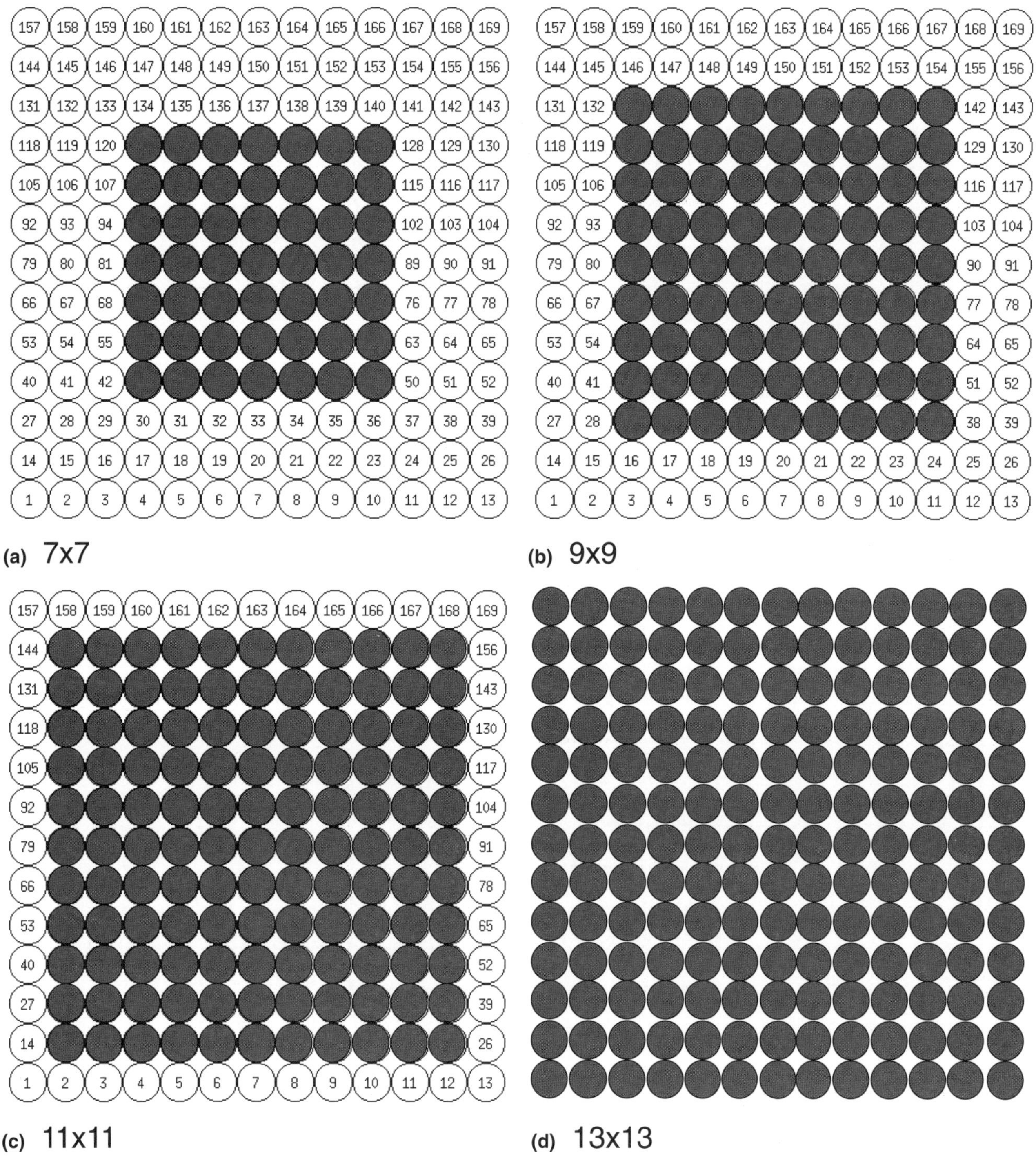

Fig. 5. The elements shaded in each of the grids depicted here were given the appropriate Chebychev weights corresponding to β = 9 and the value of *n* indicated. If this is done, the attenuation levels in Figure 6 are achieved in both the in-line and cross-line directions.

dominate the record of Figure 8 a suggests that they are merely scattered surface waves from the cross-line direction. Further evidence of this can be found in Figure 9 , in which the traces from shot record 728 only are ordered in both the in-line (Figure 9a) and cross-line (Figure 9b) directions. By measuring the apparent velocity of an event in orthogonal directions, the true horizontal velocity of the event can be calculated. When this process is performed, all events on this record are seen to have the same horizontal velocity as the surface wave.

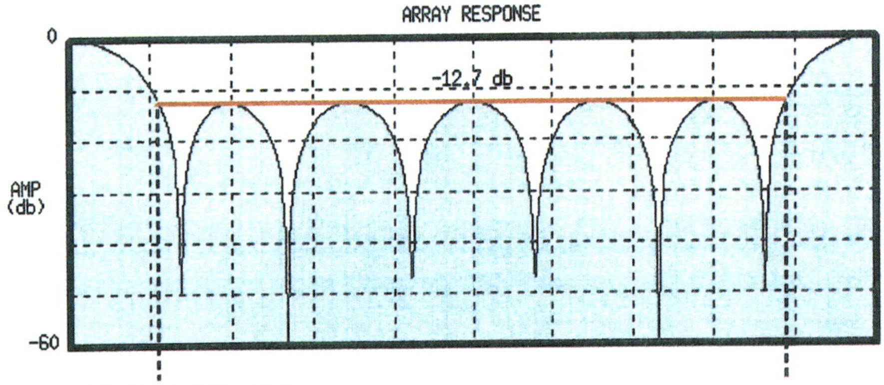

(a) n=7, β=9 (13 dB)

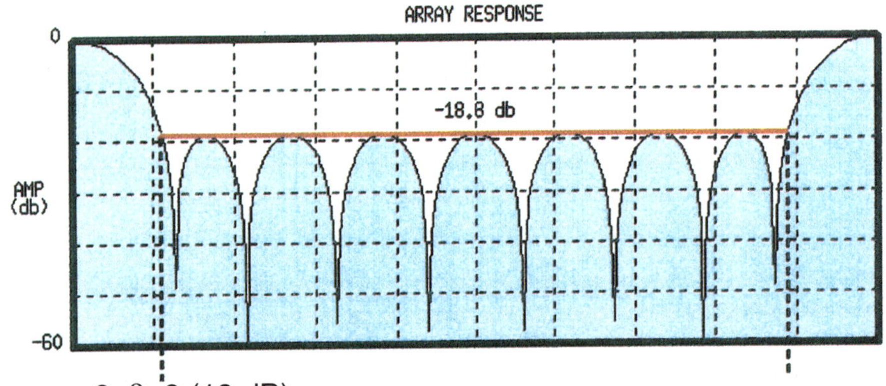

(b) n=9, β=9 (19 dB)

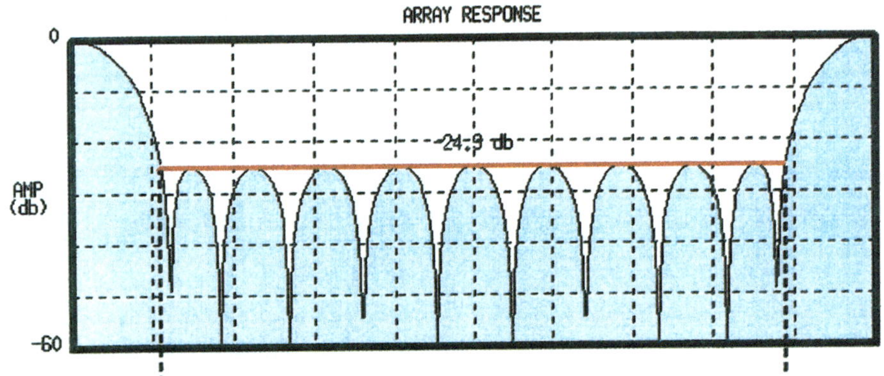

(c) n=11, β=9 (25 dB)

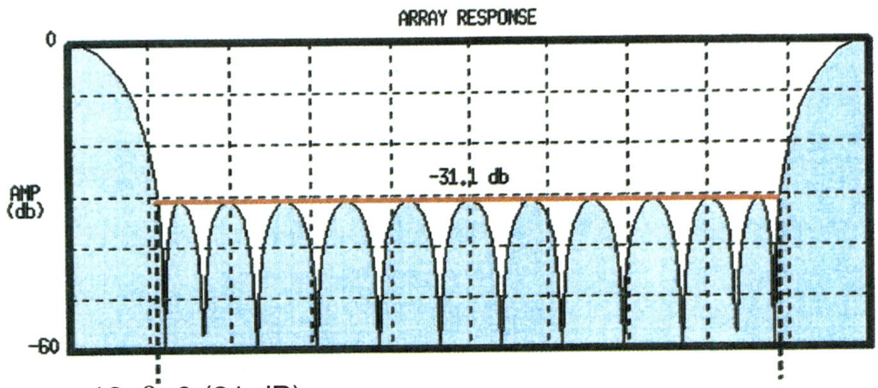

(d) n=13, β=9 (31 dB)

Fig. 6. Array response plots showing the attenuation levels achieved in both the in-line and cross-line directions by the areal Chebychev arrays formed by the summation of the shaded elements in Figure 5. Note that since β is constant, the width of the reject zone is the same in every plot. Thus, with constant β, increasing the number of elements, *n*, increases the attenuation level but does not increase the effective array length.

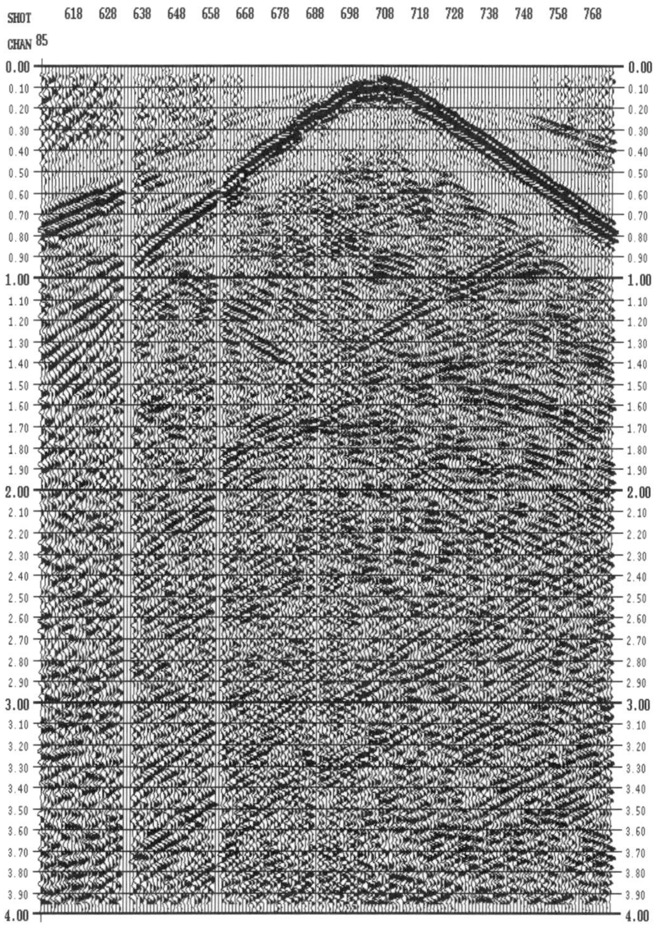

Fig. 7. The seismic record obtained by extracting only the center element of the grid (channel 85) for each source point. Each trace in this plot comes from a different source point. A 10–40 Hz bandpass temporal frequency filter and a 1000 ms AGC have been applied to the data. In this frequency band, the field array present on each trace has a negligible effect on the surface waves.

If there is signal present in the record, it is buried too deeply in the noise to be seen in these displays.

Another method for identifying the various noise events is shown in Figure 10. These images were generated by an interactive graphic program called Radar because of its radar-like displays. This program, developed by Jerry Ehlers, works on one shot record at a time and uses a simple delay and sum approach to effectively beam the grid through a range of dips and azimuths for any given arrival time of interest. In each of these displays, slowness (V^{-1}) is plotted against azimuth. The range of velocities displayed is determined by the choice of maximum slowness. For these displays, I chose a value corresponding to a velocity of 1500 m/s. Thus, events seen at the outer edge of the display would have a horizontal velocity of 1500 m/s. Three concentric circles appear in each display. The outermost circle corresponds to a velocity of 2000 m/s, the second to 3000 m/s, and the innermost to 6000 m/s; the center point of the display corresponds to infinite horizontal velocity.

The seismic record in Figure 9 was used to generate the displays in Figure 10. In Figure 10a, the arrival time chosen for analysis was 300 ms. From Figure 7, it is apparent that the dominant event occurring at a time of 300 ms for shot 728 is the direct arrival surface wave. Figure 10a shows this direct arrival surface wave as a single event at a velocity of 2000 m/s that has the same azimuth as a line from the source point to grid center. In Figure 10b, the arrival time chosen for analysis was 800 ms. With one notable exception, all the events in this display occur at a velocity of 2000 m/s and arrive at the grid from a wide range of azimuths. These are scattered surface waves. The notable exception is the event near the center of the display. This is a primary reflection that cannot be seen in Figures 7, 8, or 9 because it is masked by scattered surface waves. Figure 10c and 10d show more scattered surface waves arriving at the grid from a wide range of azimuths at arrival times of 1600 ms and 2000 ms, respectively.

Similar analyses performed on other shot records support the contention that nearly all the events seen on Figure 7 are scattered surface waves. Where is the signal? How deeply is it buried beneath the noise?

Quantifying the Signal-to-Noise Ratio

The main approach used to quantify the S/N is the formation of a nested series of high-performance arrays that was described earlier. The shaded elements of the grid shown in Figure 5a were given the Chebychev weights corresponding to $\beta = 9$ and $n = 7$ and were then summed to produce one output trace for each shot record. If the resulting traces are plotted adjacent to one another, the seismic record in Figure 11a is produced. In a similar manner, summing the shaded elements in Figures 5b, 5c, and 5d produces the seismic records in Figures 11b, 11c, and 11d. As the attenuation level increases, the events caused by scattered surface waves diminish and some reflection events appear.

Using this approach, one can identify the attenuation levels needed to cause the reflection events of interest to appear. For this example, some reflections appear at the 13 dB attenuation level (Figure 11a), while others are not discernible until the 31 dB attenuation level (Figure 11d) is reached. The acquisition and processing schemes chosen for a production seismic survey must achieve these levels of noise attenuation before the desired reflections can be seen.

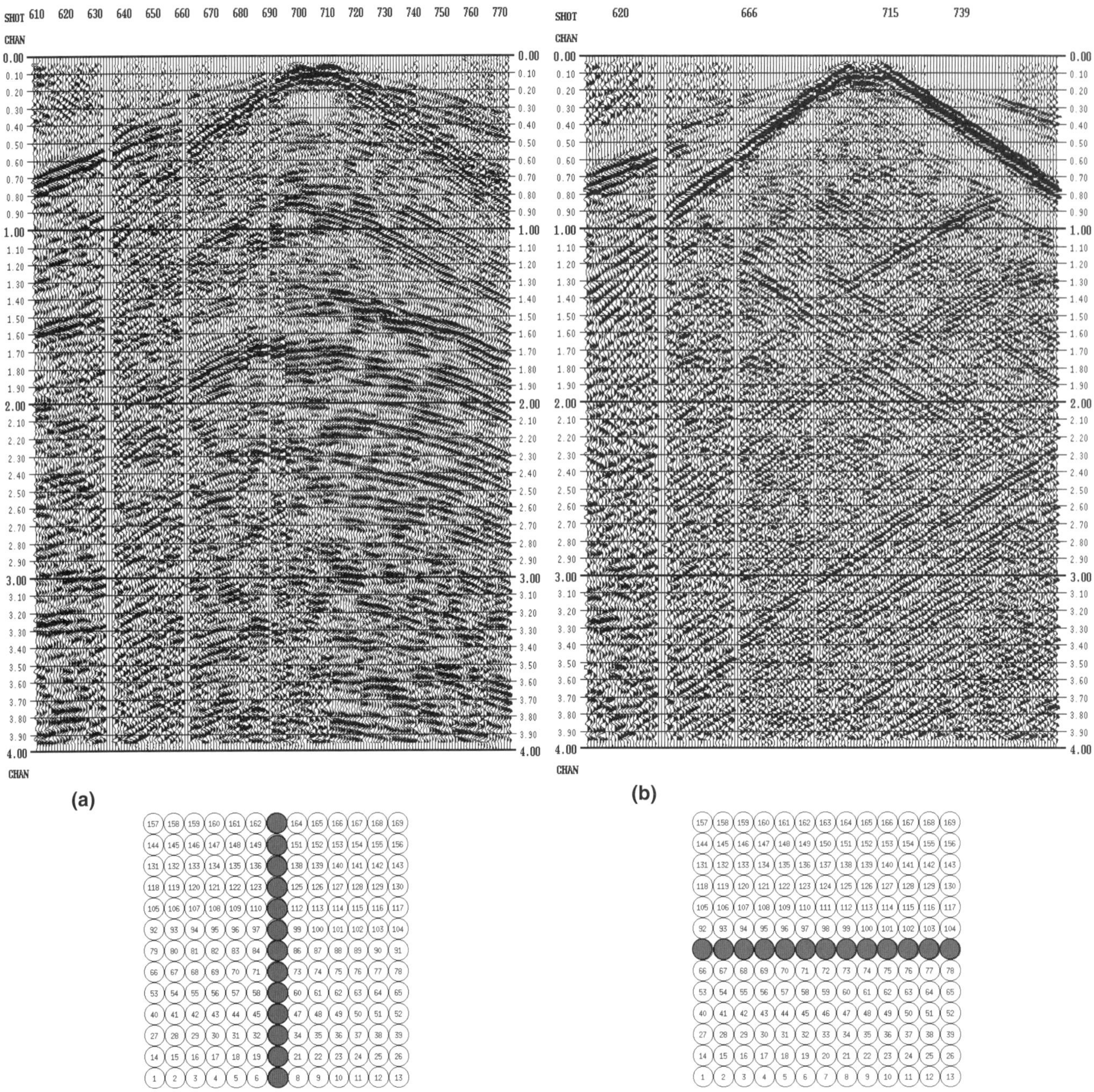

Fig. 8. The seismic records corresponding to (a) an in-line only and (b) a cross-line only Chebychev array formed by weighting and summing the elements of the grid indicated by shading. For each shot record, the grid produces one output trace per array type. This makes it possible to distinguish events scattered in the in-line direction from those scattered in the cross-line direction.

Another method that can be used to quantify the S/N is the Radar-type analysis shown earlier. An event in a Radar display is the result of a delay and sum corresponding to a particular velocity, azimuth, and arrival time. If the amplitude of each event is normalized relative to the total energy impacting the grid at that arrival time, the S/N can be read directly from the Radar display. In Figure 11d, four reflection events have been marked by arrows. Radar analyses performed at these four locations are shown in Figure 12. In each case, the S/N designated in the figure reflects the amplitude value displayed by the program for the reflection events located near the center of each display. The remarkable agreement between these values and those interpreted

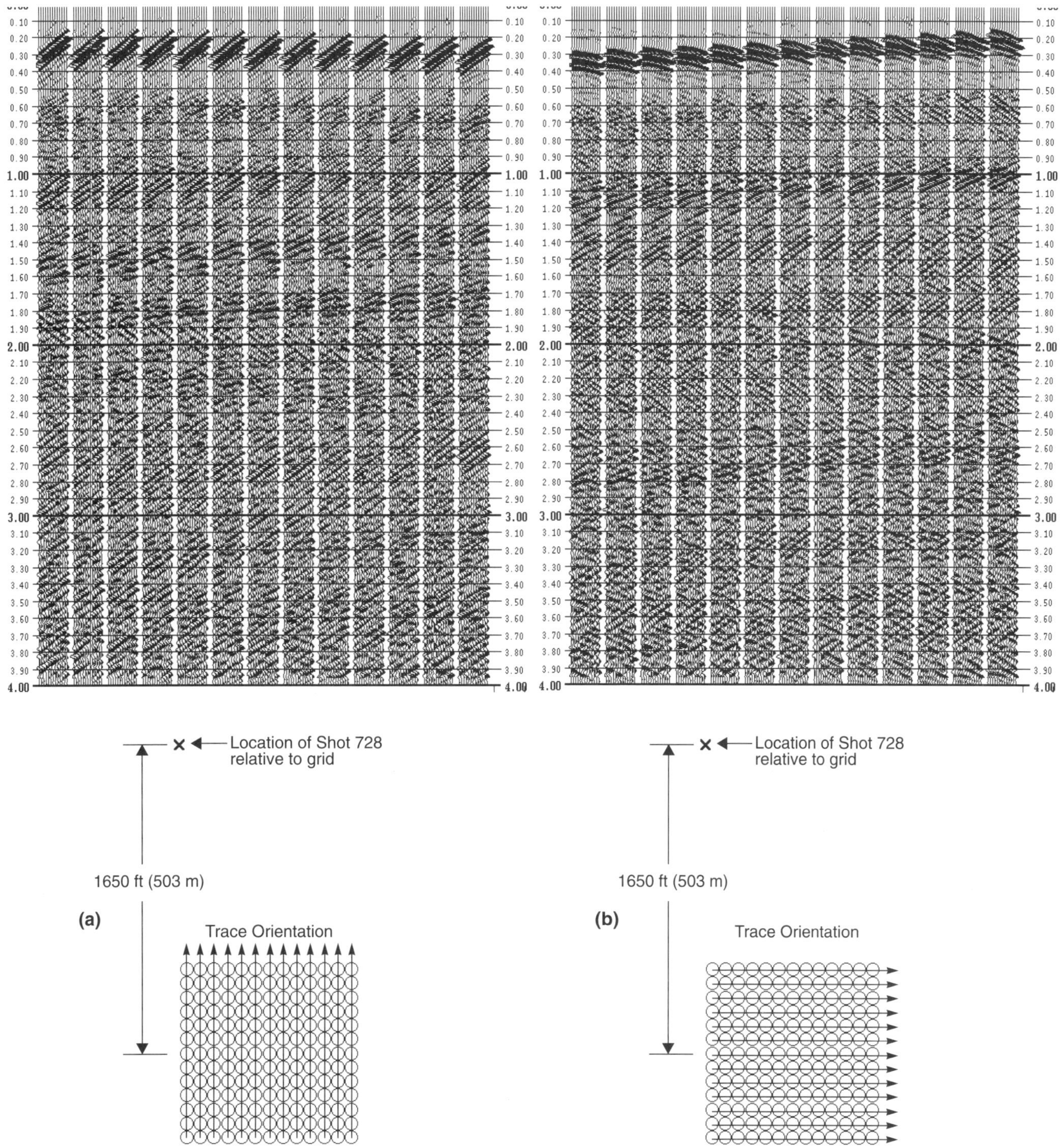

Fig. 9. In-line ordered and cross-line ordered displays for the traces from shot 728. By measuring the apparent velocity of an event in orthogonal directions, the true horizontal velocity of the event can be calculated.

from Figure 11 for the same events suggests that it may be possible to quantify the S/N for an area with just a few isolated shots and a Radar-type analysis method rather than an entire source line as was needed for the common-receiver records of Figure 11.

Example from the Former Yugoslavia

The second field example was recorded in the former Yugoslavia in 1989. The surface geology consisted largely of karstic limestone. A 19 × 19 grid was used

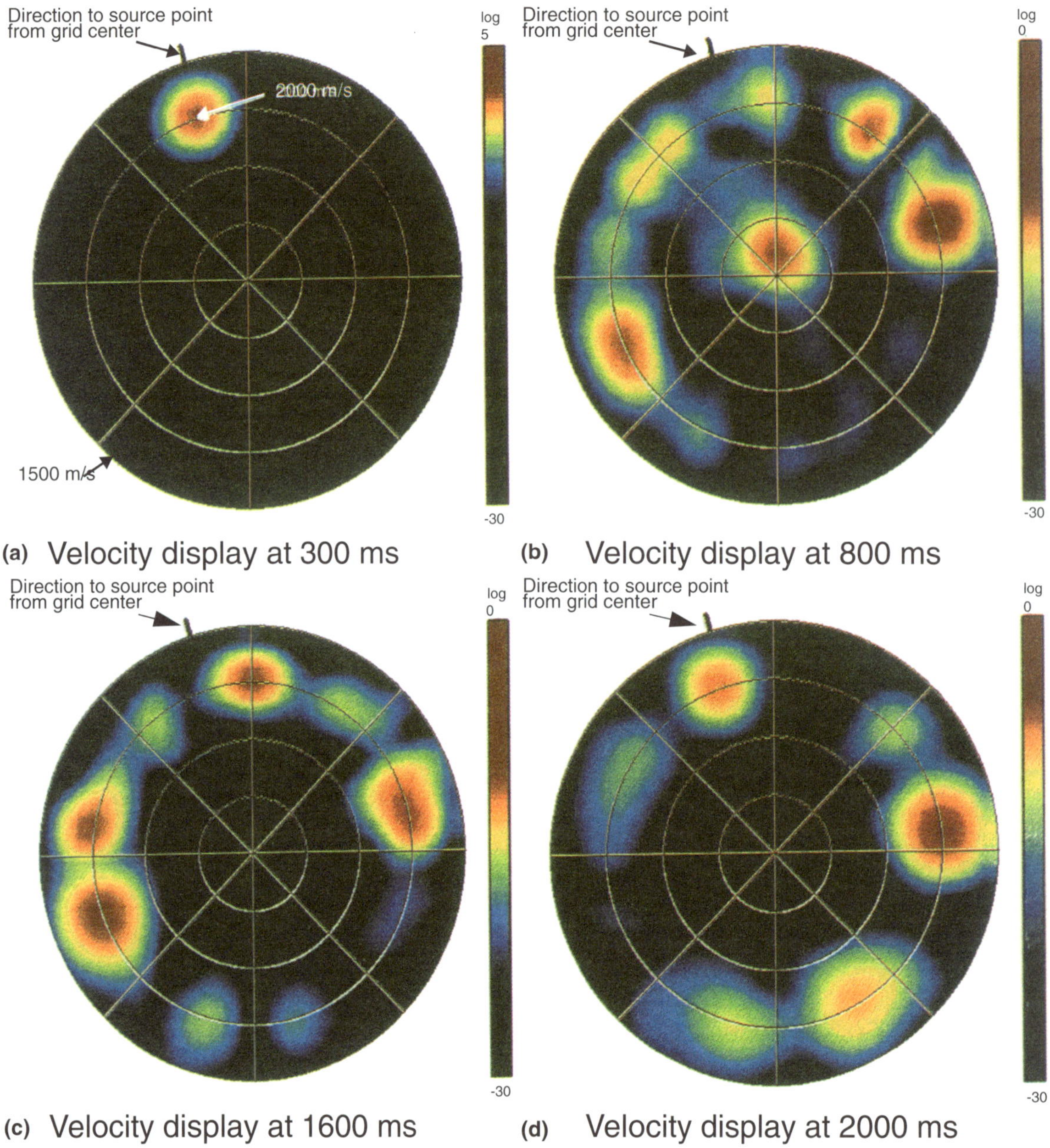

Fig. 10. A display of horizontal velocity as a function of azimuth for the source points and arrival times indicated by the arrows in Figure 11d. Desired reflection events (identified by their high horizontal velocity) are visible near the center of each display. The S/N value given is generated by the program and represents the amplitude of the signal event relative to the total energy impacting the grid at that time.

for the box wave test, and the source type was vibroseis. The field layout is shown in Figure 13.

If the center trace from the grid for each source location is extracted, the result is the common-receiver record shown in Figure 14. For the noise wavelengths present in the data, the 30-m field array used to record each trace in this record has a negligible effect and essentially all noise events are passed. A complex noise field is present in this seismic record and very little signal is apparent.

A Radar-type analysis (Figure 15) gives a clearer picture of the makeup of the noise field. In this figure, a velocity versus azimuth display is shown for shot 175 and an arrival time of 1850 ms. The source was located to the south of the grid (indicated by the tick mark near the bottom of the display in Figure 15). Some of the

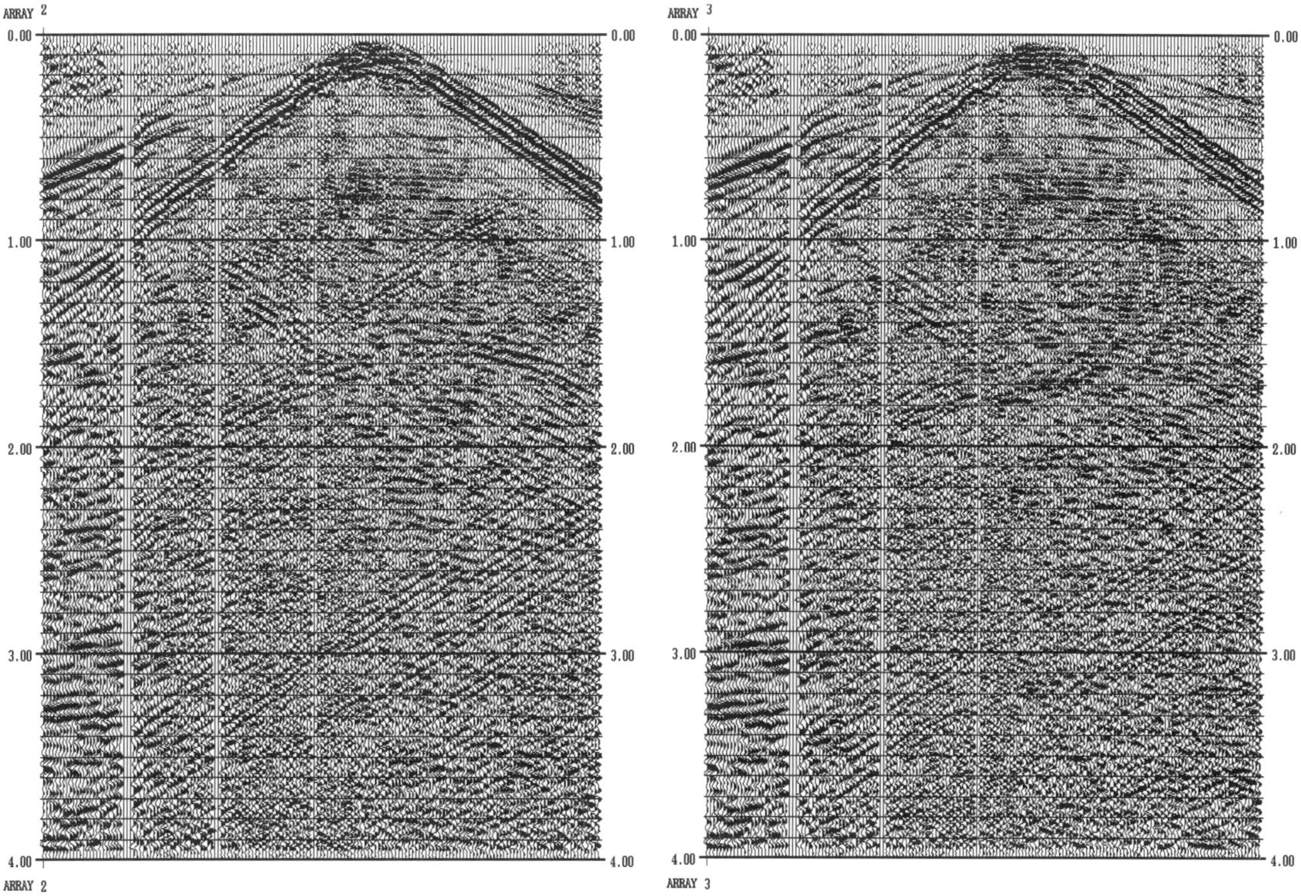

(a) 7x7 Chebychev array (13 dB) (b) 9x9 Chebychev array (19 dB)

Fig. 11. The seismic records resulting from the Chebychev-weighted summing of the shaded elements of the grid shown in Figure 5. A 10–40 Hz bandpass filter followed by a 1000 ms AGC was applied before the weighted summation, and another 1000 ms AGC was applied afterward. The array responses are shown in Figure 6. The arrows in part (d) correspond to the source points and arrival times used to create the velocity versus azimuth displays shown in Figure 12. ***(see parts c and d on next page)***

many events readily seen here are labeled for clarity. From this display, it can be seen that most of the energy present in this shot record at this recording time consists of scattered ground roll events traveling with a horizontal velocity of about 2000 m/s. Numerous scattered head wave events can be seen as well. Near the center of the display, a primary reflection event is apparent. The indication from the Radar analysis is that this reflection event is about 14 dB below the total energy impacting the grid at this arrival time. When this process was repeated for many more shots and arrival times, it was observed that the ground roll velocities were consistently around 2000 m/s, the head wave velocities were widely varied but less than 4000 m/s, and the signal was rarely more than 20 dB below the total energy.

Using the velocity knowledge gained from the Radar analysis, we could then synthesize appropriately sized arrays for both the ground roll and head wave events. The record shown in Figure 16 is the result of combining the center most 81 traces from the grid for each source point in such a way as to form a 9 × 9 Chebychev-weighted areal array with $\beta = 7$. This results in an effective array length in both the in-line and cross-line directions of 210 m and an attenuation level of 26.3 dB. A 10–40 Hz temporal filter and a 1000 ms automatic gain control (AGC) were applied before array synthesis, and a 1000 ms AGC was applied afterward. With these choices, the ground roll was attenuated fairly strongly but much head wave energy remained. Nevertheless, many reflection events can now be seen that were formerly masked by the scattered noise fields.

To determine the effect of the remaining scattered head wave energy, a larger areal array needed to be synthesized. With head wave velocities as high as 4000 m/s and frequencies as low as 10 Hz, a 400 m effective array length was needed. For this grid geometry, a 400 m

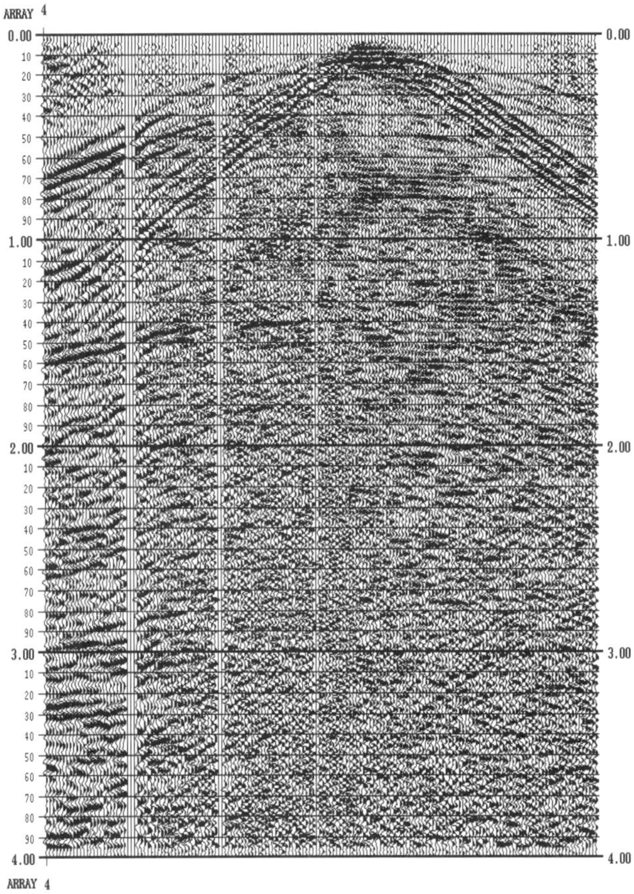

(c) 11x11 Chebychev array (25 dB)

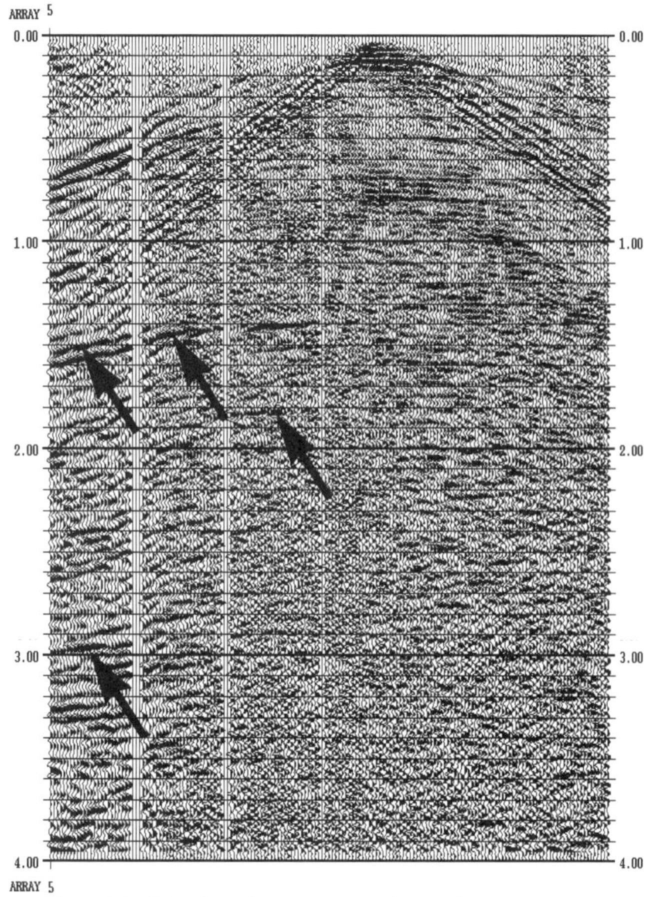

(d) 13x13 Chebychev array (31 dB)

effective array length in both in-line and cross-line directions and an attenuation level of 27 dB were achieved by forming a 17 × 17 Chebychev array with $\beta = 13.3$. Figure 17 shows the seismic record resulting from that array synthesis with the same temporal filter and AGC application used in Figure 16. Comparing Figures 16 and 17 shows that a substantial improvement in data quality has been achieved by attenuating both head waves and ground roll rather than just ground roll.

A robust method for attenuating both scattered head waves and ground roll in a production scheme is to use a form of wide-line profile like the one shown in the third data example in this paper. This approach makes it possible to synthesize the large areal arrays necessary for scattered head wave attenuation without sacrificing the ability to correct for static and cross-line dip of the signal. If one desires to attenuate only the scattered ground roll, it is often possible to record only a 2-D line with areal field arrays. This may be considerably cheaper than recording a wide-line profile. In this example, the data quality shown in Figure 16 was deemed adequate by the exploration team, and the additional improvement suggested by Figure 17 for a wide-line profile could not justify the additional expense. This assumption was tested in the following manner before final production parameters were chosen.

Two separate spreads were deployed and simultaneously recorded from one source line to produce two multifold seismic test lines. As indicated in Figure 18, one of the receiver spreads had only in-line receiver arrays and the other had cross-line arrays. Because the box wave test analysis indicated that the desired reflection events were not buried very deeply beneath the scattered noise, simple unit-weighted field arrays could be used. Limitations in the number and spacing of the available geophone strings resulted in a choice of 180 m for the array length instead of the 210 m used in the array synthesis. A decision was made to keep both receiver array types identical in terms of length and number of geophones. Thus, both receiver array types consisted of 36 geophones spread evenly over a distance of 180 m. The only difference was the direction of the array. The resulting seismic sections are shown in Figures 19 and 20. The in-line only section shown in Figure 19 had *f-k* filtering in both the common-shot and common-receiver domains in an attempt to bring the data quality up to the

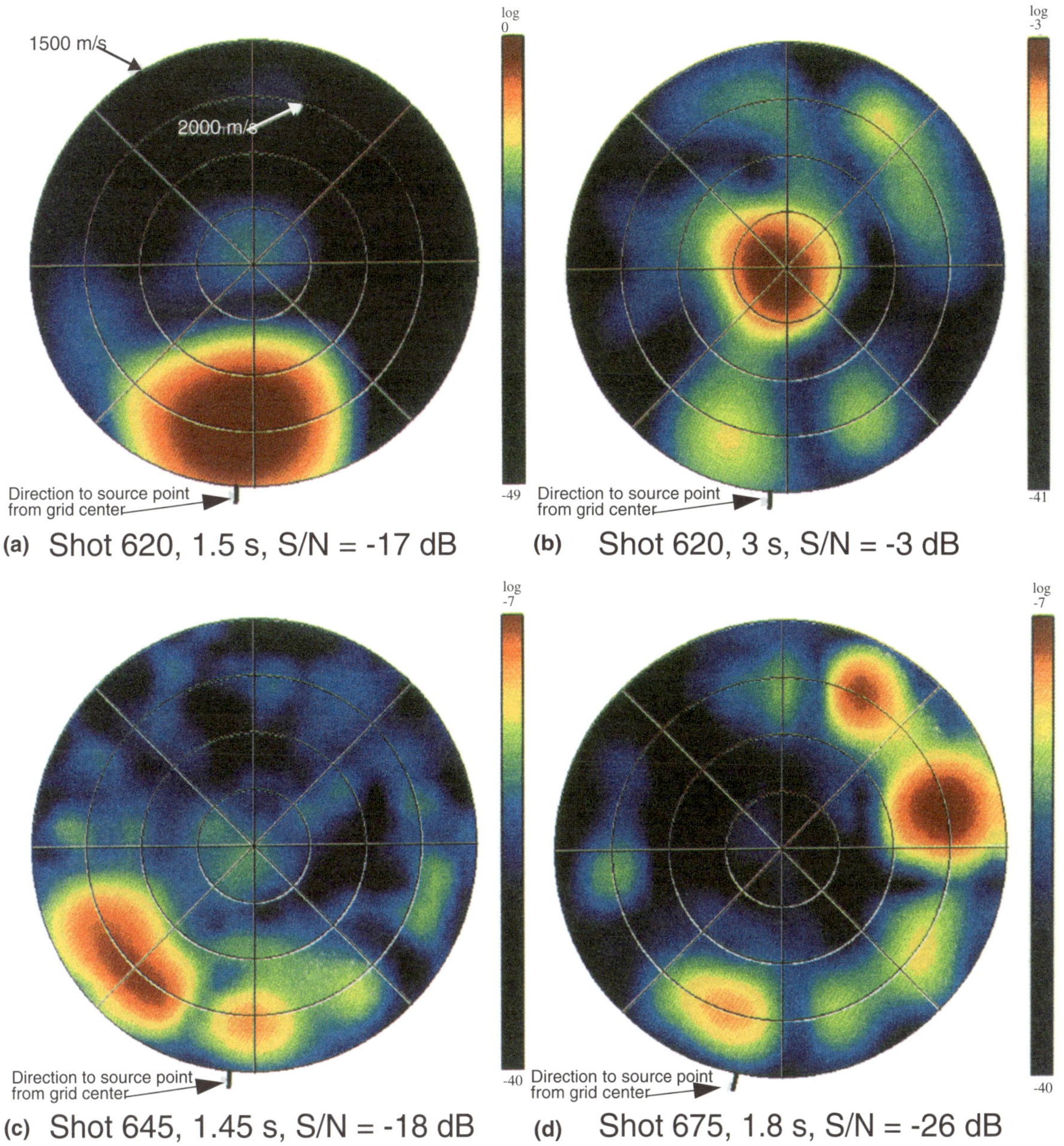

Fig. 12. A display of horizontal velocity as a function of azimuth for the source points and arrival times indicated by the arrows in Figure 11d. Desired reflection events (identified by their high horizontal velocity) are visible near the center of each display. The S/N value given is generated by the program and represents the amplitude of the signal event relative to the total energy impacting the grid at that time.

level of the cross-line array section of Figure 20, which received only the most basic processing. The significant difference between these two sections indicates that the scattered noise predictions resulting from the box wave test analysis were correct. I believe that most of the noise remaining in Figure 20 is unattenuated scattered head wave energy and that a wide-line profile would have improved the data quality even further. The next example illustrates this effect nicely by providing a comparison of a 2-D line with in-line only arrays, a 2-D line with areal field arrays of the appropriate size for attenuating ground roll, and a wide-line profile with sufficient width to attenuate scattered head waves as well as ground roll.

Example from Black Warrior Basin

This last data example comes from an area in the

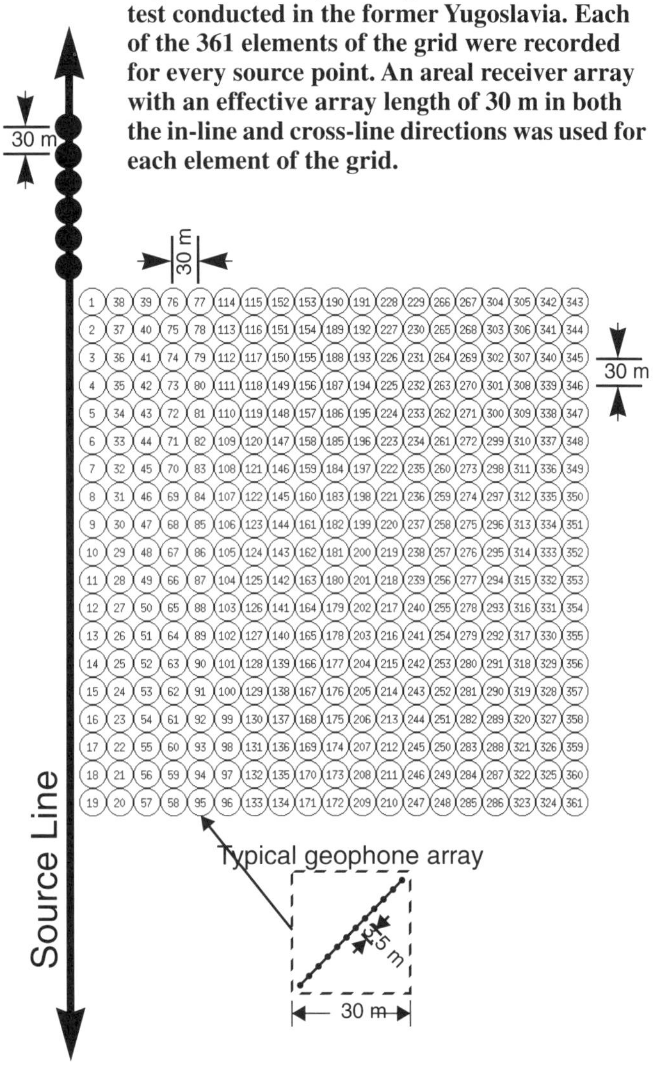

Fig. 13. Geometry description for a box wave test conducted in the former Yugoslavia. Each of the 361 elements of the grid were recorded for every source point. An areal receiver array with an effective array length of 30 m in both the in-line and cross-line directions was used for each element of the grid.

Black Warrior Basin near Meridian, Mississippi. Although this area has no surface carbonates, it illustrates that scattered noise problems also occur in areas that do not possess the obvious scatterers associated with carbonate topography and that the methods discussed in this paper have widespread application.

In this study area, existing seismic reflection data were good above a Cretaceous unconformity but poor in the geologically complex Paleozoic section below the unconformity where a possible flower structure, steep dips, and complex faulting were expected. The surface geology consisted of low-relief, poorly consolidated Tertiary clastics with no significant development.

A 2-D seismic line recorded in 1984 with only in-line arrays is shown in Figure 21. The source array consisted of a three-hole dynamite pattern with 110 ft (34 m) between holes. The charge size was 20 lbs (9 kg) per hole at a depth of 50 ft (15 m). The receiver array consisted of 12 geophones spaced 20 ft (6 m) apart. The group and shot intervals were both 220 ft (67 m), and the fold was 30. The large source effort was an attempt to obtain improved data quality by boosting what was perceived to be weak reflection energy returning from the Paleozoic section. It is apparent that this approach failed, and there are few reflection events visible beneath the Cretaceous unconformity. It was not possible to map the complex structures present in the Paleozoic section with data of this quality.

Subsequent testing revealed the presence of scattered ground roll and head waves, and in 1985, a wide-line profile consisting of five receiver lines and two shot lines was recorded directly over the 1984 2-D line. The geometry used for the wide-line profile is shown in Figure 22. Each common midpoint (CMP) line was 30-fold so that a direct comparison could be made with the previous 2-D line. For most of the line, two sources were used at each shot point. One source was a single dynamite charge of 25 lbs (11 kg) at a depth of 100 ft (30 m), and the second source was a mini-hole pattern (Figure 22). The deep-hole and mini-hole sources were fired separately so that a comparison could be made between these two extremes of source effort. The areal geophone array used at each group location consisted of 36 geophones in a 6 × 6 pattern with 25 ft (~8 m) between phones. This array was large enough to be effective at attenuating the low-velocity scattered ground roll present in the area. The final stack of CMP line 6 is shown in Figure 23. A significant improvement has been made over the 1984 seismic line shown in Figure 21. This improvement was achieved through the suppression of scattered ground roll by the areal field arrays.

When all ten CMP lines were summed together after each was brought to a final stack, an areal array was formed that was large enough to attenuate the scattered head waves. Because of the complex 3-D structure present in the Paleozoic section, it was necessary to steer the cross-line sum in the direction of the dominant dip of the subsurface reflectors in order to keep from attenuating reflection events with substantial cross-line dips. The resulting sum of all ten CMP lines is shown in Figure 24, and a simple time migration of this section is shown in Figure 25. A dramatic improvement in seismic data quality has been achieved. The flower structure can now be easily seen, and the entire section is filled with mappable events. Separate stacks made for deep-hole only and mini-hole only shots yield results comparable to Figure 24 and indicate that adequate signal from the Paleozoic section is obtained even with low source strength. One merely needs to attenuate the interfering noise.

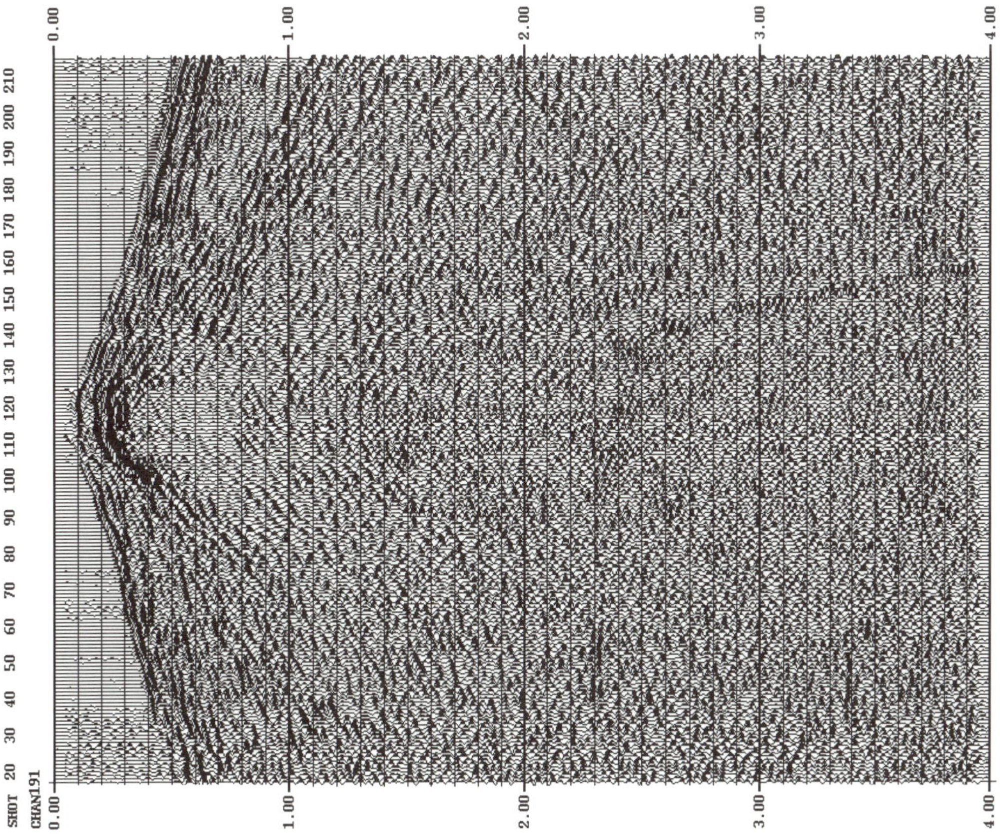

Fig. 14. The seismic record obtained by extracting from each shot record only the center trace from the grid shown in Figure 13. The noise wavelengths present in the 10–40 Hz frequency band are largely unaffected by the field array used to record these traces.

Discussion and Conclusions

It has long been known that backscattered seismic noise is a problem not only in surface carbonates but nearly everywhere. To have a successful field technique for production seismic data, one must be able to identify this noise and measure its strength relative to the desired signal. Standard field tests designed to analyze direct arrival noise events are inadequate for backscattered noise. The test procedure described here, which is based on a 3-D grid of receivers and a line of sources, has proven effective in the analysis of scattered coherent noise in many areas around the world and over many types of surface lithologies. With the new multi-streamer marine vessels being developed, this test procedure can now be performed easily even in the marine environment.

I believe that it does little good to try to predict the seismic acquisition problems one faces in a given area solely from the characteristics of the surface geology. Such attempts usually fail to produce consistent results because the subsurface reflector strengths can vary greatly from one area to the next even when the surface conditions remain unchanged. Thus, an acquisition procedure that is adequate in one area may fail completely in another area with seemingly identical surface characteristics. I believe the best approach for a difficult area is to conduct field tests, such as the box wave test described in this paper, in a number of locations throughout the problem area and actually measure the S/N.

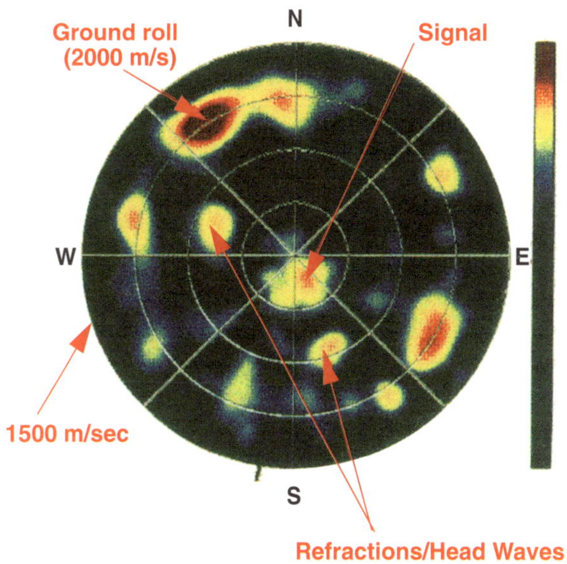

Fig. 15. A display of horizontal velocity versus azimuth for shot point 175 and an arrival time of 1850 ms. Note the presence of signal and scattered near-surface body waves in addition to the scattered surface waves.

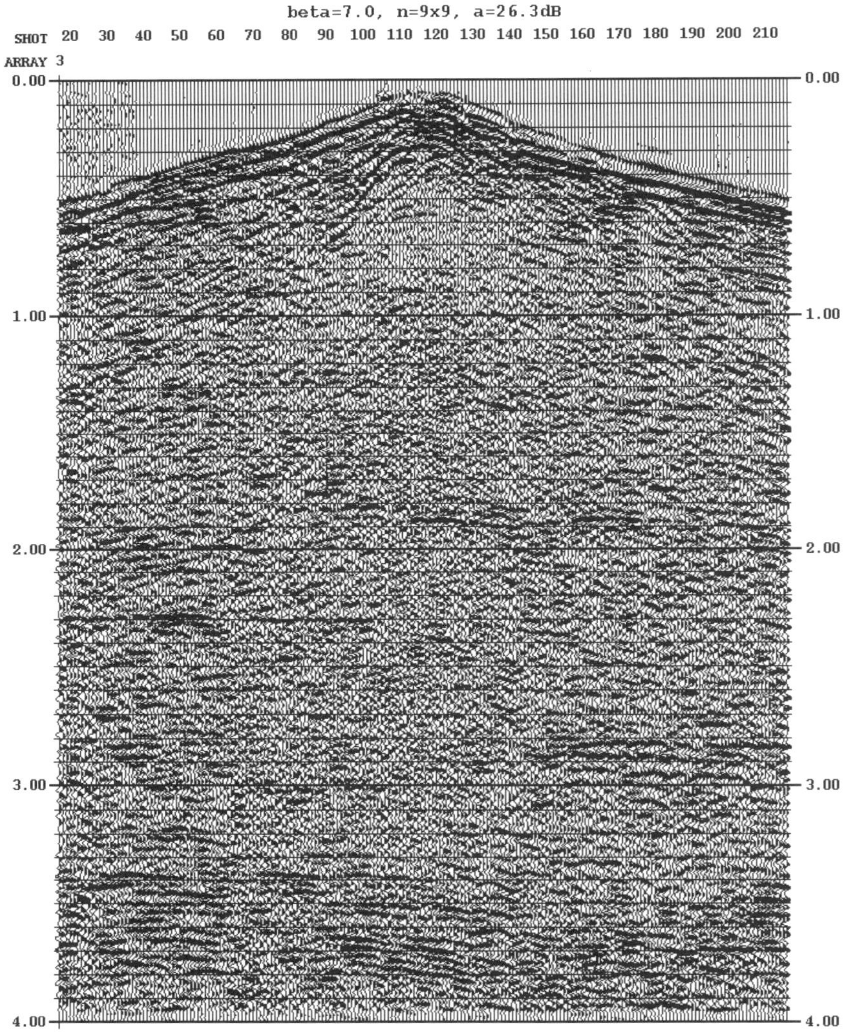

Fig. 16. The seismic record obtained by synthesizing from the grid of Figure 13—a Chebychev areal array with an effective array length of 210 m and an attenuation level of 26.3 dB.

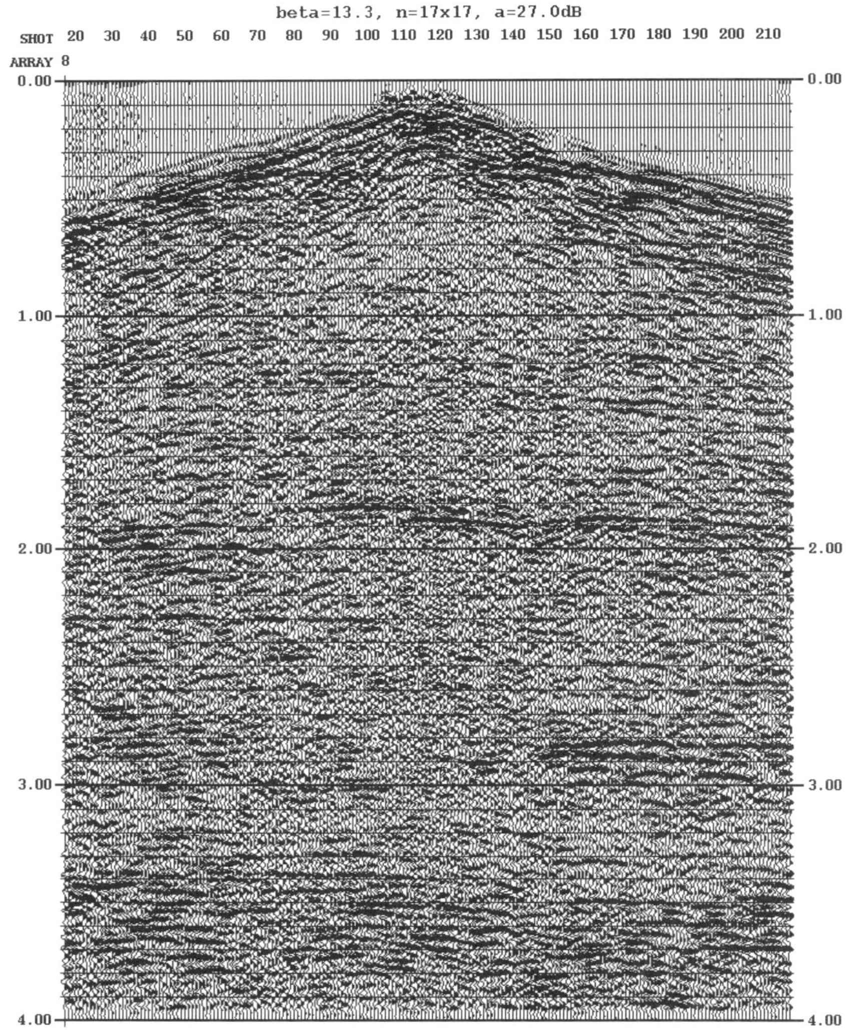

Fig. 17. The seismic record obtained by synthesizing from the grid of Figure 13—a Chebychev areal array with an effective array length of 400 m and an attenuation level of 27 dB.

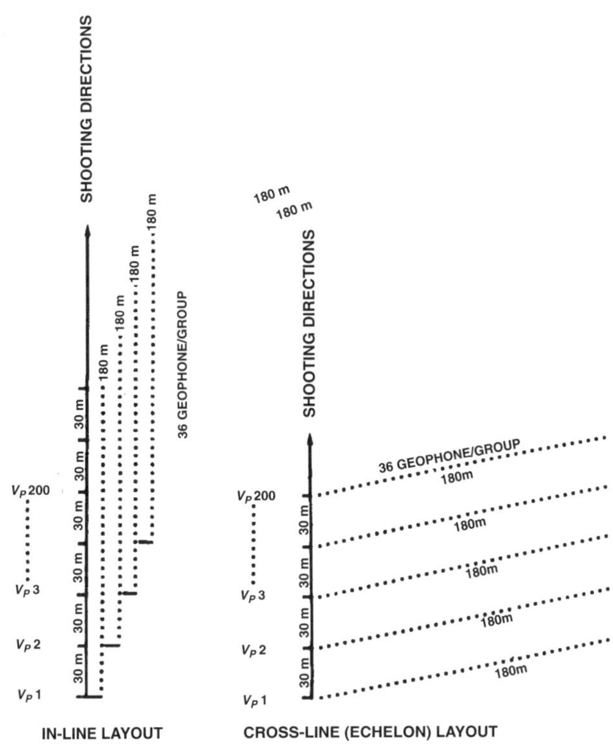

Fig. 18. The geometry used for the receiver array comparisons shown in Figures 19 and 20.

More than one test location in a problem area is desirable for the box wave test procedure because of a shortcoming with the present method. The high-performance areal arrays used to quantify the S/N do not achieve the desired levels of attenuation if strong scatterers lie within the grid itself. In these situations, the incoming coherent noise wavefronts develop significant local curvature and the plane wave assumptions of array design break down. Identification of the noise velocities is still possible, but S/N measurement may suffer. Having more than one grid present in the box wave test increases the likelihood of successful S/N measurement.

Once the identity, propagation velocity, and strength (relative to desired signal) of the problem noise events are known, one can begin to select an appropriate acquisition scheme for the production data. How to do this in the most cost-effective manner, especially with regard to 3-D surveys, is a topic of ongoing research.

References

Aki, K., and Richards, P. G., 1980, Quantitative seismology, theory and methods: W. H. Freeman & Co.

Holzman, M., 1963, Chebychev optimized geophone arrays: Geophysics, **28**, 145-156.

Hudson, J. A., and Knopoff, L., 1967, Statistical properties of Rayleigh waves due to scattering by topography: Seism. Soc. Am. Bull., **57**, 83-90.

Husseini, M. I., Glover, J. B., and Barley, B. J., 1981, Dispersion patterns of the ground roll in eastern Saudi Arabia: Geophysics, **46**, 121-137.

Larner, K., Chambers, R., Yang, M., Lynn, W., and Wai, W., 1983, Coherent noise in marine seismic data: Geophysics, **48**, 854-886.

Levander, A. R., 1990, Seismic scattering near the Earth's surface: Pure Appl. Geophys., **132**, No. 1-2, 21-47.

Newman, P., and Mahoney, J. T., 1973, Patterns with a pinch of salt: Geophys. Prosp., **21**, 197-219.

Parr, J. O., and Mayne, W. H., 1955, A new method of pattern shooting: Geophysics, **20**, 539-564.

Regone, C. J., and Rethford, G. L., 1990, Identifying, quantifying, and suppressing backscattered seismic noise: 60th Ann. Internat. Mtg., Soc. Expl. Geophys., Expanded Abstracts, 748-751.

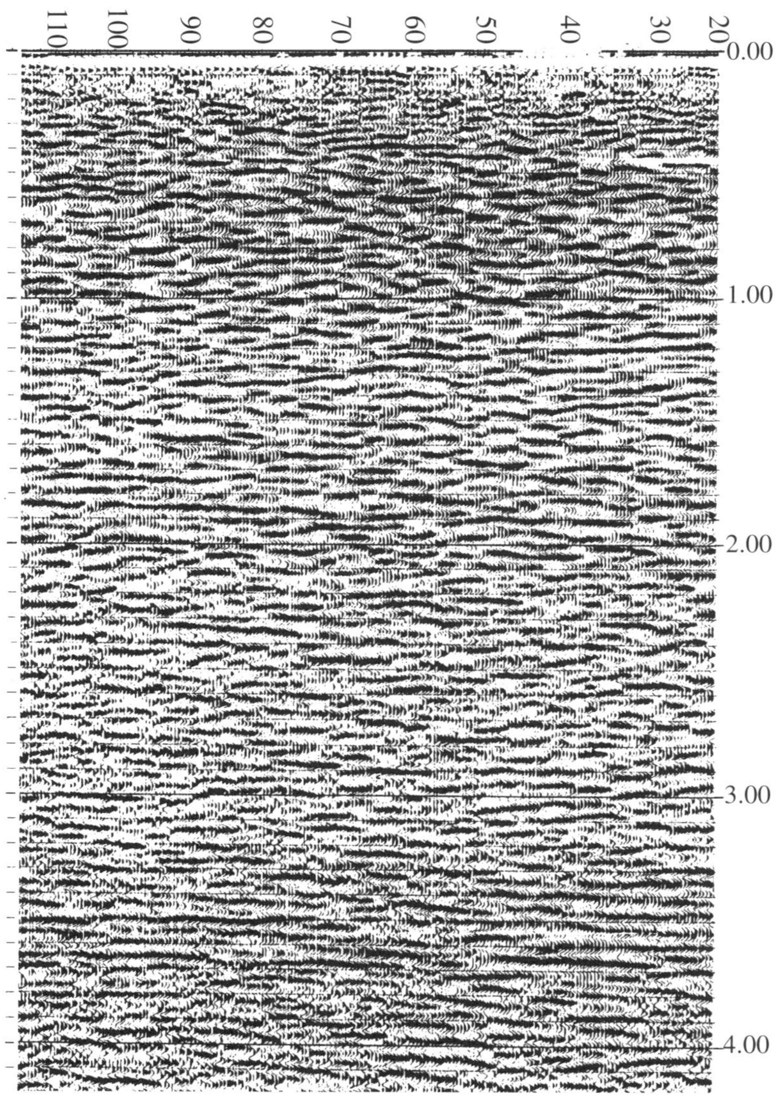

Fig. 19. The seismic line obtained from the experiment shown in Figure 18 when the geophone arrays were oriented in the in-line direction.

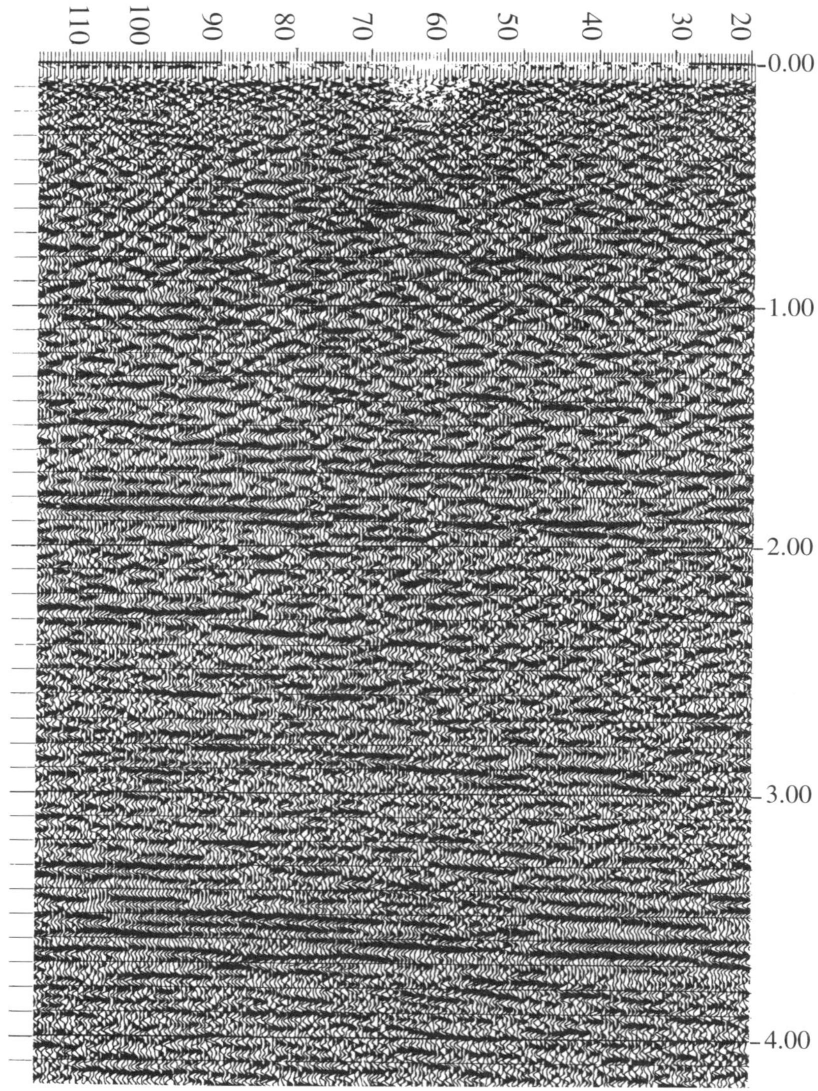

Fig. 20. The seismic line obtained from the experiment shown in Figure 18 when the geophone arrays were oriented in an *en echelon* pattern such that each array formed the diagonal of a rectangle that had an in-line dimension of 30 m and a cross-line dimension of 180 m.

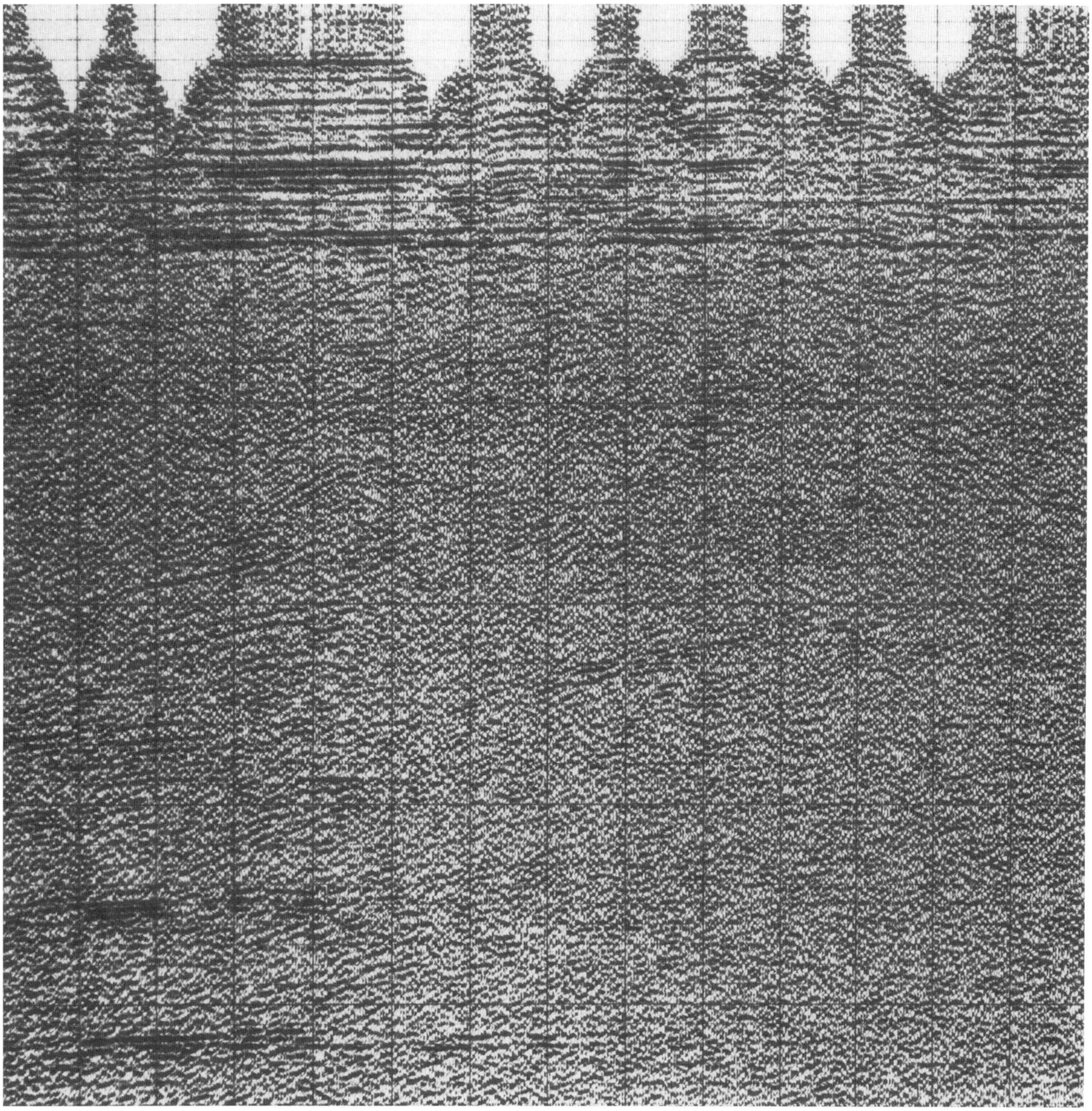

Fig. 21. A 2-D seismic line recorded in the Meridian, Mississippi, area with only in-line source and receiver arrays.

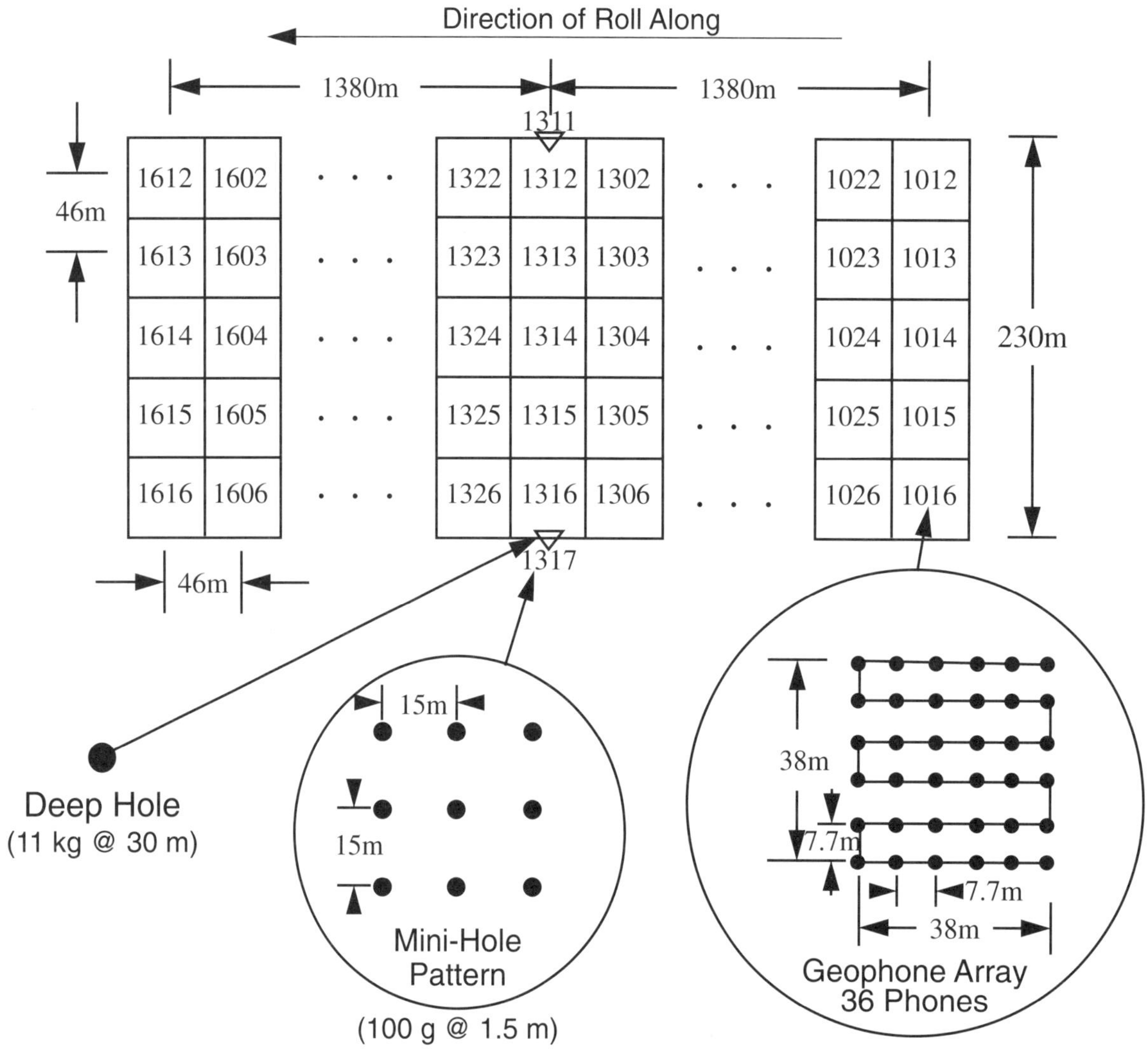

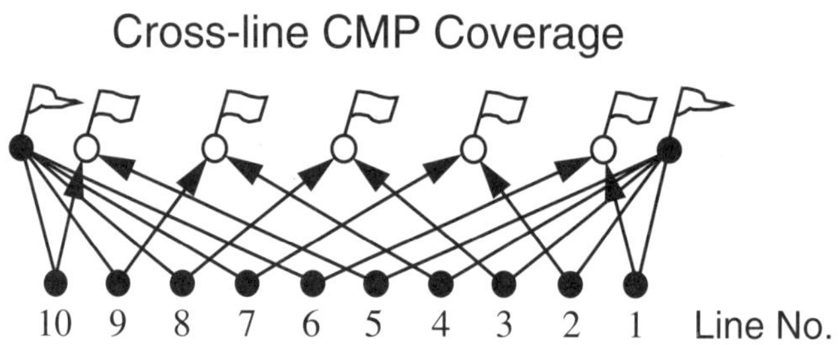

Fig. 22. The wide-line profile geometry used in the Meridian, Mississippi, area. Five receiver lines and two shot lines were used to produce ten CMP lines.

Fig. 23. The final stack of CMP line 6 from the wide-line profile used in the Meridian, Mississippi, area. This seismic line had areal field arrays to help attenuate scattered ground roll events.

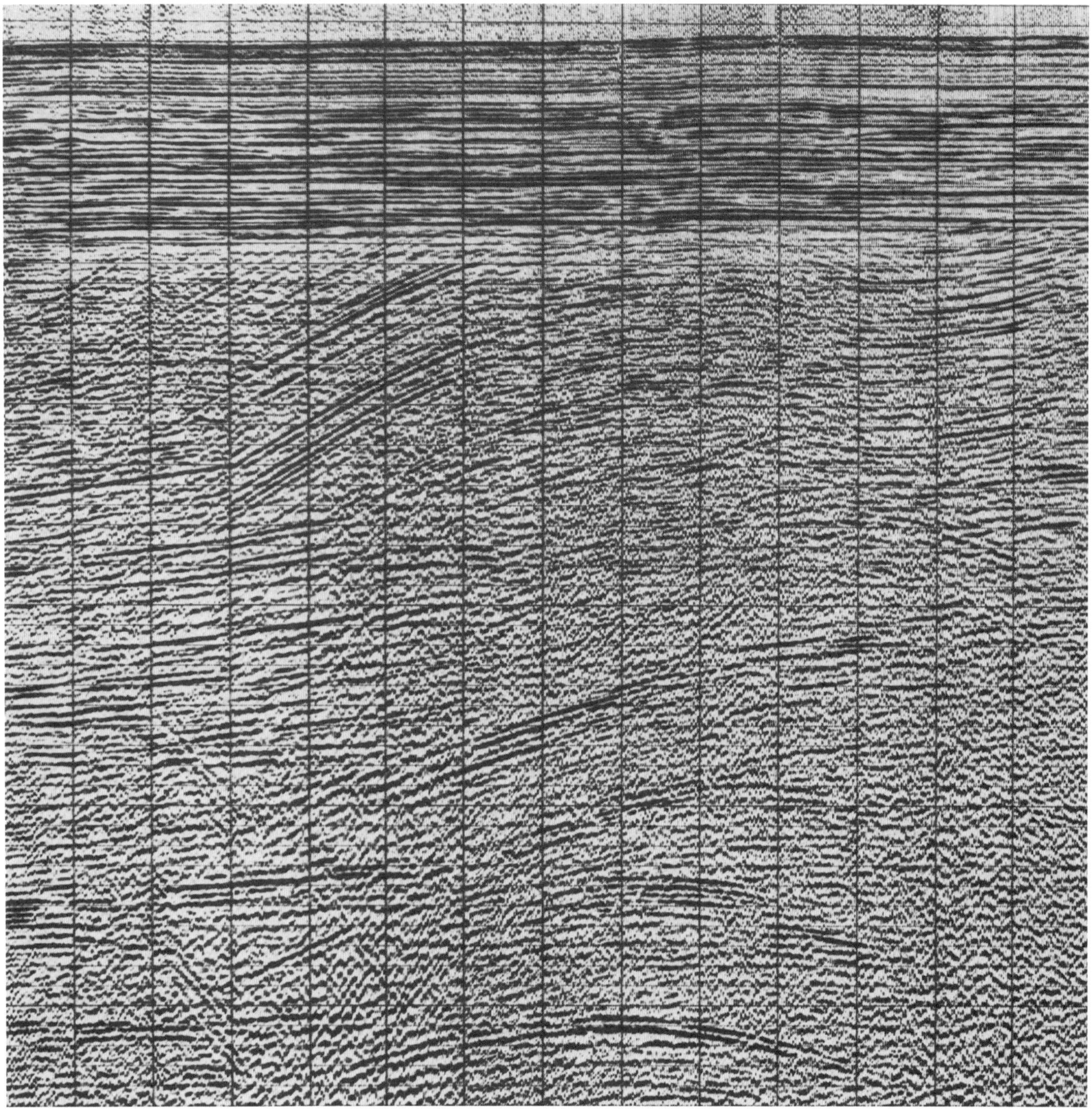

Fig. 24. The steered, dominant-dip stack of all ten CMP lines from the wide-line profile used in the Meridian, Mississippi, area.

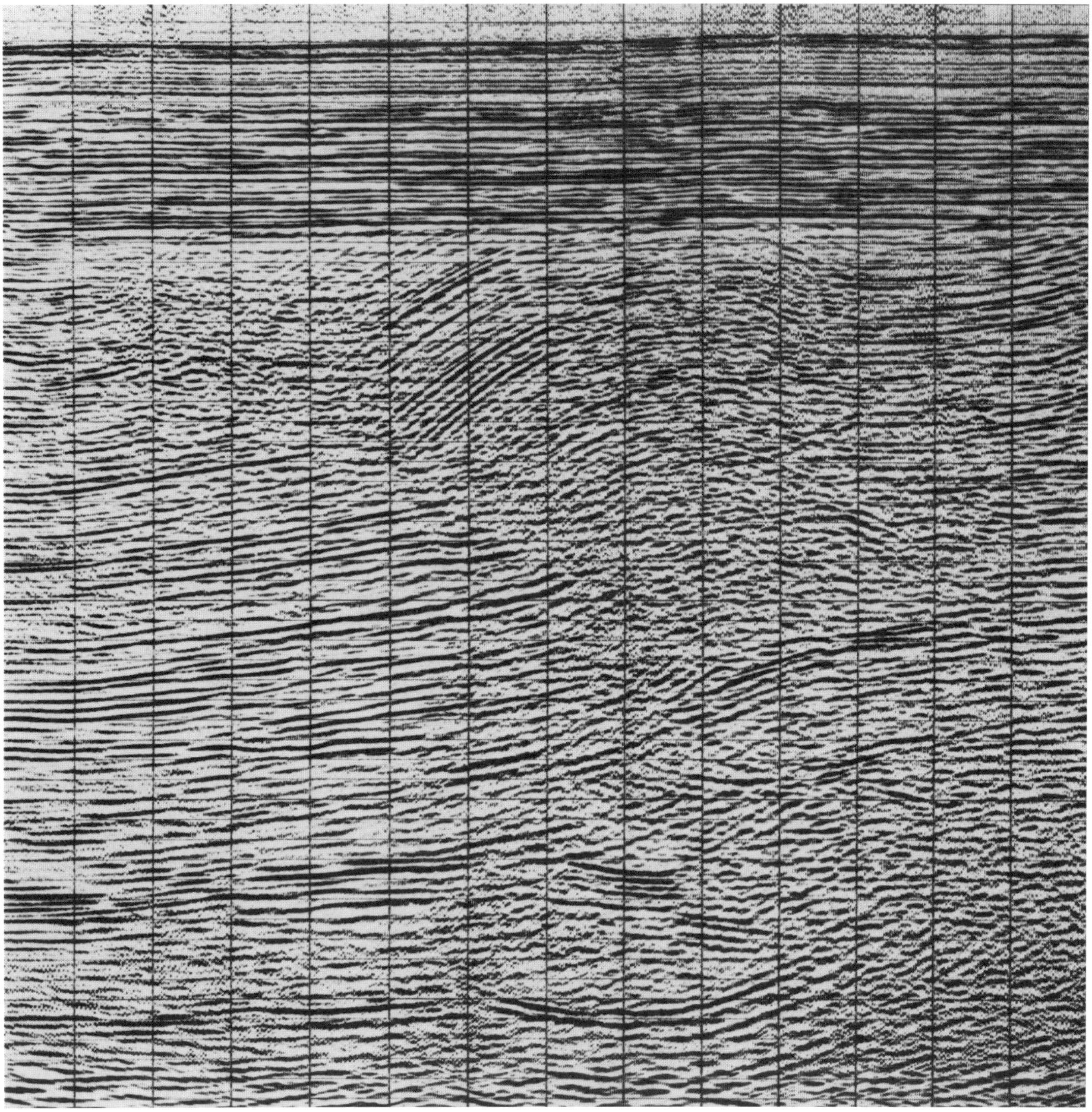

Fig. 25. The time migration of the stacked section of Figure 24.

Chapter 12

Seismic Signatures of Carbonate Diagenesis

Paul D. Wagner
Amoco EPTG
Houston, Texas, U.S.A.

Abstract

Our objective—to summarize the effects of carbonate diagenesis on seismic records—is directly linked to an industry goal of continually improving our capability of seismically imaging reservoir, seal, and hydrocarbon distributions. Carbonate diagenesis affects reservoir and seal distributions mainly through processes of differential leaching, compaction, and mineralogic replacement. The results of previous studies (both in-house and published) indicate that all processes of carbonate diagenesis are detectable by conventional seismic techniques if the combination of spatial scale and net rock property change is sufficient. The best results are seen when large changes in rock properties occur over large volumes of rock, such as in basinwide, depth-related mineralogic phase changes or in spatially extensive subsurface leaching. In many actual exploration cases, however, the seismic signatures of carbonate diagenesis are too subtle for easy detection, either because the rock-altering process acted over too small a volume or because the resultant changes in rock properties were too small. This seismic detection problem is modest at burial depths of 1000 m but is serious at burial depths ≥4000 m. Improved resolution of diagenetically altered carbonates must come through (1) increased use of higher resolution seismic techniques, such as crosswell tomography, and (2) more rigorous integration of rock, log, and seismic data in a data management environment that allows for iterative model testing, such as 3-D visualization software.

Introduction

Seismic investigation of carbonates is especially challenging because of high interval velocities and an acoustic character strongly dependent on diagenetic history. Diagenetic processes (e.g., karsting, mechanical compaction, chemical compaction, burial leaching, cementation, and dolomitization) progressively mold carbonate seismic rock properties across a continuum of scales. Reservoirs can be formed, degraded, rejuvenated, degraded, and rejuvenated again in a dynamic interplay of competing diagenetic processes that occur over the common depth range of drilling. Some basinwide diagenetic products can be imaged easily with conventional *P*-wave seismic data, but the same chemical and mechanical processes operative on a basinwide scale can form reservoirs and seals on the scales of individual fields or beds that strain the detection limits of even the best seismic techniques. At present, we do a fair job of seismically imaging important products of diagenesis at shallow burial depths. Technologic improvements in acquisition and processing are required for adequate resolution and detection in moderate to deeply buried carbonates. The best objective techniques involve calibration of surface seismic with borehole seismic, well logs, and petrologic data. Subjective appreciation of the dynamics of past diagenetic processes, and the geometries over which they may have acted, can also give valuable interpretational insight.

This paper summarizes the (1) types of diagenetic processes that affect seismic rock properties, (2) spatial scales over which the processes act, (3) types of published seismic techniques used to image diagenetic features, and (4) challenges for future interpretation.

In a general sense, the present subsurface distribution of carbonate reservoirs and seals is a function of two variables: (1) sequential depositional patterns of grain-rich and mud-rich rocks and (2) later diagenetic overprinting patterns of leaching, cementation, dolomitization, and many others. Depositional processes form large-scale morphologic features (e.g., reefs, shoaling complexes, and platform margins) that are seismically resolvable even at great depths (Figure 1) (see Wilson, 1975; Matthews, 1984). Diagenetic processes can alter the external morphology of carbonate bodies after deposition (Figure 2) and can fundamentally alter the internal distributions of reservoirs and seals.

Diagenesis is defined as the sum total of all physical and chemical changes that occur to sedimentary rocks from the time they are deposited to the time they undergo metamorphism. Emphasis here is on small-scale mechanical changes and the plethora of chemical alterations that are conventionally considered to be diagenetic. Examples of diagenetic processes are submarine cementation, pan-evaporative and reflux dolomitization, karsting, meteoric cementation, mixing zone corrosion, mechanical compaction, chemical compaction, sulfate plugging, burial leaching, hydrocarbon "poisoning" of chemical compaction in hydrocarbon pay zones, and hydrothermal mineralization (e.g., Bathurst, 1975; Hardeman, 1981; McIlreath and Morrow, 1990). These diagenetic processes can fundamentally alter the petrophysical properties of a deposited carbonate and render useless seismically defined reservoir predictions based on external morphology alone. Effects of diagenesis should be a major interpretational input into any potential seismic model of a carbonate system.

The relative importance of depositional and diagenetic processes in describing carbonate body geometries and distributions of reservoirs and seals is partially controlled by the depth of burial. At shallow depths, depositional processes are usually of overriding importance (exceptions to this are discussed later). At increasing depths, diagenetic processes command an increasingly dominant role. It is here that a large problem looms: seismic resolution degrades through the depth range where increasing detail is needed to differentiate simpler depositional geometries from increasingly complicated internal diagenetic patterns. As exploration efforts shift deeper and targets become more subtle, the cutting-edge challenge today is to refine seismic techniques to a level where we can unambiguously map diagenetic features 5–10 m thick in carbonate bodies buried to >3000–4000 m.

The intent here is not to discuss how carbonate depositional systems, such as reef outlines, platform edges, and carbonate fan edges, are defined. Seismic expression of carbonate facies are addressed elsewhere (e.g., Fontaine et al., 1987; Langdon and Malacek, 1987; Harris and Sarg, 1989; Brown et al., 1990; Sarg and Candelaria, 1992; Stafleu et al., 1995; Macurda,

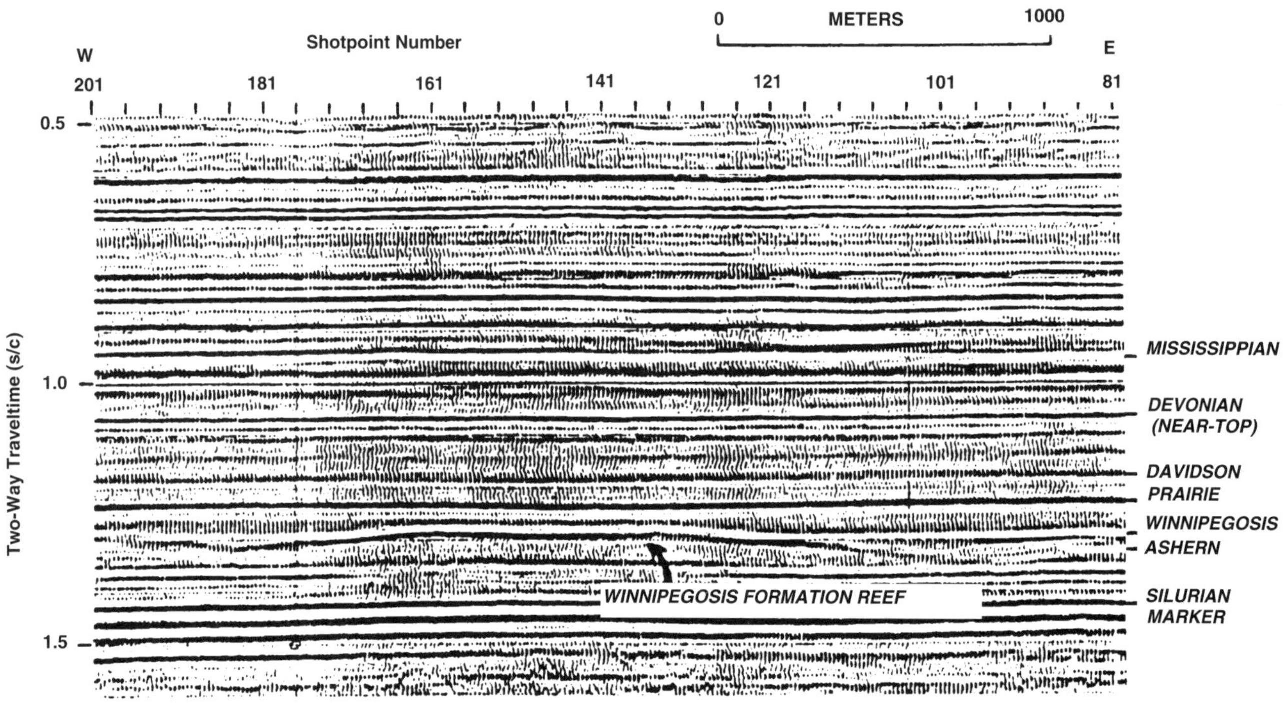

Fig. 1. Conventional *P*-wave seismic image of a carbonate reef. From Brown et al. (1990).

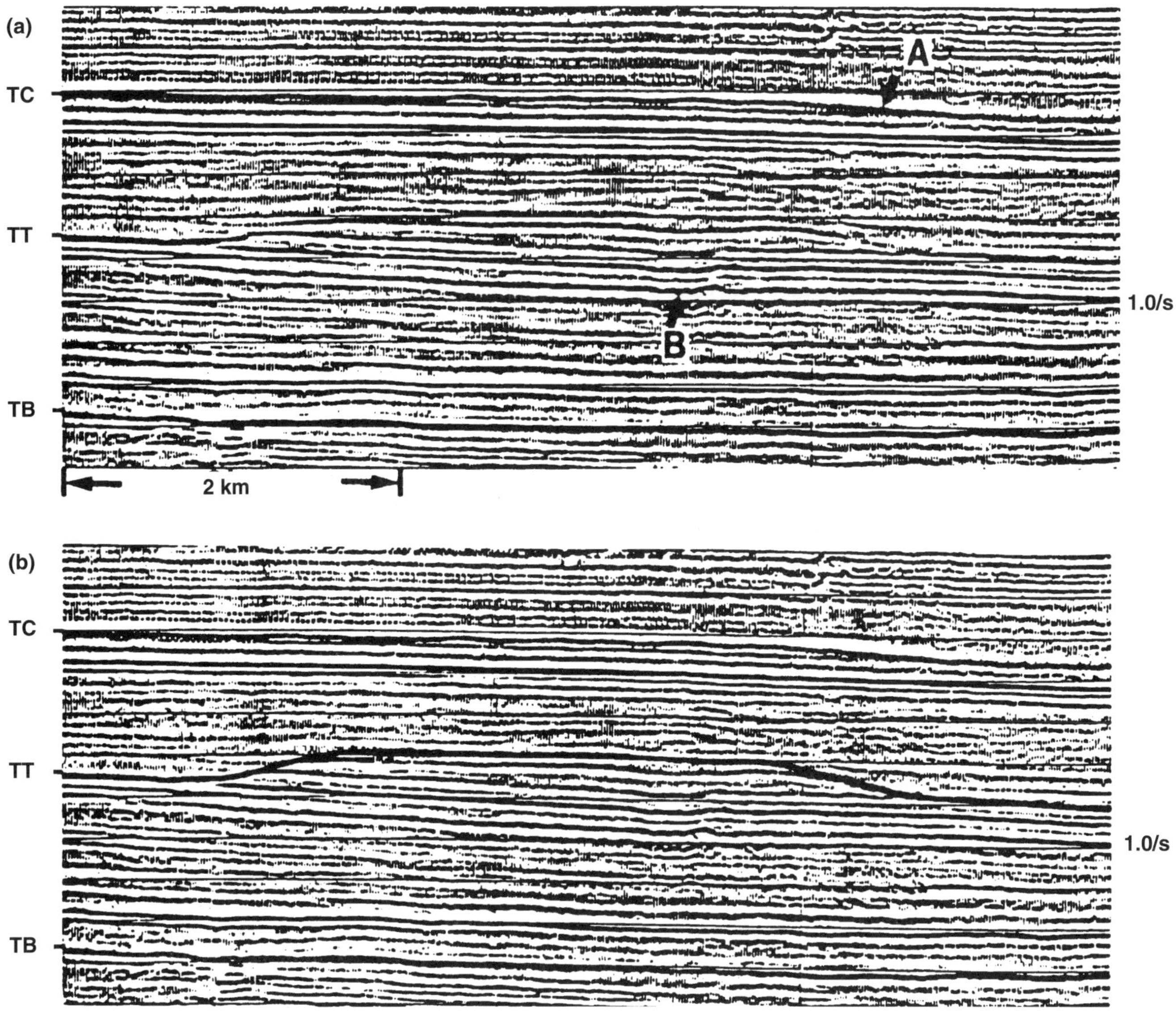

Fig. 2. (a) Uninterpreted and (b) interpreted conventional *P*-wave seismic image of a karst-related "buried hill." From Jenyon (1986).

Chapter 6; and Fitcher, Chapter 7 of this volume). However, cases where diagenesis alters the external geometry of carbonates are of interest because of the potential for associated changes in rock properties within the altered carbonate body, and these are mentioned briefly.

Spatial Scales of Diagenesis

This section is included to give the reader an idea of the spatial breadth and diversity of carbonate diagenesis. Important diagenetic processes are described that produce petrophysical changes in carbonates at different geometric scales. For example, basinwide seismic reflectors can be caused by diagenetically induced changes in mineralogic phase or cementation (e.g., Mienert, 1984; Bassinot et al., 1993; Jarrard and Symonds, 1993; Kenter and Ivanov, 1995). On scales of individual fields or beds, diagenesis can form discrete intervals of highly cemented or leached carbonate that are of great importance to hydrocarbon trap development. But these small-scale diagenetic features may not be resolvable on conventional *P*-wave seismic records. In both cases, the products of diagenesis may have formed from identical chemical and mechanical processes but the scales of operation are so different that alternate seismic techniques must be used to image them properly.

Diagenesis on a Basinwide Scale

Shallow regional reflectors can be associated with depth-related changes in carbonate bulk density. Anselmetti and Eberli (1992 and Chapter 4 of this volume) and Bassinot et al. (1993) have collected rock property data from basinal and platform carbonates of relatively shallow burial and have determined that sharp changes in degrees of carbonate lithification should correlate well with seismic amplitudes. Levels of particularly great change in acoustic impedance are associated with ooze-chalk-limestone transitions in basinal sections and with intervals of more cementation in shelf sections. Jarrard and Symonds (1993) have confirmed similar results from shelf carbonates from offshore Australia. Regional seismic stratigraphic correlation is an important use of these types of reflectors, although not all of them have time-stratigraphic significance.

More generally, diagenetic processes of *mechanical and chemical compaction* cause bulk density differences in otherwise homogeneous limestone intervals at all common drilling depths. Figure 3 shows that preferential compaction and porosity-permeability loss in muddier carbonate intervals leads to a natural layering of higher and lower acoustic impedance intervals. Lateral facies changes from grain-rich to mud-rich facies produce lateral acoustic impedance increases due to preferential compaction of mudstones and wackestones. The frequency of impedance changes is strongly linked to the degree of facies intercalation and rapidity of lateral facies changes. Variations in depositional thickness can range from centimeters to hundreds of meters. Variations in lateral depositional character can range from meters to kilometers. Regional changes in bulk density can also be caused by tectonic events that trigger burial-induced chemical compaction and associated regional loss of carbonate porosity (Wagner, 1990).

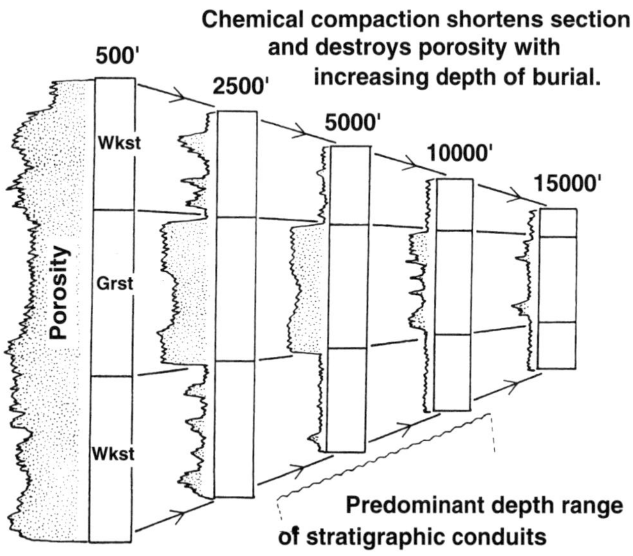

Fig. 3. Differential response of rock textures to progressive chemical compaction and associated burial cementation. Lime grainstones (Grst) compact much less readily than mud-dominated limestones such as wackestones (Wkst), leading to a natural layering of porous and nonporous intervals.

Erosional truncation is caused by a combination of chemical and mechanical "attack" on an exposed carbonate and can have seismic expression over hundreds, or even thousands, of square kilometers. Exposure can occur under either subaerial or marine conditions. If it occurs under subaerial conditions, a variety of gross morphologic changes can occur depending on climate and duration of exposure. The greatest removal of carbonate material, and thus the greatest potential for altering exterior geometries, occurs under wet climate conditions conducive to karsting. Features such as wedge-edges, rugose coral terrain, "buried hills," large sinkholes, and other solution-collapse features can form (Thornbury, 1965; Esteban and Klappa, 1983; Jenyon, 1986). Features caused by subaerial erosion can be discerned with conventional reflection seismology (Figure 2), just as normal *P*-wave seismic data can define boundaries of unaltered carbonate bodies (Figure 1). Conventional seismic records, however, may have difficulty resolving subparallel contacts. In this case, rock property changes caused by meteoric diagenesis during exposure may be the only possible cause of seismic reflection (discussed later).

If exposure of the carbonate body occurs under marine conditions, significant removal of material can occur that causes major changes in gross morphology (e.g., Campbell et al., 1993). Compared to meteoric exposures, however, submarine exposures have much less potential for producing secondary variations in porosity. This lower potential stems from the diagenetic nature of seawater as a less chemically reactive fluid than meteoric water. An exception to this rule is seen in certain aged sections (e.g., Devonian and Permian) that experienced unusually intense submarine cementation in localized areas (see next section).

Regional overpressuring can be caused by laterally extensive, diagenetically induced seal formation. The flow barrier prevents connate fluids from being expelled during continuous burial. Increased hydrostatic pressure retards mechanical and chemical compaction processes below the seal. Overpressuring can help preserve porosity in chalks (Hardeman, 1981) and presumably in any other carbonate facies. Acoustic impedance differences are greatest in all-carbonate sections and much less in shale-on-carbonate strata.

Gas hydrates are regionally extensive cements that have the seismic signature of a "bottom-simulating reflector" (Kvenvolden and Barnard, 1982; Badley, 1985). These diagenetic additions to slope sediments on the margins of continents can have the seismic signature of "dimming" reflectors by reducing acoustic impedance contrasts within the zone of cementation. Gas hydrate cements form as methane–water complexes under low-temperature, relatively high-pressure conditions. Cemented intervals can be hundreds of meters thick and can extend along hundreds of kilometers of continental slope section. Gas hydrates can theoretically seal shallow reservoirs charged with biogenic gas and late-migrated thermogenic hydrocarbons.

Permafrost is another unusual diagenetic cement with regional distribution. Intervals with ice cement generally thicken landward in high-latitude areas. In the early days of Alaskan North Slope exploration, thick permafrost complicated structural interpretations of deeper reflectors because improper interval velocity values were assigned to the cemented interval. Corrections were confirmed with the first drilling test and have been applied successfully to later wells.

Diagenesis on the Scale of Individual Fields or Beds

As mentioned previously, a first-order process of carbonate diagenesis is the differential chemical compaction of certain limestone facies during burial. Preferential compaction and porosity-permeability loss in muddier intervals leads to a natural layering of higher and lower acoustic impedance intervals in the typical limestone (Figure 3). In the absence of complicating factors, this diagenetic process closely parallels facies distributions and is operative at all scales from basin-wide to the thinnest limestone bed. Comparative resilience to compaction of grain-supported facies explains why typical limestone exploration programs are focused on facies such as reefs and grainstone shoals.

Similarly, large differences in compaction susceptibilities between dolomite and limestone allow porous dolomites to exist within compacted limestone sections at >6000 m. Dolomite is more resilient to compaction because of its lower solubility in typical basinal brines. A field composed of intercalated porous limestones and dolomites has a fairly uniform acoustic impedance character at shallow burial depths, but at greater depths and with more selective compaction of limestones, it evolves into a layered system of nonporous limestones and porous dolomites. Several fields in the deep Alberta Basin have these characteristics. Acoustic impedance differences between porous dolomites and lower porosity limestones have also been noted at shallow and intermediate burial depths (Harvey and MacDonald, 1990; Kenter and Stafleu, 1995).

Processes of burial leaching and hydrocarbon emplacement can reverse or retard effects of chemical compaction in either of the preceding cases, an example of *competitive diagenesis* (Figure 4). Individual beds composed of muddier limestone texture can be rejuvenated as reservoirs if compaction has not been too severe and if a source of burial leaching fluid is available. Emplacement of hydrocarbons in the leached intervals helps retard further compaction and burial cementation. Some Cretaceous limestone fields onshore and offshore in the U.A.E. are good examples of how these beneficial burial diagenetic processes help form and preserve porosity and permeability in carbonates (Dunnington, 1967). The chalk fields of the North Sea are also good examples of the positive effect of hydrocarbon emplacement on porosity retention with depth (Hardeman, 1981). In these cases, acoustic impedance mapping can show discrete "lozenges" of high-porosity hydrocarbon-filled carbonate positioned between the overlying seal and the underlying lower porosity, compacted aquifer rock.

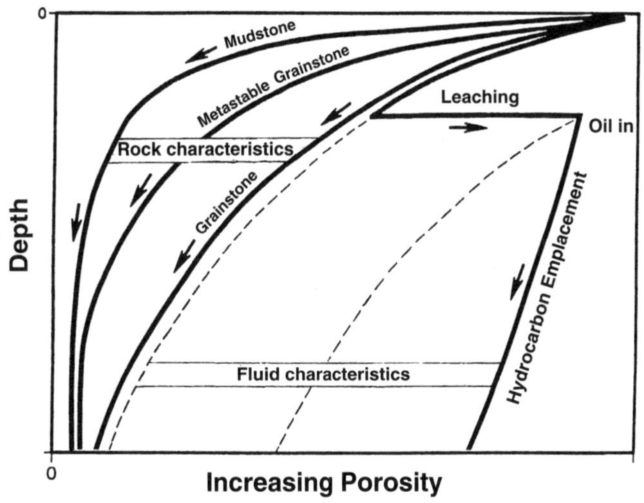

Fig. 4. Depth versus porosity diagram showing "competitive diagenesis" as a function of rock and fluid characteristics through time. The reservoir character of a limestone at any given depth depends on its rock texture and fluid event history. Leaching and hydrocarbon emplacement reverse some effects of chemical compaction and retard future degradative effects. Good deep limestone targets have a grainier rock texture that has been leached in the subsurface and charged with hydrocarbons at shallower depths.

Despite the strong affinity for grain-supported facies to survive better against the onslaught of chemical compaction, other diagenetic processes can significantly complicate the simple "facies model" of porosity distributions. Effects of diagenesis can invert reservoir characteristics in a facies assemblage, destroying porosity in grain-rich intervals and forming it in muddier intervals. Some carbonate fields in Wyoming have this characteristic. A typical producing section is composed of stacked depositional cycles of echinoderm grainstones and capping dolomitized mudstones; the pays are in the mudstones. Excellent depositional porosity and permeability in the grainstones were destroyed by meteoric cementation. Mudstones were dolomitized as the depositional system became restricted and were then extensively leached by meteoric water as the system moved from an evaporative marine to a subaerial diagenetic environment. The cycle repeated itself to produce a field where multiple pays in thin mudstone intervals are intercalated with thick intervals of highly cemented grainstones. The seismic signature of this type of reservoir distribution would be difficult to define on a discrete-bed basis, but the integrated signal could probably be effectively compared to intervals with more uniform acoustic impedance character.

As mentioned previously, subaerial erosional truncations are caused by a combination of chemical and mechanical attack on an exposed carbonate, and a variety of geometric changes can occur depending on the availability of meteoric water and the duration of exposure. Features caused by subaerial erosion can be discerned with conventional reflection seismology. What is not obvious, however, is the highly variable petrophysical nature of the remnant carbonate. In the process of being eroded by meteoric water, the matrix rock can undergo either significant leaching or cementation (Wagner et al., 1995). Hence, what may have started out as a homogeneous blanket carbonate sand may end up as a wedge-shaped body of highly variable internal porosity. Internal variations in rock properties of the body may either be below seismic resolution or require specialized reprocessing and analysis.

Figure 5 shows the seismic expression of a compartmentalized limestone reservoir formed by localized cementation under meteoric and burial conditions. The section is an acoustic impedance map of a reservoir and underlying platform located offshore of China. The backstepping nature of the platform system is clearly seen as discrete shelf edges. Hydrocarbons are trapped in the uppermost platform. Darker blue colors in the gross pay interval indicate the presence of two higher impedance intervals that correspond to cored intervals of much lower porosity. The shallower of the two intervals corresponds specifically to a meteorically cemented limestone just below an ancient subaerial exposure surface (Wagner et al., 1995). Meteoric cementation associated with the unconformity formed a tabular-shaped body of more indurated limestone that still retains an acoustic impedance signature after 1100 m of burial. The deeper of the two intervals is a low-porosity zone just below the present oil-water contact that postdates hydrocarbon emplacement. Petrographic and geochemical data indicate that late calcite spar cementation occurred in a chemical microenvironment near the hydrocarbon–water contact. The thickness of the two cemented intervals varies between 0 and 10 m.

Figure 5 also shows the seismic expression of an internal "plumbing system" in a limestone platform that has undergone extensive leaching in the burial environment (yellow color just below hydrocarbon pays in the uppermost reef). Petrographic evidence clearly shows that much of the leaching in this grain-rich interval occurred after development of stylolites (chemical compaction features produced in the subsurface by burial-induced loading). Porosity and permeability values have been significantly increased in limestones that underlie the hydrocarbon-filled section. The seismic signature of this burial leaching is seen as a lower acoustic impedance interval with a horizontal, tabular-shaped geometry of about 160 m thickness. Higher acoustic impedance streaks within the leached interval are interpreted to be muddier limestones that were significantly compacted before much of the leaching occurred.

The effects of diagenesis can cause preferential loss of porosity in certain types of grainstones. Mississippian beds containing abundant echinoderm grains are notorious for being extensively cemented, even though they were deposited as high-porosity grainstones. The reason for extensive cementation is the diagenetic potential for nucleation of cement on the original magnesium calcite mineralogy of the echinoderm grains. Nearby grainstones in an adjoining facies belt may be composed mostly of bryozoan particles (originally aragonitic) and be almost uncemented. The seismic signature of this type of mineralogy-dependent diagenesis would be lateral acoustic impedance variations within a grainstone interval where higher values represent nonporous echinoderm sands and lower values correspond to porous bryozoan sands. Grainstones are usually meters to tens of meters thick and vary widely in lateral extent depending on depositional conditions.

Another example of preferential porosity loss in grain-rich rocks is submarine cementation. This diagenetic process can completely obliterate porosity in one shoaling system, and leave porosity untouched in an adjoining one (e.g., Cretaceous oolitic sand systems of the eastern Arabian Platform area). Shoaling bodies are typically meters to tens of meters in thickness. Early submarine cementation also destroyed potential reservoirs in some Devonian pinnacle reefs in the Alberta Basin while leaving neighboring pinnacles almost entirely uncemented. Devonian reefs in the Canning Basin show similar cementation patterns, as do many Permian shelf margin carbonates in West Texas. Spatial controls on submarine cementation are not well understood but can have an overwhelming effect on the distribution of carbonate reservoirs and seals, particularly at the scale of fields and individual carbonate beds. Intervals with large amounts of submarine cements are typically several meters thick, but can be more than 100 m thick in unusual cases.

How much a carbonate is cemented and leached partially dictates its seismic character. But the style of carbonate cementation and leaching also affects seismic characteristics. Anselmetti and Eberli (1992) found that limestones with moldic porosity have higher P-wave velocities than limestones with equivalent amounts of interparticle porosity. This observation is consistent with the well-known logging technique of subtracting sonic-derived porosity from neutron- or density-derived porosity to define amounts of moldic pore space. The compressional wave moves through the rigid framework, efficiently bypassing the moldic pore spaces. This is a potentially important observation that could be used to map out areas of effective versus ineffective porosities on a field scale since some carbonates with high moldic porosity have low permeabilities. A theoretical corollary to these observations is that preferential cementation at grain contacts should increase rock rigidity and P-wave velocity. A seismic mapping tool could be used to differentiate intervals affected by cementation in the vadose diagenetic environment, where cements are concentrated at grain contacts (e.g., Longman, 1980), versus other diagenetic environments, where cements are distributed more uniformly.

Seismic Techniques Versus Scales of Diagenesis

This section is included to give a brief survey of recently published seismic techniques applicable to defining diagenetically controlled rock properties. Rock property changes caused by diagenetic processes occur over a continuum of scales. Emphasis is on those techniques effective for defining features on the scales of tens to hundreds of meters thick. Broader scale features

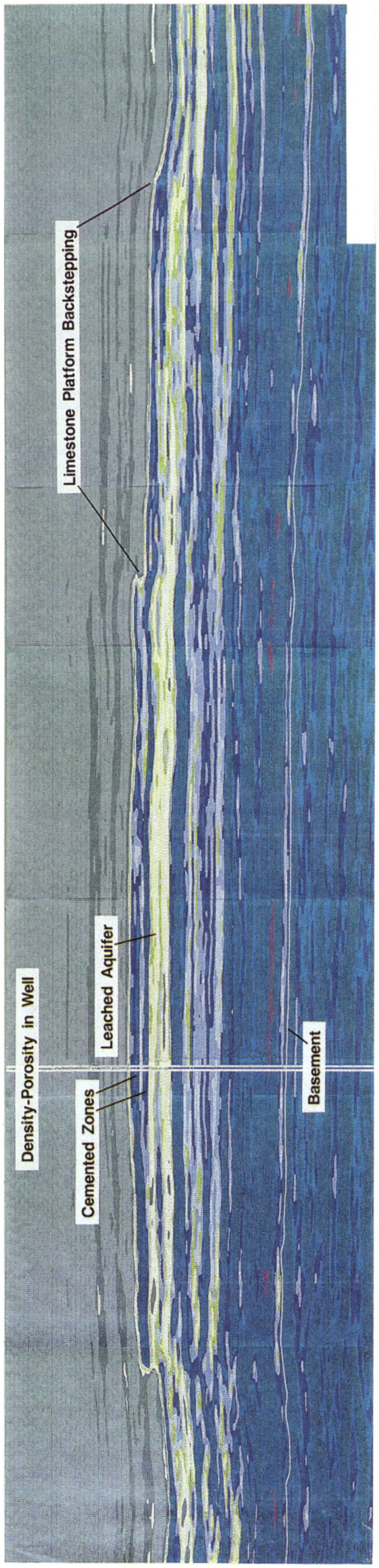

Fig. 5. An acoustic impedance cross-section off a carbonate platform offshore of China. (Overlying clastic section is shown in light blue.) Note the backstepping at the top of the limestone platform; the uppermost platform is charged with hydrocarbons. Two diagenetically cemented intervals are indicated: the uppermost is associated with an ancient subaerial exposure surface and the lower with late localized cementation near the hydrocarbon–water contact. The underlying highly leached aquifer limestone is shown in yellow, and basement is shown as a white line. Acoustic impedance maps show the internal architecture of a platform and important diagenetic features affecting subsurface oil migration, subsurface leaching, and reservoir compartmentalization.

are generally imaged with conventional *P*-wave seismic records. I first address theoretical resolution constraints then briefly discuss several published techniques.

Resolution

Vertical seismic resolution is a function of (1) bandwidth, (2) interval velocity, (3) attenuation and distortions caused by near-surface and thin-bed geology, and (4) additions caused by multiples (Waters, 1978; Badley, 1985). The resolution limit is generally taken to be 1/4 wavelength, and detection limits can be as small as 1/30 wavelength, depending on data quality (Sheriff, 1989a). Figure 6 illustrates these limiting factors for a wedge-shaped carbonate body in shale. If the carbonate has an interval velocity of 5000 m/s, then the seismic resolution of a 50-Hz wavelet would be only 25 m (1/4 wavelength) (Figure 7); for nondestructive resolution, it would be 50 m (1/2 wavelength). Deeper burial of the wedge would degrade resolution even more because of incremental high-frequency attenuation. In a natural system, an increase in depth or age of the carbonate can lead to degradation of resolution because of increased interval velocity.

Clearly, conventionally shot *P*-wave seismic is not currently adequate for resolving carbonate diagenetic features of a few meters to tens of meters in thickness at typical exploratory drilling depths of 2000–4000 m. However, the thickness of diagenetically affected intervals below tuning can be estimated from amplitude heights if proper rock calibration is available (Badley, 1985; Poster and Samwell, 1990; Stanulonis and Tran, 1992).

Three-Dimensional *P*-Wave Mapping

Three-dimensional *P*-wave mapping is the tool of choice for general definition of structure and gross depositional morphologies. Conventional 3-D surveys are especially effective in defining medium- and large-scale geometries of karsted intervals. Figure 8 shows a time slice of a karsted carbonate interval with sinkholes and rugose terrain. Definition of features is good, even though data vintage is fairly old (Brown, 1985). Karsted terrain is comparatively easy to recognize because of the multitude of features with highly identifiable geometries. The density of data in a 3-D shoot can provide necessary resolution of sinkholes, for example, where a 2-D data set might show only chaotic reflectors caused by localized solution and collapse features.

Interval Velocity and Acoustic Impedance Mapping

Pigott et al. (1991) did a detailed comparison of relative-amplitude stacking velocities to differentiate porous versus nonporous dolomitized bioherms in the Hardeman Basin, Texas. The work showed clearly lower velocities in the porous bioherm, suggesting that relative amplitude velocity work was a good tool for indicating porosity in the local area.

An example of acoustic impedance mapping inside a carbonate body was shown in Figure 5. In this example, the technique did an excellent job of imaging laterally extensive tabular bodies of highly cemented and leached carbonate. The cemented intervals segment the reservoir into compartments and partially isolate the

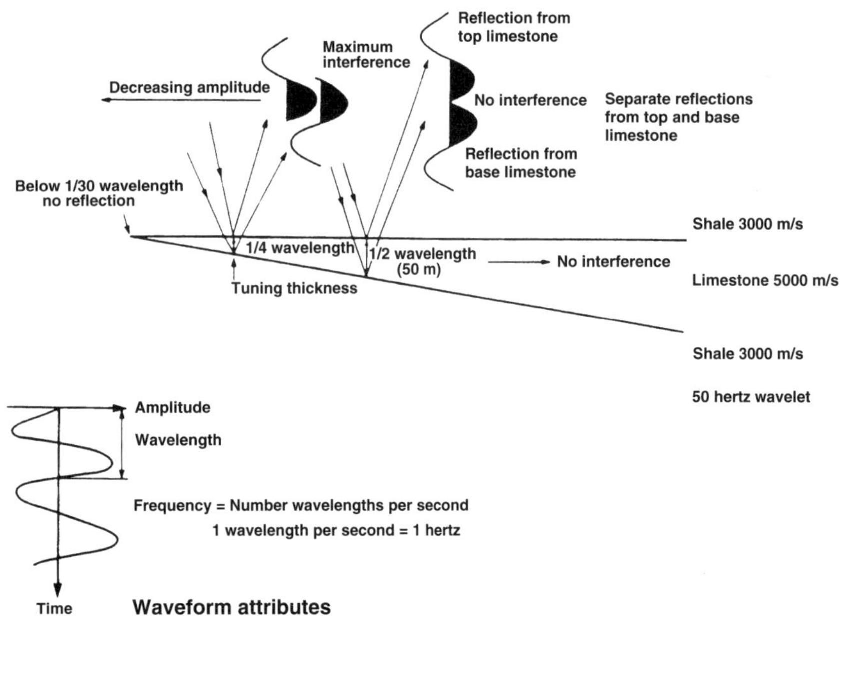

Fig. 6. A carbonate wedge model showing tuning thickness at 1/4 wavelength and no interference at 1/2 wavelength. From Badley (1985).

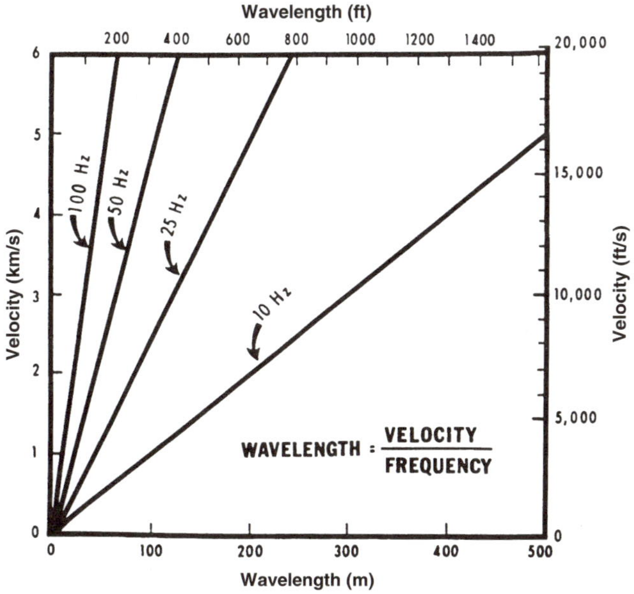

Fig. 7. Graph showing the relationship of wavelength, velocity, and frequency in carbonate rocks. From Sheriff (1989b).

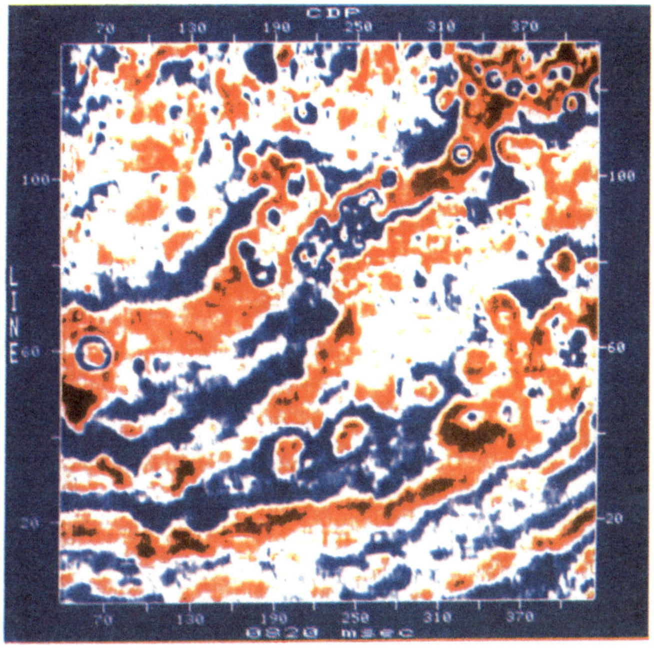

Fig. 8. Colored-enhanced 3-D time slice of a karsted limestone in the Gippsland Basin, Australia. The circular features are interpreted to be large sinkholes. Red and blue colors are used for easy visualization of peaks and troughs. From Brown (1991, courtesy of AAPG).

reservoir interval from aggressive aquifer flow. The leached interval of underlying aquifer limestone provides water drive for production and was probably the local hydrocarbon migration pathway for reservoir charging. With proper rock calibration, acoustic impedance mapping is an excellent tool for defining subsurface fluid migration pathways.

S-Wave Mapping

An important application for shear wave analysis is fracture detection (Cooper, 1994; see also Li and Müller, Chapter 14, this volume, for a fracture imaging study in the Austin Chalk of east Texas). Measured V_S is also an important control on improved inversions for lithology, porosity, and fluid types.

Inversions for Lithology and Porosity

Landro et al. (1995) used amplitude versus offset (AVO) analysis on North Sea chalk fields to define zones of discontinuous porosity. Drilling confirmed the presence of hydrocarbon-charged stratigraphic chalk traps. Their semi-quantitative results were based on transforming common midpoint (CMP) gathers from the time-offset domain to the oblique traveltime ray parameter domain.

Ho et al. (1993) described two techniques for inferring the presence of fractured dolomite in nonporous limestone. The first technique, using V_P/V_S ratios extracted from *P*-wave and *S*-wave CMP stacks, was used successfully to image the Albion Scipio dolomite field in the Michigan Basin. The authors could not specify what geologic component was contributing the most to the V_P/V_S response because the ratio is dependent on lithology, fluid type, and fracturing. The second technique, using AVO analysis of *P*-wave data only, also imaged the field, although interpretation was more subjective. Ho et al. (1993) tried three different methodologies for their AVO modeling based on previous work done by Shuey (1985), Chiburis (1987), and Chou et al. (1991). They found that the best method was the comparative technique of Chiburis (1987), which is an interactive comparison of *P*-wave AVO responses between a set reference horizon or window and individual traces in the carbonate body of interest. Chiburis (1993) claims a 96% hydrocarbon finding ratio for his technique as applied to exploration work in Central Saudi Arabia.

Stanulonis and Tran (1992) described a highly integrated technique for inferring porosity from 3-D seismic data in the Lisburne carbonate pool of Prudhoe Bay. They statistically separated the field into four regions, each having a different empirical relationship between seismic amplitude and total porosity (Figure 9). Seismic predictions of porosity before drilling have been excellent despite complicated reservoir layering of limestone, dolomite, and shale, and despite great lateral variability

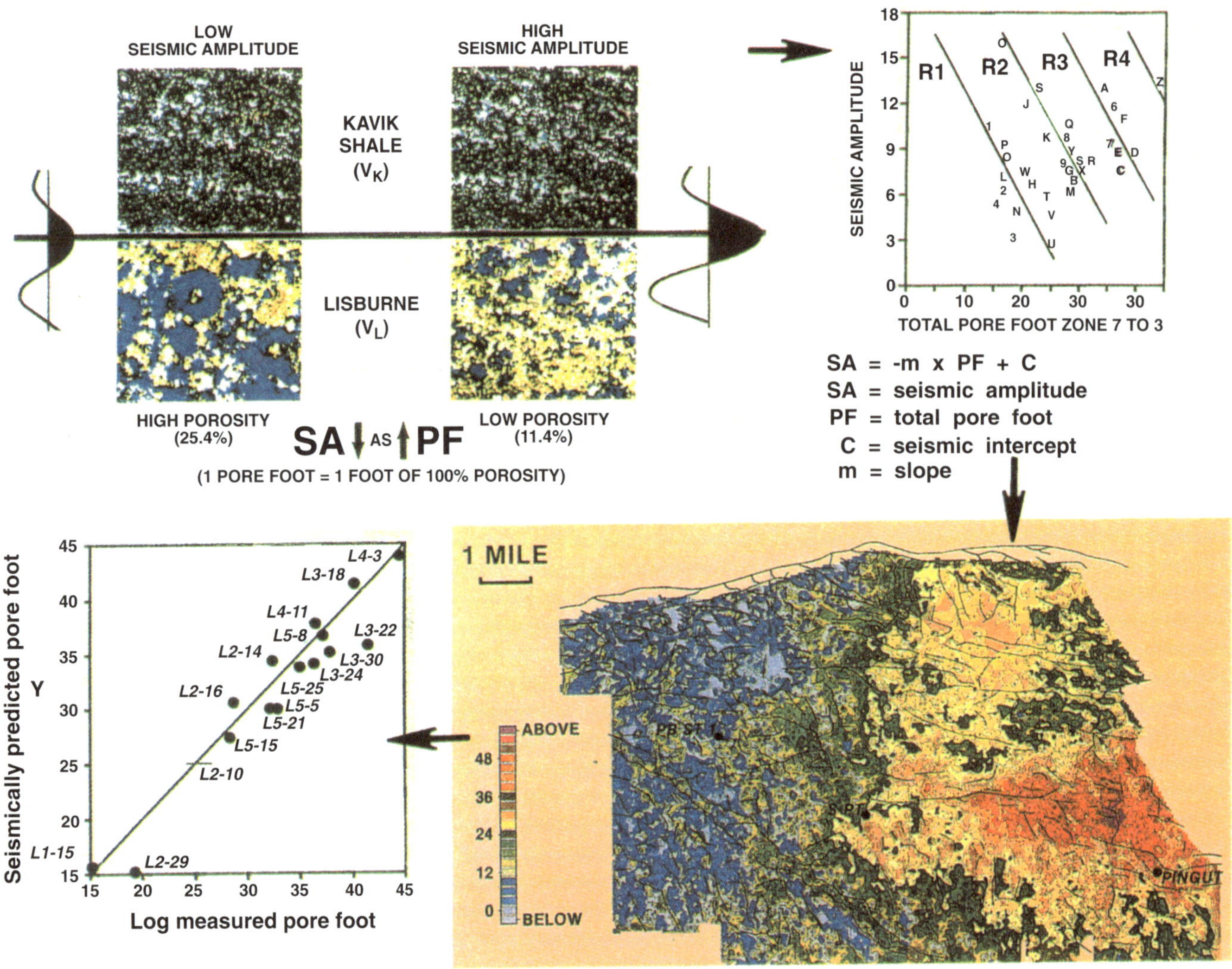

Fig. 9. Calibration of surface seismic with well log data from the Lisburne carbonate of Prudhoe Bay field, Alaska. From Stanulonis and Tran (1992). Amplitude height was correlated with total interval porosity for four regions of the field (upper right). The resulting total porosity map (lower right) was tested by drilling and found to be very good (lower left).

of porosity within individual beds. Good results are due in part to addition of as much rock control as possible. Less rigorous studies have been carried out elsewhere for carbonates of roughly similar burial depth (Krishna et al., 1990; Harvey and MacDonald, 1990).

Fraser and Jain (1988) described an integrated well log and synthetics study from a deeper section that illustrates how porosity inversions become increasingly difficult with depth. They looked at Beaverhill Lake reef porosity in the Caroline field area of Alberta, which lies at about 3700 m depth. They tested three models to infer porosity in both gas-bearing and dry wells. The first model was based on AVO, the second on amplitude versus wavelength changes in the uppermost reef, and the third on amplitude inversion of stacked traces. They found that no seismic technique alone could distinguish a porous carbonate from a shaley carbonate at target depths, although reservoir carbonates could theoretically be detected at thicknesses of as little as 10 m if porosity was great enough (18–35%). AVO modeling produced the worst results; signal strength was not high enough above local noise conditions to be observed on normal prestack data. The best porosity predictions were produced from amplitude changes on stacked sections and on velocity inversion sections (Figure 10).

Lu and Lines (1995) obtained better results with AVO modeling of deeply buried Devonian reefs. Better discriminations of porous carbonates from shaley carbonates were obtained by using range-limited stacks near well control points. Differences in AVO response were inferred to be due to large theoretical differences in Poisson's ratio between a porous carbonate and a nonporous shaley carbonate.

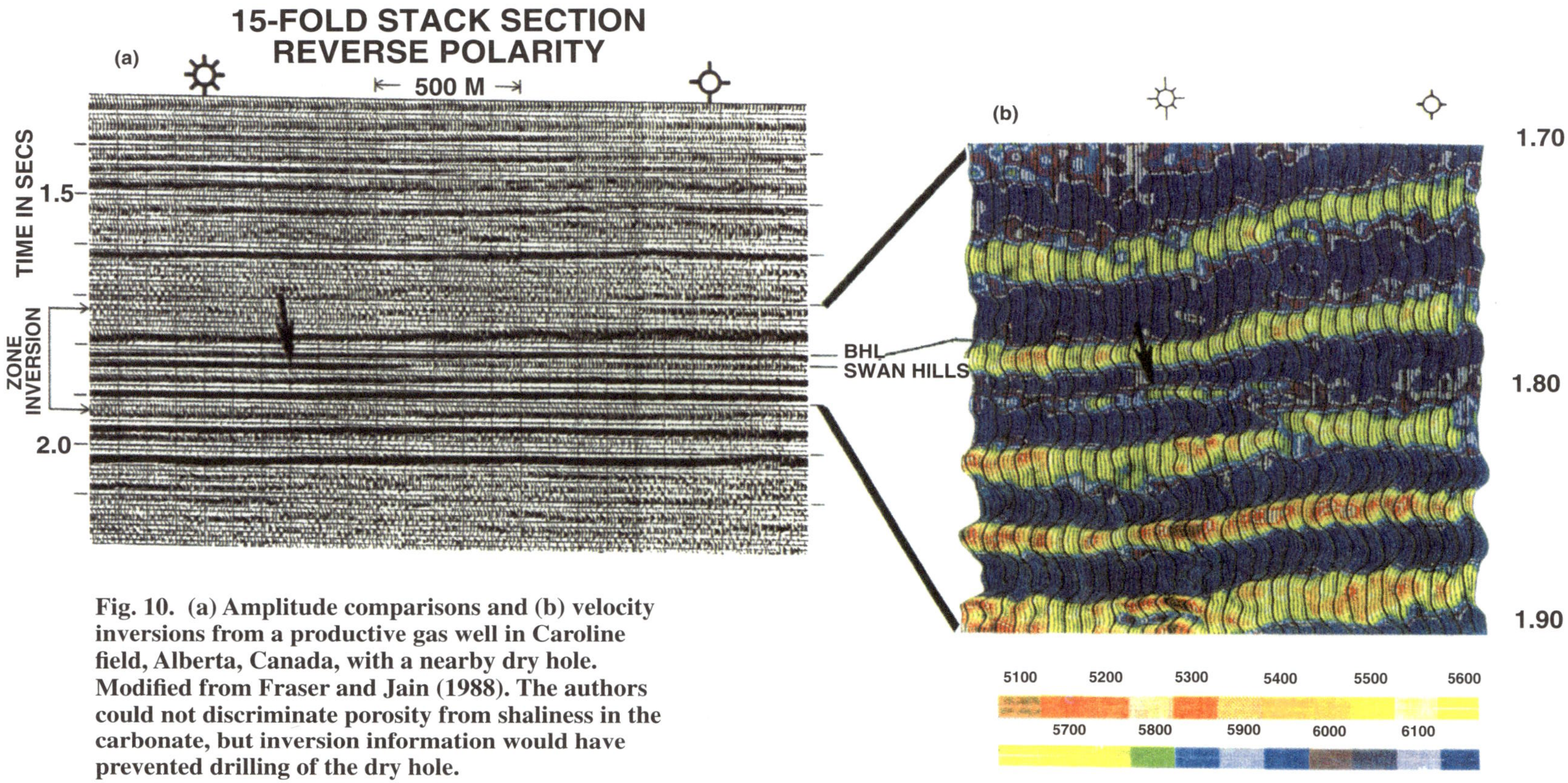

Fig. 10. (a) Amplitude comparisons and (b) velocity inversions from a productive gas well in Caroline field, Alberta, Canada, with a nearby dry hole. Modified from Fraser and Jain (1988). The authors could not discriminate porosity from shaliness in the carbonate, but inversion information would have prevented drilling of the dry hole.

Borehole Seismic Techniques

Harris et al. (1995) described excellent results of *P*-wave and *S*-wave cross-borehole studies in West Texas. They noted small-scale variations of "a few meters" in rock properties between wells 183 m (600 ft) apart. Although not specified in the article, heterogeneities in the stratigraphic intervals of interest (including the Grayburg and San Andres formations) were controlled by both facies and diagenetic factors. An integrated study involving cross-borehole seismic and available log and rock data would need to be conducted to discriminate the two. However, Harris et al.'s (1995) study demonstrates well how cross-borehole data can fill a niche between the scale of surface seismic data and that of logs and cores.

Poster and Samwell (1990) described an integrated well log and geophysical study of an unspecified offshore Dubai carbonate field. They used vertical seismic profiling (VSP) to calibrate surface seismic data for relative amplitudes, multiple attenuation, and stacking velocities. *S*-wave seismic data were also collected and integrated. They claimed a significant improvement in ability to differentiate shaliness in carbonates over earlier attempts, and an ability to correctly forward-model porosity characteristics of limestone on the scale of a large field. The paper did not specify improvements in resolution, but this integrated approach is probably continuing with some success, as Dubai Petroleum Company has scheduled the drilling of ten wells and acquisition of 1000 km^2 of 3-D seismic profiles in 1994 (World Oil, 1994). Dubai's offshore fields lie at depths of about 2800 m. A less rigorous study using VSP calibration to detect karsting and fracturing was done by Vandenberghe et al. (1986).

Cross-borehole tomography (CBT) has been used in West Texas fields to resolve diagenetically affected carbonate intervals of about 10 m thickness (Jovanovich, 1994). CBT is especially useful in deeper buried carbonates of older fields where surface seismic data are sparse or inadequate, although use on wells from waterflooded fields significantly complicates interpretations.

Challenges

One of the biggest challenges to explorationists working deeply buried carbonate plays is t2he lack of sufficient seismic resolution to image rock attributes within buildups. The offshore China line of Figure 5, although shot for a fairly shallow target carbonate, has a seismic acquisition history that is an optimistic model for future efforts in more deeply buried carbonate plays. The line was originally shot with acquisition parameters that allowed resolution of the limestone platform's exterior dimensions only. Later, the line was reshot and processed to give a 150-Hz frequency content within the limestone platform. The combination of high-frequency content and comparatively slow *P*-wave velocities at 1100 m burial depths allowed resolution of diagenetic features on the scale of <10 m. Integrated log, petrographic, and geochemical data confirmed the nature of high and low acoustic impedance intervals as diagenetically cemented and leached intervals, respectively. Field development plans were greatly aided by knowing the origin and lateral extent of cemented intervals.

These types of technologic improvements in data acquisition and processing are needed to approach adequate seismic imaging of diagenetic features in more deeply buried carbonate bodies. An excellent testing ground for new improvements is the deep Alberta Basin, where frequency attenuation and high velocities are not the only complicating factors for seismic resolution—the reservoir and seal are both predominantly carbonate. This makes it difficult to seismically image even the outer margins of reservoir-bearing carbonate bodies. The geophysicist is presently forced to infer shelf margin edges from subtly contoured acoustic impedance maps and AVO models with nonunique solutions. Other high-profile exploration areas with similar challenging geology exist in China, Venezuela, Mexico, western Africa, and the former Soviet Union.

Accurate rock property inversions from seismic data is an area of great promise for remote definition of diagenetic features. The challenge of future work is to limit inversion complications with rock calibration and other independent data. Since the only physical properties of a carbonate that affect seismic character are bulk density, bulk modulus, and shear modulus (Waters, 1978), uncalibrated inversions for bulk density and V_P can be explained with changes in porosity, lithology, organic matter content, or fluid type. Increased use of integrated seismic and petrologic data sets can improve seismic interpretations on a case-specific basis. 3-D V_S mapping, although not an independent data set, can significantly aid seismic interpretation efforts in areas with poor rock control.

Conclusions

At present, our ability to resolve important reservoir-modifying effects of carbonate diagenesis at shallow depths is fair. Specially shot seismic lines with significant high-frequency components reveal diagenetic features of exploration and exploitation significance on a scale of <10 m. At greater depths, however, seismic resolution of diagenetic features is generally poor due to high interval velocities and frequency attenuation. Wellbore seismic techniques offer a high-resolution solution to this problem, albeit one limited to fairly small volumes of rock. Continued research into improved seismic acquisition and processing and integration of petrologic data with seismic interpretation are required for the best results. Subjectively, seismic interpretations can be improved by maintaining a 4-D view of "competitive diagenesis" and its resulting spatial controls on the distribution of carbonate reservoirs and seals.

Acknowledgments

Discussions with Craig Cooper, Rick Lindsay, Chuck Skidmore, and Mike Kelly about various facets of seismic theory are gratefully acknowledged. Craig Cooper also shared an illustration from his work on high-resolution acoustic impedance mapping of a limestone platform (Figure 5). Thanks are also due to Amoco EPTG for allowing this work to be published.

References

Anselmetti, F. S., and Eberli, G. P., 1992, Sonic velocity in carbonate—a product of original composition and post-depositional porosity evolution: Ann. AAPG-SEPM-EMD-DPA-CSPG Conv. Abstracts.

Badley, M. E., 1985, Practical seismic interpretation: IHRDC.

Bassinot, F. C., Marsters, J. C., Wilkens, R. H., and Mayer, L. A., 1993, Variations of porosity in calcareous sediments from the Ontong Java Plateau (ODP Leg 130): Proc. Ocean Drill. Prog. Sci. Res., **130**, 653-661.

Bathurst, R. G. C., 1975, Carbonate sediments and their diagenesis: Springer Verlag.

Brown, A. R., 1991, Interpretation of three dimensional seismic data: Am. Assoc. Petr. Geol. Mem. **42**, 3rd Edition.

Brown, R. J., Hills, L. V., and Anderson, N. L., 1990, Seismic interpretation of Upper Elk Point (Givetian) carbonate reservoirs of western Canada: Geophys. Prosp., **38**, 719-736.

Campbell, A. E., Schlager, W., and Stafleu, J., 1993, Interpreting unconformities is seismic records and outcrop: 25th Ann. Soc Petr. Eng. Offshore Tech. Conf. Proc., **1**, 257-268.

Chiburis, E. F., 1993, AVO applications in Saudi Arabia: *in* Castagna, J. P., and Backus, M. M., Eds., Offset-dependent reflectivity—theory and practice of AVO analysis: Soc. Expl. Geophys., 211-229.

Chiburis, E. F., 1987, Studies of amplitude versus offset in Saudi Arabia: 57th Ann. Soc. Expl. Geophys. Internat. Mtg., Expanded Tech. Prog., 614-616.

Chou, T. G, Shih, T. C., and Lorenzetti, E. A., 1991, Inversion of seismic traveltime and amplitude-versus-offset information: 61st Ann. Soc. Expl. Geophys. Internat. Mtg., Expanded Tech. Prog., **2**, 1072-1075.

Cooper, C. W., 1994, Seismic technique development: World Oil, **215**, 55-56.

Dunnington, H. V., 1967, Aspects of diagenesis and shape change in stylolitic limestone reservoirs: 7th World Petr. Conf., **2**, 339-352.

Esteban, M., and Klappa, C. F., 1983, Subaerial exposure, *in* Scholle, P. A., Bebout, D. G., and Moore, C. H., Eds., Environments of carbonate deposition: Am. Assn. Petr. Geol. Mem. **33**, 1-54.

Fontaine, J. M., Cussey, R., Lacase, J., Lanaud, R., and Yapaudjian, L., 1987, Seismic interpretation of carbonate depositional environments: AAPG Bull., **71**, 281-297.

Fraser, D., and Jain, S., 1988, Exploration for porosity in carbonates: a synthetic-based study: Can. J. Expl. Geophys., **24**, 141-153.

Hardeman, R. F. P., 1981, Chalk reservoirs of the North Sea: Bull. Geol. Soc. Denmark, **30**, 119-137.

Harris, J. M., Nolen-Hoeksema, R. C., Langan, R. T., Van Schaack, M., Lazaratos, S. K., and Rector III, J. W., 1995, High resolution crosswell imaging of a West Texas carbonate reservoir, part 1—project summary and interpretation: Geophysics, **60**, 667-681.

Harris, D. G., and Sarg, J. F., 1989, Seismic modeling of a mixed carbonate-siliciclastic prograded system, middle Permian, Midland Basin, *in* Search for the subtle trap: West Texas Geol. Soc., Proc., **89-85**, 119-138.

Harvey, P. J., and MacDonald, D. J., 1990, Seismic modelling of porosity within the Jurassic aged carbonate bank, offshore Nova Scotia: Can. J. Expl. Geophys., **26**, 54-71.

Ho, S. M., Lee, S. S., and Purnell, G. W., 1993, Comparison of *P*-wave AVO (amplitude vs. offset) techniques for locating zones of fractured dolomite within nonreservoir limestone: 78th Ann. Am. Assn. Petr. Geol. Conv. Abstracts, 119.

Jarrard, R. D., and Symonds, P. A., 1993, Origin of seismic reflectors within carbonate-rich sediments from northeastern Australian margin (ODP Leg 133): Proc. Ocean Drill. Prog., Sci. Res., **133**, 649-659.

Jenyon, M. K., 1986, The seismic expression of real and apparent buried topography: J. Petr. Geol., **10**, 41-58.

Jovanovich, D. B., 1994, Seismic research continues: World Oil, **215**, 56.

Kenter, J. A. M., and Ivanov, M., 1995, Parameters controlling acoustic properties of carbonate and volcanoclastic sediments at Sites 866 and 869: Proc. Ocean Drill. Prog., Sci. Res., **143**, 287-303.

Kenter, J. A. M., and Stafleu, J., 1995, Synthetic seismograms at (ODP Leg 143) Site 866: origin of reflections and implications for recognizing the limestone/basalt transition in Cretaceous Mid-Pacific guyots: Proc. Ocean Drill. Prog., Sci. Res., **143**, 305-315.

Krishna, B. R., Rajagopalarao, M., and Subbarao, C. V., 1990, Mapping area of favorable porosity development in a carbonate reservoir from seismic data: J. Assn. Expl. Geophys., **11**, 129-136.

Kvenvolden, K. A., and Barnard, L. A., 1982, Hydrates of natural gas in continental margins, *in* Watkins, J. S., and Drake, C. L. Eds., Am. Assn. Petr. Geol. Mem. **34**, 631-640.

Landro, M., Buland, A., and D'Angelo, R., 1995, Target-oriented AVO inversion of data from Valhall and Hod fields: The Leading Edge, **14**, No. 8, 855-861.

Langdon, G. S., and Malacek, S. J., 1987, Seismic stratigraphic study of two Oxfordian carbonate sequences, eastern Saudi Arabia: AAPG Bull., **71**, 403-418.

Longman, M. W., 1980, Carbonate diagenetic textures from near-surface diagenetic environments: AAPG Bull., **63**, 461-487.

Lu, H.-X., and Lines, L., 1995, AVO and Devonian reef exploration: difficulties and possibilities: The Leading Edge, **14**, No. 8, 879-881.

Matthews, R. K., 1984, Dynamic stratigraphy: Prentice-Hall.

McIlreath, I. A., and Morrow, D. W., Eds., 1990, Diagenesis: Geosc. Can. Repr. Ser. **4**.

Mienert, J., 1984, The importance of carbonate content in the acoustic stratigraphy of Panama Basin: Mar. Geol., **54**, 237-247.

Pigott, J. D., Shrestha, R. K., and Warwick, R. A., 1991, Carbonate reservoir characterization using seismic velocity and amplitude variation with offset analysis: Hardeman Basin, Texas, test case (abs.): AAPG Bull., **75**, 654.

Poster, C. K., and Samwell, N., 1990, Integration of surface seismic and well data: case studies in carbonate reservoirs: Norwegian Petr. Soc. Develop. Geophys. Tech. Conf., Proc. **2**.

Sarg, J. S., and Candelaria, M., 1992, High-resolution seismic characterization of a Wolfcamp carbonate platform: Ann. AAPG Conv. Abstracts, 113-114.

Sheriff, R. E., 1989a, Geophysical methods: Prentice Hall.

Sheriff, R. E., 1989b, Inferring stratigraphy from seismic data, *in* Beaumont, E. A., and Foster, N. H., Eds., Geophysics III: Am. Assn. Petr. Geol. Treatise Petr. Geol. Repr. Ser. **14**, 96-110.

Shuey, R. T., 1985, A simplification of the Zoeppritz equations: Geophysics, **50**, 609-614.

Stafleu, J., Everts, A. J. W., and Kenter, J. A. M., 1995, Seismic models of a prograding carbonate platform: Vercors, SE France: Mar. Petr. Geol., **11**, 514-527.

Stanulonis, S. F., and Tran, H. V., 1992, Method to determine porosity-thickness directly from 3-D seismic amplitude within the Lisburne carbonate pool, Prudhoe Bay: The Leading Edge, **11**, No. 1, 14-20.

Thornbury, W. D., 1965, Principles of geomorphology: John Wiley & Sons.

Vandenberghe, N., Poggialiolmi, E., and Watts, G., 1986, Offset-dependent seismic amplitudes from karst limestone in northern Belgium: First Break, **4**, No. 5, 9-27.

Wagner, P. D., 1990, Geochemical stratigraphy and porosity controls in Cretaceous carbonates near the Oman Mountains, *in* Robertson, A. H. F., Searle, M. P., and Ries, A. C., Eds., The geology and tectonics of the Oman region: Geol. Soc. London Spec. Pub. **49**, 127-137.

Wagner, P. D., Tasker, D. R., and Wahlman, G. P., 1995, Reservoir degradation and compartmentalization below subaerial unconformities: limestone examples from West Texas, China and Oman, *in* Budd, D. A., Saller, A. H., and Harris, P. M., Eds., Unconformities in carbonate strata—their Recognition and the Significance of associated porosity: Am. Assn. Petr. Geol. Mem. **63**, 177-195.

Waters, K. H., 1978, Reflection seismology: a tool for energy resource exploration: John Wiley & Sons.

Wilson, J. L., 1975, Carbonate facies in geologic history: Springer-Verlag.

World Oil, 1994, 49th international outlook: Gulf Pub. Co., **215**, No. 8., 100.

Chapter 13

Porosity Detection and Mapping in a Basinal Carbonate Setting, Offshore Norway

R. M. D'Angelo
M. K. Brandal
K. O. Rørvik
Amoco Norway Oil Company
Stavanger, Norway

Abstract

The Cretaceous chalk section of offshore Norway is a prolific hydrocarbon-producing trend. The fields are typically structural traps, often with a stratigraphic trapping component. The identification and evaluation of stratigraphic traps in this chalk section require differentiation of the porous reservoir and tight nonreservoir chalk facies. Integration of geologic models, petrophysical data bases, and seismic analysis has allowed the detection and mapping of high-porosity reservoir-quality chalks along this play trend.

Sedimentologic studies of chalk deposition, burial histories, and reworking are combined with detailed petrophysical analysis of core plug samples in the lab to define rock properties that distinguish porous from nonporous chalk facies. Seismic analysis techniques, including stratigraphic processing, velocity analysis, modeling, inversion, and AVO analysis, enable the extraction of these rock properties from the seismic data, leading to prediction of the subsurface chalk porosity distribution. The results of these analyses indicate high-porosity reservoir-quality chalk. The drilling of this prospective high-porosity chalk anomaly resulted in an oil discovery.

This study demonstrates that the distribution of porosities within the Cretaceous chalk section of offshore Norway can be detected by seismic data. Furthermore, the porosity distribution in the chalk can be accurately predicted and mapped from the analysis of seismic data integrated with a detailed understanding of the petrophysical rock properties and the geologic history of the sediments.

Introduction

The South Central Trough of the North Sea contains numerous oil and gas fields in the Upper Cretaceous Chalk section. Many of these fields occur along the northwest-southeast trending Lindesnes Ridge. Amoco Norway Oil Company is the operator of the Valhall and Hod fields located on the crest of this ridge (Figure 1), where the Chalk has been fractured and deformed by uplift during the Tertiary.

Throughout 1992 and 1993, multidisciplinary studies done by Amoco Norway Oil and its partners (Enterprise Oil Norge Ltd., Elf Petroleum Norge, and Amerada Hess Norge A/S) on License PL033 uncovered potential for stratigraphic traps in the Cretaceous Chalk on the flanks of the Valhall and Hod fields. The main reservoirs in the Valhall-Hod area are the chalks of the Ekofisk (Danian), Tor (Maastrichtian–Campanian), and Hod (lower Santonian–Turonian) formations. These formations have been subdivided by the operator as shown in Figure 2 based on lithology and reservoir quality. The main oil-producing zone in both fields is Tor 1. Upper Ekofisk is also a good producer in the East Hod field, while Hod 4 is producing in both the Valhall and Hod fields. The two lowermost formations in the Norwegian North Sea Chalk, the Blodøks and Hidra formations, do not have reservoir-quality chalk.

Possible stratigraphic traps are seen as a thickening in the upper chalk sequence on 3-D seismic data and on specially acquired and processed 2-D seismic profiles. The anomalously thick chalk areas correspond to rapid accumulations of redeposited chalks that are formed

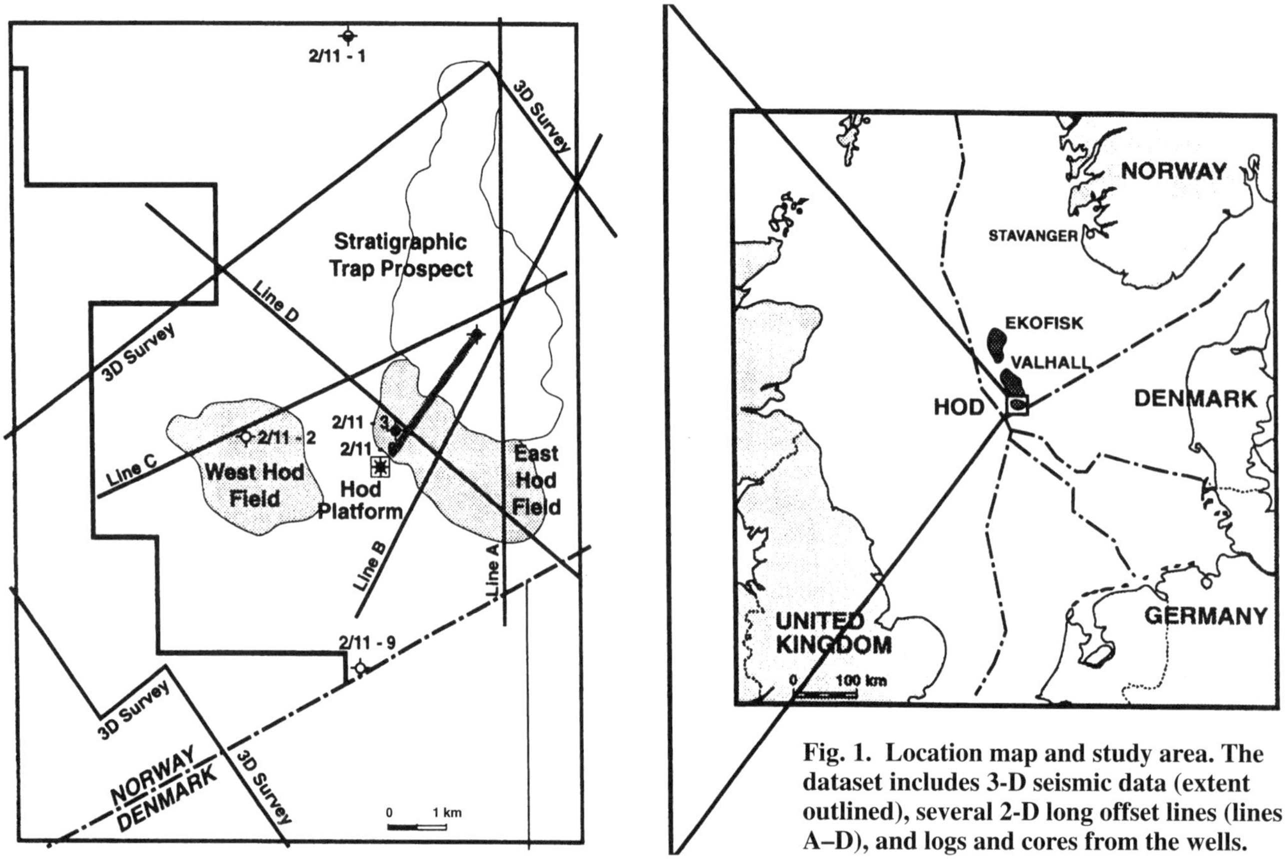

Fig. 1. Location map and study area. The dataset includes 3-D seismic data (extent outlined), several 2-D long offset lines (lines A–D), and logs and cores from the wells.

mainly by debris flows, slumps, slides, and turbidites. The redeposited chalks are surrounded by pelagic chalk facies. In addition to lateral facies changes related to chalk deposition, different degrees of diagenesis also occur in the chalk. Although the redeposited chalks usually have initial high reservoir-quality porosity compared to the pelagic chalks, porosity preservation is dependent on the timing of hydrocarbon migration and the associated degree of diagenesis. Therefore, stratigraphic traps within the chalks are formed due to lateral pinchouts, facies changes, and diagenesis, all of which provide the seal to the reservoir. Furthermore, reservoir-quality porosity is considered the greatest risk of failure.

By integrating specialized geophysical studies with rock property data from drilling operations in the fields, seismic evidence in support of the geologic model has been developed. For example, the internal seismic reflection character of the thickened chalk beds is enhanced by stratigraphic wavelet processing of the data. Furthermore, although the rock properties of the chalks have a wide range, petrophysical data from core plug measurements show a clear association of increasing porosity and decreasing *P*-wave and *S*-wave interval velocity with bulk density in the reservoir chalks of the fields. These data provide "ground truths" to guide the interpretation of special seismic study results over an interpreted stratigraphic trap prospect.

Interval velocity profiles calculated from nonroutine stacking velocity analysis show anomalously low velocity zones (including a gas chimney) at the field high, as well as in the down flank study area. Further support for the geologic model is given by forward ray-trace models of normal incidence seismic data, which agree well with the acquired data. Also, seismic trace inversion to acoustic impedance show low-impedance (high-porosity) zones in the upper chalk. Furthermore, amplitude versus offset (AVO) anomalies are observed in the prospect area. Forward AVO models indicate that these anomalies may be caused by high-porosity chalk zones in the upper chalk. Again, this interpretation supports the geologic model. Finally, an AVO inversion to elastic parameters indicates high-porosity zones in the chalk at the field and in the prospect area.

Together, the results of the geophysical analyses integrated with the petrophysical data are interpreted as evidence supporting the geologic model of a high-porosity, reservoir-quality chalk zone within the generally impermeable upper chalk section. These analyses have lowered the perceived risk of failure of the stratigraphic trap concept. Furthermore, the techniques discussed

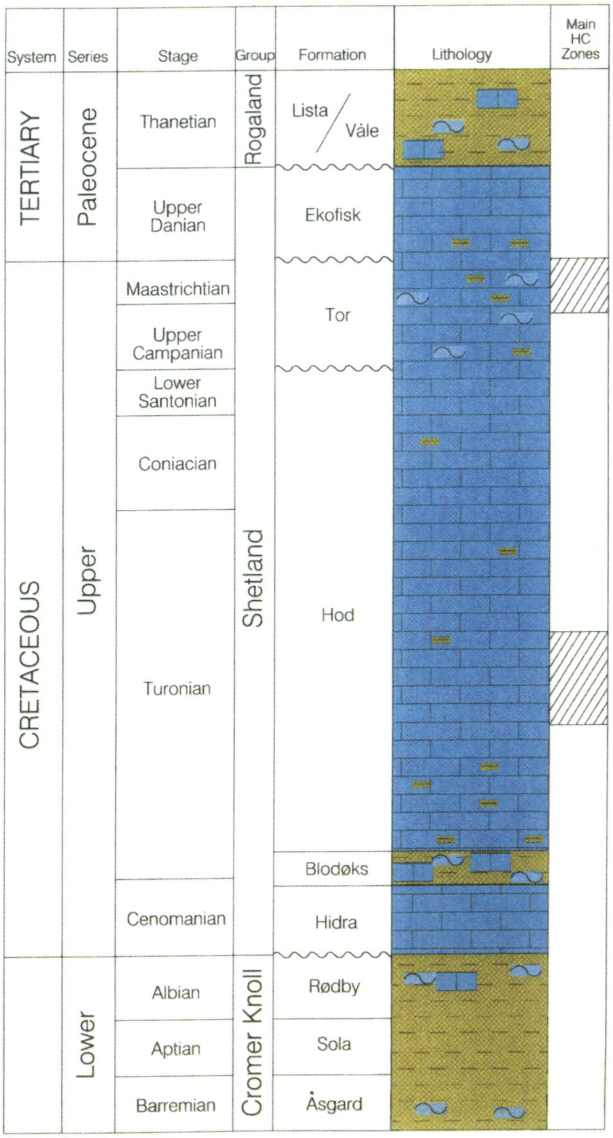

Fig. 2. Generalized stratigraphic column for the Lower Cretaceous–lower Tertiary section in the study area. Note the indicated main hydrocarbon zones, which are porous chalk intervals, for the Hod and Valhall fields in the far-right column.

here can be used as tools to explore for other stratigraphic traps along trend within the Cretaceous chalk section.

Based in-part on the results of these analyses, an exploration well was drilled in early 1994 to test the stratigraphic trap concept. The well is a discovery, proving the play concept and opening new play potential along the trend.

Geology and Stratigraphic Trap

The Cretaceous chalks of the North Sea can be categorized as either pelagic (autochthonous) or redeposited (allochthonous) based on their mode of deposition. The pelagic chalk accumulated mainly from suspension. The major part of the Hod Formation is interpreted to pelagic chalk, but the best reservoir section in this formation is thought to be both pelagic and redeposited chalk.

Redeposition of chalk is induced by basinward slopes, fault displacement along graben margins, and local halokinesis. The gravity flow mechanisms by which redeposition can occur include submarine slides, slumps, debris flows, and turbidites. The best reservoir sections in the upper chalk Ekofisk and Tor formations are interpreted to be deposited mainly by reworking of pelagic chalk. During the Late Cretaceous, inversion and fault reactivation caused an increase in chalk redeposition, which may be reflected in thickened upper chalk areas. Possibly related stratigraphic features are seen at several chalk outcrop locations in central and northern Europe, such as on the northwest coast of France.

Several authors have indicated that the mode of deposition may be the main cause of the lower porosities in the pelagic chalk and the higher porosities in the redeposited chalks (e.g., Kennedy, 1980; Hardman, 1982; D'Heur, 1984; Schatzinger et al., 1985). In the case of pelagic chalk (slow sedimentation rate), a dewatering effect can occur as a result of high bioturbation. Dewatering can prevent overpressure and can increase compaction. In addition, cement precipitation is relatively high, also due to slow sedimentation rates. Cementation can be reinforced by bioturbation, with input of additional cement through the burrows. The result is a chalk matrix with relatively low porosity and permeability.

In contrast, in redeposited chalk (fast sedimentation rate), the relative amount of cement is less and the rapid burial may promote preservation of porosity. However, the actual process is probably a much more complex interplay of mode of deposition with timing of oil migration, burial history, diagenesis, and insoluble residue concentrations (Maliva and Dickson, 1992) which affects porosity preservation in the chalks. In this paper, we present the results of geophysical analyses, constrained by petrophysical data, that are aimed at identifying and mapping porous chalk intervals that could form stratigraphic traps. These tools and methodologies may be applicable to similar chalk prospect evaluations regardless of the origin of the porosity.

An isopach map of the porous chalk that forms a stratigraphic trap prospect is shown in Figure 3. The lens trends north-south on the northeastern flank of the Hod field and is well expressed on the 3-D seismic data (Figure 3b). The trap is defined as a thickening in the upper chalk sequence between the top of the chalk and the top

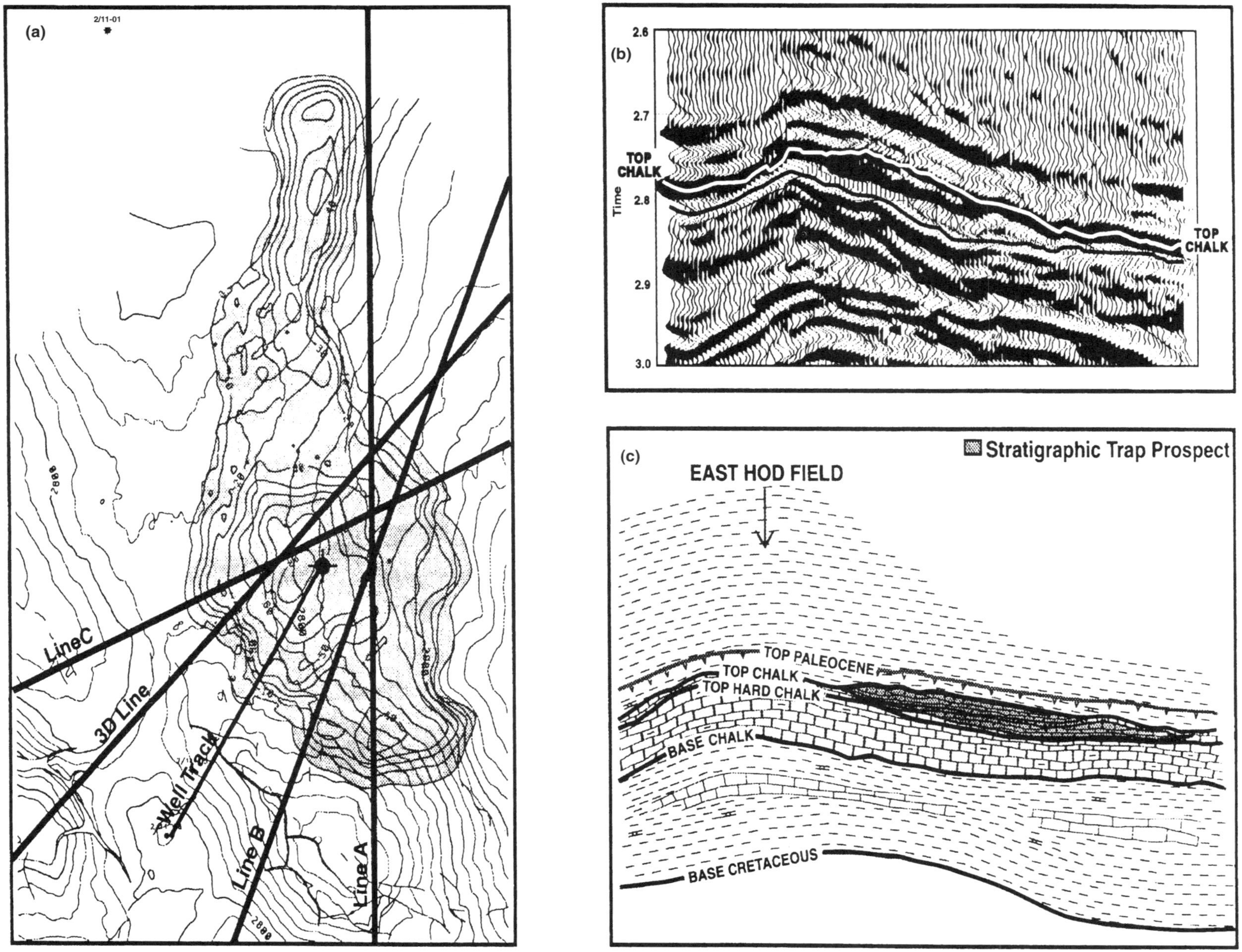

Fig. 3. (a) Isopach map of the porous chalk that forms a stratigraphic trap prospect. (b) A representative seismic line from the 3-D survey showing the appearance of the prospect in the upper chalk interval. (c) A schematic cross-section of the geologic model for the prospect along this 3-D line.

of the Hod Formation reflectors. A seismic reflector in the lens is interpreted as dividing the lens into upper and lower parts and possibly indicating an internal lithologic change. The lateral seal for the trap is provided by the low-porosity Upper Cretaceous pelagic chalk of the Hod Formation. The occurrence of anomalously high porosity within the lens is considered the greatest risk of failure for this prospect. Thus, the challenge for the explorationist is to predict and map porosity distribution directly from the seismic data.

Petrophysical Rock Properties

Extracting rock properties from seismic data is a general theme of much research and applied technology in the oil industry. In this case, integrating the rock property data with the seismic data began with a simplified but focused petrophysical study. Forty-one plugs from cores through the Cretaceous chalk interval were analyzed in Amoco's Geophysical Evaluation Module, a portable laboratory for measuring geophysical and geologic parameters such as porosity, density, mineralogy, *P*-wave velocity (V_P), and *S*-wave velocity (V_S) as a function of confining pressure and magnetic susceptibility. The cores were taken from four wells within the Valhall and Hod fields, as well as one well off-structure. Therefore, a wide range of chalk properties was sampled, including porosities from <5% up to about 40%.

Statistically significant predictive relationships are found between chalk porosity and the elastic parameters of V_P, V_S, and bulk density. Scatter plots of the data points are shown in Figure 4. Note that in the chalk, 80% of the variation in porosity can be accounted for by variation in the *P*-wave velocity ($R^2 = 0.80$). The R^2 values for the other elastic parameters are *S*-wave velocity, $R^2 = 0.77$, and bulk density, $R^2 = 0.64$. Thus, all three elastic parameters have statistically significant relationships to porosity. This knowledge is critical for a meaningful evaluation of the geologic model and porosity detection using the seismic data because the elastic parameters are the only rock properties that seismic data measure directly. Thus, the analyses presented here rely on having these meaningful correlations between the elastic parameters measured by the seismic data and chalk porosity.

The data base of petrophysical rock properties allows us to constrain forward seismic modeling reliably for various chalk porosities. Furthermore, from inversion of the seismic data to elastic parameters, we can predict porosity values within the chalk stratigraphic trap prospect. This exercise is easier said than done, however, especially because seismic data responds to contrasts in the elastic parameters. Perhaps the best that we can hope for is to predict only contrasts or relative differences in porosity. The nonuniqueness problem of seismic inversion to rock properties is another reason to submit the seismic data to a suite of rigorous analyses. Presumably, if more types of analyses confirm a geologic model, more confidence is gained in the interpretation. That is why we have applied several different approaches for extracting rock properties from the seismic data. The following sections present the results of these analyses: wavelet-shaping seismic processing, nonroutine stacking velocity analysis, 2-D normal incidence ray trace modeling, seismic trace inversion to acoustic impedance, AVO analysis and modeling, and prestack seismic inversion to elastic parameters.

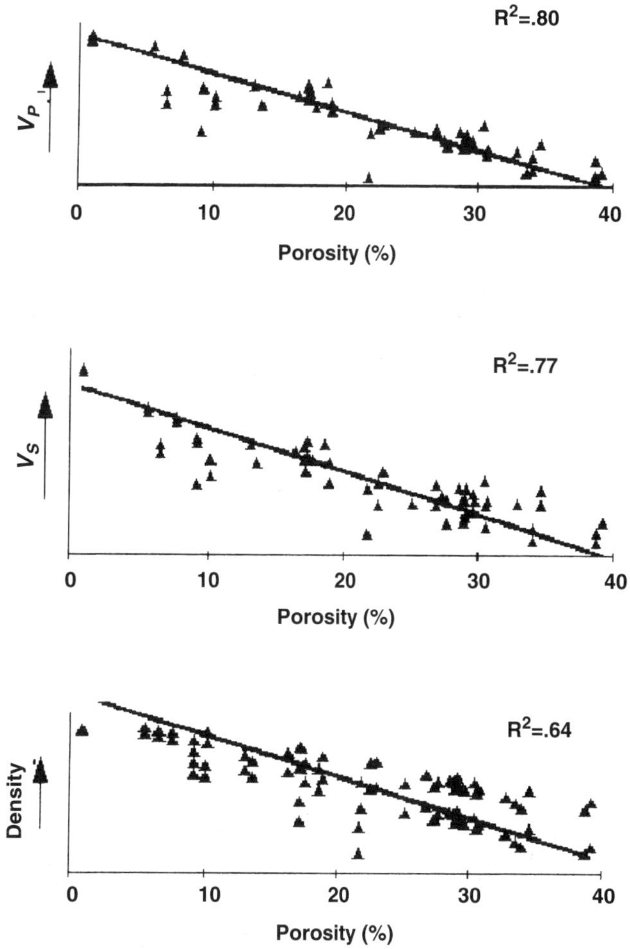

Fig. 4. Scatter plots of porosity (%) versus (a) *P*-wave velocity, (b) *S*-wave velocity, and (c) bulk density from laboratory measurements on 41 plug samples of cored Cretaceous chalk in the study area. The best-fit linear regression lines are shown, and the goodness-of-fit (R^2) values are given.

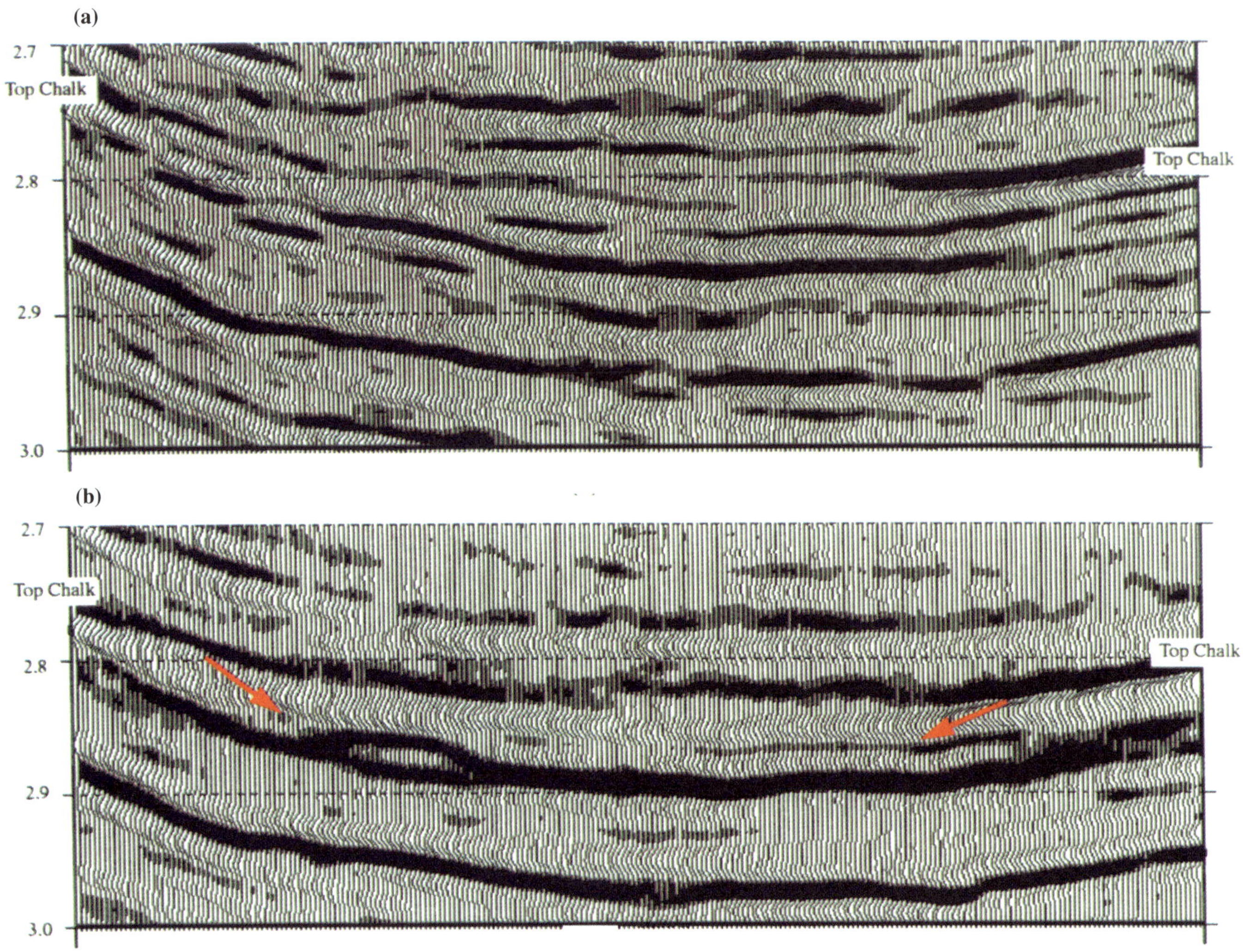

Fig. 5. (a) Seismic line A after standard contractor processing and (b) after Amoco's reprocessing. Note the enhanced detail in the upper chalk section (red arrows) on the Amoco reprocessed section.

Seismic Analyses

Seismic Acquisition, Processing, and Modeling

This section presents the results of specialized seismic acquisition and processing designed to enhance anomalous stratigraphic features in the upper chalk section. Several long offset seismic lines (4500 m cable length) were acquired over the prospective area (Figure 1) and specially processed with Cogneseis' Disco and Amoco's proprietary software. The acquisition parameters are listed in Table 1, and the generalized processing sequence is listed in Table 2. Prestack relative amplitude processing was emphasized to allow AVO analysis. The most critical step for poststack enhancement is the range of traces included in the stack. Figure 5 shows a comparison between the original contractor processed data and the reprocessed data for line A. On the reprocessed data, the upper chalk section is seen to thicken in the low and contain anomalous stratigraphic features. These features are not evident on the contractor processed data and are interpreted to be redeposited allochthonous chalk. Line B, oriented parallel to structural dip, shows more dramatically the thickening in the upper chalk section and a distinct pod (Figure 6). After applying wavelet shaping filters to this section, one can see seismic stratigraphic events that can be interpreted as downlapping progradational features. Thus, the reprocessing provides evidence that clearly fits the geologic model of redeposited chalk.

Now consider extracting just one of the elastic parameters, namely *P*-wave velocity, from the seismic data. During processing, the stacking velocities were picked every 125 m and along consistent reflection horizons across the seismic section. Figure 7 shows interval velocity fields calculated from these high-resolution

Table 1. Seismic Data Acquisition Parameters

Parameter	Description
DFS V recording	7 s at 2-ms sample rate
Navigation system	Differential GPS and Syledis
Source	5012 in.3 airgun at 2000 psi
Source depth	6 m ave. (cable depth 8 m ave.)
Source to center of first hydrophone group	113 m
Source to center of last hydrophone group	4588 m
Shotpoint interval	25 m (group interval 25 m)
Cable length	4500 m
Coverage	180 groups, 90-fold

Table 2. Generalized Processing Flow Sequence

Number	Step
1	Reformatting, resampling, and geometry indexing
2	Acquisition statics and trace editing
3	Spherical divergence correction
4	First break and direct arrivals muting
5	Designature deconvolution
6	Sort and "brute" NMO velocity analysis
7	Gapped predictive deconvolution
8	Amplitude recovery
9	Radon transform multiple removal
10	Inverse NMO and dip moveout
11	"Refined" NMO velocity analysis
12	NMO correction and stretch muting
13	Offset-limited stacking
14	Time migration
15	Coherency filter, scale, and display

"horizon-oriented" stacking velocities (Dix, 1954) for lines A and B. Lower interval velocities (high porosities) are seen in the upper chalk at the East Hod field and at the prospect location. Note the low interval velocity gas cloud in the Tertiary clastics above the field, adding credibility to the results. The low interval (P-wave) velocities in the chalk may indicate higher porosity zones. This interpretation holds true in the field where the wells have encountered chalk with porosities approaching 50%. Nevertheless, it is well documented that interval velocity calculation from stacking velocity data is subject to large errors (e.g., Al-Chalabi, 1974; Hubral and Krey, 1980; Lynn and Claerbout, 1982; Stork and Clayton, 1992; McBride et al., 1993). Errors can arise from resolution limitations of stacking velocity analysis and the effects of nonflat horizons and nonhyperbolic NMO on stacking velocity measurements. These errors carry over to the associated process of correcting the stacking velocities to rms velocities. Therefore, these data alone can not be considered strong evidence to support drilling the prospect.

A further test of the geologic model is made by forward (ray trace) modeling a normal incidence synthetic seismic section using GX Technology's GXII modeling package. A simplified 2-D model containing a high-porosity chalk lens within lower porosity chalk is used. Two elastic parameters (V_P and bulk density) are defined and constrained by the petrophysical core plug database. Normal incidence rays for P-wave reflections are traced, and the synthetic seismic section is processed to approximately match the real data in phase and frequency content. The migrated normal incidence synthetic section agrees well with the real data (Figure 8). Notice in particular the agreement in subtle details where the top of the chalk reflection begins to break up (red box in Figure 8). Also note that slight discrepancies in amplitude variations are seen, especially in the internal lens reflections (arrows). Although not a unique solution, the results of this simple 2-D modeling can be interpreted as support for the geologic model.

Seismic Trace Inversion

Seismic trace inversion is another method that considers two elastic parameters: P-wave velocity and bulk density. The seismic data should be relative amplitude, high bandwidth, zero-offset reflectivity, with a high signal-to-noise ratio. Knowledge of the seismic wavelet is also necessary. Under these conditions, running-sum integration (a recursive inversion) of the seismic data, which approximates the reflection coefficient "spike" series, results in an acoustic impedance step function series. Whereas the reflection coefficients represent lithologic boundaries (acoustic impedance contrasts), the step functions represent constant acoustic impedance values within lithologic layers.

Hampson-Russell's STRATA software (bandlimited inversion mode) is used to extract an acoustic impedance section from the near-range migrated stack of line C. Higher porosity chalks are detected as relatively low acoustic impedance zones within the upper chalk on the basis of the drop in V_P and bulk density, with increasing porosity observed in the petrophysical data. Figure 9 contains the calculated acoustic impedance for line C from the STRATA inversion analysis. Note that below the interpreted top of chalk in Figure 9, there is a laterally isolated low acoustic impedance anomaly (high-porosity) zone. This zone conforms to the upper part of the mapped stratigraphic trap prospect and is interpreted as additional evidence supporting the geologic model.

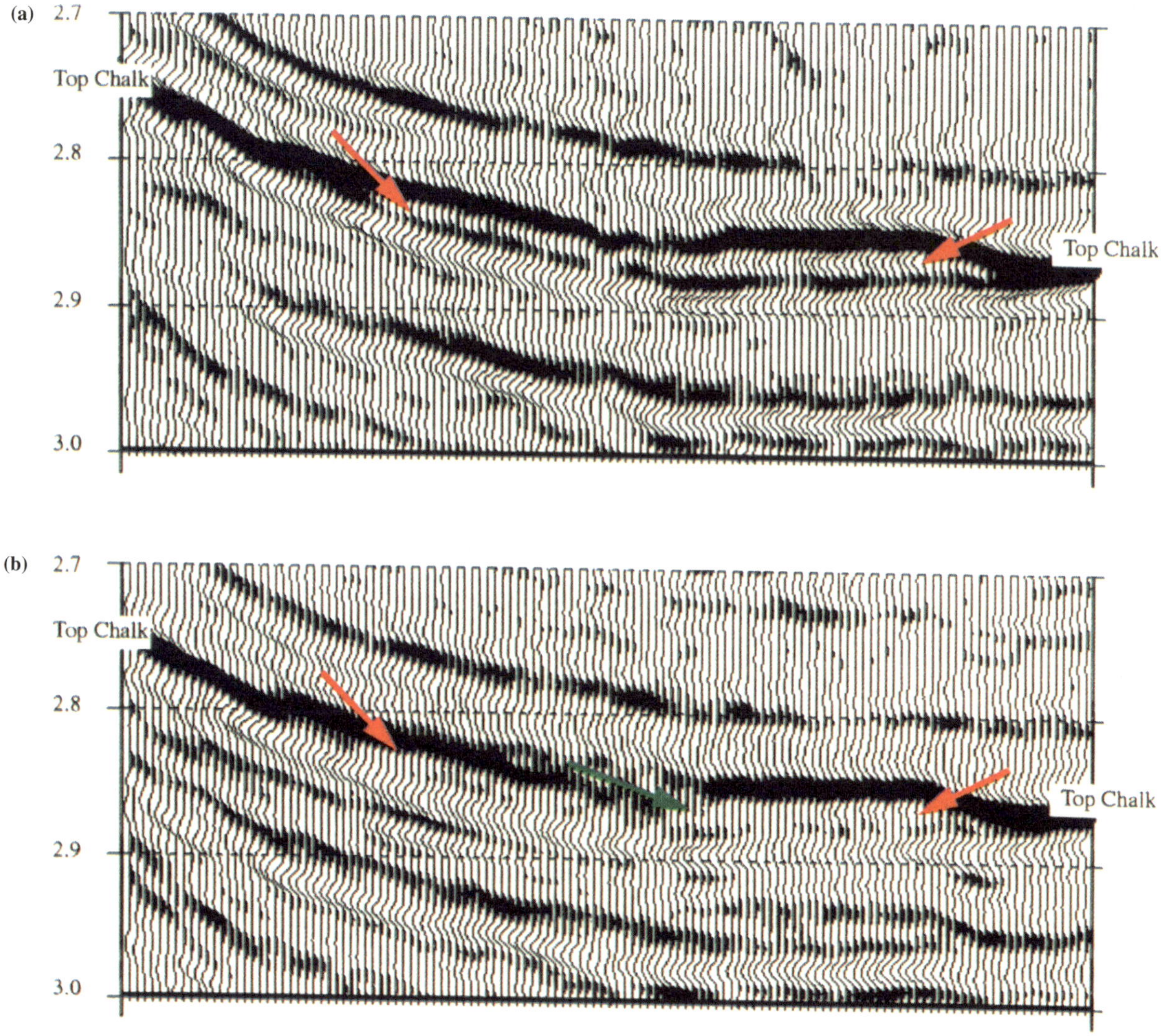

Fig. 6. (a) Seismic line B after Amoco's processing and (b) with Amoco's wavelet shaping. Note the pod-shaped thickening in the upper chalk (red arrows) and the internal reflection character in this pod that is evident after wavelet shaping (green arrow).

However, this evidence is dependent on the interpreted top of chalk reflector. Notice that if the top of chalk reflector had been interpreted to be on the lower peak (between the arrows in Figure 9a), then there would be no anomalous low acoustic impedance zone in the upper chalk. Thus, one criticism of this technique could be that it serves as a self-fulfilling prophecy. However, it is a relatively fast and efficient poststack analysis that can be applied inexpensively to an entire 3-D data volume. This was done here, with the results shown in Figure 10. The average porosities calculated from the acoustic impedances over the interval from the top of the chalk horizon to 20 ms below are displayed in color on a time-structure relief map of the top of chalk. The high-porosity zones in the East Hod field, which is on the time high, and in the off-structure stratigraphic trap prospect are easily identified.

AVO Analysis

Amplitude versus offset (AVO) analysis involves interpreting the prestack seismic data, which contain reflection energy from a range of reflection angles. Seismic reflection amplitudes from nonnormal angles of incidence are influenced by all three elastic parameters (V_P, V_S, and bulk density). Recall that distortion and contamination of the prestack relative amplitude information was avoided during processing in this study (Table 2). For example, simple spherical divergence corrections are based on a generalized velocity field for the

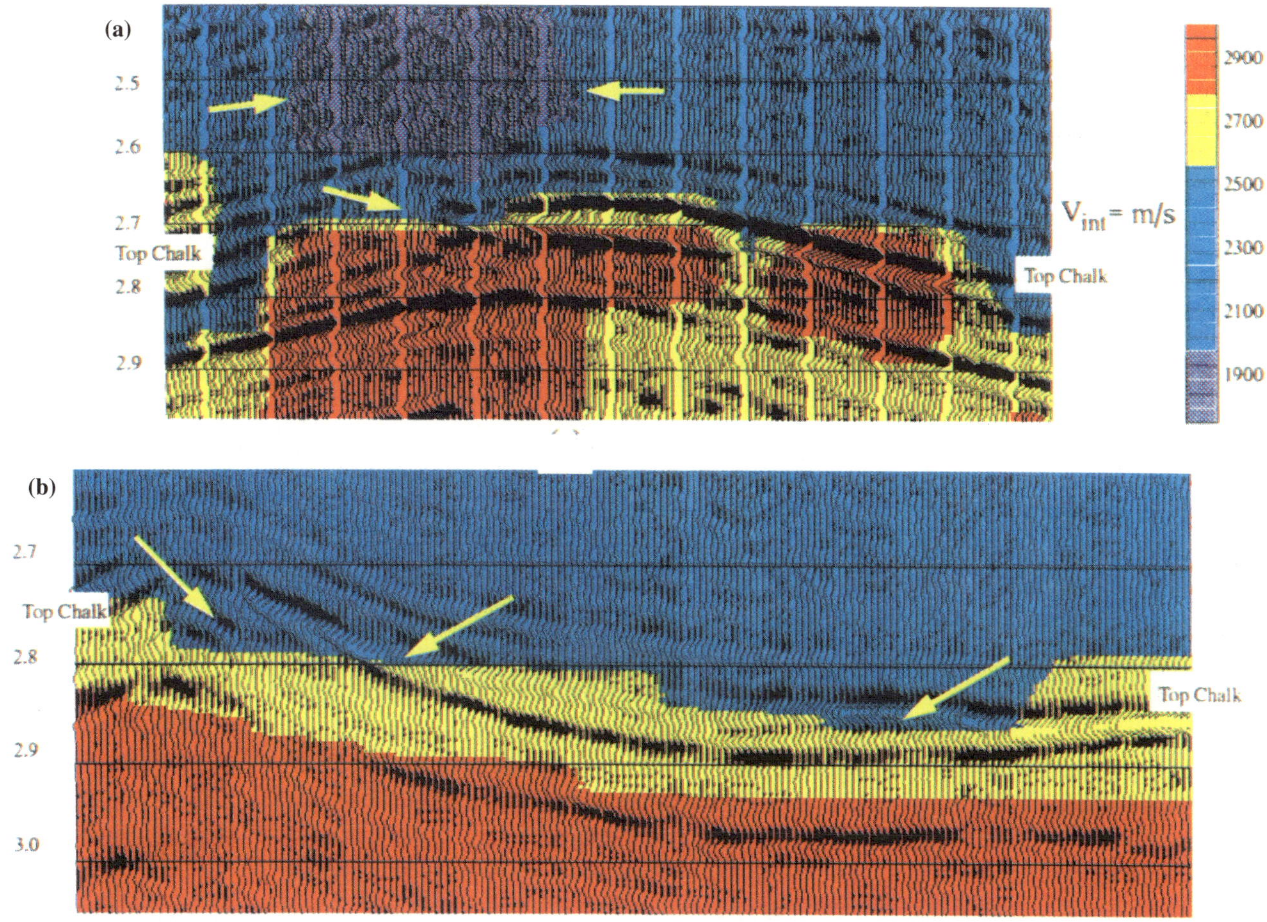

Fig. 7. Seismic data overlain on interval velocity fields calculated from stacking velocities for (a) line A and (b) line B. Higher interval velocities are shown in red and yellow and lower interval velocities in blue. The lowest calculated interval velocities (purple) in part (a) indicate a "gas cloud" effect over the East Hod field high. Note the lower velocities (blue) extending below the top of chalk (yellow arrows) in part (b).

area, and deconvolution is predictive with a gap large enough to avoid wavelet distortion. Also, a radon transform multiple removal technique and dip moveout algorithm were tested and applied to the data where necessary, with careful quality control for relative amplitude preservation.

A representative prestack CDP gather is shown in Figure 11. An obvious AVO anomaly is visible in the upper chalk interval at 2.8 s. Because of the choice of conservative processing parameters, especially in offset amplitude balancing, it is assumed that the anomaly is not a processing artifact. On line C, this strong AVO anomaly occurs at the base of the upper lens within the thickened upper chalk sequence and dies out laterally (Figure 12). Although this AVO anomaly corresponds to a stratigraphic anomaly and is quite obvious, its geologic meaning is still problematic.

First, this AVO anomaly is different than the conventional pore fluid related anomalies in Gulf Coast and North Sea Tertiary clastics (Allen and Peddy, 1993; Castagna and Backus, 1993; D'Angelo, 1994; Reilly, 1994; Barrett et al., 1995). In the classic case, changes in pore fluids affect changes in V_P/V_S ratio (or Poisson's ratio) in reservoir sandstones. This effect causes the hydrocarbon sand–shale interface to produce an AVO signature different than that of the water filled sand–shale interface. For the chalk lithologies in this case study, the plug-measured petrophysical data indicate that porosity is the dominant influence on the elastic parameters and the pore fluid effect is secondary and minor. Here, the AVO anomaly is interpreted to be caused by contrasts in porosity and not by pore fluid contrasts (oil or gas versus brine). Demonstrating this interpretation is based on forward offset modeling using the rock properties from the plug measurements.

The same basic geologic model is used for offset modeling as was used for the normal incidence modeling, that is, an anomalously porous chalk in the upper

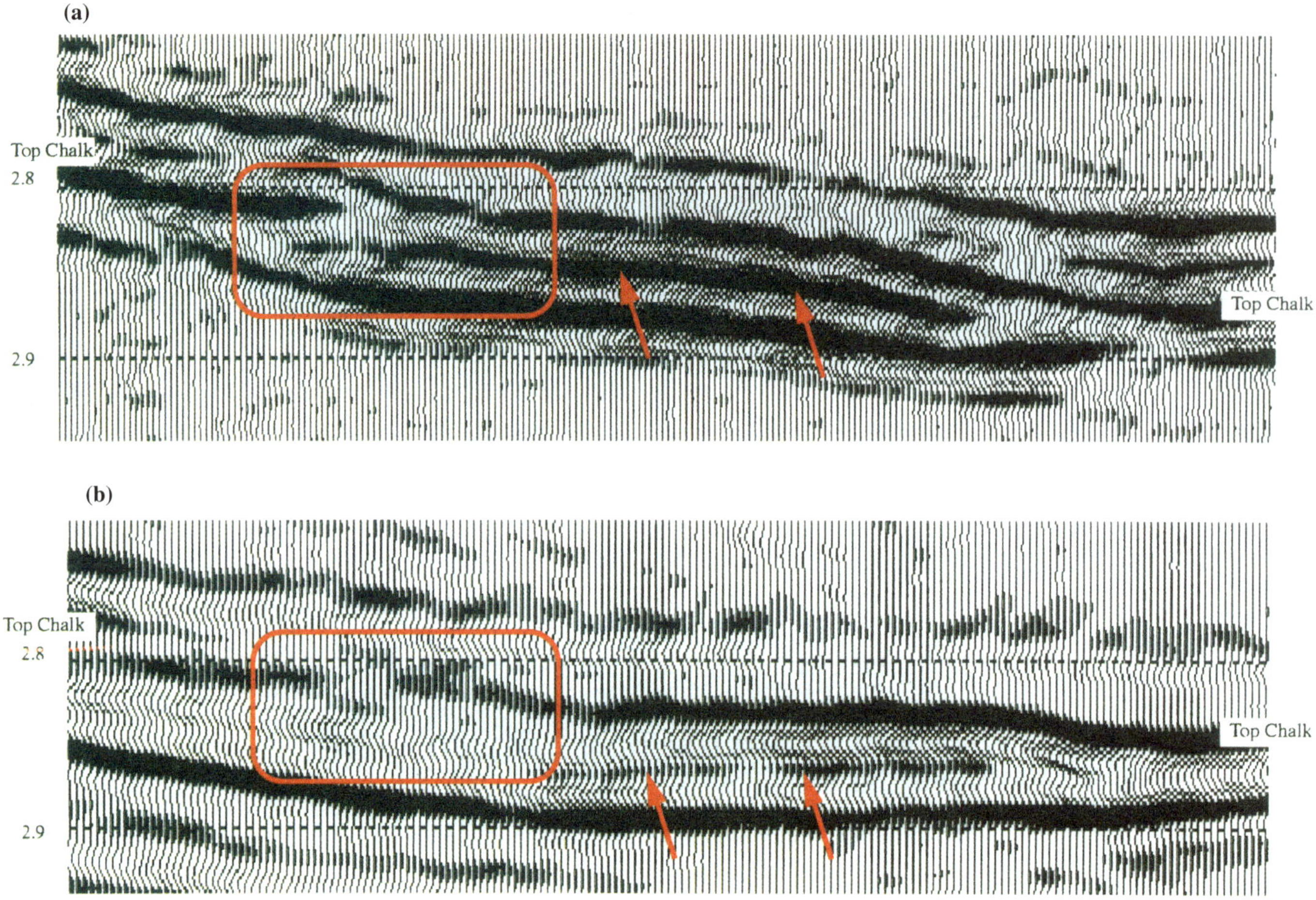

Fig. 8. (a) A synthetic normal-incidence seismic section, migrated and filtered to match (b) the real seismic section. Note the subtle similarities in the change of reflection character at the top of chalk (red box) and the mismatch of amplitudes of events (arrows).

chalk section. Synthetic *P*-wave velocity, *S*-wave velocity, and density logs are created from area type logs edited with rock properties from the plugs. A full solution to the Zoeppritz equations is used to calculate synthetic CDP gathers which are filtered to the same frequency content and phase as the real seismic data. This offset modeling approach neglects the effects of converted waves and may break down in the presence of thin layers (Simmons and Backus, 1994).

Qualitatively, the synthetic CDP shows the same AVO anomalies as the real CDP (Figure 11). Anomalously strong far-range amplitudes are seen in the upper chalk interval. Notice that the strongest anomaly is from the interface between the base of the porous chalk lens and the top of the harder chalk. The seismic response from this interface is a peak at near-normal incidence that first decreases with offset and then increases again at the farthest offsets, always remaining a peak (Figure 13). This response is interpreted here to be precritical angle energy caused by the large contrast in *P*-wave velocities across the interface.

Sheriff (1982) has stated that the reflection amplitude may be exceptionally large in the vicinity of critical reflection (precritical angle reflection energy). He defines the critical angle (ϕ) as

$$\phi = \sin^{-1}(V_1/V_2) \text{ or } \sin\phi = V_1/V_2 ,$$

where V_1 is the *P*-wave velocity in the upper layer and V_2 is the *P*-wave velocity in the lower layer. In one extreme, when the velocities in the two layers are equal ($V_1 = V_2$), then $V_1/V_2 = 1$ and $\phi = 90°$. In this case, critical angle energy would not be recorded. In the case of a "well-behaved" earth model, a gently increasing velocity field, V_2 would be slightly higher than V_1. In this typical case, critical angle reflection energy would rarely be seen at normal seismic acquisition apertures (offsets up to about ~3300 m and reflection apertures up to ~45°).

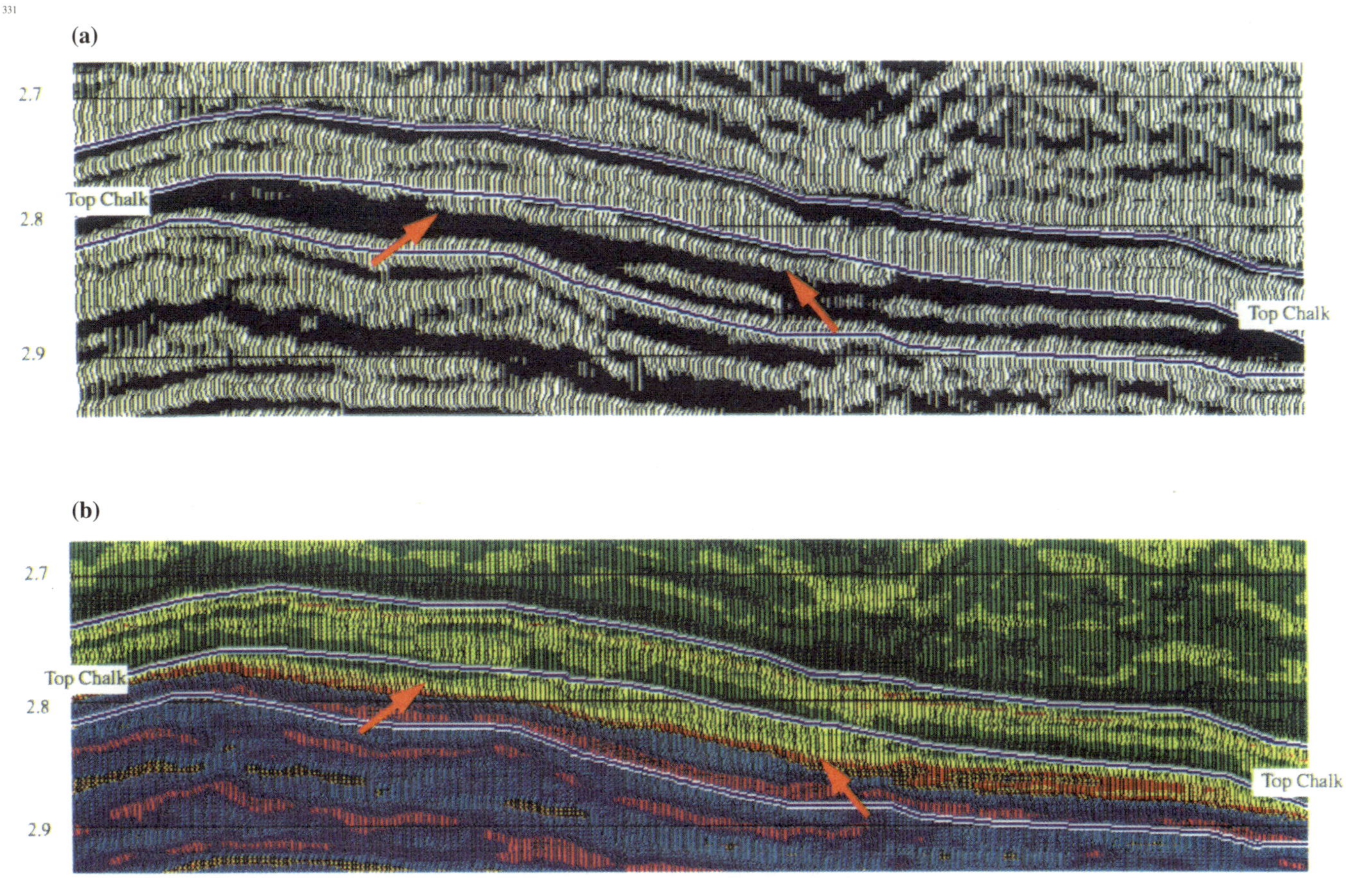

Fig. 9. Results of STRATA seismic inversion to acoustic impedance. (a) The normal-incidence seismic data input to STRATA with the interpreted horizons shown (purple). Note the upper (arrows) and lower lenses in the upper chalk section. (b) Acoustic impedance section output from STRATA. Lower acoustic impedances are shown in green and yellow and higher acoustic impedances in orange-red to blue and purple. Note the low acoustic impedance anomaly (high porosity) associated with the upper lens (arrows).

However, in the anomalous case where the velocity in the lower layer is much higher than the velocity in the upper layer ($V_2 >> V_1$), then V_1/V_2 approaches 0 and ϕ approaches 0°, another end-member case. As ϕ approaches 0°, more and more precritical angle energy is recorded and observed on the seismic data. Thus, the greater the contrast between V_1 and V_2 and the more abrupt (less gradual) the change, the stronger the precritical angle energy recorded and observed on the seismic data.

In the application to our geologic model, the lower layer is the hard, low-porosity chalk, which we assume to have a constant velocity laterally. That is, V_2 is constant, or at least approximately so, over small depth intervals. The upper layer is a chalk with laterally varying *P*-wave velocities that are related to changes in porosity. Thus, for tighter chalks in the upper layer, V_1 is about equal to V_2, but for more porous chalks in the upper layer (relative to the lower layer), V_1 is much smaller than V_2. The precritical angle energy is weak or absent in the first case of tight chalk over tight chalk. However, in the second case of porous chalk over tight chalk, precritical angle reflection energy is observed. In fact, as the strength of the observed precritical angle energy increases, the *P*-wave velocity in the upper layer decreases (assuming that V_2 is fixed) and the velocity contrast from the upper to lower layer increases.

Therefore, stronger observed precritical angle energy and AVO anomalies are used to predict relatively higher porosities in the upper chalk layer on the basis of the petrophysical relationships of velocity and porosity. However, as with the techniques discussed earlier, this interpretation also has weaknesses. Two pitfalls should be mentioned: (1) misidentification of the chalk-chalk interface (the interface may actually be clastics-chalk) and (2) violation of the assumption that V_2 is constant (V_2 may actually change laterally). In the first case, the observed precritical angle energy would not indicate porous chalk simply because it would be coming from the base of a clastic layer. In fact, it is common to see pre-

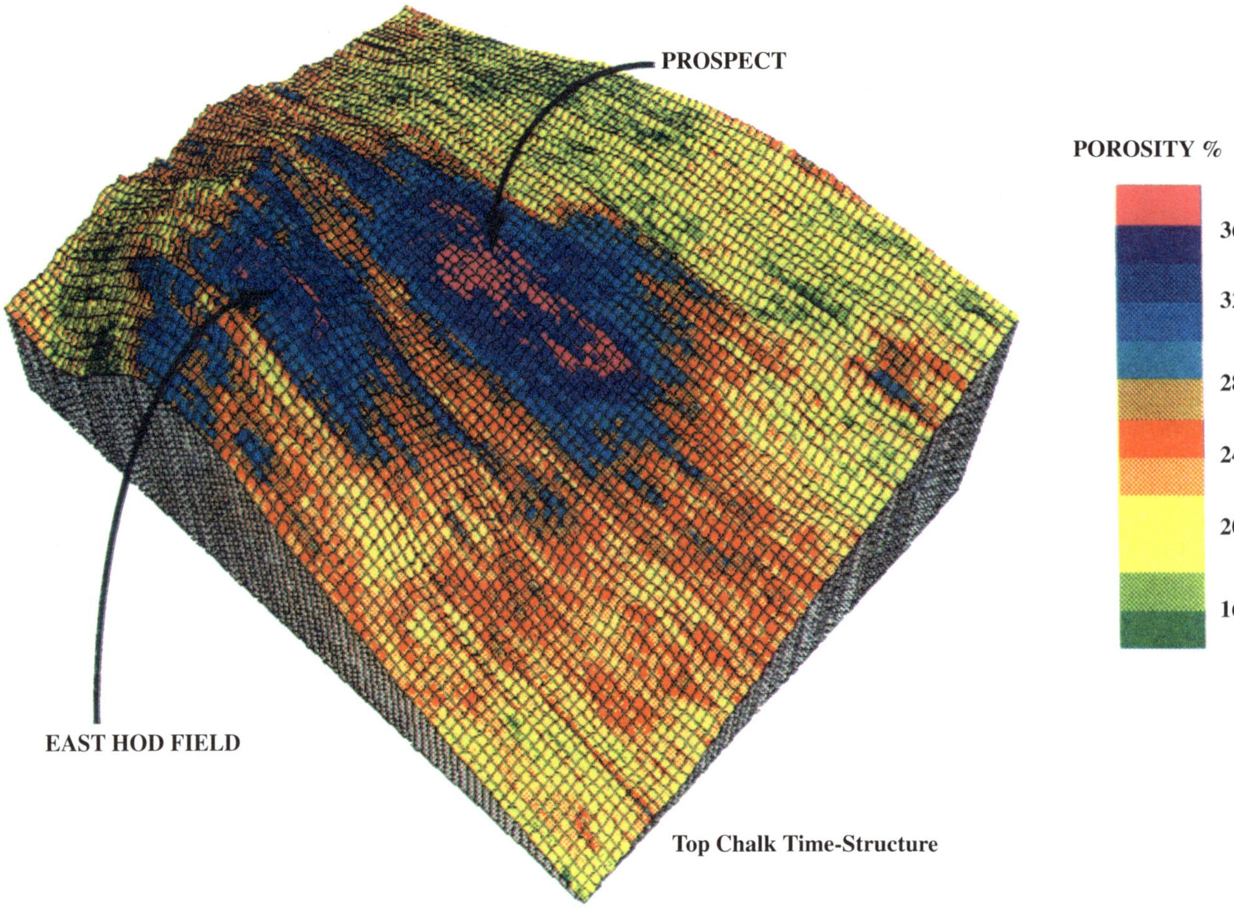

Fig. 10. Calculated porosity from a recursive inversion to acoustic impedance of a 3-D seismic data volume. Average porosities are from the top of chalk to 20 ms below and are displayed on a time-structure relief map of the top of chalk horizon. Higher porosities are shown in blue and purple. Note the high average porosity anomalies detected at the East Hod field and at the prospect.

critical angle energy from the interface at the base of the shales and top of the hard chalk. In the second case, the precritical angle energy might indicate a laterally tighter lower layer rather than a laterally more porous upper layer. As mentioned earlier, all seismic responses are to a *contrast* in rock properties, and determining the absolute values of the rock properties in the upper and lower layers is nonunique. Nevertheless, the important idea here is that, in some situations, precritical angle energy (AVO anomalies) can contain clues about geologic rock properties, even though they are often ignored or removed from the seismic data.lpq

Prestack Inversion

A separate study was completed by IKU Petroleum Research of Trondheim, Norway (Landrø et al., 1995). The study involves applying IKU's unique prestack inversion analysis to line A. The data are processed in the tau-p domain, and elastic parameters are extracted from the tau-p domain prestack data (Helgesen and Landrø, 1991). The IKU study is particularly interesting because it was undertaken independently and involved a different approach than our analyses, yet it had similar results. The IKU analysis also indicates that a high-porosity zone exists in the thickened upper chalk interval.

The IKU inversion on the prestack tau-p domain data creates three separate seismic sections, one each for *P*-wave velocity, *S*-wave velocity, and bulk density (Figure 14). Note that these results are most stable for the *P*-wave velocity parameter and least stable for the density parameter, similar to the statistical relationships of the petrophysical data base. Without the plug measurements, only these semi-quantitative estimates of V_P, V_S, and density would be possible. With the petrophysical data base, these elastic parameter data are converted to porosity by using linear regression equations.

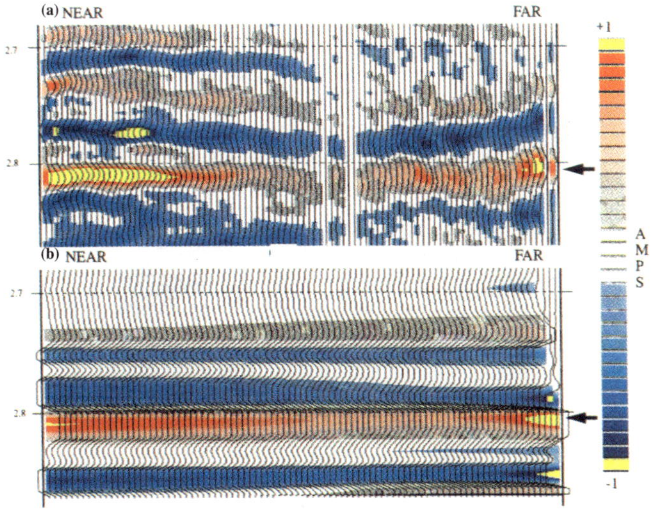

Fig. 11. Unstacked CDP gathers with near-range traces on the left and far ranges to the right. Data are relative amplitudes scaled with the strongest amplitudes in yellow. (a) A representative real CDP gather. Note the reflection from within the interpreted upper chalk just below 2.8 s that starts off as a strong peak at the near ranges, then weakens with offset, and finally increases again to a relatively strong peak at the farthest offsets. (b) A synthetic CDP gather. The event just below 2.8 s agrees qualitatively with the same event on the real CDP. This event is generated from the interface base of the porous chalk and the top of harder chalk in the model.

Figure 15 shows the calculated porosity section. Note that calculated high-porosity zones are apparent both in the low, where the prospect is located, and on the high of the East Hod field.

Conclusions

The chalk section of the Central Trough in the North Sea currently produces from several fields on structural highs along the Lindesnes Ridge. In this paper, we have presented a geologic model for additional chalk prospectivity in off-structure stratigraphic traps. The geologic model for these traps is the pinch-out and facies change of a porous, reservoir-quality allochthonous chalk facies within a sealing autochthonous chalk. The geologic model is constrained by petrophysical information and rock properties encountered in wells from the producing fields and off-structure. The biggest risk of failure of this geologic model is perceived to be the occurrence of reservoir-quality porosity within these stratigraphic traps.

Several geophysical analysis techniques, including stratigraphic processing, stacking velocity analysis, normal incidence modeling, seismic trace inversion, AVO analysis and modeling, and prestack inversion, have provided results that are interpreted to support the geologic model. All of these techniques give evidence that relatively higher porosity chalk could exist within the upper chalk stratigraphic trap prospect. Individually, all of these techniques have assumptions and weaknesses that could lead to inaccurate interpretations. The fact that all of them lead to the same conclusion, however, gives us reasonable confidence in the geologic model.

In fact, the operator was compelled to test both the stratigraphic trap concept and the viability of the geophysical methodologies. In early 1994, the stratigraphic trap prospect discussed here was drilled. The well is a discovery, proving the play concept and the technologies used to evaluate it.

Acknowledgments

The authors would like to thank the partners on Norwegian License PL033 for approval to publish these data and for their ideas and interaction, especially Nina Gravdal and Simon Campbell of Enterprise Oil. Furthermore, the contributions of scientists from the Amoco Norway Oil Company, Amoco Production Company, Amoco Research, and others, such as Claus Clausen (of Intera) and Martin Landrø (of IKU), are acknowledged. A study of this scope is always affected by the thoughts, comments, and other contributions of many more people than are possible to list.

References

Al-Chalabi, M., 1974, An analysis of stacking, RMS, average, and interval velocities over a horizontally layered ground: Geophys. Prosp., **22**, 458-475.

Allen, J. L., and Peddy, C., 1993, Amplitude variation with offset: Gulf Coast case studies: Soc. Expl. Geophys.

Barrett, R. F., Margesson, R. W., and D'Angelo, R. M., 1995, Use of rock properties and AVO in the Everest field development: Petr. Geosci., **1**, 311-317.

Castagna, J. P., and Backus, M. M., Eds., 1993, Offset-dependent reflectivity: theory and practice of AVO analysis: Soc. Expl. Geophys.

D'Angelo, R. M., 1994, An AVO study on North Sea data: North Sea Oil and Gas Reservoirs **III**, 109-114.

D'Heur, M., 1984, Porosity and hydrocarbon distribution in North Sea chalk reservoirs: Mar. Petr. Geol., **1**, 211-237.

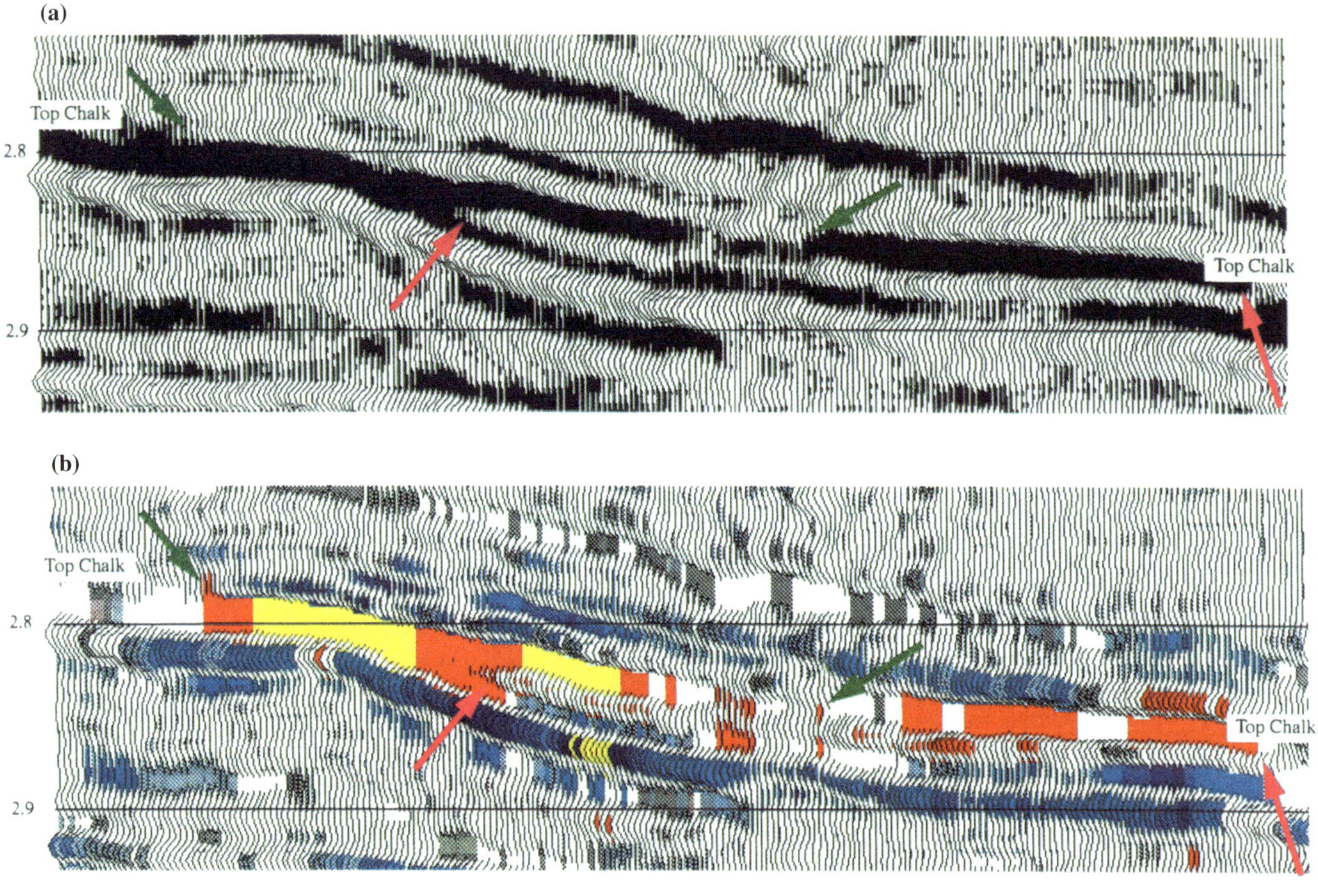

Fig. 12. (a) Normal incidence seismic section for line C. The green and pink arrows indicate the updip and downdip limits of the upper and lower parts of the prospect, respectively. (b) Same seismic wiggle data, but here the wiggle infill colors represent far-range relative amplitude information. Yellow indicates the occurrence of the highest relative amplitudes at the far ranges on the prestack CDPs (AVO anomalies shown in Figure 10). Note that these AVO anomalies occur at the base of the upper part of the prospect.

Dix, C. H., 1955, Seismic velocities from surface measurements: Geophysics, **20**, 68-86.

Hardman, R. F. P., 1982, Chalk reservoirs of the North Sea: Geol. Soc. Denmark Bull., **30**, 119-137.

Helgesen, J., and Landrø, M., 1991, Estimation of elastic parameters from AVO effects in tau-p domain: Paper presented at 53rd EAEG Meet. Tech. Exhib.

Hubral, P., and Krey, T., 1980, Interval velocities from seismic reflection time measurements: Soc. Expl. Geophys. Spec. Publ.

Kennedy, W. J., 1980, Aspects of chalk sedimentation in the southern Norwegian offshore, *in* The sedimentation of North Sea reservoir rocks: Norwegian Petrol. Soc., 29.

Landrø, M., Buland, A., and D'Angelo, R., 1995, Target-oriented AVO inversion of data from Valhall and Hod fields: The Leading Edge, **14**, 855-861.

Lynn, W. S., and Claerbout, J. F., 1982, Velocity estimation in laterally varying media: Geophysics, **47**, 884-897.

Maliva, R. G., and Dickson, J. A. D., 1992, Microfacies and diagenetic controls of porosity in Cretaceous–Tertiary Chalks, Ekofisk field, Norwegian North Sea: AAPG Bull., **76**, 1825-1838.

McBride, J. H., Lindsey, G. A., Hobbs, R. W., Snyder, D. B., and Totterdell, I. J., 1993, Some problems in velocity analysis for marine deep seismic profiles: First Break, **11**, 345-356.

Reilly, J. M., 1994, Wireline shear and AVO modeling: application to AVO investigations of the Tertiary, U.K. Central North Sea: Geophysics, **59**, 1249-1260.

Schatzinger, R. A., Feazel, C. T., and Henry, W. E., 1985, Evidence of resedimentation in the chalk from the Central Graben, North Sea, *in* Crevello, P. D., and Harris, P. M., Eds., Deep-water carbonates, buildups, turbidites, debris flows and chalks: Soc. Econ. Paleont. Mineral. Core Workshop **6**, 342-385.

Simmons, Jr., J. L., and Backus, M. M., 1994, AVO modeling and the locally converted shear wave: Geophysics, **59**, 1237-1248.

Stork, C., and Clayton, R. W., 1992, Using constraints to address the instabilities of automated prestack velocity analysis: Geophysics, **57**, 404-419.

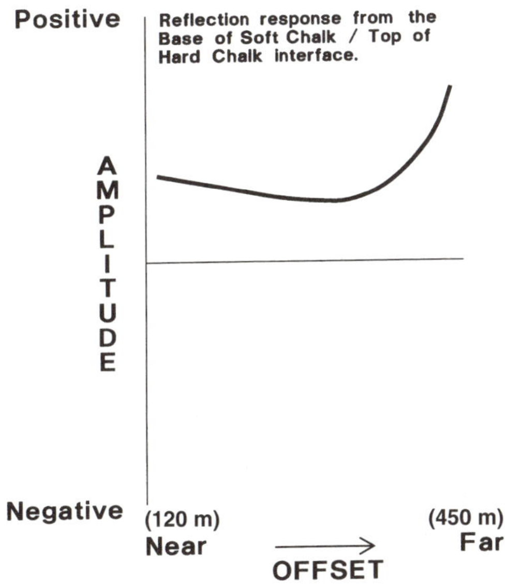

Fig. 13. **Crossplot of relative AVO from the interface at the base of the porous chalk and top of the tight chalk (see arrows in Figure 11 marking the reflection event from this interface). Note that the amplitude is a peak that decreases with offset first and then increases again at the farthest offsets, always remaining a peak.**

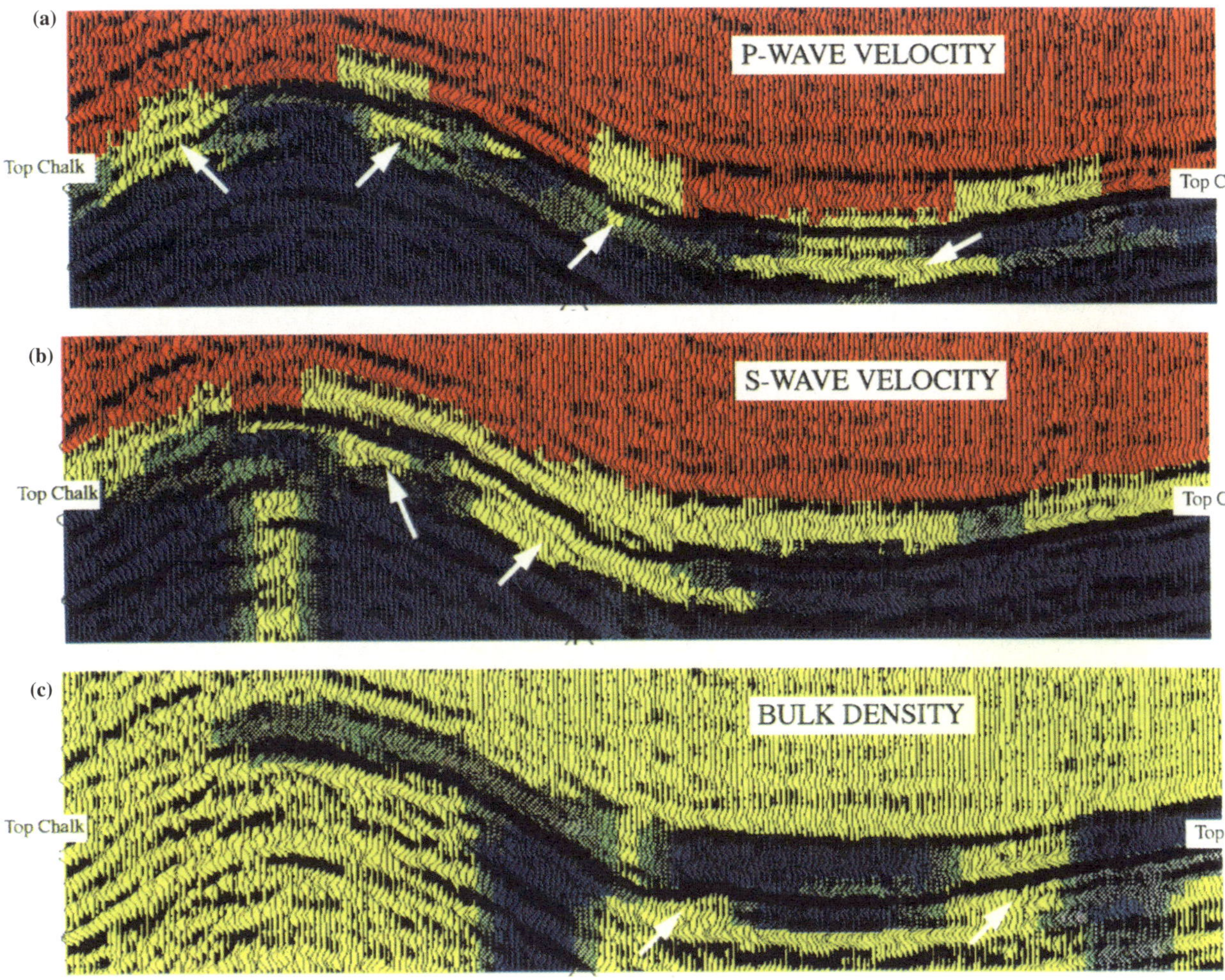

Fig. 14. **Results of IKU's prestack seismic inversion to elastic parameters. Red indicates lower values and blue higher values. Values that fall within the range of high-porosity chalk from the plug measurements are shown in yellow. Thus, below the top of chalk, blue indicates tight chalk and yellow porous chalk. (a) Inverted *P*-wave velocity section indicating several higher porosity zones (arrows) in tighter chalk. (b) Inverted *S*-wave velocity section showing some high-porosity zones (arrows) in the upper chalk. (c) Inverted bulk density section that seems to be less geologically meaningful.**

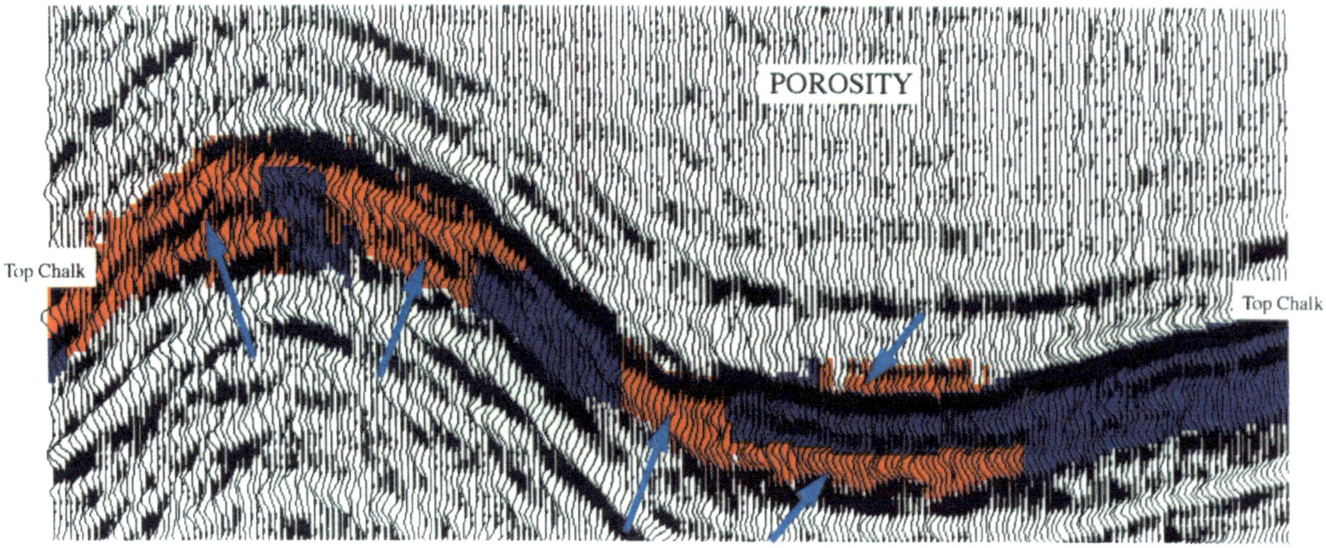

Fig. 15. A predicted porosity section calculated by combining IKU's inverted elastic parameter sections and converting to porosity based on the linear regression equations from the plug measurements that relate the elastic parameters to porosity. Blue indicates tight, lower porosity chalk, while red indicates higher porosity chalk. The plug data are appropriate only for the upper chalk interval.

Chapter 14

Case Studies of Multicomponent Seismic Data for Fracture Characterization: Austin Chalk Examples

Xiang-Yang Li
Edinburgh Anisotropy Project
British Geological Survey
Edinburgh, Scotland, U.K.

Michael C. Mueller
Amoco EPTG
Houston, Texas, U.S.A.

Abstract

Shear wave studies of multicomponent seismic data were done along the Austin Chalk trend in Texas. Six surface seismic lines of four-component shear wave data from Pearsall and Giddings fields and three zero-offset vertical seismic profiles (VSPs) from three sites with different production rates were studied to demonstrate the applications of shear wave splitting for fracture reservoir delineation. The three seismic lines (1–3) from Pearsall field formed a classic experiment for studying shear wave splitting. They have different (line) azimuth with respect to the regional fracture strike (parallel, perpendicular, and at ~40°, respectively), are in areas with different hydrocarbon production, and have different split shear wave behavior. The anisotropy along line 1 is small, which correlates with the absence of nearby commercial production. There is an increasing trend in anisotropy from line 2 to 3, which correlates with line 2 being close to and line 3 being within the producing Austin Chalk acreage. The trend of anisotropy variation along line 3 also correlates with the distribution of producing oil wells along that line. In particular, production from the three horizontal wells (W1–W3) drilled nearby correlates with the variation in shear wave polarization, time delay, and amplitude. Line 4 from Giddings field had a horizontal well drilled on it, and mud logs were obtained for identifying fracture zones. The fracture swarms identified by the mud logs are coincident with the dim spots identified from the section of the slow split shear wave. Lines 5 and 6, also from Giddings field, demonstrate classic S_2 dim spot and S_1 versus S_2 mistie behavior, respectively.

The three VSPs, from three wells with different production rates, show different shear wave responses. VSP 1, from a nonproductive well, shows minimal shear wave time delay and no amplitude anomalies. VSP 2, from a water-producing well, shows some amount of splitting but no anomalies in shear wave attributes. VSP 3, from an oil-producing well, shows clear shear wave splitting and anomalies in shear wave amplitudes. Amplitude corrections are necessary for preserving and recovering shear wave anisotropy information associated with the target, and the linear transform technique (LTT) simplifies the processing sequence for examining anisotropy in shear wave reflection data. Stacked polarization logs can be used for identifying local variations in polarization (often associated with local variations in crack geometry) and for better imaging of subsurface structure, as in line 3.

Introduction

The study of seismic waves propagating through the earth has greatly improved our understanding of the earth's interior and has developed into one of the most important tools for petroleum exploration. However, until the 1970s, most theories were developed mainly around the propagation of *P*-waves in isotropic solids (Aki and Richards, 1980; Crampin, 1981). Most applications were limited to *P*-wave energy recorded on single-component vertical geophones or pressure-sensitive hydrophones (Tatham and McCormack, 1991). Compared to shear wave recording, *P*-wave recording is cheaper and more robust and often yields better structural imaging of the subsurface. Moreover, the effects of anisotropy on *P*-wave are often small and difficult to detect in one-component *P*-wave data (Winterstein, 1990). Thus, the relevance of shear wave and seismic anisotropy to hydrocarbon exploration and development was not widely recognized.

Gardner and Harris (1968) and Tatham and Stoffa (1976) predicted significant anomalies associated with gas saturation in the velocity ratio of *P*-wave to shear wave (V_P/V_S) and identified V_P/V_S ratios as a possible indicator of lithology and even porosity variations. Keith and Crampin (1977) and Crampin (1981) showed the potential importance of shear wave anisotropy and identified shear wave splitting (birefringence) as a diagnostic feature of shear wave anisotropy, in which the polarization, time delay, and amplitude of shear waves are related to the in-situ stress orientation and the alignment and intensity of fractures and cracks. These new insights into shear wave behavior, together with the emergence of digital processing, opened new aspects in shear wave exploration. In the 1970s, this led to the development of *SH*-wave survey (one-component horizontal *SH* source and receiver) for studying V_P/V_S ratio and, in the 1980s, to the acquisition of multicomponent seismic data using both vertically and horizontally polarized sources and multicomponent receivers for studying shear wave anisotropy.

Over the past decade, the use of joint *P*- and *S*-wave surveys and the analysis of V_P/V_S ratios have been widely reported (e.g., Tatham, 1982, 1985; Ensley, 1984; Winterstein and Hanten, 1985). Tatham and McCormack (1991) have documented these applications. Observations of shear wave anisotropy have also become well-documented in multicomponent seismic surveys (e.g., Crampin, 1985; Crampin et al., 1986; Alford, 1986; Willis et al., 1986) and in earthquake recordings (Crampin et al., 1985). Crampin and Lovell (1991) have reviewed these developments. In the present study, we focus on the use of multicomponent seismic data and the analysis of shear wave anisotropy for applications to petroleum exploration.

It has long been speculated that the study of shear wave anisotropy can be used to characterize fractured reservoirs (e.g., Crampin, 1987; Lewis et al., 1991). Higher oil production in fractured reservoirs is often associated with higher fracture intensity and hence greater shear wave anisotropy. Thus, shear wave anisotropy may be directly correlated with oil production (Lewis et al., 1991; Mueller, 1991). Cliet et al. (1991) reported the correlation of shear wave anisotropy with oil production rates in a multi-offset three-component vertical seismic profiling (VSP) survey in the Romashkino field, Russia. Davis and Lewis (1990) and Lewis et al. (1991) also correlated the degree of shear wave anisotropy with oil production in a three-dimensional, three-component reflection survey in Silo field, Wyoming.

The Austin Chalk trend in Texas is a natural laboratory for testing practical applications of shear waves in anisotropic rocks. This is because the chalk has little porosity, and vertical aligned fractures provide the major pathway for oil. To analyze shear wave anisotropy in the Austin Chalk trend, Amoco Production Company acquired several lines of shear wave data, and several studies of shear wave anisotropy along the trend have been published. Alford (1986) established the presence of shear wave splitting and an algorithm for accommodating it using data from the chalk trend. Thomsen (1988) suggested that reflection amplitude anomalies (between the split shear waves) constituted a sensitive measure of local difference in anisotropy, useful for detecting fractured reservoirs. Confirming this suggestion, Mueller (1991) identified lateral fracture concentrations from the differential amplitudes of split shear waves from the Austin Chalk in the Giddings field of central Texas. Li et al. (1993a) investigated variations in shear wave attributes with varying line azimuth and compared different processing methods for preserving and enhancing shear wave splitting. Yardley and Crampin (1993) studied three VSPs from different sites along the trend and examined the different response of shear wave splitting for varying production rates.

Here we review these shear wave studies along the Austin Chalk trend. We outline the processing and interpretation methods for fracture reservoir delineation and discuss the underlying principles of shear wave splitting associated with the chalk, including its physical properties and seismic characteristics. We then present interpretations and illustrate how the fracture parameters can be deduced from variations in shear wave attributes such as polarization, time delay, and amplitude. We show that the variations in anisotropy can be correlated with oil production in the study area and that the variation in shear wave amplitude can be correlated with fracture swarms encountered from horizontal drilling, as inferred from production records and mud logs.

Overview of the Shear Wave Technique

When a shear wave propagates into an effectively anisotropic medium (e.g., resulting from aligned fractures), it necessarily splits into two modes that have different polarizations and travel with different speeds. For near-vertical propagation, the faster split shear wave is polarized along the fracture strike and the slower one is polarized nearly orthogonal to it (Crampin, 1985). Furthermore, the time delay and differential amplitude between the two shear waves are both proportional to the intensity of fracturing. This forms the basis for fracture delineation using multicomponent shear wave data. This section first reviews the shear wave information content in terms of shear wave polarization, time delay, and normal reflectivity. We then discuss the development of multicomponent acquisition and processing techniques necessary for studying shear wave splitting.

Shear Wave Attributes

The characteristics of split shear waves diagnostic of fracture distributions are referred to as *shear wave attributes*. These include the polarization of the leading shear wave, the time delay between the two shear waves, the differential reflectivity at interfaces, and the differential attenuation and scattering along the raypath. In general, polarization and time delay are quantitative attributes, but are difficult to measure in thin-layered or weakly anisotropic reservoirs. Differential reflectivity at normal incidence is a qualitative attribute, but measurable for thin-layered or weakly anisotropic reservoirs. An integration of all three attributes with other geologic information is needed for a reliable interpretation of fractured zones. Here we briefly review the three most widely used attributes: the *polarization*, the *time delay*, and the *reflectivity* at normal incidence. We discuss their relative merits and compare their potential for detecting fractures.

Yardley et al. (1991) have provided a detailed discussion on the viability of using shear wave amplitude versus offset (AVO) for fracture detection. Although the two shear waves have different AVO signatures (the reflectivity of the *SV*-type wave may have a zero crossing at relatively small offset due to conversion), it is difficult to attribute this difference to fracturing because such difference in AVO signatures also exists for *SV*- and *SH*-waves propagating in isotropic media. The pattern of the AVO curves remains the same with increasing fracturing except the reflectivity at normal incidence. Yardley et al. (1991) concluded that for fracture detection, the *differential reflectivity* at normal incidence is a first-order effect and can be measured and interpreted reliably, while the *offset-dependent reflectivity* is a second-order effect and more subtle and difficult to interpret.

Polarization

In theory, for near-vertical propagation, the leading shear wave is polarized parallel to the fracture strike of an effectively anisotropic medium when that medium possesses a single (or dominant) vertical, aligned fracture set. Thus, the polarization of the fast split shear wave can be used to infer the fracture strike of the medium through which it has propagated from seismic data recorded on the surface or in boreholes. However, the main factor affecting the use of polarization information at the target is the anisotropic effects in the overburden. A shear wave generally splits whenever it passes through an anisotropic medium. Thus, if the overburden is anisotropic, the polarization recorded at the surface is determined only by the anisotropic symmetry in the overburden; the anisotropic symmetry information in the target is then concealed by that of the overburden. Considering that anisotropy is ubiquitous (Willis et al., 1986) and often seen in the overburden (Crampin and Lovell, 1991), inferring the fracture orientation of the reservoir from surface recordings requires layer-stripping techniques (MacBeth et al., 1992; Li, 1995; Thomsen et al., 1995a). Because of these complications, Yardley and Crampin (1991) suggested that VSPs may be better used to study shear wave polarizations at target.

Note that the layer-stripping technique was first developed by Winterstein and Meadows (1991) for accommodating shallow anisotropy in shear wave VSP data; Winterstein (1991) and Meadows et al. (1994) suggested that the same techniques for VSPs could be equally applied to surface seismic data. However, to best accommodate the anisotropy effects in the overburden, a revised technique for surface data may be necessary with different time shifts for the diagonal and off-diagonal elements, as demonstrated by MacBeth et al. (1992), Li (1995), and Thomsen et al. (1995a).

Time Delay

The time delay between the fast and slow split shear waves depends on the propagation path length and direction, the symmetry of the medium, and the degree of anisotropy. Because the degree of anisotropy is mainly determined by the fracture density in the media, time delay is a valuable attribute for the inference of fracture density from a seismic section, and the fracture density can be quantitatively determined if the number of fracture sets and their orientation are known. Time delays are usually measured from the stacked sections of the fast and slow shear waves. The advantage of using the time-delay attribute lies in its simplicity. It is relatively easy to measure, and overburden anisotropy can be corrected for simply by taking the interval time delay between the top and bottom reflections of the target if the fracture orientation does not change with depth. Layer thickness can be simply handled by normalization. However, for a thin or weakly anisotropic reservoir, interval time delays may be too small to resolve reliably in the presence of noise, particularly if a significant part of the anisotropy is in the overburden and near the surface, as demonstrated by Li et al. (1993a). Also, time delays contain no information about fracture orientations, which must be deduced from other attributes. Squires et al. (1989) and Lewis et al. (1991) presented examples of interpreting interval time delays from field data.

Normal Reflectivity

Thomsen (1988) suggested that the differential amplitudes between fast and slow split shear waves can be used to identify fracture zones in stacked seismic sections if these fracture zones are terminated at reflecting boundaries such as lithologic boundaries, as commonly seen in outcrops. In such cases, the anisotropy caused by the fractures tends to affect the slow wave (with polarization perpendicular to the fracture strike) more than the fast wave (with polarization parallel to the fracture strike). The presence of the fractures lowers the velocity of the slow wave and causes a change in slow wave impedance contrast with respect to that of the fast wave, hence a difference in reflection strength in the stacked fast and slow sections occurs. This phenomenon is often referred to as *differential reflectivity at normal incidence* because stacked sections are often considered as seismic reflections at normal incidence. Note that anisotropy caused by macrofractures, whose scale length is comparable to wavelength, can affect both the fast and slow split shear waves (Liu et al., 1993).

Mueller (1991) first presented an example using normal reflectivity to identify fracture swarms in the Austin Chalk in south Texas (the results of which are included in this paper), where the fractures are confined in the brittle chalk (Corbett et al., 1987). The reflectivity attribute is of greater value in identifying the intensely fractured zones in the seismic section rather than in quantifying the exact values of the fracture intensity, although methods have been published to quantify the fracture intensity from reflectivity (Spencer and Chi, 1991; Li and Crampin, 1993c). Because this qualitative information is usually sufficient in most cases, the differential reflectivity at normal incidence is thus a useful indicator of fracture swarms. The interpretation is straightforward and similar to traditional amplitude analysis, and thin-layered or weakly anisotropic reservoirs can be resolved. The main difficulty lies in maintaining and recovering the true amplitude information during processing.

Four-Component Geometry

The use and study of shear waves requires polarized sources and receivers. In the middle 1970s, when horizontal receivers and large horizontal motion vibroseis sources became available, shear wave exploration and multicomponent surveying began to develop. Up to nine-component data can be recorded using different configurations of three-component sources and receivers. These components consist of various combinations of three mutually orthogonal source motions and receiver directions (combinations of vertical, H1 and H2, or *z*, *x*, and *y* sources and receivers) (Figure 1a). In practice, several configurations of sources and receivers have been used depending on the purpose of the survey. These include (1) three-component acquisition (vertical source recorded by vertical, H1, and H2 receivers, or *zz*, *zx*, and *zy* components) in VSPs for correlating seismic horizons (Hardage, 1991); (2) vertical and H2 (or *SH*-wave) surveying (*zz* and *yy* components) (Conoco Group Shoot, Ensley, 1984; Winterstein and Hanten, 1985; Tatham and McCormack, 1991); and (3) vertical and H1 (or *SV*-wave) surveying (*zz* and *zx* components) for mapping geologic structure and lithology (Dohr, 1985).

In the 1980s, following the completion of the Conoco Group Shoot in 1978, Amoco Production Company acquired four-component (*xx*, *xy*, *yx*, and *yy*) and five-component (*xx*, *xy*, *yx*, *yy*, and *zz*) surveys (Alford, 1986; Willis et al., 1986). The purpose of these surveys was to investigate whether shear wave splitting could be used for reservoir characterization. This configuration has several distinct features. First, under the assumption of

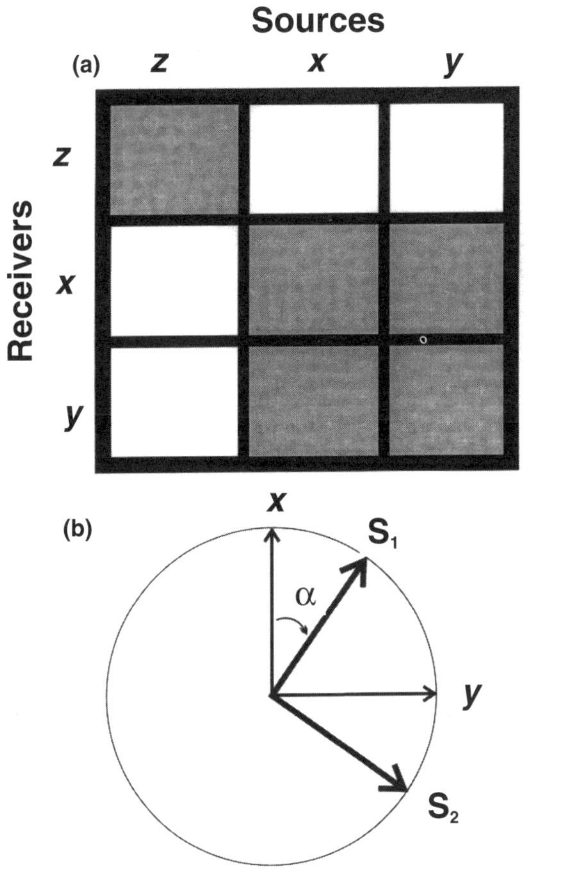

Fig. 1. Multicomponent geometry and coordinate system. (a) Full nine-component geometry: three orthogonal sources (columns) and three orthogonal receivers (rows). The shaded area is the five-component geometry (Alford, 1986) examined here. (b) Coordinate system: *x* represents the in-line source and geophone direction; *y*, the cross-line source and geophone direction; S_1, the fast split shear wave; S_2, the slow split shear wave; and α, the S_1 polarization azimuth as measured from the *x* direction.

weak anisotropy, the vertical or *P* component (*zz* component) is decoupled from the four horizontal or shear wave components (*xx*, *xy*, *yx*, and *yy*). Thus, processing of five-component data can be separated into two parts: processing of the *zz* component and processing of the horizontal or shear components. Second, the four horizontal components allow unique determination of the fast and slow split shear waves (Alford, 1986; Thomsen, 1988; Li and Crampin, 1991a, 1993a), which is discussed later in this section.

In the examples shown in this paper, all surface data were acquired using four-component geometry. The four-component shear wave survey with two horizontal sources and two horizontal receivers, as shown in Figure 1a (shaded), form a data matrix of traces $\boldsymbol{d}(t)$:

$$\boldsymbol{d}(t) = \begin{bmatrix} xx(t) & xy(t) \\ yx(t) & yy(t) \end{bmatrix}, \tag{1}$$

where the left column, $xx(t)$ and $yx(t)$, represents the in-line (*x*-axis) traces from the in-line and cross-line sources, respectively, and the right column, $xy(t)$ and $yy(t)$, represents the cross-line (*y*-axis) traces from the in-line and cross-line sources, respectively. This data matrix concept is important for processing and interpreting shear wave splitting. From here onward, the phrase *shear wave data* will refer to the data matrix unless otherwise specified.

Processing Techniques

The purpose of processing multicomponent shear wave data is to preserve and recover the three major attributes (polarization, time delay, and principal reflectivity). The shear wave data matrix represents a tensor wavefield, and conventional *P*-wave data processing methods are not directly applicable. The basic approach, as reported in the literature, is to separate the tensor wavefield into the principal vector wave fields. By restricting oneself to the shear wave plane of polarization (assuming no change along raypaths), the vector fields can then be processed separately by the conventional scalar method, including statics correction, velocity analysis, normal moveout correction, stacking, and other signal enhancing routines such as *f-k* filter. These conventional methods are well established for *P*-wave data processing (Yilmaz, 1987). The approach here is based on the assumption that polarizations do not change with depth or offset, and that once the polarizations are separated, anisotropy effects such as varying velocity with offset on the principal split shear waves are negligible over small spreads (Li, 1992; Li and Crampin, 1993b). With this approach, the separation of the split shear waves in the recorded data matrix is the crucial step in the processing sequence that allows many of the conventional scalar processing steps to be applied to each component separately. Ultimately, only a few steps, such as those associated with the accommodation of anisotropy (rotations) and near-surface distortions, are handled with a tensor or vector wavefield (multicomponent) approach.

For azimuthally anisotropic media with a uniform symmetry axis, Alford (1986) and Thomsen (1988) proposed a rotation algorithm that rotates the source and receiver axes numerically to minimize the energy in the off-diagonal elements (*xy* and *yx* components) of the data matrix. Similar results can be obtained by rotating analytically (Murtha, 1988; Lewis et al., 1991). Li and Crampin (1991a, 1993a) suggest a linear transform technique (LTT) as an alternative to the rotation technique. The LTT contains four linear transforms, which effectively recast the complicated shear wave motions in the recording coordinate system into linear motions in the transformed coordinate system. Then, the polarization estimation and split shear wave separation can be easily made from these linearized components. This is equivalent to a singular value decomposition of the data matrix. In principle, all these methods assume that the amplitudes of the two split shear waves are eigenvalues of the data matrix.

Another major challenge in processing the split shear waves involves near-surface and overburden amplitude corrections to recover the shear wave information associated with the target zone. The shear wave is inevitably affected by acquisition errors, near-surface irregularities, and overburden anomalies. Winterstein and Meadows (1991) developed a layer-stripping algorithm for the VSP context and applied it to reflection data (Winterstein, 1991). MacBeth et al. (1992), Li (1995), and Thomsen et al. (1995a) modified this stripping procedure for reflection data. Li (1994) presented another procedure for amplitude correction in multicomponent shear wave data which includes a prestack surface-consistent correction and poststack overburden scaling.

The data examples presented here have been processed using different separation algorithms. For completeness, we cover some basic equations for processing split shear waves.

Multicomponent Amplitude Corrections

According to Li (1994), factors affecting the target amplitudes in the multicomponent data can be divided into two groups: one related to the surface (or near surface) and the other to the subsurface. The surface-related group includes source and receiver distortions due to interactions with the near-surface. These effects can be corrected for by using a modified surface-consistent procedure for multicomponent seismic data before separation of split shear waves and stacking. The subsurface-related group includes attenuation, scattering, geometric spreading, anisotropy, and other undesirable wavefield properties in the overburden. These complications can be corrected by an overburden-correction scheme after separation of the split shear waves and stacking. Note that the effects of an anisotropic overburden may introduce different moveout velocities and AVO responses for the fast and slow split shear wave. Thus, different stacking velocities may be needed for stacking the fast and slow split shear waves; different amplitude scaling factors can be applied to correct the anisotropic AVO responses.

Assuming the earth is a linear system for seismic waves satisfying the convolution model (MacBeth et al., 1994), we have, in the time domain,

$$\boldsymbol{d}(t) = \boldsymbol{o}(t) * \boldsymbol{g}(t) * \boldsymbol{m}(t) * \boldsymbol{s}(t), \quad (2)$$

where $\boldsymbol{o}(t)$ is a two-by-two matrix representing the offset responses for the four components. For a data matrix containing mixed modes of split shear waves, $\boldsymbol{o}(t)$ can be assumed as a scalar $o(t)$ that is the same for all four components, and for data matrix containing separated split shear waves, $\boldsymbol{o}(t)$ is a diagonal matrix representing different offset responses for the fast and slow split shear waves. Also, $\boldsymbol{g}(t)$ is a diagonal matrix representing the receiver geophone responses, $\boldsymbol{m}(t)$ is a two-by-two matrix representing the subsurface response for shear wave splitting, and $\mathbf{s}(t)$ is a diagonal matrix representing the source responses. The symbol * denotes convolution.

Equation (2) is a direct extension of Taner and Koehler's equations for one-component *P*-wave seismic data (Taner and Koehler, 1981). In a similar way and with similar assumptions, one does not need to compute the complete frequency-dependent surface (or near-surface) response. Moreover, amplitude corrections can be implemented by multiplying the seismic trace by a scalar, which is equivalent to adding a constant value to the log of amplitude spectrum. This procedure is similar to conventional *P*-wave processing (Taner and Koehler, 1981), although there are two source scalars, two geophone scalars, and two (for separated split shear waves) or one (for mixed modes) offset scalars. Statistical solutions of $\boldsymbol{o}(t)$, $\boldsymbol{g}(t)$, and $\boldsymbol{s}(t)$ can then be obtained using the least-squares method (see Li, 1994, for details). Thus, the subsurface medium response $\boldsymbol{m}(t)$ can be obtained. We then proceed to separate the split shear waves from the medium response $\boldsymbol{m}(t)$, or the data matrix after surface-consistent deconvolution. In the following sections, we use $\boldsymbol{D}(t)$ (note capital letter) to represent the data matrix, after surface-consistent correction, that is, $\boldsymbol{D}(t)$ is the same as $\boldsymbol{m}(t)$ after surface-consistent correction.

Separation of Split Shear Waves

Assuming orthogonally polarized split shear waves, we can simulate the medium response of shear wave splitting by a two-component eigensystem, with the eigenvectors representing the polarizations of the two split shear waves and the eigenvalues representing their amplitudes. For the four-component geometry (*xx*, *xy*, *yx*, and *yy*) shown in Figures 1a and 1b, in the time domain, the medium response for shear wave splitting (or the data matrix after surface-consistent correction) can be written as

$$\boldsymbol{D}(t) = \boldsymbol{C}(\alpha)\,\Lambda(t)\,\boldsymbol{C}^T(\alpha), \tag{3}$$

where superscript T is the transpose operator, α is the polarization azimuth of the fast split shear wave (Figure 1b), $\boldsymbol{D}(t)$ is the medium response for shear wave splitting or the data matrix after surface-consistent deconvolution using equation (2), $\boldsymbol{C}$ is the orthogonal rotation matrix, and $\Lambda(t)$ is the diagonal transfer function for the two split shear waves S_1 and S_2 (also see Alford, 1986). We then have

$$C(\alpha) = \begin{bmatrix} \cos\alpha & \sin\alpha \\ -\sin\alpha & \cos\alpha \end{bmatrix}; \quad \Lambda(t) = \begin{bmatrix} S_1(t) & 0 \\ 0 & S_2(t) \end{bmatrix}. \tag{4}$$

Note that the orthogonality of split shear waves is strictly true only for phase propagation and is a good approximation only for near-vertical group propagation (Crampin, 1981; for nonorthogonal split shear waves, see Li et al., 1993b). We stress again that equation (3) holds only if the source and geophone responses have been carefully compensated by surface-consistent deconvolution and only for near-vertical propagation over small spreads. If we take the inverse of the rotation matrix, equation (3) becomes

$$\Lambda(t) = \boldsymbol{C}^T(\alpha)\,\boldsymbol{D}(t)\,\boldsymbol{C}(\alpha). \tag{5}$$

Equation (5) forms the basis of Alford (1986) rotation. The data matrix can be rotated numerically using a series of angles. The angle that minimizes the energy of the off-diagonal elements of the rotated data matrix is the polarization azimuth α.

Another way to determine the rotation angle and the amplitudes of the split shear waves is to use the LTT (Li and Crampin, 1993a). The four linear transforms are defined as

$$\begin{bmatrix} \xi(t) & \chi(t) \\ \eta(t) & \zeta(t) \end{bmatrix} = \begin{bmatrix} xx(t) - yy(t) & yx(t) - xy(t) \\ yx(t) + xy(t) & xx(t) + yy(t) \end{bmatrix}. \tag{6}$$

Substituting equation (3) into equation (6) and making some manipulations, we have

$$\begin{bmatrix} \xi(t) & \chi(t) \\ \eta(t) & \zeta(t) \end{bmatrix} = \begin{bmatrix} \cos 2\alpha & 0 \\ \sin 2\alpha & 1 \end{bmatrix} \begin{bmatrix} [S1(t) - S2(t)] & 0 \\ 0 & [S1(t) + S2(t)] \end{bmatrix}. \tag{7}$$

Thus, the time series $S_1(t)$ and $S_2(t)$ are separated from the static geometry factors α after transformation; the transformed vectors $[\xi(t)\ \eta(t)]^T$ and $[\chi(t)\ \zeta(t)]^T$ are eigenvectors representing linearly polarized motions in the time domain in the displacement plane; the rotation angle α can be determined from vector $[\xi(t)\ \eta(t)]^T$ as the *Jacobi* rotation angle:

$$\alpha = \frac{1}{4}\tan^{-1}\left[\frac{2\sum_t \xi(t+\tau)\,\eta(t+\tau)}{\sum_t [(\xi^2(\tau+t)) - \eta^2(t+\tau)]}\right]. \tag{8}$$

The summation is over a chosen time window, or the entire recorded time, and τ represents the starting point of the chosen window. If we apply equation (8) for each time sample instantaneously (e.g., the length of the time window is one sample), then we obtain one angle for each sample. This is referred to as a polarization log, or instantaneous polarization trace, which is discussed in more detail later. The fast and slow waves can also be easily determined from the transformed vectors from equation (7):

$$\begin{aligned} S1(t) - S2(t) &= \xi(t)\cos 2\alpha + \eta(t)\sin 2\alpha, \\ S1(t) + S2(t) &= \zeta(t). \end{aligned} \tag{9}$$

The simple arithmetic of equation (6) and the resulting separation of time series from geometry factors in equation (7) are the main advantages of the LTT. After separation of the split shear waves, the conventional *P*-wave stacking procedures can be performed separately for the principal wave components. Overburden correction is carried out on poststack sections.

Overburden Correction

Overburden amplitude corrections compensate for near-surface and overburden anisotropy effects, attenuation, scattering, and transmission loss. This is similar to the surface-consistent approach for prestack amplitude correction. Here a statistical approach for the overburden, assuming subsurface consistency, is taken to implement the overburden correction for surface seismic data (Li, 1994). Note that the subsurface consistency implies that the reflectivity of the overburden horizon is consistent for adjacent common depth points (CDPs). The procedure can be summarized as follows:

1. Select a quality horizon in the overburden as a reference level. An average pattern of the amplitude variation along this horizon in the overburden is calculated for each CDP location on the S_1 and S_2 stack sections.
2. Similarly, define for the target horizon an average pattern of the amplitude variation calculated for each CDP.
3. Derive a common overburden scaling factor from these two curves, assuming the scaling factor is consistent for adjacent CDPs.

This method is based on the model that the target is sandwiched between the overburden and a half space and that the overburden response can be simplified with a frequency-independent scaling factor (Li, 1994).

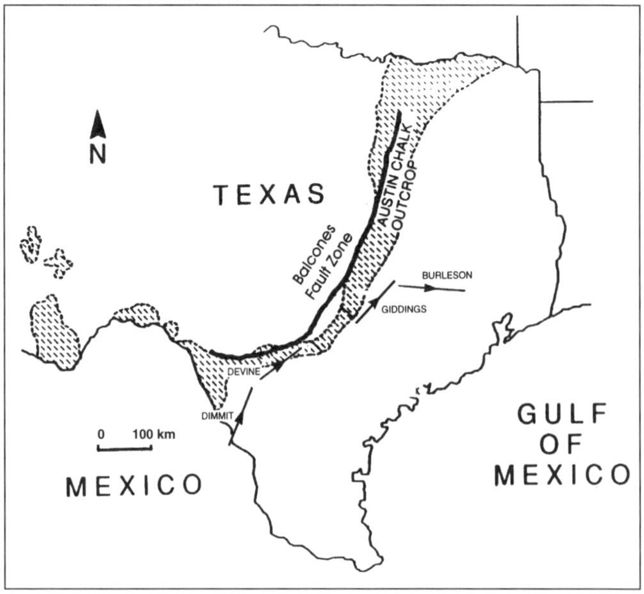

Fig. 2. Map showing the locations of the study areas. The shaded area marks the Austin Chalk outcrop, and the arrows show the regional fracture strike obtained from shear wave data. Adapted from Yardley and Crampin (1993).

Shear Wave Splitting in Austin Chalk

The multicomponent seismic data we selected in this study consist of six lines of four-component surveys from two different fields and three VSPs from three different sites over the Austin Chalk. Following a review of the geologic setting in the study area, we discuss the acquisition parameters for these data sets and their unique features. We then review the characteristics of shear wave splitting along the Austin Chalk and compare different ways to display shear wave splitting. We intend to demonstrate that clear shear wave splitting in the Austin Chalk can be observed both in multicomponent surface data and in VSPs, which makes it possible to characterize fracture swarms in the chalk.

Geologic Setting

The study area lies in an intraplate basin, and the general tectonic setting and stress regime have been described by Zoback and Zoback (1991). The outcrop pattern of the Austin Chalk is shown in Figure 2 ;sou it dips about 2° toward the Gulf of Mexico. Extension is occurring toward the Gulf of Mexico, and the maximum horizontal stress parallels the coast. The nature of the fracturing in the chalk has been described by Corbett et al. (1987), who analyzed data from cores and surface outcrops. Fracture orientations trend parallel to the regional stress directions. Studies of shear wave anisotropy in the region find that the polarization direction of the leading split shear wave parallels the stress pattern (Raikes, 1991; Mueller, 1991; Li et al., 1993a).

The area covers parts of Dimmit, Zavala, Frio, and La Salle counties in the southwest (Figure 3a) and Devine, Giddings, and Burleson counties in the northeast (Figure 2). The subsurface structures are almost horizontal layers because of the small dip. Primary oil and gas plays in the area are in Cretaceous rocks, including the Olmos Sandstone, San Miguel Sandstone, and Austin Chalk (Ames, 1990). In the southwestern part of the area, oil and gas are produced from the Olmos Sandstone; in the northeast, oil and gas are produced from the San Miguel Sandstone and the Austin Chalk.

Of these major formations, the Austin Chalk has attracted continuous interest since the initial oil discovery in the 1920s (Stapp, 1977). Interest in the Austin Chalk has been renewed with the advent of horizontal drilling technology (Kuich, 1989; Ames, 1990). In Dimmit, Zavala, Frio, and La Salle counties, the chalk is at 2000 m depth and has a thickness of ~300 m (Scott, 1977; Stapp, 1977). In Devine, Texas, which is closer to the outcrop, the chalk is at 1000 m depth and is 100 m thick. In Giddings field (in Fayette and Burleson counties), the chalk is at ~2500 m depth and is thin, with a thickness of only 30–100 m. The strike of the Austin Chalk is northeast-

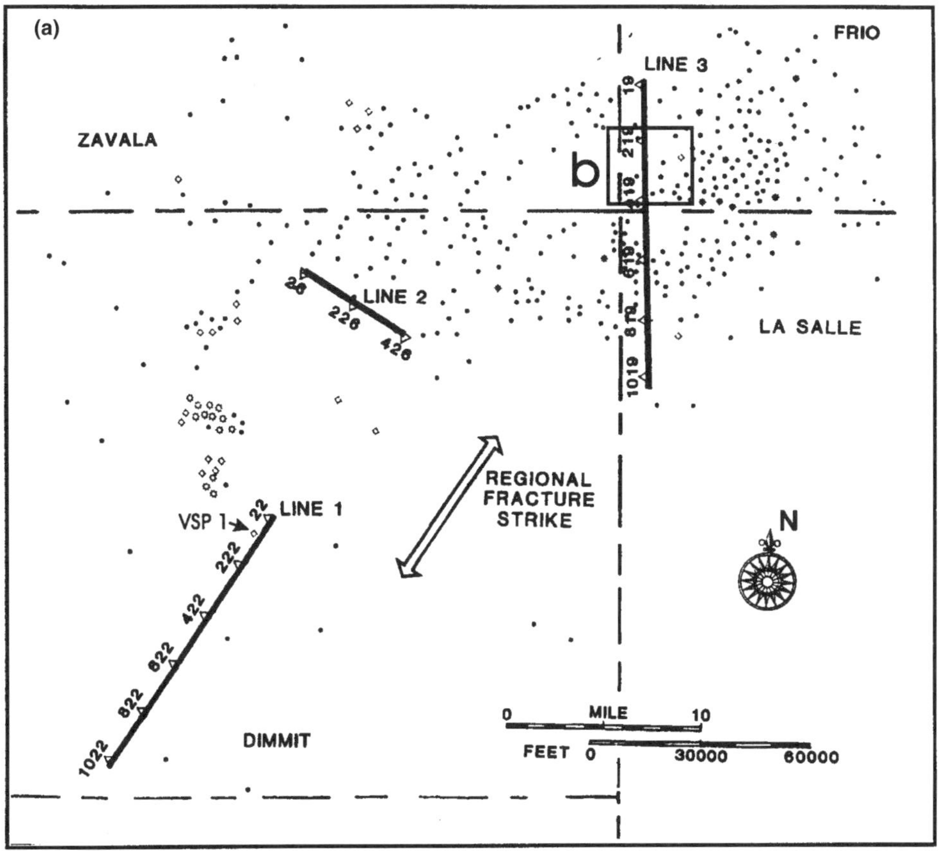

Line 1: 0.5% Line 2: 1.0% Line 3: 1.5%

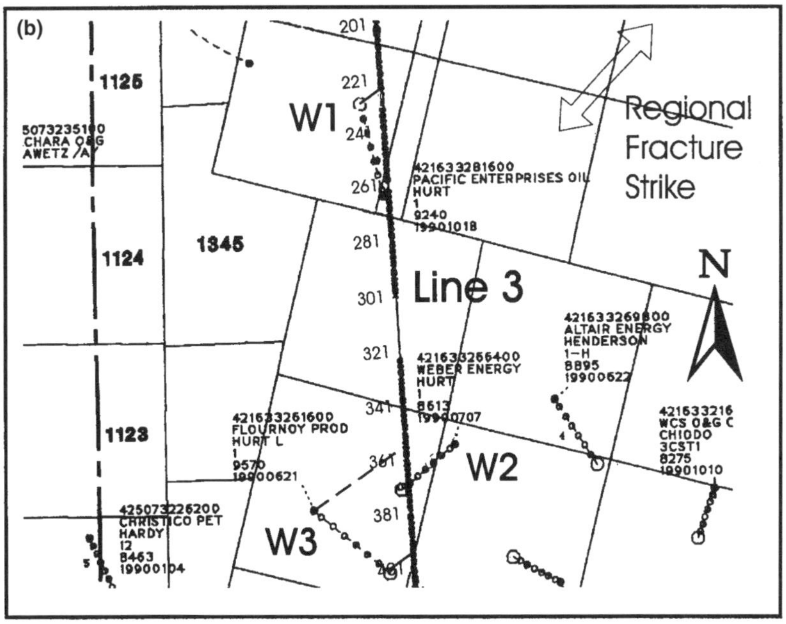

Fig. 3. (a) Locations of the three surface lines in Pearsall field in Dimmit and Frio counties. The dots mark oil wells (courtesy of Amoco Production Company). Note the location of VSP 1. (b) Map showing part of line 3 and the surrounding area. The rows of circles are surface projections of the horizontal wells; the large open circle at one end indicates the surface location of a well and the small solid circle at the other end indicates the horizontally displaced total depth of the well. W1, W2, and W3 are the three wells selected for study; the dashed lines project the horizontal wells to the survey line along fracture strike.

southwest, and the regional structural setting is parallel to that of the Gulf Coast Basin. Weeks (1945) and Dravis (1979) have reviewed the regional geology and sedimentology of the Austin Chalk.

The Austin Chalk was formed from a fine-grained carbonate mud containing skeletal remains of algae (Corbett et al., 1987). The porosity of the chalk in the subsurface is as low as 6% (Corbett et al., 1987). With such low porosity, reservoir development is almost wholly dependent on accessing fractures, and successful drilling depends on identifying and penetrating highly fractured chalk (Kuich, 1989; Mueller, 1991).

The two major fracture systems in the Austin Chalk are regional and fault-related. Both fracture systems are believed to be uniformly oriented and parallel to the strike of the structures (Stapp, 1977). There are three major reasons for this. First, according to Scott (1977), the Austin Chalk was deposited on a relatively flat surface in deep-water environments and is expected to have uniform facies. Second, the regional fractures were created by the downwarping of a flat depositional surface in response to subsequent deposition of great thicknesses of Tertiary sands and shales. This downwarping stretched the Austin Chalk uniformly along the dip direction of the structure. Third, all large subsurface faults in the Cretaceous strata are parallel to the strike of structure and have similar displacements. There is no evidence of age difference in the various fault zones (Corbett et al., 1987), and all movements are assumed to be contemporaneous and the regional stress field uniform.

Although the strike of fractures in the chalk is expected to be uniform, the fracture intensity in dipmeter logs and field mapping is not. This is probably due to minor differences in local stress and variations in the content of skeletal algae remains; thus, fracturing is more likely to occur in some areas (Scott, 1977; Stapp, 1977; Corbett et al., 1987). Thus, fractures form clusters and swarms which are the major exploration targets for horizontal drilling (Kuich, 1989; Mueller, 1991, 1994). The same uniform stress fields as in the Austin Chalk also reoriented the pore space in the Tertiary overburden along the strike of the fractures. Thus, an overall uniform anisotropy symmetry direction is expected in the basin, with possible minor local variations. This situation, together with the "layer-cake" structure, provides a favorable environment for studying shear wave splitting in seismic reflection surveys.

Data Volume and Field Acquisition

Figure 2 shows the Austin Chalk trend and the distribution of shear wave data sets. Geographically, the data are from three fields: three surface lines (lines 1–3) and one VSP (VSP 1) are from the Pearsall field in Dimmit and Frio counties in the southwestern part of the trend; one VSP (VSP 2) is from a site in Devine, Texas, in the central part of the trend; and three surface lines (lines 4–6) and one VSP (VSP 3) are from the Giddings field in Fayette and Burleson counties in the northeastern part of the trend.

The six surface seismic lines were recorded in a four-component shear wave survey design with 121 in-line and 121 cross-line receiver channels both for in-line and cross-line (horizontal) source orientations and with a station interval of 34 or 50 m (~110 or 160 ft). (Details of acquisition parameters for lines 1–3 from Pearsall field are documented by Li et al., 1993a; those of lines 4–6 from Giddings field are documented by Mueller, 1991.) The three VSPs were acquired with three orthogonal sources and three-component geophones, forming a nine-component geometry. However, only the four-component horizontal subset is analyzed here. Details of their parameters are documented by Yardley and Crampin (1993).

The three seismic lines from Pearsall field form a classic shear wave experiment for studying shear wave splitting. Three lines with different azimuths to the regional fracture strike, and located in areas with different hydrocarbon production rates, demonstrate different behavior of the split shear waves. Figure 3a shows the locations of the three survey lines and well distributions in the Pearsall field. Line 1, striking about N 40° E, is parallel to the strike of the subsurface faults and subsurface fractures. There were few exploration wells drilled in this area of southwestern Dimmit County. Line 2, trending N 50° W, is nearly perpendicular to the fracture strike, and line 3, trending about north-south, is at ~40° to the fault and fracture strike. These two lines are located near the juncture of Dimmit, Zavala, Frio, and La Salle counties, where the area has been heavily drilled (Figure 3a). Significant oil fields operated in this area are the Pearsall field in Frio County (Champion, 1936) and the Big Wells field in Dimmit and Zavala counties (Layden, 1976). Recent horizontal drilling in the Austin Chalk has also occurred here (Kuich, 1989; Ames, 1990; Mueller, 1991). Three horizontal wells (W1, W2, and W3) have been drilled along line 3.

Line 4 (Figure 4a), from the southern part of Giddings field, had a horizontal well drilled on it with supporting mud logs for identifying fracture zones. Thus, any fracture zone, as identified by shear wave attributes such as differential time delays and amplitudes between the two splits shear waves, could be carefully calibrated with the horizontal well. Figure 4a shows the local line

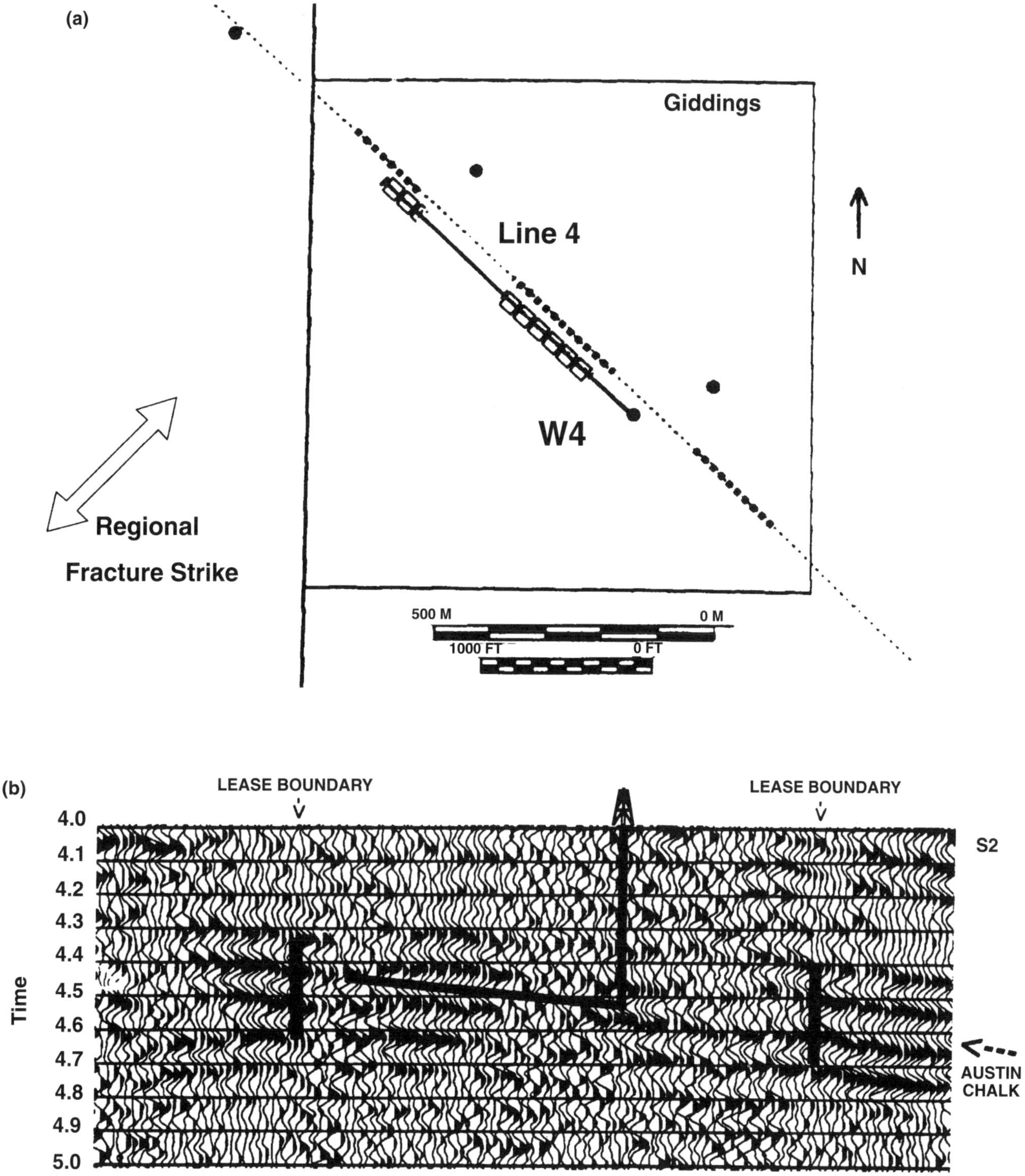

Fig. 4. (a) Location of line 4 and horizontal well W4 within the lease boundary (outer square). The light dashed line is line 4, with the small squares representing dim spots interpreted from the S_2 section (below). The solid line trending northwest represents the horizontal borehole, with the open boxes representing fracture swarms determined from mud logs. The solid dots are the surface locations of producing wells drilled to the Austin Chalk. (b) Part of the S_2 seismic section showing amplitude dim spots, with heavy vertical bars marking the lease boundary; the reversed L-shaped line marks the borehole. The seismic section has the same lateral scale as the map in part (a). Adapted from Mueller (1991).

layout for line 4. The regional fractures and fault system in this area strike southwest-northeast (Figure 2), and line 4 strikes about N 40° W, which is approximately perpendicular to the regional fracture strike. W4 marks the horizontal well with mud logs taken to verify the fractures, which are interpreted from mud logs by the presence of calcite crystals that form on the intersected fracture faces. Lines 5 and 6 are from the northern reaches of Giddings field in Burleson County (Figure 2), and VSP 3 was obtained from a well near line 5. These two lines in Burleson County strike N 40° E, and the fracture strike at this location has rotated to a east-west orientation (Figure 2).

The three VSPs also form a classic experiment. VSP 1 is from a dry hole on line 1 in Dimmit County in an area that has had little oil production (Figure 3a); VSP 2 is from a water-producing well and VSP 3 is from an oil-producing well. Of these wells, only the chalk layer at VSP 3 is producing oil. This allows comparison of the anisotropy attributes for different production rates. The three VSPs are all near-offset nine-component VSPs, thus allowing accurate determination of the anisotropy parameters. We not only use the VSPs to demonstrate different features of shear wave anisotropy as the production varies but also VSP 1 and 3 to correlate horizons with the shear wave surface data.

The six surface lines illustrate some basic features of shear wave splitting, such as coupling and decoupling of wave modes as line azimuth changes, misties of fast and slow events as the two shear waves propagate through the subsurface with different speeds, and differential amplitudes between the fast and slow shear waves as the slow wave attenuates more than the fast wave.

Shear Wave Splitting

Recording shear waves with horizontal sources and receivers is generally less robust than recording *P*-waves with vertical sources and receivers. The coupling of a horizontal source with the surface may be weak, and the horizontal phones may be more open to various noise that often propagates horizontally, such as wind and surface waves. Furthermore, because of the difficulties associated with the near surface, such as strong attenuation and scattering, low-velocity weathered layers, and irregular topographic variations, there have been serious concerns in the industry as to whether quality shear wave data can be acquired and whether shear wave splitting can be observed. Here, we use field data to examine these concerns.

Pearsall Field

Figures 5a, 5b, and 5c show one shot-gather data matrix each from lines 1, 2, and 3 in the Pearsall field. These figures show how field data change for different line azimuths. Coherent primary reflection events can be clearly identified in all three data matrices. Comparison of the diagonal elements (*xx* and *yy*) with the off-diagonal elements (*xy* and *yx*) of the data matrix shows that both line 1 and line 2 (Figures 5a and 5b) have strong, coherent shear wave events in the diagonal elements but almost no coherent signals in the off-diagonal elements. In contrast, line 3 (Figure 5c) has strong, coherent shear wave events with approximately equal energy in both diagonal and off-diagonal elements. These figures illustrate the effects of shear wave splitting in a basin having azimuthal anisotropy with uniform symmetry directions. Lines 1 and 2 are parallel to the vertical planes of symmetry, so that the in-line source excites the in-line receiver, and the cross-line source excites the cross-line receiver, and there is little cross-coupling. Line 3, in an intermediate direction, results in the in-line and cross-line source orientations each exciting both split shear waves, and there is strong cross-coupling. Comparison of the in-line source components (*xx* and *xy*) with the cross-line source components (*yx* and *yy*) shows that all three lines have larger coherent noise (the *S*-to-*P* converted arrivals) in the in-line source components than the cross-line source components above 3.0 s. This coherent noise is mainly *S*-to-*P* converted waves because the cross-line source does not favor *S*-to-*P* conversions. The *xx* components show the largest coherent noise, while the *yy* components show almost no coherent noise.

Thus, although there is noise in the recorded data matrix, quality primary shear waves can be recorded, and shear wave splitting is observed, as shown in Figure 5c. Note that here we display data in a matrix format and compare the distributions of energy among the different components for interpreting shear wave splitting. Color displays of instantaneous polarization (Li and Crampin, 1991b), which can be calculated using equation (8), are another way of displaying this information.

Figures 6a and 6b show polarization logs after LTT has been applied to the data matrices of line 1 (Figure 5a) and line 3 (Figure 5c), respectively. The polarization log of line 2 shows similar features to that of line 1 and is not shown here. Note that polarization filtering flags the noise and incoherent polarizations as un-

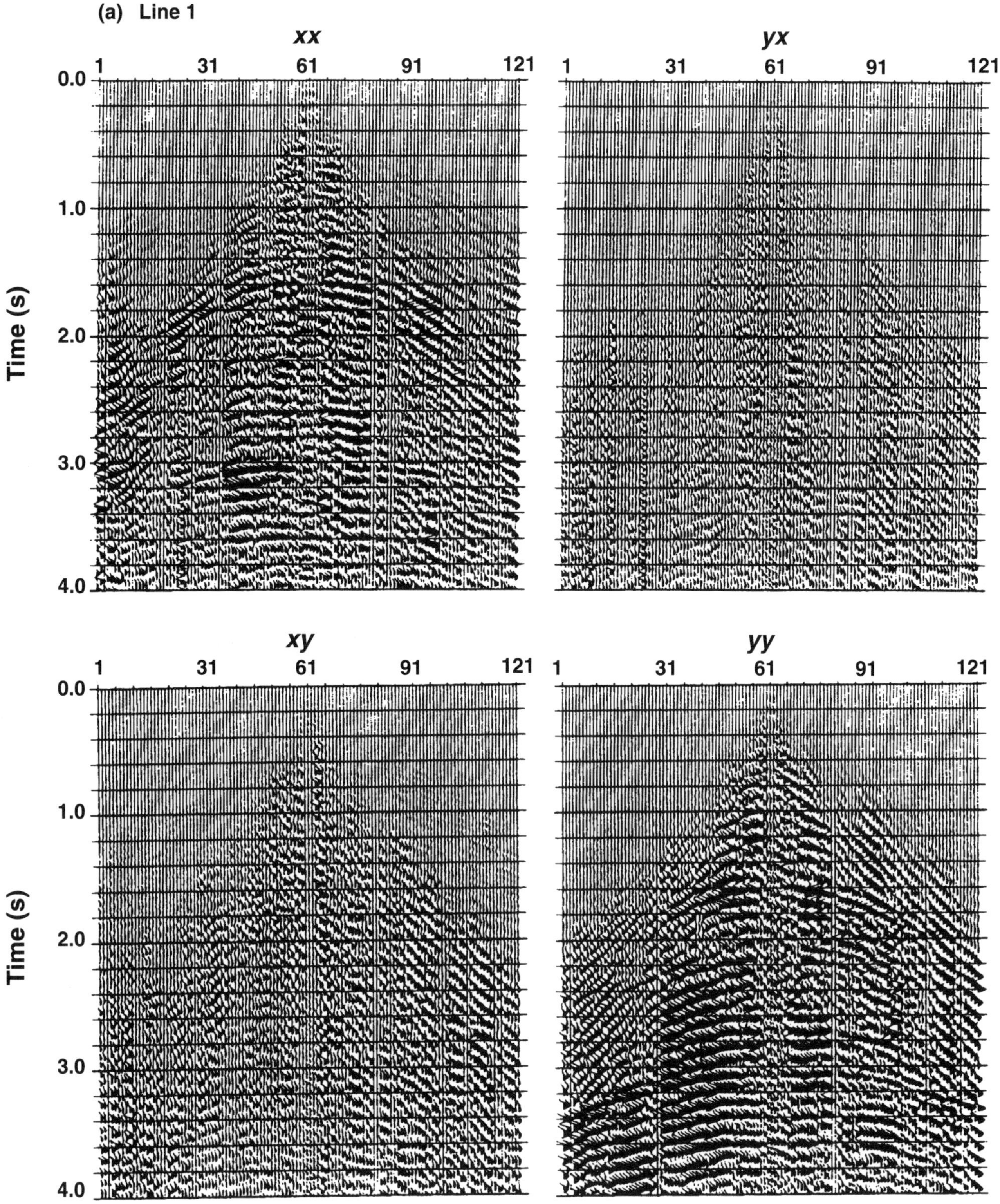

Fig. 5. Comparison of shot data matrix selected from the three lines in Pearsall field: (a) line 1, (b) line 2, and (c) line 3. Rows are common geophone orientations, and columns are common source orientations for in-line (*x*) and cross-line (*y*) horizontal components. The station interval is 50 m (165 ft), and the far offset is 3000 m (9900 ft). Relative amplitudes among the components are preserved by applying the same gain function to all four components. Note that the shear wave first breaks are weak and incoherent, likely caused by the attenuation due to source and receiver groups (Li et al., 1993a). ***(see parts b and c on next two pages)***

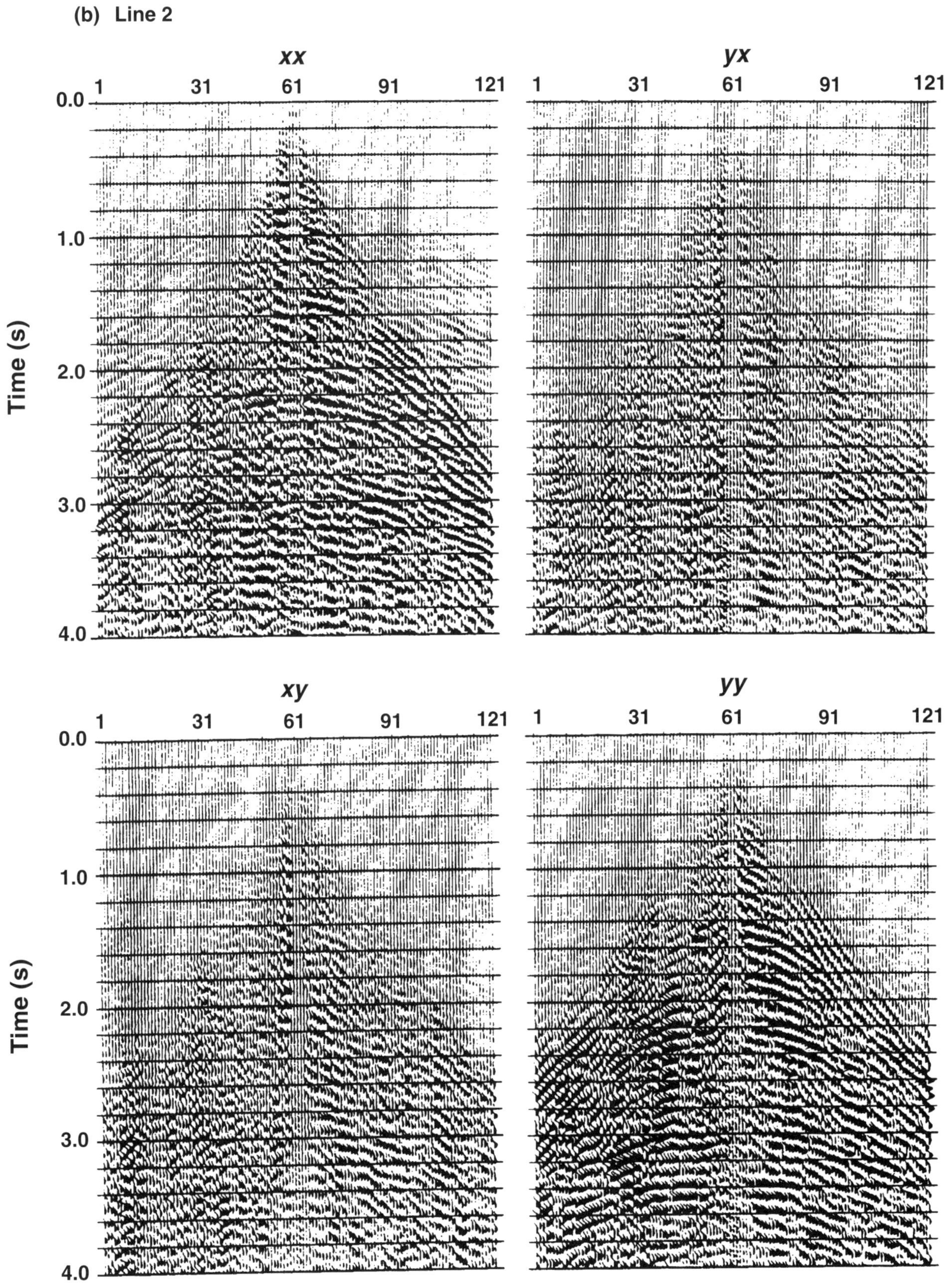
(b) Line 2
xx
yx
xy
yy
1
31
61
91
121
0.0
1.0
2.0
3.0
4.0
Time (s)

(c) Line 3

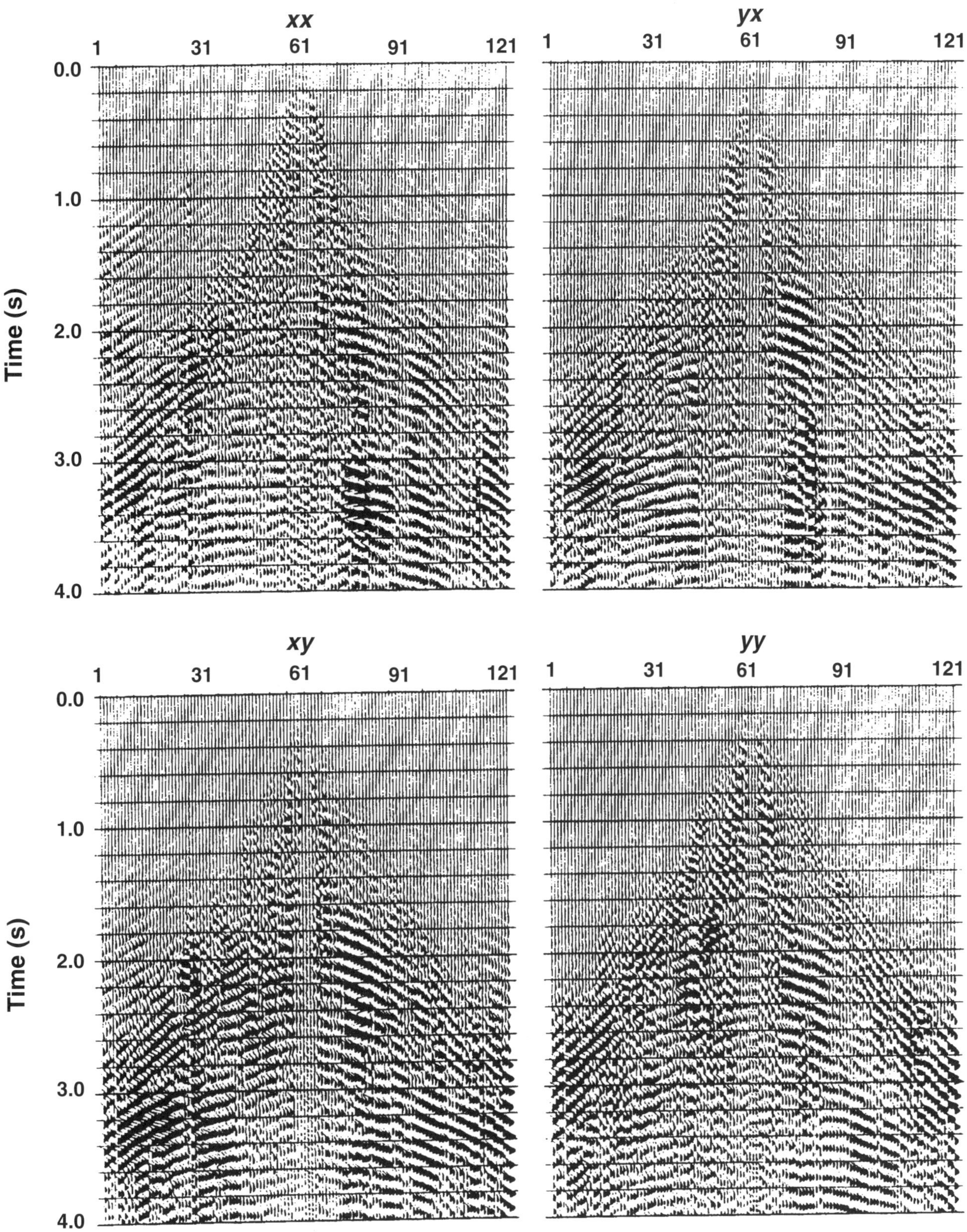

specified polarizations represented by the white background. In Figure 6a, the incoherent polarizations between the in-line (*xx* and *xy*) and cross-line source components (*yx* and *yy*) are muted (Li and Crampin, 1991b), while in Figure 6b, polarizations of 0°–15° are muted. The polarization log of line 1 (Figure 6a) shows two events: green (in-line polarization of 0°) and red (transverse horizontal cross-line polarization of 90°), indicating that the in-line and cross-line polarizations from the source are preserved. This can imply either a plane-layered isotropic structure or, as here, an anisotropic structure in which the reflection lines are along directions of vertical symmetry planes. In contrast, the polarization log of line 3 (Figure 6b) shows blue and orange events with polarizations of –50° and 40° from the in-line direction, counterclockwise for negative angles and clockwise for positive angles. This is a typical occurrence when source polarization lies between the symmetry planes of an azimuthally anisotropic medium. Comparison of Figure 5 with Figure 6 shows that color displays of instantaneous polarization are highly efficient for analyzing shear wave polarizations.

Giddings Field

Thus far, we have discussed one feature of recording shear waves in anisotropic rocks using the three lines in the Pearsall field: the variation in event coherence on the four components as the line azimuth varies in an azimuthally anisotropic medium. Using lines 5 and 6 from the Burleson County part of Giddings field, we now show two other features of recording shear waves in anisotropic rocks: mistie of events and differential amplitudes between fast and slow shear waves as they travel at different speeds.

Figure 7 shows two stack sections from line 5 representing the two split shear waves. The fast shear wave is polarized east-west, and the slow shear wave north-south. From top to bottom, there is an increasing mistie, indicating that anisotropy is present not only in the near-surface layer but also in deeper layers. At the level of the Austin Chalk (just below 4 s two-way traveltime) a cumulative mistie of 70 ms is observed, representing 2% average shear wave anisotropy from the surface to the top of the chalk (just below 4 s). Note that there is no increasing mistie from 3 to 4 s, which confirms that the thick Tertiary sandstone above the Austin Chalk is isotropic.

Figure 8 shows two stack sections from line 6 for the two split shear waves. Similar to line 5, the fast shear wave is polarized east-west and the slow shear wave north-south. At the Austin Chalk level, a mistie of 70 ms and a differential amplitude between the fast and slow shear waves can be observed. The reflections from the chalk formation in the S_1 section are consistent in strength and continuity, while those in the S_2 section vary strongly. There is a prominent weak amplitude anomaly in the center of the event that corresponds to a large fracture swarm.

Data Processing

Because the data sets showed different responses to anisotropic rocks, slightly different processing sequences were applied to different survey lines. We discuss the processing sequences separately.

Lines 1 and 2

Lines 1 and 2 are parallel and perpendicular to the presumed crack strike, respectively. Figures 5a and 5b show that the two split shear waves along lines 1 and 2 are separated so that one-component *P*-wave processing techniques can be applied to each component separately (Lynn and Thomsen, 1990; Li et al., 1993a). Because the off-diagonal elements contain little signal energy (Figures 5a and 5b), they can be omitted in an early stage of processing. Table 1 summarizes the processing sequence used for lines 1 and 2. The main points to note are as follows:

1. Even when the polarizations are separated, the velocities of the two shear waves are different and vary with offset. Li and Crampin (1993b) suggested velocity and moveout equations for CDP gathers in anisotropic media. For small spreads, the effects of varying velocities with offset are thought to be negligible, and the wide offset traces must be carefully muted, as done here. However, the difference between the stacking velocities of the two shear waves is not necessarily negligible even though it may be small, and different stacking velocities may be needed to stack the two shear waves.
2. Careful amplitude balancing is usually necessary in processing vibrator data. Traces at small offsets typically have large amplitudes that decrease sharply with increasing offset; sources at different orientations and locations may vary in strength and coupling. Thus, prestack multicomponent surface-consistent corrections are usually necessary to condition the data matrix, as indicated by equation (2).

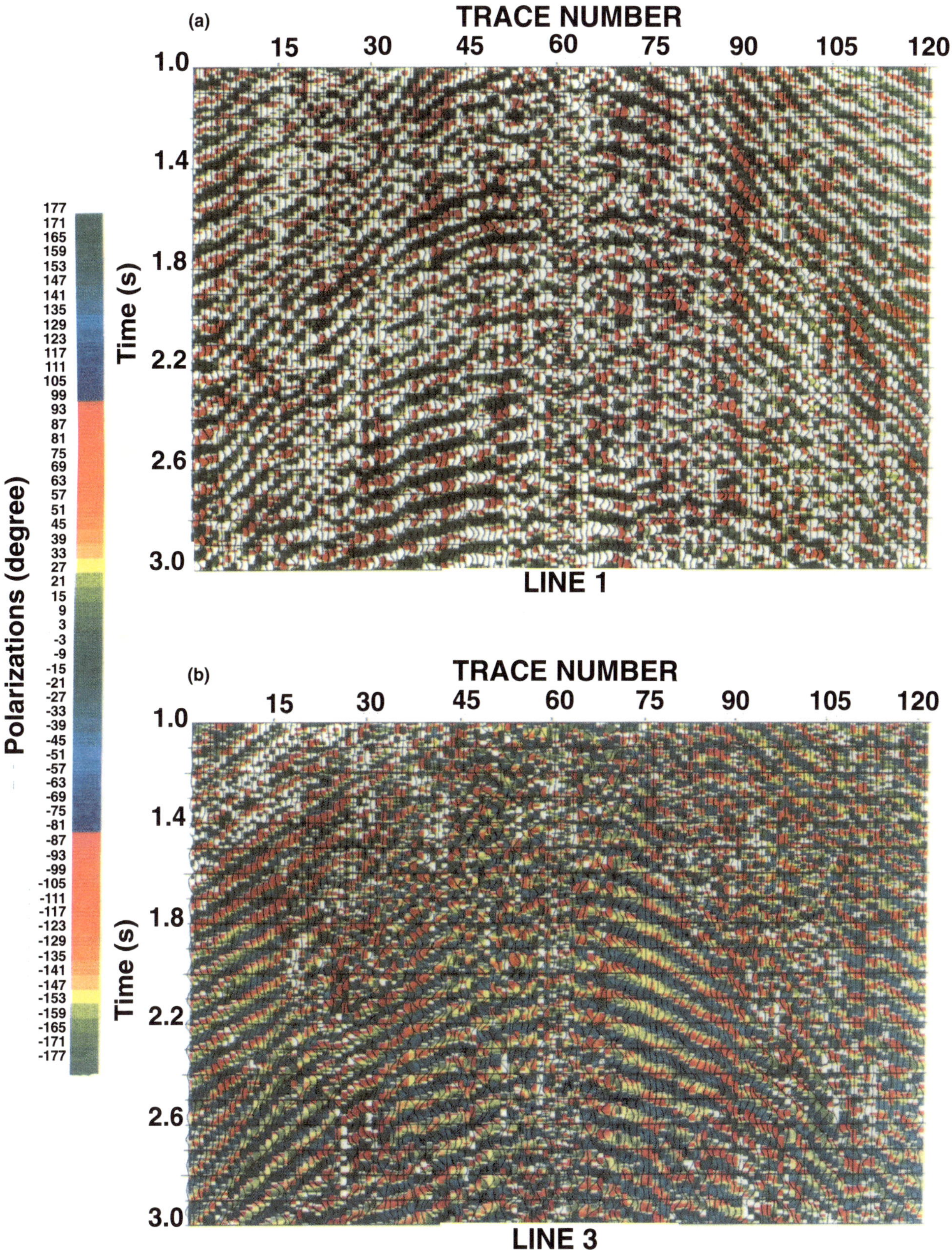

Fig. 6. Polarization logs after applying the linear transform technique (LTT) to (a) the shot data matrix in Figure 5a, line 1, and (b) the shot data matrix in Figure 5c, line 3. Positive angles in the color code indicate polarizations measured clockwise from the in-line direction, and negative angles indicate polarizations measured counterclockwise.

Table 1. Conventional Processing Sequences

Procedure	Sequence of Steps
Data editing	Geometry input Trace editing Data statics using velocity 3000 ft/s Multicomponent surface consistent amplitude correction
Velocity and residual statics	CDP sort Velocity analysis NMO correction Four-component scaling: balancing trace by off-diagonal records Stacking: form pilot traces for statics estimation Residual statics picking Solve for surface consistent residual statics Iterate above sequence
Final Stacking	CDP sort NMO correction Residual statics application Stack Overburden amplitude correction Display

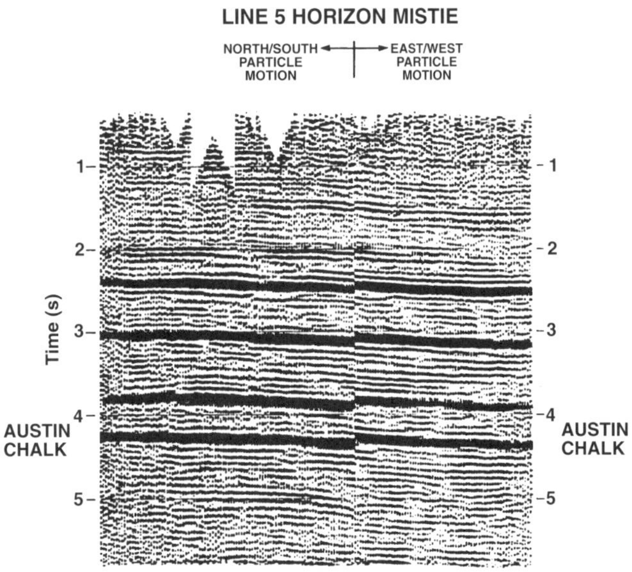

Fig. 7. A natural polarization direction comparison for line 5 in Giddings field. On the left side of the plot, the north-south particle motion is displayed, and on the right side, the east-west particle motion. The sections are joined at the same surface point. A dynamic mistie of events can be seen where the sections join. After Mueller (1991).

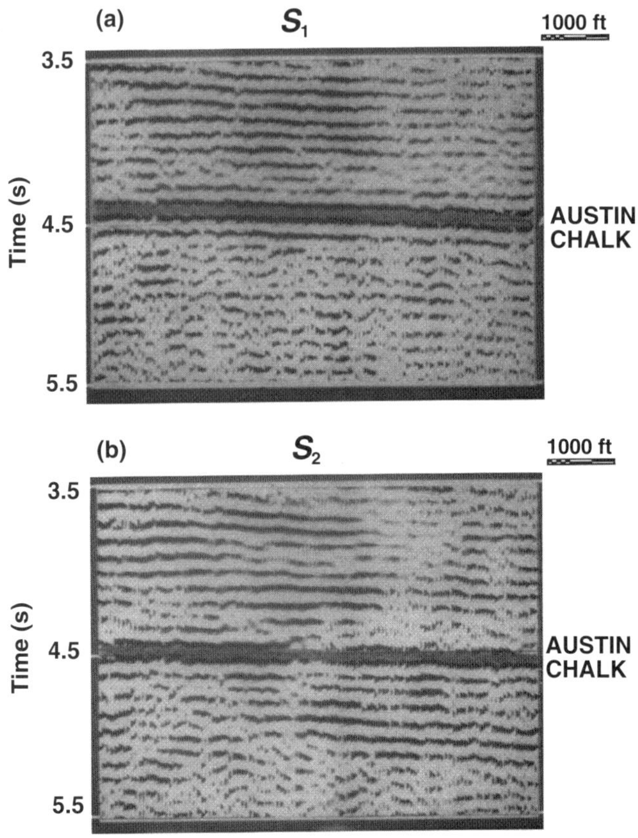

Fig. 8. Final stacked (a) S_1 and (b) S_2 sections of line 6 in Giddings field using the processing sequence in Tables 1 and 2. The S_2 section shows an amplitude dim spot along the Austin Chalk.

3. In practice, it is often assumed that the fast and slow split shear waves have the same field statics and the same residual statics. However, because shear wave splitting is due to the polarizations reacting differently to the anisotropic structure, this assumption can only be a first-order approximation. Assuming equal statics, the problem is to ensure that the same statics are applied to all components. The fast and slow split shear waves may have different stacking velocities (Thomsen, 1988; Li, 1992), and it may be necessary to carry out two separate passes of velocity analysis for S_1 and S_2.

Line 3

Line 3 shows strong energy on cross components. Separating split shear waves provides an extra challenge in addition to the problems encountered in processing lines 1 and 2. Before one can proceed to separate the split shear waves, one has to compensate for source, receiver, offset, and other near-surface factors. The data matrix after compensation may then be considered as a proper representation of the medium response for shear wave splitting. However, amplitude corrections are sensitive and can be tricky; thus, careful procedures have to be taken to ensure minimum distortion and artifacts to the data. Always check the quality control (QC) plots of the scaling factors before applying the corrections. The source and receiver factors should only show short-wavelength and minimum corrections as implied by the surface consistent criteria; the offset factor should show a smooth and almost exponential variation with offset. One can also check the observation logs to confirm any significant abnormality in the source and receiver scaling factors. In practice, we find that these scaling factors, particularly the offset scaling factor, provide an optimum amplitude balance between the near- and far-offset traces but preserve the relative amplitude between the components and CDP locations. Without such corrections, the trace amplitudes are not balanced, and the S/N ratio of the stacking section may be very low (see Li, 1994). A conventional amplitude scaling may balance the trace amplitude, but it may also distort the relative amplitudes between the components.

Here the multicomponent surface consistent correction, as shown in equation (2), and the LTT, as described in equations (6)–(9), are used for processing line 3. The multicomponent surface-consistent correction involves Gause-Seidal iteration to decompose the logarithm amplitude spectrum based on equation (2) (Li, 1994). The LTT processing is also straightforward. First, LTT is applied to the shot data matrix to transform the four components (*xx*, *xy*, *yx*, and *yy*) into the three LTT components: S_1, S_2, and the polarization log. Second, the conventional processing sequence in Table 1 is applied separately to the S_1, S_2, and polarization components to give the final stacked S_1, S_2, and polarization sections. Note that the processing of polarization logs is slightly different and usually involves a polarization filter to remove unwanted polarizations (Li and Crampin, 1991b).

To show the effect of LTT, we have included Figure 9a which is the output after applying LTT to the shot data matrix in Figure 5c. Comparing Figure 9a with Figure 5c shows that the energy in the off-diagonal panels is significantly reduced. The polarization log after applying LTT to Figure 5c was shown and discussed previously in Figure 6b. Note that Figure 9a is not as good as Figures 5a and 5b in terms of residual energy in the cross components, which highlights the limits of the data processing technique for separating split shear waves, as compared with physical separation of the two shear waves.

To compare the LTT results with the rotation technique, Figure 9b shows the output after rotating the shot data matrix in Figure 5c by 40° (angle derived from Figure 6b; see earlier discussion). Comparing Figure 9a with 9b shows that the energy in the cross components obtained by LTT is less than that obtained by conventional rotation. This indicates that the split shear waves are optimized better by LTT than by rotation. This is because in the implementation of the rotation used here, the rotation angle is constant, whereas LTT allows a sample-by-sample and trace-by-trace variation in the polarization angle (Li and Crampin, 1993a). Li et al. (1993a) explain in more depth why LTT performs better than simple rotation.

Lines 4, 5, and 6

Line 4 is perpendicular to the fracture strike and is processed as outlined for line 2.

Lines 5 and 6 also show significant shear wave splitting, as demonstrated in Figure 7 (for line 5) and Figure 8 (for line 6). The separation of shear waves in lines 5 and 6 is performed using the rotation method, as a comparison with LTT in line 3. The process of rotation analysis can be summarized as follows. First, the data editing sequence in Table 1 is applied to all four components. Again note that the editing sequence in Table 1 contains the procedure of amplitude corrections which are essential for balancing the trace amplitudes while preserving the characteristics of shear wave splitting, such as the differential amplitudes of the two shear waves. This requires careful quality control to ensure that only the short-wavelength and minimum source and receiver corrections are applied (see section on line 3 for more discussion). Second, the velocity and statics sequence in Table 1 is applied to the *yx* or *xy* component to determine the stacking velocities and residual statics. Note that the *yx* and *xy* components are expected to be about equal in noise-free azimuthal anisotropic media containing uniform symmetry (same crack orientation) throughout the depth range. In such cases, a single pass of the velocity and statics sequence is sufficient to choose off-diagonal records for preparing stacked data matrices for rotation. Third, the preprocessing sequence for rotation in Table 2 is applied to all four components separately and a stacked data matrix is obtained. This is immediately followed by the rotation sequence in Table 2. At this stage, the stacked data matrix can be used to determine the rotation angle, and this angle used to rotate the shot data matrix to separate the four-component data into two-component (S_1 and S_2) data sets. The last stage is to apply the velocity and statics sequence and the final stacking sequence in Table 1 to the S_1 and S_2 components separately to obtain the final stacked S_1 and S_2 sections.

Vertical Seismic Profiles

The LTT (from equations 6 to 9) is used to determine the fast direction in the VSPs at the three sites, and the data are then rotated to give the fast and slow sections. The time delays are calculated using cross-correlation of the fast and slow section. To compare the differential amplitude of reflected fast and slow shear waves, quadrilateral *f-k* filters are used to separate the upgoing and downgoing wavefields. Before the *f-k* filters, a depth-dependent amplitude scaling factor is applied to the data to balance the amplitudes between different depth levels; after the *f-k* filters, this scaling factor is removed to recover the amplitude variations. Before the amplitude ratio of the upgoing arrivals on the slow and fast sections can be calculated, the amplitude of the downgoing arrival for the fast and slow waves at each geophone level is used as a scaling factor to normalize the amplitude of the upgoing arrival for the corresponding fast and slow waves at the same geophone level. This means that changes in source strength between the in-line and cross-line sources and spherical divergence effects can be neglected. The amplitude ratio is then calculated at each level for the reflections from the Austin Chalk for all three VSPs.

Interpretation

The interpretation of fracture zones is based on the analysis of shear wave attributes, including polarizations, time delays, and amplitudes (dim spots) at normal incidence. Wherever possible, other well information such as production figures and mud logs are used to calibrate the results. We first discuss the results of the three lines in Pearsall field, then the results of line 4 in Giddings field, and compare the three VSPs from the three sites with different production rates. Finally, we give a regional overview of all the results. Note that we emphasize the use of shear wave attributes at normal incidence as interpreted from stacked sections. This is because the variations in shear wave attributes at normal

Table 3. Processing Sequences for Separating Split Shear Waves: Comparison of Rotation Technique with the Linear Transform Technique (LTT)

Procedure	Sequence of Steps
Preprocessing for rotation	CDP sort; NMO correction Robust average scaling Residual statics application Stack
Rotation sequence	Rotation angle estimation: rotation analysis, or other techniques Apply rotation angle to shot data Separate data into S_1 and S_2 data sets
LTT	Apply LTT to shot data Separate data into S_1, S_2, and polarization components

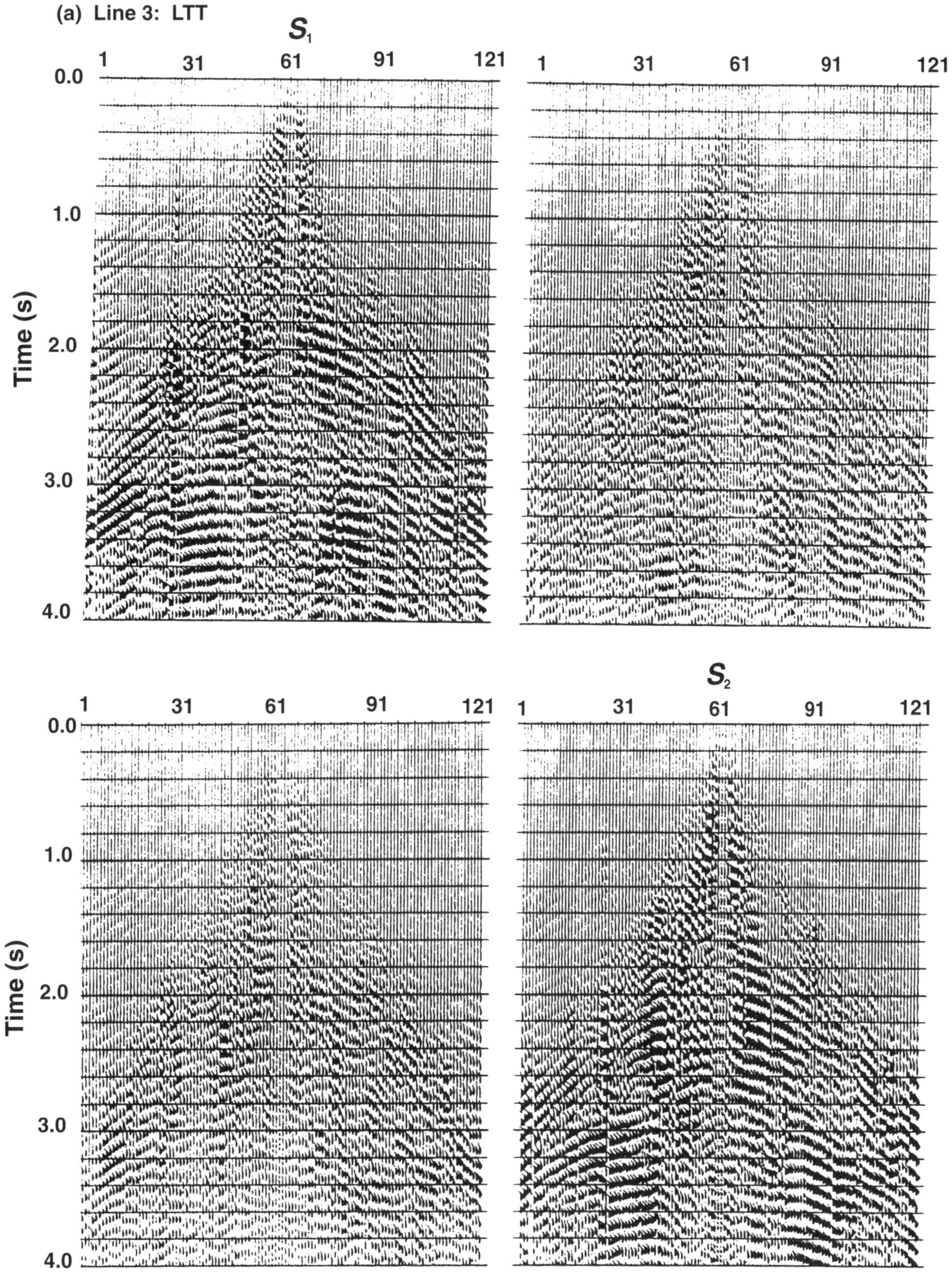

Fig. 9. (a) The shot data matrix of Figure 5c after applying LTT and (b) after applying a 40° rotation. The coherent off-diagonal energy in both is less than in Figure 5c, but the off-diagonal energy in (a) is substantially less than in (b). *(see part b on next page)*

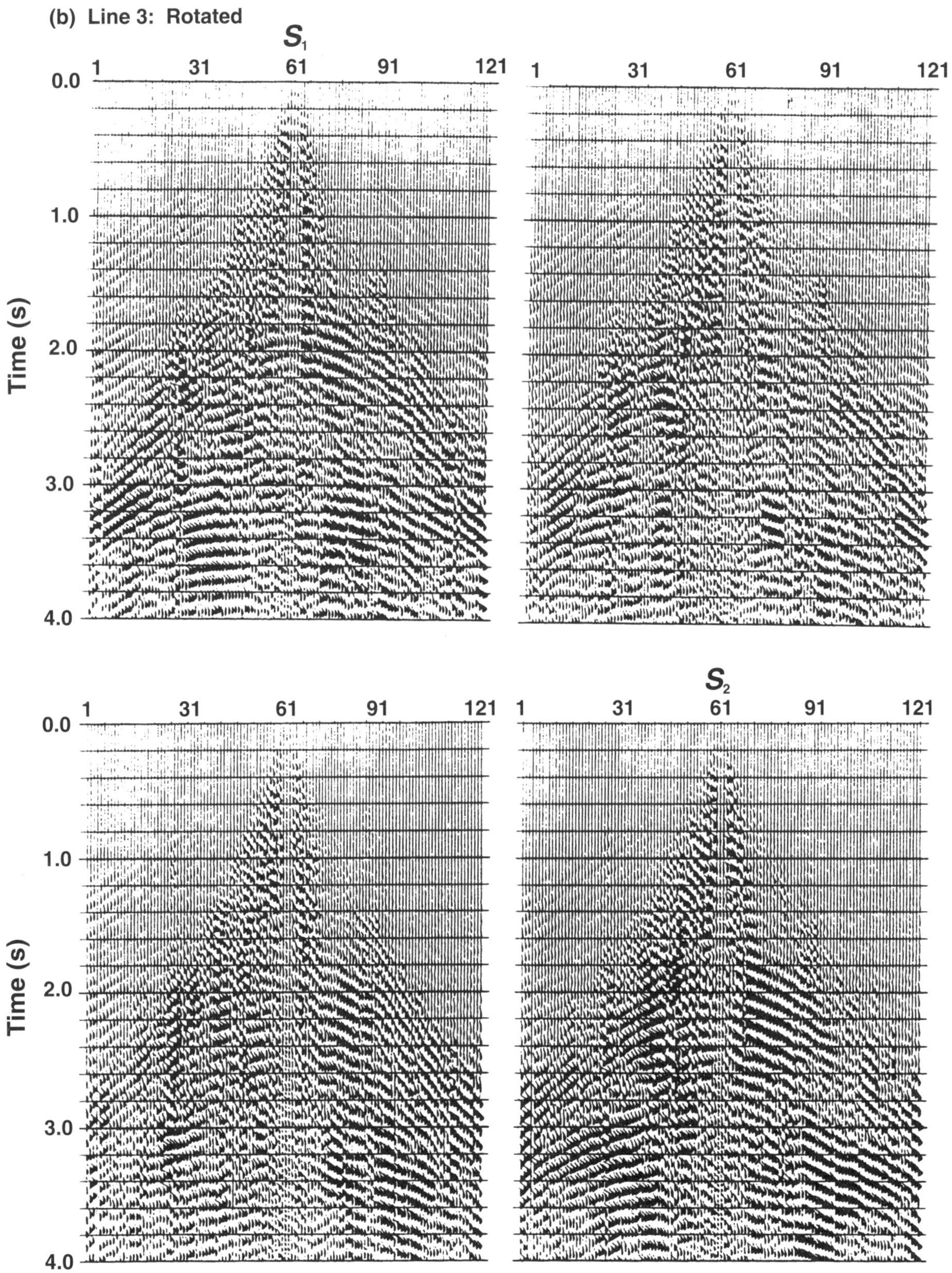
(b) Line 3: Rotated
S_1
1 31 61 91 121
Time (s)
0.0
1.0
2.0
3.0
4.0
S_2

incidence are first-order effects, as compared to their offset-dependent variations which are more subtle and difficult to interpret for fracture characterization.

Pearsall Field: Lines 1, 2, and 3

This section presents the final stacks of the three lines for S_1 and S_2 waves and the stack polarization log of line 3. We demonstrate a positive correlation of the overall anisotropy with production and a correlation of amplitude dim spots in the stacked section with fracture swarms encountered by horizontal drilling as inferred from production records.

Variations in Anisotropy

Figures 10, 11, and 12 show parts of the final stack sections of lines 1, 2, and 3 for the fast and slow waves. Figure 13 shows the stack polarizations of line 3. One can see that the stacks are "ringy" (particularly line 3 of Figures 12a, 12b), which is caused by the narrow bandwidths of the field data. As shown in Figure 5, the dominant frequency is 10 Hz with a bandwidth of less than 15 Hz. For such narrow band data, predictive deconvolution does not function properly for reducing reverberations. Note that to improve the reliability of event correlation in ringy (narrow bandwidth) data, such as in line 3, it is essential to integrate all available information including local VSP corridor stack and sonic synthetics, local *P*-wave survey lines, and other regional survey lines.

Figure 14a shows the cumulative time delays of the Georgetown Formation below the Austin Chalk, which is calculated from the cross-correlation of S_1 and S_2 sections for a 400 ms window from one cycle above the top of the chalk. Figure 14b shows the interval time delays between the Olmos Sandstone and the Austin Chalk for all three lines. These results show clear variations in anisotropy among the three lines. The average anisotropy in line 1 is small, less than 0.5%. This is calculated from the interval time delay in Figure 14b and normalized by the traveltime. Note that the interval time delays for line 1 are 0–10 ms (Figure 14b), and the Georgetown Formation appears at 2.5 s (Figure 10). This is consistent with variations in other attributes such as small differential amplitude variations and the small mistie of events in Figure 10. Similarly, the average anisotropy for lines 2 and 3 can be calculated from Figure 14. Line 2 possesses about 1% interval anisotropy and line 3 about 1.5%. This is also consistent with the greater variations in amplitude and mistie in Figures 11 and 12. Also, anisotropy increases from south to north on lines 2 and 3.

Correlation of Anisotropy with Production

Variation in anisotropy on the three lines can be roughly correlated with oil production. Figure 3a shows well locations in the study area. Few wells have been drilled near line 1, and it is more than 10 mi (16 km) from neighboring production. Line 2 is at the edge of the Pearsall and Big Wells fields in Frio, northern Dimmit, and Zavala counties, and there are about 10 wells within 1 mi (0.6 km) of the line. Line 3 is through the center of the Pearsall field, and many producing wells have been drilled along or near it. The percentages at the bottom of Figure 3a summarize the overall anisotropy in all three lines, as calculated from Figure 14: line 1, 0.5%; line 2, 1%; and line 3, 1.5%. This shows that the overall oil production along or near the three reflection lines approximately correlates with the anisotropy (time delays) observed along the three lines, where line 1 with the smallest time delays has little production nearby, line 2 with intermediate time delays shows intermediate production, and line 3 with the largest time delays has the greatest production nearby.

Variation of anisotropy along line 3 can also be correlated to oil production along or near the line. Figure 14c shows the distribution of horizontal wells within 1 mi (0.6 km) on either side of the line. Most wells are distributed in the northern part of the line (station numbers <700). There are only two wells at the northern end (station numbers <100), and only one well at the southern end of the line (station numbers >800). The time delays along line 3 in Figure 14a and 14b show wide variation. In Figure 14a at the northern end, the delay rapidly builds up to 55 ms, remains at about the same level until station 400, and then uniformly decreases to about 35 ms at station 1000 toward the south. Thus, the trend of variation of time delays broadly correlates with the overall trend of well distribution along the northern part of the line. The interval time delay in Figure 14b shows a similar trend.

The overall trend of the variation in the stacked polarization logs of line 3 in Figure 13 can also be correlated with the oil production along the line. Note that we displayed 400 traces here to show the coherency of the polarizations, while in Figure 12, we displayed only 200 traces to highlight the variations in amplitudes. Corresponding to a high production area from stations 399 to 449, the polarizations are more scattered (more white background in the lateral variations), and the polariza-

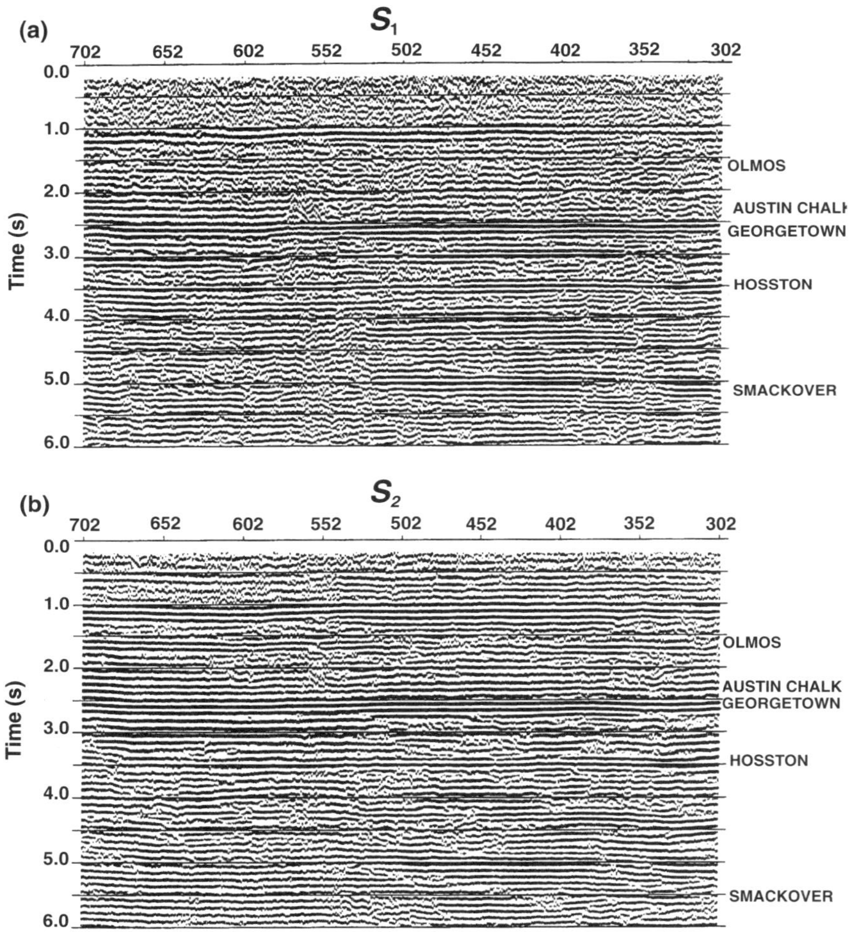

Fig. 10. Final stacked S_1 and S_2 sections of line 1 using the processing sequence in Table 1: (a) S_1 section (*xx* component) and (b) S_2 section (*yy* component).

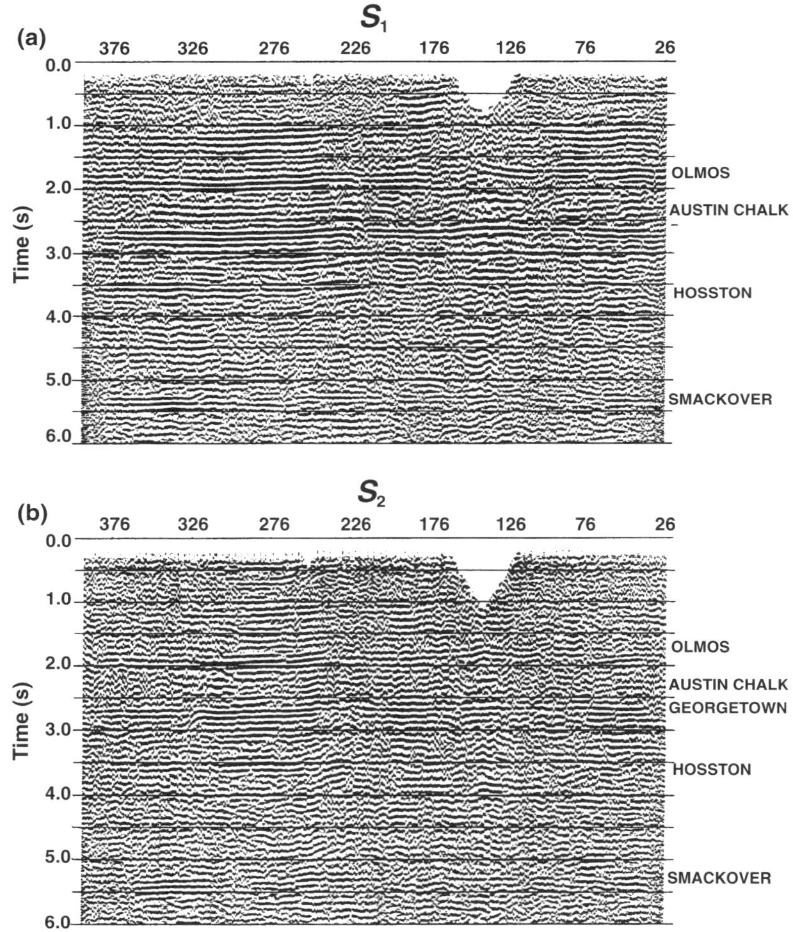

Fig. 11. Final stacked S_1 and S_2 sections of line 2 using the processing sequence in Table 1: (a) S_1 section (*yy* component) and (b) S_2 section (*xx* component).

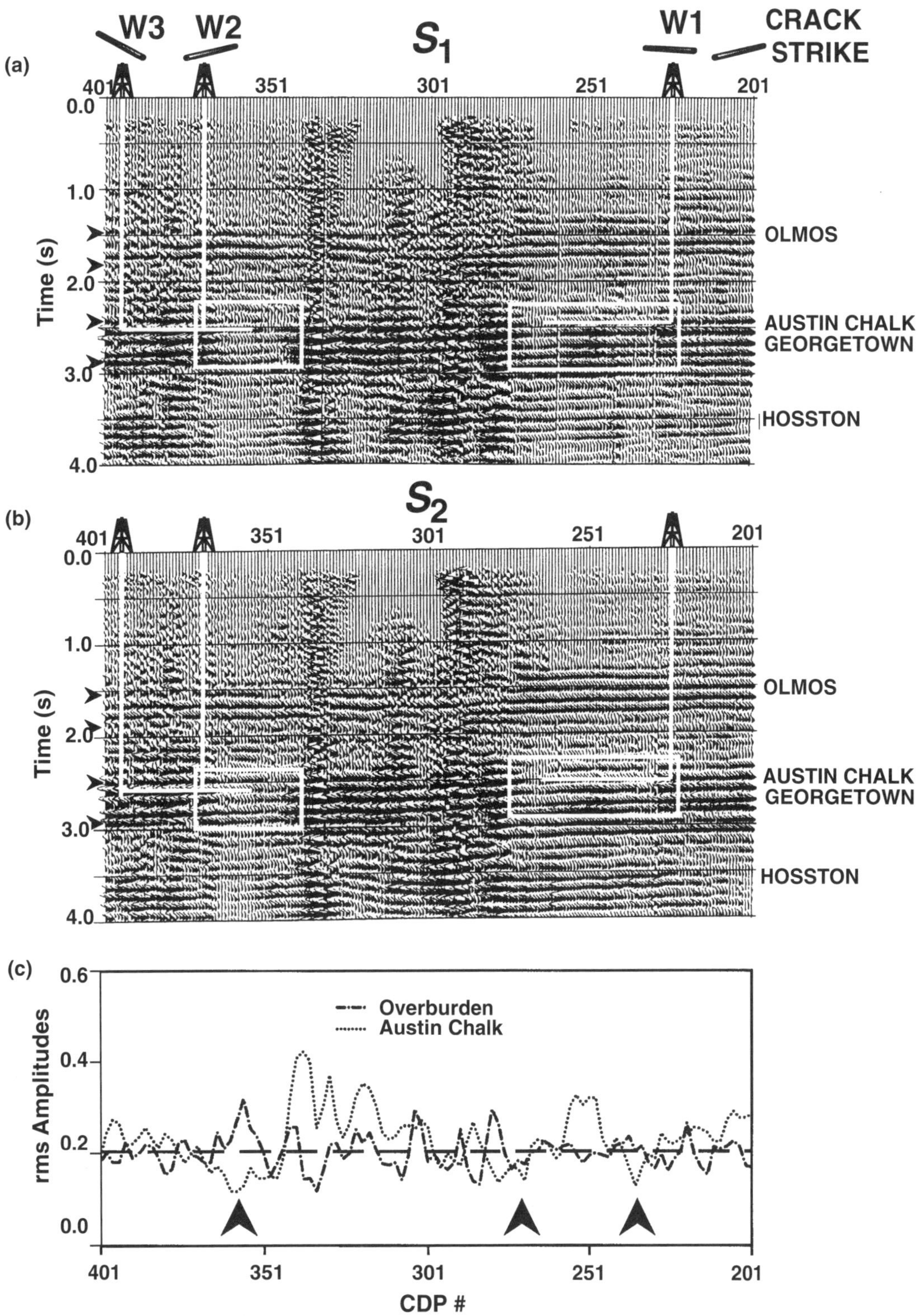

Fig. 12. Stacked (a) S_1 and (b) S_2 sections of line 3 using the processing sequence in Tables 1 and 2. W1, W2, and W3 are horizontal wells; the bars at the top represent the well azimuths, and the bar at the far right shows the crack (fracture) strike. The white boxes mark the amplitude dim spots, and the two pairs of arrows on the left mark two time windows for amplitude analysis (top = overburden, bottom = Austin Chalk). (c) Comparison of windowed rms amplitudes in the overburden and target; windows are shown by two pairs of arrows in parts (a) and (b). The dot-dashed line is the rms amplitude in the overburden window, and the dotted line is the amplitude in the Austin Chalk window. The dashed straight line represents mean amplitude; the arrows mark significant dim spots.

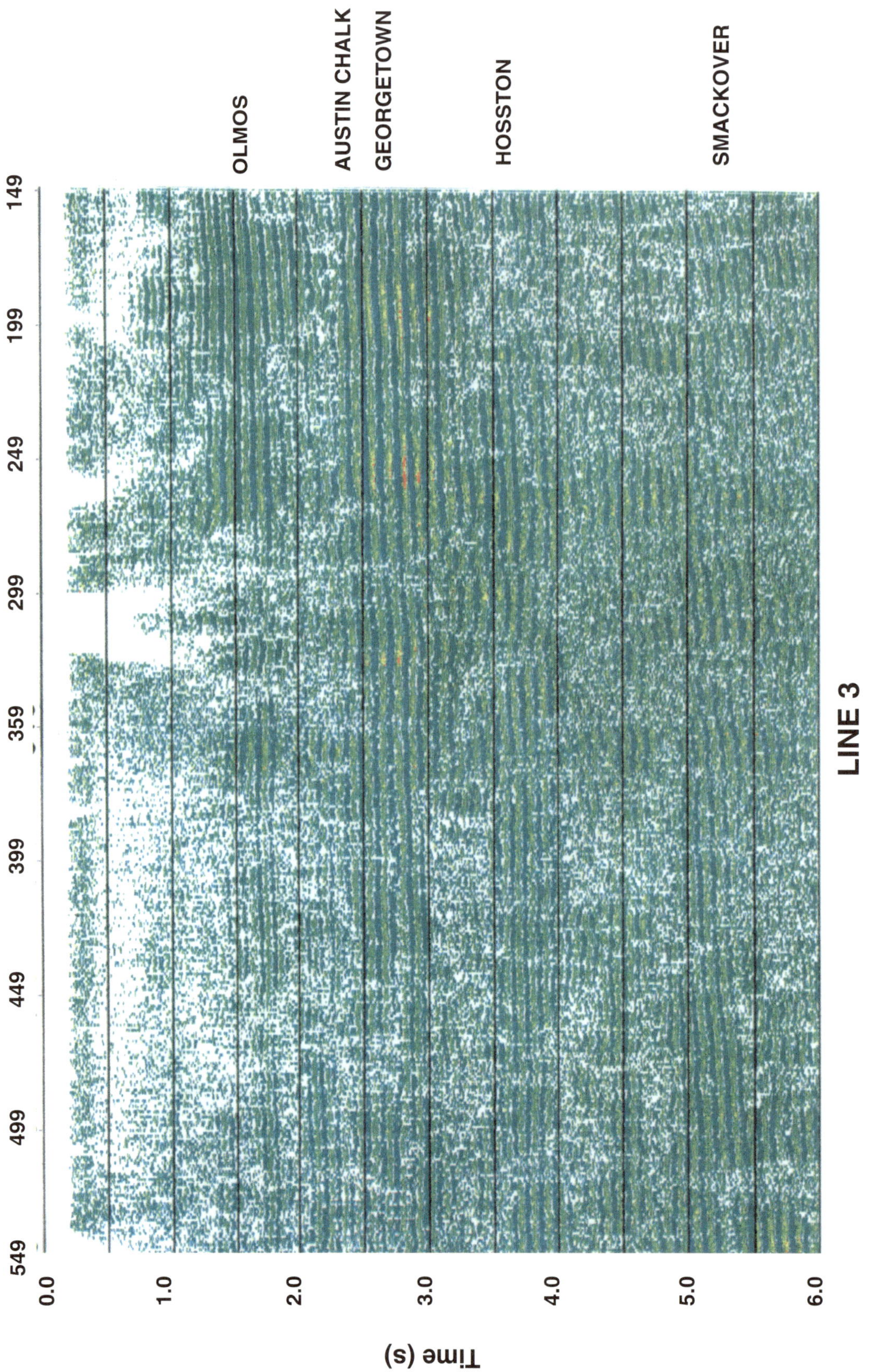

Fig. 13. Final stacked polarization logs of line 3. Color scale and definition are the same as in Figure 6.

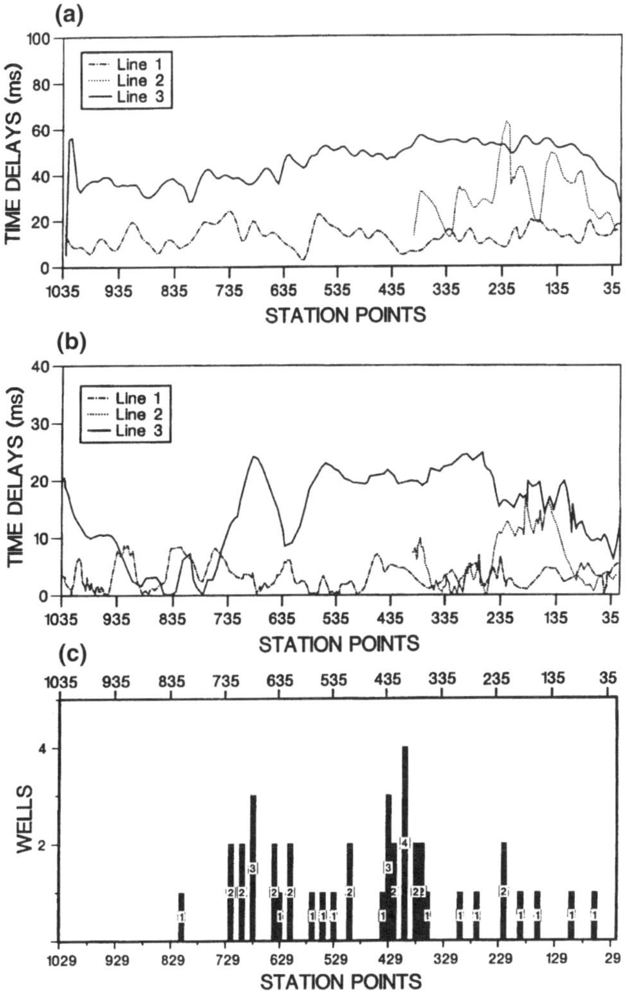

Fig. 14. (a) Comparison of cumulative time delays above the Georgetown strata between the fast and slow shear waves measured from the stacked sections of lines 1, 2, and 3 (Figures 10, 11, and 12) from the Pearsall field. (b) Comparison of interval delays between the Olmos Sandstone and the Austin Chalk. (c) Distribution of horizontal producing wells within 1.6 km (1 mi) of line 3.

tion events from the Austin Chalk are less continuous and often broken into vertical variations. Note that the white background is associated with noise and noncoherent polarization events of 0° to 15°. Dim and broken S_2 events can cause broken and noncoherent polarization events which, after stacking, are attenuated toward 0°. Clearly, identification of broken and scattered polarization events in color-coded sections is easier than identification of dim and broken S_2 events in the amplitude sections. Note that such variations are subtle and complicated, and only when the nature of such variations of polarization are better understood, from more case studies, can such interpretation be confidently made.

In summary, shear waves in the three reflection lines show typical features of shear wave splitting, although the anisotropy along line 1 is weak and marginal. The variation in overall anisotropy between the three lines can be correlated roughly with commercial oil production in the study area. Variations in anisotropy along line 3 can also be correlated with oil production along that line. High production areas are found to be associated with scattered polarizations and broken polarization events in the stacked polarization logs.

Correlation of Amplitude Dim Spots with Fracture Swarms

We selected line 3 and the three horizontal wells drilled nearby for a detailed analysis of amplitude variations and their correlation with fracture swarms, as encountered by horizontal drilling near line 3. Figure 3b is an enlarged map from Figure 3a showing the location of the survey line and the distribution of the horizontal wells. Figure 3b shows a number of horizontal wells in the area (indicated by rows of circles), all drilled in the Austin Chalk. The three wells close to the line and with similar horizontal distances—W1, W2, and W3—were selected for study. Well W1 trends northwest–southeast, at about 60° from the regional fracture strike; W2 trends southwest–northeast, parallel to the fracture strike; and W3 is southeast–northwest and perpendicular to strike. The production rates of the three wells are shown in Table 3, with W1 the most productive, W2 the least productive, and W3 moderately productive.

For similar completion programs in similarly drilled horizontal wells, the production rate (flow rate) of a horizontal well is mainly determined by the amount of fractures encountered. This is in turn determined by the length and azimuth of the horizontal well and the fracture intensity of the zone penetrated by the well. With this in mind, and by comparing Figure 3b and Table 3, one can infer that (1) local fracture swarms are present at all three sites, (2) wells parallel to fracture strike intercept fewer fractures than wells 60° or more from the fracture strike, and (3) the zone penetrated by W1 may be more intensively fractured than the zone penetrated by W3, or W1 penetrates more fracture swarms than W3 (Li, 1996). Now we examine how the variation in amplitudes can be correlated with these fracture swarms.

Figure 12 shows two groups of dim spots (white boxes) that are caused by shear wave differential reflectivity at normal incidence (and not by AVO effects). Figure 12c compares the windowed amplitude variations in the Austin Chalk with those in the overburden. (Windows are marked by two pairs of arrows on the left side of Figures 12a and 12b.) The right-hand boxes con-

Table 3. Production Records of Three Horizontal Wells W1, W2, and W3[1]

Well Name	Azimuth	Horizontal Distance (ft)	Maximum Production (bbl/day)	Cumulative Production (bbl)	Period (months)
W1	N 25° W	3000	600	108,000	14
W2	N 40° E	2000	300	37,000	19
W3	N 50° W	3000	370	74,600	21

[1]Regional fracture strike is N 40° E. Data supplied by Amoco Production Company

tain two dim spots extending from CDP 235 to 245 and from 260 to 275, and those on the left contain one dim spot from 345 to 365. To see how the horizontal wells are intercepting these dim spots, we projected the horizontal wells to the survey line along the fracture strike (Figure 3b); the projected wells were then superimposed on the seismic sections (Figures 12a and 12b). Comparison of Figures 3b and 12 shows that W1 extends from 225 to 265 and intercepts two dim spots, W2 is at the edge of the third dim spot (CDP 370), and W3 extends from CDP 355 to 395 and intercepts part of the third dim spot (355–365). Thus, W1 is likely to intercept many more fractures per unit of drilling distance than the other two wells since it intercepts more dim amplitudes than the others. This may be one of the reasons that W1 yields more than W2 and W3, demonstrating the correlation between the dim spots in the seismic sections and the local fracture swarms encountered by the horizontal wells. Dim spots appear in both the S_1 and S_2 sections here, implying the presence of macrofractures that attenuate both the fast and slow split shear waves, as demonstrated by Liu et al. (1993) at the Conoco Borehole Test Facility.

We believe that the dim spots in the S_1 section in Figure 12a are more likely to be genuine rather than artifacts of data processing, such as the overburden correction. There are two reasons for this. First, after examining the original data, we find that dim spots are also present in the stacked S_1 section before the overburden correction, but less obvious when compared to the section after the correction. Thus, the correction applied has preserved and enhanced the original features in the data. Second, both micro- and macrofractures are likely present in the Austin Chalk, and in certain areas, one set of fractures may be more developed than the others (Scott, 1977; Stapp, 1977; Corbett et al., 1987). Note that a fracture is classified as a *microfracture* if its length is less than 0.1 of the shear wavelength; otherwise it is considered to be a *macrofracture.* A medium containing aligned microfractures is equivalent to a homogeneous anisotropic medium, according to the equivalent medium theory (e.g., Hudson, 1981). However, for macrofractures, the equivalent medium theory break downs (Ebrom et al., 1990). For the seismic data studied here, the shear wavelength is 200–250 m.

Giddings Field: Lines 4, 5, and 6

In the previous example, amplitude dim spots were correlated with fracture swarms inferred from production records in horizontal wells. In the present example, a more precise correlation of amplitude dim spots with fracture swarms is made by examining the mud logs from a test horizontal well drilled on line 4.

The structure in the Giddings field can be roughly separated into four anisotropic layers. As shown in line 5 of Figure 7, the first layer (from the surface to 2.4 s two-way traveltime) has a mistie of 50 ms representing 2% shear wave anisotropy. The second layer (2.4–3.0 s) has an increasing mistie of 20 ms (interval delay) representing 3% anisotropy. The third layer (3.0–3.8 s) has no increasing mistie and is isotropic, while the fourth layer (Austin Chalk interval, 3.8–4.3 s) has an increasing mistie of 20 ms (interval delay) representing 4% anisotropy. A strong anisotropic target with an isotropic cap layer creates an ideal situation for studying the fractures in the Austin Chalk.

From line 6 of Figure 8, a cumulative mistie of 70 ms at the Austin Chalk level can also be observed. Furthermore, there is a prominent dim spot in the S_2 section. The S_1 *section* at the chalk level shows strong and consistent amplitudes, while the S_2 section shows a weak amplitude anomaly in the center of the event.

Similar dim spots can be seen in the S_2 section of

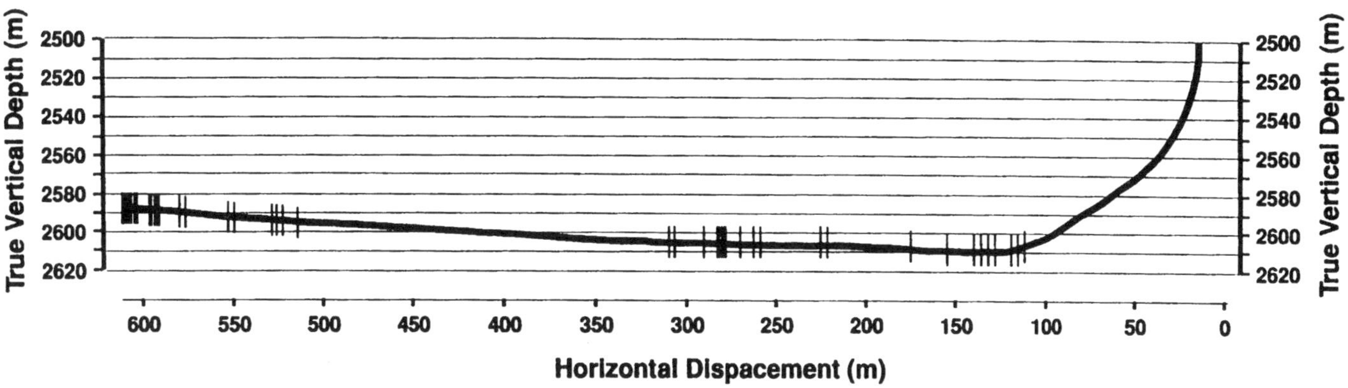

Fig. 15. A display adapted from the horizontal mud log for the well drilled to test the S_2 amplitude anomalies in Figure 4b. The short vertical lines intercepting the borehole are fractures partially filled with calcite. After Mueller (1991).

line 4 (Figure 4b) which correspond to fracture swarms confirmed by horizontal drilling. A test horizontal well was drilled toward the anomaly, and mud logs were taken to verify the fractures by the presence of calcite crystals in the mud system which often grow on the fracture faces of the Austin Chalk trend. The trajectory of the horizontal well is shown in Figure 15; the vertical lines intersecting the borehole represent fractures identified from the mud log. Figure 4b shows a part of the S_2 section, which highlights the amplitude anomaly and its interception with the horizontal well. The fractures and the amplitude anomalies are all projected onto the surface and shown in Figure 4a, where the solid line is the surface projection of the horizontal well and the parallel light dashed line is line 4. The black boxes along the survey line are amplitude dim spots interpreted from the S_2 section in Figure 4b. The boxes along the well projection are fractures identified from the mud logs (Figure 15). A good correlation clearly exists between the dim spots in the S_2 section and the fracture swarms identified by the mud logs. Note that dim spots appear only in the S_2 section here, which implies the dominance of microfractures (with length less than 20 m) in this area.

Comparison of the Three Vertical Seismic Profiles

We now examine the shear wave response in the three multicomponent VSPs, which are from three different sites: Dimmit (VSP 1), Devine (VSP 2) and Burleson (VSP 3) (see Figure 2). The three wells are dry, water producing, and oil producing, respectively.

Figures 16a, 16b, and 16c show the four-component data matrices of the three VSPs. Note that they are all near-offset VSPs and that the orientation of the downhole geophone can change direction during acquisition. Reorientation is often needed in subsequent processing, although in Figure 16, the tools have not been reoriented. VSP 1 was acquired without supporting offset VSPs for tool orientation purposes. Because of tool rotation, the energy of the direct arrival switches from one component to another as depth varies. Some bad traces were caused by loose coupling of the geophone in VSP 3 (Figure 16c).

For the three VSPs, anisotropy parameters were calculated using the linear transform technique (LTT), which allows for unknown geophone orientation. Figure 17 shows polarization and time delay estimates from these sites. The top row shows the polarization measurements, and the bottom row shows the corresponding time delay estimates. Polarization angle estimates of the leading split shear wave were about N 30° E for the Dimmit VSP (which is close to the surface measurement of N 40° E), N 55° E for the Devine VSP, and N 100° E for the Burleson VSP (which is also close to the surface measurement of an east-west orientation for line 5). Note that there is a 10° difference in measurements of the polarization between the surface and VSP results and that the measurements were constant with depth. However, no observable increases in time delay were associated with the Austin Chalk layer, and any increase in time delay must be less than one sample interval (2 ms) (bottom row of Figure 17). The chalk is about 300 m thick at the Dimmit VSP and 100 m at Devine. No buildup of time delay occurs in the chalk, suggesting that there is little birefringence in the Austin Chalk in the Dimmit and Devine VSPs and that the anisotropy observed must be attributed to the overburden. This is in agreement with the fact that the Dimmit well is a dry hole and that the Devine well has only produced minor rates of water.

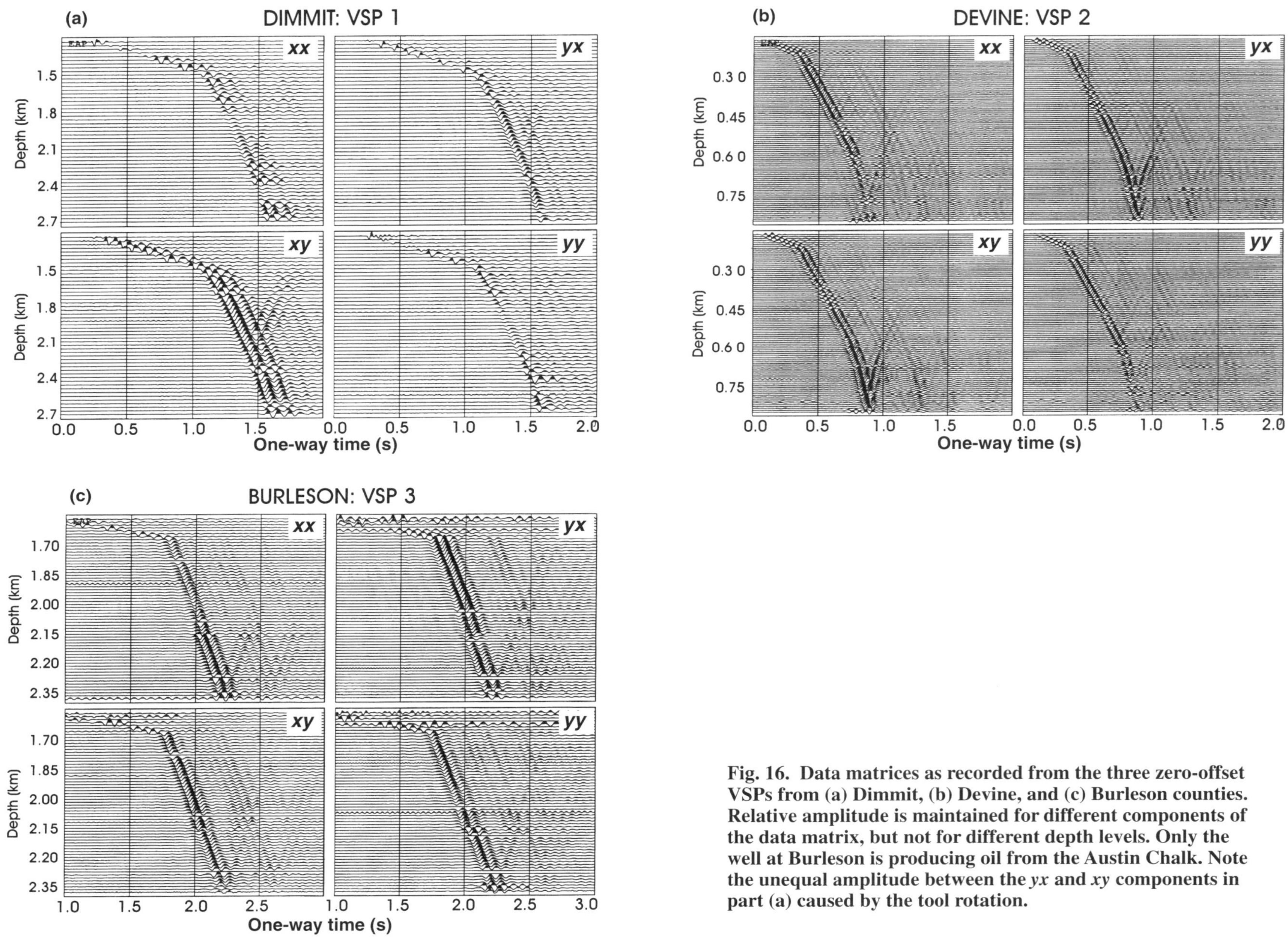

Fig. 16. Data matrices as recorded from the three zero-offset VSPs from (a) Dimmit, (b) Devine, and (c) Burleson counties. Relative amplitude is maintained for different components of the data matrix, but not for different depth levels. Only the well at Burleson is producing oil from the Austin Chalk. Note the unequal amplitude between the *yx* and *xy* components in part (a) caused by the tool rotation.

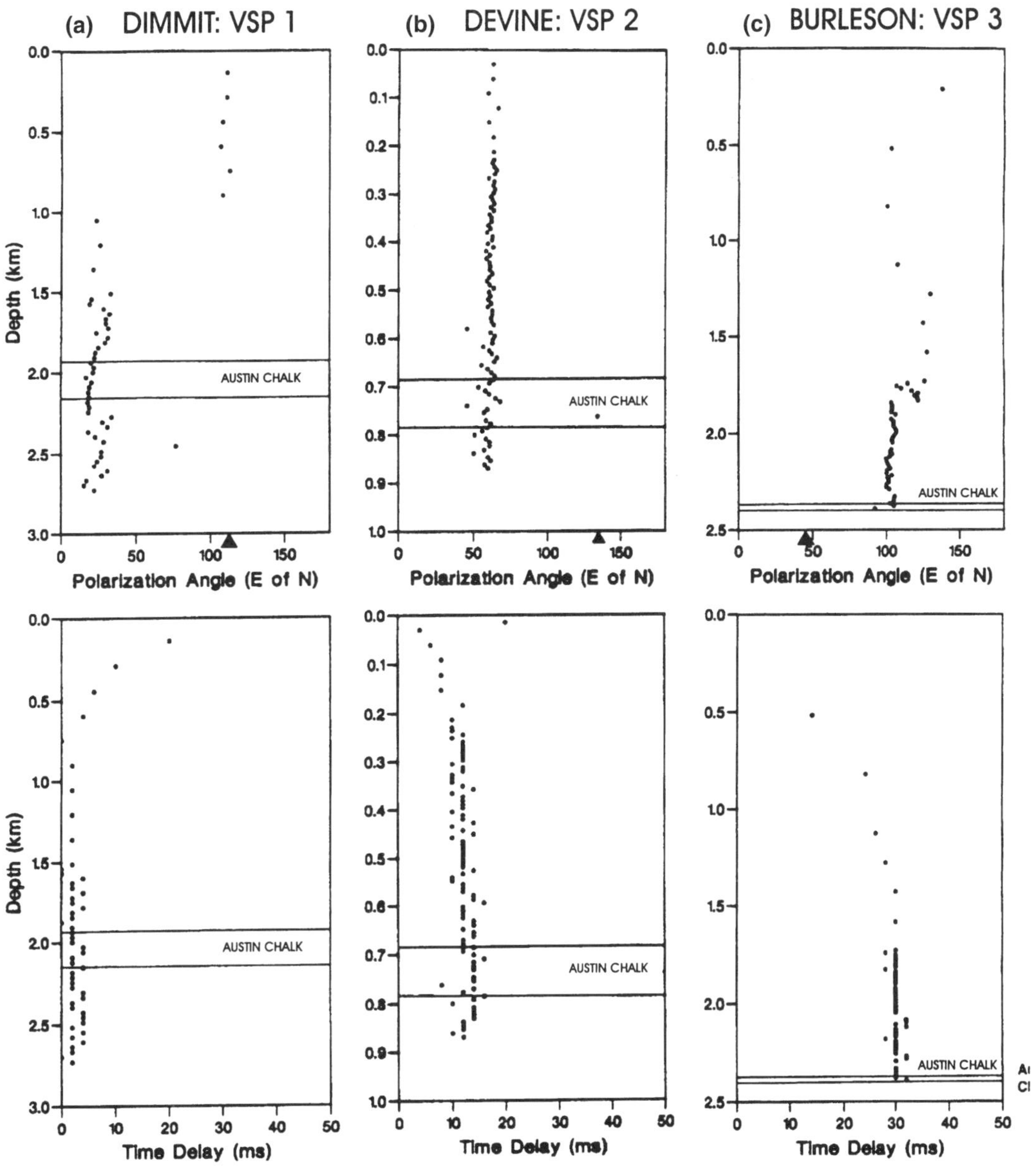

Fig. 17. Comparison of polarizations (top row) and time delays (bottom row) estimated from the three VSPs: (a) VSP 1 from Dimmit, (b) VSP 2 from Devine, and (c) VSP 3 from Burleson. The spurious results around 115° in the polarizations in the Dimmit VSP (top left) correspond to a 90° shift due to multiple solutions of the data matrix. The high time delay values for the shallow geophone levels in the Dimmit VSP are due to a 13 m difference in the in-line and cross-line source positions leading to an apparent time delay between arrivals from each source. Compared to the two-way-traveltime delay in line 1 (where the Dimmit VSP is located; Figure 14a), the time delay in the Dimmit VSP is smaller and a picking error of one or two samples is likely in the surface seismic data. Adapted from Yardley (1992).

It is not possible, however, to determine if the chalk at Burleson is anisotropic using transmitted shear waves because the chalk here is thin (only about 30 m), even though VSP 3 was acquired in a producing well. Field studies show that thin layers are more likely to be heavily fractured (Barthelemy et al., 1992). The study of reflected arrivals may provide the best way of studying anisotropy in thin layers (Thomsen, 1988; Yardley et al., 1991). Reflected amplitudes in the zero-offset VSP data were analyzed to see if they could give a more accurate determination of the anisotropy in the Austin Chalk. Figure 18 shows a comparison of the reflected amplitude ratios from the three VSPs, calculated using the processing sequence previously described.

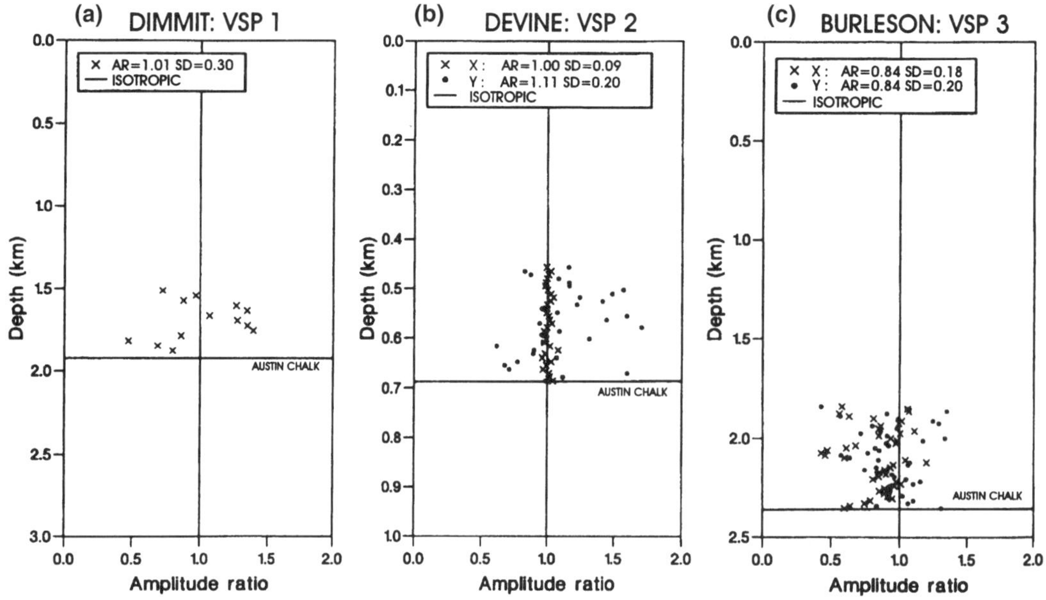

Fig. 18. Amplitude ratios (ARs) calculated for the Austin Chalk reflector at each VSP site: (a) VSP 1 from Dimmit, (b) VSP 2 from Devine, and (c) VSP 3 from Burleson. The ARs were calculated at each geophone level, and all values were averaged to give a single number for each reflector (given as AR values in the figure). Although the results are scattered because of weak reflectors, they cluster around the 1.0 line, as expected for isotropic reflectors. The exception is at Burleson, where both the in-line (*x*) and cross-line (*y*) data give amplitude ratios below unity, implying an anisotropic reflector. After Yardley and Crampin (1993).

Figure 18 shows that the average amplitude ratios for the chalk reflectors for the Dimmit and Devine VSPs are close to unity, indicating that the chalk layer is isotropic as expected from the analysis of transmitted waves. The amplitude ratios at Burleson are less than unity, implying an anisotropic chalk layer. The Burleson VSP is the only one of the three data sets to be collected in a producing well. Because the hydrocarbon reservoirs in the Austin Chalk are fractured, they are expected to be anisotropic to shear wave propagation, which is confirmed by Figure 18c. Analysis of the reflected amplitudes at the three VSP sites indicates that only the Burleson well penetrates an anisotropic chalk layer, which again correlates with the fact that the Burleson well is oil producing. Without detailed knowledge of the velocity structure and pulse shape it is not possible to quantify the amount of anisotropy in the reflecting layer. However, for these sites, a positive correlation is seen between anisotropy and productivity in the chalk layer.

Discussion

Through shear wave studies along the Austin Chalk trend, we have demonstrated that the overall degree of anisotropy, as quantified by differential shear wave traveltime reflection surveys, can be broadly correlated with regional oil production. Dim spots identified from variations in differential shear wave amplitude can be exactly correlated with the fracture swarms encountered by horizontal drilling, as inferred from production records and mud logs.

The thickness of Austin Chalk varies along the trend. If the interval of a target is too small, it is impossible to measure interval polarization and time delays reliably from surface seismic data. Thus, polarization and time delay measurements often give only an estimate of the anisotropy with low spatial resolution. In contrast, variations in amplitude can give a better resolution, as shown for lines 3 and 4.

Both the surface data and VSPs show that anisotropy occurs both above and within the Austin Chalk (Figures 7 and 17). We have assumed here that these two kinds of anisotropy have the same symmetry axis and the same strike, as suggested by the geologic evidence and as indicated by the polarization measurements from the three VSPs (Figure 17). With the uniform, nearly layer-cake structure seen here, we can reliably identify the orientations of the fractures in the chalk by the polarizations of split shear waves observed at the surface.

In the presence of near-surface complications, shear wave splitting in reflection surveys may be subjected to distortions where raypaths pass twice through the surface layers (Yardley and Crampin, 1991). These distortions are much less in VSPs, where the raypaths pass only once through surface layers, and the recorded polarizations of split shear waves are in the undisturbed rock mass within a few wavelengths of the geophone. Despite this, our examples have shown that surface recordings of split shear waves possess sufficient subsurface information for the purpose of delineating fracture reservoirs. However, as the use of shear wave technology expands, it is becoming important to extend this methodology to cases where fracture orientation may change with depth. For VSP surveys, Winterstein and Meadows (1991) demonstrated vertical variations in fracture azimuth. Meadows et al. (1994), Thomsen et al. (1995a), and Chaimov et al. (1995) demonstrated a method and case history of vertical fracture azimuth variation for the surface seismic reflection case. Studies of surface geometry sensitivity and processing methods include MacBeth et al. (1992) and Li (1994, 1995); we believe that more studies should be done in this direction. Another approach may be based on data matrix asymmetry studies (Li and MacBeth, 1995).

Line 3 shows variations in both the fast and slow sections, while the data of line 6 (Figure 8) shows variations only in the slow section. It is possible that this difference is caused by the length of the fractures involved. For microfractures, the resulting anisotropy modifies only the reflectivity of the slow wave, and only the slow wave can then show anomalies in amplitude variations. In the case of large (formation scale) fracturing and sufficiently developed orthogonal fracture sets, the resulting anisotropy may affect both the fast and slow waves. In both cases, the ability to observe the differential amplitude contrasts or the amplitude dim spots requires an appropriate impedance contrast at the interface (Yardley et al., 1991; Li and Crampin, 1993c). With large impedance contrasts, the difference in the amplitudes of shear wave reflections from fractured and unfractured rock could be much less, and other methods may prove more diagnostic of anisotropy (e.g., Thomsen et al., 1995b). Processing procedures can also introduce side effects that degrade the phenomenon. By the use of multicomponent amplitude correction and the LTT, the information contained in stacked amplitude sections is preserved and enhanced. In some cases, broken and scattered polarization events on the stacked polarization section are diagnostic (line 3, Figure 13).

Conclusions

The shear waves in the three reflection lines from Pearsall field show typical characteristics of shear wave splitting in both shot data matrix and color displays of polarization logs. The average fracture orientation of N 40° E in this area agrees with other geologic and geophysical results. In areas where commercial oil production is absent, the average cumulative time delay of the split shear waves is less than 10 ms, corresponding to about 0.5% differential shear wave anisotropy or less. In areas close to major production fields, the average delay is about 30–40 ms, corresponding to about 1.5% shear wave anisotropy. In major production areas, the delay is 50–60 ms, corresponding to about 2.5% anisotropy. Three horizontal wells drilled near line 3 were selected for a more detailed correlation of production with anisotropy. Well W1, drilled about 60° to the fracture strike in an area with relatively high anisotropy, was most productive. Well W2, parallel to the fracture strike and drilled in an area with less anisotropy, was least productive. Well W3, drilled in the same area as W1 but perpendicular to the fracture strike, was moderately productive.

The shear waves in the surface lines from Giddings field show classic amplitude dim spots between the fast and slow waves. The fast wave, polarized parallel to fracture strike, show strong continuous events, while the slow waves, polarized perpendicular to fracture strike, possess weak and dim spots associated with fracture swarms. For line 4, the existence of fracture swarms is indicated by mud logs taken from a line-parallel horizontal well. The regional fracture strike in this area changes from southwest-northeast at the southern edge of the field to east-west in the north.

The VSPs show that Austin Chalk anisotropy may not be detected by polarization and time delay variation in the transmitted downgoing wavefield from the VSPs, but may be detected by variations in amplitude on the reflected upcoming wavefield. The polarization measurements from the VSPs agree with those from the surface data within ~10° error.

The use of multicomponent surface consistent correction and overburden correction helps in preserving and recovering the shear wave anisotropy information associated with the Austin Chalk. The use of the LTT allows the generation of both stacked amplitude sections and stacked polarization logs of the split shear waves. Lateral variations in polarization (often associated with lateral variations in crack geometry) can be identified from the polarization logs, and subsurface structures can also be better imaged than with conventional stack sections. Areas of high production are found to be associated with scattered polarizations and broken polarization events in the stacked polarization logs, which correspond to dim amplitude spots in the stacked fast and slow shear wave sections.

Acknowledgments

We thank Amoco Exploration and Production and British Petroleum for contributing to the Edinburgh Anisotropy Project, and for permission to publish reflection and VSP data from Pearsall and Giddings fields (Amoco) and VSP data from Devine, Texas (BP). We also thank Gareth Yardley and Stuart Crampin for permission to include their VSP study in this review. We thank David Booth and Enru Liu for their comments on the manuscript, and Don Winterstein for his detailed review of the paper. The Edinburgh-based work on this project was supported by the sponsors of the Edinburgh Anisotropy Project and the Natural Environment Research Council. It is published with the approval of the EAP Sponsors and the Director of the British Geological Survey (NERC).

References

Aki, K., and Richards, P. G., 1980, Quantitative seismology: theory and method, I: W. H. Freeman and Co.

Alford, R. M., 1986, Shear data in the presence of azimuthal anisotropy: 56th Ann. Internat. Mtg., Soc. Expl. Geophys., Expanded Abstracts, 476-479.

Ames, G. L., 1990, Oil and gas developments in south Texas in 1989: AAPG Bull., **74**, 144-148.

Barthelemy, P., Tremolieres, P. and Andrieux, J., 1992, Fracture spacing related to bed thickness: geological or mechanical bed limits?: Presented at 4th EAPG Conf., 156-157.

Chaimov, T. A., Beaudoin, G. J., Haggard, W. W., Mueller, M. C., and Thomsen, L. A., 1995, Shear wave anisotropy and coalbed methane productivity: 65th Ann. Internat. Mtg., Soc. Expl. Geophys., Expanded Abstracts, 305-308.

Champion, O. R., 1936, The stratigraphy and structure of the Pearsall field, Frio County, Texas: AAPG Bull., **20**, 1514.

Cliet, C., Brodov, L., Tikhonov, A., Marin, D., and Michon, D., 1991, Anisotropy survey for reservoir definition: Geophys. J. Int., **107**, 417-427.

Corbett, K., Friedman, M., and Spang, J., 1987, Fracture development and mechanical stratigraphy of Austin Chalk, Texas: AAPG Bull., **71**, 17-28.

Crampin, S., 1981, A review of wave motion in anisotropic and cracked elastic media: Wave Motion, **3**, 343-391.

Crampin, S., 1985, Evaluation of anisotropy by shear wave splitting: Geophysics, **50**, 142-152.

Crampin, S., 1987. Geological and industrial implications of extensive dilatancy anisotropy: Nature, **328**, 491-496.

Crampin, S., and Lovell. J. H., 1991, A decade of shear wave splitting in the Earth's crust: what does it mean, what use can we make of it, and what should we do next?: Geophys. J. Int., **107**, 387-407.

Crampin, S., Bush, I., Naville, C., and Taylor, D. B., 1986, Estimating the internal structure of reservoirs with shear wave VSPs: The Leading Edge, **5**, No. 11, 35-39.

Crampin, S., Evans, R., and Üçer, S. B., 1985, Analysis of records of local earthquakes: the Turkish Dilatancy Projects (TDP1 and TDP2): Geophys. J. R. Astr. Soc., **83**, 1-16.

Davis, T., and Lewis, C., 1990, Reservoir characterization by 3-D, 3-C seismic imaging, Silo field, Wyoming: The Leading Edge, **9**, No. 11, 22-25.

Dohr, G. P., 1985, Seismic shear waves, part A: theory: Geophysical Press.

Dravis, J. J., 1979, Sedimentology and diagenesis of the Upper Cretaceous Austin Chalk formation, south Texas and northern Mexico: Ph.D. thesis, Rice Univ.

Gardner, G. H. F., and Harris, M. H., 1968, Velocity and attenuation of elastic waves in sands: Soc. Prof. Well Log Analysts, 9th Ann. Logging Symp., M1-M19.

Ebrom, D., Tatham, R., Sekharan, K. K., McDonald, J. A., and Gardner, G. H. F., 1990, Dispersion and anisotropy in laminated versus fractured media: an experimental comparison: 60th Ann. Internat. Mtg., Soc. Expl. Geophys., Expanded Abstracts, **II**, 1416-1419.

Ensley, R. A., 1984, Comparison of *P*- and *S*-wave seismic data: a new method for detecting gas reservoirs: Geophysics, **49**, 1420-1431.

Hardage, B. A., 1991, Vertical seismic profiling, part A: principles: Pergamon Press.

Hudson, J. A., 1981, Wave speed and attenuation of elastic waves in a material containing cracks: Geophy. J. R. Astr. Soc., **64**, 133-150.

Keith, C. M., and Crampin, S., 1977, Seismic body wave in anisotropic media: Geophys. J. R. Astr. Soc., **49**, 181-243.

Kuich, N., 1989, Seismic fracture identification and horizontal drilling: keys to optimizing productivity in a fractured reservoir, Giddings field, Texas: Trans. Gulf Coast Ass. Geol. Soc., **39**, 153-158.

Layden, R. L., 1976, Big Wells field, Dimmit and Zavala counties, Texas, *in* Braunstein J., Ed,. North American oil and gas fields: Am. Assn. Petr. Geol. Mem. **27**, 145-156.

Lewis, C., Davis, T. L., and Vuillermoz, C., 1991, Three-dimensional multicomponent imaging of reservoir heterogeneity, Silo field, Wyoming: Geophysics, **56**, 2048-2056.

Li, X.-Y., 1992, Shear wave splitting in reflection surveys: theory, methods, and case studies: Ph.D. thesis, Univ. of Edinburgh.

Li, X.-Y., 1994, Amplitude corrections for multicomponent surface seismic data: 64th Ann. Internat. Mtg., Soc. Expl. Geophys., Expanded Abstracts, 1505-1508.

Li, X.-Y., 1995, Detecting polarization changes with depth in multicomponent reflection surveys: 57th Ann. Internat. Mtg., Euro. Asso. Expl. Geophys., Expanded Abstracts, C034.

Li, X.-Y., in press, Fracture reservoir delineation using multicomponent data: Geophys. Prosp., **44**.

Li, X.-Y., and Crampin, S., 1991a, Linear transform technique for analyzing shear wave splitting in four-component seismic data: 61st Ann. Internat. Mtg., Soc. Expl. Geophys., Expanded Abstracts, **1**, 51-54.

Li, X.-Y., and Crampin, S., 1991b, Complex component analysis of shear wave splitting: case studies: Geophys. J. Int., **107**, 605-613.

Li, X.-Y., and Crampin, S., 1993a, Linear transform techniques for processing shear wave anisotropy in four-component seismic data: Geophysics, **58**, 240-256.

Li, X.-Y., and Crampin, S., 1993b, Approximations to shear wave velocity and moveout equations in anisotropic media: Geophys. Prosp., **41**, 833-857.

Li, X.-Y., and Crampin, S., 1993c, Variations of reflection and transmission coefficients with crack strike and crack density in anisotropic media: Geophys. Prosp., **41**, 859-882.

Li, X.-Y., Mueller, M. C., and Crampin, S., 1993a, Case studies of shear wave splitting in reflection surveys in south Texas: Can. J. Expl. Geophys., **29**, 189-215.

Li, X.-Y., Crampin, S., and MacBeth, C., 1993b, Interpreting non-orthogonal split shear waves in multicomponent VSPs: 55th Ann. Internat. Mtg., Euro. Asso. Expl. Geophys., Expanded Abstracts, C009.

Li, X.-Y., and MacBeth, C., 1995, Interpreting data matrix asymmetry and polarization changes with depth in multicomponent reflection surveys: 65th Ann. Internat. Mtg., Soc. Expl. Geophys., Expanded Abstracts, 719-722.

Liu, E., Crampin, S., Queen, J. H., and Rizer, D. W., 1993, Velocity and attenuation anisotropy caused by microcracks and macrofractures in a multiazimuth reverse VSP: Can. J. Expl. Geophys., **29**, 177-188.

Lynn, H. B., and Thomsen, L. A., 1990, Reflection shear wave data collected near the principal axes of azimuthal anisotropy: Geophysics, **55**, 147-156.

MacBeth, C., Li, X.-Y., Crampin, S., and Mueller, M. C., 1992, Detecting lateral variability in crack parameters from surface data: 62nd Ann. Internat. Mtg., Soc. Expl. Geophys., Expanded Abstracts, 816-819.

MacBeth, C., Zeng, X., Yardley, G., and Crampin, S., 1994, Interpreting data matrix asymmetry in near-offset, shear wave VSP data: Geophysics, **59**, 176-191.

Meadows, M. A., Winterstein, D. F., and Szymanski, D. L., 1994, New observations of shear wave polarizations from surface and VSP data at Lost Hills field, California: Presented at the 6th Internat. Workshop on Seismic Anisotropy.

Mueller, M. C., 1991, Prediction of lateral variability in fracture intensity using multicomponent shear wave surface seismic as a precursor to horizontal drilling in the Austin Chalk: Geophys. J. Int., **107**, 409-415.

Mueller, M. C., 1994, Characterization of fracture swarms in chalked reservoirs: Abstracts of the AAPG-EAPG Joint Conf. on Chalk Reservoirs, Copenhagen.

Murtha, P. E., 1988, Estimation of the rotation transformation angle for shear wave data acquired in azimuthally anisotropic regions: Presented at the 3rd Internat. Workshop on Seismic Anisotropy.

Raikes, S. A., 1991, Shear wave characterization of BP Devine test site, Texas: 61st Ann. Internat. Mtg., Soc. Expl. Geophys., Expanded Abstracts, 65-68.

Scott, R. J., 1977, The Austin Chalk–Buda trend of south Texas: Trans., Gulf Coast Assoc. Geol. Soc., **27**, 164-168.

Spencer, T. W., and Chi, H. C., 1991, Thin-layer fracture density: Geophysics, **56**, 833-843.

Squires, S. G., Kim, C. D., and Kim, D. Y., 1989, Interpretation of total wave field data over Lost Hills field, Kern County, California: Geophysics, **54**, 1420-1429.

Stapp, W. L., 1977, The geology of the fractured Austin and Buda formations in the subsurface of south Texas: Trans., Gulf Coast Assoc. Geol. Soc., **27**, 208-229.

Taner, M. T., and Koehler, F., 1981, Surface consistent corrections: Geophysics, **46**, 17-22.

Tatham, R. H., 1982, V_P/V_S and lithology: Geophysics, **47**, 336-344.

Tatham, R. H., 1985, Shear waves and lithology, *in* Dohr, G., Ed., Seismic shear waves, part B: applications: Geophys. Press, 87-133.

Tatham, R. H., and Stoffa, P. L., 1976, V_P/V_S—a potential hydrocarbon indicator: Geophysics, **41**, 837-849.

Tatham, R. H., and McCormack, M. D., 1991, Multicomponent seismology in petroleum exploration: Soc. Expl. Geophys.

Thomsen, L. A., 1988, Reflection seismology over azimuthally anisotropic media: Geophysics, **53**, 304-313.

Thomsen, L. A., Tsvankin, I., and Mueller, M. C., 1995a, Layer stripping of azimuthal anisotropy from reflection shear wave data: 65th Ann. Internat. Mtg., Soc. Expl. Geophys., Expanded Abstracts, 289-292.

Thomsen, L. A., Tsvankin, I., Mueller, M. C., 1995b, Adaptation of split shear wave techniques to coalbed methane exploration: 65th Ann. Internat. Mtg., Soc. Expl. Geophys., Expanded Abstracts, 301-304.

Weeks, A. W., 1945, Balcones, Luling, and Mexia fault zones in Texas: AAPG Bull., **29**, 1733-1737.

Willis, H. A., Rethford, G. L., and Beilanski, E., 1986, Azimuthal anisotropy: occurrence and effect on shear wave data quality: 56th Ann. Internat. Mtg., Soc. Expl. Geophys., Expanded Abstracts, 479-481.

Winterstein, D. F., 1990, Velocity anisotropy terminology for geophysicists: Geophysics, **55**, 1070-1088.

Winterstein, D. F., 1991, Method of layer stripping to predict subsurface stress regimes: U.S. Patent 5 060 203.

Winterstein, D. F., and Hanten, J. B., 1985, Supercritical reflections observed in *P*- and *S*-wave data: Geophysics, **50**, 185-195.

Winterstein, D. F., and Meadows, M. A., 1991, Changes in shear wave polarization azimuth with depth in Cymric and Railroad Gap oil fields: Geophysics, **56**, 1349-1364.

Yardley, G. S., 1992, Identification and characterization of fractured reservoirs by analysis of shear wave anisotropy: Ph.D. thesis, Univ. of Edinburgh.

Yardley, G., and Crampin, S., 1991, Extensive dilatancy anisotropy: relative information in VSPs and reflection surveys: Geophys. Prosp., **39**, 337-355.

Yardley, G., and Crampin, S., 1993, Shear wave anisotropy in the Austin Chalk, Texas, from multioffset VSP data: Case studies: Can. J. Expl. Geophys., **29**, 163-176.

Yardley, G., Graham, G., and Crampin, S., 1991, Viability of shear wave amplitude versus offset studies in anisotropic media: Geophys. J. Int., **107**, 493-503.

Yilmaz, O., 1987, Seismic data processing: Soc. Expl. Geophys., **2**.

Zoback, M. D., and Zoback, M. L., 1991, Tectonic stress field of North America and relative plate motions, *in* Slemmons, D. B., Engdahl, E. R., Zoback, M. D., and Blackwell, D. D., Eds., Neotectonics of North America: Geol. Soc. Am., 339-366.

Chapter 15

Seismic Modeling of a Carbonate Platform Margin (Montagna della Maiella, Italy): Variations in Seismic Facies and Implications for Sequence Stratigraphy

Flavio S. Anselmetti
Swiss Federal Institute of Technology ETH
Zürich, Switzerland

Present address:
University of Miami
Rosentiel School of Marine and Atmospheric Science/MGG
Miami, Florida, U.S.A.

Gregor P. Eberli
University of Miami
Rosentiel School of Marine and Atmospheric Science/MGG
Miami, Florida, U.S.A.

Daniel Bernoulli
Swiss Federal Institute of Technology ETH
Zürich, Switzerland

Abstract

Synthetic seismic sections across the exposed Cretaceous–Miocene carbonate platform margin of the Montagna della Maiella (central Italy) explain the seismic facies of a carbonate platform margin system and show the limitations of relating seismic sequences to depositional sequences. To define a layered impedance model, velocities and densities of 186 minicores from all major outcropping lithologies were determined. The impedance model was converted to synthetic seismic data by applying a computer-simulated model that uses the normal incidence ray-tracing method at variable frequencies, amplitude gains, and noise levels. The resulting synthetic seismic sections show a mostly transparent platform that is onlapped along the escarpment by a succession of high-amplitude slope reflections. The different reflectivities of platform and slope can be explained by their differences in impedance contrasts. The small impedance contrasts within platform carbonates results in weak reflections nearly indistinguishable from noise, whereas the large impedance contrasts within the slope and basin carbonates yield coherent high-amplitude reflections. The seismic image with incoherent to transparent platform, high-amplitude slope reflections, and recognizable prograding units is similar to observed seismic data across other steep carbonate platform margins (e.g., Great Bahama Bank and Adriatic Sea).

In outcrop, seven unconformity-bounded supersequences were mapped. Comparison with the synthetic seismic section shows that, at a frequency of 20 Hz, only five of these depositional supersequences can be recognized using seismic unconformities. With an increase in frequency, an increasing number of unconformities become visible, and at a frequency of 60 Hz, all seven are imaged. The synthetic seismic sections also reveal that some of the seismic unconformities are pseudo-unconformities—they do not exist in outcrop, but the seismic image shows erroneous or nonexistent geometric patterns. These are a result of the thinning of layers below seismic resolution. These observations document the problem of seismically imaging depositional sequences. Depending on the dominant frequency, an erroneous number of sequences might be interpreted. This limitation must be taken into account when making sequence stratigraphic interpretations based solely on seismic information.

Introduction

The relationship between the geologic rock record and its seismic response is not simple and straightforward. Seismic profiles are purely physical images of rock sections and do not necessarily image geologically relevant characteristics of the rocks. In a limited way, seismic data can be directly compared to the rock record if cored drill holes are available, but these provide only a 1-D view of the geology. The only procedure to evaluate the relationship between rock bodies and their seismic response is the technique of seismic modeling.

The purpose of seismic modeling is to evaluate the seismic response of a geologic setting by computer-simulation of a seismic survey. Basic principles, techniques, and applied case studies of seismic modeling are discussed by Neidell and Poggiagliolmi (1977) and Fagin (1991). Several examples of modeling describe the relationships between observed outcrop geology and its synthetic seismic response and discuss applications of seismic modeling in the context of seismic stratigraphy (Galloway et al., 1977; Meckel and Nath, 1977; Rudolph et al., 1989; Schlager et al., 1991; Biddle et al., 1992; Campbell and Stafleu, 1992; Stafleu and Schlager, 1993; Stafleu and Sonnenfeld, 1994; Stafleu et al., 1995) or model subsurface data so as to better interpret existing seismic lines (Middleton, 1987; Rudolph and Greenlee, 1991). In addition, seismic modeling can improve the interpretation of real seismic lines. For example, 3-D modeling can show the origin and position of out-of-plane reflections so that they can be identified on 2-D seismic sections (Fagin, 1991; Stäuble and Pfiffner, 1991).

In this study, seismic modeling is performed in 2-D to simulate the seismic response of the transition from an isolated carbonate platform to the adjacent pure carbonate slope of the Montagna della Maiella in central Italy. The seismic modeling is a combined result of fieldwork, laboratory analyses, and computer simulations. Fieldwork was performed to obtain information on the lithologic content and geometric stratal pattern of the depositional sequences (Vecsei, 1991; Eberli et al., 1993; Sanders, 1994). Velocity data from nearly 200 minicores provide the petrophysical base to establish an impedance model that can be converted into synthetic seismic data.

The Maiella platform, ranging in age from Jurassic to Miocene, provides excellent outcrops on a seismic scale, which make the area suitable for approaching two main objectives. First, we wish to evaluate the seismic facies of escarpment-bounded platforms and their adjacent slope and basin deposits. To date, it is not well known why many horizontally layered carbonate platforms show incoherent to chaotic seismic facies. It has been speculated that pervasive dolomitization or strong karstic overprints may be responsible for their low seismic reflectivity (Popenoe et al., 1984; Fontaine et al., 1987). Our modeling will show that the low range in impedance measured from platform samples is not sufficiently high to result in high seismic reflectivity. The characteristics of the seismic image of an escarpment, as seen in the modeled synthetic seismic sections, may help to interpret whether such escarpments are fault-controlled or growth structures (G. P. Eberli and J. Hurst, pers. comm., 1995).

Our second objective is to test the concept of seismic sequence stratigraphy in a carbonate platform-to-basin setting. The concept of sequence stratigraphy performed on seismic sections is often used in sedimentologic and stratigraphic interpretations of seismic data. This study of modeling addresses the problems caused by limited seismic resolution that may not be sufficient to correctly image depositional sequences. We show that several misinterpretations can be made by performing sequence stratigraphic analyses solely with seismic data, particularly at low seismic frequencies.

Methods

Synthetic seismic sections are calculated by establishing a petrophysical model of the rock section, which consists of several layers with characteristic geometries and impedances. This model is then converted to seismic data using computer simulations by applying a commercially available software package (Figure 1).

Impedance Measurements

Velocities of outcrop samples were measured on 1 or 1.5 in. diameter plugs at ultrasonic frequencies (600–1200 kHz) with the pulse-transmission technique (Birch, 1960; Wang, Chapter 3 of this volume) under variable confining and pore fluid pressures to simulate in-situ conditions of a buried sediment. The samples were collected on several vertical measured sections and cover all major units. Dry bulk densities were determined by weighing the oven-dried plugs and by measuring the diameter and length of the miniplugs. Saturated bulk densities were calculated from dry bulk density by adding the weight of the water-filled pore space. No corrections were made to convert the measured ultrasonic velocities into velocities at seismic frequencies. The velocity dispersion is assumed to be insignificant compared to the wide range of measured velocities, with the

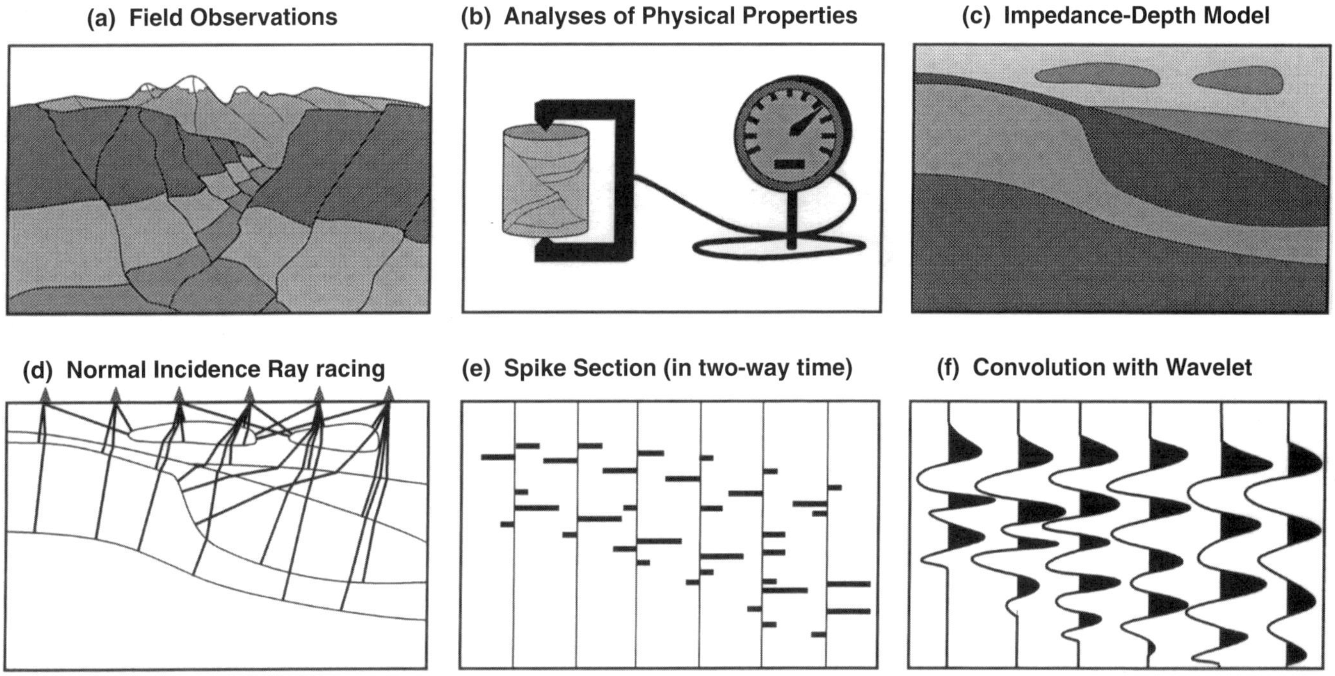

Fig. 1. Schematic steps of a study in seismic modeling combining (a) observations and sampling in the field with (b) laboratory analyses of physical properties and (c–f) computer simulations.

highest velocities being more than twice as fast as the lowest (Anselmetti, 1994). A general correction for dispersion would lower all velocity values in the model by about 5–10% (Winkler, 1986; Wang and Nur, 1990) and would thus have only a minor impact on the seismic reflection pattern. This pattern is influenced mainly by impedance contrasts between layers rather than by absolute impedance values. Furthermore, the decision to take the velocities at ultrasonic frequencies results in a negligible error compared to other assumptions that are made, such as sample selection and layer definition.

Seismic Modeling Technique

Normal Incidence Ray Tracing

Synthetic seismic sections are calculated by simulating the refraction and reflection of seismic signals initiated at the top surface of the model. The raypaths are constructed through the layers of the model using the normal incidence ray-tracing method (Fagin, 1991), which uses only rays that arrive and are reflected at a 90° angle to the reflecting target surface (Figure 1d). Downgoing and upcoming rays are thus identical, and the resulting seismic data represent a stacked but unmigrated section. This normal incidence ray-tracing method is effective in simulating traces of a common midpoint (CMP) stacked seismic survey. For the purposes of this study, normal incidence ray tracing results in reliable results even though no time-consuming prestack modeling was performed. Unlike the often used *vertical* incidence method, which only uses vertical rays, the *normal* incidence ray-tracing method calculates the true trace of a ray, which is influenced by the 2-D geometries of the layers. The signals of all rays that arrive at one station are added to the final seismogram of that station. In some cases, several reflections from the same target horizon are recorded at the same geophone station resulting in a focusing of rays, which is a characteristic pattern of irregularly dipping surfaces (see Figure 12).

Virtual geophones record the upcoming signal as reflection events (spikes) as a function of two-way traveltime at previously defined CMPs. A reflected signal is detected at the surface of a model if it arrives within a capture radius around the virtual geophone. As a default, the modeling software uses a defined capture radius equal to 15% of the CMP spacing. This value ensures the separation of rays into adjacent groups and produces reasonably accurate traveltimes, while preserving the execution speed of the simulation. After ray tracing, the program creates a spike section for every CMP (Figure 1e). For every trace, this spike seismogram includes information on the two-way traveltime, reflection coefficient, and polarity of the reflection events.

Wavelet Type and Noise

The spike seismograms are convolved with seismic wavelets of 20, 30, 40, 50, and 60 Hz dominant frequencies. The lower frequencies of 20 and 30 Hz are common in hydrocarbon exploration, whereas the higher frequencies of up to 60 Hz represent high-resolution surveys. A zero-phase Ricker wavelet was chosen to allow the highest possible resolution (Schönberger, 1974). This wavelet type has an amplitude peak in the center of the wave train (Figure 2). The resulting reflected wavelet from a layer with an impedance contrast has its main peak centered on the two-way traveltime to that reflector.

To make the synthetic seismic section and the interpretation more realistic, the seismograms were disturbed by adding background noise at 10% of the average registered amplitudes. Unlike *structural modeling* studies (Fagin, 1991), in which the main purpose is the identification and origin of out-of-plane reflections, *stratigraphic modeling* can be improved by added noise, as shown in this study. The addition of noise is useful because the main purpose in this case is the realistic imaging of reflection patterns from sedimentary sections. Such imaging depends on the interference of several reflections having variable amplitudes. Low-amplitude reflections are sensitive to the signal-to-noise ratio (S/N) and might only be visible on clean, undisturbed sections, but they remain undetected in more realistic sections with added background noise (see Figure 11). These more realistic synthetic sections are needed for testing seismic sequence stratigraphy, which is particularly dependent on the resolution and amplitude of reflections. Added background noise can thus be significant.

Gain Control

The final seismic image also depends on the type and amount of amplitude gain applied to the seismograms. In addition to the automatically adjusted amplitude loss due to geometric spreading, the software offers two options to control gain at increasing traveltimes. First, an automatic gain control (AGC) amplifies the signal to keep the sum of amplitudes constant within a time window of desired length. This window is moved along the seismograms with the result that low-amplitude events are amplified. Second, time-ramp scaling has a power function that amplifies the amplitudes at longer traveltimes. (See Figure 11 for differences between these two methods.) For interpreting the synthetic seismic sections in this study, we chose the time-ramp

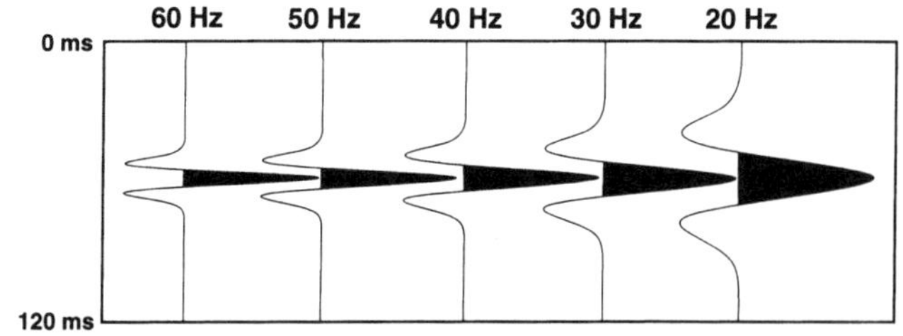

Fig. 2. Zero-phase Ricker wavelets of different frequencies used in this study. With positive polarities, a boundary at an increase in impedance results in a reflection with a black peak centered on the reflecting surface.

scaling method with an exponent of 1.2. This method produces pseudo-true amplitudes and seems more realistic for simulating the pattern of low-amplitude reflections from interfaces with low impedance contrasts.

Geologic Setting of Montagna Della Maiella

The Montagna della Maiella in central Italy, part of the Apulian carbonate platform (Figure 3), is the edge of an isolated Bahama-type platform that was situated along the southern margin of the Mesozoic Tethys ocean. Its seismic-scale exposures in a frontal anticline of the Apennine thrust belt cover an area 28 km long by 10–15 km wide with a relief of more than 2500 m. It offers excellent insights into depositional geometries and stratigraphic controls across a carbonate platform margin (Accarie, 1988; Vecsei, 1991; Bernoulli et al., 1992; Eberli et al., 1993; Sanders, 1994). Our modeled cross-section covers 14 km of this platform margin (Figure 4).

The Jurassic–Cretaceous evolution of the carbonate platform was characterized by a steep, nondepositional escarpment that separated shallow-water platform carbonates from onlapping slope sediments. Progradation of carbonates in the Late Cretaceous transformed this steep relief between platform and adjacent slope into a more gradual ramp-like margin. After emersion in latest Maastrichtian–earliest Paleocene time, subsequent deepening resulted in backstepping submarine erosion and redeposition in an upper slope regime that was established along the platform margin. In the late Eocene–early Oligocene, skeletal sands and reefal units prograded over the former slope sediments and again shifted the platform margin basinward. In the Miocene, the whole system was characterized by a wide, shallow shelf that was dominated by bioclastic sedimentation. Finally, the whole platform-basin system was uplifted during the Apennine orogeny.

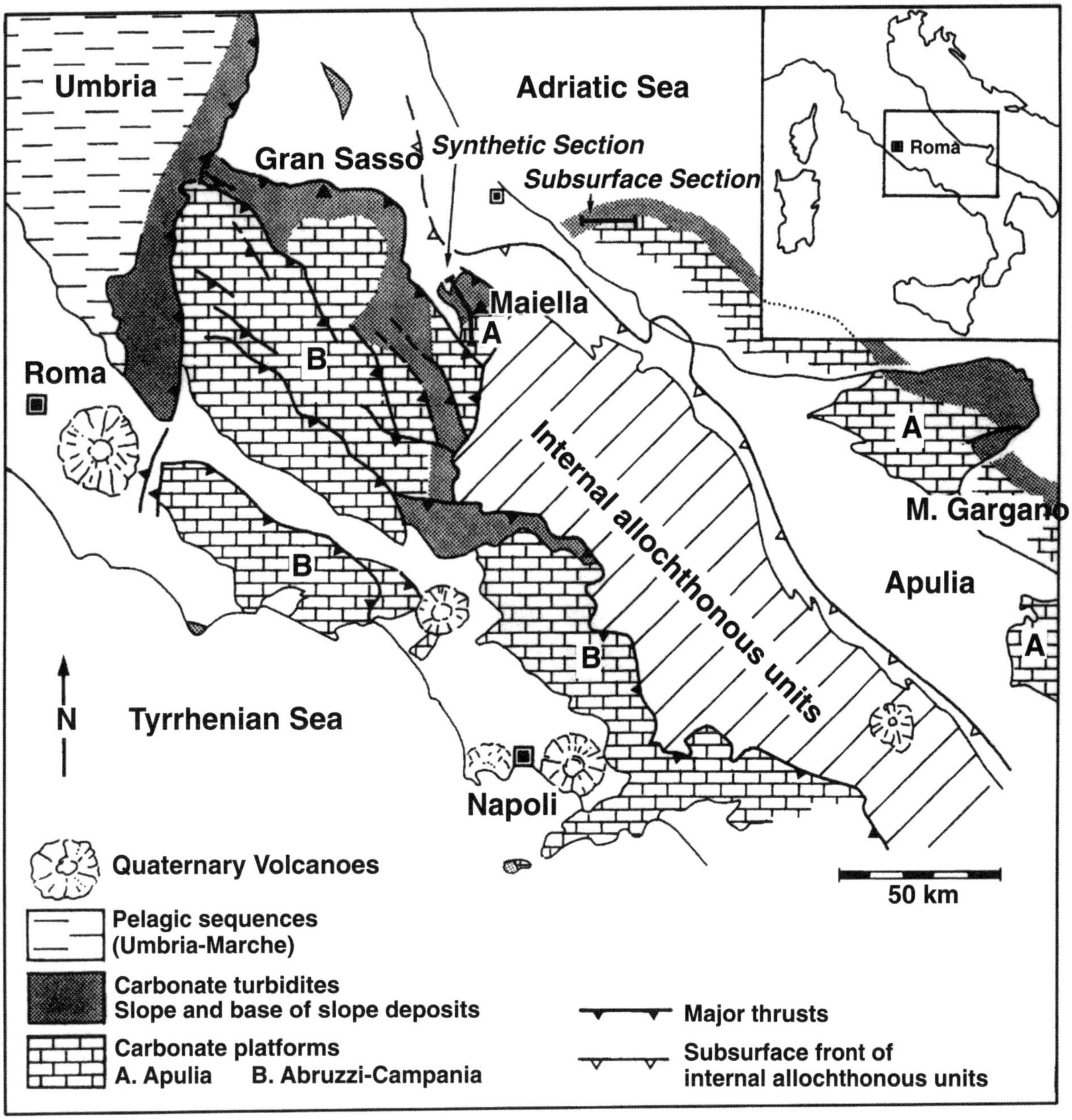

Fig. 3. Location map of the Montagna della Maiella in central Italy and the relative positions of the carbonate platforms of Apulia (A) and the Abruzzi–Campanian limestones of the southern Apennines (B). The Apulian platform is partially covered by foreland deposits and allochthonous units of the internal southern Apennines and only marginally involved in Tertiary thrusting in the Montagna della Maiella. The seismic section of the Adriatic subsurface images the continuation of the Maiella platform margin.

Depositional Supersequences

Seven depositional supersequences (including their substrate with an unknown lower boundary) can be distinguished on the basis of field observations of geometric and lithologic characteristics (Figure 5) (Vecsei, 1991; Bernoulli et al., 1992; Eberli et al., 1993; Sanders, 1994; Vecsei et al., in press). The six sequence boundaries that partition the rocks into these seven supersequences are recognized in the field by geometric unconformities and typically by prominent changes in the sedimentary systems. In proximal areas, signatures of subaerial exposure are an additional characteristic of these boundaries. The duration of observed sequences varies between 7 and 21 m.y. (Vecsei, 1991; Vecsei et al.,

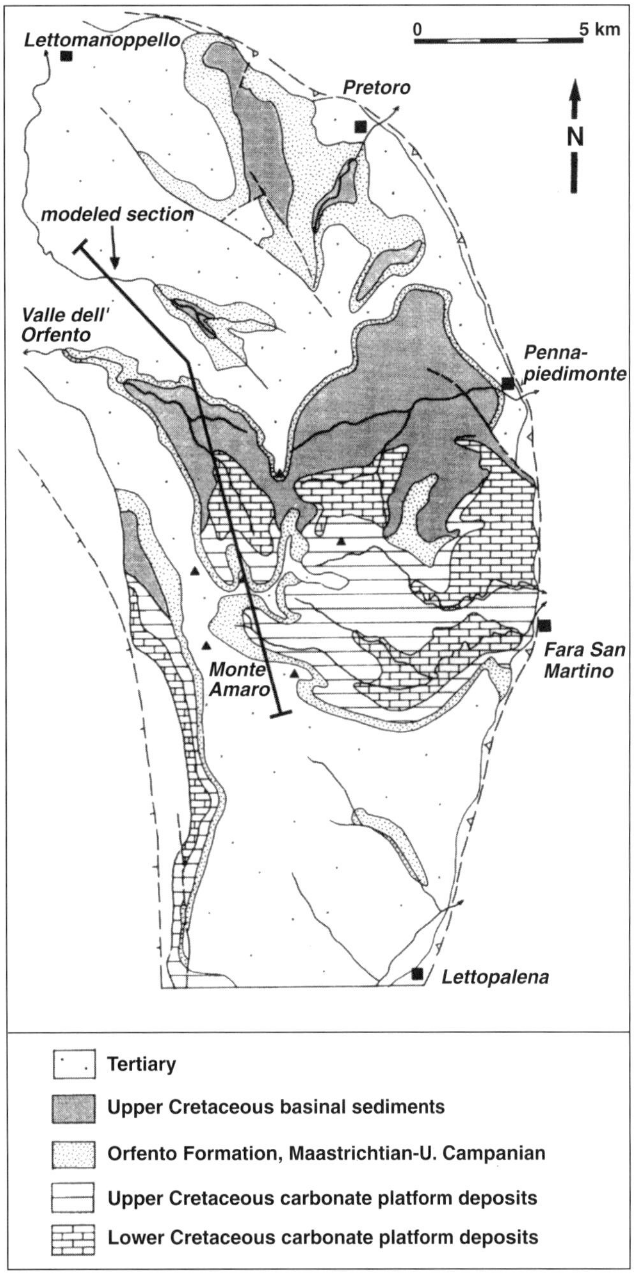

Fig. 4. Simplified geologic map of the Montagna della Maiella with the modeled cross-section. Modified from Carta Geologica d'Italia (1970) and Eberli et al. (1993).

1995), suggesting that the sequences are second-order and are the equivalent of supersequences (Haq et al., 1988). These supersequences can be subdivided into several smaller scale sequences, possibly third-order, which themselves consist of smaller scale fourth-order parasequence sets.

The following description of formations and supersequences is based on Vecsei (1991). Figure 5 shows a general cross-section of all depositional supersequences, which can be recognized in a panoramic view of the western Valle dell'Orfento (Figure 6).

Upper Jurassic–Lower Cretaceous (Supersequence 0)

Upper Jurassic–Lower Cretaceous platform rocks form the oldest outcropping formations of the Montagna della Maiella (Morrone di Pacentro Formation). Although its sequential development and lower boundary are not known, this substrate, which underlies the other six supersequences, is named "supersequence 0" because its mapped upper boundary marks the top of this oldest succession. In the modeled section, supersequence 0 consists of Lower Cretaceous carbonates that were deposited in restricted shallow water. On the platform, this supersequence is truncated by the middle Cretaceous unconformity and is cut basinward by the escarpment and onlapped by slope sediments of supersequence 1.

Supersequence 1

On the platform, the base of supersequence 1 (the lower Cenomanian–lower Campanian Cima delle Murelle Formation) is the middle Cretaceous unconformity, a distinct exposure horizon. This irregular surface is characterized by numerous karstic features such as speleothems, breccias, and bauxitic accumulations. The overlying successions of platform sediments are mainly peritidal carbonates and grainstones that alternate with rudist biostromes. The abundance of biostromes increases upward and toward the escarpment, but as in proximal supersequence 0, no massive rim facies or reefs have been observed.

During the Late Cretaceous, the platform escarpment reached a height of 1000 m with an average inclination of about 30–35°. It contains several niches that form concave structures in the platform wall that were probably inherited features from the Early Jurassic rifting phase of the continental margin, such as edges of fault blocks (Bernoulli and Jenkyns, 1974; Bice and Stewart, 1990; Eberli et al., 1993). Sediments produced on the platform bypassed the upper escarpment and were deposited as an onlapping wedge at its foot (Figures 5, 6, and 7d). The irregular and undulating surface of the escarpment and the lack of a pronounced rim facies in the preserved platform sediments indicate that the escarpment was periodically eroded and dismantled by gravitational collapse. This hypothesis is supported by the appearance of older cemented components of the platform in the onlapping megabreccias.

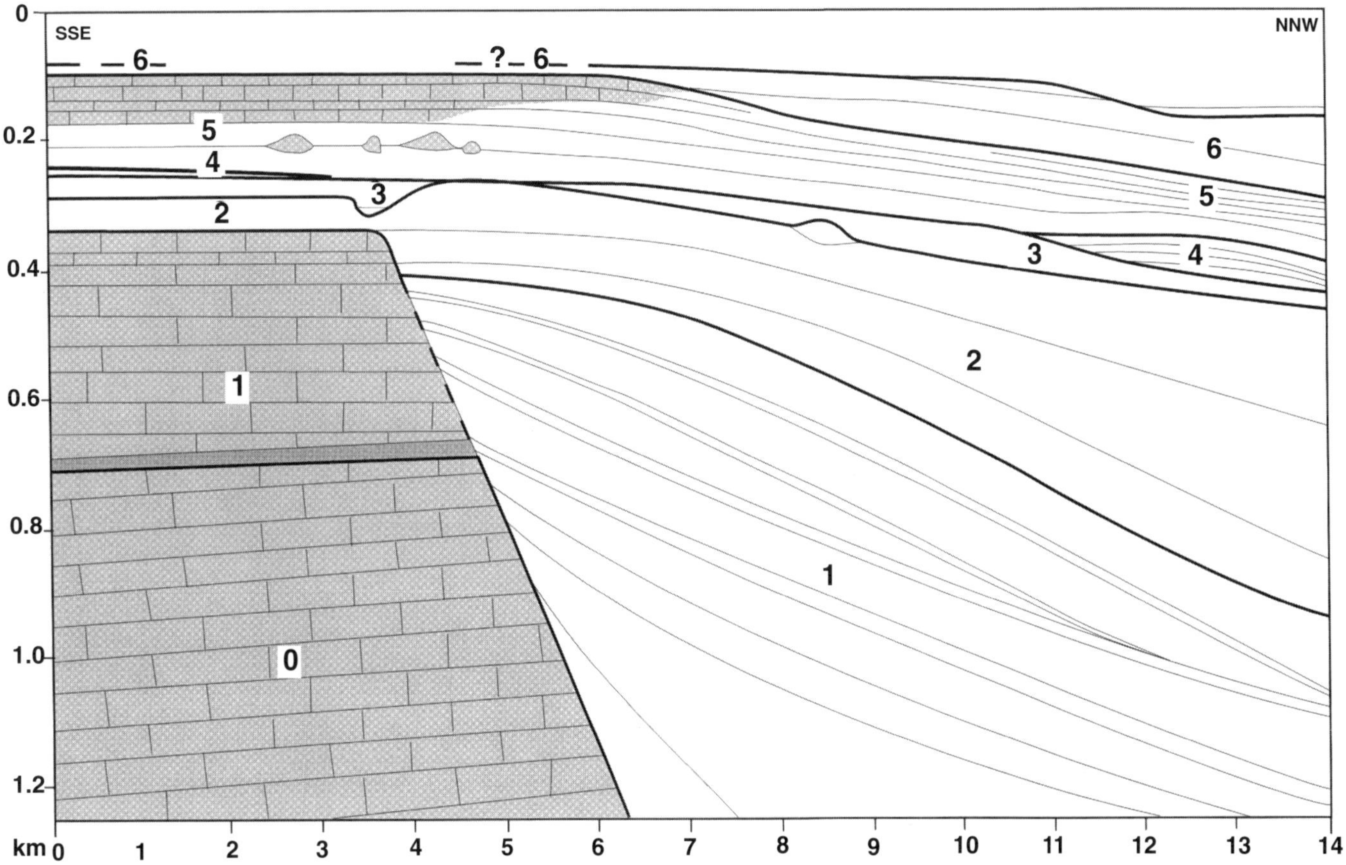

Fig. 5. Schematic cross-section of the Maiella carbonate platform. Seven second-order supersequences (0–6) can be distinguished (Vecsei, 1991). See text for discussion. Bold lines are supersequence boundaries; shaded areas indicate shallow-water carbonates.

The lower basinal unit of supersequence 1, the Valle del Inferno Unit (middle Cenomanian–Turonian), is a succession of breccias and bioclastic grainstones that form an internally cross-bedded onlapping wedge at the base of the escarpment. The upper unit is the Turonian–upper Campanian Tre Grotte Formation, an alternation of carbonate turbidites and periplatform ooze ("scaglia" of Italian authors) with intercalated megabreccias. The breccias onlap onto the platform sediments along the escarpment; they are often amalgamated and contain lithified blocks and pebbles eroded from the shallow-water platform. They are interpreted to be deposited during lowstands of sea level (Vecsei, 1991). Consequently, each megabreccia is considered to be the base of a third-order sequence.

Supersequence 2

The escarpment separating the platform from the adjacent basinal sediments was buried during deposition of supersequence 2 (upper Campanian–Maastrichtian Orfento Formation). This unit is characterized by aggrading carbonates that are overlain by prograding sediments of the upper slope, which formed the first aggradation. Then prograding units deposited on the upper slope, which finally led to a gentle ramp-like platform-to-basin transition. The most typical lithology of supersequence 2, called "Calcari Cristallini," is a well-sorted bioclastic sand of mainly rudist debris with high porosities and almost no cementation. These sands overlay a karstic surface at the top of the platform (Mutti, 1995). Within this supersequence, several horizontally and vertically stacked breccia bodies can be recognized in proximal positions that define lower order depositional sequences (Eberli et al., 1993; Mutti et al., 1994). Unlike the underlying megabreccias, these breccias do not extend far into the basin. The distal sediments of the Orfento Formation are well-sorted, fine-grained rudist sandstones that have an average porosity of about 30%.

Supersequences 3–5 (Slope)

The slope deposits of these supersequences (Paleocene–middle Eocene Santo Spirito Formation) consist of

Fig. 6. (a) View of the western side of Valle dell'Orfento showing major features of the Maiella impedance model (Figure 10). (b) Same view with interpreted geologic units. In the foreground is the Cretaceous platform, which is onlapped along the escarpment by slope sediments. The overlying uppermost Cretaceous sediments level the relief between platform and slope. The mountain top is formed by upper Eocene slope and lower Oligocene reef deposits, which prograde over slope and shelf units. Vertical relief from escarpment at lower right to mountain top is ~1200 m.

Fig. 7. Outcrop photographs of various architectural elements of the Maiella impedance model (see Figure 10). Parts a, b, and d show formations with clear seismic-scale geometries and characteristic impedance distributions, while the section in part c is difficult to model because thin layers form no impedance pattern at a seismic scale. (a) Succession of cliff-forming bryozoan limestones (supersequence 6) with intercalated hemipelagic marls in Valle San Bartolomeo. The low-velocity marls (40 m thick) form the grassy slope between the vertical cliffs (arrows) of lower (50 m thick) and upper bryozoan limestones (20 m thick). (b) Patch reefs (supersequence 5) at Pesco Falcone. Two high-velocity coral buildups are surrounded by low-velocity carbonate sands (Figure 9g). The height of the larger patch reef is ~40 m. (c) Slope sediments (supersequences 4 and 5) in Valle dell'Orfento near Decontra (height of cliff ~100 m) consisting of alternating pelagic limestones, turbidites, and breccias. The breccias cut the underlying beds (arrows) and mark the base of higher order depositional sequences. The small-scale layering provides no clear impedance distribution. (d) Platform escarpment (from lower left to upper right) with onlapping megabreccias at Monte d'Ugni (arrows). Megabreccias and platform are cliff-forming, whereas background slope sediments are tree-covered. Horizontal width of picture is ~1.5 km.

alternations of thin-bedded, fine-grained pelagic mudstone to wackestone with abundant planktonic foraminifera ("scaglia"), turbidites, and breccias (Figure 7c). The supersequences are separated using geometric unconformities that coincide with biostratigraphically dated hiatuses (Vecsei, 1991; Vecsei et al., in press). The breccias appear in bundles and may be amalgamated. Lithoclasts in many of these breccias are the products of erosion of the exposed platform top and slope. Meteoric cements in these clasts suggest that they were eroded during sea level lowstands (Sanders, 1994). Thus, they are interpreted to be the base of third-order sequences (Vecsei, 1991). The turbidites consist mainly of skeletal grains with abundant nummulitids in a micritic or sparitic matrix.

Supersequences 3 and 4 (Platform)

On the platform, supersequences 3 and 4 occur only as relics (lower part of the Paleocene–middle Eocene Santo Spirito Formation). The weakly indurated rudist sandstones of supersequence 2 are locally covered by massive breccias of supersequence 3. These breccias are intercalated with pelagic limestones and were deposited in channels that incised the platform. The breccias were deposited mainly by mass flow processes during different phases in the late Paleocene–middle Eocene (Sanders, 1994; Moussavian and Vecsei, 1995). In-situ reef and shelf limestones occur only as local relics.

Supersequence 5 (Platform)

During the late Eocene, shallowing resulted in deposition of progradational units of skeletal sands in which several small patch reefs developed (Figure 7b). They show convex-upward shapes and contain large corals and hydrozoans. The surrounding bioclastic sands are grainstones-packstones that contain benthic foraminifera, algae, bryozoans, and other skeletal fragments (Vecsei, 1991; Sanders 1994).

In the early Oligocene, coral reefs (Pesco Falcone Formation) bloomed and eventually prograded about 4 km basinward over these fine sands (Eberli et al., 1993). South of the former escarpment, where relics of younger sediments from supersequence 6 are only locally preserved (Sanders, 1994), the reefal units of the Pesco Falcone Formation usually form the modern mountain crests and thus the upper limit of the outcropping section (Figure 6).

Supersequence 6

In the Miocene, the Maiella was characterized by a wide, shallow shelf on which supersequence 6 was deposited (Bolognano Formation). Rocks belonging to this supersequence are found mainly overlying the former slope. On top of the prograding reefs of sequence 5, they are only occasionally preserved. The lower part of supersequence 6 is formed by two higher order sequences. Both contain one bioclastic grainstone unit (bryozoan limestone), overlain by marly limestones indicating deeper water deposition (Figure 7a). The top of the supersequence is not continuously preserved. Where present, it is made up of *Lithothamnium* limestones and a marly interval (Vecsei, 1991). This supersequence is onlapped by evaporitic sediments of the Gessoso-Solfifera Group of late Messinian age (Crescenti et al., 1969). These sediments were deposited during the evaporitic drawdown of the Mediterranean Sea and are part of the next supersequence.

Physical Rock Properties

The link between geology and seismic images is the physical properties of the rocks, in particular the acoustic impedance. Several published studies have investigated acoustic properties in carbonate sediments and rocks (Rafavich et al., 1984; Wang et al., 1991; Anselmetti and Eberli, 1993; Kenter and Ivanov, 1995) and provide the basic concepts for relating sonic velocity to rock lithology. We analyzed the sonic velocities and densities of 186 drilled minicores of outcrop samples in the laboratory (Figures 8, 9 and Table 1) to establish an impedance model (Figure 10 and Table 2).

The measured velocities range from 3100 to 6500 m/s, and saturated bulk densities vary from 2.0 to 2.7 g/cm^3 (Figures 8 and 9a). The variation in sonic velocity is a result of the combined effect of variations in depositional lithology and diagenetic alterations (Anselmetti and Eberli, 1993), and the range of velocity is thus dependent on the depositional and diagenetic regime. Carbonates deposited on the shallow-water platform show only a narrow range in velocity, mainly caused by a high diagenetic potential that results in densely cemented rocks and thus uniformly fast velocities (Figure 8). Carbonates from the slope and basin have a much wider range of velocity values, reflecting the variable lithologies with partially lower diagenetic potentials. This difference in velocity ranges, and thus impedance ranges, explains the different seismic facies of the different carbonate environments. The next section gives examples of velocity distributions within the various carbonate settings. We describe how the different carbonate units can be grouped into layers with characteristic acoustic impedance values and geometries so as to define an impedance model.

Impedance Model

The most crucial step in seismic modeling is to establish a realistic petrophysical model of the geometric distribution of acoustic impedance. It is evident that such a model can represent only a simplified version of the true impedance distribution of the rock section, which in reality consists of countless fine layers with distinct impedances. The seismic resolution, which is limited to about one-fourth of the signal's wavelength (Sheriff, 1977, 1985), allows us to simulate the physical properties with a simplified model on a seismic scale.

Methods

The impedance model used to calculate the synthetic sections measures 14 km wide by 1.5 km deep and consists of 75 layers with characteristic geometries (Figure 10) and impedances (Table 2). The overall geometries of the layers are based on the observed geometries of the seven supersequences of Vecsei (1991). The

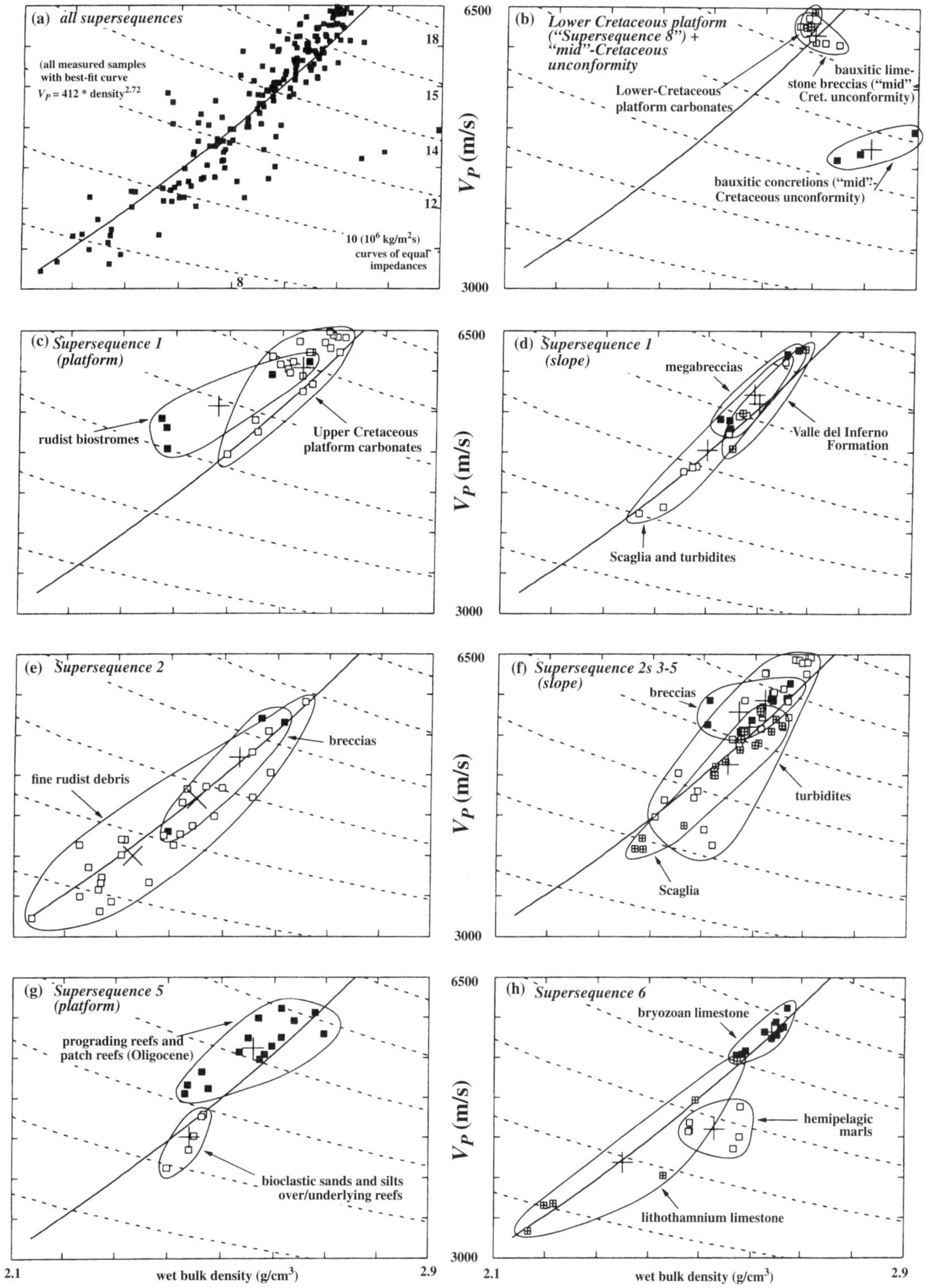

Fig. 8. (a) Velocity-density plot of 186 measured Maiella samples from all supersequences showing best-fit equation (solid line). The dashed lines mark curves of equal impedances (calculated for the density of calcite) that are the basis for defining the velocity-density values for the layers of the impedance model. (b–f) The velocity-density pairs are grouped into lithologic categories within the individual supersequences. The plus marks are the statistically averaged velocity–density values used to define the impedance value of the individual lithologies in the impedance model (Table 2 and Figure 10). The two Xs in part (e) are the arbitrarily chosen impedance values for the fine-grained Calcari Cristallini and represent the wide range of measured samples. The Lower Cretaceous platform is represented in the model by two arbitrarily chosen values that lie close to each other; only their average value is marked by a plus in part (b).

Table 1. Average Values of Porosity, Density, and V_P for Various Lithologies

Lithology[1]	Super-sequence	No. of Samples	Average porosity (%)	Average density (g/cm^3)	Average V_P (m/s)
Lower Cret. platform (Morrone di Pacentro Fm)	0	4	1.6	2.69	6315
"mid"-Cretaceous unconformity (breccias)	1	6	3.7	2.70	6160
"mid"-Cretaceous unconformity (bauxites)	1	3	12.5	2.81	4754
Up. Cretaceous platform (Cima delle Murelle Fm)	1	21	4.6	2.64	6044
Up. Cret. rudist biostromes (Cima delle Murelle Fm)	1	5	14.2	2.48	5566
Valle del Inferno Unit	1	3	6.7	2.60	5600
Up. Cretaceous megabreccias (Tre Grotte Fm)	1	5	7.5	2.59	5713
Up. Cretaceous turbidites and Scaglia (Tre Grotte Fm)	1	9	12.7	2.50	5021
Calcari Cristallini (Orfento Fm)	2	27	19.6	2.38	4283
Calcari Cristallini breccias (Orfento Fm)	2	3	11.2	2.53	5217
Paleogene reefs and patch-reefs (Pesco Falcone Fm)	5	15	9.0	2.56	5618
Paleogene nummulitid sands/silts (Santo Spirito Fm)	5	5	'6.1	2.44	4509
Paleogene breccias (Santo Spirito Fm)	3–5	12	5.9	2.62	5929
Paleogene turbidites (Santo Spirito Fm)	3–5	26	7.2	2.59	5602
Paleogene Scaglia (Santo Spirito Fm)	3–5	19	10.0	2.55	5128
Bryozoan limestones (Bolognano Fm)	6	10	4.4	2.64	5817
Hemipelagic marls (Bolognano Fm)	6	6	10.6	2.53	4598
Lithothamnium limestone (Bolognano Fm)	6	6	21.9	2.35	4186
Total samples	0–6	186	10.1	2.55	5321

[1] Formation names after Vecsei (1991).

selected cross-section represents a cut through the platform–basin transition that parallels the western side of Valle dell'Orfento (Figures 4 and 6). Several profiles were projected orthogonally into the plane of the cross-section. Faults and tectonic dips were corrected so that the resulting geometry represents the geometric relationships at the end of deposition.

The layers were defined petrophysically by grouping rocks with characteristic velocity and density pairs. Ideally, the thicknesses of the layers should be comparable to, or thinner than, the resolution of the seismic section. However, due to the variety of impedances and impedance ranges in the different parts of Montagna della Maiella, the thicknesses of the layers are highly variable. Homogeneous parts of the rock column can be simulated by fewer layers than parts that are physically more variable. Due to the large scattering of velocity and density data within a given unit, a statistical averaging of measurements within each unit was necessary to define the acoustic properties of the layers in the model. The average velocity and density values of the mappable lithologic units are given in Table 1 and Figure 9. These values provided the basis for assigning velocity-density pairs to the individual layers of the impedance model. Table 2 gives the chosen velocity-density pairs for the 75 layers of the final impedance model. The following paragraphs describe four examples of how the observed field geometries (Figure 7) and measured impedance values from the laboratory (Figure 9) were used to define elements of the impedance model.

Elements of Impedance Model

Example 1: Interbedded Bryozoan Limestone with Hemipelagic Beds (Supersequence 6)

The succession of cliff-forming, well-cemented bryozoan limestones intercalated with two intervals of hemipelagic beds (Figure 7a) is an example how geologic elements can be easily incorporated into an impedance model. This double succession of limestone and hemipelagic marls provides distinct signatures in impedance distribution. The coarse, bioclastic bryozoan limestones have significantly higher velocities than the samples collected from the intercalated hemipelagic beds (Figure 9h), resulting in a drastic impedance contrast be-

Table 2. Lithology, Velocity-Density pairs, Impedances, and Reflection Coefficients of Layers in Maiella #29 Model[2]

Layer	Supersequence	Lithology	V_P (km/s)	Density (g/cm^3)	Impedance 10^6kg/m^{2}s	Refl. Coeff.
1		top of model (=average values of all samples)	5.32	2.55	13.57	0.132
2		evaporites (gypsum)	6.10	2.90	17.69	-0.285
3	6	lithothamnium limestone	4.19	2.35	9.85	0.083
4		upper marl unit	4.60	2.53	11.64	0.138
5		upper bryozoan limestone	5.82	2.64	15.36	-0.138
6		lower marl unit	4.60	2.53	11.64	0.138
7		lower bryozoan limestone	5.82	2.64	15.36	-0.033
8	5	reefs/build-ups	5.62	2.56	14.39	0.004
9		turbidites	5.60	2.59	14.50	0.0035
10		breccias	5.93	2.62	15.54	-0.086
11		scaglia	5.13	2.55	13.08	0.086
12		breccias	5.93	2.62	15.54	-0.171
13		nummulite packstone	4.51	2.44	11.00	0.134
14		(patch) reefs/breccias	5.62	2.56	14.39	-0.134
15		nummulite packstone	4.51	2.44	11.00	0.171
16	4	breccias	5.93	2.62	15.54	-0.086
17		scaglia	5.13	2.55	13.08	0.086
18		breccias	5.93	2.62	15.54	-0.035
19		turbidites	5.60	2.59	14.50	0.035
20		breccias	5.93	2.62	15.54	0.086
21	3	scaglia	5.13	2.55	13.08	0.064
22		breccia	5.79	2.57	14.88	-0.128
23	2	calcari cristallini (rudist sand)	4.70	2.45	11.51	0.069
24		calcari cristallini (rudist sand)	5.22	2.53	13.21	-0.173
25		calcari cristallini (rudist sand)	4.00	2.33	9.32	0.173
26		calcari cristallini (rudist sand)	5.22	2.53	13.21	-0.069
27		calcari cristallini (rudist sand)	4.70	2.45	11.51	0.069
28		calcari cristallini (rudist sand)	5.22	2.53	13.21	-0.173
29		calcari cristallini (rudist sand)	4.00	2.33	9.32	0.173
30		calcari cristallini (rudist sand)	5.22	2.53	13.21	-0.026
31	1 (basin)	scaglia/turbidite	5.02	2.50	12.55	0.082
32		megabreccias	5.71	2.59	14.79	-0.082
33		scaglia/turbidite	5.02	2.50	12.55	0.082
34		megabreccias	5.71	2.59	14.79	-0.082
35		scaglia/turbidite	5.02	2.50	12.55	0.082
36		megabreccias	5.71	2.59	14.79	-0.082
37		scaglia/turbidite	5.02	2.50	12.55	0.082
38		megabreccias	5.71	2.59	14.79	-0.082
39		scaglia/turbidite	5.02	2.50	12.55	0.082
40		megabreccias	5.71	2.59	14.79	-0.082
41		scaglia/turbidite	5.02	2.50	12.55	0.082
42		megabreccias	5.71	2.59	14.79	-0.082
43		scaglia/turbidite	5.02	2.50	12.55	0.082
44		megabreccias	5.71	2.59	14.79	-0.082
45		scaglia/turbidite	5.02	2.50	12.55	0.012
46		Valle del Inferno unit	5.04	2.55	12.85	0.046
47		Valle del Inferno unit	5.48	2.57	14.08	-0.046
48		Valle del Inferno unit	5.04	2.55	12.85	0.136
49		Valle del Inferno unit	6.28	2.69	16.89	-0.029
50	1 (platform)	platform limestones	6.04	2.64	15.95	-0.072
51		rudist biostromes	5.57	2.48	13.81	0.072
52		platform limestones	6.04	2.64	15.95	-0.072
53		rudist biostromes	5.57	2.48	13.81	0.072
54		platform limestones	6.04	2.64	15.95	-0.072
55		rudist biostromes	5.57	2.48	13.81	0.072
56		platform limestones	6.04	2.64	15.95	-0.072
57		rudist biostromes	5.57	2.48	13.81	0.072
58		platform limestones	6.04	2.64	15.95	-0.072
59		rudist biostromes	5.57	2.48	13.81	0.072
60		platform limestones	6.04	2.64	15.95	-0.072
61		rudist biostromes	5.57	2.48	13.81	0.072
62		platform limestones	6.04	2.64	15.95	-0.089
63		bauxite concretions and sand	4.75	2.81	13.35	0.109
64		bauxitic limestone breccias	6.16	2.70	16.63	0.019
65	"0"	platform limestones	6.40	2.70	17.28	-0.016
66		platform limestones	6.25	2.68	16.75	0.016
67		platform limestones	6.40	2.70	17.28	-0.016
68		platform limestones	6.25	2.68	16.75	0.016
69		platform limestones	6.40	2.70	17.28	-0.016
70		platform limestones	6.25	2.68	16.75	0.016
71		platform limestones	6.40	2.70	17.28	-0.016
72		platform limestones	6.25	2.68	16.75	0.016
73		platform limestones	6.40	2.70	17.28	-0.016
74		platform limestones	6.25	2.68	16.75	0.016
75		platform limestones	6.40	2.70	17.28	0.009
HD	halfspace	Model base = platform limestones	5.50	2.71	17.61	

[2] Reflection coefficients were calculated with respect to the underlying layers (= layer with next higher number), which may not necessarily be an impedance contrast in the model, due to variable bed terminations.

tween the layers. In addition, the units vary in thickness from 30 to 50 m and thus have vertical extents of seismic scale. As a consequence, this succession is easily modeled by four layers that are characterized by the observed field geometry of each unit and by the measured and averaged impedance values of the bryozoan limestones and hemipelagic rocks (average values marked by crosses in Figure 9h).

Example 2: Fine-Grained Bioclastic Grainstones with Isolated Patch Reefs (Supersequence 5)

Another example of a straightforward model definition is the patch reefs of supersequence 5 that are over- and underlain by little-cemented, fine-grained carbonate grainstones (Figure 7b). Similar to example 1, the measured velocity and density values plotted in a density–velocity diagram mark two clusters of the two lithologies that do not overlap (Figure 9g). All reefal rocks

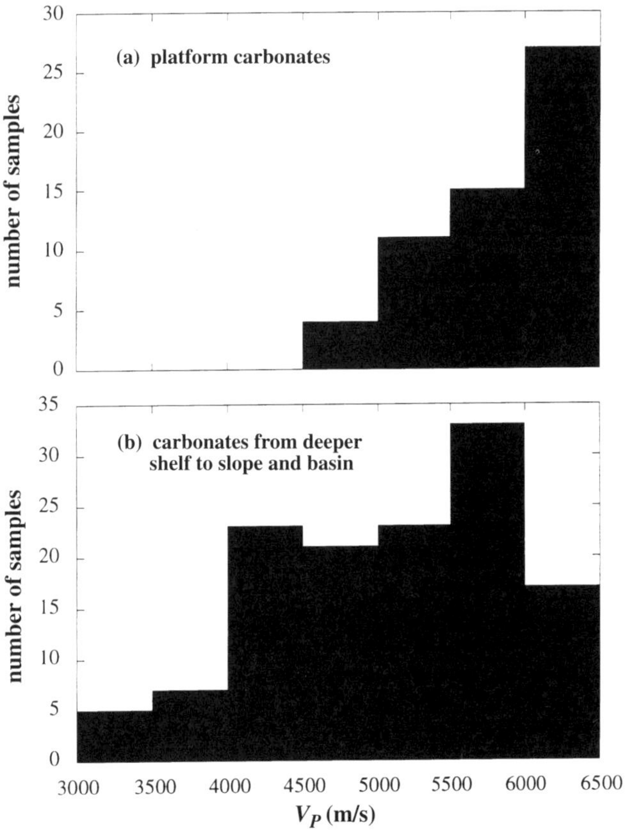

Fig. 9. Ranges of V_P (at 8 MPa effective pressure) of different depositional environments from the Maiella. (a) Carbonates deposited on the shallow-water platform have a narrow range with generally high velocities. (b) Sediments from the deeper shelf, slope, and basin show a higher variability toward lower V_P, but maximum velocities are the same as in the platform rocks.

have higher velocities than the fine-grained carbonates and thus have a significant higher average impedance value. These high velocities are not only a product of the lower porosity but also reflect the frame-forming fabric of the reefal sediments that enhances the elastic properties of the rock (Anselmetti and Eberli, 1993). In the model, these patch reefs (up to 50 m thick) were modeled by one layer with variable thickness. Because of software requirements, this reef layer is still part of the model between the outcropping patches, but it is characterized here by "zero thickness" and is thus seismically nonexistent (Figure 10).

Example 3: Alternations of Breccias, Turbidites, and Pelagic Carbonates (Supersequences 3–5)

This example shows the limitations in creating a layered impedance model from a geologic cross-section because the layers do not have seismic-scale thickness, nor do they have impedance values that can be easily grouped. The succession of slope carbonates, consisting of lithoclast breccias, fine-grained turbidites, and pelagic background sediments, forms a thin-layered unit with individual beds with an average thickness of <1 m. Because software capabilities are limited, the 100 m high cliff shown in Figure 7c cannot be modeled by a hundred or so layers. Consequently, some drastic simplifications must be made in modeling this thin-bedded alternation of different lithologies. Thus, the model for this unit consists only of a schematic succession of some layers that reflect the general bedding geometry but not the true layer or impedance distribution. In addition to the complicated geometric relationships, the velocity measurements show an overlap of values among the three lithologies (Figure 9f). Despite this overlap of petrophysical data, the average velocity of the breccias is higher than the average velocity of the turbidites. Scaglia samples revealed on average the lowest velocities of this succession and are modeled with a lower impedance value than the other two lithologies.

Example 4: Megabreccias Onlapping the Escarpment (Supersequence 1)

The megabreccias that onlap the platform escarpment can be recognized from far distances because their cliff-forming appearance contrasts sharply to the tree-covered slopes formed by the background deposits (Figure 7d). The cluster in the velocity–density diagram of these megabreccias (Figure 9d) partially overlies the data of the background sediments (Scaglia and turbidites). However, the mean velocities of the two groups

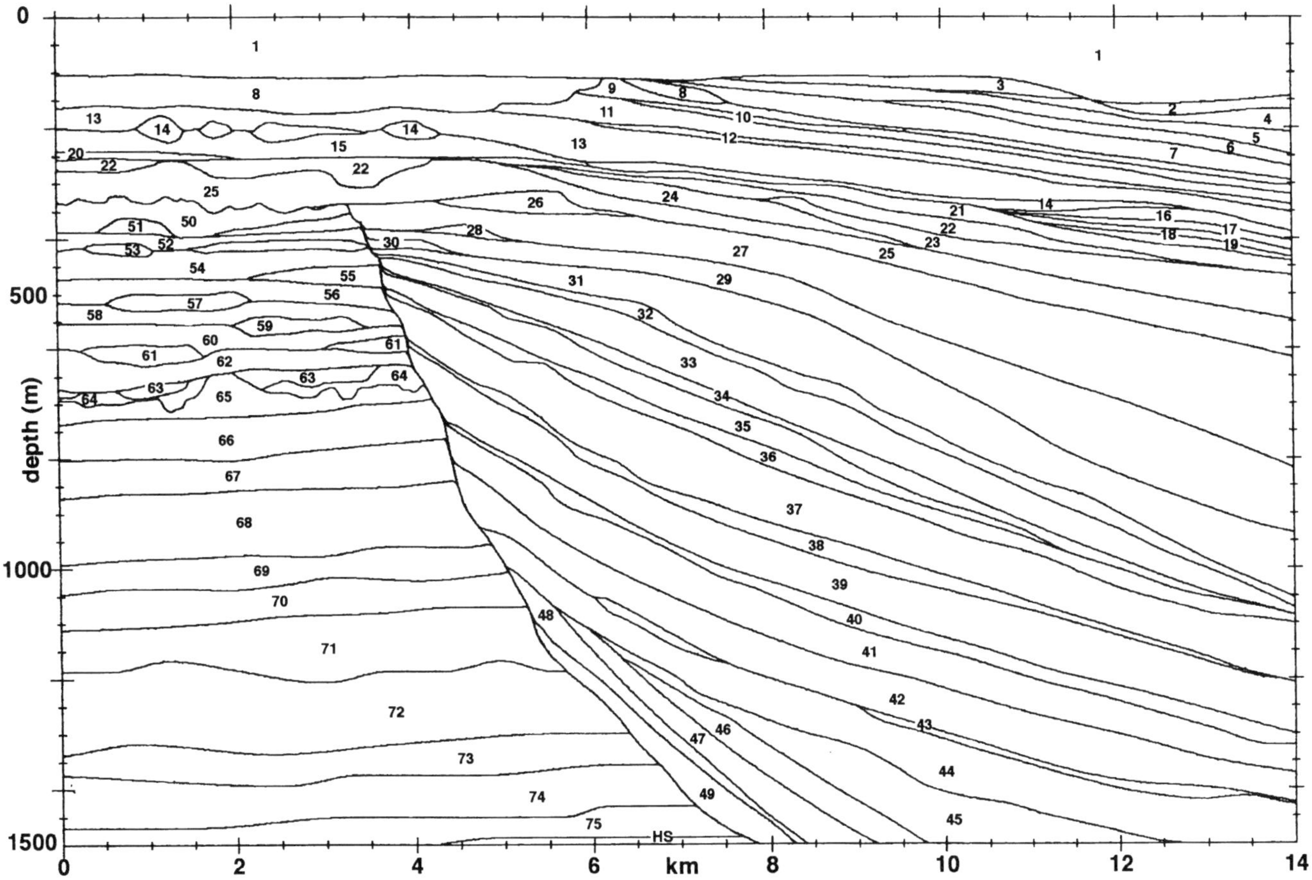

Fig. 10. Plot of impedance model (6× vertical exaggeration). This model, 14 km wide by 1.5 km deep, consists of 75 layers and the half-space below it. All layers have characteristic geometries and impedance values. The layer numbers refer to Table 2, which gives detailed information on the individual layers. Major elements of the model can be seen in the photo of western Valle dell'Orfento (Figure 6) and in the schematic cross-section (Figure 5).

show a clear impedance contrast. The irregular geometries of the megabreccias, particularly the cut-and-fill structures at the base of the breccias, are of seismic scale and were used in defining the layer geometries in the impedance model.

Summary of Examples

These four examples show that different modeling approaches must be performed to incorporate the different geologic elements in an impedance model. In the modeled section of the platform-basin transition of the Maiella, most of the section is characterized by a clear impedance distribution with mappable geometries, as described in examples 1, 2, and 4. However, some parts, such as the platform sequence of supersequence 0 and 1 and the slope portion of supersequence 2, had to be simplified and were modeled in similar approaches, as described in example 3. This was done because neither the geometries nor the impedance distribution allowed a straightforward modeling technique.

Synthetic Seismic Sections

The impedance model (14 km wide by 1.5 km deep) was covered by a virtual survey of 400 shot points that are identical to 400 CMPs and 400 receiver points. The resulting CMP distance equals 35 m, which is a reasonable value for true seismic lines. The time sample interval for all synthetic sections is 2 ms.

Seismic Imaging Capabilities

The two most important factors for the recognition of a geologic feature on a seismic section are (1) the contrast in acoustic impedance relative to the surrounding rocks and (2) the vertical and lateral extent of the geologic element. These two factors, together with the processing and display parameters (amplitude gain, S/N, and frequency), determine which parts of the rock record are imaged on a seismic section.

Required Impedance Contrast

Recognition of a reflection through the noise requires a certain amplitude level. Patterns of reflections with high coherency can be recognized at lower amplitudes than reflections with low coherency. We showed earlier that the variation of velocity in the modeled layers is remarkably high. This variability causes the carbonates to have impedance contrasts between the individual layers that result in high reflection coefficients with an average value of ~0.1 (Table 2). Jones and Nur (1982) assumed that a reflection coefficient higher than ±0.05 yields a reflection with an amplitude that can be seen on most seismic sections. Modern seismic techniques with high S/Ns might even image layer boundaries with a coefficient of only ±0.02.

Most layers in the Maiella model have reflection coefficients that are far higher than required. Only some layers do not have sufficiently high impedance contrasts to create reflections (Table 2). For example, the Lower Cretaceous platform section shows a succession of layers with velocity values that vary between 6250 and 6400 m/s, resulting in reflection coefficients of ±0.016. These low impedance contrasts and reflection coefficients cause reflections with very low energy; consequently, the layers cannot be detected when noise is present on the synthetic sections. The result is a seismically transparent Lower Cretaceous platform (Figures 11b and 11c), which demonstrates how layers with considerable vertical and lateral extent may not be resolved on a seismic section due to insignificant contrast in acoustic impedance. A modeling procedure without added noise and with an applied automatic gain control would have distorted the seismic image by showing a platform with high seismic reflectivity (Figure 11d). In contrast to the often-made assumption that added noise does not improve seismic modeling (Stafleu et al., 1995), this comparison shows that artificial noise is useful to investigate low-amplitude reflection events.

Horizontal and Vertical Resolution

The limit of vertical resolution is determined by the wavelength, which is a function of velocity and frequency (wavelength = velocity/frequency). Every layer produces reflections at its top and base, which can be superimposed to create the resulting signal. These reflections may have opposite polarities resulting in destructive interference if the layer has a thickness of less than one-fourth wavelength (Sheriff, 1977). A layer thicker than this produces two reflections that interfere to create two separate and thus detectable signals on a seismic section.

Horizontal resolution on unmigrated sections is limited by the Fresnel zone (Sheriff, 1977). The size of the Fresnel zone depends on (1) frequency, (2) depth of reflector, and (3) velocity of the layers. For a two-way traveltime of 0.5 s, a frequency of 40 Hz, and a velocity of 4000 m/s, the diameter of the Fresnel zone is ~220 m. Elements that are smaller than this can still be recognized on a seismic section but by a diffraction rather than a reflection. On migrated sections, horizontal resolution is limited more by other factors, such as noise and spatial sampling, than by the diameter of the Fresnel zones (Sheriff, 1985). In our model, most layers have lateral extents that are far larger than the Fresnel zones. Only the patch reefs have widths that are close or below its size. However, the method of normal incidence ray tracing does not simulate the effects of Fresnel zones. The main limiting factor of horizontal resolution in this study is the spacing of the detectors, which is 35 m. A layer with little horizontal extent nevertheless produces a reflected signal when it is "hit" at a normal angle by a synthetic ray that arrives at the surface within the capture radius of a virtual geophone.

Because most of the modeled layers are characterized by thicknesses close to or below the seismic resolution and by widths that are considerably larger than the geophone spacing, the vertical resolution in this study is much more critical than the horizontal one. In reality, a rock section is made up of countless physical layers that have a wider lateral extent than vertical thickness, resulting in a reflection pattern that is a product of interference from several thin layers.

Synthetic Seismic Expression and Facies of Carbonate Environments

Seismic Image of Platform Escarpment

Despite its irregular and steeply dipping geometry, the platform escarpment is imaged clearly on all synthetic sections. Regardless of the chosen frequency, polarity, or gain, the escarpment produces its own reflection and is also recognized by the termination of onlapping slope reflections. It separates high-velocity platform rocks from lower velocity slope sediments and is thus characterized by a high-amplitude reflection.

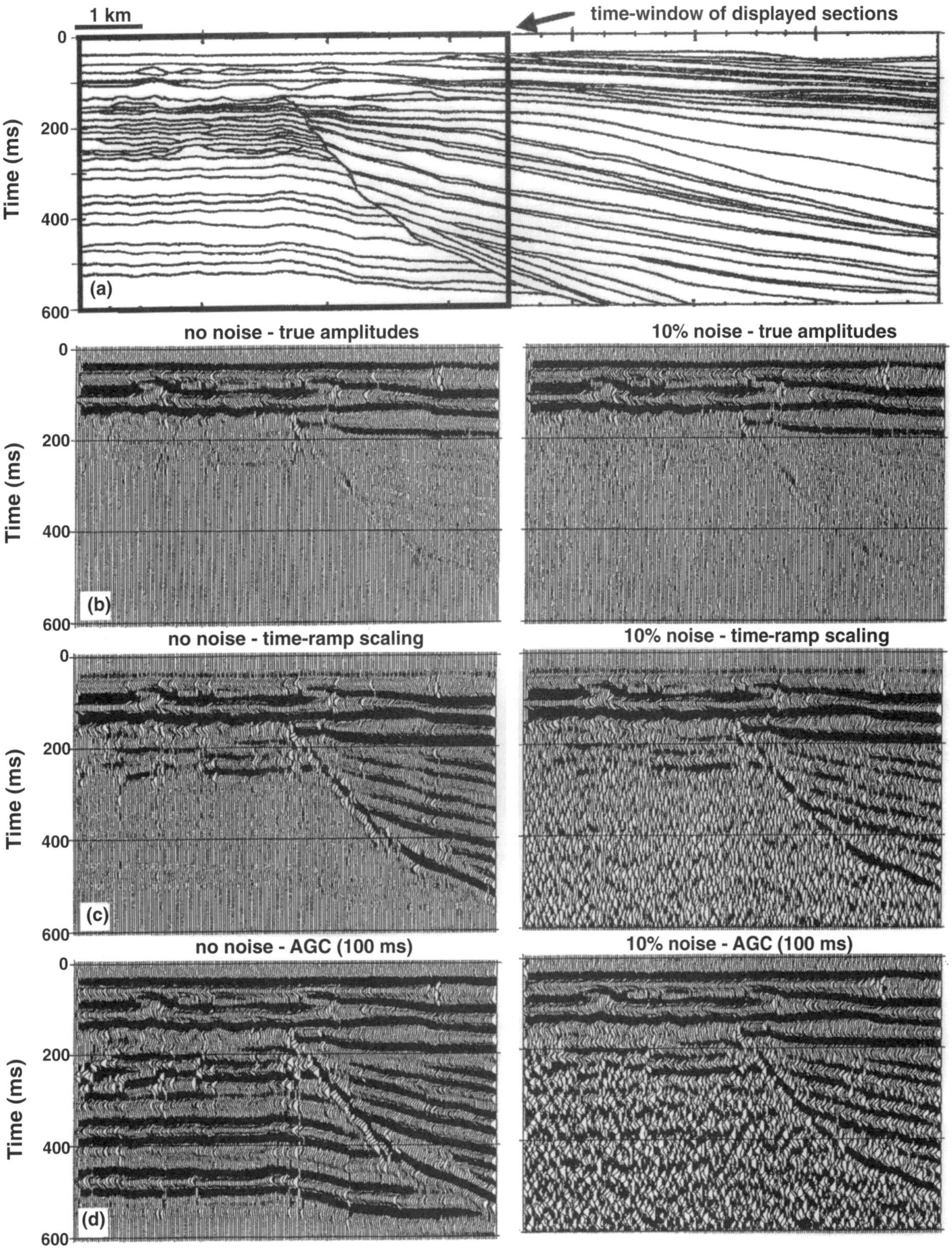

Fig. 11. Comparison of various noise and amplitude gain options. The sections on the left are seismograms with no noise, and those on the right are disturbed by randomly added noise (10% of average amplitude). (a) Impedance model converted to time. (b) True amplitude section without amplitude gain; only shallow reflections can be recognized. (c) Section with time-ramp scaling. The high-amplitude reflections on the slope are visible, but the Lower Cretaceous platform is transparent, with or without noise. The Upper Cretaceous platform shows some internal structure (rudist biostromes). (d) Sections with applied automatic gain control (AGC) of 100 ms. Without noise, the low impedance contrasts of the Lower Cretaceous platform ($R = \pm 0.016$) are amplified, distorting the seismic image.

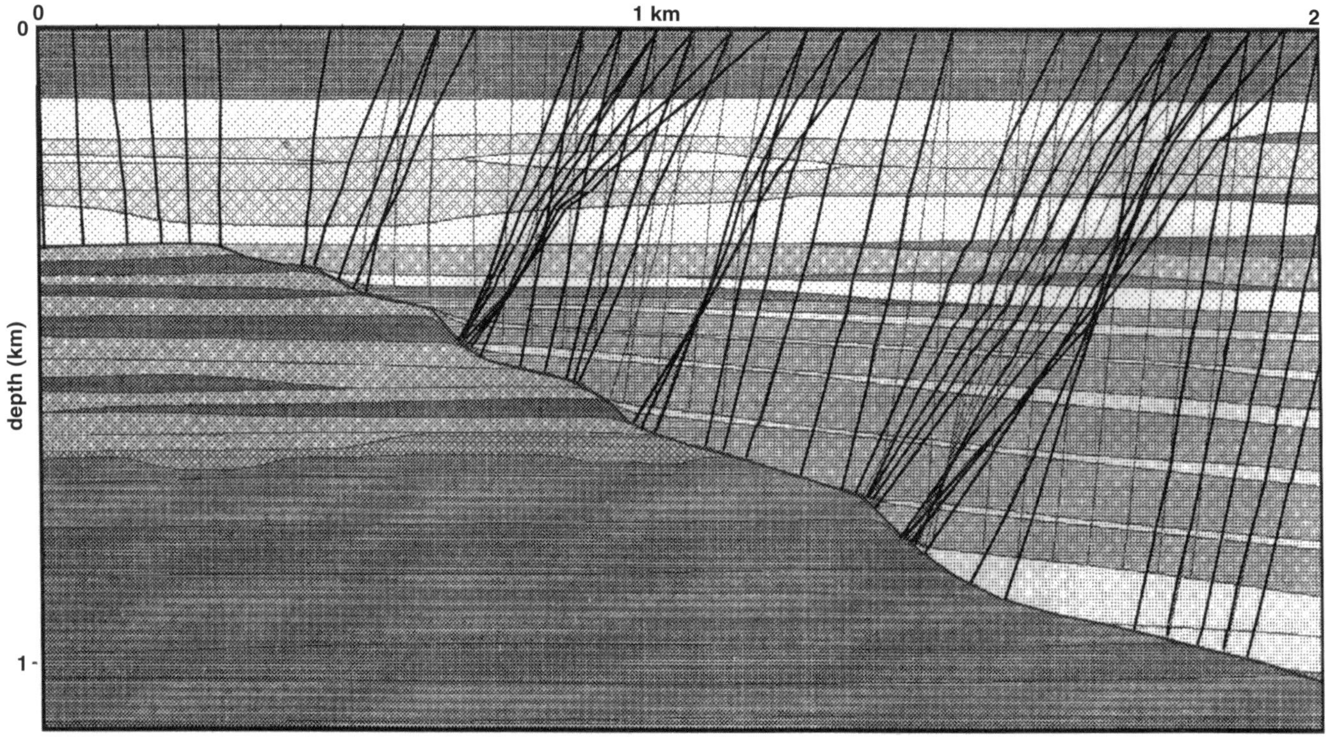

Fig. 12. Ray tracing of reflections originating from the platform escarpment. The impedance model is not vertically exaggerated and shows true angles. The rays are calculated by the normal incidence ray-tracing method, which only considers rays that arrive with a normal angle on the target surface. Downgoing and upcoming rays are thus identical. To separate individual rays, a wider CMP spacing of 60 m was used. The irregular surface of the escarpment (average dip 30°) forms steeper niches that result in bundling and divergence of the rays.

The ray tracing of all reflections from the escarpment is shown in Figure 12. A CMP spacing of 60 m was chosen for this plot so that the individual rays could be distinguished. For the synthetic seismic sections in Figures 13b–f, the CMP spacing was 35 m, thus, more rays were actually used to calculate the synthetic sections than are shown in Figure 12. In this plot, which shows true angles, most of the escarpment is imaged by rays that have angles of incidence 10–45° to the vertical. The irregular surface produces bundling and divergence effects, especially at concave and convex niches in the platform wall. In reality, the escarpment is likely to be the origin of numerous diffractions, which are not simulated properly with the modeling program. The oblique nature of the raypaths results in a relatively large offset between the position of the escarpment on the unmigrated seismic section (Figures 13b–f) and its true position on the time–depth plot (Figure 13a). This true position of the escarpment is calculated using a vertical depth-to-time conversion of the individual layers and, unlike on the synthetic sections, represents a perfectly migrated section. In reality, correct migration of such a setting would be difficult because the relief of the escarpment is complicated. It is a steep, discontinuous surface that separates horizontal to gently dipping beds that are imaged close to their true positions on both migrated and unmigrated sections. However, the offset between the reflection of the escarpment and its true position does not change the general picture and is not a major source of error in seismic interpretation.

Seismic Facies of Cretaceous Platform and Middle Cretaceous Unconformity

The top of the platform, which is overlain by the low-velocity Orfento Formation (Calcari Cristallini), forms a major contrast in acoustic impedance. This contact yields a high-amplitude reflection on the synthetic sections. The modeled karstic nature of this contact is visible on the seismic section by a wavy but continuous reflection (Figures 13b–f, at CMP 1–100, 0.15 s). Across this surface, little acoustic energy penetrates to the underlying layers of platform rocks because the high impedance contrast results in most of the energy being reflected back to the surface.

Compared with the seismic facies of the slope, the platform has only moderate seismic reflectivity and is imaged by low-amplitude reflections that barely exceed the noise levels. The Upper Cretaceous platform, which has few higher impedance contrasts due to the rudist biostromes, produces some low-coherent reflections that are partially visible through the randomly generated noise. The middle Cretaceous unconformity, marked by pockets of low-velocity rocks, is also imaged by non-continuous reflections that have a reflection pattern similar to that of the biostromes above (Figure 13b, at CMP 1–100, 0.25 s). At low frequencies, the middle Cretaceous unconformity cannot be distinguished from the biostromes. At dominant frequencies above 40 Hz, the reflection from this unconformity has a significantly higher amplitude than reflections from the rudist biostromes (Figures 13e–f, at CMP 1–100, 0.25 s) and can therefore be recognized. The transparent seismic facies of the platform, with only few low-coherent reflections, correlates well with the seismic facies of platforms on observed seismic lines across the Bahamian (Figures 14 and 15c) and Apulian platform margins (Figure 15b).

Seismic Facies of Slope Carbonates

Unlike the platform carbonates, the slope deposits show excellent reflectivity on the seismic sections due to their wide range of acoustic impedances (Figure 8). This is seen clearly on the modeled synthetic sections, where the slope reflections onlap the transparent core of the platform (Figures 13b–13f).

Measurements of physical properties show that breccias have significantly higher impedances than the pelagic background sediments and turbidites. The resulting synthetic sections image these breccias because their thickness and lateral distribution are above the limits of seismic resolution so that they can control the reflection pattern. The unconformities observed in the modeled slope sequences are not necessarily real because several pseudo-unconformities (Schlager et al., 1991; Biddle et al., 1992) are created when breccia beds or intercalated background sediments thin below seismic resolution. The observed seismic facies of slope carbonates is in good accordance with real seismic lines, which shows that slope carbonates have similar reflectivity on the modeled synthetic sections.

Seismic Image of Prograding Reefs and Patch Reefs

Using a signal with a dominant frequency of 20 Hz, we could recognize only two of the four modeled patch reefs clearly on the synthetic seismic section (Figure 13b, at CMP 40–120, 0.1 s). Dominant frequencies of 30 and 40 Hz (Figures 13c and 13d) allowed all reefs to be seen because the better vertical resolution allowed the reefs to be imaged as separate reflections that do not overlap with the strong reflections from the underlying top of the platform. A further increase in the dominant frequency did not change this pattern significantly. In contrast to the isolated patch reefs above the platform (Figures 7b and 10), the most distal reef was not seen clearly on all the sections (Figures 13b–13f, at CMP 200, 0.05 s). This invisibility is mainly the result of the small thickness of the reef and its lower impedance contrast relative to surrounding rocks.

Comparison of Synthetic Section with Real Seismic Data

A good way to test the credibility of seismic models is to compare the synthetic data with real seismic sections from the same or similar geologic environments. Such a comparison can refine and improve the interpretation of observed seismic data (e.g., Middleton, 1987; Rudolph and Greenlee, 1991; Stäuble and Pfiffner, 1991). Likewise, the synthetic sections of outcropping rocks units can be compared with true seismic data from a similar geologic environment to see the seismic response of a particular geologic setting (Stafleu and Sonnenfeld, 1994). These comparisons are qualitative only, and no attempts to quantify the reflection records of real data were intended. For the comparison with modeled outcrops of the Maiella, two seismic data sets were available that both illustrate similar carbonate platform-to-basin environments: (1) the Western seismic line from Great Bahama Bank and (2) a seismic section of the eastward continuation of the Maiella escarpment, which plunges into the subsurface beneath the Adriatic Sea (Figure 3). Although the late Cenozoic evolution of the Maiella is different from that of both compared analogs, these sections show a striking resemblance to the modeled section.

Seismic Line from the Subsurface of the Adriatic Sea

Several published seismic sections across the margin of the Apulian platform in the subsurface of the Adriatic Sea show reflection patterns similar to the modeled Apulian platform margin in the Montagna della Maiella (De Alteriis and Aiello, 1993). In particular, an available multichannel seismic line offshore in the Adriatic Sea (seismic data courtesy of Elf Italiana) (Figures 3

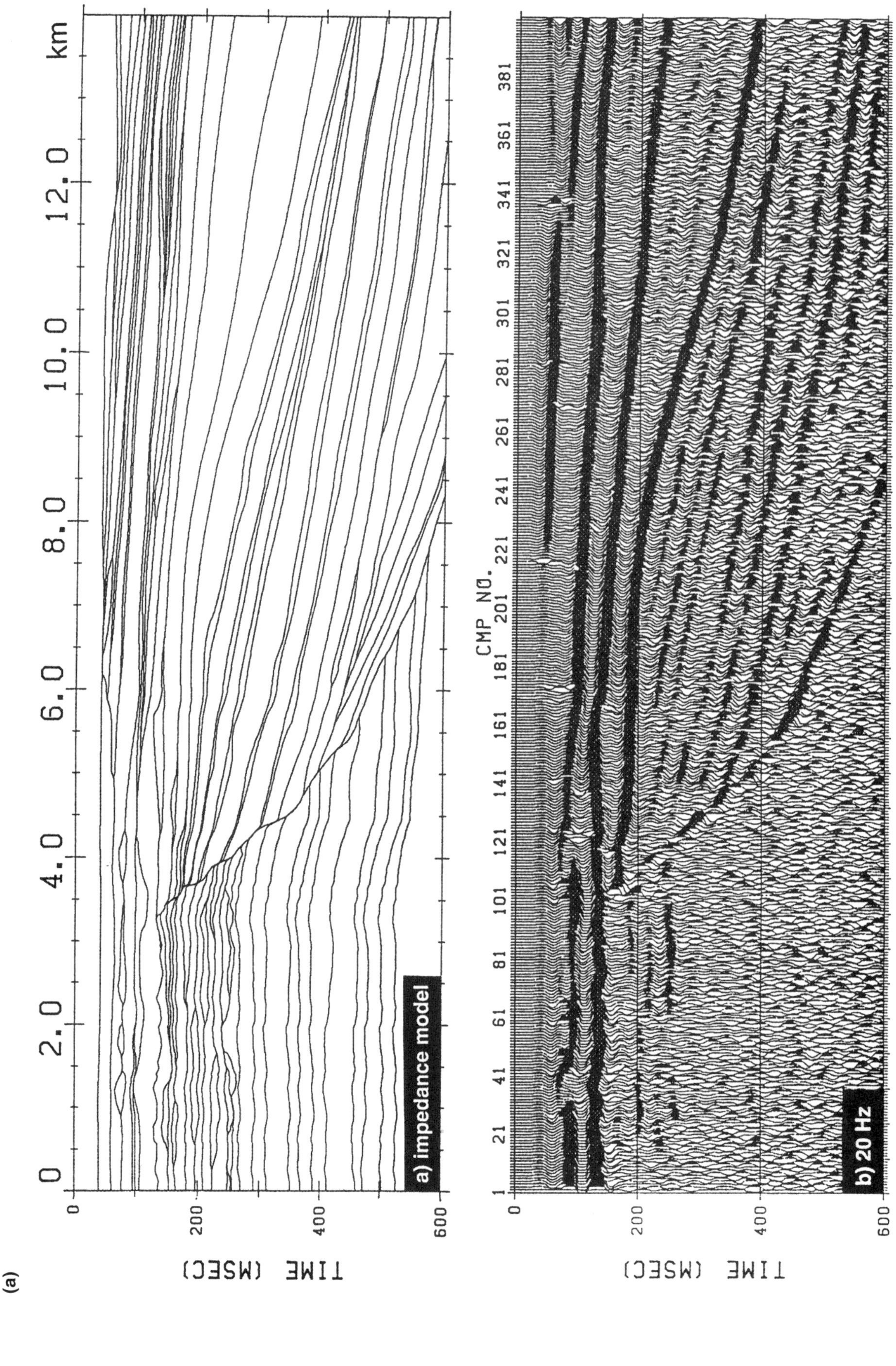

Fig. 13. (a) Impedance model converted to two-way time and shown at the same scale as the synthetic seismic sections in parts b–f. This conversion is computed only in the vertical direction and represents a migrated section; it does not consider dipping layer geometry. Note the effects of velocity pull-up below patch reefs and push-down in the platform below the escarpment. (b–f) Synthetic seismic sections of the impedance model at dominant frequencies of 20–60 Hz and positive polarity. Background noise (10% of calculated amplitudes) was added. A time-ramp power function with an exponent of 1.2 was used as the amplitude gain. *(see parts c, d, e and f on next two pages)*

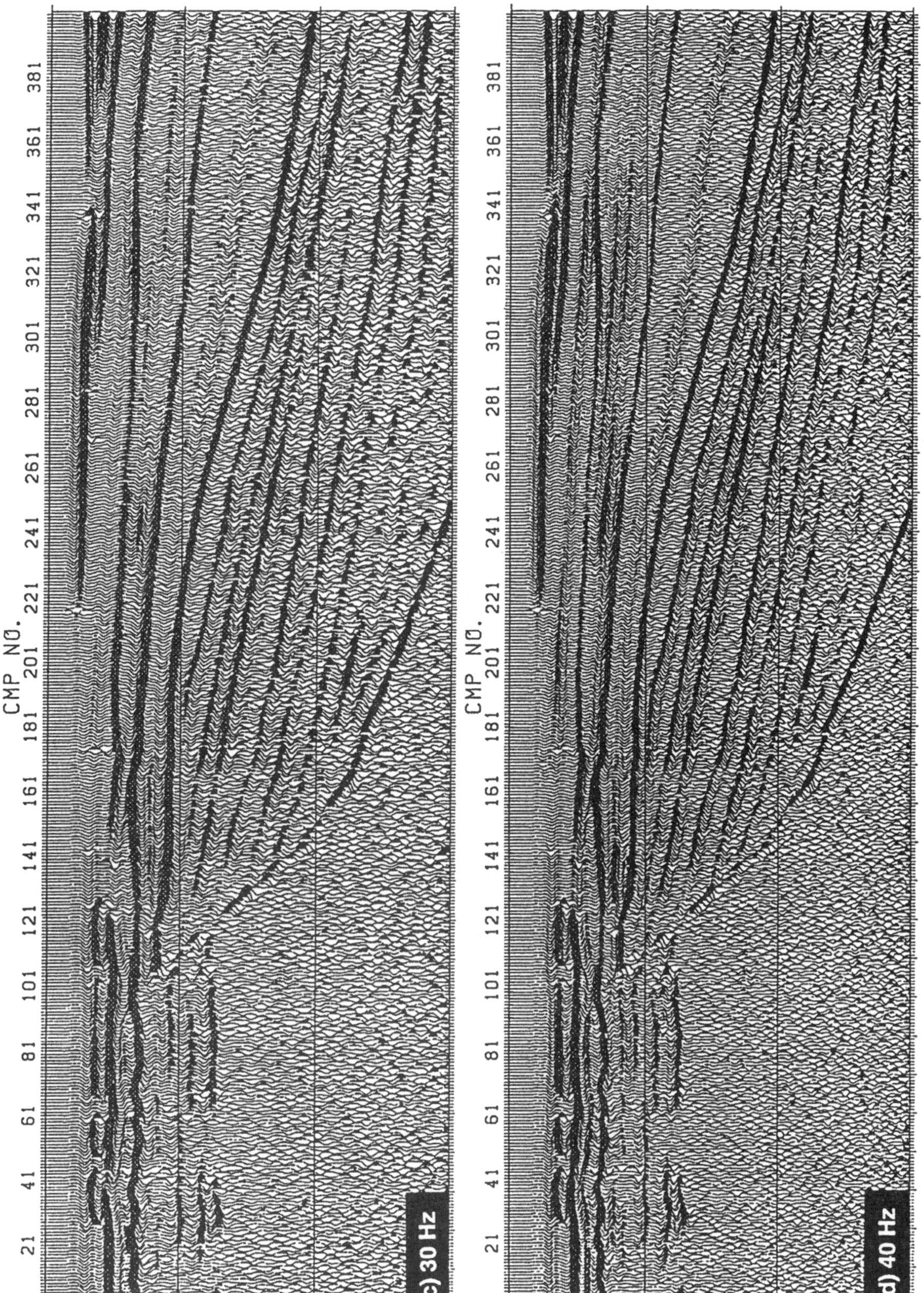

(c) (d)

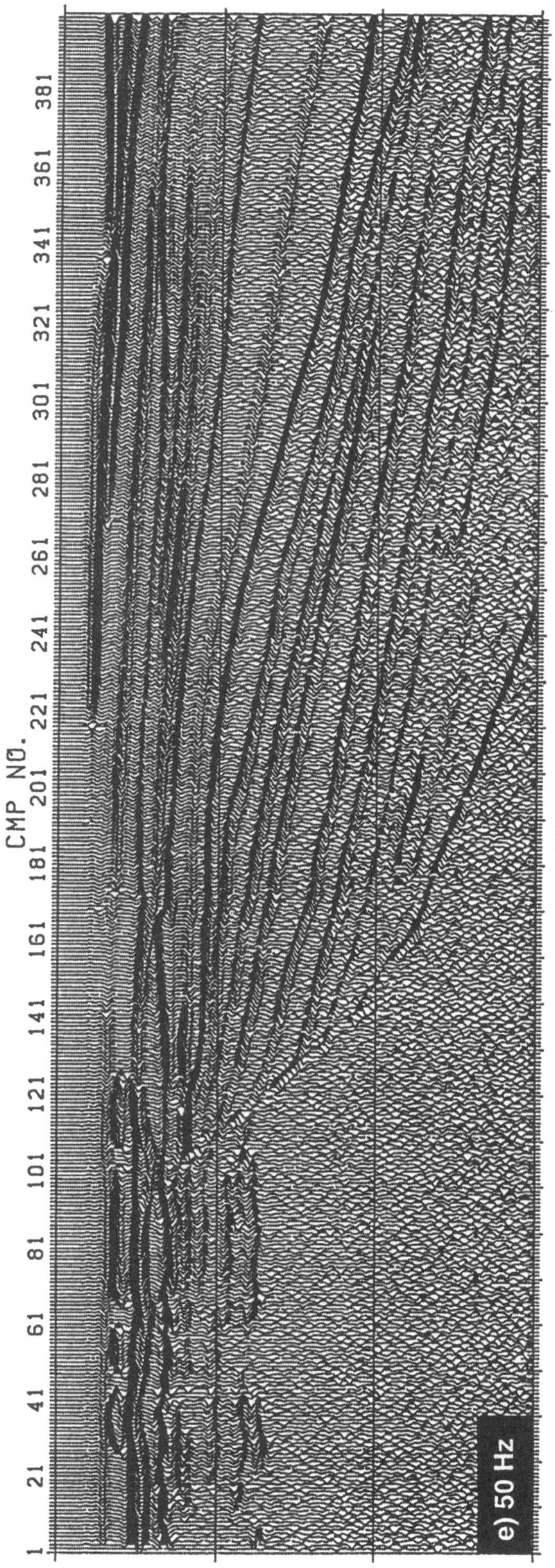

(e)

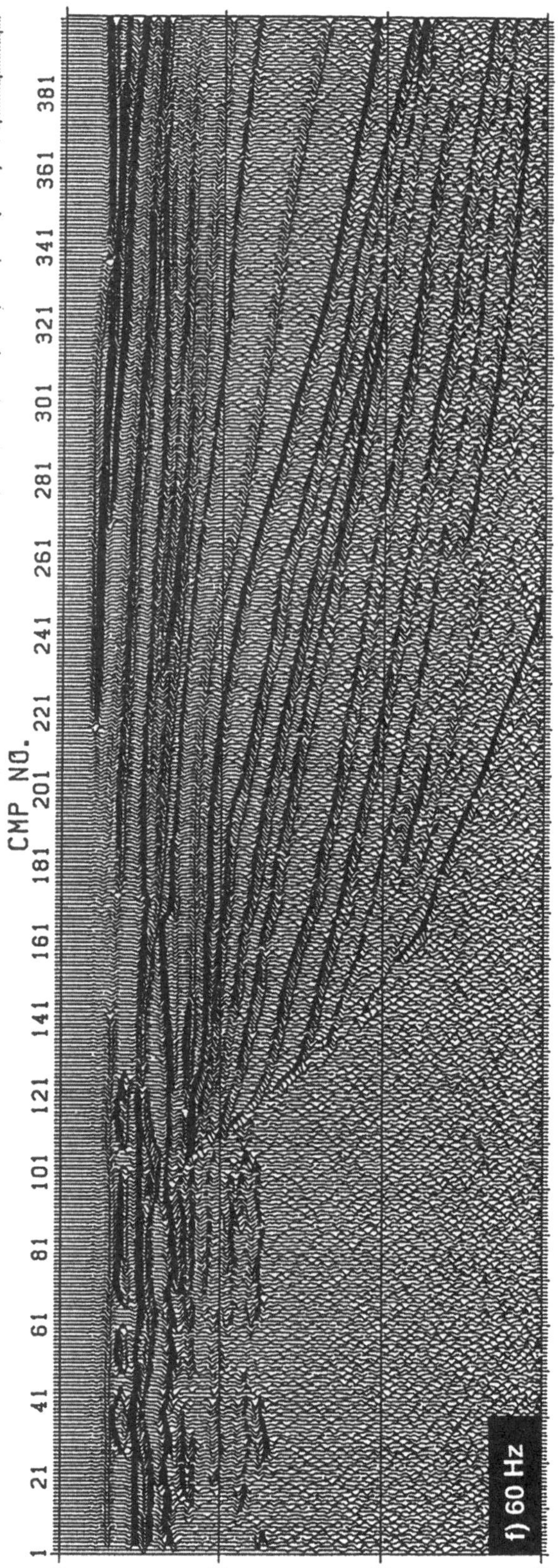

(f)

and 15b) images the buried continuation of the Apulian platform margin only 40 km away from the Maiella outcrops (Zappaterra, 1994). Figure 15b shows a window from this line that has the same dominant frequency (~30 Hz) as in the synthetic section of the Maiella (Figure 15a) and the same horizontal (distance) and vertical (two-way time) scales. Consequently, the seismic images and geologic structures (such as dip and size) can be directly compared.

Similar to the Maiella and the Bahamas sections (Figures 15a and c), the Adriatic line (Figure 15b) shows a transparent platform core that is onlapped by high-amplitude slope reflections. Due to deeper burial, the thick Pliocene–Pleistocene slope-to-basin sediments between 1.6 and 2.2 s might have higher velocities than the slope sediments of the Maiella and may thus represent a thicker rock section than the modeled slope. The buried platform section cannot be significantly faster than the Maiella since the Maiella platform was modeled with velocities of 5500–6500 m/s. Steeper inclined reflections of the slope in the Adriatic section could be explained by a velocity push-down of the overlying thick basinal sediments.

Both the Maiella and Adriatic sections (Figures 15a, 15b) show sharp boundaries between the transparent platform zone and the high-reflectivity slope section. These boundary surfaces are steep at the top and become gradually less steep at the base of the platform core. These similarities suggest that the nondepositional escarpment, as seen in the outcrops of the Montagna della Maiella, might continue into the subsurface of the Adriatic Sea. Another similarity of both sections is the burial of the escarpment and subsequent formation of a gentle platform-to-basin transition. Despite these similarities in their seismic patterns, the two areas are characterized by substantial differences in recent geologic history. In the Pliocene, the Maiella became incorporated into the nappe formation of the Apennine orogeny, while the Adriatic area to the east became the foredeep and was covered with a thick Pliocene–Pleistocene sequence overlying the Messinian evaporite deposits. On the seismic section of the Adriatic, these evaporites yield the most prominent reflection with the highest amplitude.

Unlike the synthetic Maiella section, the offshore seismic section from the Adriatic Sea displays no signs of major progradational units. This difference in geologic evolution could be a result of different subsidence history. It also could be related to a laterally variable growth potential and sedimentation rate of the carbonate platform. Such a variable growth potential could be explained by a change in orientation of the platform margin with respect to the main wind direction, which determines the direction of off-bank shedding (Hine and Neumann, 1977).

The similar seismic images suggest that both areas underwent a similar evolution with a steep bypass escarpment being gradually covered by slope sediments. This comparison shows how synthetic sections can be used to interpret seismic reflection patterns of real seismic lines.

Western Seismic Line from Great Bahama Bank

The subsurface of Great Bahama Bank, which was imaged by the Western seismic line (Eberli and Ginsburg, 1989) (Figures 14 and 15c), represents a similar geologic environment to that of the Montagna della Maiella. The multichannel seismic line across Great Bahama Bank images a buried modern platform margin and a succession of aggrading and prograding carbonate sequences similar to that of the Maiella.

Below the modern flat-top platform, the Western seismic line clearly images a buried slope section prograding into the Straits of Florida. An infilled intraplatform seaway, the Straits of Andros, can be seen in the eastern part of the section (Figure 14) (Eberli and Ginsburg, 1989). The slope carbonates can be recognized on the seismic section by their inclined geometry, showing the depositional slope angle, and by a succession of high-amplitude reflections that contrast sharply with the seismically transparent platform cores.

The seismic facies related to the platform margin is similar to that of the modeled Maiella sections. In Figure 15c, the buried platform margin of the Western seismic line is plotted with the same scale as the synthetic section of the Maiella. Despite the moderate quality of the available paper copy of the Western seismic line, the general pattern of inclined high-amplitude reflections that onlap a transparent platform is seen clearly. In addition, both sections image the geometries of the prograding units by coherent high-amplitude reflections, so that several prograding sequences can be recognized. Thus, the Maiella model and the Western seismic line show a similar seismic pattern of an aggrading and subsequent prograding carbonate platform margin, and the platform cores show a nearly identical seismic image. However, the geometries of the slope sediments of Great Bahama Bank appear to be more complicated than the slope geometries on the synthetic section of the Maiella. The slope reflections on the Bahamas line show irregularities in dip that can be partially explained by velocity pull-

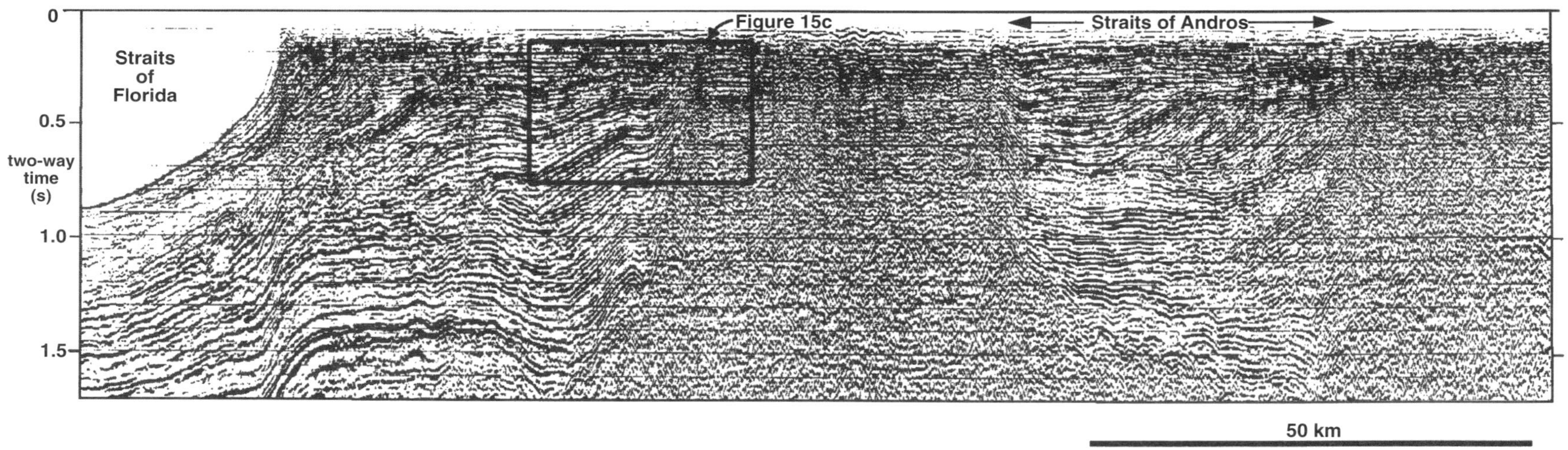

Fig. 14. The Western seismic line from Great Bahama Bank (east is to the right). The modern Great Bahama Bank consists of old platform cores, characterized by seismic transparency, and slope carbonates with inclined high-amplitude reflections. An infilled seaway, the Straits of Andros, can be recognized in the right third of the section by a succession of inclined high-amplitude reflections (Eberli and Ginsburg, 1989).

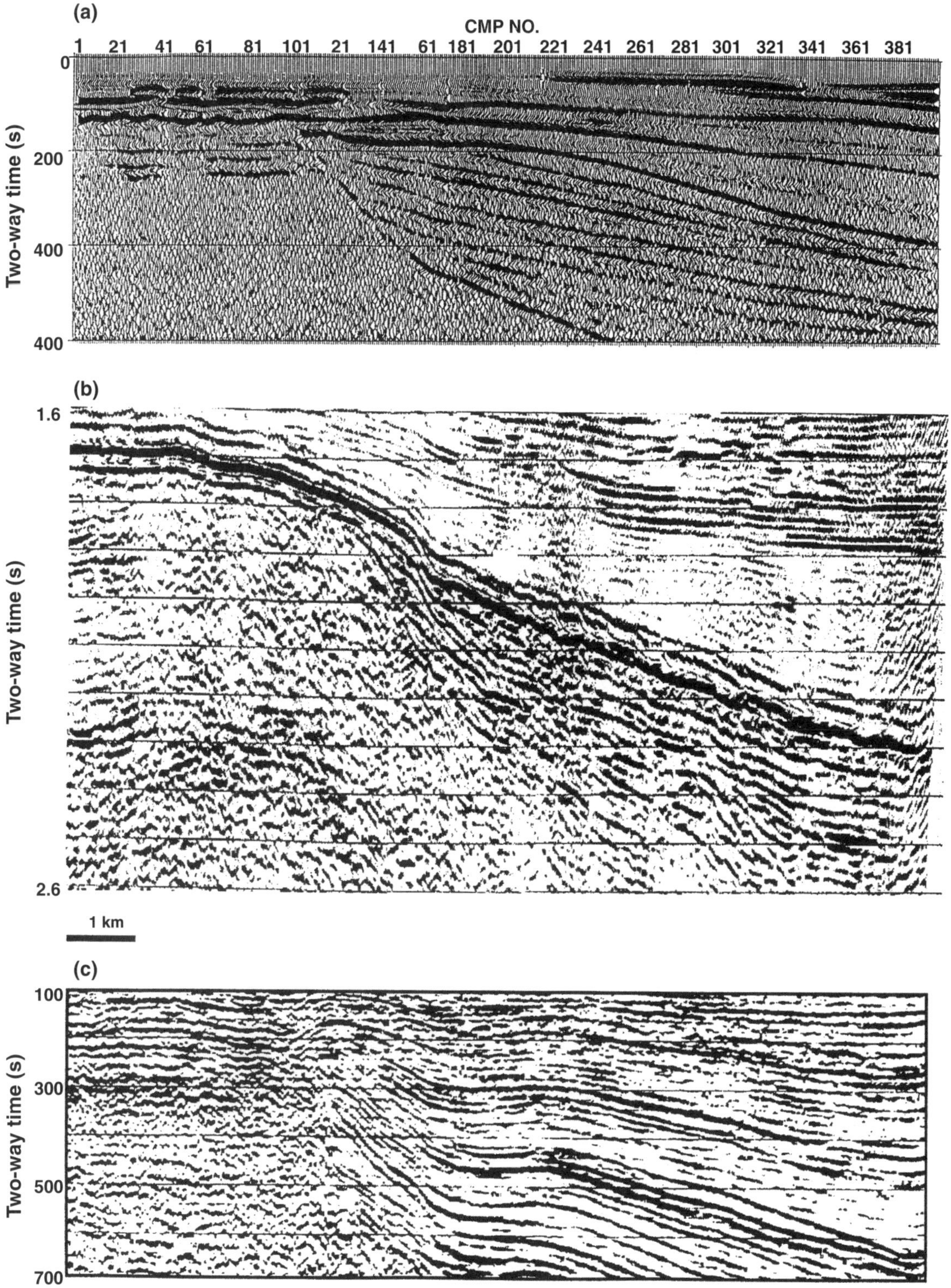

Fig. 15. Comparison of seismic sections from similar geologic environments plotted at equal horizontal and vertical scales (total width of sections is 14 km). (a) Synthetic seismic section across the platform-to-basin transition in the Montagna della Maiella (30 Hz). (b) Seismic section of the eastward continuation of the Maiella platform margin (Apulian platform) located ~40 km east of the Maiella where the margin plunges into the subsurface beneath the Adriatic Sea (Figure 3). The reflection pattern is characterized by a transparent platform onlapped by a wedge-like succession of high-amplitude reflections. (c) Part of the Western seismic line from Great Bahama Bank (Eberli and Ginsburg, 1989). The reflection pattern is similar to that above, confirming the differences in seismic facies of platform and slope.

ups of the overlying and prograding high-velocity reefal sections.

Comparison of Sequence Stratigraphy in Outcrop and on Synthetic Sections

In addition to investigating the seismic facies of the various carbonate environments, one of the main goals of seismic modeling of the Maiella outcrops is to compare the seismic sequences from synthetic seismic sections with the depositional sequences determined in the field. This allows the testing of the concept of seismic stratigraphy in a pure carbonate setting (Mitchum et al., 1977; Vail et al., 1977a; Kendall and Schlager, 1981; Sarg, 1988; Schlager, 1992). A good agreement between seismic sequences and depositional sequences would support the assumption that seismic sequence analyses can be used for stratigraphic and sedimentologic interpretations.

The model represents an attempt to image seismically the seven depositional supersequences and related geologic features. It tests whether the observed unconformities and bedding planes can be seen on synthetic seismic data. The third-order depositional sequences were not completely simulated in the petrophysical model because the thickness of these sequences is mostly beyond seismic resolution and their inclusion would have increased the size of the computer model beyond the limitations of the modeling software. In other sedimentary environments, specifically in siliciclastic rocks, third-order sequences, which are the basis of the global sea level curve (Haq et al., 1988), seem to have generally higher sedimentation rates and therefore greater seismic significance. Under favorable conditions, even fourth-order sequences can be recognized on seismic sections (Fulthorpe, 1991). However, the modeled platform margin-to-slope transition of the Maiella is characterized by relatively low bulk sedimentation rates, and the third-order sequences are therefore thin and not modeled.

Sequence stratigraphic analysis has been performed on the synthetic sections following the procedure described by Mitchum and Vail (1977). The first step of such an analysis is to determine where reflections terminate or where two reflections merge into one. In the next step, unconformities are defined by the positions of these reflection terminations. The geometric characteristics of these unconformities are examined (surfaces of onlap, downlap, toplap, and truncation). On a seismic section, the presence of an unconformity is the only criteria to define a seismic sequence boundary. The results of the seismic sequence analyses are shown in Figures 16 and 17 and discussed below.

Chronostratigraphic Significance of Reflections

An important assumption in applying the concept of seismic sequence stratigraphy is that seismic reflections do not cross stratigraphic time lines, and thus they have chronostratigraphic significance (Vail et al., 1977b). Although other seismic modeling studies have indicated that this assumption is not automatically fulfilled (Tipper, 1993), a comparison of the Montagna della Maiella impedance model with the synthetic sections shows that no reflections significantly cut the stratigraphic time lines. The escarpment causes problems with assigning time lines to seismic reflections because the time lines merge along the escarpment. The upper part of the Upper Cretaceous platform belongs to the same supersequence as part of the slope sediments that onlap the escarpment. The correlation of sequences and time lines between the platform and slope across the escarpment unconformity cannot be resolved due to the lack of physical connections across this by-pass margin. The onlap surface of the escarpment is indicated as a seismic sequence boundary, even where it separates sediments belonging to the same supersequence (dashed line in Figure 17a). This apparent seismic sequence boundary seems to crosscut time lines. However, this is not a result of the limits in seismic resolution of the various wavelets but rather a geometric characteristic of escarpments. A seismic interpreter who is aware of the nature of platform escarpments will know that the seismic section provides no information on how much the time lines merge along the unconformity of an escarpment.

Effect of Frequency

The interpreted seismic sections with dominant frequencies of 20, 30, 40, 50, and 60 Hz and normal polarity are shown in Figures 16a–e. Figure 17 shows a simplification of the interpretation that characterizes the observed seismic sequence boundaries. The seismic sequences are labeled s0 through s6 (Figures 16 and 17), corresponding to their equivalent depositional supersequences.

At a frequency of 20 Hz, only five of the seven supersequences identified in the field are recognized on the synthetic section (Figures 16a and 17b). The oldest sequence that can be separated by seismic unconformities is the Cretaceous platform core (s0). Its apparent upper boundary is marked by the onlap of slope superse-

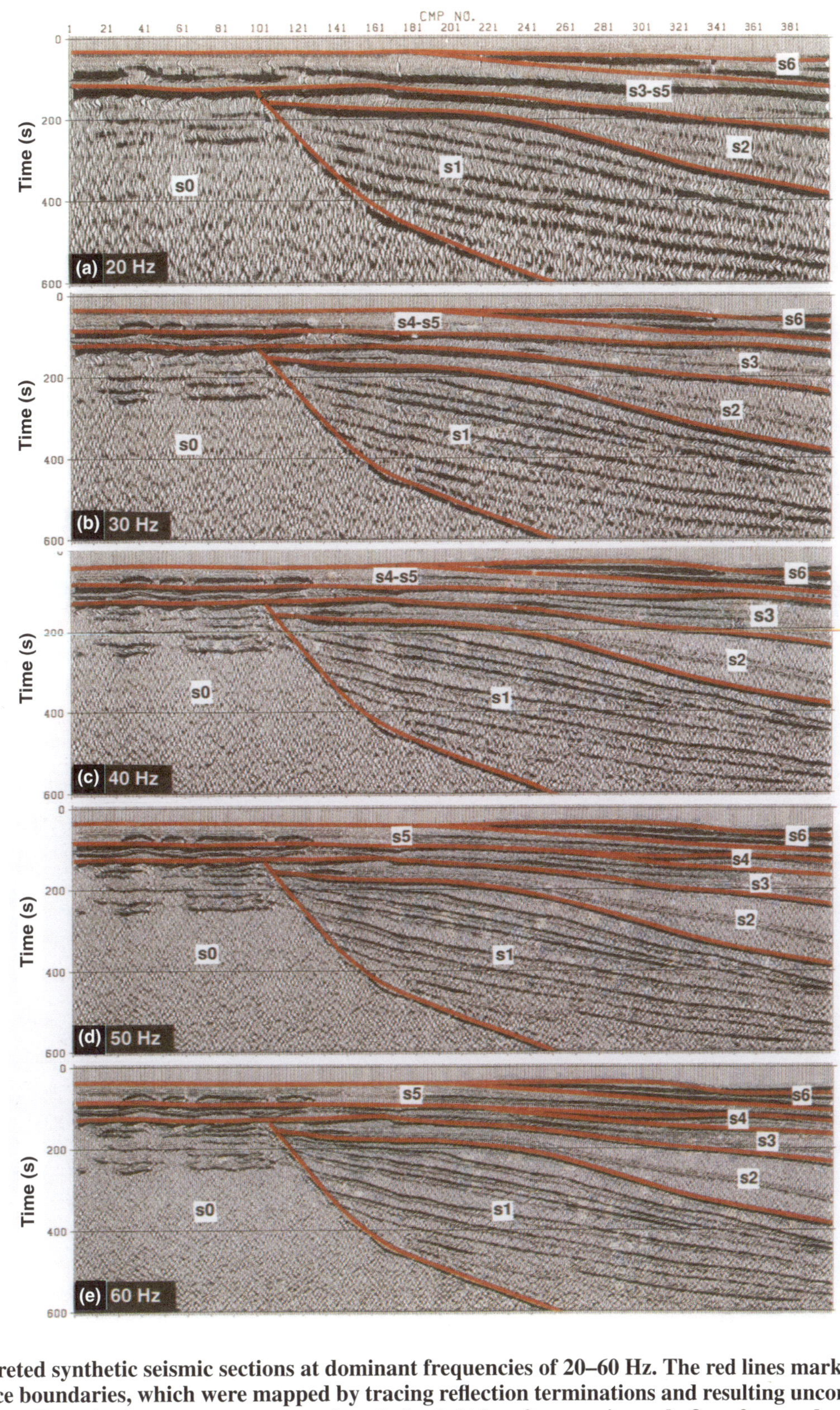

Fig. 16. Interpreted synthetic seismic sections at dominant frequencies of 20–60 Hz. The red lines mark interpreted seismic sequence boundaries, which were mapped by tracing reflection terminations and resulting unconformities on the synthetic sections. The seismic sequences are numbered s0–s6. At low frequencies, only five of seven depositional supersequences are imaged by an equivalent seismic sequence. Also, pseudo-toplaps (e.g., top of s1) and pseudo-onlaps (e.g., base of s2) are seen on low-frequency images. Only at 60 Hz does the synthetic section image the correct geometries and number of sequences.

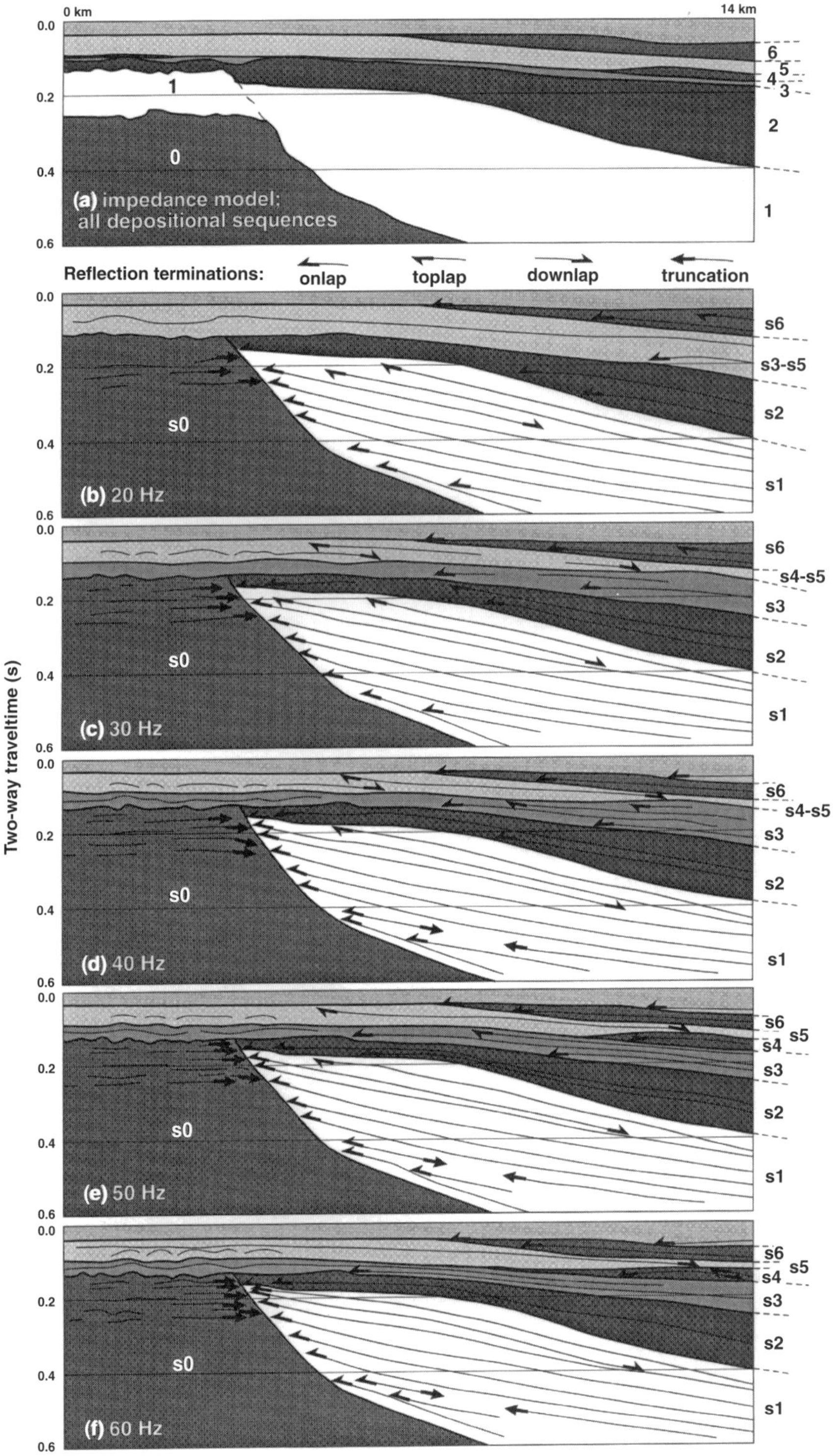

Fig. 17. Comparison of depositional supersequences 0–6, which are the basis of the impedance model, with seismic sequences s0–s6 recognized on the synthetic seismic sections (Figures 16a–e). (a) Plot of the input impedance model in time with seven depositional supersequences. (b–f) Seismic line drawings derived from the synthetic seismic sections at positive polarity and various frequencies (Figures 13 and 16). The black arrows mark reflection terminations with their termination character (onlap, toplap, downlap, and truncation); these were used to constrain the seismic sequence analyses. Unconformities define seismic sequence boundaries (bold lines), which can be compared with the depositional supersequences in part a. At 20 Hz, only five of the seven supersequences are recognized, while at 60 Hz, all seven can be mapped.

quence 1 along the escarpment (s1, Figure 16a, at CMP 100–250, 0.15–0.6 s). Seismic sequence s1 represents the true depositional supersequence 1 on the slope, whereas its platform equivalent cannot be separated from the Lower Cretaceous platform. The lower boundary of s1 is characterized by a real onlap on the escarpment, whereas at this low frequency, its upper boundary is seen as a toplap surface (Figure 16a, at CMP 150–220, 0.2 s) that is neither part of the model nor observed in reality. As shown in other models (Schlager et al., 1991), these pseudo-surfaces are the result of layers that thin below the limits of seismic resolution so that the reflections appear to terminate far from the true termination of the layers. In our model, the two uppermost megabreccias of supersequence 1 merge toward the platform and become thinner, so that their reflections terminate far from the escarpment; however, in reality, the thinned breccias onlap the escarpment. The toplap unconformity is an artifact and is thus called a pseudo-toplap.

The third seismic sequence that can be distinguished at 20 Hz is the equivalent of supersequence 2 (s2). The lower boundary of this seismic sequence is imaged by an onlap surface that is also a pseudo-unconformity. Supersequence 2 forms an overall wedge with layers thinning toward the platform. Thus, the layer thicknesses reach the limits of resolution, and their reflections terminate with an onlap geometry above the pseudo-surface of s1 (Figure 16a, at CMP 220–340, 0.2–0.3 s). The upper boundary of s2 is onlapped by the overlying sequence that is the equivalent of the three depositional supersequences 3, 4, and 5 (s3–s5). Although an onlap occurs in reality by wedge-out of some of the breccias, the surface appears too low on the profile and marks a horizon within supersequence 2 (Figure 16a, at CMP 300–350, 0.18–0.22 s). This low horizon is an effect produced by the extremely low velocity of the Calcari Cristallini of the Orfento Formation, which forms a negative impedance contrast that is too low in the positive polarity section due to the reversed shape of the reflected signal. The thin platform equivalents of supersequences 3 and 4 are not detected at 20 Hz because they are far below the vertical resolution at this frequency. Similar to the seismic image of supersequence 1, the platform part of supersequence 2 cannot be distinguished from the underlying platform core because the whole platform succession forms a concordant stack of layers that lacks geometric unconformities and is detached from the basinal time equivalents across the escarpment.

The boundaries between depositional supersequences 3, 4, and 5 on the slope cannot be resolved using the 20 Hz signal because they are too thin to produce a succession of reflections; the vertical resolution at 20 Hz is 50–75 m. The next seismic sequence above onlaps the top of s3–s5 and matches depositional supersequence 6 (s6, Figure 16a, at CMP 200–360, 0.05–0.1 s). This onlap of bryozoan limestones and hemipelagic marls onto the underlying slope sediments is real. The succession of two high-velocity bryozoan limestones with the two interbedded low-velocity marls forms a package with sufficient impedance contrast and layer thickness to be imaged as reflections, so that this onlapping sequence can be recognized at the lowest frequency of 20 Hz.

With increasing frequencies, the appearance of the lowest two sequences (s0 and s1) on the seismic sections does not change significantly (Figures 16b–e). The pseudo-toplap at the upper boundary of s1 disappears, but a true onlap appears along the escarpment. The pseudo-toplap terminations of the two upper megabreccias move from distal positions at low frequencies toward the escarpment with increasing frequencies. At 60 Hz, both reflection terminations are true onlaps (Figure 16e, at CMP 110, 0.17–1.2 s). This evolution demonstrates how increasing resolution at higher frequencies changes not only the appearance but also the type of a seismic unconformity.

The positions of the two reflection terminations that define the pseudo-onlap at the base of s2 (Orfento Formation) also move updip with increasing frequencies, but continue to form an onlap geometry (Figures 16b–e, at CMP 120–250, 0.18–0.22 s). At higher frequencies, these two reflections merge to one because they are formed only by the negative impedance contrast of a single horizon. A wavelet without the two negative lobes of the zero-phase Ricker wavelet (Figure 2) would not have produced two reflections, but the onlap geometry is independent of the chosen wavelet.

The main differences at increasing frequencies concern seismic sequences s3–s5. At 30 Hz, a low-amplitude downlap is observed within s3–s5 (Figure 16b, downlaps at CMP 170, 0.08 s and CMP 340, 0.12 s). This surface is not necessarily a sequence boundary, but could represent a maximum flooding surface.

No additional unconformities can be recognized at 40 Hz. However, a subtle reflection on the lower slope seems to form an onlap just below the sequence defined on the 30 Hz plot (s4–s5, Figure 16c, at CMP 340, 0.15 s). At 50 Hz, this onlap appears stronger and a new seismic sequence (s4) can be distinguished. The upper boundary of this sequence is now the downlap surface of the next overlying sequence s5. This new sequence represents depositional supersequence 4, which forms a limited onlapping wedge with breccias, turbidites, and pelagic carbonates on the lower slope. At 60 Hz, s4 is imaged clearly with an additional subtle intrasequence reflection.

The onlap of the Gessoso-Solfifera evaporites on the top of supersequence 6 is imaged at frequencies higher than 30 Hz (Figures 16 b–e, at CMP 370, 0.06 s). This sequence is not included in the interpretation because it was not part of the analysis of physical properties and was only modeled with a single layer as the upper limit of the modeled section.

The succession of sequence analyses at different frequencies clearly documents that seismic images do not simply mimic true geologic conditions and demonstrates the limitations of performing stratigraphic sequence analyses on seismic sections with low dominant frequencies. Only five of the seven supersequences were recognized at low frequencies. Using signals with increased frequencies allows the remaining two supersequences to be resolved. In addition, the geometric criteria used to discern individual sequences are often artifacts produced by pseudo-unconformities. Furthermore, the position of seismic sequence boundaries may be different from the positions of geologic sequence boundaries, especially if such boundaries are characterized by negative impedance contrasts and if positive polarity plots are used.

The correctly scaled overlay (Figure 18) of the synthetic seismic section over a photograph of the valley flank outcrop (Figure 6) demonstrates that, at low seismic frequencies, all major elements of the modeled rock section are well imaged by the seismic survey. Nevertheless, the overlay in Figure 18 reminds any field geologist that geologic information observed in outcrop on a microscale has drastically higher resolution than macroscale seismic images.

Conclusions and Implications

The seismic response of the exposed carbonate platform margin in the Montagna della Maiella can be in-

Fig. 18. Plot of synthetic seismic section (30 Hz, Figure 13c) over the modeled valley flank outcrop (Valle dell'Orfento, Figure 6). This overlay, distorted and plotted at approximately the same scale and position, show that all major elements of the geologic section (e.g., escarpment, megabreccias, and reefs) are well imaged. However, seismic resolution is not comparable to detailed geologic information obtained from outcrops.

vestigated by combining the technique of seismic modeling with laboratory measurements of physical rock properties. The applied normal incidence ray-tracing method gives reliable results, such that no time-consuming prestack processing has to be performed. The calculated synthetic seismic sections reveal that various depositional environments can be imaged by characteristic seismic facies. These different seismic responses can be related to differences in acoustic impedance. Prominent geologic boundaries and layers, such as the platform top, platform escarpment, megabreccias, reefal units, and patch reefs, generally have sufficient impedance contrasts and lateral and vertical extents above seismic resolution to allow them to be imaged on the synthetic sections as individual reflections. The narrow range of impedance values within the platform sediments yields a low-reflectivity to transparent seismic facies for shallow-water carbonates. The much larger range in impedance measured in slope deposits causes the slope carbonates to be seismically imaged with a succession of high-amplitude reflections. This pattern seen in the synthetic sections can be confirmed on real seismic sections from similar platform margin environments.

In addition to the distribution of physical rock properties, the seismic image is affected by several seismic parameters that vary during seismic modeling. These technical parameters influence the number and characteristics of reflection terminations and related seismic unconformities, which is especially important for testing the concept of seismic sequence stratigraphy. Seismic frequency is the most important parameter because it determines the wavelength of the acoustic signal and thus the limits of seismic resolution. Increasing frequency improves resolution, so that more geologic details can be recognized. Modeling shows that the number and positions of the reflection terminations change at different frequencies. At low frequencies, some unconformities that are part of the model are not recognized or are recognized only with incorrect characteristics (pseudo-unconformities). They can be resolved correctly with increasing frequency. The threshold frequency at which the unconformities are correctly imaged is variable and depends on the geometric characteristics of the unconformity. In the models, several pseudo-unconformities observed with 20 Hz dominant signals shows their true patterns when the signal frequencies are increased to 40 and 60 Hz.

In addition to signal frequency, different amplitude scales and different levels of background noise can mask low-amplitude reflections or make them visible on a seismic section. Therefore, the S/N of the seismic data is crucial for recognition of geologic features, in particular for low-coherent reflections. Platform carbonates with a narrow range in impedance, such as the Lower Cretaceous platform section in the Montagna della Maiella, are not resolved by reflections in the synthetic sections if noise is added, even though they form layers that have lateral and vertical extents far above the limits of seismic resolution.

An intercalation of layers with high impedance ranges but with vertical extents below seismic resolution, such as the modeled slope sediments, can still be imaged by a succession of reflections. Seismic modeling shows that the observed reflections are the result of interference from several thin layers rather than just the effect of one particular reflector. Because velocities are relatively homogeneous within an individual layer, the variability in velocity or impedance is much higher in a direction vertical to bedding than in a direction parallel to bedding. Consequently, the reflections can still image the sedimentary bedding direction and do not significantly cross-cut time lines.

In our models, the seismic stratigraphic requirement that reflections should have chronostratigraphic significance is thus fulfilled. Only along the reflection of the platform escarpment do time lines merge because the escarpment physically separates time-equivalent platform from onlapping slope layers, both belonging to the same supersequence. However, this is not a problem of seismic imaging but a characteristic of the original depositional geometry.

Our synthetic sections document that the number of recognizable seismic sequences separated by unconformities is frequency dependent. In the Maiella example, all seven modeled supersequences can be recognized with the highest frequency signal of 60 Hz, whereas two sequences are missed using the lowest frequency of 20 Hz. In general, the concordant platform parts of the sequences cannot be distinguished when the platform deposits are detached from their basinal equivalents across an escarpment. Several observed pseudo-unconformities may result in a misleading interpretation of depositional environments or rock type, but not necessarily in erroneous stratigraphic correlations obtained by seismic sequence analyses. The use of seismic sequences for stratigraphic purposes is more suspect because not all supersequences are recognized at low frequencies. This means that information concerning long time periods is missing on the section or sequences are wrongly correlated when the section is compared with other sections or with the global cycle chart of Haq et al. (1988).

In addition, the modeled second-order supersequences are only partially recognized on low-frequency seismic data, which demonstrates that high sedimentation rates are required for sequence analysis. In the studied case of a pure carbonate platform margin, the smaller scale third-order depositional sequences are mostly not discernible on the synthetic section and cannot be used to perform seismic stratigraphic correlations. Only in the Cretaceous slope sediments with onlapping megabreccias are sedimentation rate and geometric characteristics sufficient to result in third-order sequences that can be seen on seismic sections. Despite all these limitations, we show that carbonate rocks, due to their diagenetic reactivity and related high impedance range, have a high potential to be imaged correctly by seismic survey methods.

In summary, the modeled synthetic sections enable us to relate carbonate depositional facies to seismic facies and to explain several characteristics of seismic images of carbonate platform-to-basin settings. The most critical factors for a misleading seismic image are (1) insufficient impedance contrasts, (2) thin-bedded impedance patterns below seismic resolution that can result in pseudo-unconformities, (3) low seismic frequencies causing low seismic resolution, and (4) low S/Ns. An understanding of the seismic imaging process and its limitations is essential for any seismic interpreter to avoid misinterpretations of seismic data. Geologic boundary conditions, such as expected sedimentation rates (which influence sequence thickness) and expected lithologies (which determine impedance variations), should thus be incorporated into any seismic interpretation.

Acknowledgments

We would like to thank Adam Vecsei, Diethard Sanders, Maria Mutti, and Johannes Pignatti for providing their field data and inspiring discussions. Alan Green and Heinrich Horstmeyer from the reflection seismology group of the Geophysical Institute of the ETH offered to use their seismic modeling program and discussed the results. The Swiss National Science Foundation is acknowledged for financial support of the Maiella project over several years. The first series of velocity measurements were performed at the petrophysics laboratory at the University of Geneva, where Souad Sellami and Catia Moschella provided access and help. ELF Italiana S.P.A. in Pescara was so nice to release a copy of their offshore Adriatic seismic section. The manuscript was improved by helpful comments of the editors and a SEG reviewer.

References

Accarie, H., 1988, Dynamique sédimentaire et structurale au passage plateforme/bassin. Les faciés carbonatés Crétacés et Tertiares: Massif de la Maiella, Abruzzes Italie: École de Mines Paris, Mém. Sci. de la Terre **5**, 162.

Anselmetti, F. S., 1994, Physical properties and seismic response of carbonate sediments and rocks: Ph.D. thesis, Eidgenössische Technische Hoshschule, Zürich.

Anselmetti, F. S., and Eberli G. P.,1993, Controls on sonic velocity in carbonates: Pure Appl. Geophys., **141**, 287-323.

Bernoulli, D., and Jenkyns, H. C., 1974, Alpine, Mediterranean, and central Atlantic Mesozoic facies in relation to the early evolution of the Tethys, *in* Dott, Jr., R. H., and Shaver, R. H., Eds., Modern and ancient geosynclinal sedimentation: Soc. Econ. Paleont. Mineral. Spec. Publ. **19**, 129-160.

Bernoulli, D., Eberli, G. P., Pignatti, J. S., Sanders, D., and Vecsei, A., 1992, Sequence stratigraphy of Montagna della Maiella: Quinto simposio di ecologia e paleoecologia delle comunitá bentoniche, Paleobenthos V, libro-guida delle escursioni: Universitá "La Sapienza," Rome, 85-109.

Bice, D. M., and Stewart, K. G., 1990, The formation and drowning of isolated carbonate seamounts: tectonic and ecological controls in the northern Apennines, *in* Tucker, M. E., Wilson, J. L., Crevello, P. D., Sarg, J. R., and Read, J. F., Eds., Carbonate platforms: facies, sequences, and evolution: Internat. Assn. Sed. Spec. Publ. **9**, 145-168.

Biddle, K. V., Schlager, W., Rudolph, K. W., and Bush, T. L., 1992, Seismic model of a progradational carbonate platform, Picco di Vallandro, the Dolomites, northern Italy: AAPG Bull. **76**, 14-30.

Birch, F., 1960, The velocity of compressional waves in rocks to 10 kilobars, part 1: J. Geophys. Res., **65**, 1083-1102.

Campbell, A. E., and J. Stafleu, 1992, Seismic modeling of an Early Jurassic drowned platform: The Djebel Bou Dahar, High Atlas, Morocco: AAPG Bull., **76**, 1760-1777.

Carta Geologica d'Italia, 1970, Foglio 147, Lanciano, 1:100,000: Servizio Geologico d'Italia.

Crescenti, U., Crostella, A., Donzelli, G., and Raffi, G., 1969, Stratigrafia della serie calcarea dal Lias al Miocene nella regione Marchigiano-Abruzzese, parte II—Litostratigrafia, Biostratigrafia, Paleogeografia: Mem. Soc. Geol. Italiana **8**, 343-420.

De Alteriis, G., and Aiello, G., 1993, Stratigraphy and tectonics offshore of Puglia, Italy, southern Adriatic Sea: Mar. Geol., **113**, 233-253.

Eberli, G. P., and Ginsburg. R. N., 1989, Cenozoic progradation of northwestern Great Bahama Bank, a record of lateral platform growth and sea-level fluctuations, *in* Crevello, J. L., Wilson, J. L., Sarg, J. F., and Read, J. F., Eds., Controls on carbonate platform and basin development:

Soc. Econ. Paleon. Mineral. Spec. Publ. **44**, 339-351.

Eberli, G. P., Bernoulli, D., Sanders, D., and Vecsei, A., 1993, From aggradation to progradation: the Maiella platform, Abruzzi, Italy, *in* Simo, J. T., Scott, R. W., and Masse, J.-P., Eds., Cretaceous carbonate platforms: Am. Assn. Petr. Geol., Mem. **56**, 213-232.

Fagin, S. W., 1991, Seismic modeling of geologic structures: Soc. Explor. Geophys.

Fontaine, J. M., Cussey, R., Lacaze, R., Lanaud, R., and Yapaudjian, L., 1987, Seismic interpretation of carbonate depositional environments: AAPG Bull., **71**, 281-297.

Fulthorpe, C. S., 1991, Geologic control on seismic sequence resolution: Geology, **19**, 61-65.

Galloway, W. E., Marshall, S. Y., and Whipple, A. P., 1977, Seismic stratigraphic model of depositional platform margin, eastern Anadarko Basin, Oklahoma, *in* Payton, C. E., Ed., Seismic stratigraphy—applications to hydrocarbon exploration: Am. Assn. Petr. Geol. Mem. **26**, 439-449.

Haq, B. U., Hardenbol, J., and Vail, P., 1988, Mesozoic and Cenozoic chronostratigraphy and cycles of sea-level change, *in* Wilgus, C. K., Hastings, B. S., Kendall, C. G. St. C., Posamentier, H. W., Ross, C. A., and Van Wagoner, J. C., Eds., Sea-level changes—an integrated approach: Soc. Econ. Paleont. Mineral. Spec. Publ. **42**, 71-108.

Hine, A. C., and Neumann, C., 1977, Shallow carbonate-bank-margin growth and structure, Little Bahama Bank, Bahamas: AAPG Bull., **61**, 376-406.

Jones, T., and Nur, A., 1982, Seismic velocities and anisotropy in mylonites and the reflectivity of deep crustal fault zones: Geology, **10**, 260-263.

Kendall, C. G. St. C, and Schlager, W., 1981, Carbonates and relative changes in sea level: Mar. Geol., **44**, 181-212.

Kenter, J. A. M., and Ivanov, M., 1995, Parameters controlling acoustic properties of carbonate and volcaniclastic sediments at sites 866 and 869: Ocean Drill. Proj., Sci. Res., **143**, 287-303.

Meckel, Jr., L. D., and Nath, A. K., 1977, Geologic considerations for stratigraphic modeling and interpretation, *in* Payton, C. E., Ed., Seismic stratigraphy—applications to hydrocarbon exploration: Am. Assn. Petr. Geol. Mem. **26**, 417-438.

Middleton, M. F., 1987, Seismic stratigraphy of Devonian reef complexes, northern Canning Basin, Western Australia: AAPG Bull., **71**, 1488-1498.

Mitchum, Jr., R. M., and Vail, P. R., 1977, Seismic stratigraphic interpretation procedure, *in* Payton, C. E., Ed., Seismic stratigraphy—applications to hydrocarbon exploration: Am. Assn. Petr. Geol. Mem. **26**, 117-133.

Mitchum, Jr., R. M., Vail, P. R., and Thompson, S., 1977, The depositional sequence as a basic unit for stratigraphic analyses, *in* Payton, C. E., Ed., Seismic stratigraphy—applications to hydrocarbon exploration: Am. Assn. Petr. Geol. Mem. **26**, 53-62.

Moussavian, E., and Vecsei, A., 1995, Paleocene reef sediments from the Maiella carbonate platform, Italy: Facies, **32**, 213-222.

Mutti, M., 1995, Porosity development and diagenesis in the Orfento supersequence and its bounding unconformities, Upper Cretaceous, Montagna della Maiella, Italy *in* Budd, D., Saller, A., and Harris, M., Eds., Unconformities in carbonate strata: their recognition and the significance of associated porosity: Am. Assn. Petr. Geol. Mem. **63**, 141-158.

Mutti, M., Bernoulli, D., and Eberli, G. P., 1994, Progradation of the Late Cretaceous sequences of Montagna della Maiella: the importance of changes in sea-floor topography and sediment reworking in sequence organization: 15th Regional Mtg., Internat. Assn. Sed., Abstracts, 300-301.

Neidell, N. S., and Poggiagliolmi, E., 1977, Stratigraphic modeling and interpretationÑgeophysical principles and techniques, *in* Payton, C. E., Ed., Seismic stratigraphy—applications to hydrocarbon exploration: Am. Assn. Petr. Geol. Mem. **26**, 389-416.

Popenoe, P., Kohout, F. A., and Manheim, F. T., 1984, Seismic reflection studies of sinkholes and limestone dissolution features on the northeastern Florida shelf: 1st Multidiscipl. Conf. on Sinkholes, Proc., 43-57.

Rafavich, F., Kendall, C. H. St. C., and Todd, T. P., 1984, The relationship between acoustic properties and the petrographic character of carbonate rocks: Geophysics, **49**, 1622-1636.

Rudolph, K. W., Schlager, W., and Biddle, K. T., 1989, Seismic models of a carbonate foreslope-to-basin transition, Picco di Vallandro, Dolomite Alps, northern Italy: Geology, **17**, 453-456.

Rudolph, K. W., and Greenlee, S. M., 1991, Seismic modeling of a pinnacle reef: an example from the Williston Basin, *in* Fagin, Ed., Seismic modeling of geologic structures: Soc. Expl. Geophys., 197-208.

Sanders, D. G. K., 1994, Carbonate platform growth and erosion: the Cretaceous–Tertiary of Montagna della Maiella, central Italy: Ph. D. thesis, Eidgenössische Technische Hoshschule, Zürich.

Sarg, J. F., 1988, Carbonate sequence stratigraphy, *in* Wilgus, C. K., Hastings, B. S., Kendall, C. G. St. C., Posamentier, H. W., Ross, C. A., and Van Wagoner, J. C., Eds., Sea-level changes—an integrated approach: Soc. Econ. Paleont. Mineral. Spec. Publ. **42**, 155-188.

Schlager, W., 1992, Sedimentology and sequence stratigraphy of reefs and carbonate platforms: Am. Assn. Petr. Geol. Continuing Ed. Course Notes **34**, 7.

Schlager, W., Biddle, K. T., and Stafleu J., 1991, Picco di Vallandro, Dürrenstein—a platform-basin transition in outcrop and seismic model: Dolomieu Conf. Carbonate Platforms and Dolomitization, Guidebook Excursion D, 22.

Schönberger, M., 1974, Resolution comparison of minimum-phase and zero-phase signals: Geophysics, **39**, 826-833.

Sheriff, R. E., 1977, Limitations on resolution of seismic reflections and geologic detail derivable from them, *in* Payton, C. E., Ed., Seismic stratigraphy—applications to hydrocarbon exploration: Am. Assn. Petr. Geol. Mem. **26**, 3-14.

Sheriff, R. E., 1985, Aspects of seismic resolution, *in* Berg, O. R., and Woolverton, D. G., Eds., Seismic stratigraphy II—an integrated approach: Am. Assn. Petr. Geol. Mem. **39**, 1-34.

Stafleu, J., and Schlager, W., 1993, Pseudo-toplap in seismic models of the Schlern-Raibl contact, Sella platform, northern Italy: Basin Res., **5**, 55-65.

Stafleu, J., and Sonnenfeld, M. D., 1994, Seismic models of a shelf-margin depositional sequence: upper San Andres Formation, Last Chance Canyon, New Mexico: J. Sed. Res., **B64**, 481-499.

Stafleu, J., Everts, A. J. W., and Kenter, J. A. M., 1995, Seismic models of a prograding carbonate platform: Vercors, SE France: Mar. Petr. Geol., **11**, 514-517.

Stäuble, M., and Pfiffner, O. A., 1991, Processing, interpretation, and modeling of seismic reflection data in the Molasse Basin of eastern Switzerland: Eclogae Geol. Helv., **84**, 151-175.

Tipper, J. C., 1993, Do seismic reflections necessarily have chronostratigraphic significance?: Geol. Mag., **130**, No. 1, 47-55.

Vail, P. R., Mitchum, Jr., R. M., and Thomson, S., 1977a, Relative changes of sea level from coastal onlap, *in* Payton, C. E., Ed., Seismic stratigraphy—applications to hydrocarbon exploration: Am. Assn. Petr. Geol. Mem. **26**, 63-81.

Vail, P. R., Todd, A. G., and Sangree, J. B., 1977b, Chronostratigraphic significance of seismic reflections, *in* Payton, C. E., Ed., Seismic stratigraphy—applications to hydrocarbon exploration: Am. Assn. Petr. Geol. Mem. **26**, 99-116.

Vecsei, A., 1991, Aggradation und Progradation eines Karbonatplattform-Randes: Kreide bis Mittleres Tertiär der Montagna della Maiella, Abruzzen: Mitteilungen aus dem Geologischen Institut der Eidgenössischen Technischen Hochschule und der Universität Zürich, Neue Folge **294**, 170.

Vecsei, A., Sanders, D. G. K., Bernoulli, D., Eberli, G. P. and Pignatti, J. S., in press, Sequence stratigraphy and evolution of the Maiella carbonate platform margin, Cretaceous to Miocene, Italy: Soc. Econ. Paleont. Mineral. Spec. Publ. **57.**

Wang, Z., and Nur, A., 1990, Dispersion analysis of acoustic velocities in rocks: J. Acoust. Soc. Am., **87**, 2384-2395.

Wang, Z., Hirsche, W. K., and Sedgwick, G., 1991, Seismic velocities in carbonate rocks: J. Can. Petr. Tech., **30**, 112-122.

Winkler, K. W., 1986, Estimates of velocity dispersion between seismic and ultrasonic frequencies: Geophysics, **51**, 183-189.

Zappaterra, E., 1994, Source rock distribution model of the Periadriatic region: AAPG Bull., **78**, 333-354.

Chapter 16

Development of an Exploration Concept for the Shuaiba Formation Using Seismic Sequence and Facies Analysis with Forward Modeling

K. C. Fischer
U. Möller
Schlumberger GeoQuest
Hannover, Germany

R. Marschall
Schlumberger Geco-Prakla
Hannover, Germany

Abstract

The Lower Cretaceous Shuaiba Formation, along with its Bab Member (upper Thamama Group, Aptian), is one of the most important hydrocarbon reservoirs in Abu Dhabi. It is unconformably overlain by the shales of the Nahr Umr Formation in the Wasia Group, which form the regional seal. The Shuaiba Formation represents a shelf carbonate sequence deposited in a shallow marine environment, with several lithologic units comprising lateral and vertical variations in reservoir development. The remaining potential of the Shuaiba Formation is likely to be restricted to combined structural-stratigraphic or stratigraphic traps.

In development of an exploration concept and evaluation of the prospect potential, seismic sequence and facies analysis is an important and necessary tool. Seismic stratigraphic interpretation of the Shuaiba Formation, along with core descriptions, results in a detailed lithologic model (grid size 2 m × 2 m) that is verified by forward modeling (verification loop). Here 3-D techniques can also be applied, especially 3-D amplitude versus offset (AVO). At a later stage (during ongoing production), 4-D seismics (front tracking) can be applied to optimize hydrocarbon production. Modern techniques such as horizontal drilling especially benefit from the detailed reservoir knowledge obtainable in this manner.

Low relief structures in the area are sensitive to lateral velocity variations caused by lithologic variations in the overburden; time–depth conversion should thus be verified using the wavefront method.

Geologic Setting

The Lower Cretaceous Shuaiba Formation and its Bab Member (upper Thamama Group, Aptian age) consist of a complex carbonate section that extends throughout Abu Dhabi, where it represents one of the most important hydrocarbon reservoirs. It is unconformably overlain by the Nahr Umr Shale of the Wasia Group, which represents the regional seal for the Shuaiba Formation.

The shallow-water carbonates of the Shuaiba Formation were deposited on a vast shelf covering the eastern Arabian Peninsula. The formation of intrashelf basins significantly influenced the facies distribution within the Shuaiba Formation by localizing the deposition of basinal and shallow shelf carbonates (Figure 1). A buildup-prone belt interpreted to be present in this area consists of a band of scattered mounds having a wide range of sizes. It partially rims the intrashelf basin centered in Abu Dhabi.

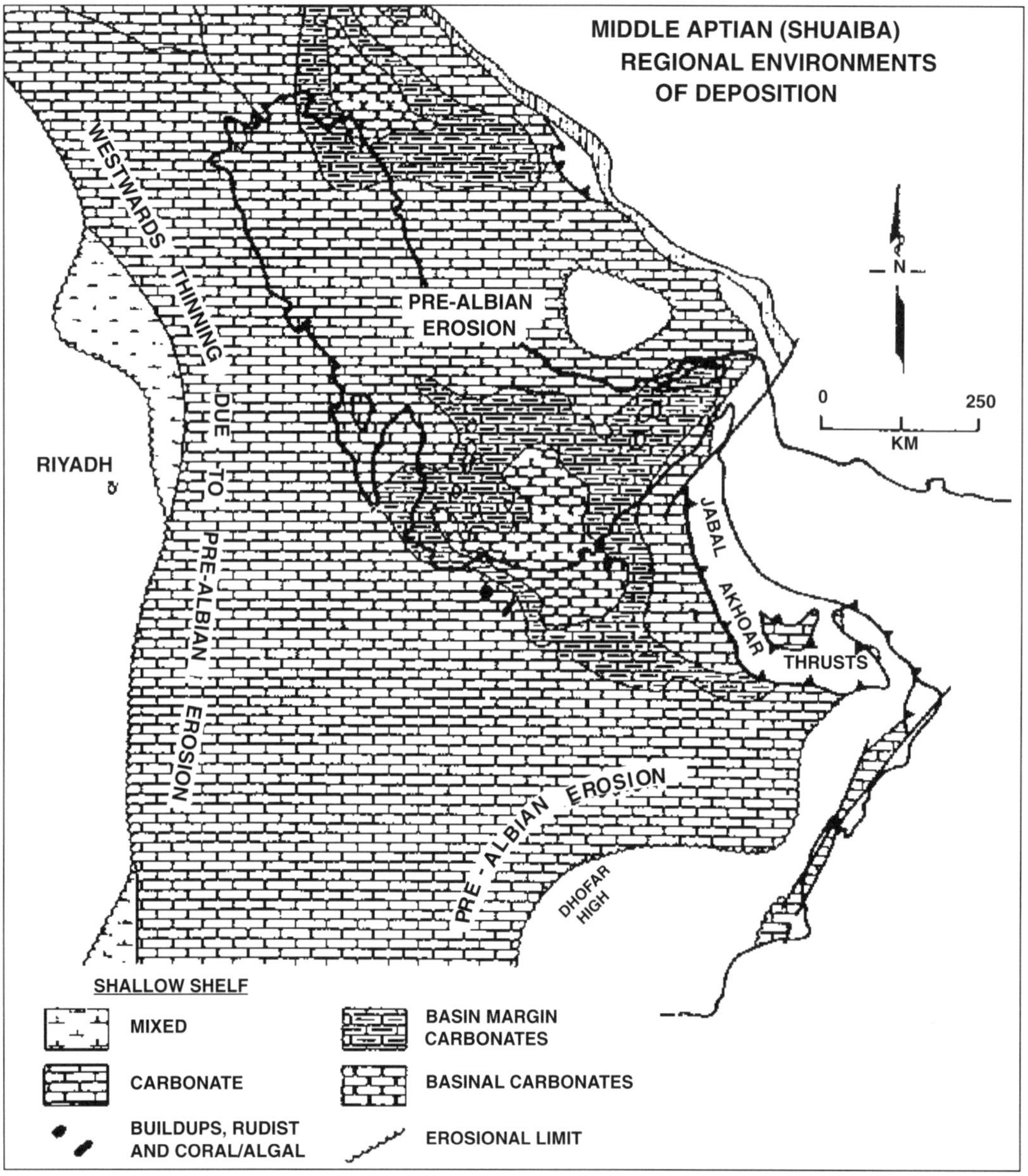

Fig. 1. Map of depositional facies of the Shuaiba Formation, on the eastern Arabian Peninsula. After Frost et al. (1983).

Shelf margin buildups, mainly formed by rudists, today form the hydrocarbon reservoirs and are the main targets of exploration. The rudist buildups were formed during the Cretaceous from the interaction of three main factors (Frost et al., 1983):

- The development of the Abu Dhabi intrashelf basin established deep, open-marine conditions on what had formerly been a shallow carbonate platform and localized a buildup-prone belt around the basin.
- Rising sea level in the early Aptian (Kauffmann, 1977; Vail et al., 1977) gave a vertical component to the buildups which contrasts with the usual sheet-like biostromes of rudists and coral-algal communities.
- The stratigraphic position of the buildups in the upper Shuaiba favorably positioned them for fresh-water leaching of the rudists and other skeletal fragments during subaerial exposure in the middle Aptian.

Stratigraphic Sequences and Play Concepts

In onshore Abu Dhabi, the Shuaiba Formation comprises a typical shelf limestone group and a basinal shale-limestone group. The shelf group consists of several stratigraphic units, and the basinal group includes a lowermost porous limestone, which is equivalent to the lowermost units of the shelf group. These units are followed by the undifferentiated interval of the remaining shelfal rock units and are topped by a shale-limestone section called the Bab Member (Figure 2).

The Shuaiba Formation is conformably underlain by the dense Hawar limestones of the Kharaib Formation and disconformably overlain by the shales of the Nahr Umr Formation. The Shuaiba Formation can be divided into three stratigraphic sequences (Abou-Choucha et al., 1990) (Figure 2):

1. A highstand sequence was deposited during a period of sea level highstand prior to the 112 Ma unconformity formed by a drop in sea level. Shelfward of the shelf margin related to this highstand, this sequence comprises the entire Shuaiba interval. The highstand sequence rapidly thins basinward and mainly contains nonporous carbonates. Within this sequence, the shelf, slope, and basin differentiation of the late highstand, with its rapid lateral facies changes, developed from the low-relief ramp setting of the early highstand, characterized by gradual facies changes.
2. A lowstand sequence was deposited after the 112 Ma drop in sea level. It comprises the basin-restricted lower Bab Member with its calcareous shales and fine-grained limestones. The sequence onlaps the lower slope of the previous highstand sequence.
3. The 108 Ma restricted highstand sequence comprises the upper Bab Member. This sequence is again basin restricted, but it onlaps the upper slope of the previous highstand. A gradation from argillaceous lime mudstones at the base upward to coarse-grained porous packstones and grainstones characterizes this sequence.

There are two main stratigraphic play concepts related to the Shuaiba Formation (Abou-Choucha et al., 1990). The primary one is Shuaiba prograding highstand multiple reef trends with their shelf margin buildups. These trends have stratigraphic relief and contain porous shoal-water carbonates. Adjacent carbonates, both shelfward and basinward, are less porous. In a strike direction from the trend, structural assist is required to provide a strike closure. The secondary play is associated with basinward and shelfward pinchouts of porosity within the upper Bab Member, and structural assist is again needed for strike closure.

Exploration Concepts

Available Data

The data available for our investigation included a poststack, time-migrated seismic line tying in two wells, from which we obtained a time–depth curve, a sonic log,

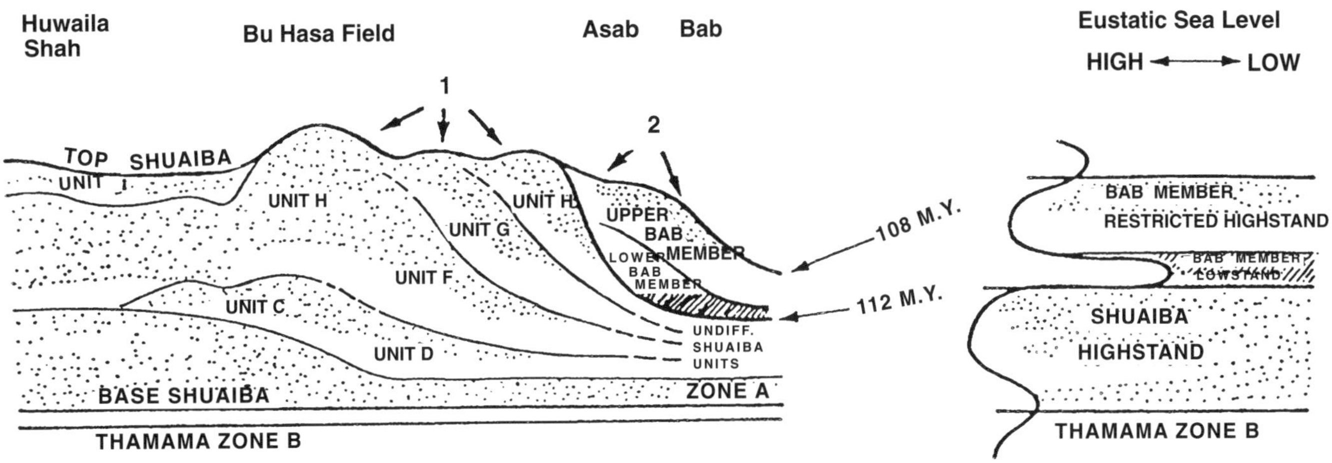

1. Shuaiba thick porous reef trends (Shelf Margin Buildups) with stratigraphic relief equivalent to less porous carbonates shelfward and basinward
2. Shuaiba Bab Member porosity pinching out basinward and/or shelfward

Porous Carbonate

BAB MEMBER Shale

Fig. 2. Schematic cross-section (left) and sequence stratigraphy (right) showing the units of the Shuaiba Formation. After Abou-Choucha et al. (1990).

a density log, a porosity log, and core descriptions. In additional, we had some general information about the wells. Well A found oil in the reefal development of the Shuaiba Formation, and drilling of well B was planned to identify the extension of the reservoir. The target was expected in a structurally higher level than in well A. However, the new well found the target in a structurally deeper position and encountered only low-porosity development in the Shuaiba.

For future activities, an exploration concept as proposed by Fischer et al. (1993, and Chapter 9, this volume) is recommended. The following steps should be taken to generate a well-defined result:

- Optimum processing of the given seismic data, including inversion
- Structural interpretation, followed by stratigraphic interpretation
- Verification of the resulting lithologic model by forward modeling
- Verification of the resulting velocity model by velocity checking by wavefronts
- Appropriate decompaction
- Definition of geophysical risk factors, which take into account the verification of the velocity field and the lithologic model

Structural Interpretation

The available seismic data showed a low-relief culmination between the two wells for the main unconformities and regional markers. However, the top Shuaiba marker did not parallel the main top Kharaib and top Nahr Umr markers. The exact definition of the top of the Shuaiba is not possible on a single line. However, a change in both the reflection pattern and the continuity in the interval of interest can be observed (Figure 3a).

A prograding reflection pattern can be recognized in the overlying intervals which indicates lateral variations in lithology and thus velocity. Simple interpolation of velocities between the two wells by itself cannot show these subtle velocity variations. Thus, the velocity model must be updated and corrected by the "velocity checking by wavefronts" method (Marschall, 1991). This method uses prestack traveltimes from constant-offset sections to verify and correct the assumed velocity model by depth migrating these traveltimes. At least two offsets with corresponding traveltimes per common midpoint (CMP) are needed here. Note that only measured quantities (traveltimes) are needed to achieve the necessary laterally varying correction. This results in an updated velocity field for time–depth conversion. The lateral velocity variations can result in a change in the structure with depth, such as a lateral shift or an opening of the culmination after time–depth conversion.

Stratigraphic Interpretation

The seismic data were reprocessed for the stratigraphic interpretation. The reprocessing resulted in an overall improvement in vertical and lateral resolution. The top of the Shuaiba, which is the target, was identified more easily and took into account the reflection terminations (toplaps) at the upper sequence boundary. A prograding internal reflection pattern also became visible. Based on these data, a seismic sequence and facies analysis was performed (see Chapter 6, this volume). The first sequence was characterized by a parallel reflection configuration directly above the base of the Shuaiba. This sequence was present along the whole section. The second sequence was characterized by a prograding internal reflection pattern with some mounds. The upper sequence boundary was marked by toplaps, and the lower was identified from downlap reflection configurations. In the northeastern part of the section, this seismic facies changed into a parallel reflection configuration combined with thinning of the sequence. The youngest sequence identified was restricted to the northeastern part of the section corresponding to the thin part of sequence 2. It onlapped onto the preceding sequence and mainly showed a parallel reflection configuration except near the southwestern termination, where mounds could be identified (Figures 3b and c).

This resulted in a sequence stratigraphic model for the Shuaiba Formation that corresponded to its general depositional model. The first sequence was deposited in a shelfal environment with little topography, resulting in roughly parallel layering. With rising sea level and establishment of the intrashelfal basin, a shelf margin began to develop. The shelf margin prograded basinward, which was reflected by the progradational reflection pattern. Several prograding cycles occurred. Subsequently, dropping sea level caused emergence of the shelf margin and deposition was restricted to the former basinal area. During a restricted sea level highstand, another shelf margin of minor extent developed basinward of the previous shelf margin. Sea level then dropped again, and after subsequent sea level rise, the Nahr Umr shale was deposited (Figures 3b and 3c). This interpretation was confirmed by the available well data.

Forward Modeling: Verification Loop

The essential step in verification is to forward model

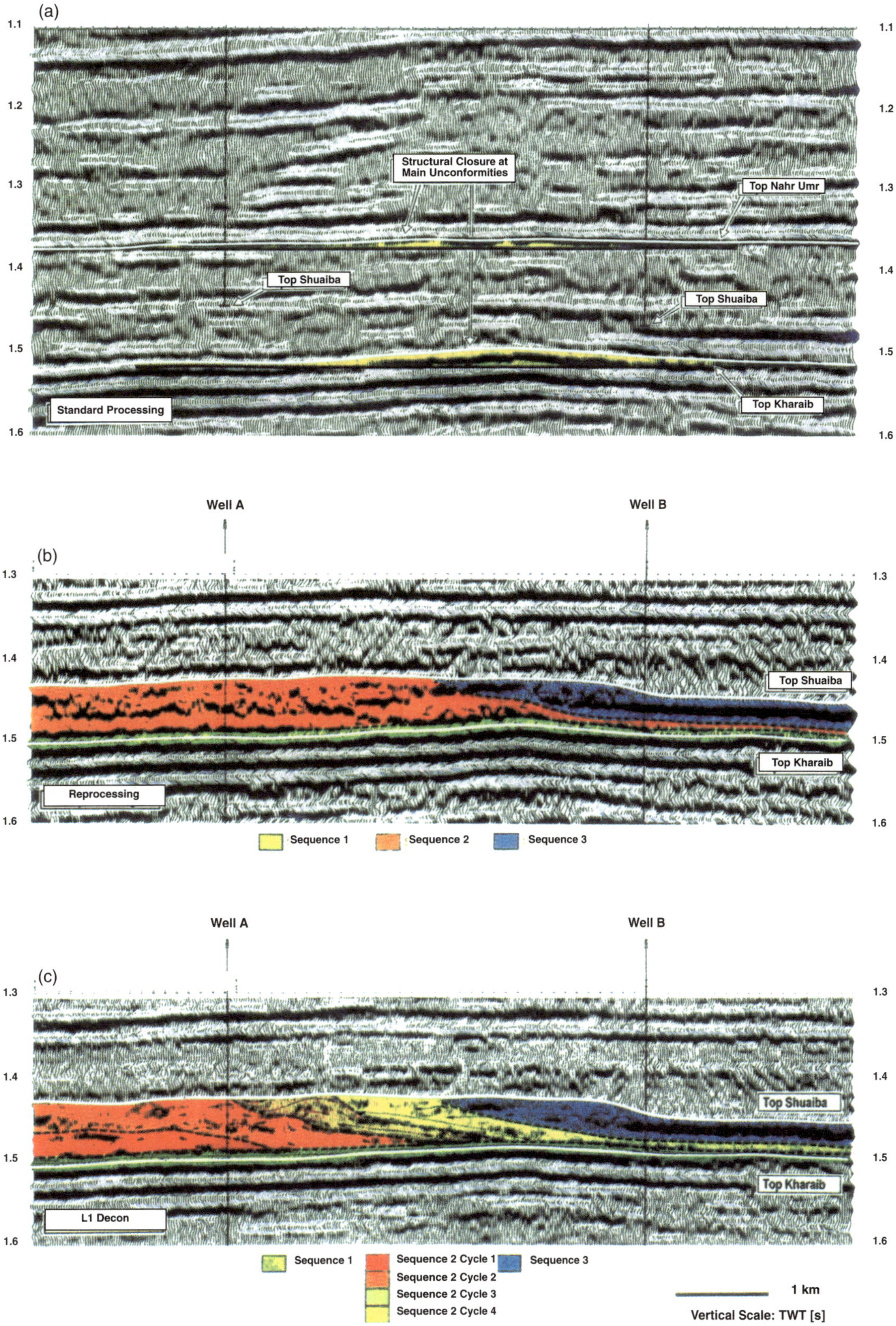

Fig. 3. (a) Seismic profile showing standard processing with structural interpretation of main unconformities (yellow) and well locations. (b) Portion of same profile reprocessed with sequence stratigraphic interpretation. (c) Same portion of profile as above but showing sequence stratigraphic interpretation based on L1-deconvolution.

an established model in terms of its acoustic and elastic response in either prestack or poststack mode to validate or correct the interpretation. The complete verification loop was composed of the following steps:

The complete verification loop of our interpretation of our seismic field data was composed of the following steps:

- Definition of the initial geologic-lithologic model
- Forward modeling, either acoustic or elastic, either in prestack or in poststack mode
- Seismic migration of the synthetic seismic data
- Comparison of the real and synthetic data
  - Visual comparison of synthetic and seismic data (in time or depth domain)
  - Numerical comparison, i.e., sample-by-sample division of synthetic and seismic data. In this case we are comparing impedances, i.e., both seismic and synthetic data have to be inverted prior to sample-by-sample division
- Reinterpretation and redefinition of the geologic-lithologic model, followed by
  - Further iterations, if necessary

For this example, the poststack or exploding reflector option (i.e., modeling of stacked sections) was applied for forward modeling.

To generate a detailed lithologic model, L_1-Deconvolution was applied to the reprocessed seismic section. This resulted in an overall higher resolution of the given data (Figure 3c). An optimum match was achieved with synthetic seismograms, calculated by applying a zero-phase 10–60 Hz wavelet (as compared to the conventionally processed data, which is 14–50 Hz).

According to the available core descriptions, seven different lithologic units were defined and assigned to the relevant facies (Table 1). The values for velocity, density, and porosity were derived from crossplots of the well logs (Figure 4). Note that porosities were derived from neutron measurements, which causes a shift of lithologic unit 7 (shale) to high porosities due to high neutron readings in shaly intervals. These shales should have almost 0% effective porosity.

The L_1-Deconvolution version of the seismic data allowed us to be able to generate a more detailed seismic facies analysis, thereby providing a higher resolution lithologic model. The lithologic units derived from the well data were assigned according to the relevant paleoenvironments (Figure 5).

In a next step, the impedance logs of the wells were blocked according to the lithologic unit boundaries identified on the seismic section corresponding to the core descriptions and log data. The resulting synthetic seismogram was compared to the original synthetics to check whether the simplification of the model had an impact on the resulting synthetic trace. According to the resulting good match, the lithologic model established was used as a first approximation (Figure 6).

The synthetic section matched the original section except for a few discrepancies. Most of these were corrected by changing the external geometry of the lithologic units. A major discrepancy occurred in the northeastern part of the section. Here lithologic unit 6 was split into two units to account for the wide spread of ve-

Table 1. Lithologic Units Used for Forward Modeling

Litho. Unit	Lithology	Paleoenvironment	Velocity (m/s)	Density (avg) (g/cm^3)	Porosity (avg) (%)
1	Bioclastic packstone-grainstone	Reef	4233	2.300	20
2	Peloidal packstone-grainstone	Reef debris	3907	2.275	23.5
3	Packstone	Upper–middle slope	3763	2.425	25.5
4	Packstone-wackestone	Middle–lower slope	4323	2.300	19
5	Mudstone	Lower slope, basin, lagoon	5166	2.650	5.5
6	Grainstone-packstone-wackestone-mudstone	Mixture	3615	2.450	—
6A	Grainstone-packstone-wackestone-mudstone	Mixture, upper ramp	3615	2.450	21.5
6B	Mudstone-wackestone-packstone-grainstone	Mixture, lower ramp	4997	2.475	9
7	Shale	Basin, transgressive systems tract	3175	2.500	—

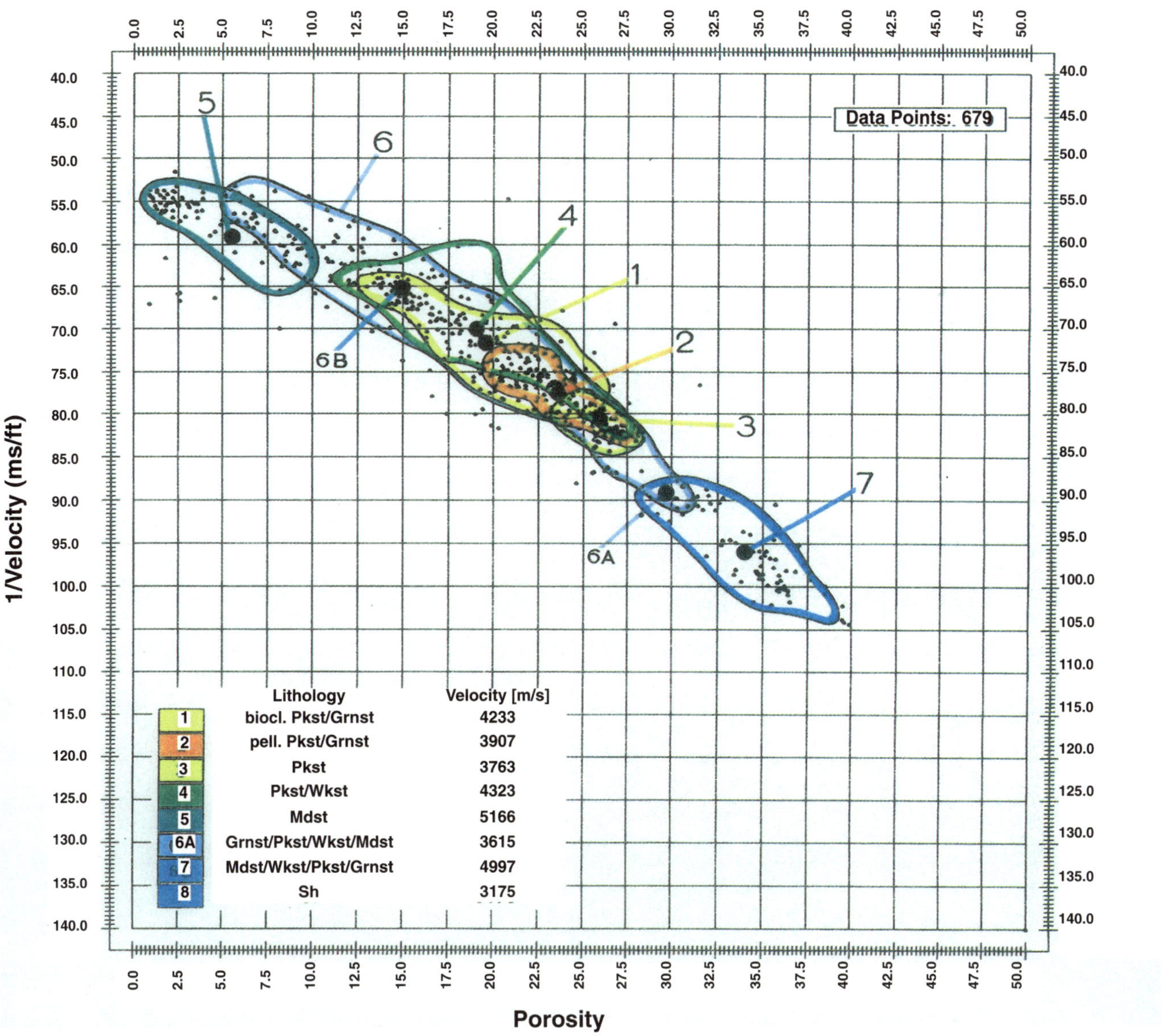

Fig. 4. Sonic versus porosity crossplot for all interpreted lithologic units of the Shuaiba Formation.

locities in this area (see units 6A and 6B in Table 1). The southwestern part remained as in the first approximation, whereas higher velocities were assigned in the northeastern part. The lateral change in lithology of the early highstand was caused by different sediments in upper and lower carbonate ramp environments.

The resulting synthetic section of the first iteration fit the original section and thus supported the lithologic model (Figure 5). The areas of good porosity were delineated in detail using the lithologies and relevant porosities assigned. As a result, recommendations for future exploration could be given.

Additional information about the target was gained from the application of amplitude versus offset (AVO) techniques (Figure 7). This figure serves two purposes. First, on the right side, it shows the synthetic AVO response of the final model at the location of well A, which was used to calibrate the AVO results by using the amplitude versus offset variation. This was obtained from the forward modeling results by interpreting the shot gather as the corresponding CMP-AVO gather. This proved to be especially useful when extended AVO techniques (resulting in an instantaneous pseudo V_P/V_S section) were applied (Nieuwland and Marschall, 1993).

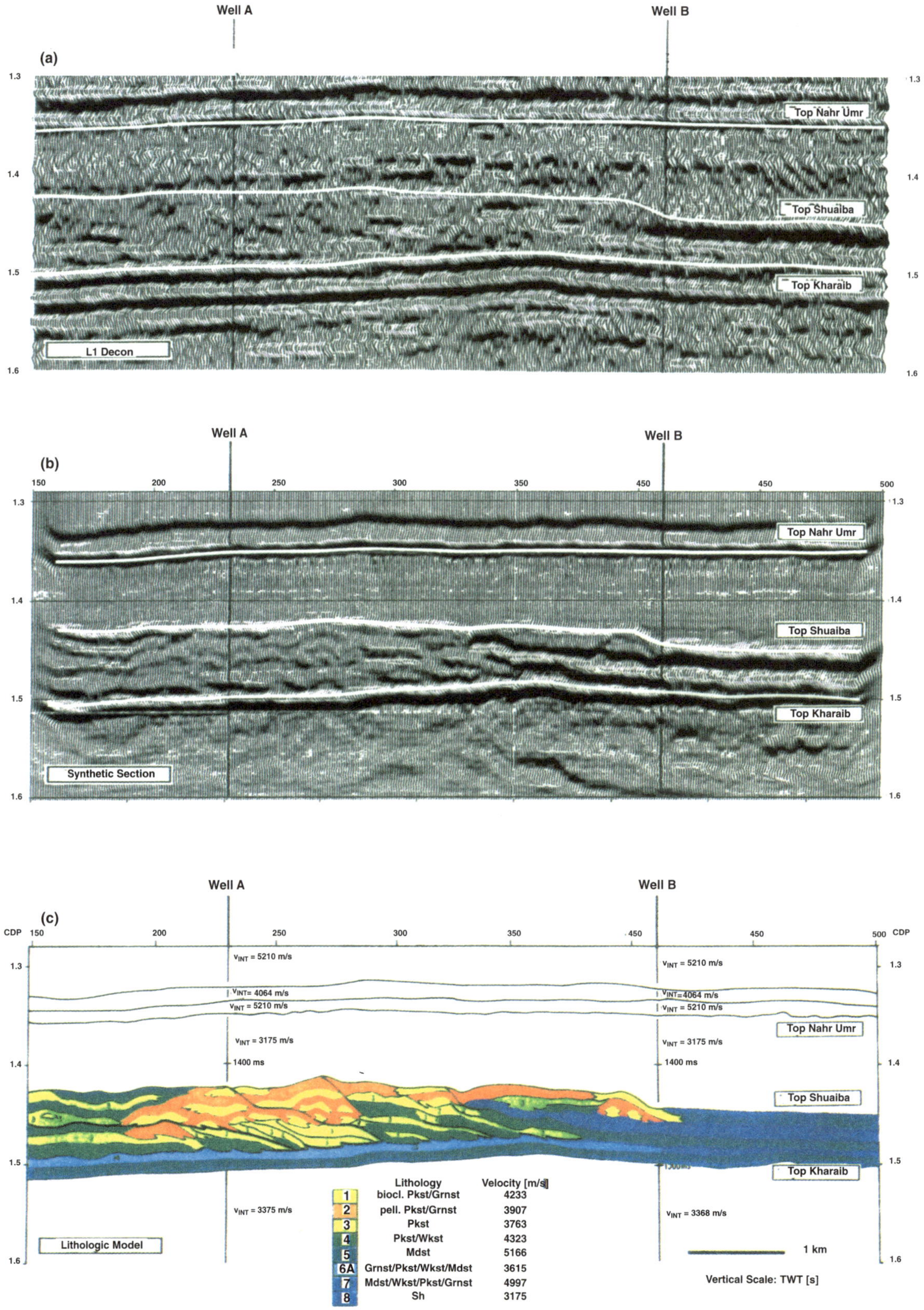

Fig. 5. Comparison of (a) L_1-deconvolution section with (b) synthetic seismic section and (c) lithologic model, first iteration. (Sections are oriented southwest-northeast.)

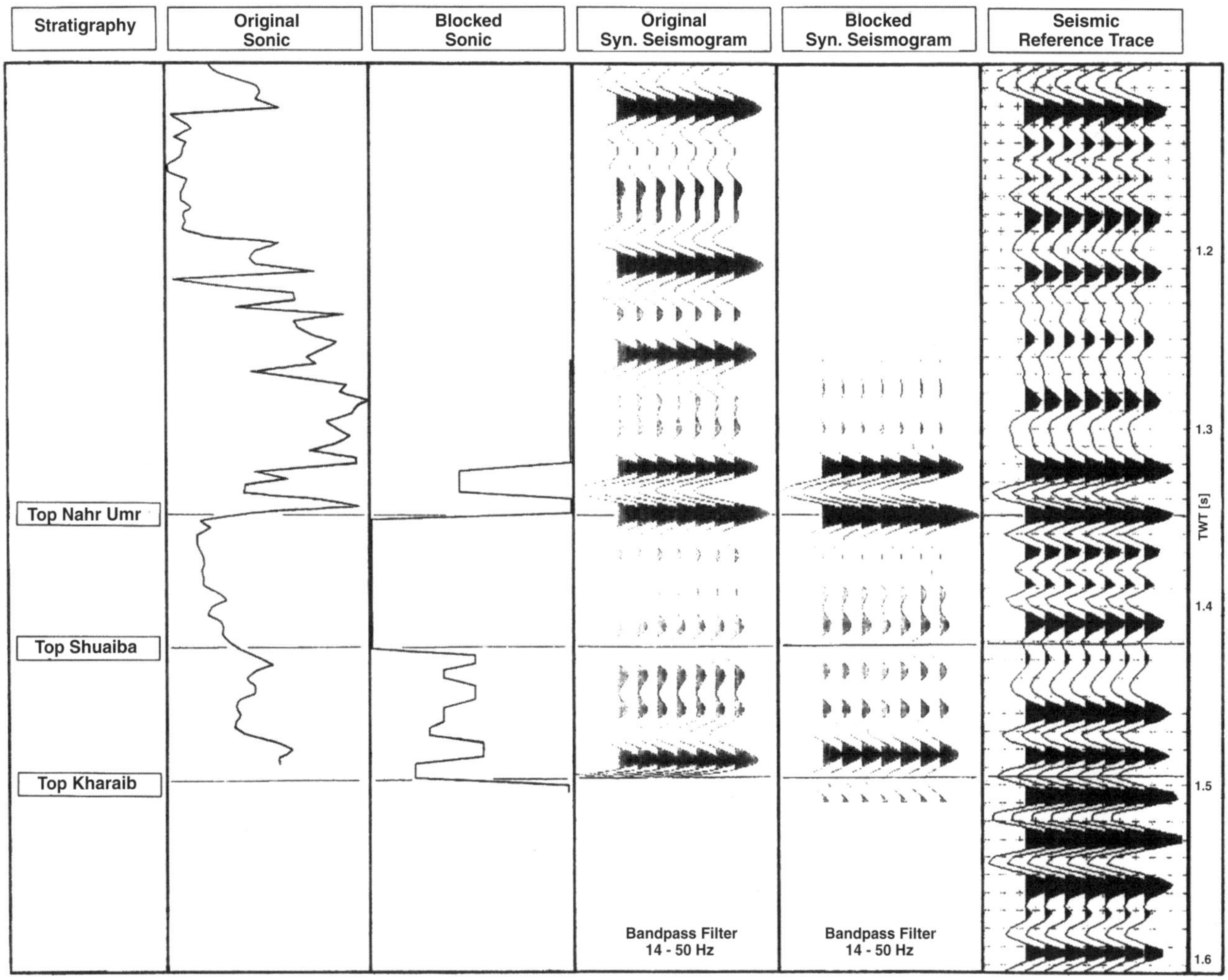

Fig. 6. Comparison of original blocked sonic and relevant synthetic seismograms with the seismic reference trace for well A.

The other purpose of Figure 7 is that it shows (on the left side) a comparison of the conventional normal incidence plane wave synthetic seismogram with the more appropriate reflectivity method synthetic. This is computed by elastic forward modeling, followed by NMO correction using *P*-wave velocities and stacking within the appropriate offset range. We believe that in the future such multi-offset elastic synthetics, based on the reflectivity method, will become the new standard, replacing the normal incidence acoustic plane wave synthetics.

The methodology applied to 3-D surveys (including 3-D AVO) enhances our knowledge of facies variations within the target and thus provides important input for further field development. The 4-D seismic method involves repeated 3-D seismic profiles over the production area which highlight areas of change in pore fill due to production. The application of this method in a later stage of ongoing production would also test the model in terms of favorable porosity and permeability trends and would thus provide additional information to optimize the exploitation of hydrocarbons.

Conclusions

A structural interpretation alone could not explain the structural and geologic setting of the target level (the Shuaiba Formation). Proper reprocessing of seismic data in close cooperation with the interpretation team improved the data quality and allowed for a stratigraphic interpretation of the data. This resulted in a structural and stratigraphic model that was consistent with the given well information and the geologic and sequence stratigraphic models. The lithologic model derived from the results was confirmed by applying a verification loop, which was done in two iterations. Additional investigations that could be applied are AVO and 3-D seismic surveys, followed later by 4-D seismic methods (or

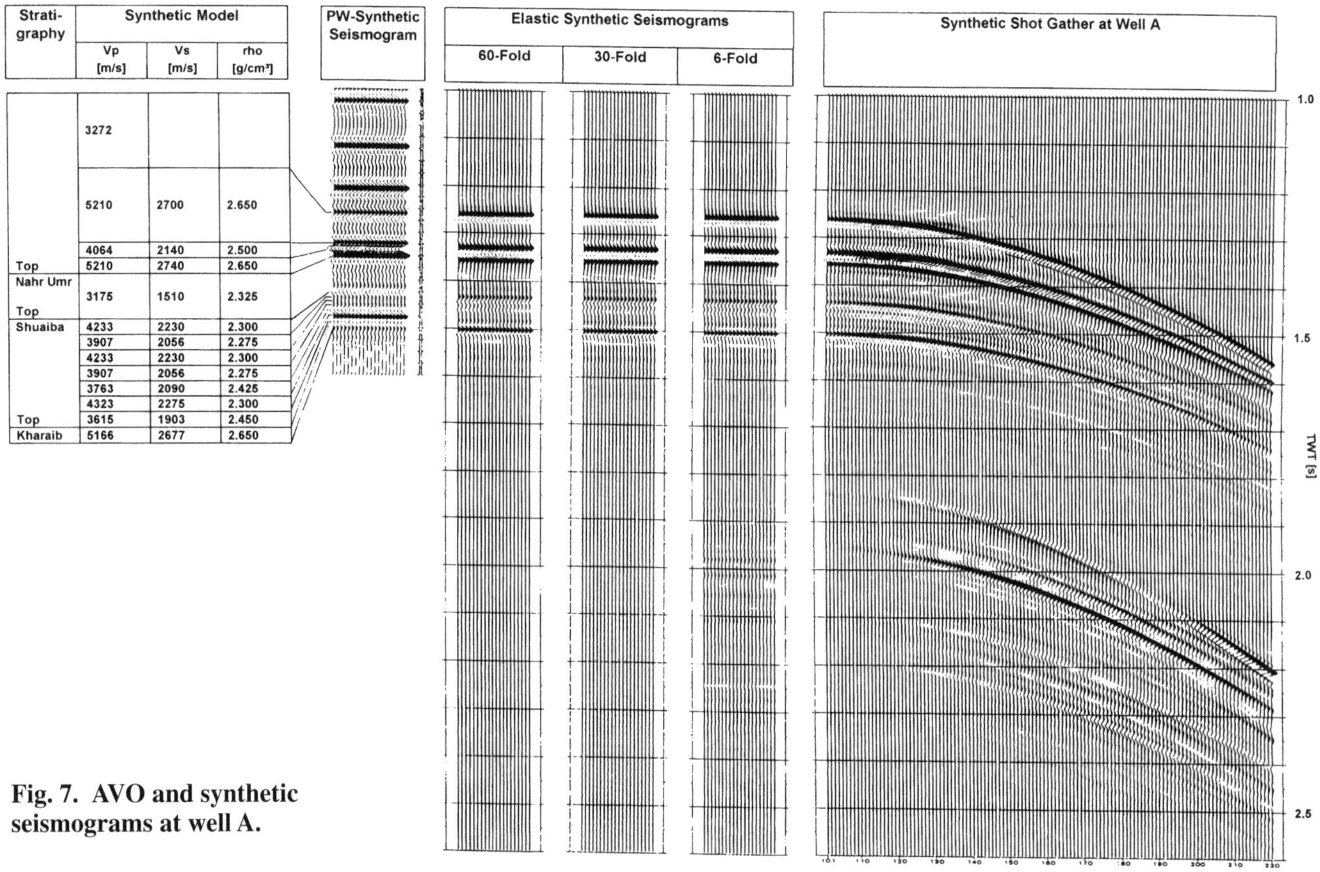

Fig. 7. AVO and synthetic seismograms at well A.

repeated data acquisition). These methods optimized the model for subsequent reservoir simulation purposes.

Acknowledgments

The authors wish to thank the Management of Abu Dhabi National Oil Company (ADNOC) for permission to publish this paper.

References

Abou-Choucha, M. K. A., and Ennadi, I. M., 1990, Possible stratigraphic traps in the Shuaiba Formation, onshore Abu Dhabi: Presented at the 4th Abu Dhabi Petr. Conf., Soc. Petr. Eng., **303**.

Fischer, K., Holling, M., Marschall, R., and Mau, J., 1993, Remarks on exploration tools: integrated exploration strategy applied to carbonate environments: Presented at the 8th Middle East Oil Show, Soc. Petr. Eng. **25579**.

Frost, S. H., Bliefnick, D. M., and Harris, P. M., 1983, Deposition and porosity evolution of a Lower Cretaceous rudist buildup, Shuaiba Formation of eastern Arabian Peninsula, *in* Harris, P. M., Ed., Carbonate buildups—a core workshop: Soc. Econ. Paleont. Mineral. Core Workshop **4**, 381-410.

Kauffman, E. G., 1977, Geological and biological overview, Western Interior Cretaceous basin: Mount. Geol., **14**, 75-99.

Marschall, R., 1991, Macro model optimization and depth migration: Proc. 2nd Intl. Congr. Braz. Geoph. Soc., Paper WOR **1.6**.

Murris, R. J., 1980, Middle East stratigraphic evolution and oil habitat: AAPG Bull., **64**, 597-618.

Nieuwland, F., and Marschall, R., 1993, From AVO to shear, a data example: Presented at the Soc. Expl. Geophys. Intl. Conf. & Exhib.

Vail, P. R., Mitchum, R. H., and Thompson, S., 1977, Seismic stratigraphy and global changes of sea level, pt. 4, global cycles of relative changes of sea level, *in* Payton, C. E., Ed., Seismic stratigraphy—applications to hydrocarbon exploration: Am. Assn. Petr. Geol. Mem. **26**, 83-97.

Chapter 17

Imaging of a Stratigraphically Complex Carbonate Reservoir with Crosswell Seismic Data

R. T. Langan
Chevron Petroleum Technology Co.
La Habra, California, U.S.A.

S. K. Lazaratos
TomoSeis, Inc.
Houston, Texas, U.S.A.

J. M. Harris
Stanford University
Stanford, California, U.S.A.

A. A. Vassiliou
Amoco EPTG
Tulsa, Oklahoma, U.S.A.

T. L. Jensen
Chevron U.S.A.
Midland, Texas, U.S.A.

J. W. Fairborn
Well Seismic Computing Services
Newport Beach, California, U.S.A.

Abstract

We conducted a crosswell survey to evaluate the stratigraphy and structure in and near a stratigraphically complex limestone reservoir in the Permian Basin of West Texas. Although the reservoir was detected with surface seismic data and was successfully drilled, the surface data had insufficient resolution to image its internal structure and to determine its relationship with nearby sedimentary rocks. This is a common problem in surveys of carbonate strata, where large seismic velocities result in long characteristic wavelengths. The thickness of the reservoir in our example is ~75 m, whereas the shortest wavelength in the surface seismic data is ~60 m. The high-resolution crosswell seismic data, which contain frequencies more than 10 times greater than those of the surface data, clearly delineate the lateral and vertical extent of the reservoir unit between the wells. The data also reveal the inner stratification of the reservoir and show the stratigraphy of units adjacent to, above, and below the reservoir. Anisotropy in the compressional velocities obtained from the crosswell data is an indicator of the facies present. Our analysis of the *P*-wave direct arrivals suggests that the reservoir itself has little velocity anisotropy but that a thinly interbedded facies of mixed siliciclastics and carbonates has a horizontal velocity of ~5% lower than its vertical velocity. The shale units have an intermediate degree of anisotropy of ~3%.

Introduction and Data Acquisition

Modern 3-D surface seismic data collected in the Midland Basin portion of the Permian Basin of West Texas (Figure 1) have revealed subtle amplitude and traveltime anomalies in the carbonate-rich Wolfcamp Formation. These anomalies are at the threshold of resolution of the seismic data. When the anomalies are drilled, they often contain reservoir-quality carbonates. Information from wells that either penetrate or narrowly miss these anomalies gives only a partial explanation. In one location, we had two closely spaced wells available for a study: one that penetrated an anomaly and another one 640 ft (195 m) away that had missed. This is shown diagramatically in Figure 2. The well that penetrated an anomaly encountered a massive limestone unit, while the well only a few hundred feet away encountered siliciclastic-rich carbonates at the same depth, which one could interpret as debris.

A crosswell survey was done between these wells to gain a greater understanding of the reservoir. Of interest is the internal structure of these anomalies, their stratigraphic relationships with adjacent strata, and the depositional environment. Because some of the operational aspects of this survey are proprietary, we can discuss the imaging results and their interpretation only in a general sense.

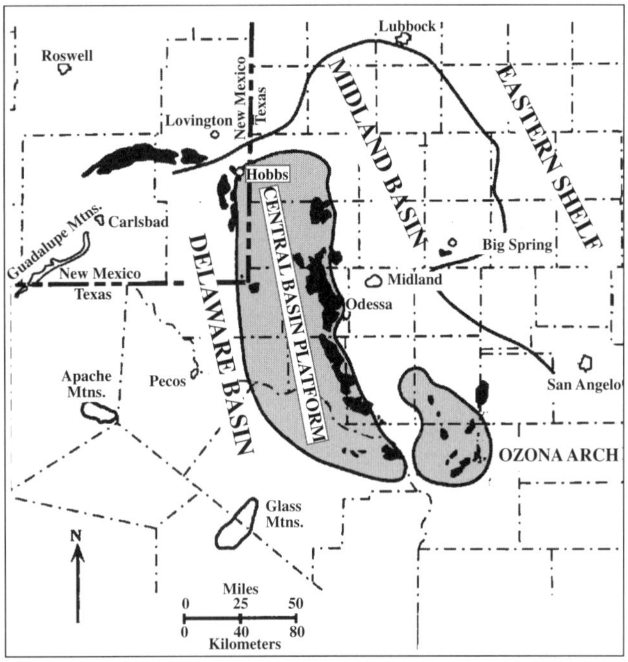

Fig. 1. Map of the eastern edge of the Central Basin platform in the Midland Basin showing the major provinces of the Permian Basin of West Texas. The seismic anomaly is located east of the eastern edge of the Central Basin platform. Oil fields producing from the Permian Grayburg Formation are highlighted in black. After Harris and Walker (1988) and Ward et al. (1986).

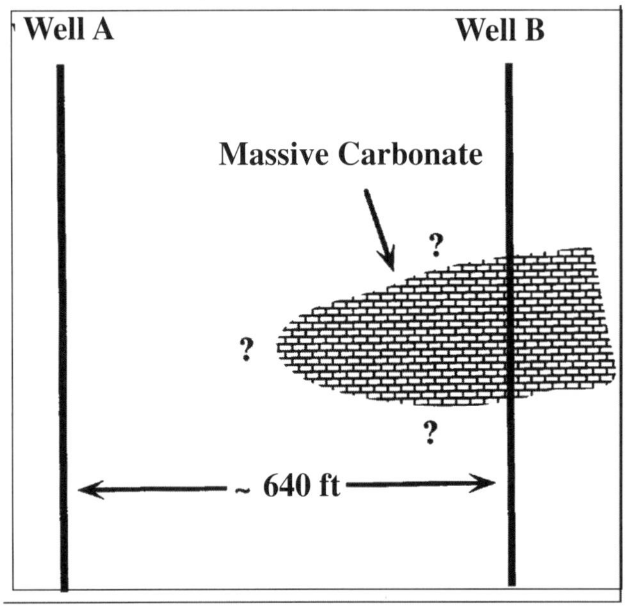

Fig. 2. Schematic diagram showing the anomalous zone of this study, known to exist from surface seismic data. Well data indicate that the zone contains massive reservoir-quality carbonate. The surface seismic data have insufficient resolution to determine the full lateral and vertical extent of the zone.

The crosswell data were collected with a piezoelectric cylindrical bender source and hydrophone receivers in the depth range of 8200–9200 ft (~2500–2800 m). The survey consisted of 201 source levels spaced at 5 ft (1.5 m) depth intervals and 203 receiver levels, also spaced 5 ft apart. This resulted in a raw data set with about 40 000 traces. The input signal was a 250–1250 Hz linear upsweep. See Harris et al. (1995) for details on the acquisition system and the techniques employed.

The data were correlated and edited in the field. The signal-to-noise ratio of the correlated data was generally good, allowing clear observation and identification of the wave modes of interest (direct *P*-wave arrivals and *P*-to-*P* reflections). We have not attempted to extract shear wave information from the data, although there is considerable *S*-wave energy present.

Data Processing

Traveltime Tomography: Isotropic Medium

We computed a *P*-wave traveltime velocity image (tomogram), assuming an isotropic medium, with an algorithm described by Harris (1994) that uses an adaptive-gridding technique. The basis of this algorithm is the definition of velocity through the use of a sparse grid of nodes. Formulation of the inversion problem in terms of node-type models provides the capability to use adaptive gridding to address the problems of nonuniform ray coverage and inhomogeneous resolution. The spacing and density of nodes are adaptively selected to provide more uniform ray density per node or to match the geometric pattern of the geologic structure being imaged. In this way, reconstruction artifacts can be reduced, while velocity estimates can be made more reliably. Also, "unknowns" are not wasted on homogeneous zones but can be concentrated in heterogeneous regions of the image. Values of velocity at any point can be derived by interpolating the values at nearby nodes.

The velocity tomogram is shown in Figure 3, along with the sonic logs from each well. The anomaly in the 3-D surface data corresponds to a depth of 8700–9000 ft (2650–2740 m) in depth in well B. The tomogram delineates an oval-shaped low-velocity zone extending from well B to about the midpoint between the wells. This low-velocity zone is only partially apparent from the sonic logs and has a velocity at its center of about 15,000 ft/s (~4600 m/s). The surrounding regions have velocities approaching or exceeding 20 000 ft/s (~6100 m/s).

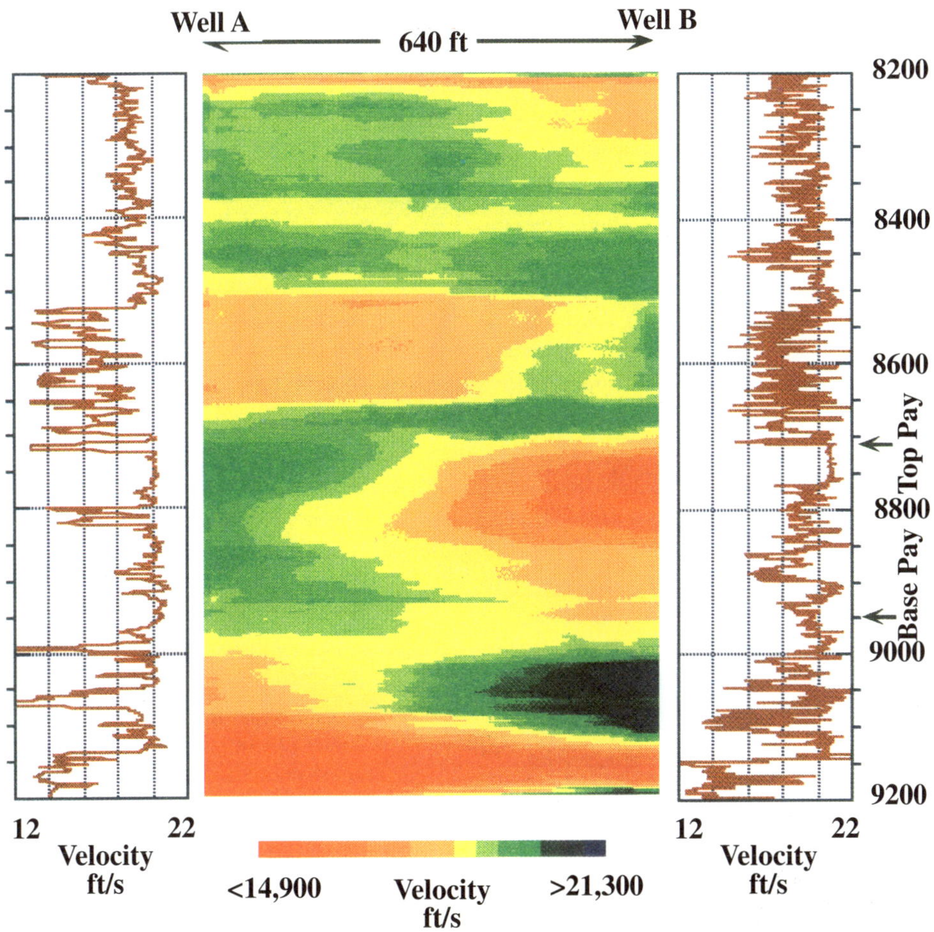

Fig. 3. The velocity image obtained using traveltime tomography for an isotropic model reveals a distinctive low-velocity zone in well B that correlates with the massive reservoir-quality carbonates. The low-velocity zone extends laterally ~350 ft (~100 m) between the wells.

Lower velocities in the carbonates of this region often correlate with higher porosities. In our example, this low velocity corresponds with a massive reservoir-quality limestone. In well A, at the same depth as the massive limestone, a mixed carbonate-siliciclastic facies occurs, with some thin beds of shale. Just above the massive carbonate is a shale unit that is about 125–150 ft (38–46 m) thick, which is expressed in the tomogram as a tabular low-velocity layer at about 8525–8675 ft (~2600–2640 m) depth. The highest velocities are associated with a mix of siliciclastic and carbonate facies rocks.

The apparent resolution in this image approaches 25 ft (~7.5 m) vertically in some places and 75 ft (~23 m) laterally. The spacing of the source and receiver levels was 5 ft (~1.5 m), which gives a spatial Nyquist of 10 ft (~3 m), so any attempt to resolve finer velocity details in the velocity structure than the Nyquist limit is improper.

Traveltime Tomography: Weak Elastic Anisotropy

If *P*-wave velocity anisotropy is present in these rocks, our velocity tomogram in Figure 3 will have errors and be biased toward the horizontal component of velocity, which is the dominant direction of energy propagation. It is possible to estimate the degree of anisotropy by using traveltime tomography and a model based on weak elastic anisotropy for *P*-wave energy (Thomsen, 1986; Vassiliou et al., 1994). Each point in the model now has three variables associated with it (instead of the single velocity variable in an isotropic medium). These three variables are denoted as V_{pv}, ε, and δ. V_{pv} is the vertical *P*-wave velocity and ε is given by

$$\varepsilon = (V_{pv} - V_{ph}) / V_{pv} ,$$

where V_{ph} is the horizontal velocity and ε is a measure of the difference between the horizontal and vertical velocities (which is one way of evaluating the degree of anisotropy present). The final variable, δ, is a measure of how the velocity distribution differs from an elliptical one. If $\delta = \varepsilon$, the distribution is double elliptical. (See Thomsen, 1986, for further the details of this model.)

Because we are solving for three variables at each point rather than one, our model must have a coarser grid than was used in our isotropic inversion, thereby reducing the apparent resolution. However, if we find that the anisotropy is relatively small, the isotropic result will be a reasonably good one. In addition, any anisotropy we derive from this inversion may be an indicator of the lithology present.

Another concern we have when trying to obtain anisotropic measures from traveltimes alone is that the system of equations for this problem can be ill-conditioned. We control part of this problem by having a wide aperture in the data acquisition. Any remaining problem with ill-conditioning is regularized through the use of the larger node spacing and by "damping" the second-order spatial derivatives of the field of velocity variables (e.g., see Pratt and Chapman, 1992; Vassiliou et al., 1994).

The vertical velocity obtained from the inversion is shown in Figure 4. Note that this velocity distribution is similar to that in Figure 3, which is the isotropic result (although the color scales are slightly different). The image in Figure 4 is also smoother because of the coarser grid spacing and the degree of "damping" necessary to obtain a stable solution.

The measure of anisotropy, ε, is shown in Figure 5. The values range from a little less than 1% (red) to slightly more than 5% (blue). The region associated with the anomaly has the lowest value, while those regions associated with known mixtures of siliciclastic rocks and carbonates have the highest values. Intermediate values of ε correlate with shale units. Anisotropy is often associated with alternating thin layers of lithologies with contrasting velocities. The fact that we obtained greater anisotropy in mixed siliciclastic-carbonate facies suggests that there is more of this kind of interbedding present in the mixed facies than in the shale units. Both the core samples and the well logs indicate that there is sufficient interbedding present in the mixed facies to give rise to this anisotropy. The low values in the anomaly area are consistent with our knowledge of it being a massive reservoir-quality carbonate. There is

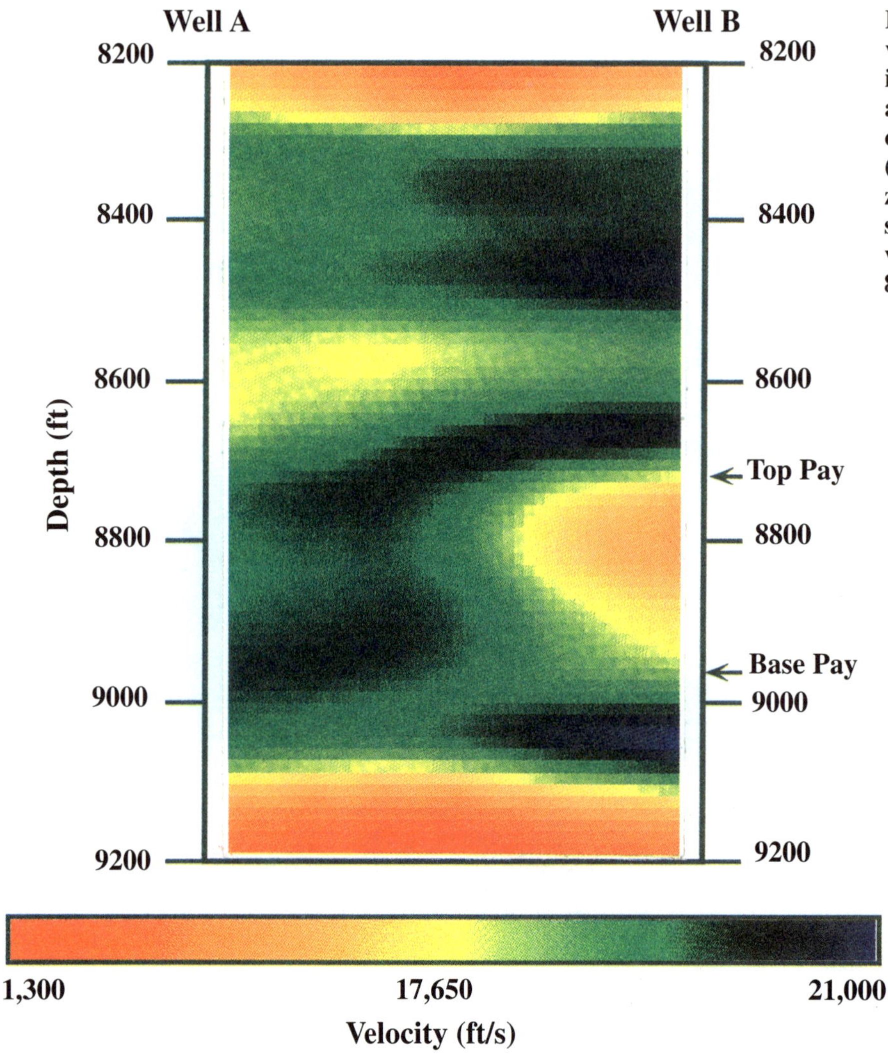

Fig. 4. Image of the vertical velocity obtained from a traveltime inversion of the direct *P*-wave arrivals. We have assumed a model of weak elastic anisotropy (Thomsen, 1986). The low-velocity zone causing the anomaly in the surface seismic data is clearly visible in well B at a depth range of 8700–9000 ft (~2650–2740 m).

fracturing present in the reservoir zone, which enhances the permeability. The low (but nonzero) values for ε are consistent with the presence of some fractures. (See the section on joint interpretation for more details of the facies present.)

Our solution for δ, the measure of deviation from a double elliptical model, is similar to that for ε in Figure 5, and therefore is not shown. This indicates that a double elliptical model is a satisfactory one for our example. It is interesting that the degree of anisotropy, ε, correlates with the magnitude of the velocity (higher velocities have a higher ε), and one might be tempted to infer that this is an artifact of ill-conditioning. However, when we have used this algorithm on other crosswell data sets, we usually do not obtain this kind of correlation.

Reflection Imaging

The crosswell data were first wavefield separated through the use of a series of filters designed to extract *P*-to-*P* primary upgoing reflections from undesirable elastic modes that interfere with them. The design of the wavefield separation sequence was based on the principles described by Rector et al. (1994, 1995). The wavefield separation sequence consisted of filtering steps aimed at attenuating the following modes:

1. Tube waves in the source and receiver wells
2. *P*-wave direct arrival
3. *S*-wave direct arrival
4. *P*-to-*S* converted waves (reflected and transmitted)
5. Downward propagating events

The wavefield-separated reflections were subsequently imaged in depth using the VSP-CDP mapping algorithm described by Lazaratos et al. (1995). This algorithm can be viewed as a special case of a Kirchhoff depth migration whereby each point in the time domain maps to a single point in the depth domain. The velocity model used for the imaging was a flat-layer model that was determined by interactively matching the direct arrival times obtained from ray tracing with those picked from the data. Although the reservoir is complex, as evidenced by the tomographic images of velocity, we found that a flat-layered model was a sufficient approximation

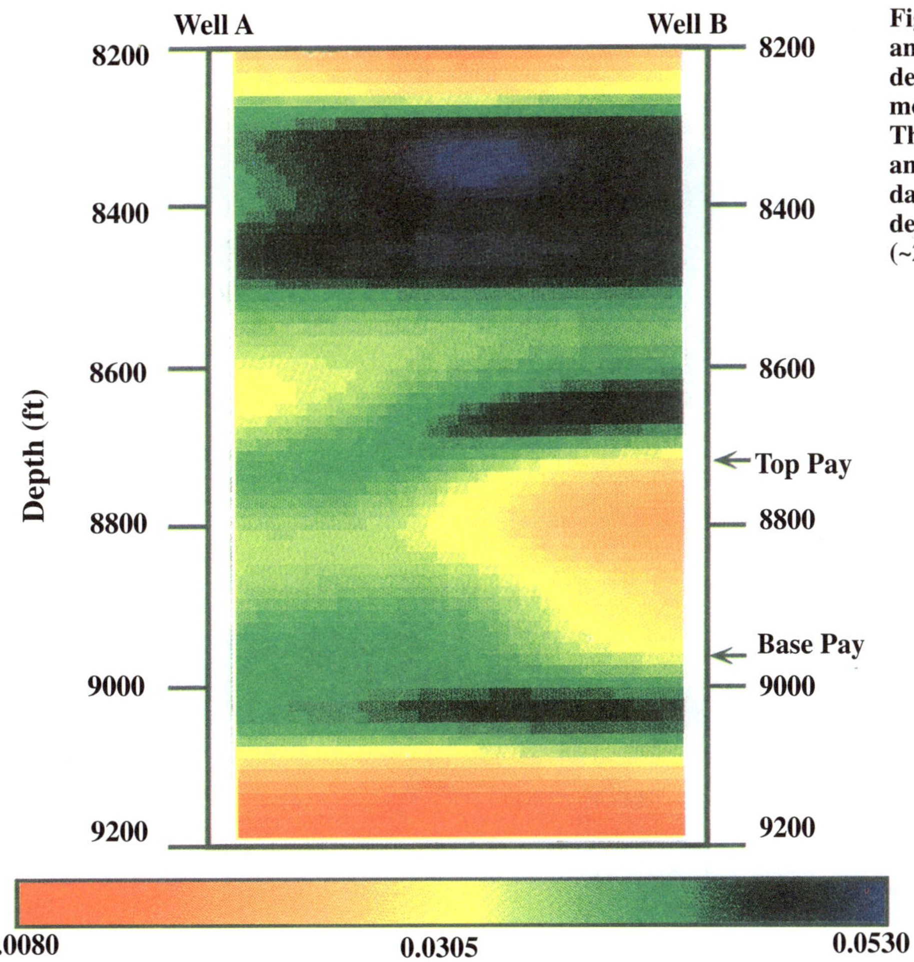

Fig. 5. Image of the measure of anisotropy called epsilon (ε), as defined by Thomsen's (1986) model of weak elastic anisotropy. The low-velocity zone causing the anomaly in the surface seismic data is clearly visible in well B at a depth range of 8700–9000 ft (~2650–2740 m).

to allow stacking of reflection events. This is partially because at depths shallower than the top of reservoir (~8700 ft, or ~2650 m), the interwell region is essentially horizontally layered in velocity.

The depth-imaged data were stacked after removing wide-angle reflections that were excessively distorted by "stretching" effects. The final stack is shown in Figure 6. A 200 ft (60 m) long automatic gain control (AGC) was applied to enhance the reflections in the reservoir zone, which were relatively weak compared to the strong events above it. The reflections within the reservoir interval are of two types: (1) high-frequency flat-lying events that do not extend from well to well and (2) lower frequency sinuous events that are also discontinuous in nature. The reflections above the reservoir are relatively flat and continuous. The spatial wavelengths in this image are about 15–30 ft (~4.5–9 m), at least an order of magnitude superior to surface seismic data.

Joint Interpretation

Figure 7 shows our joint interpretation of the *P*-wave crosswell velocity tomogram (isotropic), the *P*-wave crosswell reflection image, the well log data, and core samples that were taken in certain locations. The tomographic images for velocity anisotropy had not been computed at the time we made this interpretation, but they are consistent with it. The reservoir facies (massive limestone) is delineated by both the reflection image and the tomogram. The edge of the reservoir facies grades laterally into a talus facies and then into the mixed carbonate-siliciclastic facies seen in well A. In addition, some thin shale layers break up the mixed carbonate-siliciclastic facies.

Conclusions

Crosswell seismology is similar to other seismic techniques in that it includes both velocity estimation (tomographic imaging) and reflection imaging (VSP-CDP mapping and migration). Reflection imaging provides more resolution and is finding applications in reservoir delineation and characterization.

These images fill a resolution gap between surface seismic and well log data. The surface seismic data can have a great deal of volumetric coverage of a reservoir but relatively poor resolution. Well log and core data have a high degree of resolution but relatively poor volumetric coverage. One of our hopes for crosswell imaging is that by filling this resolution gap, it will help us tie together the wide range of more traditional data types while giving us an understanding of reservoir behavior at a unique scale of resolution.

Crosswell reflection imaging is in its infancy. The first reflection image obtained from data collected in the Permian Basin of West Texas dates back to only 1992, and significant strides have since been made in the imaging technology. The images shown in this paper may evolve as they benefit from improvements in processing technology. The interpretation of some of the features we see in these reflection images may be speculative be-

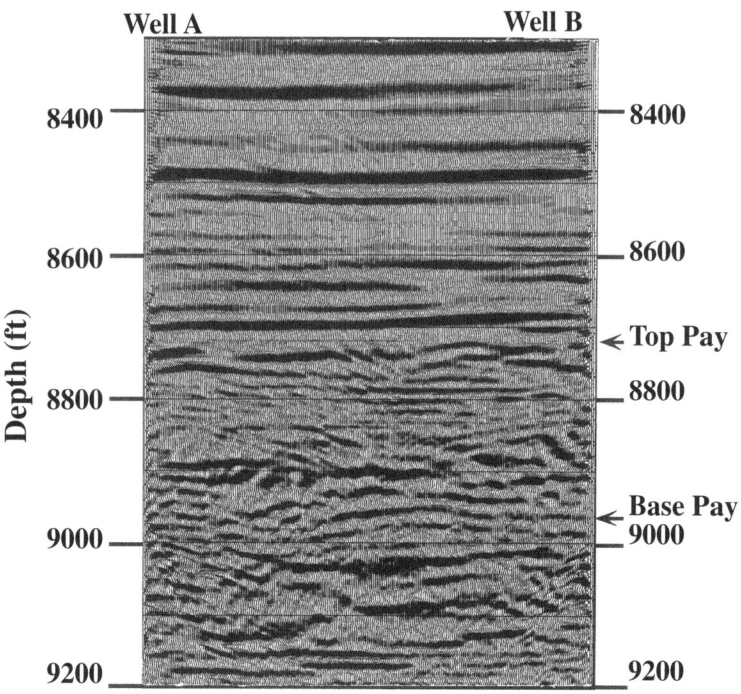

Fig. 6. A crosswell reflection image for the *P*-to-*P* reflections obtained using a variant of the VSP-CDP transform. There are numerous reflections in the reservoir zone defined by the low-velocity region on the tomogram. The base of the reservoir zone has short, sinuous events and the center has thin, flat laminations that appear to terminate at the edges of the low-velocity zone.

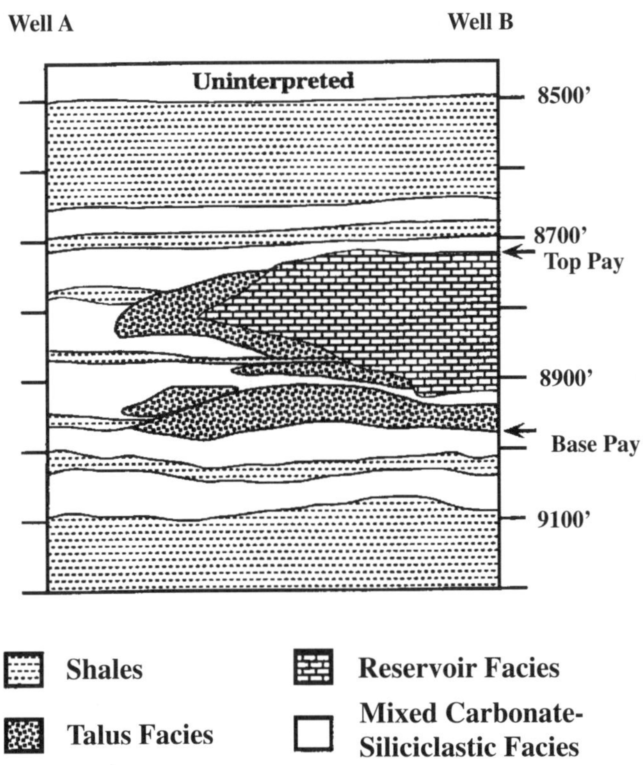

Fig. 7. A stratigraphic interpretation based on the *P*-wave tomogram, the crosswell reflection image, well logs, and core data. The massive reservoir-quality limestone extends laterally about 350 ft (~100 m) and grades into a talus facies and then into a mixed carbonate-siliciclastic facies.

cause of our limited experience with the interpretation of crosswell reflection images. However, it is unlikely we could obtain these subtle features from more traditional data types such as 3-D surface seismic data or well logs.

More powerful sources with greater bandwidths have been deployed since we collected the data for this case history. These new tools will result in higher quality data sets and greater well spacings than we show here. In addition, multiple hydrophone strings are now being used to collect several profiles simultaneously, which has increased our acquisition efficiency.

References

Harris, J. M., 1994, An approach to adaptive gridding for traveltime tomography: Ann. Rep., Seismic Tomography Project, Stanford Univ.

Harris, P. M., and Walker, S. D., 1988, McElroy field, Central Basin platform, U.S. Permian Basin, *in* Beaumont, E.A., and Foster, N. H., Eds., Atlas of oil and gas fields: Am. Assn. Petr. Geol., Treatise Petr. Geol. **1**.

Harris, J. M., Nolen-Hoeksema, R. C., Langan, R. T., Van Schaack, M. A., Lazaratos, S. K., and Rector, J. W., 1995, High-resolution crosswell imaging of a west Texas carbonate reservoir: part 1—project summary and interpretation: Geophysics, **60**, 667-681.

Lazaratos, S. K., Harris, J. M., Rector, J. W., and Van Schaack, M. A., 1995, High-resolution crosswell imaging of a west Texas carbonate reservoir: part 4—reflection imaging: Geophysics, **60**, 702-711.

Pratt, R. G., and Chapman, C. H., 1992, Traveltime tomography in anisotropic media: part II—applications: Geophys. J. Int., **109**, 20-38.

Rector, J. W., Lazaratos, S. K., Harris, J. M., and Van Schaack, M. A., 1994, Multidomain analysis and wavefield separation of crosswell seismic data: Geophysics, **59**, 27-35.

Rector, J. W., Lazaratos, S. K., Harris, J. M., and Van Schaack, M. A., 1995, High-resolution crosswell imaging of a west Texas carbonate reservoir: part 3—wavefield separation of reflections: Geophysics, **60**, 692-701.

Thomsen, L., 1986, Weak elastic anisotropy: Geophysics, **51**, 1954-1966.

Vassiliou, A. A., Thomsen, L., Treitel, S., Queen, J., and Harlan, W. S., 1994, Crosswell tomography in anisotropic media—case studies: Ann. Mtg., 1994 Eur. Assn. Expl. Geophys., Expanded Abstracts.

Ward, R. F., Kendall, C. G. St. C., and Harris, P. M., 1986, Upper Permian (Guadalupian) facies and their association with hydrocarbons—Permian Basin, West Texas and New Mexico: AAPG Bull., **70**, 239-262.

Chapter 18

Carbonate Platform Seismic Sequence Attributes, Maracaibo Basin, Venezuela

T. Chris Stiteler
Amoco Overseas Exploration Co.
Houston, Texas, U.S.A.

Fernando J. Chacartegui
Maraven S.A.
Caracas, Venezuela

Orlando J. Ortega
University of Texas
Austin, Texas, U.S.A.
and Maraven S.A., Caracas, Venezuela

Robert W. Scott
Precision Stratigraphy Associates
Tulsa, Oklahoma, U.S.A.

Izaskun Azpiritxaga
Maraven S.A.
Caracas, Venezuela

Chris K. Taylor
Geco-Prakla/Schlumberger
on assignment to Maraven S.A., Caracas, Venezuela

Abstract

A sequence stratigraphic evaluation of Cretaceous platform carbonates was conducted in the Lake Maracaibo region of Venezuela. The study was part of a larger multidisciplinary project in which cores, logs, a 3-D seismic survey, and well test data from the Cretaceous carbonate section were analyzed. Carbonate platform stratigraphy was interpreted to be tabular shaped and consists of alternating cycles of subparallel strata with intermittent shale beds at maximum-flooding intervals. Primary porosity is poorly developed in grain-rich facies, but secondary porosity is well developed in grain-rich facies in the vicinity of fault-associated fractures.

The seismic interpretation was conducted on a 3-D survey acquired in Lake Maracaibo. The seismic responses of the platform carbonates were divided into two types: a shallower section character in which some of the larger scale cycles/sequences and lateral changes in rock properties were interpreted with reasonable constraint, and a deeper section response composed mainly of "false stratigraphy" that *appears* to be stratigraphically controlled but is actually caused by side lobes and subtle changes in structure. The interpretation was also complicated by the weak seismic response of the platform carbonate section. A systematic interpretation method was proposed to help the interpreter distinguish between seismic events due to stratigraphic changes and those due to other effects. Keys to successful interpretation involve use of core data, seismic modeling, and 3-D seismic data that are calibrated to well control.

Introduction

The interpretation of seismic records of carbonates at shelf margins and where buildups are developed is well established by numerous case studies (Bubb and Hatlelid, 1977; Jansa, 1993; Scott, 1993; Yurewicz et al., 1993). At the shelf margin, the geometry of seismic reflectors mimics the stratal geometry, although stratal convergence may be indistinguishable from stratal pinchout (Campbell and Stafleu, 1992; Stafleu et al., 1994). However, on the platform interior, continuous parallel reflectors are characteristic of carbonate platforms. They may "represent interference patterns caused by small-scale cyclicity in the beds of the platform interior" (Campbell and Stafleu, 1992, p. 1776). Alternatively, they may result from side lobes, multiples, and complex interference distributions and not provide clues to bedding thickness and lithologic changes (Galloway et al., 1977). The challenge is to identify criteria to be applied to the interpretation of seismic characteristics in platform carbonates.

Stratigraphic Setting

The Maracaibo Basin in northwestern South America is a large Tertiary structure filled with Cretaceous carbonates and clastics and Tertiary clastics (Ross and Scotese, 1987) (Figure 1). Cretaceous carbonates of the Cogollo Group form a major hydrocarbon reservoir with ultimate recovery on the order of 1 billion bbl of oil (Reijers and Bartok, 1985). During Cretaceous time, the Maracaibo Basin subsided at approximately the same rate as sediment supply so that water depth remained shallow (Bartok et al., 1981). Local positive and negative structures influenced stratal patterns and facies during deposition of the lower Cogollo Group, but the upper Cogollo was relatively unaffected. Deposition of the Aptian–Albian Cogollo began on a ramp that sloped into a deep basin north of present-day Lake Maracaibo (Figure 2) (Azpiritxaga, 1991; Piccoli, 1993; Vahrenkamp et al., 1993). The platform changed into a carbonate-rimmed platform near the end of Cogollo deposition. Shallow platform deposition ended during the latest Albian–Cenomanian drowning and deposition of the deep-shelf La Luna Formation.

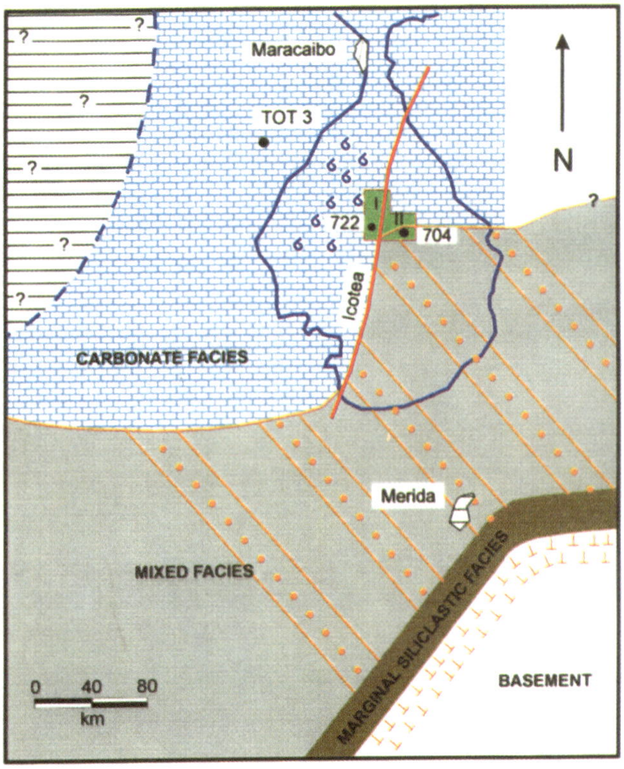

Fig. 2. Paleogeographic map of the Lisure Formation during Albian time, Maracaibo platform, northwestern Venezuela. Adapted from Vahrekamp et al. (1993).

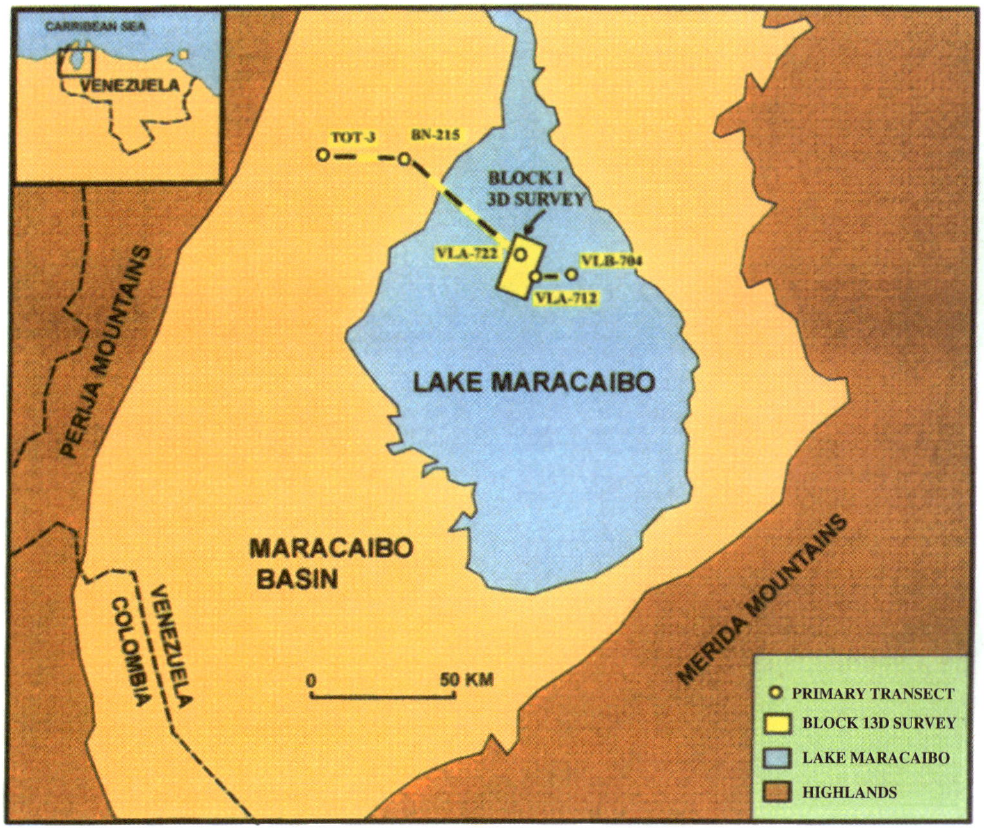

Fig. 1. Regional map of Maracaibo Basin study area in Venezuela.

The Cogollo Group is locally more than 450 m (1500 ft) thick. In the study area, it is divided into five lithostratigraphic units (from oldest to youngest): the Tibu Formation, the Machiques and Piche members of the Apon Formation, the Lisure Formation, and the Maraca Formation (Figure 3).

The stratigraphic nomenclature in the Lake Maracaibo area differs locally (Vahrenkamp et al., 1993). In the outcrops west and south of the lake, complex facies changes are formalized by the nomenclature (Renz, 1982). The Cogollo is underlain by the Rio Negro Formation consisting of fluviodeltaic clastics of variable thickness. It is overlain by the La Luna Formation comprised of thin-bedded limestones, which contain organic matter, phosphate, and planktonic foraminifera. The La Luna varies in thickness from 25 to 100 m (80 to 330 ft) (Renz, 1982) and is an effective hydrocarbon source rock. The age of the basal La Luna varies from earliest Albian to Turonian and can be as young as Coniacian (Renz, 1982). The La Luna is overlain by the Colon Formation, a thick, overpressured marine shale, the lowermost strata of which is the Socuy Member, a thin, phosphatic outer shelf carbonate with a thickness of 10–15 m (30–50 ft).

Structural Setting

This study was conducted on Block I where a faulted anticline is bisected by the large Icotea fault (Figure 4). The shallower upthrown east flank of the fault is at a depth of 2700–3400 m and is a relatively simple structure cut by small faults with little throw. The deeper west flank is at a depth of 3600–4600 m and has a thrusted section that varies in throw and trends parallel to the Icotea fault. Major structural development took place during the Tertiary, dividing the subtly dipping shallow-carbonate platform into a large structure with both shallow and deep components. Our comparison of the contiguous shallow and deep sections led to solutions in distinguishing stratigraphic changes from seismic events that are not related to stratigraphy.

Methods

This study of shallow platform strata was conducted in a 24,000-km² area from the western margin of Lake Maracaibo across the main part of the lake. In this area, two cores of the complete section of the Cogollo Group were described from the Totumos-3 onshore well and the VLB-704 offshore well (Figure 5). In addition, between these wells on Block I, cores and well logs from the VLB-722, VLA-711, VLA-712, and VLA-978 wells were described. Lithology and microfacies descriptions were based on petrographic analyses of several hundred thin sections. Also, shales from key stratal contacts were analyzed by x-ray diffraction for mineralogy and by mass spectrometer for carbon and oxygen isotopes. Micropaleontologic analyses of foraminifera and calcareous algae were based on examination of the thin sections. Palynologic analyses were conducted on carbonates and shales from the Totumos-3 core by J. A. Stein (Amoco, Houston). Ages were based on a graphic correlation of the cored wells with a Cretaceous composite standard. Sequence stratigraphy analysis was extended from six key wells with core to 13 wells in the vicinity by interpretation of log cross-sections.

Seismic studies were based primarily on the Block I 3-D survey in Lake Maracaibo that covers a 10 km × 25 km area over the carbonate structure (Figure 4). Seismic data were processed and correlated with seven synthetic seismograms. Synthetic seismograms were created from sonic logs, density logs, velocity surveys, and pseudo-sonic logs derived from cross-plotting of resistivity, gamma ray, and density logs. Seismic and synthetic seismic data were correlated, which confirmed that the wavelet from the 3-D survey was zero-phase. Also, a stratigraphic seismic model was created from a linear interpolation of a regional cross-section of 13 well synthetic seismograms that encompassed the study area. The 3-D survey was interpreted by studying seismic versus synthetic seismic data, models and analyses of sequence stratigraphy, petrophysics, log interpretations, and well tests from wells on and near the survey.

The process of integrating core and seismic interpretations began with a carbonate stratigraphic view of the core. Major sequence boundaries and associated log responses were identified, with particular attention paid to sonic and density log changes that could affect the seismic data. Carbonate sedimentologists refined the sequences and conducted tests to calibrate rock data to composite standards, while geophysicists created synthetic seismograms and models, calibrated well data to the seismic data, and began interpreting reliable sequence boundaries. The data were then synthesized with particular attention given to intervals where the smaller wavelength scale of the sequence stratigraphy derived from core data matched the observations from larger-scale seismic cycles.

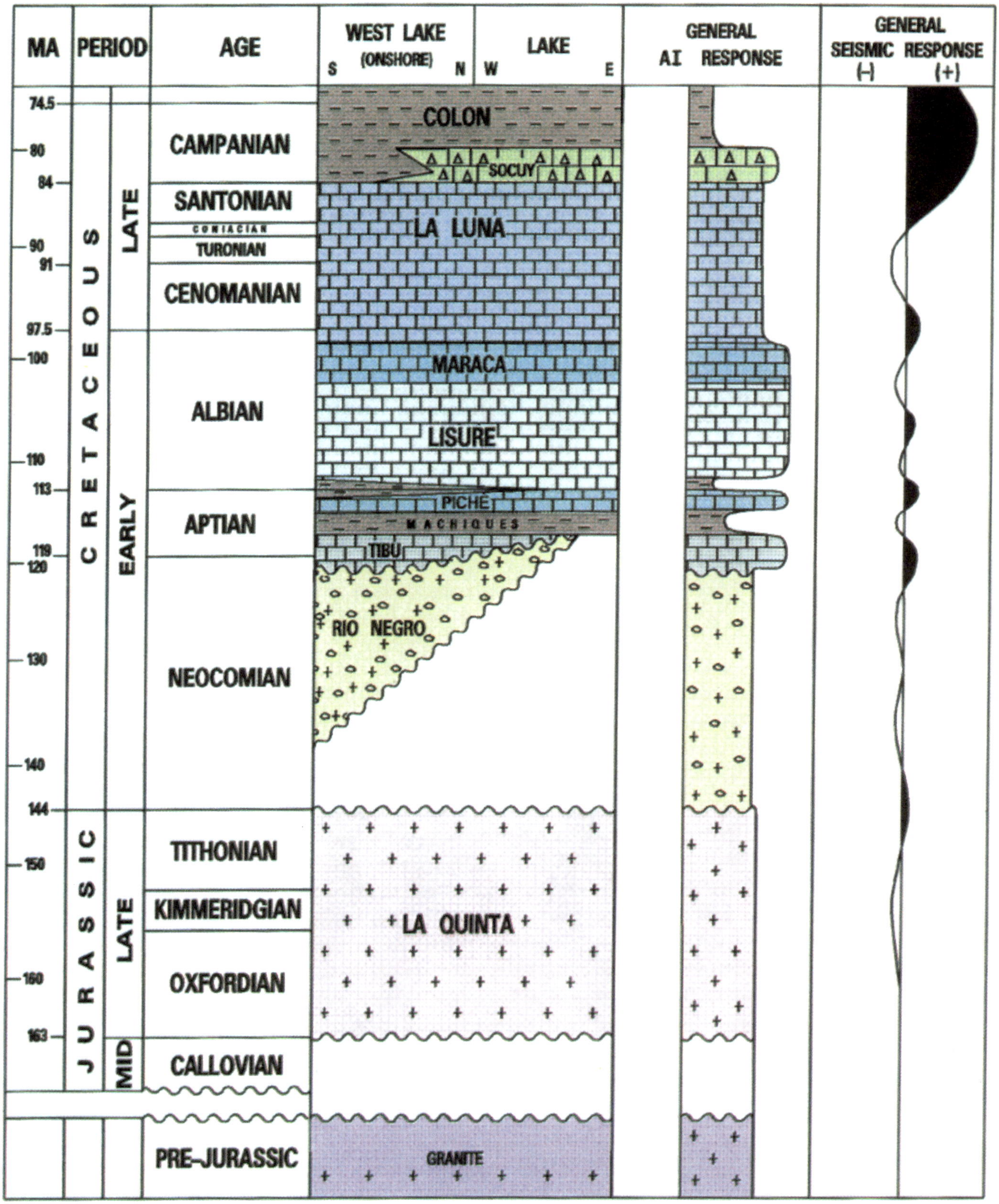

Fig. 3. Stratigraphic column for the Lake Maracaibo area showing general acoustic impedance and seismic responses.

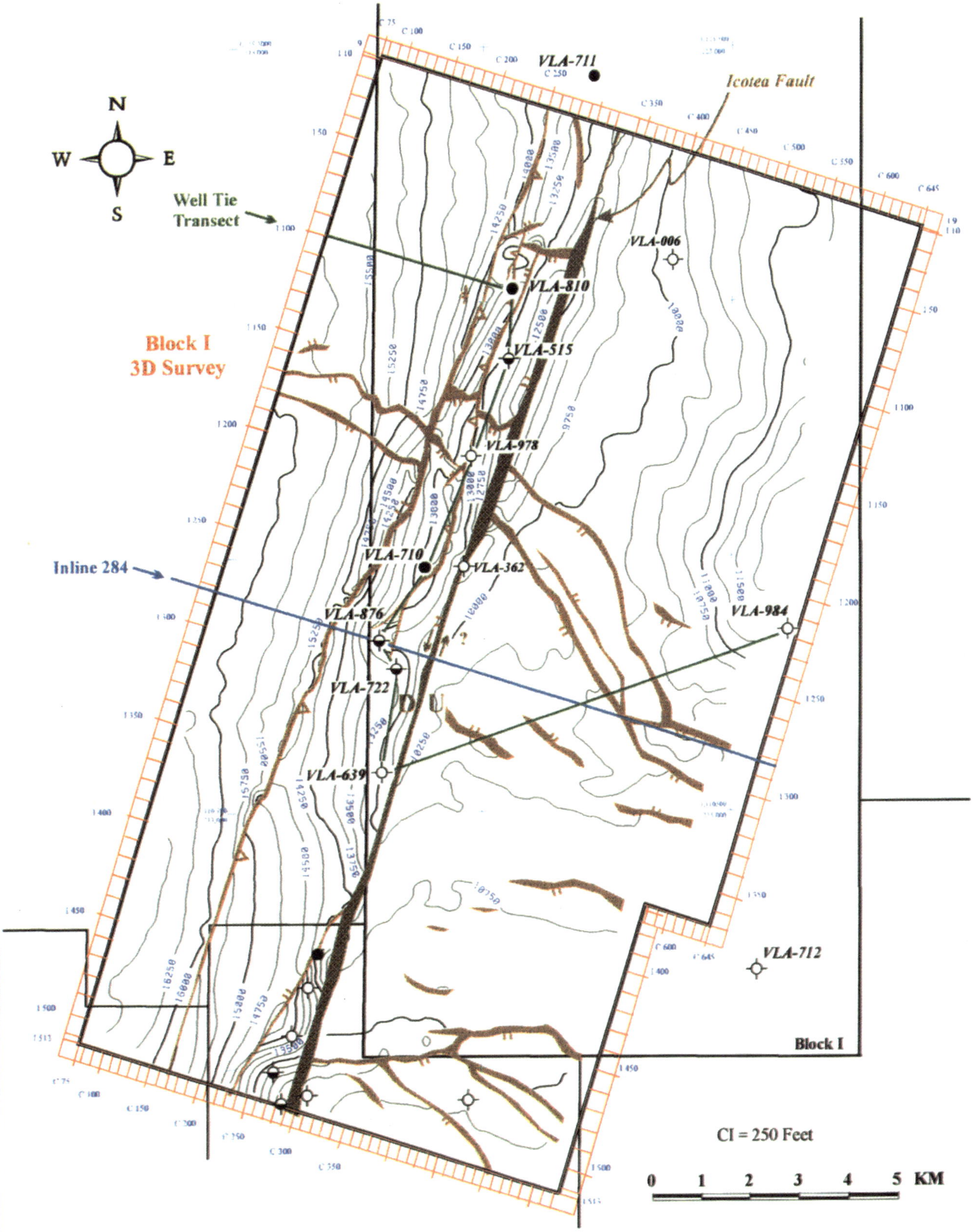

Fig. 4. Depth structure map of the top of the Socuy Member in Block I, showing the 3-D survey area, well control, and seismic transects.

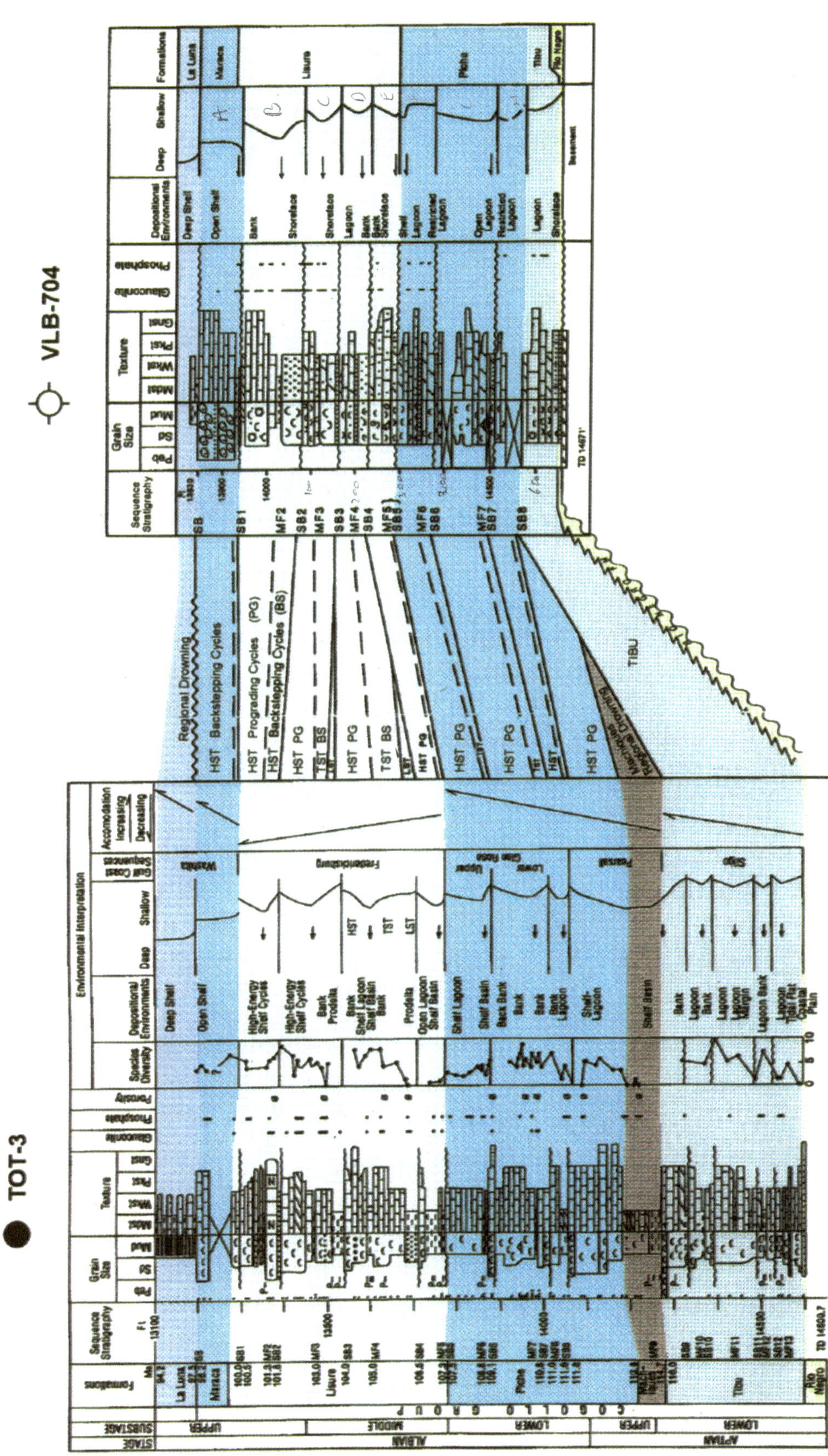

Fig. 5. Stratigraphic cross-section of the TOT-3 and VLB-704 wells, Maracaibo platform.

Results of Modeling Seismic Attributes of Platforms

Synthetic Seismograms, Models, and Seismic Data

The detection and resolution of the stratigraphic dimensions of thickness and lateral continuity are difficult in platform interior settings where tabular, subparallel stratal patterns dominate and there is little change in the gross acoustic properties. Also, seismic events from side lobes and subtle structural events are often mistaken for stratigraphy, and even worse, these patterns can change over the area due to differences in frequency and depth.

These concerns were managed by first creating synthetic seismograms and models and calibrating seismic data to well control. Synthetic ties and wavelet extraction from six wells indicate a good calibration between seismic and well data. Wavelet extraction confirms that the seismic data are zero-phase (using the convention where a positive peak represents an increase in acoustic impedance). The frequency band of the Cretaceous carbonate was measured as 5–50 Hz in the shallow area at 2700–3400 m and as 5–30 Hz in the deep area at 3600–4600 m. Also, a 2-D section tied to well control was extracted from the Block I 3-D survey (Figure 6a), flattened at the top of the Maraca (Figure 6b) and compared with modeled data (Figure 7).

Practical Limits in Seismic Interpretation

Comparison of synthetic seismograms, models, and seismic profiles and a study of resolution led to a list of practical limits and guidelines to aid the interpreter in distinguishing carbonate stratigraphic responses from other events not caused by stratigraphy. A number of factors were found to limit the reliability of platform carbonate interpretations:

1. Small changes in acoustic rock properties yield weak reflections that are lost in noise. Seismic profiles depict *changes* in velocity and density along lithologic interfaces. Where the changes in acoustic properties are small, the reflections are weak in amplitude and are lost in normal background noise. For example, the regional model (Figure 7a) shows that wells exhibit little change in sonic logs over the Cogollo section, resulting in low seismic amplitudes associated with this phenomenon (Figure 7b). The regional seismic model with 10% added noise (Figure 7c) shows that small reflections from homogeneous stratigraphy are lost in background noise that is comparable to the noise level observed on actual seismic profiles. Also, the reflection at the top of the Piche is reliable when thick shales are present above the top of the Piche, as indicated in wells TOT-3, BN-215, and VLA-712 (Figure 7).
2. Frequency attenuation causes shallower data to have better resolution. Frequency attenuation describes how the earth filters out high frequencies at increasing depths (Figure 8a). As a result, the ability to detect and resolve thin beds diminishes with depth (Figure 8b). Another lesser known phenomenon is that the ability to identify lateral changes in rock properties also diminishes with increasing depth and lower frequency. The limiting area is referred to as the Fresnel zone (Figure 8c).
3. Another factor that decreases the reliability of platform carbonates is the limited vertical resolution of beds. Given the limited seismic frequency content of the shallow section, one cannot resolve bed thicknesses less than 45 ± 6 m (150 ± 20 ft). Changes in character can be detected in beds thicker than ~25 m (~75 ft), but the *thickness* of the changed zone cannot be predicted reliably below the resolvable limit of 45 ± 6 m. The seismic response of the carbonate section in depth is indicated in a log track for the TOT-3 in Figure 9. Note the long wavelengths of the seismic events compared to the fine details of the well log.
   Resolution also varies across the survey transect area. For example, the filter tests from Block I, Inline 284 (Figures 10 and 11) show higher frequency in the shallow east flank of the Block I Cretaceous structure than in the deeper west flank. Normal and flattened sections tying Block I wells also show this effect across the Icotea fault (Figures 6 and 12). This is due to the attenuation of higher frequencies through differing thicknesses of overburden. For example, seismic waves do not travel as far through the thinner, shallower east flank overburden as through the thicker, deeper west flank, so the high frequencies are not attenuated as much.
4. Transmission loss at the top of the Socuy reflection is another factor limiting platform carbonate interpretations. The large change in acoustic impedance at the contact of the Colon shale and Socuy carbonate creates a strong reflection and limits the amount of energy transmitted into the underlying section.
5. Side lobe interference from the strong La Luna

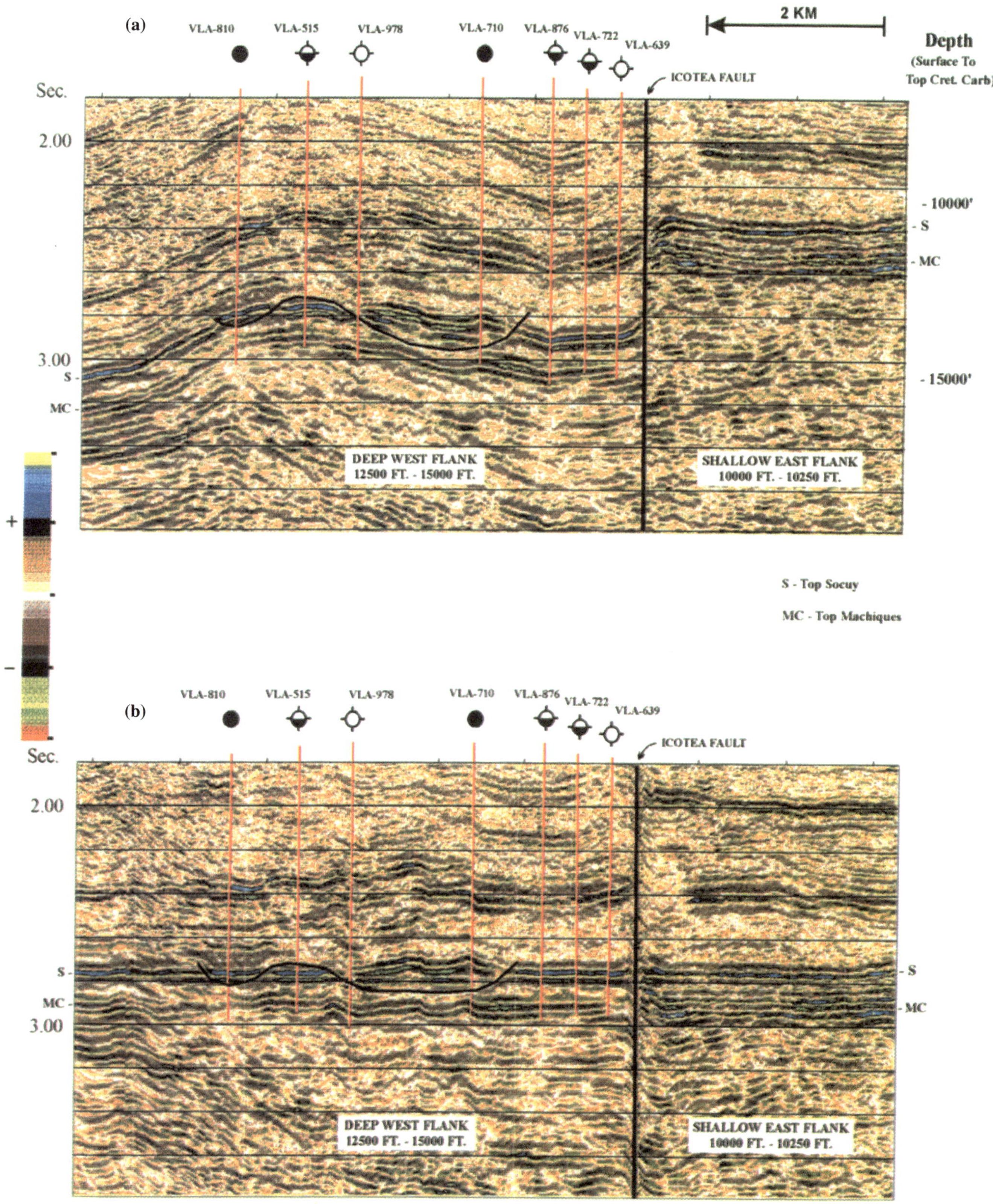

Fig. 6. Seismic profiles showing (a) the normal well-tie transect and (b) the well-tie transect flattened on the top of Maraca reflection.

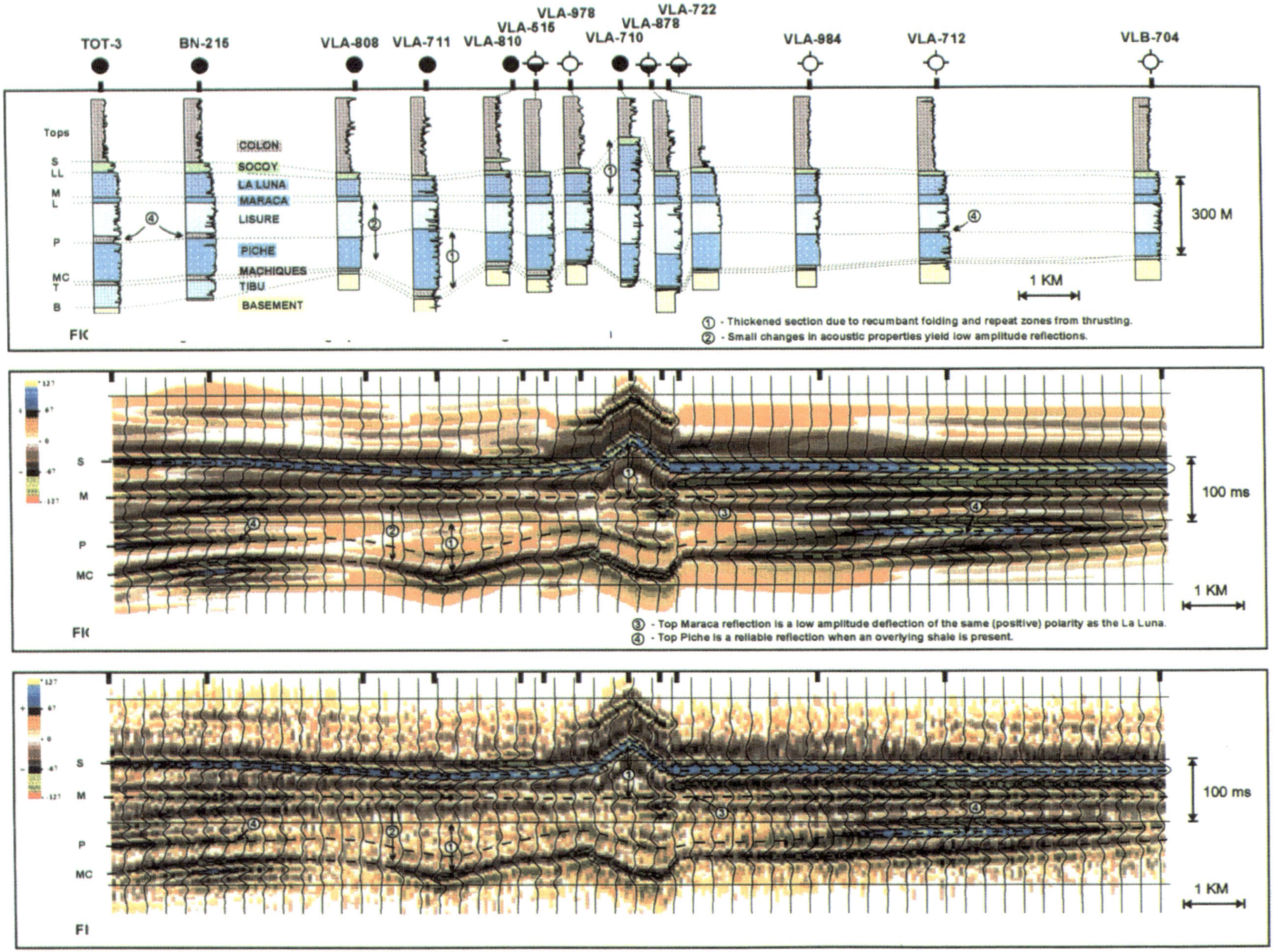

Fig. 7. Regional seismic stratigraphic model for Maracaibo platform area. (a) Sonic log cross-section. (b) Seismic model at 5-10-15-50 Hz. (c) Seismic model with 10% added noise at 5-10-15-50 Hz.

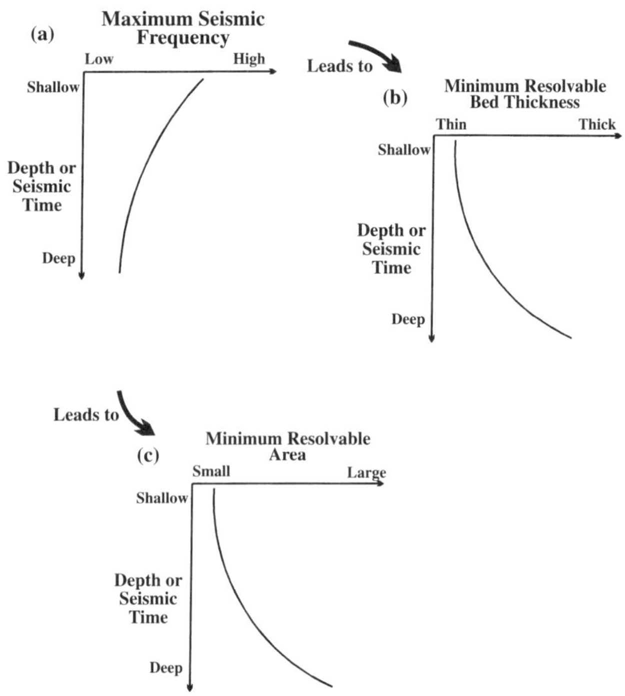

Fig. 8. Frequency attenuation and resolution. Loss of frequency content with increasing depth limits the bedding thickness and the area that can be resolved on seismic data. (a) Frequency attenuation versus depth. (b) Vertical bed resolution versus depth. (c) Fresnel zone or horizontal area resolution versus depth.

wavelet can mask the top of the Maraca reflection. The weaker top of Maraca reflection is 50 ± 10 ms from the strong top of Socuy reflector. When the frequency content is low, as in the deeper west flank, the subtle top of Maraca reflection is lost in destructive interference with the side lobes of the top of Socuy reflection. This is evident from a typical tie of a synthetic seismogram to the seismic section (Figure 13).

6. Amplitude dimming is associated with faults and fractures. The top of Socuy reflection has been observed to dim and form low-amplitude "halos" in the vicinity of mapped faults in this data set (Figure 14). This may be due to geometric distortion in close proximity to faults or porosity development in fracture zones extending away from the faults. As such, amplitude dimming clarifies fault interpretation and gives an indication that being in the vicinity of a "fault halo" may be good by increasing the likelihood of enhanced porosity and permeability.
7. A final factor limiting the reliability of platform carbonate interpretations is that subtle structural changes may be confused with stratigraphic changes. For example, the well-tie line extracted from the 3-D survey (Figure 6) and seismic model of the same wells (Figure 7) obliquely transects a thrust fault, resulting in a section that *appears* to thicken and develop mounds and other stratigraphic features. Actually, the stratigraphy is relatively consistent over this area, and the unusual features are caused by a sinuous fault trace, a thickened section where the carbonate section repeats, and spurious fault trace reflections. Also, the thrust fault is confirmed by fault cuts with slickensides from cores and by thickened and repeated sections on logs.

Summary of Interpretation Guidelines

Interpretation of platform carbonates involves many factors, but a few deserve emphasis. Careful interpretation of stratigraphy between wells is essential, along with an eye for pitfalls, to distinguish between geologic reflections and false images that are a natural result of the seismic process. This is especially true where a bright overlying reflection occurs with parallel bedding below. In this case, the interpreter has the tendency to pick side lobes from the bright reflector and interbed multiples as primary reflections.

Modeling and synthetic seismogram ties to seismic profiles are essential to distinguish the seismic response of the strata from the "masking" effect of the strong top of Socuy reflection. Away from well locations, one can distinguish Cogollo Group geology from seismic artifacts by recognizing that side lobes and interbed multiples tend to mirror amplitude changes in the top of Socuy reflection, whereas geology exhibits changes in frequency and amplitude independently. For example, the seismic response of the Cogollo is characterized by a series of parallel, discontinuous reflections below the top of Socuy reflection that do not appear to be interbed multiples because they dim and change character, while the overlying Socuy reflection remains relatively constant in amplitude and character.

In areas where well control is not available, the ability to distinguish changes in stratigraphy from subtle structural changes is also difficult. This problem must be addressed by completing both a structural and stratigraphic interpretation in three dimensions, taking care to identify the causes of observed reflections and terminations. Mapping of horizon amplitude (Figure 14) and other attributes may help in recognizing linear trends due to faulting and associated fracturing and may aid in distinguishing these features from geometries that are

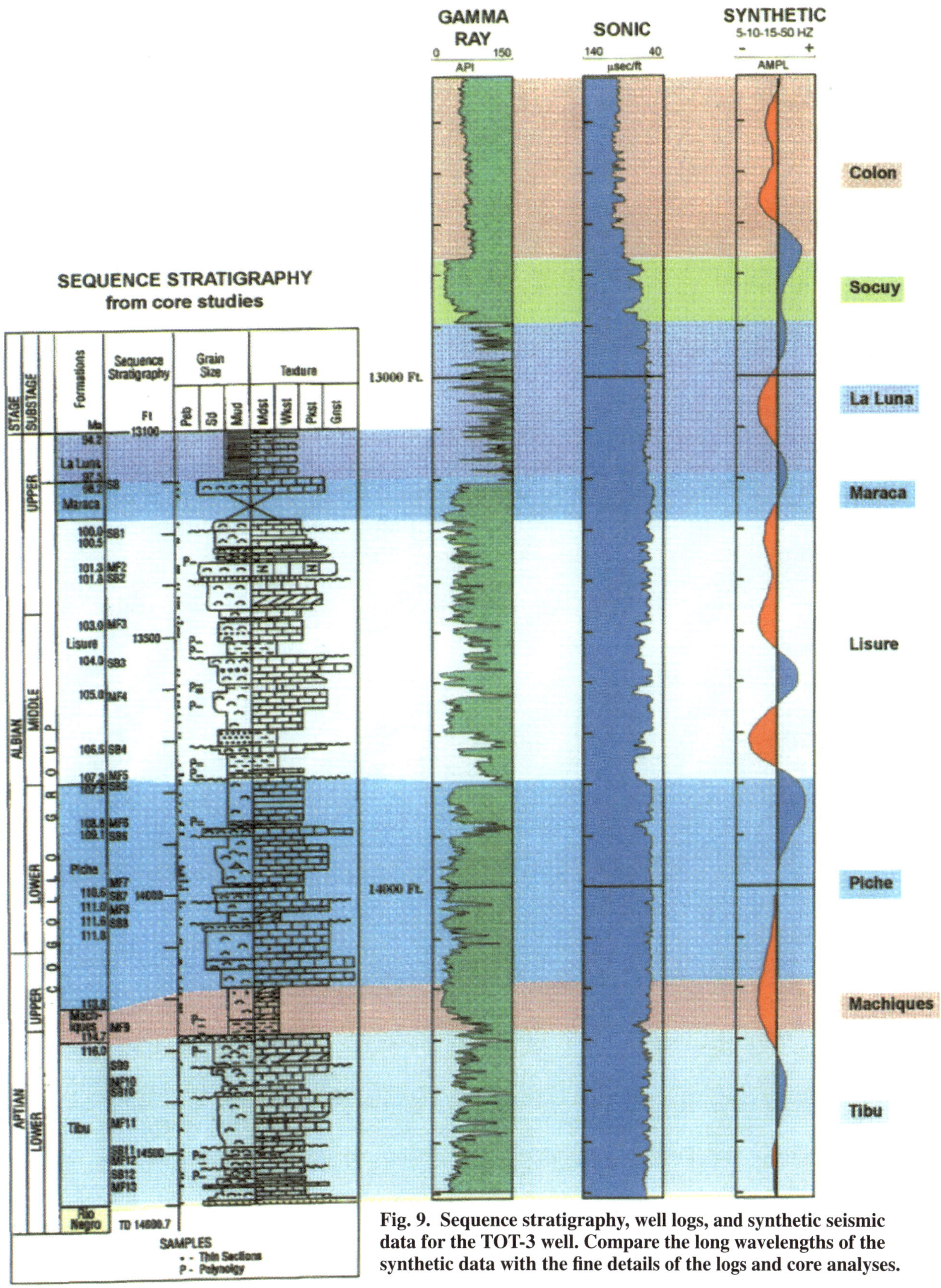

Fig. 9. Sequence stratigraphy, well logs, and synthetic seismic data for the TOT-3 well. Compare the long wavelengths of the synthetic data with the fine details of the logs and core analyses.

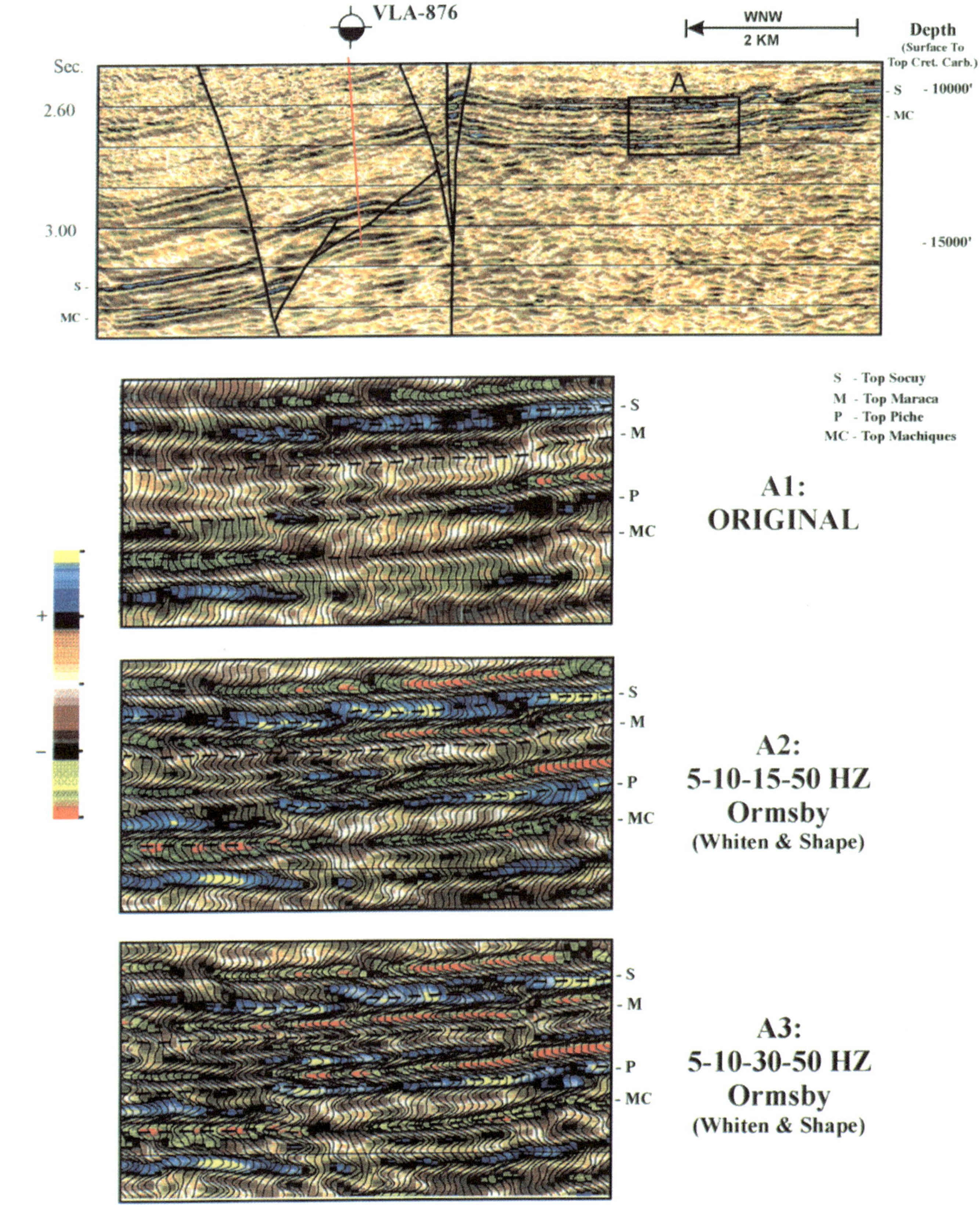

Fig. 10. Shallow section filter test "A" for the Inline 284 3-D seismic profile. Note that the top of the Maraca (M) is a fairly reliable, low-amplitude, positive reflection. The top of Piche (P) reflection is also reliable.

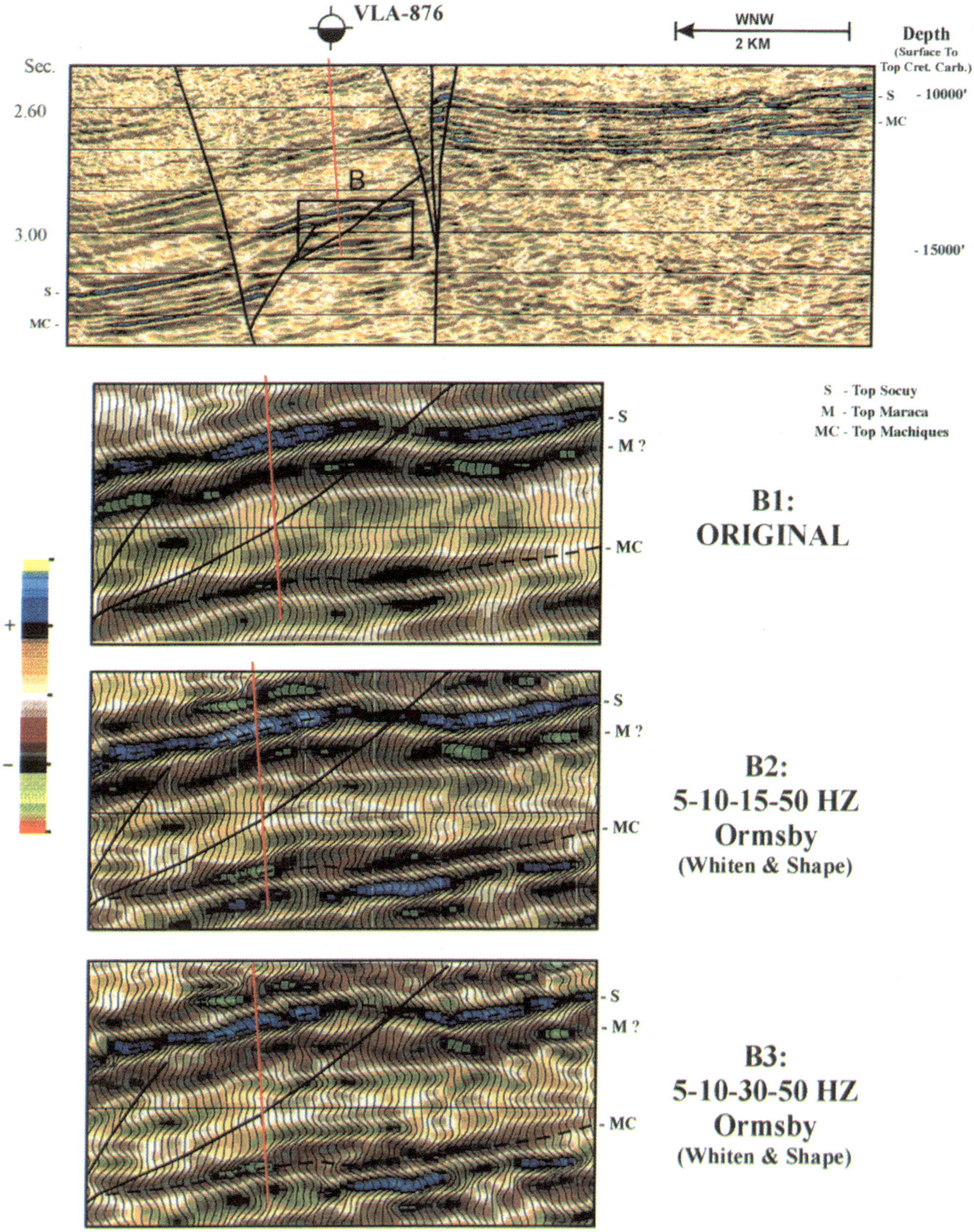

Fig. 11. Deep section filter test "B" for the Inline 284 3-D seismic profile. High frequency content and resolution are less than in the shallow filter test "A" (Figure 10) and are not recovered with filtering.

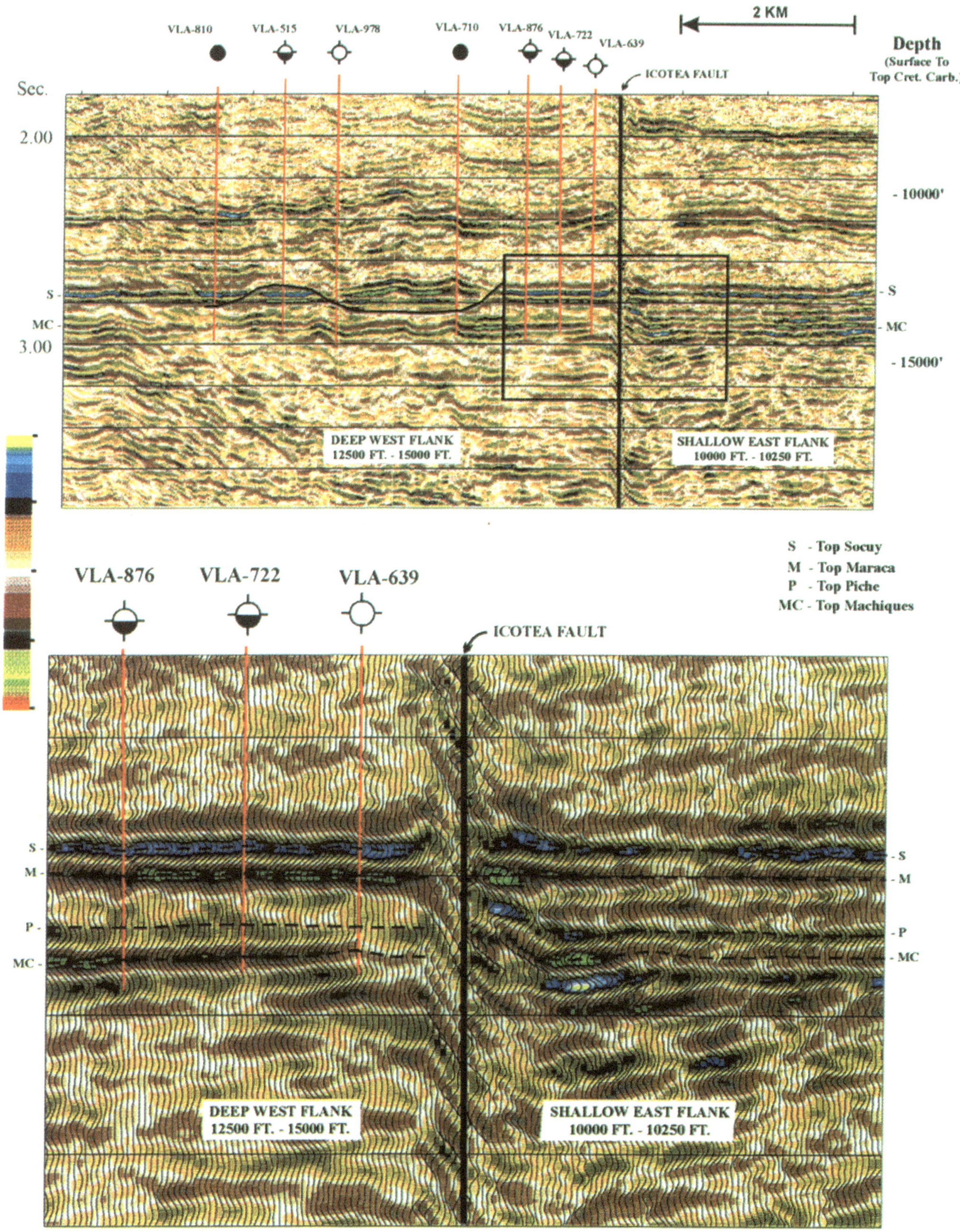

Fig. 12. Well tie seismic line flattened on the top of the Maraca. Note the differences between the shallow and deep data.

more likely to be associated with carbonate depositional or erosional patterns.

Stratigraphy of the Lake Maracaibo Carbonate Platform

Sequence Boundaries Identifiable on Seismic Profiles

We applied the methods and guidelines discussed in the previous section to the evaluation of Lake Maracaibo platform carbonates. The most striking contrasts in this area are between the relatively continuous, parallel, higher frequency response of the shallow section (Figure 10) and the homogeneous, low-amplitude, mounded, discontinuous response of the deeper section (Figure 11).

Seismic characteristics of some of the identifiable sequence boundaries are as follows:

1. The top of the Socuy is a reliable reflection. The overlying Colon shale is homogeneous, and amplitude changes are generally caused by changes in the underlying Socuy and La Luna section. However, it has been noticed in other areas that overpressured zones in the Colon shale can yield amplitude changes.
2. The sequence boundary at the top of the Maraca is marginally reliable over part of the shallow section where the frequency content is high and the top of the Maraca appears as a low-amplitude positive peak. This reflection is unreliable in the deep section and in some of the shallow section where the frequency is low. Model and synthetic ties confirm that the top of Maraca reflection is due to a positive impedance contrast representing an increase in velocity from the slower velocity, organic-rich La Luna carbonate section to the faster velocity Cogollo carbonate (Figure 7). However, the reflection is weak and generally falls in a trough on seismic profiles because it is masked by the large negative side lobe of the nearby top of Socuy reflection (Figure 13). In many cases, a phantom top of Maraca event was picked on a negative event near the well tie with the understanding that this event indicates a position from well control and is not a reliable indicator of changes in stratigraphy.
3. The top of the Piche near the MF-5 sequence boundary is a marginally reliable reflection where shales greater than ~20 m in thickness overlay this boundary. Model and synthetic ties indicate that the top of Piche reflection is reliable where thick shales are present above it, as indicated in wells TOT-3, BN-215, and VLA-712 (Figure 7).
4. The top of the Machiques is a reliable reflection over most of the Block I 3-D survey area due to the presence of the thick Machiques shale. Model and synthetic studies indicate a weak, unreliable reflection at the base of the Cretaceous carbonate when the Machiques shale is absent.
5. The base of the Cogollo is locally detected outside of the 3-D survey area as a low-amplitude reflection that is not easy to pick because the acoustic impedance contrast is small between the Cogollo carbonates and the underlying Rio Negro conglomerates and basement metamorphics. In the Block I 3-D survey area, it is difficult to distinguish the basal Cogollo reflection from the Machiques reflection because the Tibu Formation is thin or absent over much of area.

Platform Sequence Stratigraphy

In the interior of the Maracaibo carbonate platform, subhorizontal tabular-shaped units were deposited during cycles of relative rise and fall of sea level, probably driven by a combination of local tectonic and eustatic processes. At certain times, the platform was widely exposed and weathered, while at other times, it was drowned by a shallow sea from the proto-Caribbean Ocean. The resulting carbonate layers vary in facies and porosity from place to place. Porosity was also affected after burial by diagenetic processes (Bartok et al., 1981; Reijers and Bartok, 1985).

Consequently, the rock properties of each layer vary laterally in relation to facies and postdepositional structures. The challenge is to identify which of these translate to a seismic expression or response. Fracture porosity is higher near many faults, and a low-amplitude seismic expression has been observed. Matrix porosity is generally only a few percent in thin intervals 0.3–2.4 m thick (1–8 ft). Porosity is also developed within sequences in the transgressive facies and is not related to sequence boundaries. The presence of dolomite in the highstand facies can also alter the density and velocity of the rocks in unpredictable ways. These layers often directly underlie sequence boundaries and may add to the density and velocity contrasts, but in this case, the small contrasts and limited affected volume are often not large enough to be noticed on seismic data.

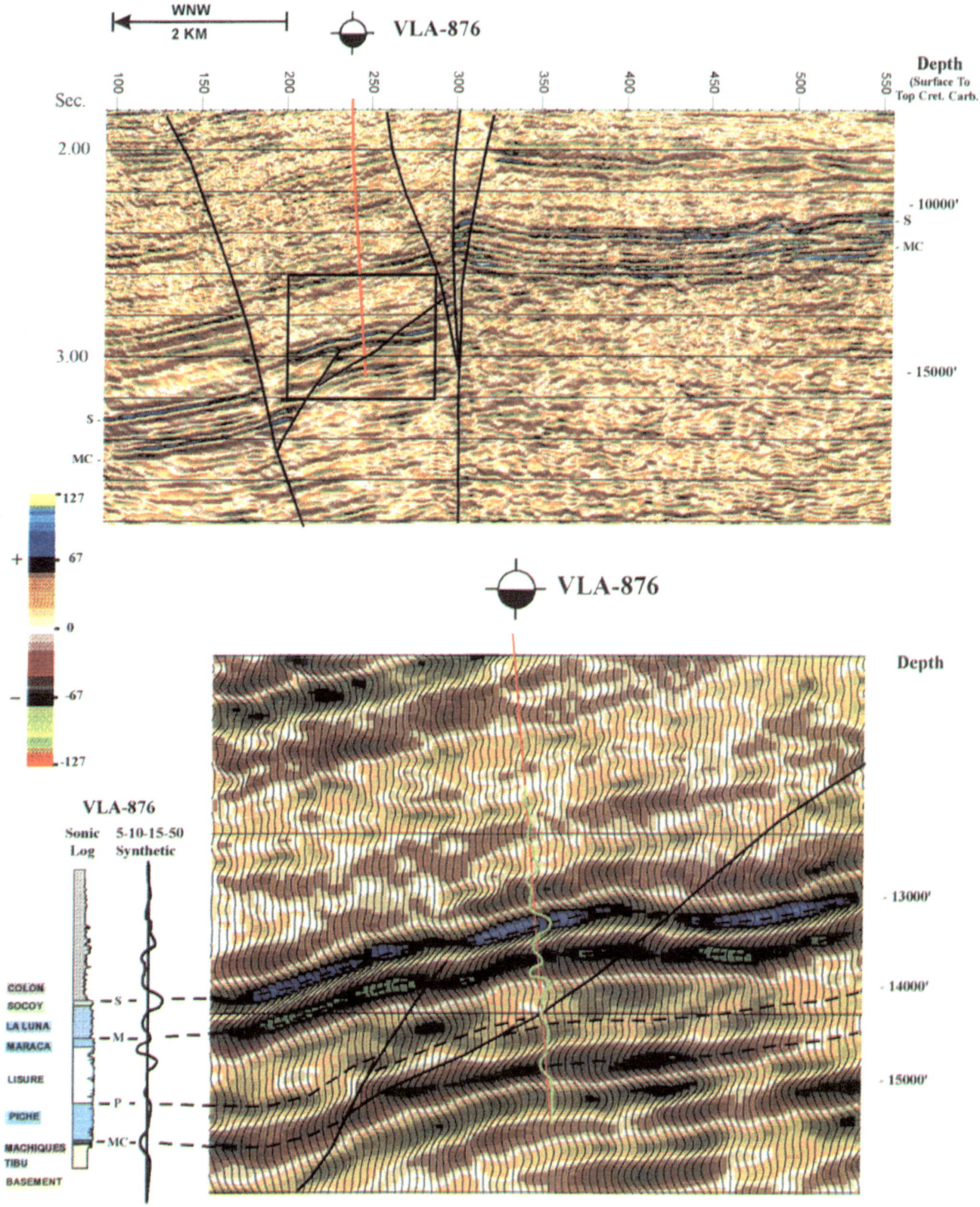

Fig. 13. Tie of synthetic seismogram between VLA-876 and Inline 284 seismic profile. Note that the top of the Maraca (M) is not detectable as a low-amplitude, positive reflection because it is masked by the strong negative side lobe of the top of the Socuy (S).

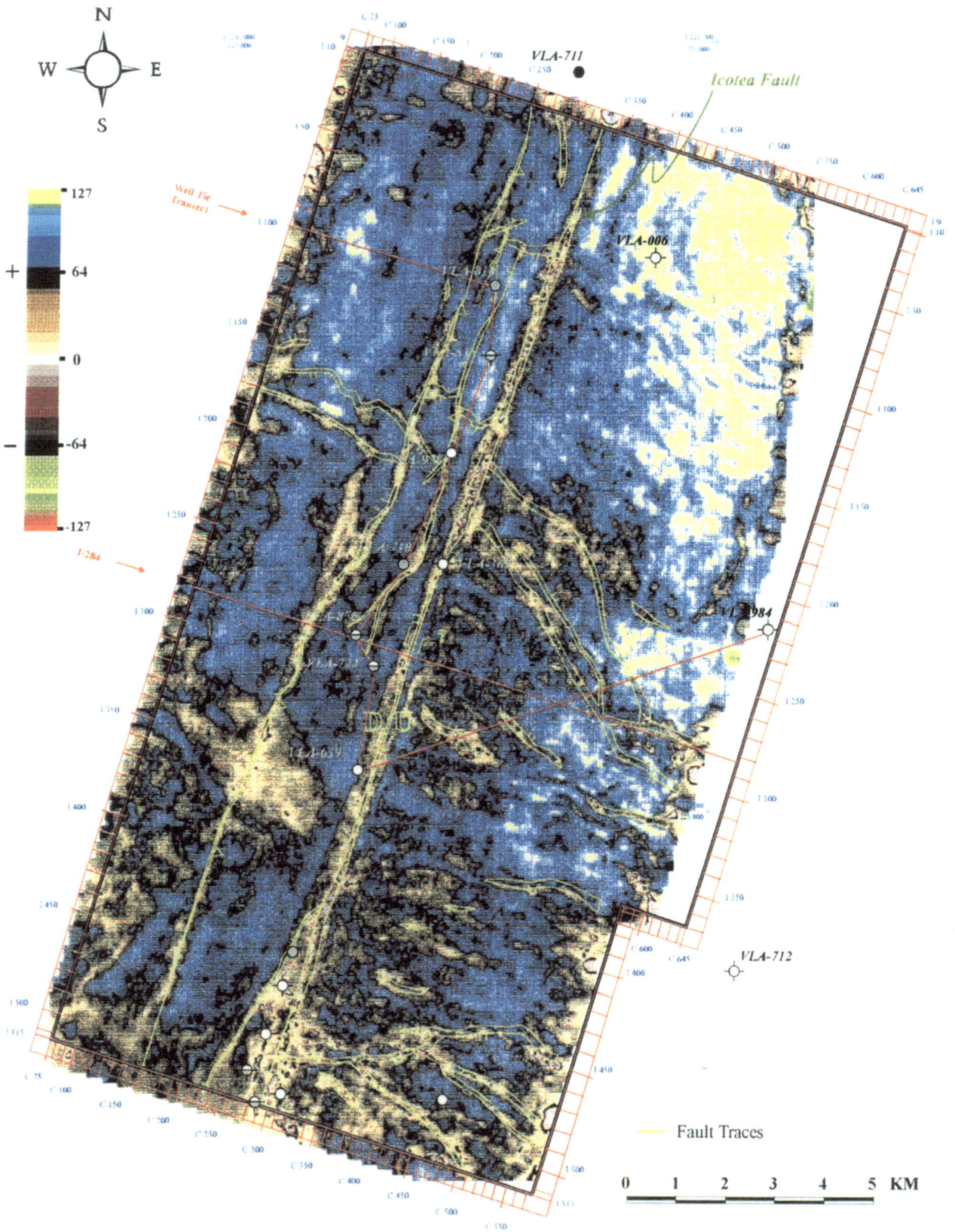

Fig. 14. Horizon amplitude map of the top of the Socuy Member from the Block I 3-D survey, showing low-amplitude fracture "halos" in the vicinity of faulting.

Are All Reflectors Sequence Boundaries?

Depositional sequences consist of facies sets that are bounded by erosional or nondepositional surfaces, and they record cycles of shoreline advance and retreat (Galloway, 1989). On carbonate platforms, sequence boundaries are not necessarily erosional exposure surfaces controlled by eustasy, nor are they necessarily time lines, although they separate younger from older strata.

We recognize three scales of depositional sequences that form the stratigraphic framework of the Cogollo Group in the study area: long-term cycles that are identified on shallow, high-frequency seismic profiles and the middle- and short-term cycles that are not resolvable on seismic data in this area. Three longer term transgressive-regressive cycles are separated by regional drowning events on the Maracaibo platform (Azpiritxaga and Murat, 1993). The Tibu-Machiques, Piche, and Lisure-Maraca lithostratigraphic units were deposited by these three cycles, each about 8–9 m.y. in duration. These larger scale sequences range in thickness from about 30 to 175 m (100–580 ft).

Portions of these sequences can be detected in the shallower section of Block I where (1) the top of the Maraca is recognized as a positive amplitude event and is unmasked by side lobes of the top of Socuy reflection; (2) the top of Piche event is evident from the presence of maximum-flooding shales; and (3) the Tibu is thick enough to be resolved. The top of Lisure event is not detected in this area due to low acoustic impedance contrast. Also, the Tibu is missing over the Block I area, so the resolvable longer term cycles on seismic profiles here are limited to the Machiques-Piche and Lisure-Maraca. Shorter cycle interpretation must be inferred from log and core analysis. In the deeper section, even the large-scale cycles are unreliable because the weak internal reflections of the Cogollo are unresolved and overridden by side lobes, seismic responses to structure, and other nonstratigraphic events.

The middle-order cycles are the carbonate sequences separated by regional-scale surfaces of nondeposition and even subaerial exposure. These cycles range in thickness from 10 to 45 m (30 to 150 ft) and are too thin to be resolved in the seismic data. The Tibu Formation consists of five sequences of about 1.6 m.y. duration each. The Piche Formation consists of four cycles that average 2.25 m.y. in duration beginning with the Machiques drowning. The Lisure Formation has four sequences of about 1.5 m.y. duration, and the Maraca Formation contains two sequences about 1 m.y. each.

The middle-order sequences and their contacts can be defined in cores and well logs by the vertical succession of sedimentary and paleoecologic features. They were not first identified by means of seismic features. The boundaries between these middle-scale sequences are sharp, generally nonerosional contacts between lagoonal mudstones and high-energy shell beds (Figure 5). The facies shifts across these contacts represent changes from regression to transgression. The maximum flooding intervals are generally deeper lagoonal mudstones. The gamma-ray characteristics are subtle fluctuations reflecting clay-rich carbonates, but they can be traced regionally by use of well logs. The boundaries between these middle-scale sequences may directly underlie high-gamma, deeper water marine mud rocks, or they may be picked in a negative-gamma trough at the base of shallow-water, transgressive beds. In a few places, the high-gamma kick defines a soil horizon or a phosphate-rich fill into the eroded surface of the underlying sequence, so the boundary is placed above the gamma peak. The presence of some sequence boundaries may be detected if thick maximum-flooding surface shales are present, but the thicknesses of the units are below seismic resolution and detection.

The shortest term cycles are the meter-scale, coarsening- and fining-upward facies sets that are separated by sharp lithologic contacts. Azpiritxaga (1991) defined 55 cycles in the entire Cogollo section that spans from upper Barremian to upper Albian, about 25 m.y. duration. Thus, the mean cycle length is about 450 k.y. This is close to one of the longest climatic Milankovitch orbital cycles (Schwarzacher, 1993). These events are well below the limits of seismic resolution.

Conclusions

In the interior of carbonate platforms, the relationships between sequence boundaries and reflectors may be direct and strong, or diffuse and indeterminate. Not all subparallel seismic reflectors correspond to significant stratal boundaries. There must be enough acoustic impedance contrast at the sequence boundaries to yield a reflection that is strong enough to be detected above background noise and seismic events caused by other factors. Also, the frequency content must be high enough to allow resolution and detection of thin beds. Maximum-flooding marine units, where they are at least 10 m (30 ft) thick, commonly relate to distinct seismic reflectors. However, depositional sequences and sets of sequences thinner than 30–60 m (100–200 ft) generally cannot be resolved on the standard seismic sections currently in use in this area.

The minimum lateral resolution within the Lake Maracaibo carbonate platform using current seismic data is about 300 m (1000 ft), measured as a radius from the seismic trace. This should be sufficient to resolve most lateral facies changes that are related to acoustic impedance contrasts and are thick enough to be detected and resolved. This resolution may also allow detection of fractured zones within wide fault zones.

The identification of carbonate platform sequences in the Block I area is limited mostly to large-scale cycles because of (1) the limited high-frequency content that is naturally attenuated with increasing depth and (2) the lack of sufficient acoustic impedance contrasts between beds required to cause strong reflections. In cases where sequence boundaries can be detected, there is a method to distinguish these reflections from those that are not caused by changes in stratigraphy.

Also, this study concludes that 3-D seismic data are essential in identifying subtle changes in the Cogollo Group carbonate section, clarifying complex structure, and following fault traces. Three-dimensional seismic data also provides critical spatial data needed for establishing repeatability of observations and identification of pitfalls. Synthetic seismic analyses are essential to tie the geology to seismic data and to avoid serious pitfalls in sequence stratigraphic interpretations.

It is also important to note that close work among geologists, geophysicists, petrophysicists, fracture experts, and engineers was critical to this study. The cross-discipline teamwork yielded a good understanding of the interdependencies and gaps in the data, a reasonable understanding of what results could be achieved with the available data, and a strong mutual respect for the people involved.

References

Azpiritxaga, I., 1991, Carbonate depositional styles controlled by siliciclastic influx and relative sea level changes, Lower Cretaceous, central Lake Maracaibo, Venezuela: M.Sc. thesis, Univ. of Texas.

Azpiritxaga, I., and Murat, B., 1993, Types of Cretaceous carbonate platforms in western Venezuela: Annual Mtg., Am. Assn. Petr. Geol., Program with Abstracts, 70.

Bartok, P., Reijers, T. J. A., and Juhasz, I., 1981, Lower Cretaceous Cogollo Group, Maracaibo Basin, Venezuela: Sedimentology, diagenesis, petrophysics: Am. Assn. Petr. Geol. Bull., **65**, 1110-1134.

Bubb, J. N., and Hatlelid, W. G., 1977, Seismic stratigraphy and global changes of sea level, part 10: seismic recognition of carbonate buildups, *in* Payton, C. E., Ed., Seismic stratigraphy—applications to hydrocarbon exploration: Am. Assn. Petr. Geol. Mem. **26**, 185-204.

Campbell, A. E., and Stafleu, J., 1992, Seismic modeling of an Early Jurassic drowned carbonate platform: Djobel Bou Dahar, High Atlas, Morocco: AAPG Bull., **76**, 1760-1777.

Galloway, W. E., 1989, Genetic stratigraphic sequences in basin analysis 1: architecture and genesis of flooding-surface-bounded depositional units: AAPG Bull., **73**, 125-142.

Galloway, W. E., Yancey, M. S., and Whipple, A. P., 1977, Seismic stratigraphic model of depositional platform margin, eastern Anadarko Basin, Oklahoma, *in* Payton, C. E., Ed., Seismic stratigraphy applications to hydrocarbon exploration: Am. Assn. Petr. Geol. Memoir **26**, 439-449.

Jansa, L. F., 1993, Early Cretaceous carbonate platforms of the northeastern American margin, *in* Simo, J. A., Scott, R. W., and Masse, J.-P., Eds., Cretaceous carbonate platforms: Am. Assn. Petr. Geol. Mem. **56**, 111-126.

Mayer, L. A., 1979, Deep sea carbonates: acoustic, physical, and stratigraphic properties: J. Sed. Pet., **49**, 819-836.

Piccoli, L., 1993, Microfacies and depositional environments of Lower Cretaceous carbonates in western Venezuela: Ann. Mtg., Am. Assn. Petr. Geol., Program with Abstracts, 165.

Reijers, T. J. A., and Bartok, P., 1985, Porosity characteristics and evolution in fractured Cretaceous carbonate reservoirs, La Paz field area, Maracaibo Basin, Venezuela, *in* Roehl, P. O., and Choquette, P. W., Eds., Carbonate petroleum reservoirs: Springer-Verlag, 407-423.

Renz, O., 1982, The Cretaceous ammonites of Venezuela: Birkhauser Verlag.

Ross, M. I., and Scotese, C. R., 1987, A hierarchical model of the Gulf of Mexico and the Caribbean region: Tectonophysics, **155**, 139-168.

Schwarzacher, W., 1993, Cyclostratigraphy and the Milanokovitch theory, *in* Developments in Sedimentology, **52**: Elsevier Sci. Publ. B.V.

Scott, R. W., 1993, Cretaceous carbonate platform, U.S. Gulf Coast, *in* Simo, J. A., Scott, R. W., and Masse, J.-P., Eds., Cretaceous carbonate platforms: Am. Assn. Petr. Geol. Mem., **56**, 97-109.

Stafleu, J., Everts, A. J. W., and Kenter, J. A. M., 1994, Seismic models of a prograding carbonate platform: Vercors, southeast France: Mar. Petr. Geol., **11**, 514-527.

Vahrenkamp, V. C. R., Franseen, R. C. W. M., Grotsch, J., and Munoz, P. J., 1993, Maracaibo platform (Aptian–Albian), northwestern Venezuela, *in* Simo, J. A., Scott, R. W., and Masse, J.-P., Eds., Cretaceous carbonate platforms: Am. Assn. Petr. Geol. Mem., **56**, 25-33.

Yurewicz, D. A., Marier, T. B., Meyerholtz, K. A., and Siroky, F. X., 1993, Early Cretaceous carbonate platform, north rim of the Gulf of Mexico, Mississippi and Louisiana, *in* Simo, J. A., Scott, R. W., and Masse, J.-P., Eds., Cretaceous carbonate platforms: Am. Assn. Petr. Geol. Mem., **56**, 81-96.